ASTRONOMY AND ASTROPHYSICS LIBRARY

Kenneth R. Lang

The Sun from Space

Second Edition

Springer

Kenneth R. Lang
Department of Physics and Astronomy
Tufts University
Medford MA 02155
USA
ken.lang@tufts.edu

Cover image: **Solar cycle magnetic variations**. These magnetograms portray the polarity and distribution of the magnetism in the solar photosphere. They were made with the Vacuum Tower Telescope of the National Solar Observatory at Kitt Peak from 8 January 1992, at a maximum in the sunspot cycle (*lower left*) to 25 July 1999, well into the next maximum (*lower right*). Each magnetogram shows opposite polarities as darker and brighter than average tint. When the Sun is most active, the number of sunspots is at a maximum, with large bipolar sunspots that are oriented in the east–west (*left–right*) direction within two parallel bands. At times of low activity (*top middle*), there are no large sunspots and tiny magnetic fields of different magnetic polarity can be observed all over the photosphere. The haze around the images is the inner solar corona. (Courtesy of Carolus J. Schrijver, NSO, NOAO and NSF.)

ISBN: 978-3-662-49597-1   ISBN: 978-3-540-76953-8 (eBook)
DOI: 10.1007/978-3-540-76953-8

© Springer-Verlag Berlin Heidelberg 2009
Softcover re-print of the Hardcover 2nd edition 2009

Cover design: eStudio Calamar S.L.

Printed on acid-free paper

9 8 7 6 5 4 3 2 1

springer.com

This book is dedicated to everyone who is curious, imaginative, and intelligent, thereby strengthening the human spirit.

Preface to the Second Edition

The *First Edition* of *The Sun from Space*, completed in 1999, focused on the early accomplishments of three solar spacecraft, *SOHO*, *Ulysses*, and *Yohkoh*, primarily during a minimum in the Sun's 11-year cycle of magnetic activity. The comprehensive *Second Edition* includes the main findings of these three spacecraft over an entire activity cycle, including two minima and a maximum, and discusses the significant results of six more solar missions. Four of these, the *Hinode, RHESSI, STEREO*, and *TRACE missions* were launched after the *First Edition* was either finished or nearly so, and the other two, the *ACE* and *Wind* spacecraft, extend our investigations from the Sun to its varying input to the Earth.

The *Second Edition* does not contain simple updates or cosmetic patch ups to the material in the *First Edition*. It instead contains the relevant discoveries of the past decade, integrated into chapters completely rewritten for the purpose. This provides a fresh perspective to the major topics of solar enquiry, written in an enjoyable, easily understood text accessible to all readers, from the interested layperson to the student or professional.

The main scientific accomplishments of the *ACE, RHESSI, SOHO, TRACE, Ulysses*, and *Wind* missions, which are included in their 2005 Senior Review Proposals, have been described in Chapter 1 and included in greater detail in the relevant chapters. (*Yohkoh* completed its decade-long scrutiny of the X-ray Sun in 2001, while *Hinode* and *STEREO* were not launched until late 2006.)

Members of the Solar Physics Community were consulted about key discoveries or important reviews during the past decade. Persons who provided important review articles or other significant new information include Loren Acton, Markus J. Aschwanden, Eugene H. Avrett, Sarbani Basu, Paul M. Bellan, Benjamin D. G. Chandran, James Chen, Steven R. Cranmer, George A. Doschek, Murray Dryer, Peter Foukal, Joseph V. Hollweg, Gordon D. Holman, Stephen W. Kahler, James A. Klimchuk, Jun Lin, Donald Liebenberg, Noé Lugaz, Ward Manchester, Scott W. McIntosh, Mark Miesch, Ronald L. Moore, Judit Pap, Eugene Parker, Alexei A. Pevtsov, Arthur I. Poland, Arik Posner, Ilia Roussev, Wilfred Schröder, Leonard Strachan, Yi-Ming Wang, David F. Webb, Thomas N. Woods, Jie Zhang, and Thomas H. Zurbuchen.

Draft chapters were then sent to experts in the field, who have provided important suggestions for changes, deletions, and omissions. They are: Loren Acton, Markus

Aschwanden, George Doschek, Leon Golub, Bernhard Fleck, Arik Posner, Takashi Sakurai, Carolus Schrijver, and Saku Tsuneta for the introductory Chapter 1; each commenting on the spacecraft they are most familiar with; Steven Cranmer for Chapter 2; Thomas Duval, Bernhard Fleck, John Harvey, Rachel Howe, Mark Miesch, and Junwei Zhao for Chapter 3; Loren Acton, Markus Aschwanden, and Steven Cranmer for Chapter 4; Steven Cranmer, John Kohl, Edward Stone, Yi-Ming Wang, and Thomas Zurbuchen for Chapter 5; Markus Aschwanden, Arnold Benz, Richard Canfield, Terry Forbes, Stephen Kahler, Ronald Moore, Alexei Pevtsov, and Jie Zhang for parts of Chapter 6, and Peter Foukal, Stephen Kahler, Alexei Pevtsov, David Webb, and Thomas Woods for Chapter 7.

The combination of Senior Reviews, advice from the Solar Physics Community, and careful reading of individual chapters by experts in the field have assured that the text is fully up to date and complete.

The First Chapter of the *Second Edition* describes the scientific objectives of each of the nine solar missions, together with the instruments that are being used to accomplish these objectives. The Second Chapter retains the historical perspective to studies of the Sun and heliosphere found in the *First Edition*, including seminal contributions to our understanding of the Sun during the past century. The next five chapters present key improvements in our understanding of the solar interior; the heating of the million-degree outer atmosphere of the Sun, known as the solar corona; the origin and nature of the Sun's winds; the cause, prediction, and propagation of explosive solar flares and coronal mass ejections; and all aspects of space-weather interactions of the Sun with either the Earth or with unprotected astronauts and spacecraft in outer space or on the Moon or Mars.

Each of these five chapters ends with Summary Highlights of key ideas and fundamental discoveries in capsule form, followed by a time line of significant events in our understanding of the Sun, updated to include recent results. An Appendix provides Internet Addresses for the spacecraft and the topics of each book chapter. The *Second Edition* of *The Sun from Space* ends with comprehensive references to more than 2,500 fundamental research papers and review articles.

Since the publication of the *First Edition*, the *Ulysses* spacecraft has completed a second orbit over the solar poles, during a maximum in the solar activity cycle. After more than 17 years of pioneering solar science, the *Ulysses* mission ended on 1 July 2008. As shown in this *Second Edition*, *Ulysses* has therefore determined the distribution of solar wind velocities and, together with other solar spacecraft, helped identify the wind sources at both activity maximum and minimum. The *Second Edition* also presents the results of years of observations of solar oscillations acquired by *SOHO*, which probe the rotation and motions of the solar interior, as well as extensive new *SOHO* data on the magnetic structure of the solar wind sources, and the powerful, explosive solar flares that are triggered and powered by magnetic reconnection in the low solar corona.

The main results of the *Transition Region And Coronal Explorer*, abbreviated *TRACE* and launched on 1 April 1998, are also included in this *Second Edition* of *The Sun from Space*. It has obtained significant new information about the nature of the magnetic loops that thread the solar corona and the sources of heating the million-degree solar corona.

The new edition additionally includes the seminal findings of the *Ramaty High Energy Solar Spectroscopic Imager*, or *RHESSI* for short, launched on 5 February 2002. It has used images and spectroscopy of the high-energy radiation from solar flares to explore the basic physics of particle acceleration and explosive energy release, including new information about magnetic reconnection and the emission of the neutron capture and pair-annihilation lines during solar flares.

The *Second Edition* of *The Sun from Space* provides an account of the initial discoveries of the Japanese *Hinode* (formerly known as *Solar-B*) spacecraft launched on 23 September 2006. It consists of a coordinated set of optical, EUV, and X-ray instruments that study the interaction between the Sun's magnetic field and its high-temperature, ionized atmosphere. They measure the detailed density, temperature, and velocity structures in the visible solar disk, the photosphere, the low corona, and the transition region between them with high spatial, spectral, and temporal resolution, resulting in new information about the Sun's varying magnetism and its relationship to solar flares and the expansion of the solar corona into the Sun's winds.

Also included are significant new findings of NASA's *Solar TErrestrial RElations Observatory*, or *STEREO* for short, launched on 25 October 2006. It is a pair of spacecraft, leading and lagging the Earth in its orbit, which investigates the origin, evolution, and interplanetary consequences of the billion-ton solar eruptions known as coronal mass ejections. The combined observations from the two *STEREO* spacecraft provide a three-dimensional view of these outbursts, from their onset at the Sun to the orbit of the Earth, improving our understanding of these solar explosions and our ability to predict their trajectories and consequences.

This *Second Edition* of *The Sun from Space* emphasizes the human impact of NASA solar missions, with a full discussion of their implications for Sun-driven space weather. Our technological society has become increasingly vulnerable to these storms from the Sun, both on Earth and in the human and robotic exploration of the heliosphere. Solar flares or coronal mass ejections may affect the health and safety of travelers in space, influence the habitability of the Moon or Mars, and damage or disable spacecraft both near the Earth and in outer space.

Instruments aboard *RHESSI, SOHO*, and *Hinode* are, for example, helping us understand and forecast solar flares. These catastrophic outbursts can suddenly flood the solar system with intense radiation and energetic particles, releasing energy equivalent to millions of 100-megaton hydrogen bombs exploding at the same time. The X-rays from a solar flare modify our atmosphere, disrupt radio communications, and alter satellite orbits. The energetic particles that are hurled out into interplanetary space during solar flares can threaten the safety of unprotected astronauts traveling beyond the safety of the Earth's magnetic field to the Moon or Mars. These flare particles can also damage or destroy satellites used for communication, navigation, and military reconnaissance and surveillance.

The instruments aboard *SOHO* and *STEREO* are additionally providing new insights to the mechanisms and prediction of coronal mass ejections; giant magnetic bubbles that expand as they propagate outward from the Sun to rapidly rival it

in size. These violent eruptions throw billions of tons of material into interplanetary space, and their associated shocks accelerate and propel vast quantities of
high-speed particles ahead of them. They can produce spectacular auroras, create intense geomagnetic storms, and disrupt satellites, radio communications, and Earth's
power systems. Energetic particles associated with coronal mass ejections can also
be hazardous to spacecraft and astronauts traveling to the Moon or exploring its
surface.

Special thanks are extended to Joe Bredekamp and Bill Wagner at NASA Head-
quarters. They have actively encouraged the writing of this book and helped fund
it through NASA Grant NNX07AU93G entitled NASA'S COSMOS, with funding
evenly split between NASA's Solar and Heliospheric Physics SR&T Program and
NASA's Applied Information Systems Research Program.

Tufts University and Anguilla,   *Kenneth R. Lang*
British West Indies
New Year's Eve, 2008

Preface to the First Edition

Our familiar, but often inscrutable, star exhibits a variety of enigmatic phenomena that have continued to defy explanations. Our book begins with a brief account of these unsolved mysteries. Scientists could not, for example, understand how the Sun's intense magnetism is concentrated into dark sunspots that are as large as the Earth and thousands of times more magnetic. Nor did they know exactly how the magnetic fields are generated within the Sun, for no one could look inside it.

Another long-standing mystery is the million-degree solar atmosphere, or corona, that lies just above the cooler, visible solar disk, or photosphere. Heat should not emanate from a cold object to a hotter one anymore than water should flow up hill. Researchers have hunted for the elusive coronal heating mechanism for more than half a century.

The Sun's hot and stormy atmosphere is continuously expanding in all directions, creating a relentless solar wind that seems to blow forever. The exact sources for all the wind's components, and the mechanisms for accelerating it to supersonic velocities, remained perplexing problems as well.

The relatively calm solar atmosphere can be violently disrupted by powerful explosions, filling the solar system with radio waves, X-rays, and gamma rays, and hurling charged particles out into space at nearly the speed of light. Other solar explosions, called Coronal Mass Ejections, throw billions of tons of coronal gases into interplanetary space, creating powerful gusts in the solar wind. Yet, we have only just begun to understand the detailed causes of the Sun's explosive outbursts, and no one has been able to predict exactly when they will occur.

In less than a decade, three pioneering spacecraft, named the *SOlar and Heliospheric Observatory*, or *SOHO* for short, *Ulysses*, and *Yohkoh*, have transformed our perception of the Sun. They are also described in the introductory chapter, together with their principal scientific goals. This scientific troika has examined the Sun with exquisitely sensitive and precise instruments that have widened our range of perception, giving us the eyes to see the invisible and the hands to touch what cannot be felt. They have extended our gaze from the visible solar disk, down to the hidden core of the Sun and out in all directions through the Sun's tenuous, expanding atmosphere. *SOHO, Ulysses*, and *Yohkoh* have together provided insights that are vastly more focused and detailed than those of previous solar missions, providing clues to many of the crucial, unsolved problems in our understanding of the Sun.

Scientific discoveries are usually not isolated "eureka" moments, occurring in a mental or experiential vacuum. They are built upon a foundation of previous investigations and extrapolated into the future. Our recent accomplishments in solar physics are the culmination of a long history of prior research, from the earliest optical and radio telescopes to rockets and then full-fledged spacecraft such as *Skylab, Helios 1* and *2*, and the *Solar Maximum Mission.*

The second chapter of our book describes the discovery of space, providing the historical framework needed to understand subsequent results. Here we also introduce ideas and terms that have become so common to our space-age vocabulary that even many practicing scientists have forgotten their origin and meaning.

As mentioned in Chapter 2, just half a century ago most people visualized our planet as a solitary sphere traveling in a cold, dark vacuum around the Sun. Inquisitive spacecraft have now shown that the space outside the Earth's atmosphere is not empty, and demonstrated that the Earth and other planets are immersed within an eternal, stormy wind. The solar wind is mainly composed of electrons and protons, set free from hydrogen atoms, the most abundant element in the Sun, but it also contains heavier ions and a magnetic field.

The second chapter also describes the discovery that intense magnetism pervades the Sun's atmosphere, creating an 11-year variation in the level of explosive solar activity, as well as the amount of cosmic rays reaching the Earth from the space outside our solar system. X-ray photographs from rockets and the *Skylab Space Station* demonstrated that the Sun's magnetic fields mold and constrain the million-degree gases found in the low solar atmosphere. Long-lasting coronal holes, without any detectable X-ray radiation in *Skylab's* photographs, were also compared with other spacecraft measurements to show that these places expel a high-speed solar wind.

Contextual and historical preludes are also woven into the texture of subsequent chapters, and usually presented in chronological order. Every chapter, except the first, contains a concluding time-line table that highlights key events and discoveries in the area under consideration. Complete references to seminal articles are given at the end of this book. The reader can consult these fundamental research papers for further details on important scientific results, from centuries ago to the present day.

Set-aside focus elements are also inserted throughout the text, which enhance and amplify the discussion with interesting details or fundamental physics. They will be read by the curious person or serious student, but do not interfere with the general flow of the text and can be bypassed by the educated reader who wants to follow the main ideas.

The discovery of how the Sun shines by thermonuclear reactions within its hot, dense core is spelled out in the beginning of the third chapter of our book. It then describes how the unseen depths of the Sun are being probed using the new science of helioseismology. Today, we can peel back the outer layers of the Sun, and glimpse inside it by observing widespread throbbing motions in the Sun's visible material. Helioseismologists decipher these oscillations, caused by sounds trapped within the Sun, to reveal its internal constitution. This procedure resembles Computed Axial Tomography (CAT) scans that derive views inside our bodies from numerous readings of X-rays that cross them from different directions.

Contents

the nanoTesla (nT) unit of magnetic flux density, where $1\,nT = 10^{-9}$ Tesla $= 10^{-5}$ Gauss, the electron volt (eV) unit of energy, with $1\,eV = 1.6 \times 10^{-19}$ Joule, and the ton measurement of mass, where 1 ton $= 10^3$ kilograms. The accompanying Table P.2 provides numerical values for solar quantities and fundamental constants.

Table P.2 Solar quantities and fundamental constants*

Symbol	Name	Value
$L_\odot$	Luminosity of Sun	3.854×10^{26} Joule s^{-1}
$M_\odot$	Mass of Sun	1.989×10^{30} kilograms
$R_\odot$	Radius of Sun	6.955×10^8 meters
AU	Mean distance of Sun	1.496×10^{11} meters
$T_{e\odot}$	Effective temperature of photosphere	5780 Kelvin
c	Velocity of light	2.9979×10^8 m s^{-1}
G	Gravitational constant	6.6726×10^{-11} N m^2 kg^{-2}
k	Boltzmann's constant	1.38066×10^{-23} J K^{-1}
h	Planck's constant	6.6261×10^{-34} J s
a	Radiation density constant	7.5659×10^{-16} J m^{-3} K^{-4}
m_e	Mass of electron	9.1094×10^{-31} kg
e	Charge of electron	1.6022×10^{-19} C
m_H	Mass of hydrogen atom	1.673534×10^{-27} kg
m_p	Mass of proton	1.672623×10^{-27} kg
$\varepsilon_\odot$	Permittivity of free space	$10^{-9}/(36\pi) = 8.854 \times 10^{-12}$ F m^{-1}
$\mu_\odot$	Permeability of free space	$4\pi \times 10^{-7} = 12.566 \times 10^{-7}$ N A^{-2}

*Adapted from Lang (1991). One Joule per second $= 1\,J\,s^{-1}$ is equal to one Watt $= 1\,W$. The unit symbols are J for Joule, s for second, kg for kilogram, m for meter, K for degree Kelvin, C for Coulomb, N for Newton, and A for ampere.

luminous intensity is nevertheless expected to boil our oceans away in about 3 billion years; and the Sun will expand into a giant star in another 4 billion years. It will then engulf Mercury, melt the Earth's surface rocks, and turn the frozen moons of the distant giant planets into globes of liquid water.

I am very grateful to my expert colleagues who have read portions of this book, and commented on its accuracy, clarity, and completeness, substantially improving the manuscript. They include Loren W. Acton, Markus J. Aschwanden, W. Ian Axford, Arnold O. Benz, Richard C. Canfield, Edward L. Chupp, Edward W. Cliver, George A. Doschek, A. Gordon Emslie, Bernhard Fleck, Peter V. Foukal, Claus Fröhlich, John W. Harvey, Hugh S. Hudson, Stephen W. Kahler, Mukul R. Kundu, John W. Leibacher, Michael E. Mann, Richard G. Marsden, Ronald L. Moore, Eugene N. Parker, John C. Raymond, Andrew P. Skumanich, Charles P. Sonett, Barbara Thompson, Virginia L. Trimble, Bruce T. Tsurutani, Yi-Ming Wang, and David F. Webb.

Locating quality figures is perhaps the most time-consuming and frustrating aspect of producing a volume like this, so I am especially thankful for the support of ESA, ISAS, and NASA for providing them. Individuals that were especially helpful in locating and providing specific images include Loren W. Acton, David Alexander, Cary Anderson, Frances Bagenel, Richard C, Canfield, Michael Changery, David Chenette, Fred Espenak, Bernhard Fleck, Eigil Friis-Christensen, Claus Fröhlich, Bruce Goldstein, Leon Golub, Steele Hill, Gordon Holman, Beth Jacob, Imelda Joson, Therese Kucera, Judith Lean, William C. Livingston, Michael E. Mann, Richard Marsden, Michael J. Reiner, Thomas Rimmele, Kazunari Shibata, Gregory Lee Slater, Barbara Thompson, Haimin Wang, and Joachim Woch.

In conclusion, this book uses the International System of Units (Systeme International, SI) for most quantities, but the reader should be warned that centimeter-gram-second (c.g.s.) units are employed in the nearly all of the seminal papers referenced in this book. Moreover, many solar astronomers still use the c.g.s. units. The following Table P.1 provides unit abbreviations and conversions between units. Some other common units are the nanometer (nm) with $1\,\text{nm} = 10^{-9}$ meters, the Angstrom unit of wavelength, where 1 Angstrom $= 1\,\text{Å} = 10^{-10}$ meters $= 10^{-8}$ centimeters,

Table P.1 Principal SI units and their conversion to corresponding c.g.s. units

Quantity	SI units	Conversion to c.g.s. units
Length	Meter (m)	100 centimeters (cm)
	Nanometer (nm) $= 10^{-9}\,\text{m}$	$10^{-7}\,\text{cm} = 10$ Angstroms $= 10\,\text{A}$
Mass	Kilogram (kg)	1,000 grams (g)
Velocity	Meter per second (m s^{-1})	100 centimeters per second (cm s^{-1})
Energy*	Joule (J)	10,000,000 erg
Power	Watt (W) $=$ Joule per second (J s^{-1})	$10,000,000\,\text{erg s}^{-1}$
Magnetic flux density	Tesla (T)	10,000 Gauss (G)

*The energy of high-energy particles and X-ray radiation are often expressed in units of kilo-electron volts, or keV, where $1\,\text{keV} = 1.602 \times 10^{-16}$ Joule, or MeV $= 1,000\,\text{keV}$.

transmission lines on Earth. These solar ejections travel to the Earth in a few days, so there is some warning time.

Intense radiation from a solar flare reaches the Earth's atmosphere in just 8 min, moving at the speed of light – the fastest thing around. Flaring X-rays increase the ionization of our air, and disrupt long-distance radio communications. A satellite's orbit around the Earth can be disturbed by the enhanced drag of the expanded atmosphere.

The Earth's magnetic field shields us from most of the high-speed particles ejected by solar flares or accelerated by CME shock waves, but the energetic charged particles can endanger astronauts in space and destroy satellite electronics. Some of the particles move at nearly the speed of light, so there is not much time to seek protection from their effects. If we knew the solar magnetic changes preceding these violent events, then spacecraft could provide the necessary early warning. As an example, when the coronal magnetic fields get twisted into an S, or inverted S, shape, they are probably getting ready to release a mass ejection.

As our civilization becomes increasingly dependent on sophisticated systems in space, it becomes more vulnerable to this Sun-driven space weather, which is tuned to the rhythm of the Sun's 11-year magnetic activity cycle. It is of such vital importance that national centers employ space-weather forecasters, and continuously monitor the Sun from ground and space to warn of threatening solar events.

As also described in the seventh chapter, solar X-ray and ultraviolet radiation are extremely variable, changing in step with the 11-year cycle of magnetic activity. The fluctuating X-rays produce substantial alterations of the Earth's ionosphere, and the changing ultraviolet modulates our ozone layer. The varying magnetic activity also changes the Sun's total brightness, but to a lesser degree. During the past 130 years, the Earth's surface temperature has been associated with decade-long variations in solar activity, perhaps because of changing cloud cover related to the Sun's modulation of the amount of cosmic rays reaching Earth. Observations of the brightness variations of Sun-like stars indicate that they are capable of a wider range of variation in total radiation than has been observed for the Sun so far.

Radioactive isotopes found in tree rings and ice cores indicate that the Sun's activity has fallen to unusually low levels at least three times during the past 1,000 years, each drop corresponding to a long cold spell of roughly a century in duration. Further back in time, during the past one million years, our climate has been dominated by the recurrent ice ages, each lasting about 100,000 years. They are explained by three overlapping astronomical cycles, which combine to alter the distribution and amount of sunlight falling on Earth.

Our book ends with a description of the Sun's distant past and remote future. The Sun is gradually increasing in brightness with age, by a startling 30% over the 4.5 billion years since it began to shine. This slow, inexorable brightening ought to have important long-term terrestrial consequences, but some global thermostat has kept the Earth's surface temperature relatively unchanged, as the Sun grew brighter and hotter. A powerful atmospheric greenhouse might have warmed the young Earth, gradually weakening over time, or plants and animals might have beneficially controlled their environment for the past 3 billion years. The Sun's steady increase in about 750 km s^{-1} and a slow one with about half that speed. *Ulysses* provided the first measurements of the solar wind all around the Sun, conclusively showing that much of the steady, high-speed wind squirts out of polar coronal holes. A capricious, gusty, slow wind emanates from the Sun's equatorial regions near the minimum in the 11-year solar activity cycle when the *SOHO* and *Ulysses* measurements were made. The high-speed wind is accelerated very close to the Sun, within just a few solar radii, and the slow component obtains full speed further out.

SOHO instruments have unexpectedly demonstrated that oxygen ions move faster than protons in coronal holes. Absorbing more power from magnetic waves might preferentially accelerate the heavier ions, as they gyrate about open magnetic fields. *Ulysses* has detected magnetic waves further out above the Sun's poles, where the waves apparently block the incoming cosmic rays. Instruments aboard *SOHO* have also pinpointed the source of the high-speed wind; it is coming out of honeycomb-shaped magnetic fields at the base of coronal holes.

Our sixth chapter describes sudden, brief, and intense outbursts, called solar flares, which release magnetic energy equivalent to billions of nuclear bombs on Earth. The Sun's flares flood the solar system with intense radiation across the full electromagnetic spectrum from the longest radio wavelengths to the shortest X-rays and gamma rays. The radio bursts provide evidence for the ejection of very energetic particles into space, either as electron beams moving at nearly the speed of light or as the result of shock waves moving out at a more leisurely pace.

The *Solar Maximum Mission*, abbreviated *SMM*, and *Yohkoh* spacecraft have shown that solar flares also hurl high-speed electrons and protons down into the Sun, colliding with the denser gas and emitting hard X-rays and gamma rays. Soft X-ray observations indicate that Earth-sized regions can be heated to about ten million degrees during the later stages of a solar flare, becoming about as hot as the center of the Sun.

Magnetic bubbles of surprising proportion, called Coronal Mass Ejections, or CMEs for short, are also discussed in the sixth chapter. They have been routinely detected with instruments on board several solar satellites, most recently from *SOHO*. The CMEs expand as they propagate outward from the Sun to rapidly rival it in size, and carry up to ten billion tons of coronal material away from the Sun. Their associated shocks accelerate and propel vast quantities of high-speed particles ahead of them.

The sixth chapter additionally describes how explosive solar activity can occur when magnetic fields come together and reconnect in the corona. Stored magnetic energy is released rapidly at the place where the magnetic fields touch. Here we also discuss how high-speed particles, released from solar flares or accelerated by CME shock waves, have been directly measured in situ by spacecraft in interplanetary space. These spacecraft also detect intense radiation, shock waves and magnetic fields.

The seventh chapter describes how forceful coronal mass ejections can create intense magnetic storms on Earth, trigger intense auroras in the skies, damage or destroy Earth-orbiting satellites, and induce destructive power surges in long-distance

The Sun's internal sound waves have been used as a thermometer, taking the temperature of the Sun's energy-generating core and showing that it agrees with model predictions. This strongly disfavors any astrophysical solution for the solar neutrino problem in which massive subterranean instruments always come up short, detecting only one-third to one-half the number of neutrinos that theory says they should detect. The ghostly neutrinos seem to have an identity crisis, transforming themselves on the way out of the Sun into a form that we cannot detect and a flavor that we cannot taste.

SOHO's helioseismology instruments have shown how the Sun rotates inside, using the Doppler effect in which motion changes the pitch of sound waves. Regions near the Sun's poles rotate with exceptionally slow speeds, while the equatorial regions spin rapidly. This differential rotation persists to about a third of the way inside the Sun, where the rotation becomes uniform from pole to pole. The Sun's magnetism is probably generated at the interface between the deep interior, which rotates at one speed, and the overlying gas that spins faster in the equatorial middle.

Internal flows have also been discovered by the *SOHO* helioseismologists. White-hot currents of gas move beneath the Sun we see with our eyes, streaming at a leisurely pace when compared to the rotation. They circulate near the equator, and between the equator and poles.

In the fourth chapter, we describe the tenuous, million-degree gas called the corona that lies outside the sharp apparent edge of the Sun. The visible solar disk is closer to the Sun's core than the million-degree corona, but is several hundred times cooler. This violates common sense, as well as the second law of thermodynamics, which holds that heat cannot be continuously transferred from a cooler body to a warmer one without doing work.

Attempts to solve the Sun's heating crisis are also discussed in the fourth chapter. The paradox cannot be solved by sunlight, which passes right through the transparent corona, and spacecraft have shown that sound waves cannot get out of the Sun to provide the corona's heat. Moreover, when *SOHO* focuses in on the material just above the photosphere, it all seems to be falling down into the Sun, so nothing seems to be carrying the heat out into the overlying corona.

Yohkoh and *SOHO* images at X-ray and extreme-ultraviolet wavelengths have provided some solutions to the coronal heating problem, in which the million-degree outer solar atmosphere overlies the Sun's cooler visible disk. They have shown that the hottest, densest material in the low corona is concentrated within thin, long magnetized loops that are in a state of continual agitation. Wherever the magnetism is strongest, the coronal gas is hottest. These magnetic loops are often coming together, releasing magnetic energy when they make contact. This provides a plausible explanation for heating the low corona.

The hot coronal gases are expanding out in all directions, filling the solar system with a ceaseless flow – called the solar wind – that contains electrons, protons, and other ions, and magnetic fields. The fifth chapter provides a detailed discussion of this solar wind, together with recent investigations into its origin and acceleration. Early spacecraft measurements showed that the Sun's wind blows hard and soft. That is, there are two kinds of wind, a fast one moving at

List of Figures

List of Focus Elements

List of Tables

Chapter 1
Instruments for a Revolution

1.1 Solar Mysteries

From afar, the Sun does not look very complex. To the casual observer, it is just a smooth, uniform ball of gas. Close inspection, however, shows that the star is in constant turmoil. The seemingly calm Sun is a churning, quivering, and explosive body, driven by intense, variable magnetism.

The ultimate power source for the Sun's relentless activity lies deep down in its energy-generating core, where nuclear fusion releases radiation that works its way out and eventually drives an overturning, convective motion of the solar gas just below its visible disk, the photosphere (Fig. 1.1). The photosphere is the source of all our visible light, the part of the Sun we see each day and the lowest, densest level of the solar atmosphere. The overlying gas includes the thin chromosphere, with a temperature of about 10,000 K, and the hot, extended corona at temperatures of a few million Kelvin. Both the chromosphere and corona are so rarefied that we look right through them, just as we see through the Earth's clean air. They are separated by a narrow transition region, a highly variable, corrugated zone which is dynamically modulated by brief, jet-like spicules that can shoot up for thousands of kilometers.

Although we cannot see it with our eyes, the diffuse solar gas extends all the way to the Earth and beyond. The gas is so hot, and moving so fast, that it overcomes the Sun's gravity and perpetually expands out into surrounding space. The Sun's hot and stormy atmosphere is carried throughout the solar system by these solar winds. Contemporary solar spacecraft can observe the winds at their origin on the Sun and sample its ingredients in situ, or in place, near the Earth's orbit.

The entire solar atmosphere is permeated by magnetic fields that are generated inside the Sun and rise up through the photosphere into the overlying atmosphere, carrying energy with them. Dark islands on the Sun's visible disk, called sunspots, mark intense concentrations of magnetism, which can be as large as the Earth and thousands of times more magnetic. Invisible magnetic arches loop between regions of opposite magnetic polarity in the underlying photosphere (Fig. 1.2). Many of these invisible coronal loops are filled with high-temperature, ionized gas, which is detected by its extreme-ultraviolet or X-ray radiation (Fig. 1.3).

K.R. Lang, *The Sun from Space*, Astronomy and Astrophysics Library,
© Springer-Verlag Berlin Heidelberg 2009

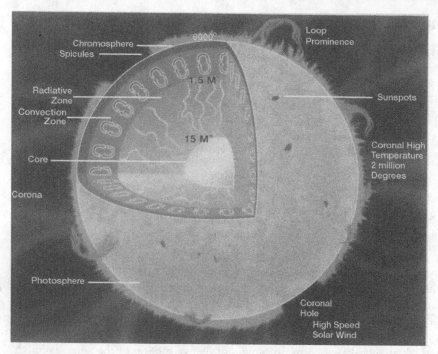

Fig. 1.1 Anatomy of the Sun. The Sun is an incandescent ball of ionized gas powered by the fusion of hydrogen in its core. As shown in this interior cross-section, energy produced by nuclear fusion is transported outward, first by countless absorptions and emissions within the radiative zone, and then by convection. The visible disk of the Sun, called the photosphere, contains dark sunspots, which are Earth-sized regions of intense magnetic fields. A transparent atmosphere envelops the photosphere, including the low-lying chromosphere with its jet-like spicules and the million-degree corona that contains holes with open magnetic fields, the source of the high-speed solar wind. Loops of closed magnetic fields constrain and suspend the hot million-degree gas within coronal loops and cooler material in prominences

Driven by motions inside the Sun, and rooted in the photosphere, the coronal loops rise, fall, move sideways, and interact, carrying energy with them. Their concentration varies with the 11-year solar activity cycle, delineated by the number of sunspots and bright coronal loops joining them. Some of the magnetism is even carried into interplanetary space by the Sun's intense winds, providing channels for the flow of energetic particles and modulating the effects of space weather.

This new perception of a volatile, ever-changing Sun has resulted in several fundamental, unsolved mysteries. Scientists could not, for example, understand how the Sun generates its magnetic fields, which are responsible for most solar activity. Nor did they know why some of this intense magnetism is concentrated into sunspots. Furthermore, they could not explain why the Sun's magnetic activity varies dramatically, waning and intensifying again every 11 years or so. The answers to these puzzles have been hidden deep down inside the Sun where its powerful magnetism is generated.

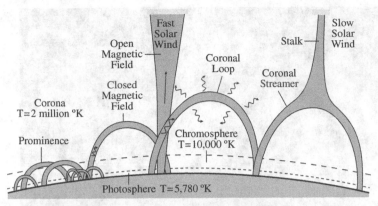

Fig. 1.2 Coronal loops. The corona is stitched together with the ubiquitous coronal loops that are created when upwelling magnetic fields generated inside the Sun push through the photosphere into the overlying chromosphere and corona. These closed magnetic structures are anchored in the photosphere at footpoints of opposite magnetic polarity. Coronal loops can be filled with hot gas that shines brightly at extreme ultraviolet and X-ray wavelengths. Driven by motions in the underling photosphere and below, the coronal loops twist, rise, shear, writhe, and interact, releasing magnetic energy that can heat the solar corona and power intense solar flares or coronal mass ejections. Large coronal loops are found in the bulb-like base of coronal steamers, whose long, thin stalks extend out into space. Magnetic fields that are anchored in the photosphere at one end can also be carried by the solar wind into interplanetary space, resulting in open magnetic fields and a channel for the fast solar wind

Further out, the normally invisible solar atmosphere, the corona, presents one of the most puzzling paradoxes of modern science. It is unexpectedly hot, reaching temperatures of a few million Kelvin, which are several hundred times hotter than the Sun's underlying visible disk or photosphere. Heat simply should not flow outward from a cooler to a hotter region. It violates the second law of thermodynamics and common sense as well. After more than a half-century of speculations, scientists still could not explain precisely how the million-degree corona is heated.

The high-temperature atmosphere creates an outward pressure that enables its outward expansion in regions of open magnetic flux, filling the solar system with a ceaseless flow – called the solar wind – that contains electrons, ions, and magnetic fields. Early spacecraft obtained important information about the speed and particle content of the solar wind, but the exact sources of all the wind's components and the mechanisms for accelerating it to high velocities remained longstanding mysteries.

Without warning, the relatively calm solar atmosphere can be torn asunder by sudden, catastrophic outbursts of awesome scale. These short-lived solar flares can flood the solar system with intense radiation and energetic particles, releasing energy equivalent to millions of 100-megaton hydrogen bombs exploding at the same time.

The X-rays from a solar flare modify our atmosphere, disrupt radio communications, and alter satellite orbits. The energetic particles that are hurled out into interplanetary space during solar flares can threaten unprotected astronauts traveling beyond the safety of the Earth's magnetic field to the Moon or Mars. These flare

Fig. 1.3 The ultraviolet Sun. This composite image, taken by two instruments aboard the *SOlar and Heliospheric Observatory*, abbreviated *SOHO*, reveals the ultraviolet light of the Sun's atmosphere from the base of the corona to millions of kilometers above the visible solar disk. The region outside the black circle, obtained by the UltraViolet Coronagraph Spectrometer, abbreviated UVCS, instrument, shines in the ultraviolet light emitted by oxygen ions flowing away from the Sun to form the solar wind. The inner image, obtained by the Extreme-ultraviolet Imaging Telescope, or EIT for short, shows the ultraviolet light emitted by iron ions at a temperature near 2×10^6 K. The overall structure is controlled by the solar magnetic field. Dark regions, known as coronal holes, have open magnetic fields that extend into interplanetary space; they are the source regions of the fast solar wind. Active regions are closed magnetic structures with very strong magnetic fields, and they mark the location of explosive solar outbursts known as solar flares. The largest closed-field structures are found at the base of coronal steamers, shown here extending out near the solar equator; they are related to the gusty, slow wind. [Courtesy of the *SOHO* UVCS consortium (*outer region*) and the *SOHO* EIT consortium (*inner region*). *SOHO* is a project of international cooperation between ESA and NASA.]

particles can also damage or destroy Earth-orbiting satellites used for communication, navigation, and military reconnaissance and surveillance.

Arching magnetic fields also suspend masses of relatively cool and dense material within the million-degree corona. These filaments are huge structures, bigger than active-region coronal loops and sometimes large enough to stretch from the Earth to the Moon. A quiescent filament can remain for weeks or even months at a time, never falling down, and then suddenly lose its equilibrium as the result of some sort of magnetic or kink instability, and erupt, carrying vast quantities of material into space.

Other explosive outbursts from the Sun hurl billions of tons of coronal gases into interplanetary space, producing strong gusts in the steady outward flow of the relentless solar wind. These giant magnetic bubbles, called coronal mass ejections, expand as they propagate outward from the Sun to rapidly rival it in size. They are often associated with the eruption of an underlying filament and are frequently accompanied by solar flares. Their associated shocks accelerate and propel vast quantities of high-speed particles ahead of them. When impacting the Earth with the right magnetic orientation, coronal mass ejections can produce spectacular auroras, create intense magnetic storms on Earth, and disrupt satellites, radio communications, and Earth's power systems.

Energetic particles associated with coronal mass ejections can also be hazardous to spacecraft and astronauts traveling to the Moon or Mars and exploring their surfaces. Yet, no one has been able to predict exactly when either solar flares or coronal mass ejections will occur or exactly how they are created. So, despite their vital importance for space weather, scientists have not yet been able to reliably warn and protect astronauts and spacecraft from these powerful outbursts from the Sun.

In summary, there were many perplexing enigmas that resulted from our new perception of the Sun. These unsolved puzzles include the internal generating mechanism for the Sun's all-important magnetic field, the heating of the million-degree corona, the origin and driving force of the solar wind, and the triggering and energy source of the Sun's unpredictable explosions, the solar flares and coronal mass ejections. While these mysteries were quite well established empirically, very little was understood about the underlying physical causes. To clarify and help solve many of these outstanding mysteries – and to better predict the Sun's impact on our planet – nine solar missions have been employed. Listed in the order of the date of their first solar observations, they are named *Yohkoh, Ulysses, Wind, SOHO, ACE, TRACE, RHESSI, Hinode,* and *STEREO*, with launch dates and references to instrument papers given in Table 1.1.

As illustrated in Fig. 1.4, these nine solar missions have been launched near either the minimum or the maximum of the Sun's 11-year solar activity cycle, sometimes viewing an entire subsequent cycle. This cycle is illustrated by the changing number of sunspots, but all forms of solar activity vary in step with it, including solar flares, coronal mass ejections, and even the total luminous output of the Sun.

·The extraordinary results of the *Yohkoh, Ulysses, Wind, SOHO, ACE, TRACE, RHESSI, Hinode,* and *STEREO* spacecraft are due to observations at wavelengths that do not reach the ground, such as extreme-ultraviolet or soft X-ray radiation, by direct detection of solar wind particles, and by observing visible sunlight in ways that are impossible to achieve on the ground, such as high-resolution observations unimpeded by the Earth's obscuring atmosphere, precise detection of small variations in the Sun's luminous output, and stereoscopic observations of coronal mass ejections.

Virtual observatories now provide a way to find and access the data from all of these missions, with tools for analyzing their results. The Virtual Heliospheric Observatory is at http://vho.nasa.gov, and the Virtual Space Physics Observatory can be found at http://vspo.gsfc.nasa.gov. The Virtual Solar Observatory has three

Table 1.1 Launch dates and instrument papers for nine solar missions

Name	Launch date	Instrument papers
Yohkoh ("*sunbeam*") (Formerly *Solar-A*)	30 August 1991	Y. Ogawara et al., T. Kosugi et al., S. Tsuneta et al., M. Yoshimori et al., and J. L. Culhane et al., *Solar Physics* **136**, 1–104 (1991). Also see L. W. Acton et al., *Science* **258**, 618–625 (1992).
Ulysses	6 October 1990 (First polar passage 13 September 1994)	*Astronomy and Astrophysics Supplement Series, Ulysses Instruments Special Issue* **92**, No. 2, 207–440 (1992).
Wind	1 November 1994	M. H. Acuña et al., *Space Science Reviews* **71**, 5–21 (1995).
SOHO (*SOlar and Heliospheric Observatory*)	2 December 1995	V. Domingo et al., and numerous others, *Solar Physics* **162**, 1–531 (1995).
ACE (*Advanced Composition Explorer*)	25 August 1997	E. C. Stone, and nine co-investigators their colleagues, *Space Science Reviews* **86**, No. 1–4 (1998).
TRACE (*Transition Region and Coronal Explorer*)	1 April 1998	B. N. Handy et al., *Solar Physics* **187**, 229–260 (1999). Also see C. J. Schrijver et al., *Solar Physics* **187**, 261–302 (1999).
RHESSI (*Ramaty High Energy Solar Spectroscopic Imager*)	5 February 2002	R. P. Lin et al., D. M. Smith et al., G. J. Hurford et al., *Solar Physics* **210**, 3–32, 33–60, 61–86 (2002).
Hinode ("sunrise") (Formerly *Solar-B*)	23 September 2006	T. Kosugi et al. (overview), J. L. Culhane et al. (EIS), L. Golub et al. (XRT), *Solar Physics* **243**, 3–17, 19–61 and 63–86 (2007), and S. Tsuneta et al. (SOT), *Solar Physics* **249**, 167–196 (2008). Also see K. Shibata et al., *Astronomical Society of the Pacific Conference Series* **369**, 1–593 (2007).
STEREO (*Solar TErrestrial RElations Observatory*)	25 October 2006	M. L. Kaiser et al. (introduction), J. L. Bougeret et al. (SWAVES), H. B. Galvin et al. (PLASTIC), R. A. Howard et al. (SECCHI), and J. G. Luhmann et al. (IMPACT), *Space Science Reviews* **136**, Issues 1–4, 5–16, 487–586, 437–486, 67–115, 117–184 (2008).

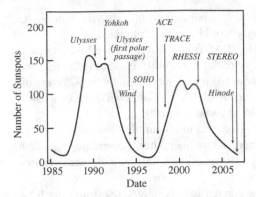

Fig. 1.4 Spacecraft launch dates and the Sun's activity cycle. The launch dates of nine solar missions are shown on this plot of the international sunspot number, recorded at monthly intervals and smoothed. Two 11-year solar cycles of magnetic activity are included, each from a maximum in sunspot numbers to a minimum number and back to a maximum. The acronyms are *SOHO* for the *SOlar and Heliospheric Observatory*, *ACE* for the *Advanced Composition Explorer*, *TRACE* for the *Transition Region And Coronal Explorer*, *RHESSI* for the *Ramaty High Energy Solar Spectroscopic Imager*, and *STEREO* for the *Solar TErrestrial Relations Observatory*. Although *Yohkoh* was shut down in December 2001, all of the other eight spacecraft were still observing the Sun throughout 2007

components, at the Goddard Space Flight Center, http://umbra.nascom.nasa.gov/vso, at the National Solar Observatory, http://vso.nso.gov, and at Stanford University http://vso.stanford.edu.

We will next give an account of the scientific objectives of each spacecraft, together with an outline of some of their significant discoveries and descriptions of their instruments. The main part of this book will then explain their fascinating results in greater detail.

1.2 *Yohkoh* Detects Unrest on an Awesome Scale

Scientists in Japan, the land of the rising Sun, have been particularly interested in sudden, powerful outbursts called solar flares. These dynamic, short-lived events involve the sudden release of enormous amounts of energy, both as intense radiation in all regions of the electromagnetic spectrum and as accelerated particles whose speeds approach the velocity of light. Such flares are best studied during the maximum in the 11-year solar activity cycle, when they occur most frequently.

Following its successful *Hinotori*, the Japanese word for "fire-bird," solar flare mission, launched on 21 February 1981, near a solar maximum, the Institute of Space and Astronautical Science, abbreviated ISAS, in Japan organized a new mission, initially called *Solar-A*. It was designed to investigate high-energy X-ray and gamma ray radiations from the Sun. Solar flares would be studied with high angular and energy resolution during the next maximum in solar activity; quiescent, non-flaring structures and pre-flare activity would also be scrutinized.

The 390-kg spacecraft was launched by ISAS from the Kagoshima Space Center on 30 August 1991, into a 96-min, nearly circular Earth orbit. Following launch, the mission was renamed *Yohkoh*, the Japanese word for "sunbeam." After a decade of studying high-energy processes on the Sun, *Yohkoh* was shut down due to a loss of control during an annular solar eclipse on 14 December 2001.

The primary scientific objective of *Yohkoh* was to obtain high-resolution spatial and temporal information about high-energy flare phenomena, permitting detailed scrutiny of where and how flare energy is released and flare particle acceleration takes place.

To accomplish its objectives, *Yohkoh* carried four co-aligned soft and hard X-ray imaging and spectrometry instruments contributed by Japan, the United Kingdom, and the United States. The X-ray radiation resembles light waves, except with shorter wavelengths and greater energy. A soft X-ray is one of relatively long wavelength and low energy; this form of electromagnetic radiation has energies of 1–10 keV and wavelengths between 10^{-9} and 10^{-10} m. Hard X-rays have shorter wavelengths, between 10^{-10} and 10^{-11} m, and higher energy, between 10 and 100 keV. Gamma rays are even shorter and more energetic. (The kilo-electron volt, abbreviated keV, is a unit of energy. For conversions, $1 \text{ keV} = 1.602 \times 10^{-16}$ J and the photon energy, E, associated with a wavelength λ in nm is $E = (1.986 \times 10^{-16}/\lambda)$ J, where 1 nm is 10^{-9} m.)

The *Yohkoh* telescopes had full-Sun fields of view with then unparalleled angular and spatial resolution. The Hard X-ray Telescope, abbreviated HXT, imaged hard X-rays emitted by high-speed electrons accelerated in impulsive flares. Both the flaring and the quiescent, or non-flaring, Sun were detected at soft X-ray wavelengths with a rapid, uniform rate using the Soft X-ray Telescope, or SXT for short. It routinely imaged high-temperature gas, above $2–3 \times 10^6$ K, across the solar disk. The SXT observed the Sun with an accuracy and steadiness that permitted images at invisible X-ray wavelengths which are as sharp and clear as pictures made in visible wavelengths from telescopes on the ground.

Yohkoh also contained four Bragg Crystal Spectrometers, abbreviated BCS, which were capable of measuring violent gas motions and upflowing gas at multi-million degree temperatures in solar flares.

The soft X-ray and hard X-ray instruments on *Yohkoh* have been used to compare the location and geometry of flaring sources to the topology of the solar magnetic field. They have shown that solar flares and coronal mass ejections can be triggered by magnetic reconnection, where oppositely directed magnetic fields merge together, releasing the necessary energy at the place where they touch. This site of magnetic energy release and particle acceleration is often just above low-lying coronal loops, and attributed to the interaction of such loops. *Yohkoh's* Soft X-ray Telescope has additionally shown that solar flares or coronal mass ejections may be triggered when the bright X-ray emitting coronal loops are twisted into a helical, kinked, and twisted S-shape.

The Soft X-ray Telescope has obtained more than six million images that have been used to study how the hot gases evolve, interact, and change their magnetic structure when a flare is not in progress, showing that the quiet Sun within and away from active regions is in a continued state of change on both small and global scales. Dynamic processes and transient events, such as bright points, faint nanoflares, and jets, have been discovered and monitored. The intrepid spacecraft has shown that the ever-changing corona has no permanent features and is never still, always in a continued state of metamorphosis.

Yohkoh was also the first spacecraft to continuously observe the Sun in X-rays over an entire solar activity cycle. Since the million-degree corona emits soft X-rays, its Soft X-ray Telescope has provided information on the heating and expansion of the high-temperature gas and tracked its dramatic year-to-year evolution over the 11-year cycle of solar magnetic activity (Fig. 1.5).

The Bragg Crystal Spectrometer's aboard *Yohkoh* detected gas heated from 10,000 K in the chromosphere up to 20×10^6 K, flowing into flaring coronal loops at typical speeds of 350 km s^{-1} and producing the copious soft X-ray emission seen during flares. These motions were studied and compared with the prediction of theoretical models of flares. Turbulent motions in flaring gas of unknown origin on the order of 150 km s^{-1} and decaying to 60 km s^{-1} or less were also observed from the onset of flares into the decay phase. The BCS instruments were able to show

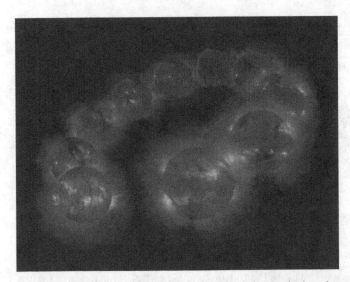

Fig. 1.5 X-ray view of the solar cycle. Dramatic changes in the solar corona are revealed in this 10-year montage of images from the Soft X-ray Telescope, abbreviated SXT, aboard *Yohkoh*. The images are spaced in 4-month intervals from the time of the satellite's launch in August 1991, at the maximum phase of the 11-year solar activity cycle (*left*), through the minimum phase (*center*) and on to the next maximum (*right*). The bright glow of X-rays near activity maximum comes from very hot, million-degree coronal gases that are confined within the powerful magnetic fields of active regions. Near the cycle minimum, the active regions have almost disappeared, and there is an overall decrease in X-ray brightness by a factor of 100. (Courtesy of JAXA, NASA and LMSAL.)

that the bulk of the gas in solar flares reaches a maximum temperature of about 23×10^6 K, or hotter than the center of the Sun. Some stellar flares can be much hotter, and the reason for the solar limit is still unknown.

Yohkoh provides an excellent example of international cooperation. Even with the challenge of a formidable language barrier, the *Yohkoh* team of scientists from Japan, America, and England has collaborated and shared resources in a collegial and trusting atmosphere, working shoulder-to-shoulder operating the spacecraft, analyzing data and publishing results in an extremely austere staffing environment.

Yohkoh's principal telemetry and operation control center was provided by ISAS near Tokyo, Japan, while NASA obtained the telemetry data captured during passes over its Deep Space Network ground stations in Goldstone, California, Canberra, Australia, and Madrid, Spain. The operation of the *Yohkoh* mission has been the responsibility of the *Yohkoh* science team. Interested scientists throughout the world have shared data from all four instruments. All of the useable scientific data obtained from all of the *Yohkoh* instruments are available at http://solar.physics.montana.edu/ylegacy/, in forms convenient for use by non-experts. This *Yohkoh* legacy archive includes flare catalogs, images and movies from the Soft X-ray Telescope, archival documentation, and weekly science highlights from 1997 to 2002.

Detailed descriptions of the *Yohkoh* instruments are provided in Table 1.2. The Principal Investigators of these instruments and their institutions are given in

Table 1.2 *Yohkoh's* instruments arranged alphabetically by acronym[a]

Instrument	Measurement
BCS	The Bragg Crystal Spectrometer measured X-ray spectral lines of highly ionized iron, Fe XXV and Fe XXVI, calcium, Ca XIX, and sulfur, S XV, between 0.18 and 0.51 nm with a time resolution as short as 0.125 s and across the full solar disk. Subtract one from the ion symbol to get the number of missing electrons.
HXT	The Hard X-ray Telescope imaged flare radiation with energy from 20 to 80 keV in four channels, with an angular resolution of about 7 arcseconds, a time resolution of 0.5 s and a field of view that includes the entire visible disk.
SXT	The Soft X-ray Telescope achieved 4 s of arc spatial resolution and 2 s temporal resolution, detecting radiation with energy between 0.6 and 4 keV (2.07–0.31 nm) with a wide dynamic range (up to 200,000). It rendered images taken at a uniform rate across the full solar disk for both the faint, quiescent and intense flaring X-ray structures with temperatures above 2–3×10^6 K.
WBS	The Wide Band Spectrometer measured X-rays and gamma-rays from 3 keV to 30 MeV during solar flares, and was also sensitive to neutrons emitted during flares; it observed the full solar disk with a time resolution as short as 0.125 s.

[a]The hardware for BCS was provided by the United Kingdom, and the U.S. Naval Research Laboratory funded the structure and crystals for the BCS, and also data analysis software later. The HXT and WBS were contributed by Japan's ISAS and the SXT was prepared by the Lockheed-Martin Solar and Astrophysics Research Laboratory, the National Astronomical Observatory of Japan and the University of Tokyo with the support of NASA and ISAS. The mission overview is given by Y. Ogawara et al. (1991), and the BCS, HXT, SXT and WBS instruments, respectively, described by J. L. Culhane et al. (1991), T. Kosugi et al. (1991), S. Tsuneta et al. (1991) and M. Yoshimori et al. (1991).

Table 1.3 *Yohkoh* principal investigators and their institutions

Instrument	Principal investigator	Institution
BCS	Eijiro Hiei	National Astronomical Observatory of Japan, Mitaka, Tokyo, Japan
	J. Leonard Culhane	Mullard Space Science Laboratory, University College, London, United Kingdom
	George A. Doschek	Naval Research Laboratory, Washington, District of Columbia, United States
HXT	Kazuo Makishima	University of Tokyo, Japan
	Takeo Kosugi	University of Tokyo, Japan
SXT	Loren W. Acton	Montana State University, Bozeman, Montana, United States
	Tadashi Hirayama	National Astronomical Observatory of Japan, Mitaka, Tokyo, Japan
	Saku Tsuneta	University of Tokyo, Japan
WBS	Masato Yoshimori	Rikkyo University, Tokyo, Japan
	Jun Nishimura	Institute of Space and Astronautical Science, Japan

Table 1.3. The Project Scientist for *Yohkoh* was Yutaka Uchida of the University of Tokyo. *Yohkoh's* accomplishments are described in greater detail within subsequent chapters of this book.

1.3 *Ulysses* Moves into Unexplored Territory

Ionized gases expand away from the Sun in all directions, carving out a bubble in the space around our star, which extends far beyond the planets. This vast region, known as the heliosphere, is filled with the Sun's tenuous, expanding atmosphere and dominated by the continuous outflow of its winds. They are composed of electrons, protons, and heavier ions, technically called plasma, and magnetic fields. Yet, until quite recently, most of our knowledge of the immense heliosphere came from interplanetary spacecraft that traveled within – or very close to – the ecliptic, the plane of the Earth's orbital motion about the Sun. This is because spacecraft often rendezvous with another planet whose orbit lies near that plane, and also because their launch vehicles obtain a natural boost by traveling in the same direction as the Earth's spin and in the plane of the Earth's orbit around the Sun.

Because the solar equator nearly coincides with the ecliptic, scientific probes have only sampled a thin plane that intersects the Sun at its midsection. They have been unable to directly sample any regions more than 16° north or south solar latitude, confining their perspective to a narrow, two-dimensional, "flat-land" slice of the heliosphere. (The solar latitude is the angular distance from the plane of the Sun's equator.) Then in the 1990s a single craft, *Ulysses*, ventured out of this thin zone into uncharted interplanetary space, moving up to 80° north and south of the solar equator, and probing the third dimension of the heliosphere for the first time.

The *Ulysses* mission is a joint undertaking of the European Space Agency, abbreviated ESA, and NASA. Dornier Systems of Germany built the spacecraft for ESA, while NASA provided the launch via the *Space Shuttle Discovery* and the U.S. Department of Energy supplied a radioisotope thermoelectric generator that powers the spacecraft. European and United States investigators provided science instruments, and *Ulysses* was operated from the Jet Propulsion Laboratory in California by a joint team from ESA and NASA.

The spacecraft is named *Ulysses* in honor of the mythical Greek adventurer Ulysses, whose exploits and long wanderings are recounted in Homer's *Iliad* and *Odyssey*. In the 26th Canto of Dante Alighieri's *Inferno*, Ulysses recalls his restless desire:

To venture the uncharted distances.

Exhorting his friends to embark on one last occasion.

Of feeling life and the new experience.
Of the uninhabited world behind the Sun.
To follow after knowledge and excellence.

Like the legendary explorer, the *Ulysses* spacecraft has traveled into previously unexplored regions, improving out knowledge of them.

The thrust and high speed needed to send a probe directly on a trajectory over the poles of the Sun are beyond the capability of today's most powerful rockets. To move out of the ecliptic, *Ulysses* therefore had to travel out to Jupiter, using the planet's powerful gravity like a slingshot, to accelerate and propel the spacecraft into an inclined orbit that sent it under the Sun.

Launched by NASA's *Space Shuttle Discovery* on 6 October 1990, *Ulysses* encountered Jupiter on 8 February 1992, which hurled the spacecraft into an elliptical orbit about the Sun with a period of 6.3 years. Its distance from the Sun varies from 5.4 AU, near Jupiter's orbit, to 1.3 AU at its closest approach to the Sun; the distance over the Sun's polar regions ranges from 2.0 AU (north) to 2.3 AU (south). The astronomical unit, 1.0 AU, is the mean distance from the Sun to the Earth, or about 150×10^6 (1.5×10^8) km. Thus, at the time of polar passage, *Ulysses* was not close to the Sun; it was more than twice as far from the Sun as the Earth.

After traveling about 3×10^9 km for nearly 4 years since leaving Earth, brave stalwart *Ulysses* reached the summit of its trajectory beneath the south polar regions of the Sun on 13 September 1994. In response to the Sun's gravitational pull, *Ulysses* arched up toward the solar equator and crossed over the north polar regions on 31 July 1995. The spacecraft's trajectory took it back toward Jupiter's orbit, and then on a return trip over the solar poles, passing beneath the Sun's southern polar regions on 27 November 2000, and over its north polar regions on 13 October 2001. *Ulysses* then continued on its epic journey, traveling away from the Sun and back again to once more go where no other spacecraft has gone before, swooping under the Sun's south polar regions on 17 February 2007, and over its north polar regions on 14 January 2008. The *Ulysses* mission ended on 1 July 2008.

The primary scientific objective of the *Ulysses* spacecraft is to explore and define the heliosphere in three dimensions, characterizing it as a function of solar latitude and time. The solar wind speed and magnetic field strength have been measured from pole to pole, compositional differences have been established for the expanding solar atmosphere, and the flux of galactic cosmic rays has been measured as a function of solar latitude. Interplanetary dust, solar energetic particles, plasma waves, and solar radio bursts have also been measured.

In more than 17 years of *Ulysses* operations, the spacecraft has completed three polar orbits, providing a unique perspective from which to study the Sun and its effect on surrounding space. It has followed the complete course of the 11-year solar activity cycle, first during a minimum in the number of sunspots (in 1994–1995), then when the sunspot number was at its maximum (2000–2001).

Ulysses thus measured the distribution of solar wind velocities at both the minimum and maximum of the solar cycle (Fig. 1.6). During its first polar orbit, at sunspot minimum, it found fast wind over the poles, and slower, variable wind confined near the solar equator. The second polar orbit, performed near solar maximum, revealed variable solar wind at all latitudes including near the poles. During this second polar orbit at sunspot maximum, the Sun's polar fields disappeared and then reappeared with the opposite polarity or direction. The third orbit over the solar poles is also at an activity minimum (2007–2008), but under very different circumstances after a reversal of the magnetic poles of the Sun, permitting studies of the changed magnetic field and its effect on the solar wind, galactic cosmic rays, and solar energetic particles.

Ulysses has also provided new insight to the composition of the solar wind, showing that it contains ions originating from both the Sun and the interstellar gas.

As the solar wind streams away from the Sun, it pulls the Sun's magnetic field radially outward, and the Sun's rotation bends and coils the radial pattern into a spiral shape within the plane of the solar equator. *Ulysses* found indications for deviations from this simple model, related to the fact that one end of the solar magnetic field remains firmly rooted in the ever-changing solar photosphere and below, while the other end is extended and stretched out into space by the solar wind. Random motions and systematic differential rotation of the magnetic anchoring points influence the three-dimensional structure of the heliospheric magnetic field, perhaps by causing direct magnetic connections at different latitudes far out into interplanetary space.

The instruments aboard *Ulysses* have additionally enhanced and extended previous investigations of interstellar atoms that have moved into the heliosphere, become ionized there, and then picked up by the solar wind and entrained in its flow. These pick-up ions have a different, low state of ionization from more highly ionized elements coming from the Sun. The pick-up helium ions have been carefully scrutinized; interstellar pick-up hydrogen was discovered; and the oxygen, nitrogen, and neon pick-ups observed. These investigations permitted study of the physical and

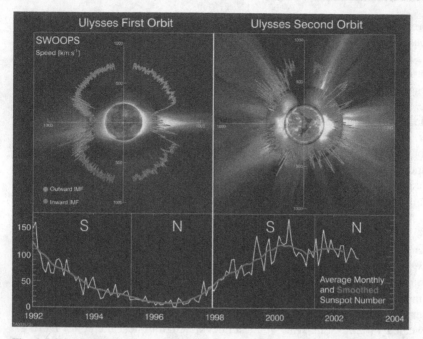

Fig. 1.6 Distributions of solar wind speeds at solar minimum and maximum. Plots of the solar wind speed as a function of solar latitude, obtained from two orbits of the *Ulysses* spacecraft (*top panels*). The north and south poles of the Sun are at the top and bottom of each plot, the solar equator is located along the middle, and the velocities are in units of kilometers per second, abbreviated km s^{-1}. The sunspot numbers (*bottom panel*) indicate that the first orbit (*top left*) occurred near a minimum in the 11-year solar activity cycle, and that the second orbit (*top right*) spanned a maximum in activity. Ultraviolet images of the solar disk and white-light images of the inner solar corona form a central backdrop for the wind speed data, and indicate probable sources of the winds. Near activity minimum (*top left*), polar coronal holes, with open magnetic fields, give rise to the fast, low-density wind streams, whereas equatorial streamer regions of closed magnetic fields are associated with the slow, gusty, dense winds. At solar maximum (*top right*), small, low-latitude coronal holes give rise to fast winds, and a variety of slow-wind and fast-wind sources resulted in little average latitudinal variation of solar wind speed. (Courtesy of David J. McComas and Richard G. Marsden. The *Ulysses* mission is a project of international collaboration between ESA and NASA. The central images are from the Extreme-ultraviolet Imaging Telescope and the Large Angle Spectrometric Coronagraph aboard the *SOlar and Heliospheric Observatory*, abbreviated *SOHO*, as well as the Mauna Loa K-coronameter.)

chemical properties of the nearby interstellar medium. In addition, singly ionized carbon was discovered, which has to originate very close to the Sun rather than from interstellar space.

As the result of coronal expansion and solar rotation, the fast and slow wind flows intersect, forming structures that co-rotate with the Sun. *Ulysses* has provided new insight to the three-dimensional nature of these co-rotating interaction regions, including their production of energetic particles and shocks and their relationship to the motion of open magnetic fields connected to the Sun.

Ulysses' formidable accomplishments have been achieved using four European and five American instruments that make the measurements of fundamental parameters as a function of distance from the Sun and solar latitude. They include devices that measure the solar wind speed, a pair of magnetometers that measure magnetic fields, and several instruments that study electrons, protons, and heavier ions that come both from the quiescent and active Sun and from the interstellar space, the galactic cosmic rays. The spacecraft radio is used to study the Sun's outer atmosphere, in addition to its communication functions. The acronyms and measurements of all of the instruments on *Ulysses* are given in Table 1.4, while the Principal

Table 1.4 *Ulysses'* instruments arranged alphabetically by acronym[a]

Instrument	Measurement
COSPIN	The COsmic ray and Solar Particle INstrument records ions with energies of 0.3–600 MeV per nucleon and electrons with energies of 4–2000 MeV.
DUST	Measures interplanetary dust particles with mass between 10^{-19} and 10^{-10} kg.
EPAC/GAS	The Energetic PArticle Composition and interstellar neutral GAS instrument records energetic ion composition for energies of 80 keV–15 MeV per nucleon and neutral helium atoms.
GRB	The Gamma Ray Burst instrument measures solar flare X-rays and cosmic gamma ray bursts with energies of 15–150 keV.
GWE	The Gravitational Wave Experiment records Doppler shifts in the satellite radio signal that might be due to gravitational waves.
HISCALE	The Heliosphere Instrument for Spectra, Composition, and Anisotropy at Low Energies instrument measures ions with energies of 50 keV–5 MeV and electrons with energies of 30–200 keV.
SCE	The Solar Corona Experiment uses the radio signals from the spacecraft to measure the density, velocity, and turbulence spectra in the solar corona and solar wind.
SWICS	The Solar Wind Ion Composition Spectrometer records elemental and ionic-charge composition, temperature, and the mean speed of solar wind ions for speeds of 145 (H^+) to 1350 (Fe^{+8}) km s^{-1}.
SWOOPS	The Solar Wind Observations Over the Poles of the Sun instrument measures ions from 237 eV to 35 keV per charge and electrons from 1 to 860 eV.
URAP	The Unified Radio And Plasma wave experiment records plasma waves at frequencies from 1 to 60 kHz, remotely senses traveling solar radio bursts exciting plasma frequencies from 1 to 940 kHz, and measures the electron density.
VHM/FGM	A pair of magnetometers, the Vector Helium Magnetometer and the scalar Flux Gate Magnetometer, that measure the magnetic field strength and direction in the heliosphere and the Jovian magnetosphere from 0.01 to 44,000 nT or from 10^{-7} to 0.44 G.

[a]The *Ulysses* instruments are described in the *Astronomy and Astrophysics Supplement Series* **92**, No. 2, 207–440 (1992). See for example the articles by A. Balogh et al. (1992), S. J. Bame et al. (1992), J. A. Simpson et al. (1992), and R. G. Stone et al. (1992). Particle energies are often expressed in electron volts, or eV, where $1 \, eV = 1.602 \times 10^{-19} \, J = 1.602 \times 10^{-12}$ erg. High-energy particles can have energies expressed in kilo-electron volts, or keV, and mega-electron volts, or MeV, where $1 \, keV = 10^3 \, eV = 1.602 \times 10^{-16} \, J$ and $1 \, MeV = 10^6 \, eV = 1.602 \times 10^{-13} \, J$. The magnetic field strength is given in units of Tesla, where $1 \, T = 10,000 \, G$ or $1 \, nT = 10^{-9} \, T = 10^{-5} \, G$

Table 1.5 *Ulysses* principal investigators and their institutions

Instrument	Principal investigator	Institution
COSPIN	R. Bruce McKibben	University of New Hampshire, Durham, New Hampshire, United States
DUST	Eberhard Grün	Max-Planck-Institut für Kernphysik, Heidelberg, Germany
EPAC	Erhard Keppler	Max-Planck-Institut für Sonnensystemforschung Katlenburg-Lindau, Germany
GRB	Kevin C. Hurley	University of California, Berkeley, United States
GWE	Bruno Bertotti	Universitá di Pavia, Italy
HISCALE	Louis J. Lanzerotti	New Jersey Institute of Technology, Newark, New Jersey, United States
SCE	Mike K. Bird	Universität Bonn, Germany
SWICS	Johannes Geiss	International Space Sciences Institute, Bern, Switzerland
	George Gloeckler	University of Michigan, Ann Arbor, Michigan, United States
SWOOPS	Dave J. McComas	Southwest Research Institute, San Antonio, Texas, United States
URAP	Robert J. MacDowall	Goddard Space Flight Center, Greenbelt, Maryland, United States
VHM/FGM	Andre Balogh	Imperial College, London, United Kingdom

Investigators and their institutions are tabulated in Table 1.5. The Project Scientists are Edward J. Smith at the Jet Propulsion Laboratory and Richard Marsden of the European Space Agency. Fundamental information about the various experiments can be found on the *Ulysses* home page (http://ulysses.jpl.nasa.gov/) or (http://helio.estec.esa.nl/ulysses/). *Ulysses'* accomplishments are discussed in subsequent chapters of this book. Not included here are *Ulysses'* study of cosmic gamma-ray bursts and its investigations of the Jovian magnetosphere obtained during the Jupiter flyby.

1.4 *Wind* Investigates the Sun's Varying Input to Earth

Light and heat are not the Sun's only contribution to our home planet Earth. There is the rarefied solar wind of charged particles, carrying the Sun's magnetic fields with them, and the much more energetic electrons, protons, and heavier nuclei hurled into interplanetary space by explosive outbursts on the Sun, the solar flares, and coronal mass ejections. We are protected from these invisible, subatomic particles by our planet's magnetic field, which deflects them around the Earth. So we reside in a magnetic cocoon, called the magnetosphere, which shields the Earth from the full force of the Sun's stormy weather and tempestuous behavior.

NASA's *Wind* mission was designed to make accurate measurements of particles and magnetic fields upstream from a place where the solar wind first meets the Earth's magnetosphere. The 1250-kg spacecraft was carried into space aboard a Delta II rocket from the Cape Canaveral Air Station, Florida, on 1 November 1994. It was initially sent into a lunar swing-by orbit in which the Moon's gravity helped propel *Wind* through the sunlit, dayside of the Earth's magnetosphere. The spacecraft was then placed just outside the Earth's magnetosphere on its sunward side, at a special location called Lagrange L_1 where the Sun and Earth's gravitational pull cancel out allowing the spacecraft to hover, rather than orbit around Earth. Thus, *Wind* is capable of providing nearly continuous, direct, in situ measurements of the solar wind, magnetic fields, and energetic particles arriving at the magnetosphere, which it has been doing for more than a decade. (The spacecraft was redirected into the Earth's deep magnetotail on the planet's night-side for about 2 months in late 2003 and early 2004, before resuming its long vigil on the dayside.)

These *Wind* observations complement those made by the *Polar* spacecraft, which looked down at the Earth's magnetic polar cusps where charged particles can penetrate into the Earth's atmosphere, and by the *Geotail* spacecraft, which skimmed the outer edge of the magnetosphere and dove deep into the magnetotail, more than a million kilometers from Earth. And the *SOHO* spacecraft was also included, to identify and analyze the solar events that contribute to these terrestrial effects. It was all part of an International Solar-Terrestrial Physics, abbreviated ISTP, Program in the 1990s that combined measurements from all four spacecraft to determine how the Sun generates energy and feeds it into the space near Earth, and how energy is coupled back and forth between the gusty, incident solar wind and the magnetosphere on one hand and between the magnetosphere and the Earth's upper atmosphere on the other hand.

Since launch, *Wind* has also made important measurements that supplement the mission's original plan. It has investigated shocks generated by coronal mass ejections, using their radio signals to track them from launch in the corona through interplanetary space to the Earth, and showed that the most powerful shocks produce a wider wavelength range of radio emission. Radio signals produced by these shocks are triangulated using instruments aboard the *Wind, Ulysses, Cassini*, and the twin *STEREO* spacecraft, permitting a three-dimensional determination of their trajectory.

The characteristics of magnetic clouds, the special subclass of coronal mass ejections that preserve their tubular magnetic flux rope geometry all the way to the Earth and beyond, have been delineated, including their magnetic connections back to the Sun. And *Wind* has also made in situ measurements of impulsive bursts of energetic electrons and ions that reach the Earth's orbit following solar flares, while tracking the radio emission of these electrons as they traveled through space.

One recent very intriguing discovery, resulting from measurements by both the *Wind* and *ACE* spacecraft, was solar-wind magnetic fields that merge and join together near the Earth's orbit, in long, steady reconnection layers that stretch out for hundreds of Earth radii. This magnetic reconnection was previously thought to occur only within very small regions in a very short, patchy manner. The discovery of the longer, steady reconnection process, which is thought to be also in operation in the solar corona, has significant implications for our understanding of how the Sun can initiate and launch coronal mass ejections.

One of *Wind's* instruments has additionally shown that the slowest solar winds contain the greatest amount of helium, while the faster winds have the least abundance of helium. This gives clues to the source regions of the slow solar wind, which is currently not known in sufficient detail.

Table 1.6 *Wind's* instruments arranged alphabetically by acronym[a]

Instrument	Measurement
EPACT	The Energetic Particles: Acceleration, Composition, and Transport instrument makes comprehensive observations of solar and interplanetary particles over wide ranges of charge, mass, and intensity, but particularly during the more intense solar outbursts. http://lheawww.gsfc.nasa.gov/docs/gamcosray/lecr/EPACT/epact.html
KONUS	The KONUS instrument provides continuous monitoring of cosmic gamma-ray bursts and solar flares in the energy range 10 keV–10 MeV. http://heasarc.gsfc.nasa.gov/docs/heasarc/missions/wind.html
MFI	The Magnetic Field Experiment measures the interplanetary vector magnetic field. http://lepmfi.gsfc.nasa.gov/mfi/windmfi.html
SMS	The SMS instrument consists of three different sensors: the Solar Wind Ion Spectrometer (SWICS), the High Mass Resolution Spectrometer (MASS), and the Supra-Thermal Ion Composition Spectrometer (STICS). Their combined data determine the composition of the solar wind and of solar and interplanetary energetic particles. http://solar-heliospheric.engin.umich.edu/wind/
SWE	The Solar Wind Experiment measures the velocity distribution functions of electrons, protons and helium ions within the solar wind. http://web.mit.edu/afs/athena/org/s/space/www/wind.html
3DP	The Three-Dimensional Plasma experiment is designed to measure the full three-dimensional distribution of suprathermal electrons and ions at energies from a few eV to over several hundred keV. Its high sensitivity, wide dynamic range, and good energy and angular resolution make it especially capable of detecting and characterizing the numerous populations of particles that are present in interplanetary space at energies above the bulk of the solar wind particles and below the energies typical of most cosmic rays. http://sprg.ssl.berkeley.edu/wind3dp/
TGRS	The Transient Gamma-Ray Spectrometer makes a high-resolution spectroscopic survey of cosmic gamma-ray bursts, but also detects gamma-ray lines emitted during solar flares. http://heasarc.gsfc.nasa.gov/docs/heasarc/missions/wind.html
WAVES	The WAVES instrument provides comprehensive coverage of radio and plasma wave phenomena in the frequency range from a fraction of a Hertz up to about 14 MHz for the electric field and 3 kHz for the magnetic field. It provides two-dimensional radio tracking of shocks driven by coronal mass ejections, and when combined with *Ulysses* and/or *Cassini* radio observations determines the three-dimensional trajectory of these shocks. http://lep694.gsfc.nasa.gov/waves/waves.html

[a]*Wind* instrument descriptions are given at http://pwg.gsfc.nasa.gov/wind_inst.shtml

Wind continues with its direct measurements of the incoming solar wind plasma, its magnetic fields and energetic particles, but as a result of the spacecraft's recent accomplishments, the mission's scientific objectives now include studies of the interplanetary manifestations of coronal mass ejections, with their associated shocks and magnetic clouds, investigations of large-scale solar-wind structures including interplanetary magnetic reconnection events and turbulent structures, and measurements of energetic particles that arrive at the Earth's orbit, accelerated by a rich variety of shocks throughout the heliosphere.

Driven by coronal mass ejections, magnetic clouds, and/or interacting solar wind streams, interplanetary shocks pump up a particle's energy before arriving at Earth. And when they reach the planet, *Wind* detects them. Such energized electrons, protons, and other nuclei have been dubbed solar energetic particles, to distinguish them from more energetic and less abundant cosmic rays that arrive from interstellar space. Scientists are now using *Wind* data to help determine where the solar energetic particles come from, how they are accelerated, and how they escape and propagate from solar flares or coronal mass ejections to Earth. This is accomplished by combining *Wind* measurements with those from the *ACE, STEREO* and/or *Ulysses* spacecraft at the Earth's orbit or beyond, and with those of *RHESSI, TRACE*, or *SOHO* that look back to the Sun, the source of it all.

All of *Wind*'s accomplishments are discussed in greater detail in subsequent chapters of this book. The acronyms and measurements of its instruments are given in Table 1.6, while the instrument Principal Investigators and their institutions are provided in Table 1.7. The NASA Project Scientist for *Wind* is Adam Szabo, and Mario H. Acuña et al. have described the Global Geospace Science Program and the *Wind* instruments in *Space Science Reviews* **71**, 5–21 (1995). Additional

Table 1.7 *Wind* principal investigators and their institutions

Instrument	Principal investigator	Institution
EPACT	Tycho T. von Rosenvinge	Goddard Space Flight Center, Greenbelt, Maryland, United States
KONUS	Evgenii P. Mazets	Ioffe Institute, St. Petersburg, Russia
	Thomas L. Cline	Goddard Space Flight Center, Greenbelt, Maryland, United States
MFI	Ronald P. Lepping	Goddard Space Flight Center, Greenbelt, Maryland, United States
SMS	George Gloeckler	University of Michigan, Ann Arbor, Michigan, United States
SWE	Keith W. Ogilvie	Goddard Space Flight Center, Greenbelt, Maryland, United States
3DP	Robert P. Lin	University of California, Berkeley, United States
TGRS	Bonnard J. Teegarden	Goddard Space Flight Center, Greenbelt, Maryland, United States
WAVES	Jean-Louis Bougeret	Observatoire de Paris, Meudon, France
	Michael L. Kaiser	Goddard Space Flight Center, Greenbelt, Maryland, United States

information about the *Wind* mission and its data may be found at http://wind.nasa.gov. Near real-time *Wind* data are available at http://pwg.gsfc.nasa.gov/windnrt/.

1.5 The Sun Does Not Set for *SOHO*

A great observatory of the Sun, the *SOlar and Heliospheric Observatory*, or *SOHO* for short, has stared the Sun down with an unblinking eye for more than a decade. This 1850-kg spacecraft was launched from Cape Canaveral, Florida, aboard an Atlas II-AS vehicle on 2 December 1995 and reached its permanent position on 14 February 1996, beginning normal operations in May 1996. The one-billion-dollar mission is a project of international cooperation between the European Space Agency and NASA.

SOHO is located sunward at about 1.5×10^6 km out in space, or at about one percent of the way to the Sun. At this place, only about 1/100th of the distance to the Sun, the combined gravity of the Earth and Sun keep *SOHO* in an orbit locked to the Earth–Sun line. Such a position is known in astronomy as the inner Lagrangian, or L_1, point after the French mathematician Joseph Louis Lagrange, who first calculated its location near the end of the 19th century. From this strategic vantage point, *SOHO* can monitor the Sun with a continuous, uninterrupted view 24 h a day, every day, all year round.

The spacecraft has been keeping watch over the Sun through all its tempestuous seasons – from a lull in its 11-year magnetic activity cycle at the time of launch to an activity maximum near the turn of the century and on into the next activity minimum. Except, that is, for a 3-month interlude when controllers lost contact with the spacecraft (Focus 1.1).

Focus 1.1

SOHO Lost in Space and Recovered from Oblivion

After more than 2 years of uninterrupted views of the Sun, and completing its primary mission with unqualified success, *SOHO's* eyes were abruptly closed. During routine maintenance maneuvers and spacecraft gyroscope calibrations on 25 June 1998, the spacecraft spun out of control and engineers could not re-establish radio contact with it. *SOHO* was no longer pointing at the Sun and therefore losing electrical power normally supplied by its solar panels.

A group of experts, with the ponderous title "The *SOHO* Mission Interruption Joint ESA/NASA Investigation Board," was assembled to find out what went wrong. Their post-mortem found no fault with the spacecraft itself; a sequence of operational mistakes added up to its loss. A combination of human error and faulty computer command software, which had not been previously used or adequately tested, disabled some of *SOHO's* stabilizing gyroscopes. Continuous firings of its jet thrusters failed to bring the spacecraft into balance, and instead sent it spinning faster.

For nearly a month, the crippled spacecraft failed to respond to signals sent daily. Scientists feared that *SOHO* might be drifting away from its expected orbit, gone forever and never to be heard from again. However, the wayward satellite was found nearly a month after it was lost, by transmitting a powerful radar pulse to it from the world's largest radio telescope located in Arecibo, Puerto Rico. The faint radio echo indicated that *SOHO* was close to its predicted location with its solar panels turned edge-on toward the Sun, so its internal power had to be draining away and the spacecraft was unable to receive or send communications.

Electrical power might nevertheless be regained as the panels slowly turned to a more favorable alignment with the Sun during *SOHO*'s annual orbit around the star. Antennas in NASA's Deep Space Network therefore continued to send the satellite wake-up messages, asking it to call home.

After nearly 6 weeks of silence, a feeble and intermittent response was received from the dormant spacecraft, like the faint, erratic heartbeat of a patient in a coma or the worn-out, distressed cries of a tired, lost child. The elated European engineers, who built the spacecraft, knew that *SOHO* was alive and immediately began regaining control of it.

It was a long, slow recovery. The onboard batteries had to be recharged, and the inner workings had to be warmed up after an enforced period of deep freeze. Almost all of the propellant in its tanks had been frozen solid, and the pipes that carry fuel to the craft's jet thrusters also had to be thawed out. Fortunately, the fuel, named hydrazine, does not expand when it freezes, so the fuel pipes did not crack open as sometimes happens when a building's water pipes freeze during the loss of heat in a severe winter storm. *SOHO* operations eventually returned to normal on 25 September 1998, so altogether it took 3 months from the initial loss of radio contact to recovery.

As luck would have it, *SOHO*'s tribulations were not yet over, since its gyroscopes acted up just a few months after recovery. The satellite had to constantly fire its onboard jets to keep it balanced and pointed toward the Sun, and this was rapidly exhausting its fuel supply. Ingenious engineers fixed the problem by instructing the spacecraft to bypass the gyroscopes and use stars to determine its position, somewhat like the ancient mariners who navigated by the stars. All of *SOHO*'s sensitive instruments were restored to full health, showing no signs of damage from their unexpected ordeal, and the mission continued uninterrupted for many years.

The fantastic rescue from deep space was an intense, unprecedented drama, and a remarkable achievement by a heroic international team working together in the face of overwhelming obstacles. In the end, it was human ingenuity, hope, and perseverance that brought *SOHO* back to life.

If *SOHO* were positioned exactly at the Lagrangian point, NASA's tracking telescopes would look directly at the Sun. The intense radio interference generated by the Sun would then severely limit satellite communications with Earth and the bright glare might even burn terrestrial telescope receivers up. As a result, *SOHO* flies as close to the Lagrangian point as possible, in an elliptical orbit whose plane is perpendicular to the line joining the Sun to the Earth and whose radius is about 600,000 km.

When viewed from Earth, the spacecraft forms a closed curve around the Sun, like a halo, so its trajectory is often called a halo orbit. As it loops about the Lagrangian point, *SOHO* also orbits the Sun in step with Earth, experiencing perpetual day and continuously gazing at the Sun with an unobstructed view.

All previous observatories with telescopes that look at the Sun have either been on the solid Earth, and spinning with it, or in orbit around our planet. Terrestrial telescopes are limited by inclement weather conditions and atmospheric distortion of the Sun's signal, and of course they cannot observe the Sun at night. Although the weather problem has been removed for solar telescopes orbiting the Earth above its atmosphere, the observations are still periodically interrupted when the spacecraft enters our planet's shadow. So, *SOHO* was the first solar observatory to look directly at the Sun nonstop.

The *SOHO* spacecraft was designed and built in Europe by an industry team led by Matra Marconi Space, Toulouse, France. NASA launched *SOHO* and operates the satellite from the Experimenters' Operations Facility at the Goddard Space Flight Center in Greenbelt, Maryland. Large radio telescopes around the world, which form NASA's Deep Space Network, are used to track the satellite and retrieve its data. Over a thousand images have been routed to the Experimenters' Operation Facility every day for more than 10 years. Here solar physicists from around the world continue to work together, watching the Sun night and day from a room without windows.

The *SOHO* mission has three principal scientific goals: to measure the structure and dynamics of the solar interior; to gain an understanding of the heating mechanisms of the Sun's million-degree atmosphere, or solar corona, and related processes in the chromosphere and the transition region to the corona; and to determine where the solar wind and solar energetic particles originate and how they are accelerated. *SOHO* has accomplished many of these objectives, providing an unparalleled breadth and depth of information about the Sun from its interior, through the hot and dynamic atmosphere, and out to the solar wind. In addition, the spacecraft has provided near real-time solar data for space weather prediction.

Analysis of global helioseismology data from *SOHO* has shed new light on structural and dynamic phenomena in the solar interior. It has shown that differential rotation, in which equatorial regions rotate faster than the regions closer to the poles, persists throughout the convective zone. At greater depths, in the radiative zone, the rotation speed becomes uniform from equator to pole. The Sun's magnetism is probably generated at the interface between the two zones, where the outmost part of the radiative zone, rotating at one speed, meets the bottom of the convective zone, which spins faster in its equatorial middle.

SOHO temperature measurements of the solar core have placed significant constraints on the amount of neutrinos generated there, suggesting that the paucity of observed neutrinos must be due to changes in neutrino type or flavor, which subsequent experiments have demonstrated. Evidence for the possible existence of gravity waves, located deep within the Sun, has also been provided, for the first time, by an instrument aboard *SOHO*.

A *SOHO* helioseismology instrument has made pioneering observations of large-scale streams inside the Sun, discovering sub-photospheric zonal and meridional flows, transient storms, high and low pressure zones, and swirling flows near active regions, which vary like weather patterns in the Earth's atmosphere.

SOHO has also helped pioneer local helioseismology on finer scales, obtaining the first three-dimensional structure and flow images of the interior of a star. These acoustic tomography results have revealed the internal structure of sunspots, showing that they are relatively shallow and that strong converging downflows hold sunspots together, explaining why they can last for weeks without breaking apart.

The *SOHO* helioseismologists have also demonstrated that sound waves traveling through the Sun can be used to estimate the location and magnitude of large active regions on the invisible backside of the Sun well before they rotate into view on the solar disk. This technique has been refined to give daily images of these unseen active regions from both *SOHO* and the ground-based Global Oscillation Network Group, abbreviated GONG.

SOHO's instruments have monitored ever-changing magnetic forces as they interact to help shape, mold, heat, and constrain the solar atmosphere and modulate conditions in the solar wind. They have shown that there is significant, variable magnetism dispersed all across the photosphere outside of sunspots. Tens of thousands of small magnetic loops are constantly being generated, rising up and out of the photosphere, interacting, fragmenting and disappearing within hours or days. The entire magnetic content of the so-called quiet Sun outside active regions is continuously replenished every 40 h or so, transferring magnetic energy from the visible disk toward the overlying corona. When magnetic fields of adjacent magnetic loops meet, they can break apart and reconnect with each other into simpler magnetic configurations, releasing energy that can provide heat to the quiet corona.

The solar magnetic field that has been drawn out into space by the Sun's wind evolves in response to the changing photospheric field at it base. This evolution, together with rotation of the Sun, helps drive space weather through continuously changing conditions in the solar wind and the magnetic field embedded within it.

SOHO instruments have also helped identify source regions and acceleration sites for the fast and slow components of the solar wind. Regions of open magnetic fields, known as coronal holes, give rise to fast, low density winds, both near the solar poles and sometimes closer to the solar equator. The fast winds are accelerated very close to the Sun. The slower, dense wind puffs out from areas of closed magnetic fields near equatorial regions, and takes a longer time to get up to speed. One interesting proposal for the origin of the fast solar wind invokes open magnetic funnels in the low corona, which might be fed by closed magnetic loops, swept by convection into the funnel regions.

Moreover, heavier ions in the polar coronal holes move faster than light ones, so something is unexpectedly and preferentially energizing the more massive ions. Magnetic waves might provide the extra boost that pushes the heavier particles to higher speeds. Instruments aboard *Ulysses* have detected magnetic fluctuations attributed to such Alfvén waves moving far above the Sun's poles, and they have been observed closer to the Sun, in the chromosphere and corona, by instruments aboard *Hinode*.

Through high-resolution images and movies, the *SOHO* mission has also revealed how the Sun's global magnetic fields interact and restructure to release energy on incredible scales, forming gigantic magnetic bubbles, the coronal mass ejections, which can be hurled toward the Earth with potentially dangerous consequences. One of *SOHO's* instruments has provided over 10 years of measurements of their physical properties, such as their mass, speed, and energy, with fascinating images that display their various magnetic structures including twisted ropes, vacant cavities, and halos (Fig. 1.7). Such observations have provided the groundwork for more elaborate, three-dimensional observations of coronal mass ejections from the *STEREO* mission described in Sect. 1.10.

Another *SOHO* instrument monitors the dimming of the Sun's extreme ultraviolet radiation that occurs when coronal mass ejections tear out part of the solar corona. When these awesome bubbles lift off, the closed magnetic fields are ripped open at their tops, causing them to shake and reverberate at lower levels and producing waves in the extreme ultraviolet radiation.

SOHO's numerous discoveries have been achieved with a dozen instruments; European scientists provided nine of them and United States scientists a further

Fig. 1.7 Coronal mass ejection. A huge coronal mass ejection is seen in this coronagraph image, taken on 5 December 2003, with the Large Angle and Spectrometric COronagraph, abbreviated LASCO, on the *SOlar and Heliospheric Observatory*, or *SOHO* for short. The black area corresponds to the occulting disk of the coronagraph that blocks intense sunlight and permits the corona to be seen. An image of the singly ionized helium, denoted He II, emission of the Sun, taken at about the same time, has been appropriately scaled and superimposed at the center of the LASCO image. The full-disk helium image was taken at a wavelength of 30.4 nm, corresponding to a temperature of about 60,000 K, using the Extreme-ultraviolet Imaging Telescope, or EIT for short, aboard *SOHO*. (Courtesy of the *SOHO* LASCO and EIT consortia. *SOHO* is a project of international cooperation between ESA and NASA.)

three. International consortia involving scientific institutes in 15 countries developed them. These instruments have examined the Sun from its deep interior, through its million-degree atmosphere and ceaseless wind, to our home planet, Earth. Three devices probe the Sun's internal structure and dynamics; six measure the solar atmosphere; and three keep track of the star's far-reaching winds.

The hidden interior of the Sun is illuminated by its in-and-out, heaving motions. These oscillations arise from sounds that move inside the Sun. On striking the Sun's apparent edge and rebounding back inside, the sound waves cause the visible gas to move up and down. The oscillating motions these sounds create are imperceptible to the naked eye, but *SOHO* instruments routinely pick them out.

The Sun's hot atmosphere is probed by observing ultraviolet, extreme-ultraviolet, or X-ray radiation. This is because hot material emits most of its energy at these wavelengths. Also, the underlying visible disk of the Sun is too cool to emit intense radiation at these wavelengths, so it appears dark under the hot gas.

To map out structures across the solar disk, ranging in temperature from 6000 to 2×10^6 K, the *SOHO* instruments tune into ultraviolet or extreme ultraviolet spectral features emitted by different ions at various definite wavelengths. Atoms in a hotter gas lose more electrons through collisions and so they become more highly ionized. The wavelength-specific radiation of different ions therefore serves as a kind of thermometer, and *SOHO* telescopes transform the emission from a given ion into high-resolution images of the structures formed at the relevant temperature. The speed of the material moving in these regions is inferred from the Doppler wavelength changes of the spectral features that *SOHO* records.

Other *SOHO* instruments directly measure energetic solar electrons, protons, and heavier ions in situ as they sweep past the satellite. By analyzing these elements, scientists reach conclusions about conditions in the Sun's atmosphere where its winds originate. *SOHO* and other spacecraft near Earth also detect very energetic particles accelerated by coronal mass ejections and solar flares.

SOHO's instruments uniquely examine many different levels in the solar atmosphere simultaneously, including for the first time the locations where the million-degree gases are heated and the Sun's winds are accelerated. Moreover, many of the instruments complement each other and are operated together, establishing connections between various phenomena at different places in the volume of the Sun and in the interplanetary medium. Their combined data can link events in the Sun's atmosphere and solar wind to the changes taking place within the solar interior. In effect, *SOHO* can combine data from its instruments to look all the way from deep inside the Sun to our planet. It provides a completely new perspective of how agitation originating inside the Sun and transmitted through the solar atmosphere directly affects us on Earth.

SOHO has made pioneering contributions to space weather investigations. Its continued stream of extreme-ultraviolet images has been used to describe changes in the Sun's dynamic atmosphere and to forecast disruptive activity there, while another *SOHO* instrument monitors the Sun for coronal mass ejections. *SOHO* helioseismology observations detect powerful active regions anywhere on the far side of the Sun, providing up to 14 days warning before they rotate to face and

Table 1.8 *SOHO's* instruments arranged alphabetically by acronym within three areas of investigation[a]

Instrument	Measurement
Helioseismology instruments	
GOLF	The Global Oscillations at Low Frequencies device records the velocity of global oscillations within the Sun. http://golfwww.medoc-ias.u-psud.fr/
MDI	The Michelson Doppler Imager measures the velocity of oscillations, produced by sounds trapped inside the Sun, and obtains high-resolution magnetograms. http://soi.stanford.edu/science/obs_prog.html/
VIRGO	The Variability of solar IRradiance and Gravity Oscillations instrument measures fluctuations in the Sun's brightness, as well as its precise energy output. http://www.ias.u-psud.fr/virgo/
Coronal instruments	
CDS	The Coronal Diagnostic Spectrometer records the temperature and density of gases in the corona. http://solar.bnsc.rl.ac.uk/
EIT	The Extreme-ultraviolet Imaging Telescope provides full-disk images of the chromosphere and the corona at wavelengths of 17.11, 19.51, 28.42 and 30.38 nm, corresponding to radiation produced by highly ionized iron, Fe XI and Fe X, Fe XII, and Fe XV, and helium, He II, respectively, formed at temperatures of 1.0×10^6, 1.5×10^6, 2.5×10^6 and 60,000 K. http://umbra.nascom.nasa.gov/eit/
SUMER	The Solar Ultraviolet Measurements of Emitted Radiation instrument gives data about the temperatures, densities, and velocities of various gases in the chromosphere and corona. http://www.mps.mpg.de/en/projekte/soho/sumer/
LASCO	The Large Angle and Spectrometric COronagraph provides images that reveal the corona's activity, mass, momentum, and energy, including coronal mass ejections and comets that come near the Sun. http://lasco-www.nrl.navy.mil/
UVCS	The UltraViolet Coronagraph Spectrometer measures the temperatures and velocities of hydrogen atoms, oxygen ions and other ions in the low corona between 1.5 and 10 solar radii. http:/cfa-www.harvard.edu/uvcs/
SWAN	The Solar Wind ANisotropies device monitors latitudinal and temporalvariations in the solar wind. http://www.fmi.fi/research_space/space_7.html
"In-situ" instruments	
CELIAS	The Charge, ELement and Isotope Analysis System quantifies the mass, charge, composition, and energy distribution of particles in the solar wind. http://sci.esa.int/science-e/www/object/index.cfm?fobjectid=30956/
COSTEP	The COmprehensive SupraThermal and Energetic Particle analyzer determines the energy distribution of protons, helium isotopes, and electrons. http://www.ieap.uni-kiel.de/et/ag-hebe r/costep/
ERNE	The Energetic and Relativistic Nuclei and Electron experiment measures the energy distribution and isotopic composition of protons, other ions, and electrons. http://www.srl.utu.fi/projects/erne/index_english.html/

[a]The *SOHO* instruments are described in *Solar Physics* **162**, 3–531 (1992). For an overview see article by V. Domingo et al. (1995); for specific instruments see GOLF: A. H. Gabriel et al. (1995), VIRGO: C. Fröhlich et al. (1995), MDI: P. H. Scherrer et al. (1995), SUMER: K. Wilhelm et al. (1995), CDS: R. A. Harrison et al. (1995), EIT: J.-P. Delaboudinière et al. (1995), UVCS: J. L. Kohl et al. (1995), LASCO: G. E. Brueckner et al. (1995), SWAN: J. L. Bertaux et al. (1995), CELIAS: D. Hovestadt et al. (1995), COSTEP: R. Müller-Mellin et al. (1995), and ERNE: J. Torsti et al. (1995).

threaten the Earth. Short-term forecasting of explosive solar outbursts, which can send lethal energetic particles toward our planet, may be achieved by detecting high-speed electrons that arrive near the Earth about an hour ahead of the more dangerous ions. *SOHO* instruments have also refined measurements of the Sun's total radiative output, the so-called solar constant, helping to monitor nearly three decades of its variation. This unparalleled record indicates that brightening of the Sun is unlikely to significantly affect future global warming by humans.

Table 1.9 *SOHO* principal investigators and their institutions

Instrument	Principal investigator	Institution
Helioseismology instruments		
GOLF	Alan H. Gabriel	Institut d'Astrophysique Spatiale, Orsay, France
MDI	Philip H. Scherrer	Stanford University, Stanford, California, United States
VIRGO	Claus Fröhlich	Physikalisch-Meteorologisches Observatorium, World Radiation Center, Davos, Switzerland
Coronal instruments		
CDS	Richard A. Harrison	Rutherford Appleton Laboratory, Chilton, Oxfordshire, United Kingdom
	Andrzej Fludra	Rutherford Appleton Laboratory, Chilton, Oxfordshire, United Kingdom
EIT	Jean-Pierre De-laboudinière	Institut d'Astrophysique Spatiale, Orsay, France
	Frédéric Auchère	Institut d'Asrophysique Spatiale, Orsay, France
SUMER	Klaus Wilhelm	Max-Planck-Institut für Aeronomie, Katlenburg-Lindau, Germany
	Werner Curdt	Max-Planck-Institut für Aeronomie, Katlenburg-Lindau, Germany
LASCO	Guenter Brueckner	Naval Research Laboratory, Washington, District of Columbia, United States
	Russell Howard	Naval Research Laboratory, Washington, District of Columbia, United States
UVCS	John L. Kohl	Smithsonian Astrophysical Observatory, Cambridge, Massachusetts, United States
SWAN	Jean-Loup Bertaux	Service d'Aéronomie, Verrieres-le-Buisson Cedex, France
	Eric Quémearis	Service d'Aéronomie, Verrieres-le-Buisson Cedex, France
"In-situ" instruments		
CELIAS	Peter Bochsler	University of Bern, Switzerland
	Berndt Klecker	Max-Planck-Institut für Extraterrestrische Physik, Garching, Germany
COSTEP	Horst Kunow	University of Kiel, Germany
	Bernd Heber	University of Kiel, Germany
ERNE	Jarmo Torsti	University of Turku, Finland
	Eino Valtonen	University of Turku, Finland

All of *SOHO's* magnificent accomplishments are discussed in the subsequent chapters of this book. The acronyms and measurements of its instruments are given in Table 1.8; the Principal Investigators and their institutions are provided in Table 1.9. The Project Scientist for *SOHO* is Bernhard Fleck of ESA. Many of the unique images and movies that *SOHO* detects move nearly instantaneously to the *SOHO* home page (http://sohowww.nascom.nasa.gov) on the World Wide Web.

1.6 *ACE* Measures the Composition of High-Energy Particles Bombarding Earth

Our planet is immersed within a cosmic shooting gallery of protons, other atomic nuclei, and ions coming from the Sun and interstellar space. The solar wind supplies a steady stream of them. These charged particles originate near the Sun and can be additionally accelerated on their way to our planet. Energetic outbursts on the Sun, the solar flares and coronal mass ejections, expel brief pulses of energized atomic nuclei, producing powerful gusts in the solar wind. And extraordinarily energetic nuclei, the comic rays, rain down on the Earth, coming in at all directions from interstellar space and traveling at nearly the velocity of light.

So there is danger blowing in the Sun's winds, with its gusts and squalls, and the less abundant but more energetic cosmic rays also pose a threat. Energetic nuclei from both sources can injure and even kill unprotected astronauts traveling in deep space, repairing spacecraft or a space station, or visiting the Moon or Mars. This space weather can also endanger Earth-orbiting satellites, disrupt communications on Earth, and produce geomagnetic storms that can overload power grids.

NASA's *Advanced Composition Explorer*, abbreviated *ACE*, spacecraft keeps a careful watch on this vast and shifting web of subatomic particles as they bombard the Earth, measuring their properties in order to determine their origin and subsequent transformations. The 785-kg spacecraft was built at the Applied Physics Laboratory of the John Hopkins University and launched by a Delta II rocket from the Cape Canaveral Air Station, Florida, on 25 August 1997. *ACE* is located at the inner Lagrangian, or L_1, point of gravitational equilibrium between the Earth and the Sun, just outside our planet in the direction of the incoming solar wind just 1.5×10^6 km from the Earth and about 150×10^6 km from the Sun. From this vantage point, *ACE* has monitored the solar wind, with its charged particles and magnetic fields, and observed high-energy particles accelerated at the Sun, within the solar wind, or in the galactic regions beyond the heliosphere. It has additionally provided nearly a decade of continuous, real-time, space-weather observations, forecasting advance warnings of hazardous events.

ACE carries a total of nine instruments. Six of these instruments measure with unprecedented precision the composition of energetic nuclei and ions from low solar-wind energies to high cosmic ray ones. The other three instruments provide the interplanetary context for these studies, including the interplanetary magnetic field and the velocity, density, and temperature of the solar-wind protons.

The primary scientific goal of *ACE* is to measure and compare the composition of nuclei originating at the Sun or in the local interstellar medium, and to identify how these nuclei are subsequently transformed as they make their way through interplanetary space. The mission's objectives include the specification of the elemental and isotopic composition of energetic nuclei at their place of origin, and the identification of processes that subsequently accelerate the nuclei to higher energies, alter their composition, and affect their motion.

ACE has accomplished many of its scientific objectives during recent years. It has provided new insights to the acceleration of nuclei by impulsive flares and by interplanetary shocks associated with either coronal mass ejections or the interaction of fast and slow steams in the solar wind. When solar energetic particles arriving at *ACE* and *Wind* are, for example, compared with *RHESSI* observations of particles accelerated in large flares, it is found that the flares are not responsible for the most energetic solar particles arriving at Earth. They are most likely accelerated by interplanetary shocks driven by fast coronal mass ejections. Data from instruments on *ACE* have indeed shown that the energetic solar electrons and ions arriving at Earth are usually released after flare onset. In some situations, however, there can be difficulty in distinguishing between direct flare acceleration and acceleration by the coronal-mass-ejection shocks.

Solar energetic particles observed with the ULEIS instrument on *ACE* show enrichments in the rare isotope of helium, designated ^{3}He rather than the usual, more abundant ^{4}He. Although discovered in the 1970s and subsequently associated with impulsive events and radio bursts from the Sun, their source remained unknown. Then comparisons with *SOHO* images indicated that every one of the *ACE* events, observed over a 6-year period, originates from small, flaring active regions that reconnect with open magnetic field lines in nearby coronal holes. The ^{3}He events and associated jet-like ejections are attributed to footpoint exchanges between closed and open field lines.

ACE instruments have additionally demonstrated that the magnetic fields entrained in the solar wind can be reconfigured when oppositely directed magnetic fields merge and join together, reconnecting with each other and even disconnecting from the Sun. The spacecraft has provided the first direct evidence that such magnetic reconnection occurs in the solar wind near the Earth's orbit, producing magnetic reversals and releasing magnetic energy with an associated blast of accelerated ions. Instruments aboard *ACE* have observed numerous instances of such magnetic reconnections within interplanetary space, many of them also observed with instruments aboard the *Wind* spacecraft. Multi-spacecraft observations have demonstrated that the magnetic reconnections are prolonged and steady, extending for hundreds of Earth radii along the X-line where oppositely directed magnetic field lines meet.

The SWICS instruments on both *ACE* and *Ulysses* have made comprehensive and continuous observations of singly ionized, abundant elements with just one electron missing. They are largely of interstellar origin due to atoms that have drifted into the heliosphere to become ionized there and picked-up by the magnetic field entrained in the solar wind. Measurements of these pick-up ions have been used to infer the abundance, chemical and physical properties of the local interstellar cloud, and the extent of the heliosphere.

Measurements of the velocity distribution of both solar-wind and pick-up ions, with instruments aboard *ACE* and *Ulysses*, reveal unexpectedly rapid ion motions of up to at least 50 times the speed of other ions in the solar wind. Since these exceptionally fast ions have energies larger than those typically found in the hot, expanding corona, they have been dubbed suprathermal ions. By measuring the composition and time variations of these previously unknown ions, the instruments on *ACE* were able to show that shocks driven by coronal mass ejections and transient interplanetary shocks preferentially accelerate the suprathermal nuclei, which have already been pre-accelerated by some currently unknown process. That is, solar energetic particles arriving at the Earth are not accelerated from the bulk solar wind, but instead consist of suprathermal ions that have been further accelerated to higher energies by interplanetary shocks.

All of *ACE*'s accomplishments are discussed in greater detail in subsequent chapters of this book. The acronyms and measurements of its instruments are given in Table 1.10, and described in instrument papers by Edward C. Stone and others in a special issue of *Space Science Reviews* **86**, No. 1–4 (1998). The *ACE* Principal Investigator is Edward C. Stone at the California Institute of Technology, abbreviated Caltech. The NASA Project Scientist for *ACE* is Tycho T. von Rosenbinge, and the Caltech *ACE* Mission Scientist is Richard A. Mewaldt. There are 18 scientific co-investigators of the *ACE* mission, whose names and institutions can be found on the *ACE* Home Page at http://www.srl.caltech.edu/ACE/.

1.7 *TRACE* Focuses on Fine Details of the High-Temperature Gas

When a telescope in space zooms in to take a close look at the Sun's detailed behavior, which cannot be fully observed from the ground, the telescope can discover a host of unsuspected fine structure and rapid dynamic changes. Such information cannot be achieved from observations with coarser resolution, from space or on the ground, for they integrate or smear out the small-scale and rapidly varying phenomena. The *Transition Region And Coronal Explorer*, abbreviated *TRACE*, was the first spacecraft to achieve this high-resolution perspective for the transition region between the chromosphere and the corona, and the low corona as well.

TRACE is the fourth mission of NASA's *Small Explorer*, or SMEX, Program for focused, relatively inexpensive space science missions. SMEX spacecraft are limited in mass to between 200 and 300 kg, so they can be sent into orbit using rocket launch vehicles, and each mission is expected to cost approximately $35 million for design, development, and operations through the first 30 days in orbit. The 241-kg *TRACE* spacecraft was launched from Vandenberg Air Force Base on 1 April 1998 aboard a Pegasus XL vehicle, a three-stage rocket launched from a jet aircraft and used to deploy small payloads. It was sent into a full-Sun orbit, the first solar observatory to have that advantage.

Table 1.10 *ACE* instruments arranged alphabetically and by acronym[a]

Instrument	Measurement
CRIS	The Cosmic Ray Isotope Spectrometer observes 24 species of cosmic rays in seven energy ranges from 60 to more than 400 MeV per nucleon. http://www.srl.caltech.edu/ACE/CRIS_SIS/cris.html
EPAM	The Electron, Proton and Alpha Monitor observes ions with energies between 0.05 and 5 MeV, and electrons with 38 to 312 keV in energy. http://sd-www.jhuapl.edu/ACE/EPAM/
MAG	The Magnetometer measures the vector magnetic field. http://www.ssg.sr.unh.edu/mag/ACE.html
SEPICA	The Solar Energetic Particle Ion Composition Spectrometer measures the charge state of eight species of solar and interplanetary particles with energies between 0.3 and 5 MeV per nucleon. This instrument has not been fully operational since 2000. http://www.ssg.sr.unh.edu/tof/Missions/Ace/
SIS	The Solar Isotope Spectrometer observes solar particles and cosmic rays with energies of 6 to more than 100 MeV per nucleon. http://www.srl.caltech.edu/ACE/CRIS_SIS/sis.html
SWEPAM	The Solar Wind Electron Proton Alpha Monitor measures the solar-wind electrons, the solar-wind proton velocity, density, and temperature, and the helium to hydrogen ion ratio $^4\text{He}/\text{H}^+$ in the solar wind. http://swepam.lanl.gov/
SWICS	The Solar Wind Ion Composition Spectrometer measures the helium density, the helium isotope ratio $^3\text{He}/^4\text{He}$, the bulk and thermal ion speeds of helium, carbon, oxygen, and iron nuclei, the abundances, relative to oxygen, and charge state distributions of helium, carbon, neon, magnesium, silicon, and iron. http://solar-heliospheric.engin.umich.edu/ace/
SWIMS	The Solar Wind Ion Mass Spectrometer observes the nitrogen and sulfur abundances relative to oxygen and the isotopes of magnesium and neon. http://solar-heliospheric.engin.umich.edu/ace/
ULEIS	The Ultra Low-Energy Ion Spectrometer observes solar and interplanetary particles with energies between 0.04 and 4 MeV per nucleon, including the helium ion abundance ratio $^3\text{He}/^4\text{He}$. http://sd-www.jhuapl.edu/ACE/ULEIS/

[a]The *ACE* instrument papers are given in a special issue *of Space Science Reviews* **86**, No. 1–4, 1–22, 267–632 (1998). EPAM, MAG, SIS and SWEPAM supply *ACE* real-time solar wind data.

TRACE has provided new insights to the magnetized coronal loops that mold and constrain the million-degree gas in solar active regions, resolving them into long, thin strands or threads that stand alone and are not braided together (Fig. 1.8). The individual loops are brightest at their base, where the legs emerge from and return to the photosphere, suggesting that the loop material is heated from below. Some of the coronal loops are so unexpectedly dense, so crammed full with hot gas, that they ought to break apart unless the material is being fed into them and drained away from below. They seem to be continuously pumped up and deflated, heated from below by upward pulses of hot material in matters of minutes, almost bursting apart at the seams, but cooling down by radiation and the conductivity of heat to lower levels in the chromosphere.

High-cadence *TRACE* images have detected highly dynamic processes in the bright, non-flaring coronal loops located in solar active regions, including flows.

Fig. 1.8 Magnetic loops made visible. An electrified, million-degree gas, known as plasma, is channeled by magnetic fields into bright, thin loops. The magnetized loops stretch up to 500,000 km from the visible solar disk, spanning up to 40 times the diameter of planet Earth. The magnetic loops are seen in extreme ultraviolet radiation of eight and nine times ionized iron, denoted Fe IX and Fe X, at a wavelength of 17.1 nm, formed at a temperature of about 1.0×10^6 K. The hot plasma is heated at the bases of loops near the place where their legs emerge from and return to the photosphere. Bright loops with a broad range of lengths all have a fine, thread-like substructure with widths as small as the telescope resolution of 1 s of arc, or 725 km at the Sun. This image was taken with the *Transition Region And Coronal Explorer*, abbreviated *TRACE*, spacecraft. (Courtesy of the *TRACE* consortium, LMSAL and NASA; *TRACE* is a mission of the Stanford-Lockheed Institute for Space Research, a joint program of the Lockheed-Martin Solar and Astrophysics Laboratory, or LMSAL for short, and Stanford's Solar Observatories Group.)

These magnetic loops also oscillate, or move back and forth, with periods of between 2 and 7 min, appearing when flares excite the oscillations. Wave and oscillating activity in coronal loops can be used to probe their physical characteristics, in a relatively new investigation known as coronal seismology.

The high-resolution and rapid sequential images of the *TRACE* telescope describe how coronal loops evolve, varying with the ever-changing magnetism. When combined with other simultaneous observations, from the ground or space, they provide evidence for the mechanisms that trigger solar flares or coronal mass ejections. Driven by churning, turbulent motions at their roots in the photosphere and below, the coronal loops shear, twist, and writhe, sometimes merging together, expanding, or breaking out of their confinement. Then, in the wake of these outbursts and the release of magnetic energy, the coronal loops reform in stable configurations, often in elegant arcades (Fig. 1.9).

TRACE has also provided new characterizations of the quiet corona and transition region, away from solar active regions. Bright, extreme-ultraviolet, flare-like events, dubbed nanoflares, flash on and off, but with insufficient energy, frequency, or power

Fig. 1.9 Stitching up the wound. An arcade of post-flare loops shines in the extreme ultraviolet radiation of eight and nine times ionized iron, Fe IX and Fe X at a wavelength of 17.1 nm, formed at a temperature of about 1.0×10^6 K. This image was taken on 8 November 2000, just after a solar flare occurred in the same active region; at least one coronal mass ejection also accompanied the event. High-energy particles from the flare entered the Earth's radiation belts, and an associated coronal mass ejection produced a strong geomagnetic storm. This image was taken from the *Transition Region and Coronal Explorer*, abbreviated *TRACE*. (Courtesy of the *TRACE* consortium and NASA; *TRACE* is a mission of the Stanford-Lockheed Institute for Space Research, a joint program of the Lockheed-Martin Solar and Astrophysics Laboratory, or LMSAL for short, and Stanford's Solar Observatories Group.)

to noticeably heat the corona. Ultraviolet brightness oscillations are also observed across the Sun, with the 5-min period of internal sound waves that push the photosphere in and out from below. Some of these waves leak out into the chromosphere, powering shocks that drive upward flows along magnetic flux tubes and forming the ubiquitous, jet-like spicules. More than 10,000 spicules can be seen at any moment on the Sun, continuously rising and falling every 5 min and carrying a mass flux of 100 times that of the solar wind into the low solar corona. This *TRACE* discovery of the previously unknown cause of spicules confirms related studies from the Swedish Solar Telescope on La Palma, which uses adaptive optics to achieve high angular resolution from the ground.

When comparing *TRACE* images with in situ measurements of the solar wind obtained by the *ACE* spacecraft, scientists have shown that the speed and composition of the solar wind at the Earth's orbit have deep roots in the solar atmosphere. There is a shallow, dense chromosphere below the strong, closed magnetic regions that are related to the slow, dense solar-wind outflow. Deep, less dense chromosphere is found below the open magnetic regions, or coronal holes, where the fast, tenuous winds originate.

Detailed discussions of these and other discoveries from *TRACE* are found in the other chapters of this book. They provide the foundation for more comprehensive investigations with the *Hinode* spacecraft discussed in Sect. 1.9.

The 30-cm aperture *TRACE* telescope zeros in on specific regions of the Sun's low corona with an 8.5×8.5 arc minute field of view and an angular resolution of one second of arc, the highest resolution of coronal structures ever achieved. By way of comparison, the angular diameter of the Sun is about 30 min of arc, which is observed in its entirety by *SOHO* telescopes with less angular resolution. By observing the solar atmosphere in the ultraviolet and extreme-ultraviolet radiation of different spectral lines, the *TRACE* instrument provides detailed images at temperatures from about 10,000 to about 10×10^6 K. They are spectral lines formed by ionized iron and helium, denoted by Fe IX/X, Fe XII, Fe XV, and He II, at respective wavelengths of 17.11, 19.51, 28.52, and 30.38 nm and formation temperatures of 1.0×10^6, 1.5×10^6, 2.5×10^6, and 60,000 K. B. N. Handy et al. (1999) give the *TRACE* instrument paper; also see C. J. Schrijver et al. (1999).

Alan M. Title, of the Lockheed Martin Solar and Astrophysics Laboratory, abbreviated LMSAL, was the Principal Investigator of the *TRACE* Mission, but Carolus J. Schrijver, also of LMSAL, now replaces him. The Project Scientist of the *TRACE* Mission is Joseph B. Gurman of NASA's Goddard Space Flight Center. The *TRACE* team consists of scientists at LMSAL, the Smithsonian Astrophysical Observatory, the Montana State University, the Goddard Space Flight Center, the Rutherford Appleton Laboratory in the United Kingdom, and Stanford University. More information about these groups can be found on the World Wide Web at the sites: http://www.lmsal.com, http://hea-www.harvard.edu/SSXG/, http://solar.physics.montana.edu, http://sspg1.bnsc.rl.ac.uk/Share/sol.html, and http://sun.stanford.edu/.

1.8 High-Energy Solar Outbursts Observed with *RHESSI*

Intense X-rays and extreme-ultraviolet radiation are not the only things unleashed during a solar flare. For large flares, a significant fraction of the total explosion energy is used to accelerate electrons, protons, and other ions to speeds approaching the velocity of light. Solar flares are indeed the most powerful accelerators in the solar system, producing ions with energies up to many GeV and electrons as energetic as hundreds of MeV. (Billions and millions of electron volts, respectively abbreviated as GeV and MeV, are the units of energy used to describe very energetic particles, where $1\,\text{eV} = 1.6 \times 10^{-19}$ J and 1 J is equal to 10×10^6 erg.)

The energetic particles move rapidly away from their acceleration site in the low solar corona, traveling down into the Sun and out into space. When the energetic ions are beamed into the lower solar atmosphere, they initiate nuclear collisions that produce hard X-ray and gamma-ray radiation. The electrons also emit radiation at these wavelengths when they are hurled down into the denser parts of the Sun's atmosphere.

The *Ramaty High Energy Solar Spectroscopic Imager*, abbreviated *RHESSI*, mission provides unique observations of high-energy acceleration processes close to the Sun by observing energetic flare radiation over an enormous range in energy, from soft X-rays (3 keV) to gamma rays (17 MeV). It thereby obtains information about the location, energy spectrum, and composition of the flare-accelerated particles. This has important implications for flare particle acceleration and escape processes, as well as energy release in solar flares. *RHESSI* has observed over 12,000 flares with detectable emission above 12 keV, more than 1500 above 25 keV, and more than a dozen gamma ray flares.

RHESSI is one of NASA's *Small Explorer*, or SMEX, Missions designed to do focused space science research in a relatively inexpensive manner. It is the only solar mission named for an individual, Reuven Ramaty, a pioneer in the fields of solar physics, gamma-ray astronomy, nuclear astrophysics, and cosmic rays, and a co-investigator and one of the founding members of the *HESSI* team.

The 293-kg spacecraft was launched from the Kennedy Space Center on 5 February 2002, aboard a Pegasus XL vehicle released from a jet aircraft.

The main scientific objective of *RHESSI* is to explore the basic physics of particle acceleration and energy release in solar flares through spectroscopy and imaging of gamma-ray lines at specific wavelengths and the X-ray and gamma-ray continuum radiation emitted at all wavelengths by flare-accelerated ions and electrons. It has determined the frequency, location, and evolution of impulsive flare emission in the corona, and studied the acceleration, propagation, and evolution of flare-associated electrons, protons, and heavier ions.

Although previous solar spacecraft have detected solar gamma-ray lines, produced by nuclear collisions with flare-accelerated ions, *RHESSI* provided the first high-resolution spectroscopy and imaging of them. It provided the first ever gamma-ray line image of a solar flare, showing that the flare-accelerated ions are separated from the accelerated electrons by about 15,000 km, which was confirmed by subsequent *RHESSI* flare observations. This could be explained by different acceleration sites for ions and electrons, or possibly by a common acceleration site with a difference in the place where they lose their energy by collision. The gamma-ray images from solar flares observed so far are nevertheless located in the flare active region, showing that the responsible ions are accelerated by a flare process and not by shocks such as those generated during fast coronal mass ejections.

One of the more fascinating spectral features is the one produced when anti-matter is destroyed almost immediately after its creation during a solar flare. That is, positrons, the anti-matter counterpart of electrons, annihilate with electrons producing radiation at 0.511 MeV (Fig. 1.10), which is the energy contained in the entire mass of a non-moving electron. *RHESSI* has resolved the positron annihilation line for the first time and showed that it is unexpectedly broad, a disconcerting anomaly that no one has been able to explain so far.

RHESSI has also confirmed that large-scale magnetic reconnection in the low corona is the most likely explanation for how flares suddenly release so much

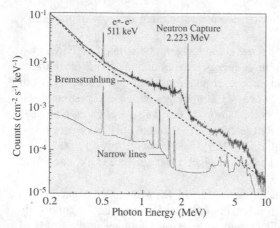

Fig. 1.10 Flare spectral lines at high energy. This energy spectrum of radiation, from a flare on 28 October 2003, exhibits two prominent spectral lines and numerous less intense, narrow ones. The line with energy of 511 keV, or 0.511 MeV, is emitted when electrons collide with their anti-matter counterparts, the positrons or "positive electrons"; both particles are destroyed while emit-ting the high-energy radiation. The neutron-capture line at 2.223 MeV, and several narrow lines from accelerated protons are also seen. Underlying the line features is the bremsstrahlung con-tinuum from accelerated electrons. These data were obtained from NASA's *Ramaty High Energy Solar Spectroscopic Imager*, abbreviated *RHESSI*. (Courtesy of Brian R. Dennis, NASA.)

energy. The spacecraft's X-ray images indicate that the coronal source of a flare can be separated from the underlying flare loops, rising to higher altitude and moving away from the reconnection site.

RHESSI observations of the particles accelerated during large flares have been compared with in situ energetic particle measurements near the Earth's orbit with the *WIND* and *ACE* spacecraft, and for the most energetic ions at the ground. Many were surprised when this comparison showed that the most energetic ions and elec-trons accelerated during solar flares are not responsible for many of the powerful solar energetic particles that arrive in the vicinity of the Earth. It appears that a different acceleration process, one associated with the shocks of fast coronal mass ejections, may often be responsible for the very energetic particles that can threaten humans and satellites near our planet, with important implications for space weather. The relative roles of flare acceleration and shock acceleration are nevertheless still controversial, and energetic particles accelerated in modest flares still appear to be similar to weaker concentrations of energetic particles arriving near Earth.

These and other significant accomplishments of *RHESSI* are discussed in greater detail in other chapters of this book.

They have been achieved with instruments that image solar flares from X-rays to gamma rays with a full Sun field of view and an angular resolution of 2 s of arc for photon energies below 100 keV to 36 s of arc for photon energies above 100 MeV, and a temporal resolution as short as tens of milliseconds. The Paul Scherrer Insti-tut in Switzerland provided the imaging telescope and optical aspect system, while NASA's Goddard Space Flight Center provided the detection grids. *RHESSI* also observes the spectral lines emitted during solar flares with a spectral resolution as fine as 1 keV for photon energies below 100 keV and just a few keV for higher MeV photon energies.

R. P. Lin et al. (2002) give the *RHESSI* instrument paper, the *RHESSI* spectrometer is described by D. M. Smith et al. (2002), and G. J. Hurford et al. (2002) delineate the *RHESSI* imaging concept.

RHESSI is the first of NASA's Small Explorers to be managed in the Principal Investigator mode, by Robert P. Lin at the Space Sciences Laboratory of the University of California, Berkeley. The *RHESSI* Mission Scientist is Brian R. Dennis of the Solar Physics Laboratory at NASA's Goddard Space Flight Center. Supporting scientists are located throughout the world at institutions such as Montana State University, the University of Alabama, the Lawrence Berkeley National Laboratory, the ETH Institute of Astronomy in Zurich, Switzerland, the University of Glasgow, Scotland, the National Astronomical Observatory of Japan, and, the Observatoire de Paris-Meudon, France. A complete list of participating scientists may be found at the *RHESSI* Web Sites http://hessi.ssl.berkeley.edu/ and http://hesperia.gsfc.nasa.gov/hessi. *RHESSI* data are available at http://rhessidatacenter.ssl.berkeley.edu and a mirrored version at http://hesperia.gsfc.nasa.gov/rhessidatacenter/.

1.9 *Hinode* Observes How Varying Magnetic Fields Heat the Corona and Power Explosive Outbursts There

Solar scientists have now found that magnetic energy is the most likely source of heating the million-degree corona and powering solar flares and coronal mass ejections. And instruments aboard the *Yohkoh, SOHO, RHESSI*, and *TRACE* spacecraft have provided abundant evidence for changing magnetic fields that interact to release magnetic energy as heat, energetic radiation, and particle acceleration. Building on these results, the *Hinode* mission uses a coordinated set of three telescopes, operating at visible-light, X-ray, and extreme-ultraviolet wavelengths, to measure the emergence, decay, structure, and motions of the Sun's ever-varying magnetic field with continuous fine detail, and to investigate how the high-temperature, ionized atmosphere responds to the changing magnetism.

The 900-kg Japanese *Solar-B* spacecraft was launched on a M-V rocket out of the Uchinoura Space Center in Kagoshima Prefecture, Japan on 23 September 2006, and renamed *Hinode*, the Japanese word for "sunrise," after launch. It has been placed in a full-Sun orbit about the Earth, permitting continuous observations of the Sun for 24 h a day during 8 months each year.

Hinode is a mission of the Japanese Aerospace Exploration Agency, abbreviated JAXA, with the National Astronomical Observatory of Japan, or NAOJ for short, as domestic partner and NASA and the Science and Technology Facilities Council, abbreviated STFC, of the United Kingdom as international partners. It is operated by these agencies in co-operation with ESA and NSC (Norway).

The Institute of Space and Astronautical Science, or ISAS/JAXA, developed and launched the spacecraft. NAOJ and ISAS/JAXA provided the visible-light Solar Optical Telescope and the camera for the X-Ray Telescope. In the United States, NASA helped with the development, funding, and assembly of the spacecraft's three science instruments. The United Kingdom has additionally supported the development and construction of the Extreme-ultraviolet Imaging Spectrometer through the Particle Physics and Astronomy Research Council, abbreviated PPARC, with contributions from Japan and the United States. The European Space Agency, in partnership with Norway, provides spacecraft tracking and data acquisition, down linking *Hinode* data during each orbit using the satellite station situated on the Norwegian Svalbard islands. Mission operations are the shared responsibility of the Japanese, United Kingdom, and United States partners.

Hinode's comprehensive set of scientific objectives includes an understanding of the creation and destruction of the Sun's magnetic fields, determining how magnetic interactions and related processes generate the solar atmosphere and solar activity, investigating the coupling between the photosphere and corona, determining how varying magnetism modulates the total output of energy from the Sun over its 11-year activity cycle, investigation of how magnetic interactions convert magnetic energy into intense ultraviolet and X-ray radiation, studies of the magnetic causes of energetic outbursts known as solar flares and coronal mass ejections, and an understanding of the origin and acceleration of the solar wind.

To accomplish these scientific objectives, *Hinode* contains a Solar Optical Telescope, abbreviated SOT, an X-Ray Telescope, or XRT for short, and an Extreme-ultraviolet Imaging Spectrometer, denoted EIS. Impressive first results were already on hand a year after *Hinode* launch, following the completion of calibration processes. Its uninterrupted high-resolution observations, from each instrument alone or combined with each other, are helping scientists unravel several longstanding solar mysteries.

The Solar Optical Telescope aboard *Hinode* is the largest visible-light solar telescope ever flown in space, with an angular resolution as good as 0.2 s of arc. It provides the first space-borne observations of the photosphere's vector magnetic field (Fig. 1.11). Although ground-based solar telescopes can resolve such fine detail under conditions of exceptionally good seeing, the Solar Optical Telescope provides it all the time. Moreover, it measures both the longitudinal and the horizontal components of the photosphere's magnetic field, while most previous instruments observed just the longitudinal one. At the center of the solar disk, the longitudinal component is the vertical one, and the horizontal component is transverse. One surprising new finding from the *Hinode* observations is the ubiquitous presence of intense horizontal magnetic fields in the quiet solar disk away from active regions; the horizontal components are on average five times stronger than the vertical ones.

The uninterrupted, high-resolution observations from *Hinode's* Solar Optical Telescope have also been used to study the emergence, motions, evolution, flows, decay, and disintegration of sunspots with unprecedented detail. And its combination of high spatial and temporal resolution have shown that the photosphere's magnetism is much more turbulent and dynamic than previously thought, continuously

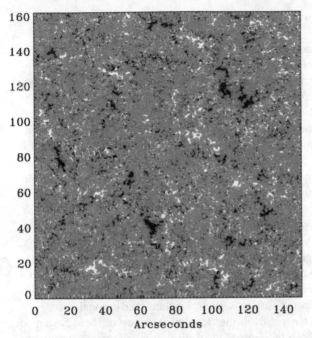

Fig. 1.11 Magnetic fields with high resolution. The longitudinal component of the magnetic field in the quiet solar photosphere, measured with the Spectro-Polarimeter, or SP for short, of the Solar Optical Telescope aboard *Hinode* on 22 November 2006. The black areas indicate magnetic fields pointing into the Sun, and the white places denote fields pointing out of the Sun. The average magnetic flux density of the longitudinal component, at the *Hinode* SP angular resolution of 0.3 s of arc, is 11.0 Maxwell per square centimeter. In contrast, the horizontal component of the photosphere magnetic field for the same area has an average apparent flux density of 55 in the same units, assuming these structures are spatially resolved. [Courtesy of Bruce Lites, the SOT consortium, NASA, JAXA and NAOJ, adapted from Lites and co-workers (2007). *Hinode* is a Japanese mission developed and launched by ISAS/JAXA, with NAOJ as domestic partner and NASA and STFC (UK) as international partners. It is operated by these agencies in co-operation with ESA and NSC (Norway).]

bringing magnetic fields together to power jets and waves that move through the overlying chromosphere and corona.

The *Hinode* X-Ray Telescope provides an unprecedented combination of spatial resolution, field of view, and image cadence. It has the broadest temperature coverage of any coronal imager to date, from 1×10^6 to 30×10^6 K. Its extremely large dynamic range permits detection of the entire corona, from coronal holes to the largest X-ray flares, and the high data rate permits observations of rapid changes in coronal magnetic and temperature structures.

The new level of detail afforded by the X-Ray Telescope, in both space and time, have provided new insights to the X-ray bright points, discovered from *Skylab*, and the X-ray jets previously investigated using the Soft X-ray Telescope aboard *Yohkoh*. The *Hinode* results have demonstrated that the bright points, which appear all over the Sun, are not points at all, but are instead resolved into small-scale magnetic loops.

High-speed jets of hot material are ejected when a closed magnetic loop emerges within a locally unipolar, or open, magnetic field region, releasing energy by means of magnetic reconnection. Observations from *Hinode's* X-Ray Telescope have shown that the jets are formed frequently in polar coronal holes, where open magnetic fields predominate, at the rate of about 60 jets per day. The large number of events, coupled with the high velocities of the apparent flows, indicate that the jets may generate magnetic waves, or Alfvén waves, which could help drive the fast solar wind into space. The X-Ray Telescope has also discovered continuous outflow of X-ray emitting gas from the edge of an active region adjacent to a coronal hole, supplying heated material into the upper corona and providing a possible source of the slow solar wind.

The X-Ray Telescope aboard *Hinode* has additionally provided new insights to the sigmoid shapes previously observed with poorer spatial resolution from *Yohkoh*. The twisted and tangled magnetic fields store large amounts of magnetic free energy, which can be released when they relax to a simpler configuration. The sigmoid shapes may therefore be used to predict when the Sun is about to release magnetic energy in powerful explosive outbursts, the solar flares and coronal mass ejections. The X-Ray Telescope has resolved the detailed structure of these S-shaped sigmoids, showing that they can be composed of discontinuous loops, strands, or threads (Fig. 1.12). Such observations are being used to test various models of how the sigmoids are formed and subsequently release magnetic energy to power solar outbursts.

The *Hinode* Extreme-ultraviolet Imaging Spectrometer measures the temperature, density, and velocity structures of the Sun's high-temperature, ionized atmosphere with high spatial and temporal resolution. It is the first extreme-ultraviolet, orbiting solar spectrometer capable of obtaining high spectral resolution data with both high spatial and temporal resolution. Electron temperature and density maps can be obtained by using the instrument to image the Sun in different spectral lines, with an angular resolution of about 2 s of arc for both quiet and active solar structures. The spectral line wavelengths and profiles can also be used to determine velocities of motions using the Doppler effect. These capabilities sharpen and refine tests of coronal heating models.

The Extreme-ultraviolet Imaging Spectrometer aboard *Hinode* is placing constraints on both static and dynamic heating mechanisms for active-region coronal loops over a broad temperature range. Loop widths and cross sections have been established for relatively cool and hot structures, and strong non-thermal velocities of about $100 \, \text{km} \, \text{s}^{-1}$ have been measured in the loop legs. In non-thermal motion, the emitting material is moving faster than expected on the basis of its temperature alone.

New and unexpected results have also been obtained from Extreme-ultraviolet Imaging Spectrometer observations of the quiet Sun away from active regions. The quiet places are highly structured, with loops at temperatures between 0.4×10^6 and 2.6×10^6 K. The cooler loops are the smaller and low-lying ones, and they are

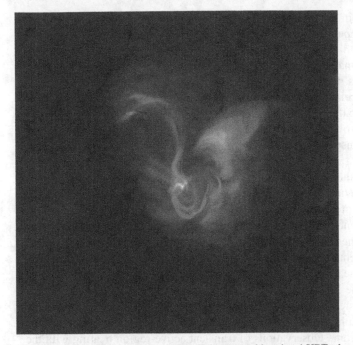

Fig. 1.12 Sigmoid details. The X-Ray Telescope, abbreviated XRT, aboard *Hinode* has resolved the S-shaped sigmoid in this active region into discontinuous, distinguishable threads highly interwoven with the surrounding material. The magnetic fields have a complex, non-potential configuration that is capable of storing free magnetic energy, which can be released to power solar flares or coronal mass ejections. This image of AR 10949 was taken by the XRT on 5 February 2007. [Courtesy of Monica Bobra, Leon Golub, Katharine Reeves, the XRT consortium, SAO, NASA, JAXA and NAOJ. *Hinode* is a Japanese mission developed and launched by ISAS/JAXA, with NAOJ as domestic partner and NASA and STFC (UK) as international partners. It is operated by these agencies, in co-operation with ESA and NSC (Norway).]

isolated from the larger, hotter, higher ones. Large outflow velocities on the order of $100 \, \text{km s}^{-1}$ have also been discovered in the quiet corona.

The three *Hinode* telescopes are operated together, continuously viewing the same structures at different heights. The Solar Optical Telescope monitors the rapidly changing photospheric magnetism; the Extreme-ultraviolet Imaging Spectrometer provides detailed temperature, density, and dynamic information in the overlying transition region; and the X-Ray Telescope provides high-resolution temporal, temperature, and context information in the low corona. The combined data from all three instruments determine how the solar magnetic field originates and changes, what magnetic configuration initiates energy release in powerful solar outbursts, what magnetic reorientation is needed to produce these outbursts over a short time, and how this variability modulates the Sun's output and creates powerful gusts in the solar wind.

The Solar Optical Telescope has, for example, been used to observe dynamic phenomena in the chromosphere that are associated with jets and waves detected by the X-Ray Telescope in the overlying corona. The chromospheric jets often exhibit an upside-down Y shape, providing evidence of ubiquitous magnetic reconnections. Energy associated with chromospheric Alfvén waves may be sufficient to heat the chromosphere or corona and accelerate the solar wind.

The twisted shapes of active regions, detected in X-rays with the X-Ray Telescope, have been correlated with sheared magnetic fields in the underlying photosphere, observed from the Solar Optical Telescope. The shear creates non-potential magnetic fields in the overlying corona, detected with the X-Ray Telescope. In one instance, colliding sunspots detected with the Solar Optical Telescope (Fig. 1.13) were associated with a major solar flare that was detected by other *Hinode* instruments, as well as by the spacecraft that observed threatening high-energy protons near Earth. By using *Hinode* to follow the evolution of solar structures that outline the magnetic field before, during, and after solar flares and coronal mass ejections, scientists hope to enhance our understanding of the way magnetic fields connect and reconfigure to cause these energetic outbursts.

The three instruments aboard *Hinode* are also being combined to improve our understanding of the heating of the solar atmosphere. Simultaneous observations of active regions with the X-Ray Telescope and the Solar Optical Telescope, for example, show that transient X-ray brightening is associated with emerging magnetic flux in the underlying photosphere, which must be contributing to the heating of the chromosphere and corona. Observations of transient active-region heating with the Extreme-ultraviolet Imaging Spectrometer and the X-Ray Telescope indicate that

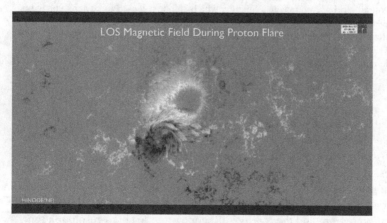

Fig. 1.13 Colliding sunspots. *Hinode's* Solar Optical Telescope detects the whirl of a new developing sunspot colliding with an existing spot. Yingna Su and co-workers (2007) have shown how the sheared magnetic fields caused by the colliding sunspots resulted in a powerful X-class flare from this active region on 13 December 2006. The flare produced high-energy protons that reached the Earth at the time of a *Space Shuttle* flight. [Courtesy of *Hinode*, JAXA, NASA, and PPARC. *Hinode* is a Japanese mission developed and launched by ISAS/JAXA, with NAOJ as domestic partner and NASA and STFC (UK) as international partners. It is operated by these agencies in co-operation with ESA and NSC (Norway).]

Table 1.11 *Hinode* instruments

Instrument	Measurement
SOT	The Solar Optical Telescope has a 50-cm aperture with an angular resolution of 0.25 arcseconds, or 175 km on the Sun, covering a visible-light wavelength range of 480–650 nm. It provides vector magneto grams, Doppler velocity, and intensity measurements of the photosphere with a temporal resolution of 5 min. S. Tsuneta et al., *Solar Physics* **249**, 167–196 (2008).
XRT	The X-Ray Telescope is a high-resolution grazing incidence telescope, a successor to the highly successful Soft X-ray Telescope on *Yohkoh*, but with a better angular resolution of about 2 arcseconds over a broad temperature range of $1–30 \times 10^6$ K with both full disk and partial fields of view and a temporal resolution as short as 2 s. L. Golub et al., *Solar Physics* **243**, 63–86 (2007) and http://xrt.cfa.harvard.edu/
EIS	The Extreme-ultraviolet Imaging Spectrometer provides images of the transition region and corona with an angular resolution of 2 arcseconds and a time resolution as short as 3 s, determining the temperature and density structure of the ionized gas and measuring the flow velocity, or speed, of the solar particles with a resolution of 3 km s^{-1}. J. L. Culhane et al., *Solar Physics* **243**, 19–61 (2007).

coronal loops at temperatures of about 1×10^6 K are not in equilibrium, and that dynamic, impulsive heating may play an important role in heating them.

The *Hinode* instruments are described in greater detail in Table 1.11. The Principal Investigators of these instruments are listed in Table 1.12, together with their

Table 1.12 *Hinode* principal investigators and their institutions

Instrument	Principal investigator	Institution
SOT	Saku Tsuneta	National Astronomical Observatory of Japan, Mitaka, Tokyo, Japan
	Ted Tarbell	Lockheed Martin Solar and Astrophysics Laboratory, Palo Alto, California, United States, succeeding Alan Title of the same institution.
XRT	Kiyoto Shibasaki	National Astronomical Observatory, Nobeyama, Radio Observatory, Nobeyama, Japan
	Edward E. DeLuca	Center for Astrophysics, Smithsonian Astrophysical Observatory, Cambridge, Massachusetts, United States, succeeding Leon Golub of the same institution.
EIS	Tetsuya Watanabe	National Astronomical Observatory of Japan, Mitaka, Tokyo, Japan
	Louise Harra	Mullard Space Science Laboratory, University College, London, United Kingdom, succeeding J. Leonard Culhane of the same institution.
	George A. Doschek	Naval Research Laboratory, Washington, District of Columbia, United States

institutions. Takeo Kosugi was the *Hinode* Project Manager through launch until his very untimely death. T. Kosugi and colleagues give an overview of the *Hinode* mission in *Solar Physics* **243**, 3–17 (2007). Ichiro Nakatani is the current JAXA Project Manager for *Hinode*; the ESA Project Scientist for the mission is Bernhard Fleck. Additional information about the mission may be found at the ISAS site http://www.isas.ac.jp/e/enterp/missions/solar-b/, the *Solar-B* project site at http://www.isas.jaxa.jp/home/solar/, the NASA site at http://solarb.msfc.nasa.gov, and the National Astronomical Observatory of Japan site http://solar-b.nao.ac.jp/index_e.shtml.

1.10 *STEREO* Observes Coronal Mass Ejections in Three Dimensions from the Sun to Earth

When impacting the Earth with the proper orientation, the billion-ton solar outbursts known as coronal mass ejections can cause powerful geomagnetic storms and intense auroras, and disrupt satellites, radio communications, and power grids. And an understanding of their trajectories and consequences is now being dramatically improved with the three-dimensional imaging provided for the first time by the twin spacecraft of NASA's *Solar TErrestrial RElations Observatory*, or *STEREO* for short; the pair of spacecraft are designated *STEREO A* and *STEREO B*, each with a mass of approximately 620 kg, which study the origin, evolution, and interplanetary consequences of coronal mass ejections.

Just as the slight offset between a person's eyes provides depth perception, the separation of the two *STEREO* spacecraft, which precede and follow the Earth in its orbit, allows a three-dimensional, stereoscopic view of coronal mass ejections from their onset at the Sun to the orbit of the Earth. This will enable scientists to forecast their arrival at the Earth within a few hours, permitting satellite and utility operators to take precautions to minimize damage. Shocks driven by coronal mass ejections play a significant role in accelerating the solar energetic particles that can damage spacecraft and harm unprotected astronauts, so *STEREO* can also be used for warnings of these threats to interplanetary spacecraft or astronauts on the Moon or Mars.

STEREO was launched on 25 October 2006, from a Delta II rocket launch vehicle from Cape Canaveral, Florida. The two spacecraft drift away from Earth at an average rate of about 22.5° per year; *STEREO A* drifts ahead of the Earth and *STEREO B* behind. The Johns Hopkins University Applied Physics Laboratory designed, built, and operates the twin observatories. NASA'S Goddard Space Flight Center's Solar Terrestrial Probes Program Office manages the *STEREO* Mission, instruments and science center.

STEREO's scientific objectives are to understand the origin, evolution, and mechanisms of coronal mass ejections and to characterize their propagation through the solar system. Its instruments are additionally investigating the mechanisms and sites of energetic particle acceleration by coronal mass ejections in the low corona and the interplanetary medium, and improving our understanding of the structure of the gusty solar wind.

Additional details about the *STEREO* Mission can be found at http://stereo. jhuapl.edu and http://stereo.gsfc.nasa.gov/, and in the book edited by Christopher Russell (2008). The instruments described in Table 1.13 are accomplishing *STEREO's* scientific objectives. They include in situ measurements of particles and magnetic fields at the location of the two spacecraft, as well as extreme-ultraviolet imagers, white-light coronagraphs, and heliosphere imagers that can follow the evolution of coronal mass ejections from the Sun through interplanetary space. They have provided the first three-dimensional reconstructions of coronal loops and the first images of coronal mass ejections outside the Sun–Earth line. The instrument Principal Investigators and their institutions are given in Table 1.14. The *STEREO* Project Scientist is Michael Kaiser at NASA's Goddard Space Flight Center. The accomplishments of the *STEREO* Mission are discussed in greater detail in other chapters of this book.

Armed with the most sophisticated detectors and telescopes ever focused on the Sun and space, this scientific armada of nine solar missions, named *Yohkoh, Ulysses, Wind, SOHO, ACE, TRACE, RHESSI, Hinode,* and *STEREO,* has vastly improved our understanding of the Sun and its interaction with the Earth. To fully interpret their results, we must first understand the landscape that they have either moved within or viewed from afar. We therefore next describe the early exploration of space, providing the historical context from which our modern discoveries came.

Table 1.13 *STEREO's* instruments arranged alphabetically by acronym

Instrument	Measurement
IMPACT	The In-situ Measurements of Particles and CME Transients samples the three-dimensional distribution and provides plasma characteristics of solar energetic particles and the local vector magnetic field. See J. G. Luhmann et al., M. H. Acuña et al., R. P. Lin et al., G. M. Mason et al., and J. A. Sauvaud et al., *Space Science Reviews* **136**, 117–184, 203–236, 241–255, 257–284, 227–238 (2008) and http://sprg.ssl.berkeley.edu/impact/ for more information.
PLASTIC	The PLAsma and SupraThermal Ion COmposition instrument provides plasma characteristics of protons, alpha particles, and heavy ions. It determines the mass and charge state composition of heavy ions and characterizes the CME plasma. Additional details are found at A. B. Galvin et al., *Space Science Reviews* **136**, 437–486 (2008) and http://stereo.sr.unh.edu/.
SECCHI	The Sun Earth Connection Coronal and Heliospheric Investigation has four instruments: an extreme ultraviolet imager, two white-light coronagraphs, and a heliospheric imager. These instruments study the three-dimensional evolution of CMEs from their origin in the low solar corona, through the expanding solar atmosphere, and interplanetary medium, to their eventual impact at Earth. Additional details of the SECCHI instruments can be found at R. A. Howard et al., *Space Science Reviews* **136**, 67–115 (2008) and http://secchi.nrl.navy.mil, http://secchi.lmsal.com/EUVI/, http://cor1.gsfc.nasa.gov/, and http://www.stereo.rl.ac.uk/science/
SWAVES	STEREO/WAVES is an interplanetary radio burst tracker that traces the generation and evolution of traveling radio disturbances from the Sun to the orbit of Earth. Additional details are available at J. L. Bougeret et al., *Space Science Reviews* **136**, 487–528 (2008) and http://swaves.gsfc.nasa.gov/.

Table 1.14 *STEREO* principal investigators and their institutions

Instrument	Principal investigator	Institution
SECCHI	Russell Howard	Naval Research Laboratory, Washington, District of Columbia, United States
SWAVES	Jean-Louis H. Bougeret	Centre National de la Recherche Scientifique, Observatoire de Paris, France
IMPACT	Janet G. Luhmann	University of California, Berkeley, United States
PLASTIC	Antoinette Galvin	University of New Hampshire, Durham, New Hampshire, United States

This will establish the framework needed for envisaging the new results, while introducing basic concepts and fundamental terms.

1.11 Summary Highlights: Modern Solar Space Missions

- The Sun is a vigorous and violent place of hot, writhing gases and powerful explosive outbursts, driven by intense variable magnetism.
- A transparent, high-temperature atmosphere, which lies above the visible solar disk or photosphere, consists of a thin chromosphere at a temperature of about 10,000 K and an extended million-degree corona.
- Magnetic fields that are generated inside the Sun loop through the photosphere and overlying solar atmosphere. These coronal loops constrain high-temperature, ionized gas that can be detected by its extreme-ultraviolet and X-ray radiation. Strong magnetic fields are often connected to sunspots within solar active regions, which are the sites of intense outbursts called solar flares. There is almost no detectable X-ray emission from coronal holes, whose open magnetic fields provide a pathway for material leaving the Sun.
- Scientists have had difficulty in explaining how the Sun's intense magnetism is generated, why it is concentrated into Earth-sized sunspots, and why solar magnetic activity varies over an 11-year cycle.
- The heating mechanism for the Sun's million-degree outer atmosphere, the solar corona, remained an unresolved paradox for more than half a century.
- The exact sources and accelerating mechanisms for the Sun's perpetual winds of electrons, ions, and magnetic fields are just beginning to be understood.
- The location, energy source, and triggering mechanism for the Sun's explosive outbursts, known as solar flares and coronal mass ejections, remained perplexing enigmas for decades.
- Eruptions on the Sun send energetic particles, intense radiation, and massive magnetic bubbles into interplanetary space. They can affect the health and safety of travelers in space, influence the habitability of the Moon and Mars, damage or disable spacecraft both near the Earth and in outer space, produce spectacular

auroras and intense geomagnetic storms, and can damage electrical transmission systems on the Earth. Scientists are not yet able to reliably predict these space weather effects and thereby provide protection from them.

- Sensitive and precise instruments aboard nine solar missions, named *Yohkoh, Ulysses, Wind, SOHO, ACE, TRACE, RHESSI, Hinode*, and *STEREO*, detect otherwise invisible particles, optical or visible, ultraviolet, and X-ray radiation, and magnetic fields, tracing the flow of energy and matter from down inside the Sun to the Earth and beyond.

- The nine solar missions have been used to solve many of the outstanding problems in our understanding of the Sun and its impact on Earth, while also revealing new unsolved dilemmas.

- The X-ray telescopes aboard the *Yohkoh* spacecraft tracked variations in the million-degree corona for 10 years, over an entire solar activity cycle, pinpointed the coronal site of magnetic energy release and particle acceleration during solar flares, and described how solar flares may be triggered by contorted magnetic fields and magnetic interactions in the low solar corona. *Yohkoh* also demonstrated that the X-ray emitting corona outside solar active regions has no permanent features, exhibiting trans-equatorial loops, bright transient jets and nanoflares.

- *Ulysses* is the first and only spacecraft to pass over the north and south polar regions of the Sun, measuring the properties of the Sun's magnetic fields and the solar wind there. It has moved across each of these polar regions three times in a 17-year period, which included two minima and one maximum in the 11-year solar activity cycle. When combined with the results of other spacecraft, instruments aboard *Ulysses* have observed dramatic changes in the solar wind structure from the minimum to the maximum in the 11-year solar activity cycle. In quiet times, near solar minimum, the high-speed solar wind rushes out of open magnetic fields in polar coronal holes and the slow-speed wind originates at low solar latitudes near equatorial coronal streamers, regions of closed magnetic fields and extended stalks. The fast and slow winds have different compositions, related to their sources, and different electron temperatures. *Ulysses* and other spacecraft have shown that the solar wind does not just flow unperturbed from the Sun to the Earth and beyond, and that it is instead altered by interplanetary magnetic reconnections, by the interactions of the fast and slow winds, and by coronal mass ejections as they travel through the vast territory between the planets. *Ulysses* has additionally provided new insights to "pick-up" ions, some of which were created when interstellar atoms penetrated the heliosphere, became ionized there, and were picked up and entrained by the solar wind.

- Instruments aboard the *Wind* spacecraft measure the solar-wind particles, magnetic fields, and other energetic particles arriving at the dayside of the Earth's magnetosphere. The spacecraft also tracks radio signals generated by solar-flare electrons and coronal-mass-ejection driven shocks, and when combined with similar observations from *STEREO, Ulysses*, and/or *Cassini* determines the three-dimensional locations of their sources. *Wind, ACE*, and other spacecraft have demonstrated that oppositely directed magnetic fields in the solar wind are joining together near the Earth's orbit, reconnecting in long, steady layers that stretch out for hundreds of Earth radii. *Wind* also measures impulsive bursts of energetic particles arriving at Earth from solar flares, as well as those more gradual but much longer lasting bursts likely coming from coronal mass ejections. These *Wind* observations help to determine the sources of solar energetic particles, the physical processes that accelerate them, and how they escape and propagate through space, all crucial information for protecting space assets and astronauts.

- *SOHO* is equipped with helioseismology instruments that measure the structure, rotation, and flows within the solar interior. They have shown that differential rotation, in which equatorial regions rotate faster than regions closer to the poles, persists to about one-third the way into the Sun, where the rotation changes and the Sun's magnetic field is probably generated. The temperature of the Sun's core, inferred from the helioseismology data, is consistent with our understanding of the Sun's internal structure, which showed that the shortfall in observed neutrinos from the Sun was due to an incomplete knowledge of neutrinos. The helioseismology results also contain hints of gravity waves that might reside in the central parts of the Sun. Internal flows have been discovered moving at different speeds in the direction of rotation and from equator to pole and back. Helioseismology instruments also routinely use sound waves to delineate active regions on the invisible backside of the Sun. The new techniques of local helioseismology have been used with *SOHO* data to obtain three-dimensional images of the structure and flows below the visible solar disk, including those below sunspots and active regions.

- *SOHO's* magnetic imager has demonstrated that tens of thousands of small magnetic loops are constantly being generated and renewed from inside the Sun, rising up and out of the photosphere to help heat the solar atmosphere. Other *SOHO* instruments have confirmed that the fast and slow solar winds respectively originate in regions of open and closed magnetic fields, which are rooted deep within the chromosphere or below, and suggested that the some of the fast winds could be fed from magnetic funnels arising from a magnetic network within polar coronal holes. Observations from *SOHO* have additionally shown that the heavier ions in the polar regions unexpectedly move faster than lighter ones, perhaps as the result of magnetic waves. Yet another *SOHO* instrument has provided more than a decade of physical information about coronal mass ejections, including their various structures, while incidentally resulting in the discovery of over 1350 comets. Dimming and waves associated with coronal mass ejections are found in extreme ultraviolet radiation detected by another *SOHO* instrument.

- Instruments aboard the *ACE* spacecraft measure the composition of energetic nuclei as they bombard the Earth, providing clues to their origin and subsequent transformations, both near the Sun and in the solar wind. Other *ACE* instruments monitor magnetic fields and shocks in the solar wind at the Earth's orbit. The source of threatening solar energetic particles arriving at Earth has been investigated by combining *ACE* observations with those of spacecraft that look back at the Sun, such as *RHESSI* and *SOHO*. Particle acceleration by interplanetary shocks, driven by coronal mass ejections, seems to be responsible for the most energetic solar particles arriving at Earth. *ACE* produced the first direct evidence that magnetic fields in the solar wind can merge and join together near the Earth's orbit, releasing magnetic energy to produce an accelerated ion flow. Observations with *ACE, Wind* and other spacecraft have demonstrated that this magnetic re-connection can be prolonged, steady and large-scale, extending for hundreds of Earth radii along the layer where oppositely directed magnetic fields meet. *ACE* and *Ulysses* observations of the velocity distributions of ions reveal a "suprather-mal" tail with up to 50 times the speed of other ions in the solar wind. They serve as a pre-accelerated source for further acceleration by interplanetary shocks asso-ciated with either coronal mass ejections or interacting fast and slow solar winds.

- Pioneering high-resolution observations with the *TRACE* telescope have been used to study the low solar corona, and the transition region between the visible-light photosphere and corona, with finer detail than ever achieved before. It has demonstrated the ubiquity of dynamic evolution of the coronal magnetic field and the intermittent heating of the solar corona; discovered the wide-spread presence of magneto-hydrodynamic, or MHD, waves, both longitudinal and transverse, many of which travel through the chromosphere with unanticipated efficiency; and resolved the structure of the corona down to the smallest observable scale, with little significant cross-sectional variation in coronal loops. *TRACE* has shown that bright arch-like magnetic structures in solar active regions are com-posed of numerous long, thin strands that are heated at their base and contain flowing material, with little observable twists or braids and generally surprisingly efficient reconnection. *TRACE* has investigated the structure and dynamic behav-ior of these coronal loops before, during and after solar flares and coronal mass ejections, providing new insights to the triggering mechanisms and magnetic en-ergy release of these explosive outbursts. It has also permitted measurements of reconnection rates, twist and writhe in filaments, and showed that active-region magnetic loops oscillate to and fro or reform in the wake of solar flares. Fine de-tails of moving structures have also been observed all across the Sun, including those that produce ubiquitous, jet-like spicules.

- *RHESSI* obtains images and spectroscopy of solar flares at hard X-ray and gamma-ray wavelengths, locating the regions of flare energy release in the low corona and studying the acceleration, propagation, and evolution of flare-accelerated electrons, protons, and heavier ions. It has provided the first ever gamma-ray line image of a solar flare and the first spectroscopic resolution of the anti-matter positron annihilation line, confirmed that solar flares can be ener-gized by large-scale magnetic reconnection in the low corona, and together with other spacecraft shown that the most energetic solar particles arriving at Earth are often accelerated by coronal mass ejections rather than flares.

- Instruments aboard *Hinode* measure the Sun's varying magnetic field with un-precedented detail, and observe how magnetic energy is ultimately dissipated to heat the million-degree corona, drive the solar wind, and energize solar flares and coronal mass ejections. It contains the largest visible-light solar telescope ever flown in space, an X-ray telescope with unprecedented spatial resolution and the broadest temperature range of any coronal imager to date, and the first extreme-ultraviolet orbiting solar spectrometer capable of obtaining simultaneous spectral data with high spectral, spatial, and temporal resolution. Impressive first results indicate that the Sun's magnetism is much more dynamic than previously known, with evidence for filamentary structures, current systems, magnetic waves, oscillations and shocks, and rapid reconfiguration and reconnection. New *Hinode* discoveries and observations include intense, ubiquitous horizontal magnetic fields in the photosphere, the detailed emergence, evolution and disintegration of sunspots, the loop structure of X-ray bright points, the magnetic reconnection that energizes high-speed jets in the chromosphere and corona, magnetic Alfvén waves and outflowing hot gas that may contribute to the solar wind, the detailed structure and underlying magnetic shear of X-ray sigmoids that precede solar outbursts, non-thermal motions in active-region coronal loops, and both cool and hot loops in the quiet Sun away from active regions.

- *STEREO* consists of two identical satellites, moving before and after the Earth in its orbit, each equipped with instruments that together permit observations of coronal mass ejections in three dimensions from the Sun to the Earth. Each *STEREO* spacecraft includes in situ particle and magnetic field detectors, an extreme-ultraviolet imager, a white-light coronagraph, a heliospheric imager, and a radio burst tracker. These instruments are poised to determine the trajectories and consequences of coronal mass ejections, and to greatly improve forecasts of their potential effects. Some of them have already been used for the first three-dimensional reconstruction of coronal loops, and to obtain the first view of a coronal mass ejection from outside the Earth–Sun line.

Chapter 2
Discovering Space

2.1 Space Is Not Empty

2.1.1 Auroras and Geomagnetic Storms from the Sun

The first suggestions that space is not a cold, empty void came from the Earth, where auroras light up the polar regions and magnetic variations make compass needles quiver. They both suggested that the Sun was sending material corpuscles into space.

The aurora borealis and aurora australis, or northern and southern lights, illuminate the Arctic and Antarctic skies, where curtains of multi-colored light dance and shimmer across the night sky far above the highest clouds (Fig. 2.1). And it is not by accident that the auroras occur near the Earth's geographic poles, for the Earth is a huge dipolar magnet, with north and south magnetic poles located near the geographic ones. That is why a compass points roughly north or south. As suggested by the great English astronomer Edmond Halley in 1716, the aurora lights are due to "magnetical effluvia" circulating poleward in the Earth's dipolar magnetic field.

In 1733, the French scientist Jean Jacques d'Ortous de Mairan asserted that the aurora is a cosmic phenomenon, arising from the entry of solar gas into the Earth's atmosphere. He also suggested a possible connection between the "frequency, the cessation and the return of sunspots and the manifestation of the aurora borealis." Nearly one and a half centuries later, Elias Loomis, Professor of Natural Philosophy and Astronomy at Yale College, demonstrated a correlation between great auroras and the times of maximum in the number of sunspots, which varies with a period of about 11 years.

At the turn of the 20th century, the Norwegian physicist Kristian Birkeland argued that the Earth's magnetism focuses incoming electrons to the polar regions where they produce auroras. This is because magnetic fields create an invisible barrier to charged particles, causing them to move along their magnetic conduits rather than across them. Birkeland demonstrated his aurora theory in laboratory experiments by sending cathode rays, or beams of electrons, toward a magnetized sphere, called a terella, with a dipolar magnetic field, using phosphorescent paint to show where the electrons struck it. The resulting light was emitted near the magnetic

K.R. Lang, *The Sun from Space*, Astronomy and Astrophysics Library,
© Springer-Verlag Berlin Heidelberg 2009

Fig. 2.1 Aurora Borealis. Swirling walls and rays of shimmering green and red light are found in this portrayal of the fluorescent Northern Lights, or Aurora Borealis, painted in 1865 by the American artist Frederic Church. (Courtesy of the National Museum of American Art, Smithsonian Institution, gift of Eleanor Blodgett.)

poles, with glowing shapes that reproduced many of the observed features of the auroras.

As Birkeland expressed it in 1913:

> It seems to be a natural consequence of our points of view that the whole of space is filled with electrons and flying ions of all kinds. We have assumed that each stellar system in evolutions throws off electric corpuscles into space. It does not seem unreasonable therefore to think that a greater part of the material masses in the universe is found, not in the solar systems or nebulae, but in "empty" space.

Another indication of the fullness of space came from watching compass needles, which fluctuate, jiggle, and do not always point in exactly the same direction. When the compass needles swing violently to and fro, with the greatest sway, the Earth's magnetic fields are being shook to their very foundations. And these powerful, global disturbances of the Earth's magnetism, the intense geomagnetic storms, are also synchronized with the Sun's 11-year sunspot cycle, suggesting that outbursts from the Sun are buffeting, compressing, and distorting the Earth's magnetism, causing at least some of the geomagnetic storms that produce large, irregular fluctuations of compass needles.

During a 9-month northern polar expedition, in 1902 and 1903, Birkeland monitored auroras and geomagnetic activity, noting that there are always detectable variations in the geomagnetic field at high latitudes and that a faint aurora borealis is also usually present. So, electrically charged particles seemed to be always flowing from the Sun to the Earth, indirectly accounting for both the nearly continuous northern lights and the geomagnetic storms, but with varying brilliance and intensity that seemed to be controlled by the Sun.

Birkeland thought in terms of electron beams and currents of electrons from the Sun. Such ideas were challenged first by Arthur Schuster in 1911 and then in 1919 by Frederick A. Lindemann, Professor of Physics at Oxford and Science Advisor to Winston Churchill. Both scientists argued that a beam of solar electrons would disperse, and effectively blow itself apart, because of the repulsive force of like charges, long before reaching the Earth. Lindemann noted, however, that a stream of solar electrons and protons in equal numbers could retain its shape and travel to the Earth to initiate magnetic disturbances there. Then in 1930–1931, Sidney Chapman and Vincent Ferraro described how sudden changes in the geomagnetic field could result when clouds of solar electrons and protons collide with the Earth's magnetic field.

But what, you might wonder, is the Sun made out of, and how could the apparently calm, serene, and unchanging Sun be sending this material into space?

2.1.2 The Main Ingredients of the Sun

The visible disk of the Sun is called the photosphere, which simply means the sphere in which sunlight originates – from the Greek *photos* for light. And when the photosphere's light is spread out into its different colors, or wavelengths, it is cut by several dark gaps, now called absorption lines, which identify the chemical ingredients of the Sun. They were first noticed by William Hyde Wollaston in 1802 and investigated in far greater detail by Joseph Von Fraunhofer in 1814–1815.

The dark spectral features are called absorption lines because they each look like a line in a spectral display of light intensity at different wavelengths, and because they are produced when atoms in a cool, tenuous gas absorb the radiation of hot, dense underlying material. Since each element, and only that element, produces a unique set of dark absorption lines, their specific wavelengths can be used to fingerprint the atom or ion from which they originated.

By comparing the solar absorption lines with those emitted by the elements vaporized in the laboratory, Gustav Kirchhoff was able to identify many of the elements in the solar spectrum in 1861. Working with Robert Bunsen, inventor of the Bunsen burner, Kirchhoff showed that the Sun contains sodium, calcium, and iron (Table 2.1). This suggested that the Sun is made out of terrestrial elements that are vaporized at the high stellar temperatures, but that is only partly true. Many of the visible spectral lines were associated with hydrogen, a terrestrially rare element.

In 1885 Johann Balmer developed a mathematical formula that described the regular spacing of the wavelengths of the Fraunhofer absorption lines of the hydrogen atom that are seen in the visible light of the Sun. His formula was used to predict hydrogen spectral lines that were subsequently observed at the invisible

Table 2.1 Prominent absorption lines in photosphere sunlight[a]

Wavelength (nm)	Fraunhofer letter	Element symbol and name
393.368	K	Ionized Calcium, Ca II
396.849	H	Ionized Calcium, Ca II
410.175	h	Hydrogen, H_δ, Balmer delta transition
422.674	g	Neutral Calcium, Ca I
431.0 ± 1.0	G	CH molecule
438.356	d	Neutral Iron, Fe I
486.134	F	Hydrogen, H_β, Balmer beta transition
516.733	b_4	Neutral Magnesium, Mg I
517.270	b_2	Neutral Magnesium, Mg I
518.362	b_1	Neutral Magnesium, Mg I
526.955	E	Neutral Iron, Fe I
587.6	D_3	Helium, He I
588.997[b]	D_2	Neutral Sodium, Na I
589.594	D_1	Neutral Sodium, Na I
656.281	C	Hydrogen, H_α, Balmer alpha transition
686.719	B	Molecular Oxygen, O_2, in our air
759.370	A	Molecular Oxygen, O_2, in our air

[a]Adapted from Lang (2001). The wavelengths are in nanometer units, where $1\,nm = 10^{-9}$ m. Astronomers have often used the Angstrom unit of wavelength, where 1 Angstrom $= 1\,Å = 0.1\,nm$. Joseph von Fraunhofer used the letters around 1814 to designate the spectral lines before they were chemically identified, but the subscripts denote components that were not resolved by Fraunhofer. A Roman numeral I after an element symbol denotes a neutral, or unionized atom, with no electrons missing, whereas the Roman numeral II denotes a singly ionized atom with one electron missing. The lines A and B are produced by molecular oxygen in the terrestrial atmosphere
[b]Fraunhofer's D line includes the two sodium lines, designated D_1 and D_2, and the helium line at 587.6 nm, designated D_3

infrared, ultraviolet and radio wavelengths. The most intense Balmer line is now known as the Balmer hydrogen alpha line, designated Hα, at a red wavelength of 656.3 nanometers.

Balmer's equation was explained by Niels Bohr's 1913 model of the hydrogen atom, in which the atom's single electron can only occupy specific orbits with definite, quantized values of energy. The observed absorption lines of hydrogen occur when an electron jumps between these allowed orbits.

Detailed investigations of the Sun's absorption-line intensities, by Albrecht Unsöld in 1928, suggested that the Sun is mainly composed of the lightest element, hydrogen, which is terrestrially rare. The observed luminosity of the Sun additionally requires that the entire star must be predominantly composed of hydrogen. In contrast, the Earth is primarily made out of heavy elements that are relatively uncommon in the Sun.

Helium is so rare on the Earth that it was first discovered in the Sun – by the French astronomer Pierre Jules César (P. J. C.) Janssen and the British astronomer Sir Joseph Norman Lockyer as an unidentified yellow emission line in the solar spectrum observed during the solar eclipse of 18 August 1868. It was probably not until the following year that Norman Lockyer convinced himself that the yellow line at 587.6 nm could not be identified with any known terrestrial element, and named the element "helium" after the Greek Sun god, *Helios*, who daily traveled across the sky in a chariot of fire drawn by four swift horses. Helium was not found on Earth until 1895, when the Scottish chemist William Ramsay discovered it as a gaseous emission given off by a heated uranium mineral clevite. Ramsay received the Nobel Prize in Chemistry in 1904 for his discovery of inert gaseous elements in the air and his determination of their place in the periodic system.

Altogether, 92.1% of the atoms of the Sun are hydrogen atoms, 7.8% are helium atoms, and all the other heavier elements make up only 0.1%. In contrast, the main ingredients of the rocky Earth are the heavier elements like silicon and iron, which explains the Earth's higher mass density – about four times that of the Sun, which is about as dense as water.

Earth has insufficient gravity to hold either hydrogen or helium in its atmosphere. Today, helium is used on Earth to inflate party balloons and in its liquid state to keep sensitive electronic equipment cold, but there is so little helium left on the Earth that we will run out of it soon.

2.1.3 The Hot Solar Atmosphere

We sometimes consider the photosphere to be the surface of the Sun, but it is not really a surface. Being entirely gaseous, the Sun has no solid surface and no permanent features. Moreover, the sharp outer rim of the Sun is illusory, for a hot, invisible atmosphere envelops it and extends all the way to the Earth and beyond. The photosphere merely marks the level beyond which the solar gases become tenuous enough to be transparent.

The atmosphere just above the round, visible disk of the Sun is far less substantial than a whisper, and more rarefied than the best vacuum on Earth. It is so tenuous that we see right through it, just as we see through the Earth's clear air. This diaphanous atmosphere of the Sun includes, from its deepest part outward, the photosphere, the chromosphere, from *chromos*, the Greek word for color, and the *corona*, from the Latin word for crown, as well as the transition region between the chromosphere and the corona.

Because of their very low densities and high temperatures, the chromosphere and corona produce bright spectral features called emission lines. Atoms and ions in a hot tenuous gas produce such emission features, heated to incandescence and shining at precisely the same wavelengths as the dark absorption lines produced by the same substance.

The chromosphere is normally invisible because of the glare of the photosphere shining through it, but it becomes briefly visible for just a few seconds at the beginning or end of a total eclipse of the Sun, as a narrow pink or rose-colored rim around the Moon's disk. The reddish hue is due to the bright red Balmer alpha transition of hydrogen, dubbed hydrogen alpha, at 656.3 nm, the wavelength of Fraunhofer's

C line. The unusual intensity of the chromosphere's hydrogen line, which is seen in emission rather than absorption, confirmed the great abundance of hydrogen in the solar atmosphere, for larger amounts of a substance tend to produce a brighter spectral line.

Other prominent emission lines of the chromosphere are the yellow line of helium at 587.6 nm and the two violet lines of calcium ions. The calcium atoms have been singly ionized, so they are missing one electron and are designated Ca II. They emit radiation at wavelengths of 393.4 and 396.8 nm, and are often called the calcium H and K lines after Fraunhofer's designation of the corresponding absorption lines in the underlying photosphere.

In 1891, the French astronomer Henri Deslandres at Meudon and the American astronomer George Ellery Hale at Mount Wilson independently invented an entirely new way of observing the chromosphere. Instead of looking at all of the Sun's colors together, they devised an instrument, called the spectroheliograph, meaning Sun spectrum recorder. It creates an image of the Sun in just one color or wavelength, without the blinding glare of all the other visible wavelengths. In a spectroheliograph, the sunlight falls on a vertical slit, and light coming through the slit is spread out in wavelength by a diffraction grating (Fig. 2.2). Light at the wavelength of one of the bright emission lines is then directed through a second slit. The two slits are moved together, with the first slit scanning the Sun from side to side, and an image of the Sun is obtained at the chosen wavelength.

By tuning in the red emission of hydrogen or a violet line of calcium, the spectroheliograph can be used to isolate the light of the chromosphere and produce photographs or digital images of it without all the rest of visible sunlight. In this way,

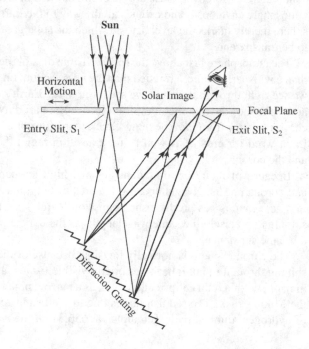

Fig. 2.2 Spectroheliograph.
A small section of the Sun's image at the focal plane of a telescope is selected with a narrow entry slit, S_1, and this light passes to a diffraction grating, producing a spectrum. A second slit, S_2, at the focal plane selects a specific wavelength from the spectrum. If the plate containing the two slits is moved horizontally, then the entrance slit passes adjacent strips of the solar image. The light leaving the moving exit slit then builds up an image of the Sun at a specific wavelength

the chromosphere can be observed across the entire disk whenever the Sun is in the sky, rather than just at the edge during a brief, infrequent solar eclipse.

Bright regions, called *plage* from the French word for "beach", glow in hydrogen-alpha light (Fig. 2.3); they are often located near sunspots in places with intense magnetism. Long, dark filaments also curl across the hydrogen-alpha Sun. They are huge regions of dense, cool gas supported by powerful magnetic forces. Indeed, the Sun's magnetism dominates the chromosphere and gives rise to its startling inhomogeneity.

The corona becomes visible to the unaided eye for only a few minutes when the Sun's bright disk is blocked out, or eclipsed, by the Moon. During such a total solar eclipse, the corona is seen at the limb, or apparent edge, of the Sun, against the blackened sky as a faint halo of white light, or all the visible colors combined. The eclipse corona contains high-density coronal streamers that are bulb shaped in

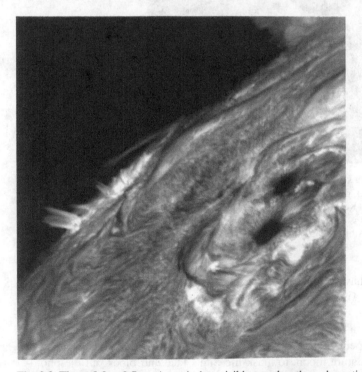

Fig. 2.3 The red-faced Sun. At optical or visible wavelengths, solar activity is best viewed by tuning to the red line of atomic hydrogen – the Balmer hydrogen alpha transition at 656.3 nm. Light at this wavelength originates in the chromospheric layers of the Sun, which lie just above the part we see with the eye. An active region, shown in the right half of this image, contains two round, dark sunspots, each about the size of the Earth, and bright plage that marks highly magnetized regions. Long, dark filaments are held in place by arched magnetic fields. This image was taken on 26 April 1978. (Courtesy of Victor Gaizauskas, Ottawa River Solar Observatory, Herzberg Institute of Astrophysics National Research Council of Canada.)

Fig. 2.4 Eclipse corona streamers. The million-degree solar atmosphere, known as the corona, is seen around the black disk of the Moon, photographed in white light, or all the colors combined, from atop Mauna Kea, Hawaii during the solar eclipse on 11 July 1991. The million-degree, electrically charged gas is concentrated in numerous fine rays as well as larger helmet streamers. (Courtesy of the High Altitude Observatory, National Center for Atmospheric Research.)

the low corona and have narrow, ray-like stalks which can extend to large distances from the Sun (Fig. 2.4). Such features are sometimes called helmet streamers, whose name is derived from spiked helmets once common in Europe.

To look further out in the corona at times other than a total eclipse of the Sun, astronomers use a special telescope called the coronagraph, which has a small occulting disk, or miniature moon, to mask the Sun's face and block out the photosphere's light. A coronagraph detects the photosphere sunlight scattered off coronal electrons, providing an edge-on, side view of the corona.

The first coronagraph was developed in 1930–1931 by the French astronomer Bernard Lyot, and soon installed by him at the Pic du Midi observatory in the Pyrenees. As Lyot realized, such observations are limited by the bright sky to high-altitude sites where the thin, dust-free air scatters less sunlight. The best coronagraph images with the finest detail are obtained from high-flying satellites where almost no air is left and where the daytime sky is truly and starkly black.

As detailed in Chaps. 4 and 5, space-borne coronagraphs can provide unique perspectives on the regions where the corona is heated, and the solar winds originate and are accelerated to high speeds. In 1979–1982, for example, John L. Kohl and his co-workers used 5-min rocket flights of an ultraviolet coronagraph spectrometer to pioneer spectroscopic observations of the low, extended corona in the absence of a total solar eclipse. They used similar observations from *Spartan 201's* for about 40 h each in the 1990s, in preparation for the UltraViolet Coronagraph Spectrometer, abbreviated UVCS, instrument aboard the *SOlar and Heliospheric Observatory*, or *SOHO*.

The solar corona presents one of the most puzzling enigmas of solar physics. It is hundreds of times hotter than the underlying photosphere. The corona's searing heat was suggested by the identification of emission lines, first observed during solar eclipses more than a century ago. And the exceptionally hot corona was fully confirmed by observations of the Sun's radio emission.

During the solar eclipse of 7 August 1869, both Charles Young and William Harkness first found that a conspicuous green emission line characterizes the spectrum of the solar corona. Within half a century, eclipse observers had detected at least ten coronal emission lines and the number doubled in a few decades more, but not one of them had been convincingly explained.

Since none of the coronal features had been observed to come from terrestrial substances, astronomers concluded that the solar corona consisted of some mysterious ingredient, which they named "coronium." Belief in the new element lingered for many years, until it became obvious that there was no place for it in the atomic periodic table, and it was therefore not an unknown element, but a known substance in an unusual state.

The solution to the coronium puzzle was provided by Walter Grotrian, of Potsdam, and the Swedish spectroscopist, Bengt Edlén, in 1939 and 1941, respectively, who attributed the coronal emission lines to familiar terrestrial elements, but in an astonishingly high degree of ionization and at an unexpectedly hot temperature. The spectral lines were attributed to relatively rare transitions in a very tenuous gas (Table 2.2). They are called forbidden emission lines because collisions can keep them from happening in even the best vacuum on Earth.

The identified features were emitted by the atoms deprived of 9–14 electrons (Table 2.2). The coronal particles would have to be moving very fast, with temperatures of millions of Kelvin, to have enough energy to rip off so many electrons during atomic collisions. Elements also move at a faster speed in a hotter gas, producing Doppler shifts in their radiation and broadening the observed spectral features. Already in 1941, Edlén noticed that the observed widths of the emission lines in the corona indicate a temperature of about 2×10^6 K.

In 1946 Australian scientists, led by Joseph L. Pawsey, used war-surplus radar (radio detection and ranging) equipment to monitor the Sun's radio emission at 1.5-m wavelength, showing that its intensity, though highly variable, almost never fell below a threshold level. Pawsey's colleague, David F. Martyn, noticed that the source of the steady radio component must be the corona, since radiation at meter

Table 2.2 Strong coronal forbidden emission lines[a]

Wavelength (nm)	Ion	Name	Wavelength (nm)	Ion
338.8	Fe XIII		670.2	Ni XV
423.2	Ni XII		789.2	Fe XI
530.3	Fe XIV	Green line	802.4	Ni XV
569.4	Ca XV	Yellow line	1074.7	Fe XIII
637.4	Fe X	Red line	1079.8	Fe XIII

[a]Adapted from Edlén (1941). The symbols Ca, Fe, and Ni denote, respectively, calcium, iron, and nickel. Subtract one from the Roman numeral to obtain the number of missing electrons. Thus, the ion Fe XIII is an iron atom missing 12 electrons. The wavelength is in units of nanometers, or 10^{-9} m. Astronomers have often used the Angstrom unit of wavelength, where 1 Angstrom = $1 \text{ Å} = 0.1 \text{ nm} = 10^{-10}$ m

wavelengths would be reflected by this outer part of the Sun and one could not use long radio waves to look past it. Moreover, the observed intensity of the radio emission from such a tenuous gas corresponded to a temperature of approximately a million Kelvin. The existence of a million-degree corona, first suggested by the corona's emission lines, was thereby confirmed from its radio emission, in papers published in the same year by Martyn, Pawsey and independently by the Russian physicist Vitalii L. Ginzburg.

The corona is so intensely hot that its abundant hydrogen is torn into numerous electrons and protons; each hydrogen atom consists of one electron moving about one central, nuclear proton. The solar atmosphere therefore consists mainly of electrons and protons, with smaller amounts of heavier ions created from the less abundant elements in the Sun. These electrons, set free from former atoms, scatter the photosphere sunlight that strikes them, providing the pearl-white coronal radiation seen during a total solar eclipse.

Although the corona's electrons, protons, and other ions are electrically charged, the overall corona is electrically neutral and has no net charge. In such an ionized gas, often called plasma, the total negative charge of all the electrons is equal to the combined positive charge of all the ions.

Somehow the corona is heated to a few million Kelvin just above the photosphere with a temperature of 5,780 K, which is completely unexpected. It is something like watching a glass of water boiling on your cool kitchen counter. Heat normally moves from a hotter to a colder region, not the other way around. Simply put, there is too much heat in the corona. Early attempts to resolve this heating paradox and modern possibilities for explaining it are given in Chap. 4.

But for now the main point, as far as interplanetary space is concerned, is that the corona is so hot that it cannot stay still. At a million Kelvin, the sizzling heat is too hot to be entirely constrained by either the Sun's inward gravitational pull or its magnetic forces. An overflow corona is forever expanding in all directions, filling the solar system with a great eternal wind of charged particles and magnetic fields that are always blowing away from the Sun.

2.1.4 Discovery of the Solar Wind

The notion that something is always being expelled from the Sun first arose from the observations of comet tails. Comets appear unexpectedly almost anywhere in the sky, moving in every possible direction, but with tails that always point away from the Sun. A comet therefore travels headfirst when approaching the Sun and tail-first when departing from it. Ancient Chinese astronomers concluded that the Sun must have a *chi*, or "life force," that blows the comet tails away. And in the early 1600s, the German astronomer Johannes Kepler proposed that the pressure of sunlight pushes the comet tails away from the Sun.

Modern scientists noticed that a comet could have two tails. One is a yellow tail of dust, which can litter the comet's curved path. The dust is pushed away from the Sun by the pressure of sunlight. The other tail is electric blue, shining in the light of ionized particles. The comet ion tails always stream along straight paths away from the Sun with velocities many times higher than could be caused by the weak pressure of sunlight.

In the early 1950s, the German astrophysicist Ludwig Biermann proposed that streams of electrically charged particles, called corpuscular radiation, poured out of the Sun at all times and in all directions to shape the comet ion tails. Summing up his work in 1957, Biermann concluded that:

> The acceleration of the ion tails of comets has been recognized as being due to the inter-action between the corpuscular radiation of the Sun and the tail plasma. The observations of comets indicate that there is practically always a sufficient intensity of solar corpuscu-lar radiation to produce an acceleration of the tail ions of at least about twenty times solar gravity.

Thus, the ion tails of comets act like an interplanetary windsock, demonstrating the existence of a continuous, space-filling flow of charged particles from the Sun. Biermann proposed that a perpetual stream of electrically charged particles pours out of the Sun at all times and in all directions, colliding with the cometary ions and imparting momentum to them. This would accelerate the comet ions and push them radially away from the Sun in straight tails.

The comet tails serve as probes of the solar particles streaming away from the Sun; it is something like putting a wet finger in the wind to show it is there. Bier-mann estimated a speed of $500-1,000\,\mathrm{km\,s^{-1}}$ from moving irregularities and the directions of comet tails.

By 1957, the English geophysicist Sydney Chapman had presented mathemat-ical arguments that seemed to show that Biermann was wrong. Chapman noticed that the free electrons in the very hot corona make it a good thermal conductor, even better than a metal. The electrons in the corona would therefore carry its intense heat far into space, somewhat like an iron bar that is heated at one end and therefore becomes hot all over. This meant that a static corona must spread out to the Earth's orbit and beyond. According to Chapman, this extended, non-expanding corona would block any outward corpuscular stream from the Sun. Biermann disagreed, arguing that the solar corpuscles would sweep any stationary gas out of the solar system.

In 1958 Eugene N. Parker, of the University of Chicago, reconciled these conflicting ideas of a static corona and a perpetual stream of charged particles. He added dynamic terms to Chapman's equations, showing how a relentless steady-state outward expansion from the Sun might work – and dubbing it the solar wind. A very hot coronal gas can create an outward pressure that becomes greater than the inward pull of the Sun's gravity at increasing distances from the Sun, where the expanding gas accelerates from slow, subsonic speeds near the Sun to fast, supersonic speeds in interplanetary space. Parker also argued that the wind would pull the Sun's magnetic field into surrounding interplanetary space, obtaining a spiral shape due to the combination of radial flow and solar rotation.

Parker's theoretical conclusions were very controversial, and were initially received with a great deal of skepticism by established scientists. Referees objected to the publication of Parker's ideas, and Sydney Chapman publicly clashed with Parker over his notion of an expanding solar wind, preferring a static, non-expanding corona. The Sun might be sporadically hurling material out from localized, explosive outbursts, but it was difficult to envisage a continual ejection over the entire Sun. Critics also wondered how the solar atmosphere could be hot enough to sustain such a powerful wind.

Even with a temperature of 10^6 K, the thermal energy of protons is several times less than the Sun's gravitational pull on them at the bottom of the corona. In scientific terms, the proton's average thermal velocity, due to its hot temperature, is still less than the escape velocity of the Sun (Focus 2.1).

Focus 2.1

Escape from the Sun, Earth, and Moon

When the kinetic energy of motion of an object or a particle of mass m moving at velocity V is just equal to the gravitational potential energy exerted on it by a larger mass M, we have the relation:

$$\text{Kinetic energy} = \frac{mV^2}{2} = \frac{GmM}{D} = \text{Gravitational potential energy,}$$

where the Newtonian gravitational potential is $G = 6.67 \times 10^{-11} \, \text{N m}^2 \, \text{kg}^{-2}$, and D is the distance between the centers of the two masses. When we solve for the velocity, we obtain the escape velocity

$$V_{\text{escape}} = \sqrt{\frac{2GM}{D}},$$

where the $\sqrt{}$ sign denotes the square root of the following term and the subscript escape has been added to show that the object or particle must be moving faster than V_{escape} to leave a larger mass M. This expression is independent of the smaller mass m.

At the solar photosphere, where D becomes the solar radius of 696×10^6 (6.96×10^8) m and the Sun's mass is 1.989×10^{30} kg, the equation gives

$$V_{\text{escape}}(\text{Sun's photosphere}) = 617 \text{ km s}^{-1}.$$

By way of comparison, the escape velocity at the Earth's surface is:

$$V_{\text{escape}}(\text{Earth's surface}) = 11.2 \text{ km s}^{-1},$$

where the mean radius of the Earth is 6.378×10^6 m and the mass of the Earth is 5.9742×10^{24} kg. The escape velocity from the surface of the Moon is:

$$V_{\text{escape}}(\text{Moon's surface}) = 2.37 \text{ km s}^{-1},$$

for the Moon's mass is 7.348×10^{22} kg and its mean radius is 1.738×10^6 m.

A rocket must move faster than 11.2 km s^{-1} if it is to move from the Earth into interplanetary space, and if it travels at a slower speed the rocket will crash back down into the Earth. A lunar craft only needs to be propelled at about one-fifth of this speed to leave the Moon. There is no atmosphere on the Moon because it has a very low escape velocity, and molecules that are heated by the Sun's radiation can therefore easily leave it.

An atom, ion, or molecule moves about because it is hot. Its kinetic temperature T is defined in terms of the thermal velocity, V_{thermal}, given by the expression equating the thermal energy to the kinetic energy of motion,

$$\text{Thermal energy} = \frac{3}{2} k T = \frac{1}{2} m V_{\text{thermal}}^2 = \text{Kinetic energy}$$

or solving for the thermal velocity:

$$V_{\text{thermal}} = \sqrt{\frac{3kT}{m}},$$

where Boltzmann's constant $k = 1.38 \times 10^{-23}$ J s, the temperature of the particle is denoted by T, and its mass by m. We see right away that at a given temperature, lighter particles move at faster speeds. Colder particles of a given mass travel at slower speed.

Anything will cease to move when it reaches absolute zero on the Kelvin scale of temperature. But everything in the Universe moves, and there is nothing completely at rest. You might say that motion seems to define existence.

For the lightest known element, hydrogen, our expression gives

$$V_{\text{thermal}}(\text{hydrogen atoms}) = 157 \, T^{1/2} \text{m s}^{-1},$$

where the mass of a hydrogen atom is $m = 1.674 \times 10^{-27}$ kg. So, hydrogen atoms move at about 12.0 and 15.7 km s^{-1} in the photosphere and the chromosphere, where the respective temperatures are 5,780 and 10,000 K. Since these velocities are way below the Sun's escape velocity, hydrogen and any other heavier element must be retained in the low solar atmosphere. Even at the corona's temperature of 2×10^6 K, the thermal velocity of a hydrogen atom is 222 km s^{-1}. To leave the Sun, hot coronal hydrogen has to be given an extra push out to a distance of a few solar radii, where the solar gravity and escape velocity have become smaller. The same conclusion applies to protons that have essentially the same mass as a hydrogen atom. Since a free electron is 1,836 times lighter than a proton, it has a thermal velocity that is 42.8 times faster at a given temperature, so the electrons in the million-degree corona can easily escape the Sun's powerful gravity.

But that's not the entire story, for if the electrons all escaped, then the protons would be left behind and a net charge would build up. The effective "escape speed" is also affected by the electric attraction between electrons and protons, and a more complete discussion has been given in Arthur Hundhausen (1972a), Nicole Meyer-Vernet (1999), and Steven Cranmer (2002).

The hot, million-degree protons are not moving fast enough, on average, to overcome the gravity in the low corona, next to the visible Sun, and move out into the solar wind. So, some additional source of energy seemed to be required to help the corona break away from the Sun's powerful gravitational grasp.

Nevertheless, in just 4 years American and Soviet scientists confirmed the solar wind's existence with direct observations of the interplanetary medium, verifying many aspects of Parker's model. Arthur Hundhausen (1972a) has told the colorful story of the turbulent years preceding this vindication of the ideas of a relatively young scientist in his book, *Coronal Expansion and the Solar Wind*. Historical recollections have also been published by Parker (1998, 2001, 2002).

The confirmation of Parker's theory did not, however, reveal how or where the electrons and protons in the solar wind are accelerated to their high speeds, or the precise place of their origin on the Sun. These topics are considered in Chap. 5.

2.2 Touching the Unseen

2.2.1 The First Direct Measurements of the Solar Wind

Our civilization was forever changed with the launch of *Prosteyshiy Sputnik*, the simplest satellite, by the Soviet Union on 4 October 1957. It began a space age that has led to our daily use of artificial satellites for communications, navigation, and weather forecasting. And it also instigated a competitive race to land the first humans on the Moon, which the United States won.

To many Americans, *Sputnik* verified the threat that the Soviet Union posed to world peace. So it is not surprising that many of the space-age improvements in our life grew out of military applications. They include reconnaissance satellites that monitor enemy activity, navigation satellites that accurately target missiles launched from ships or airplanes, and intercontinental ballistic missiles that can either carry nuclear warheads to distant countries or toss satellites into space. Yet, there has always been a strong scientific component to the American and Russian space programs from the very beginning.

Some of the earliest spacecraft reached out to touch, feel, and identify the invisible constituents of space. For instance, America's first *Satellite 1958α* – launched on 1 February 1958 and better known today as *Explorer 1*, included James A. Van Allen's instruments that detected energetic charged particles in nearby space. They are trapped by the Earth's magnetic field into donut-shaped regions that girdle our planet's equator. Other spacecraft soon characterized the material components of the solar wind outside the Earth's magnetic domain.

The first direct measurements of the solar wind's corpuscular, or particle, content were made by a group of Soviet scientists led by Konstantin I. Gringauz, using four ion traps aboard the *Lunik 2* (*Luna 2*) spacecraft launched to the Moon on 12 September 1959. Each trap contained external, charged grids that acted as gates to exclude low-energy ions and to keep energetic electrons out. Only high-speed ions could pass through the trap door. All four ion traps detected the energetic ions, leading Gringauz and his colleagues to report in 1960 that "the corpuscular emission of the Sun ... has thus been observed for the first time in the interplanetary space outside the magnetic field of the Earth."

In the following year, Gringauz reported that the maximum current in all four ion traps corresponded to a solar wind flux of two million million (2×10^{12}) ions (presumably protons) per square meter per second. This is in rough accord with all subsequent measurements. No evidence for a stationary component of the interplanetary gas was found. Moreover, since the ion traps were strategically placed around the rotating spacecraft, the group could show that the wind flows from the Sun and not toward it.

A Massachusetts Institute of Technology group first carried out solar wind measurements by American scientists with instruments aboard *Explorer 10* in 1961, but unfortunately the spacecraft did not travel completely beyond the Earth's magnetic field to pristine, "undisturbed" interplanetary space.

All reasonable doubt concerning the existence of the solar wind was removed by the measurements made on board *Mariner 2*, launched on 27 August 1962. Marcia Neugebauer and Conway W. Snyder of the Jet Propulsion Laboratory used more than 100 days of *Mariner 2* data, obtained as the spacecraft traveled to Venus, to show that charged particles are continuously emanating from the Sun, for at least as long as *Mariner 2* observed them. The key aspect of their measurements was that both the velocity and the density were measured separately while the Russian mission just measured the total particle flux, which is related to the product of the velocity and density. Only by knowing the separate values could Parker's theory be fully confirmed. The velocity of the solar wind was accurately determined from the *Mariner 2* observations, with an average speed of $500 \, \text{km s}^{-1}$, in rough accord with Biermann's and Parker's predictions. In 1997 Marcia Neugebauer published a personal history of these early direct measurements of the solar wind, as well as later ones.

The solar wind flux determined by Neugebauer and Snyder was in good agreement with the values measured with the ion traps on *Lunik 2*. The average wind ion density was shown to be five million (5×10^6 protons per cubic meter near the distance of the Earth from the Sun. We now know that such a low density close to the Earth's orbit is a natural consequence of the wind's expansion into an ever-greater volume, but that variable wind components can gust with higher densities.

The *Mariner 2* data unexpectedly indicated that the solar wind has a slow and a fast component. The slow one moves at a speed of 300–400 km s^{-1}; the fast one travels at twice that speed. The low-velocity wind was identified with the perpetual expansion of the very hot corona. Uncertainties over which type of solar wind, the fast or slow one, is "ambient" persisted for decades because the observations were always made near the ecliptic, until *Ulysses* passed over the solar polar regions in 1994 and showed that the uniform fast wind is the dominant component, at least during the minimum in the 11-year solar activity cycle, with the slow wind contributing a varying input at low solar latitudes.

The high-velocity component swept past the *Mariner 2* spacecraft every 27 days, suggesting long-lived, localized sources on the rotating Sun. (The Sun spins about its axis with a period of about 25 days at the equator, but since the Earth is orbiting the Sun in the same direction that the Sun rotates, the rotation period observed from the Earth is about 27 days.) Moreover, peaks in geomagnetic activity, also repeating every 27 days, were correlated with the arrival of these high-speed streams at the Earth, indicating a direct connection between some unknown region on the Sun and disturbances of the Earth's magnetic field.

2.2.2 Properties of the Solar Wind at Earth's Orbit

Interplanetary space probes have been making in situ (Latin for "in place") measurements for decades, both within space near the Earth and further out in the Earth's orbital plane. Unlike any wind on Earth, the solar wind is rarefied plasma or an electrically neutral mixture of electrons, protons, and heavier ions, and magnetic fields streaming radially outward in all directions from the Sun at supersonic speeds of hundreds of kilometers per second.

This perpetual solar gale brushes past the planets and engulfs them, carrying the Sun's corona out into interstellar space at the rate of almost a billion (10^9) kilograms every second. As the corona disperses, gases welling up from below to feed the wind must replace it. Exactly where this material comes from and how it is accelerated to high speeds are important subjects of contemporary space research, discussed in greater detail in Chap. 5.

Although the Sun is continuously blowing itself away, the outflow can continue for billions of years without significantly reducing the Sun's mass. Every second, the solar wind blows away a million tons (10^6 tons $= 10^9$ kg). That sounds like a lot of mass loss, but it is four times less than the amount consumed every second during the thermonuclear reactions that make the Sun shine. To supply the Sun's present luminosity, hydrogen must be converted into helium, within the Sun's energy-generating

Table 2.3 Relative importance of solar wind and radiation[a]

	Solar wind	Sun's radiation
Mass loss rate	$10^9 \, \mathrm{kg \, s^{-1}}$	$4 \times 10^9 \, \mathrm{kg \, s^{-1}}$
Total mass loss (In five billion years)	$0.00005 \, M_\odot$	$0.0002 \, M_\odot$
Energy flux (1 AU)	$0.00016 \, \mathrm{J \, m^{-2} \, s^{-1}}$	$1{,}400 \, \mathrm{J \, m^{-2} \, s^{-1}}$
Momentum flux (1 AU)	$8 \times 10^{-10} \, \mathrm{kg \, m^{-1} \, s^{-1}}$	$5 \times 10^{-6} \, \mathrm{kg \, m^{-1} \, s^{-1}}$
Despin time	10^{16} years	10^{12} years

[a]Adapted from Axford (1985). $M_\odot$ denotes the Sun's mass of $2 \times 10^{30} \, \mathrm{kg}$. The mean distance of the Earth from the Sun is one astronomical unit, or 1 AU for short, where $1 \, \mathrm{AU} = 1.496 \times 10^{11} \, \mathrm{m}$

core, with a mass loss of about 4×10^6 tons every second. It is carried away by the Sun's radiation, whose energy and momentum flux vastly exceed those of the solar wind (Table 2.3). A more significant concern is the depletion of hydrogen; the Sun will run out of hydrogen in its core in about seven billion years, when our star will expand into a giant star. By that time, the Sun will have lost only about 0.005% of its mass by the solar wind at the present rate.

The hot corona extends all the way to the Earth, where it has only cooled to a little more than 100,000 K. Even though the electrons near the Earth are awfully hot, they are so scarce and widely separated that an astronaut or satellite will not burn up

Table 2.4 Mean values of solar wind parameters at the Earth's orbit[a]

Parameter	Mean value
Particle density, N	$N \approx 10$ million particles per cubic meter (five electrons and five protons per cubic centimeter)
Velocity, V	$V \approx 600 \, \mathrm{km \, s^{-1}}$
Mass density, ρ	$\rho = 10^{-20} \, \mathrm{kg \, m^{-3}}$ (protons)
Flux, F	$F \approx 10^{12} – 10^{13}$ particles per square meter per second
Temperature, T	$T \approx 120{,}000 \, \mathrm{K}$ (protons) to $140{,}000 \, \mathrm{K}$ (electrons)
Particle thermal energy, kT	$kT \approx 2 \times 10^{-18} \, \mathrm{J} \approx 12 \, \mathrm{eV}$
Proton kinetic energy, $0.5 m_p V^2$	$0.5 m_p V^2 \approx 10^{-16} \, \mathrm{J} \approx 1{,}000 \, \mathrm{eV} = 1 \, \mathrm{keV}$
Particle thermal energy density	$NkT \approx 10^{-11} \, \mathrm{J \, m^{-3}}$
Proton kinetic energy density	$0.25 N \, m_p V^2 \approx 10^{-9} \, \mathrm{J \, m^{-3}}$
Radial magnetic field, H_r	$H_r = 2.5 \times 10^{-9} \, \mathrm{T} = 2.5 \, \mathrm{nT} = 2.5 \times 10^{-5} \, \mathrm{G}$
Alfvén velocity, V_A	$V_A = 32 \, \mathrm{km \, s^{-1}}$
Sound speed, C_s	$C_s = 41 \, \mathrm{km \, s^{-1}}$

[a]These solar wind parameters are at the mean distance of the Earth from the Sun, or at one astronomical unit, 1 AU, where $1 \, \mathrm{AU} = 1.496 \times 10^{11} \, \mathrm{m}$; the Sun's radius, $R_\odot$, is $R_\odot = 6.96 \times 10^8 \, \mathrm{m}$. Boltzmann's constant $k = 1.38 \times 10^{-23} \, \mathrm{J \, K^{-1}}$ relates temperature and thermal energy. The proton mass $m_p = 1.67 \times 10^{-27} \, \mathrm{kg}$. The flux of cosmic-ray protons arriving at the Earth, discussed in the next section, is about one ten billionth (10^{-10}) that of the solar wind at the Earth's orbit. The most abundant cosmic ray protons near Earth have an energy of $10^{-10} \, \mathrm{J}$ but a local energy density of about $10^{-13} \, \mathrm{J \, m^{-3}}$, or about one ten thousandth (10^{-4}) the kinetic energy density of protons in the solar wind

when venturing into interplanetary space. Although the velocity is high, the density of the expanding corona is so low that if we could go into space and put our hands on it, we would not be able to feel it.

The reason that space looks empty is that the solar wind is very tenuous, even when compared to our transparent atmosphere. By the time it reaches the Earth's orbit, the solar wind is diluted to about five electrons and five protons per cubic centimeter, a very rarefied gas (Table 2.4). By way of comparison, there are 25 billion, billion (2.5×10^{19}) molecules in every cubic centimeter of our air at sea level.

Still, at a mean speed of about $600 \, \text{km s}^{-1}$, the flux of solar wind particles is far greater than anything else out there in space. Between one and ten million million $(10^{12} – 10^{13})$ particles in the solar wind cross every square meter of space each second (Table 2.4). That flux far surpasses the flux of more energetic cosmic rays that enter our atmosphere, which are discussed next.

2.3 Cosmic Rays

Although the Sun's winds dominate the space within our solar system, cosmic rays form an important additional ingredient. These extraordinarily energetic charged particles enter the Earth's atmosphere from all directions in outer space.

Cosmic rays were discovered in 1912, when the Austrian physicist Victor Franz Hess, an ardent amateur balloonist, measured the amount of ionization at different heights within our atmosphere. It was already known that radioactive rocks at the Earth's surface were emitting energetic "rays" that ionize the atmosphere near the ground, but it was expected that the ionizing substance would be completely absorbed after passing through sufficient quantities of the air.

Although the measured ionization at first decreased with altitude, as would be expected from atmospheric absorption of rays emitted by radioactive rocks, the ionization rate measured by Hess from balloons increased at even higher altitudes to the levels exceeding that at the ground. This meant that some penetrating source of ionization came from beyond the Earth. By flying his balloon at night and during a solar eclipse, when the high-altitude signals persisted, Hess showed that they could not come from the Sun, but from some other source. In 1936 he was awarded the Nobel Prize in Physics for his discovery of these cosmic rays.

Further balloon observations by Werner Kolhörster of Berlin showed that the ionization continued to increase with height. By 1926 the American physicist Robert A. Millikan had used high-altitude balloon measurements to confirm that the "radiation" comes from beyond the terrestrial atmosphere, and incidentally gave it the present name of cosmic rays. Millikan was the first president of the California Institute of Technology, and the first American to win the Nobel Prize in Physics – in 1923 for his work on the elementary charge of electricity and the photoelectric effect.

Cosmic rays were initially believed to be high-energy radiation, in the form of gamma rays, but global measurements showed that they are charged particles deflected by the Earth's magnetic field toward its magnetic poles (Focus 2.2). In 1927, the Dutch physicist Jacob Clay published the results of cosmic ray measurements during ocean voyages between Genoa and the Dutch colony of Java. He found lower cosmic ray intensity near the Earth's equator, suggesting that the cosmic rays consisted, in part at least, of charged particles. Clay's results were confirmed and extended between 1930 and 1933 by Arthur H. Compton of the University of Chicago. He conclusively demonstrated an increase in cosmic ray intensity with terrestrial latitude, and also made measurements at mountain altitudes, where the latitude increase was even stronger.

Focus 2.2

Charged Particles Gyrate Around Magnetic Fields

A charged particle cannot move straight across a magnetic field, but instead gyrates around it. If the particle approaches the magnetic field straight on, in the perpendicular direction, a magnetic force pulls it into a circular motion about the magnetic field line. Since the particle can move freely in the direction of the magnetic field, it spirals around it with a helical trajectory.

The size of the circular motion, called the radius of gyration and designated by R_g, depends on the velocity $V_\perp$ of the particle in the perpendicular direction, the magnetic field strength B, and the mass m and charge Ze of the particle. That gyration radius is described by the equation:

$$R_g = [m/(Ze)][V_\perp/B],$$

provided that the velocity is not close to the velocity of light c. The period P of the circular orbit is $P = 2\pi R_g/V_\perp = 2\pi m/(eZB)$, and the frequency v_g is $v_g = 2\pi/P = eZB/m$.

This expression for the gyration radius provides a shorthand method of explaining a lot of intuitively logical aspects of the charged-particle motion. It says that faster particles will gyrate in larger circles, and that a stronger magnetic field tightens the gyration into smaller coils. The attractive magnetic force increases with the charge, also resulting in a tighter gyration.

For an electron with mass $m_e = 9.11 \times 10^{-31}$ kg and charge $e = 1.60 \times 10^{-19}$ C, with $Z = 1.0$, the corresponding gyration radius is:

$$R_g(\text{electron}) = 5.7 \times 10^{-12}[V_\perp/B] \quad \text{m},$$

where $V_\perp$ is in m s^{-1} and B is in Tesla. Since the proton has the same charge, but a mass m_p that is larger by the ratio $m_p/m_e = 1836$, the proton's gyration radius is:

$$R_p(\text{proton}) = 1.05 \times 10^{-8}[V_\perp/B] \quad \text{m}.$$

At high particle velocities approaching that of light, the radius equation is multiplied by a Lorentz factor $\gamma = [1 - (V/c)^2]^{-1/2}$, which becomes unimportant at low velocities V when $\gamma = 1$. For cosmic ray protons of high velocity and large energies, $E = \gamma mc^2$, our equation becomes:

$$R_g = 3.3\, E/B \quad \text{m},$$

when the energy E is in units of GeV and the magnetic field strength is in Tesla. Thus, the path of a 1-GeV cosmic ray proton will be coiled into a gyration radius of about 10^{10} m in the interstellar medium where $B \approx 10^{-10}\,\text{T} = 10^{-6}\,\text{G}$, which is about 15 times smaller than the mean distance between the Earth and the Sun. When that 1 GeV proton encounters the terrestrial magnetic field, whose strength is roughly $10^{-4}\,\text{T}$ or 1 G, its radius of gyration is $R_g \approx 10,000\,\text{m}$, or hundreds of times smaller than the Earth whose mean radius is 6.37×10^6 m. That means the proton will gyrate about the magnetic field, and move toward Earth's magnetic poles.

The latitude increase showed that cosmic rays had to be electrically charged, but both negative and positive charges would show a similar effect. The sign of the charge was inferred in the late 1930s from the hemispheric distribution of the cutoff energy below which no vertically arriving particles are found. Lower-energy particles of positive charge will be observed if they arrive from the west; negatively charged cosmic rays of lower energy will be found in the east. Measurements of this east–west effect showed that the most abundant cosmic rays are positively charged, and most likely protons.

By the late 1940s, instruments carried by high-altitude balloons established that the cosmic rays consist of hydrogen nuclei (protons), helium nuclei (alpha particles), and heavier nuclei (Table 2.5). Cosmic ray electrons were not discovered until 1961, mainly because they are far less abundant than the cosmic ray protons at a given energy. The presence of very energetic electrons in interstellar space had nevertheless been inferred from the radio emissions of our Galaxy. Sometimes the term galactic cosmic ray is used to distinguish between very energetic, charged particles coming from interstellar space and somewhat less energetic ones coming from the Sun, historically called solar cosmic rays. Nowadays, the term cosmic ray usually refers to those originating outside the solar system, and the most energetic particles from the Sun's outbursts are known as solar energetic particles.

Table 2.5 Average fluxes of primary cosmic rays at the top of the atmosphere[a]

Type of Nucleus	Flux (particles $m^{-2}\,s^{-1}$)
Hydrogen (protons)	640
Helium (alpha particles)	94
Carbon, Nitrogen, Oxygen	6

[a]Adapted from Friedlander (1989). The flux is in units of nuclei per square meter per second for particles with energies greater than 1.5 billion (1.5×10^9) electron volts per nucleon, denoted 1.5 GeV per nucleon, arriving at the top of the atmosphere from directions within 30° of the vertical

Since the cosmic rays enter the atmosphere with energies greater than could be produced by particle accelerators on Earth, physicists used them as colossal atom destroyers, examining their subatomic debris after colliding with atoms in our atmosphere. These studies of fundamental particles led to several other Nobel Prizes in Physics – to Charles T. R. Wilson in 1927 for his cloud chamber that makes the tracks of the electrically charged particles visible by condensation of vapor; to Carl D. Anderson in 1936 for his detection of the positron, the anti-particle of the electron, amongst the byproducts of cosmic rays colliding in the air; to Patrick M. S. Blackett in 1948 for his development of the Wilson cloud chamber method, and his discoveries therewith in the fields of nuclear physics and cosmic radiation; and to Cecil F. Powell in 1950 for his discovery of a fundamental particle, called the meson, from high-altitude studies of cosmic rays using a photographic method.

Unlike electromagnetic radiation, the charged cosmic ray particles are deflected and change direction during encounters with the interstellar magnetic field that wends its way through the stars. So, we cannot look back along their incoming path and tell where cosmic rays originate; the direction of arrival just shows where they last changed course. As suggested by Walter Baade and Fritz Zwicky in 1934, cosmic rays are most likely accelerated to their tremendous energies during the explosions of dying stars, called supernovae.

A great majority of these supernovae are formed by the core collapse of massive O and B stars, which originate in OB associations whose component supernova explosions combine to blast out a surrounding superbubble – described by Mordecai-Mark Mac Low and Richard McCray in 1988. A decade later, James C. Higdon, Richard E. Lingenfelter, and Reuven Ramaty proposed that a substantial fraction of cosmic rays have originated in these superbubbles, which could account for the abundances of cosmic ray neon isotopes observed more recently from the *ACE* spacecraft.

Although they are relatively few in number, cosmic rays contain phenomenal amounts of energy. That energy is usually measured in units of electron volts, abbreviated eV – for conversion use $1\,\text{eV} = 1.602 \times 10^{-19}$ J. The greatest flux of cosmic ray protons arriving at Earth occurs at about 1 GeV, or at a billion (10^9) electron volts of energy. This is about a million times more than the kinetic energy of a typical proton in the solar wind. At this energy, a cosmic ray proton must be traveling at 88% of the speed of light (Table 2.6) while a proton in the Sun's wind moves about 500 times slower.

Cosmic rays do not all have the same energy, and there are fewer of them with higher energy above peak energy of about a billion (10^9) electron volts or 1 GeV. The number of particles with kinetic energy E is proportional to $E^{-\alpha}$, where the index $\alpha = 2.5$–2.7 for energies from a billion to a million million $(10^9$–$10^{12})$ electron volts. At higher energies there are relatively few particles. At lower energies the magnetized solar wind acts as a valve for the more abundant lower-energy cosmic rays, controlling the amount entering the solar system.

Table 2.6 Particle speeds at different particle energies, expressed as fractions of the speed of light, c^a

Particle kinetic energy (keV)	Electron speed (times c)	Proton speed (times c)
1 keV	$0.063c$	$0.0015c$
1,000 keV = 1 MeV	$0.94c$	$0.046c$
100,000 keV = 100 MeV	$0.999987c$	$0.43c$
1,000,000 keV = 1 GeV	$0.99999987c$	$0.88c$

[a]An energy of one kilo-electron volt is $1\,\text{keV} = 1.6022 \times 10^{-16}\,\text{J}$, and the speed of light $c = 2.99792458 \times 10^8\,\text{m s}^{-1}$

2.4 Pervasive Solar Magnetism

2.4.1 Magnetic Fields in the Photosphere

To most of us, the Sun looks like a perfect, white-hot globe, smooth and without a blemish. However, detailed scrutiny indicates that our star is not perfectly smooth, just as the texture of a beautiful face increases when viewed close up. Magnetism protrudes to darken the skin of the Sun in Earth-sized spots detected in its visible disk, the photosphere (Fig. 2.5). Our understanding of what causes these dark, ephemeral sunspots is relatively recent, at least in comparison to how long they have been known. The earliest Chinese records of sunspots, seen with the unaided eye, date back 3,000 years.

In 1908 George Ellery Hale first established the existence of intense, concentrated magnetic fields in sunspots, using subtle wavelength shifts in spectral lines detected at his solar observatory on Mt. Wilson, California. Such a shift, or line splitting, is called the Zeeman effect after Pieter Zeeman who observed it with magnetic fields on the Earth in 1896; Hendrik Lorentz predicted the effect, and the two Dutch physicists shared the Nobel Prize in Physics in 1907 for their researches into the influence of magnetism upon radiation.

The size of the sunspot wavelength shifts measured by Hale indicated that they contain magnetic fields as strong as 0.3 T, or 3,000 G (Focus 2.3). That is about 10,000 times the strength of the terrestrial magnetic field, which orients our compasses. The intense sunspot magnetism acts as both a valve and a refrigerator, choking off the outward flow of heat and energy from the solar interior and keeping the sunspots cooler and darker than their surroundings.

Focus 2.3

The Zeeman Effect

When an atom is placed in a magnetic field, it acts like a tiny compass, adjusting the energy levels of its electrons. If the atomic compass is aligned in the direction of the magnetic field, the electron's energy increases; if it is aligned in the opposite direction, the energy decreases. Since each energy change coincides with a change

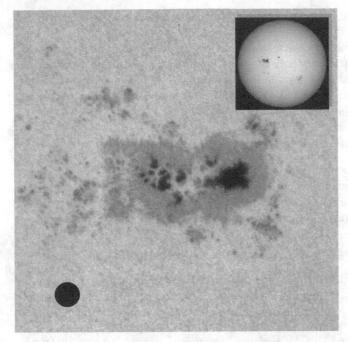

Fig. 2.5 Sunspot group. Intense magnetic fields emerge from the interior of the Sun through the photosphere, producing groups of sunspots. The sunspots appear dark because they are slightly cooler than the surrounding photosphere gas. This composite image shows the visible solar disk in white light, or all the colors combined (*upper right*), and an enlarged white-light image of the largest sunspot group (*middle*), which is about 12 times larger than the Earth whose size is denoted by the black spot (*lower left*). (Courtesy of *SOHO*, ESA and NASA.)

in the wavelength or frequency emitted by that electron, a spectral line emitted at a single wavelength by a randomly oriented collection of atoms becomes a group of three lines of slightly different wavelengths in the presence of a magnetic field. This magnetic transformation has been named the Zeeman effect, after Pieter Zeeman, who first noticed it in the terrestrial laboratory. The size of an atom's internal adjustments, and the extent of its spectral division, increases with the strength of the magnetic field.

We can understand the Zeeman effect by considering the motion of a free electron in the presence of a magnetic field. The electron will circle about the magnetic field with a radius, R_g, described in Focus 2.2, and with a period, P, given by:

$$P = 2\pi R_g / V_\perp = 2\pi m_e / (eB).$$

At the velocity $V_\perp$, the electron goes once around the circumference $2\pi R$ in the period P. The frequency, ν_g, of this motion is

$$\nu_g = 2\pi / P = eB / m_e = 1.8 \times 10^{11} B \quad \text{Hz},$$

where B is the magnetic field strength in Tesla. Here e and m_e, respectively, denote the charge and mass of the electron.

When an atom is placed in a magnetic field, a very similar thing happens to its electrons and the spectral lines they emit. A line that radiates at a wavelength λ without a magnetic field becomes split into two or three components depending on the orientation of the magnetic field. For the three component split, the shift, $\Delta\lambda$, in wavelength of the two outer components is given by:

$$\Delta\lambda = \lambda^2 \Delta v/c = 4.7 \times 10^{-8} \lambda^2 B \text{ nm,}$$

where the shift in frequency is Δv, the velocity of light is $c = 2.9794 \times 10^8 \text{ m s}^{-1}$; both $\Delta\lambda$ and λ are in nanometers, or 10^{-9} m, and B is in Tesla.

The separation is thus proportional to the magnetic field strength B. In 1908 George Ellery Hale made measurements of this Zeeman splitting in sunspots, showing that they have magnetic field strengths of about 0.3 T or 3,000 G.

Magnetic fields are described by lines of force, like those joining the north and south poles of the Earth or the opposite poles of a bar magnet. When the lines are close together, the force of the field is strong; when they are far apart, the force is weak. The direction of the lines of force, and the orientation of the magnetic fields, can be inferred from the polarization of the spectral lines. Magnetic field lines pointing out of the Sun have positive magnetic polarity, while inward-directed fields have negative polarity.

Sunspots tend to travel in pairs of opposite polarity, roughly aligned in the east west, rotational direction on the Sun. The twinned sunspots are joined by invisible magnetic loops that rise above the photosphere into the corona, like an arching bridge connecting two magnetic islands. The magnetic field lines emerge from the Sun at a spot with one polarity and re-enter it at another one of opposite polarity.

The magnetized atmosphere in, around, and above bipolar sunspot groups is called a solar active region. Active regions are places of concentrated, enhanced magnetic fields, large enough and strong enough to stand out from the magnetically weaker areas. These disturbed regions also emit intense X-rays and are prone to awesome explosions. They mark one location of extreme unrest on the Sun.

Unlike the Earth, magnetism on the Sun does not consist of just one simple dipole. The Sun is spotted all over, like a young child with the measles, and contains numerous pairs of opposite magnetic polarity. They are found everywhere on the Sun, including active regions with their sunspots and the quiet places outside of active regions.

The invisible magnetized bridges that join the ubiquitous pairs of opposite magnetic polarity have become known as coronal loops. Powerful magnetism, spawned deep inside the Sun, threads its way through the solar atmosphere, creating a dramatic, ubiquitous, and ever-changing panorama of coronal loops and arcades.

2.4.2 The 11-Year Magnetic Activity Cycle

The Sun's magnetism is forever changing and is never still, just like everything else in the Universe. Sunspots, and the active-region magnetic loops that join them, are temporary. They come and go with lifetimes ranging from hours to months. Moreover, the total number of sunspots varies periodically, from a maximum to a minimum and back to a maximum, in about 11 years (Fig. 2.6 and Table 2.7). Samuel Heinrich Schwabe, an amateur astronomer of Dessau, Germany, discovered this periodic variation in the early 1840s. At the maximum in the sunspot cycle, there may be 100 or more spots on the visible hemisphere of the Sun at one time; at sunspot minimum very few of them are seen, and for periods as long as a month none can be found.

The number of active regions, with their bipolar sunspots and coronal loops, varies in step with the sunspot cycle, peaking at sunspot maximum when they dominate the structure of the inner corona. At sunspot minimum, the active regions are largely absent and the strength of the extreme-ultraviolet and X-ray emission of the corona is greatly reduced. And since most forms of solar activity are magnetic in origin, the sunspot cycle is also called the solar cycle of magnetic activity.

The places that sunspots emerge and disappear also vary over the slow 11-year sunspot cycle (Fig. 2.6). At the beginning of the cycle, when the number of sunspots just starts to grow, sunspots break out in belts of activity at middle solar latitudes of 30°–45°, parallel to the equator in both the northern and the southern hemisphere. The two belts move to lower solar latitudes, or toward the equator, as the

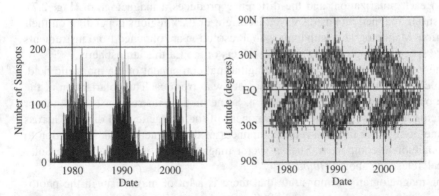

Fig. 2.6 Solar magnetic activity cycle. The 11-year solar cycle of magnetic activity plotted from 1975 to 2007. Both the positions of sunspots (*left*) and the numbers of sunspots (*right*) wax and wane in cycles that peak every 11 years. Similar 11-year cycles have been plotted for more than a century. At the beginning of each cycle, the first sunspots appear at about 30° solar latitude and then migrate to 0° solar latitude, or the solar equator, at the cycle's end. This plot of the changing positions of sunspots resembles the wings of a butterfly, and has therefore been called the butterfly diagram. The cycles overlap with spots from a new cycle appearing at high latitudes when the spots from the old cycle persist in the equatorial regions. The solar latitude is the angular distance from the plane of the Sun's equator, which is very close to the plane of the Earth's orbit about the Sun, called the ecliptic. (Courtesy of David Hathaway, NASA/MSFC.)

Table 2.7 Dates of the minimum and maximum in solar activity since 1960[a]

Cycle number	Date of minimum	Date of maximum
Cycle 20	1964.9	1968.9
Cycle 21	1976.5	1979.9
Cycle 22	1986.8	1989.6
Cycle 23	1996.9	2000.5
Cycle 24	2008.3	2011.5

[a]Courtesy of David H. Hathaway

cycle progresses and the number of sunspots increases to a maximum. The zones of sunspot activity keep on moving closer and closer to the equator, but they never reach it. The sunspots fizzle out and gradually disappear at sunspot minimum, just before coming together at the equator.

As old spots linger near the equator, new ones break out about a third of the way toward the poles (Fig. 2.6). But the magnetic polarities of the new spots are reversed with north becoming south and vice versa – as if the Sun had turned itself inside out. The magnetic field at the solar poles approximates that of a dipole; this field also reverses polarity every 11 years. Thus, the full magnetic cycle for the return of both the sunspot pair orientation in each hemisphere and the polarity of the general polar magnetic field takes an average of 22 years.

Magnetographs are now used to chart the magnetic fields running in and out of the Sun. These instruments consist of an array of tiny detectors that measure the Zeeman effect at different locations on the photosphere. Two images are produced, one in each polarization; and the difference produces a magnetogram (Fig. 2.7), with strong magnetic fields displayed as bright or dark regions depending on their direction. Magnetograms with high spatial resolution are obtained from instruments aboard spacecraft, above the obscuring effects of the Earth's atmosphere.

The Zeeman effect measures the longitudinal component of the magnetic field, or the component that is directed toward or away from us. The polarization of the radiation provides the in or out direction. Some modern instruments, called vector magnetographs, also measure the component of the magnetic field directed across our line-of-sight, or the transverse component of the magnetic field, in addition to the longitudinal component. Such a vector magnetograph is attached to the Solar Optical Telescope aboard *Hinode*.

The magnetograms demonstrate that there is a lot of magnetism in the photosphere outside sunspots. More than 90% of these magnetic fields are concentrated into intense magnetic flux tubes that appear, disappear, and are renewed in just 40 h. The individual flux tubes are a few hundred kilometers across and have magnetic fields strengths comparable to those of the much larger sunspots.

The magnetograms also indicate that at sunspot minimum there is still plenty of magnetism, even though there are no large spots on the Sun (Fig. 2.8). Small magnetized regions continuously well up all over the photosphere throughout the 11-year solar cycle. Moreover, the solar magnetic fields are always clumped together

Fig. 2.7 Magnetogram. This magnetogram was taken on 12 February 1989, close to the maximum in the Sun's 11-year cycle of magnetic activity. Yellow represents positive or north polarity pointing out of the Sun, with red the strongest fields which are within or near sunspots; blue is negative or south polarity that points into the Sun, with green the strongest. In the northern hemisphere (*top half*) positive fields lead, in the southern hemisphere (*bottom half*) the polarities are exactly reversed and the negative fields lead. (Courtesy of William C. Livingston, NSO and NOAO.)

into intense bundles that cover only a few percent of the Sun's surface, both at sunspot minimum and at sunspot maximum.

2.4.3 Magnetic Fields in the Corona

About 100 years ago Frank H. Bigelow used the shape of the corona, detected during total eclipses of the Sun, to make some very prescient speculations about solar magnetism on large scales. He supposed that rays seen near the solar poles are open "lines of force discharging coronal matter from the body of the Sun." Bigelow also concluded that the "long equatorial wings" of the corona, detected at periods of minimum activity, are "due to the closing of the lines of force about the equator."

At times of reduced magnetic activity, prominent coronal streamers, detected during a solar eclipse, are indeed restricted to near the solar equator. At activity minimum, the streamers are molded by the available magnetism into extended shapes

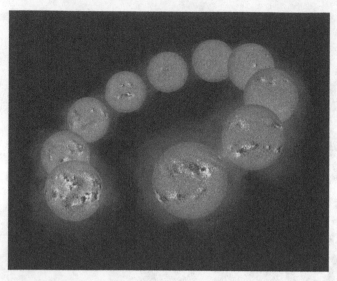

Fig. 2.8 Solar cycle magnetic variations. These magnetograms portray the polarity and distribution of the magnetism in the solar photosphere. They were made with the Vacuum Tower Telescope of the National Solar Observatory at Kitt Peak from 8 January 1992, at a maximum in the sunspot cycle (*lower left*) to 25 July 1999, well into the next maximum (*lower right*). Each magnetogram shows opposite polarities as darker and brighter than average tint. When the Sun is most active, the number of sunspots is at a maximum, with large bipolar sunspots that are oriented in the east–west (*left–right*) direction within two parallel bands. At times of low activity (*top middle*), there are no large sunspots and tiny magnetic fields of different magnetic polarity can be observed all over the photosphere. The haze around the images is the inner solar corona. (Courtesy of Carolus J. Schrijver, NSO, NOAO and NSF.)

that point along the equatorial plane, often forming a ring or belt of hot gas that extends around the Sun. The streamers have bulb shapes in the low corona near the Sun, where electrified matter is densely concentrated within closed magnetic loops or arches that straddle the equator. Farther out in the extended corona, the equatorial streamers narrow and stretch out into long, thin stalks extending tens of millions of kilometers in space.

Near a minimum in the activity cycle, the corona is relatively dim at the poles where faint plumes diverge out into interplanetary space, apparently outlining a global, dipolar magnetic field of about 0.001 T or 10 G in strength. This large-scale magnetism becomes pulled outward at the solar equator, confining hot material in the streamer belt. The streamers are then sandwiched between regions of opposite magnetic direction or polarity, and are confined along an equatorial current sheet that is magnetically neutral.

Near the maximum in the activity cycle, the shape of the corona and the distribution of the Sun's extended magnetism can be much more complex. The corona then becomes crowded with streamers that can be found closer to the Sun's poles.

At times of maximum magnetic activity, the widths of streamers are smaller and their radial extension is shorter; near minimum, they are wide and well developed

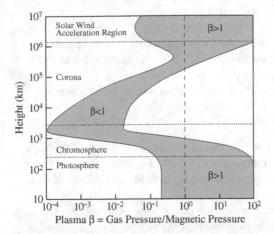

Fig. 2.9 Magnetic and gas pressure. The ratio of gas to magnetic pressure, denoted by the symbol β, plotted as a function of height above the photosphere. The magnetic pressure is greater than the gas pressure in the low corona, where β is less than one and magnetic fields dictate the structure of the corona. Large local variations of β occur in active regions because of the presence of dense loops. Further out, the gas pressure can exceed the magnetic pressure and the Sun's magnetic field is carried out into interplanetary space by the solar wind. In the photosphere, below the corona and chromosphere, the gas pressure also exceeds the magnetic pressure, and the moving gas carries the magnetic fields around. [Adapted from G. Allen Gary (2001); this survey ignores non-radial structure.]

along the equator. The shape, location, and tilt of extended sheets of opposing magnetic polarity are variable near sunspot maximum. At about this time, the Sun swaps its north and south magnetic poles, so a much more volatile corona exists at solar maximum.

Throughout the solar atmosphere, a dynamic tension is set up between the charged particles and the magnetic field (Fig. 2.9, Focus 2.4). In the photosphere and below, the gas pressure dominates the magnetic pressure, allowing the magnetic field to be carried around by the moving gas. Because the churning gases are ionized, and hence electrically conductive, they sweep the magnetic field along. The situation is reversed in the low corona within active regions. Here strong magnetism wins and the hot particles are confined within coronal loops. Nevertheless, the loops are themselves tied into the underlying photosphere, which is stirred up by mass motions. The gas pressure can become greater than the magnetic pressure further out in the corona, where the magnetic field decreases in strength and the solar wind carries the Sun's magnetism out into interplanetary space.

Focus 2.4

Magnetic and Gas Pressure

A magnetic field tends to restrain a collection of electrons and protons, called plasma, while the plasma exerts a pressure that opposes this. The pressure, P_B, produced by a magnetic field transverse to its direction is given by:

$$P_B = B^2/(2\mu_0),$$

for a magnetic field of strength B in Tesla, and the permeability of free space $\mu_0 = 4\pi \times 10^{-7}$ in units of newtons per square ampere or $N\,A^{-2}$. As expected, a stronger magnetic field applies a greater restraining pressure.

Hot plasma generates a gas pressure, P_G, owing to the motions of its particles. It is described by the expression

$$P_G = NkT,$$

where N is the particle number density, $k = 1.38 \times 10^{-23}$ joule per Kelvin, or $J\,K^{-1}$, is Boltzmann's constant, and T is the temperature in Kelvin. Hotter particles move faster and create greater pressure to oppose the magnetic field, and denser plasma also results in greater pressure.

The two kinds of pressure compete for control of the solar atmosphere. In the low solar corona, strong magnetic fields in active regions hold the hot, dense electrified gas within coronal loops. The magnetic and gas pressure become equal for a magnetic field, B, given by:

$$B = [(2\mu_0 k)NT]^{1/2} = [3.46 \times 10^{-29}NT]^{1/2} \quad T.$$

If a coronal loop contains a hot, dense plasma, with $N = 10^{17}$ electrons per cubic meter and $T = 10^6\,K$, the magnetic field must be stronger than $B = 0.002\,T = 20\,G$. The magnetic field strengths of coronal loops in active regions are therefore strong enough to hold this gas in, at least within the low corona near sunspots.

Far from the Sun, the magnetic fields of coronal loops also become too weak to constrain the outward pressure of the hot gas, and the loops might expand or break open to allow electrons and protons to escape, contributing to the solar wind and carrying the magnetic fields away. Within the solar wind, the gas pressure of the electrons and protons is roughly equal to the magnetic pressure of the interplanetary magnetic field.

2.4.4 Interplanetary Magnetic Fields

Low-energy, cosmic ray protons are significantly more numerous when solar activity is minimal. And when solar activity is at the peak of its 11-year cycle, the flux of low-energy cosmic rays detected at the top of the Earth's atmosphere is least. This unexpected anti-correlation is often called the Forbush effect, after its discovery by Scott Forbush in the early 1950s. It was explained in 1956 by Peter Meyer, Eugene Parker, and John Simpson, who proposed that enhanced interplanetary magnetism near the maximum in the 11-year solar activity cycle deflects cosmic rays from their Earth-bound paths.

Interplanetary magnetic fields act as a barrier to electrically charged cosmic rays that are coming from the depths of space, preventing them from reaching the Earth. During the maximum in the solar cycle, stronger solar magnetic fields are carried out into interplanetary space by the Sun's wind, deflecting more cosmic rays. Less extensive interplanetary magnetism, during a minimum in the 11-year cycle of magnetic activity, lowers the barrier to the cosmic particles and allows more of them to arrive at Earth.

In 1958, Parker showed that the solar wind ought to carry both the Sun's particles and the magnetic fields outward to the far reaches of the solar system. He also concluded that interplanetary magnetism ought to have a spiral shape. This results from the combined effects of solar rotation and the radial expansion of the solar wind. While one end of the interplanetary magnetic field remains firmly rooted in the Sun, the other end is extended and stretched out by the solar wind. The Sun's rotation bends this radial pattern into an interplanetary spiral shape, coiling the magnetism up like a tightly wound spring (Fig. 2.10). Because it is wrapped up into a spiral, the magnetic field's strength only falls off linearly with distance from the Sun, in contrast to the wind density that decreases much more rapidly with distance as it fills a larger volume. Of course, Parker's model was a theoretical speculation, with many skeptics. At the time, even the solar wind had yet to be directly measured, and evidence for Parker's spiral awaited the observations.

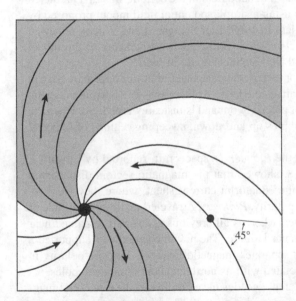

Fig. 2.10 Interplanetary magnetic sectors. If viewed from above, the magnetic field in the Sun's equatorial plane would be drawn out into a spiral pattern that is divided into magnetic sectors which point in opposite directions (*arrows, gray* and *white*). The rotating Sun sweeps these sectors of opposite magnetic polarity past the Earth, represented by the small black dot. At the Earth's location, the pulling effect of the solar wind is about equal to the twisting effect of the Sun's rotation, so the stretched-our field makes an angle of about 45° with the radial direction from the Sun. [Adapted from Lief Svalgaard and John M. Wilcox (1976).]

Studies of the interplanetary magnetic field were pioneered using America's *Interplanetary Monitoring Platform 1*, or *IMP 1* for short, launched on 26 November 1963. Its magnetic field experiment, under the direction of Norman F. Ness from NASA's Goddard Space Flight Center, measured both the strength and the direction of the magnetic field in interplanetary space near the Earth. The measurements confirmed that the interplanetary magnetic field has a spiral shape. The magnetometers aboard *IMP 1* also provided important results on the confinement and distortion of the Earth's magnetic field by the solar wind.

Ness teamed up with John M. Wilcox to investigate the relation between solar and interplanetary magnetism, discovering large interplanetary regions of alternating magnetic polarity. In 1964, they showed that the magnetic field borne by the solar wind near a minimum in the 11-year solar activity cycle is divided into several magnetic sectors directed toward or away from the Sun within the plane of the solar equator and the orbital plane of the planets, the ecliptic. As reviewed by Wilcox in 1968, the interplanetary magnetic field observed by spacecraft is thus dominated by a sector structure, co-rotating with the Sun and following an Archimedean spiral pattern (also see Fig. 2.10); as the field swings past the Earth, it points predominantly away from the Sun for several days, and then toward the Sun for several days, etc. Wilcox also suggested a connection between the fast solar wind and the open magnetic fields near the Sun's polar regions.

The polarity of the interplanetary magnetic field in the ecliptic was first predicted from photosphere magnetograms using a simple, potential field model proposed by Kenneth Schatten, Wilcox, and Ness in 1969; Lief Svalgaard and Wilcox reviewed the many successes of this model in 1978. Subsequent refinements to the model include that of Yi-Ming Wang and Neil R. Sheeley, Jr., in 1992.

A magnetically neutral layer, or electric current sheet, that lies approximately in the plane of the solar equator separates the large-scale regions of opposite magnetic polarity. This dividing sheet is not precisely flat and is instead warped. So, when the Sun rotates, the current sheet wobbles up and down, sweeping regions of opposite magnetic polarity past the Earth.

Magnetic measurements from the *Pioneer 11* spacecraft, reported by Edward J. Smith and his colleagues in 1978, showed that the magnetic sectors disappeared when the spacecraft rose above the equatorial current sheet, where they were replaced by a magnetic field of one polarity. *Pioneer 11* traveled up to solar latitude of 16° after its encounter with Jupiter, whereas all previous spacecraft remained near the ecliptic plane at less than 7° solar latitude. The new measurements supported a model in which the current sheet separates magnetic fields of opposite polarity in each hemisphere of the Sun, associated with its north and south magnetic poles.

Measurements over the subsequent decades have confirmed the equatorial magnetic sectors near activity minimum, and repeatedly demonstrated that the interplanetary magnetic field is oriented into a simple spiral pattern, if averaged over a sufficiently long time. When a spacecraft is below the equatorial current sheet, it detects magnetic fields pointed in one direction. When located above the sheet, the observed fields point in the opposite direction. The magnetic field on one side of the sheet points toward the Sun near the minimum in one 11-year activity cycle, and away from the Sun near the minimum of the next cycle, but the fields on the two sides of the current sheet always point in opposite directions.

Nevertheless, until very recently no spacecraft has ventured far out of the ecliptic to see what happens way above the equatorial plane of the Sun. As described in Chap. 5, the *Ulysses* spacecraft has now completed three orbits above the Sun's poles, observing the Parker spiral from above, and showing that it is generally preserved to the highest latitudes – but with significant departures that might be attributed to magnetic waves in the solar wind.

2.5 X-rays from the Sun

Guglielmo Marconi, an Italian electrical engineer, astonished the world on 12 December 1901 when he sent a long-wavelength radio signal across the Atlantic Ocean. Radio waves travel in straight lines, and cannot pass through the solid Earth; so many wondered how Marconi could send them half way around the world.

The radio signals got around the Earth's curvature by reflection from the ionosphere, or spherical shell of ions, an electrically charged layer in the upper atmosphere located about 100 km above your head. X-rays from the Sun create this spherical layer of ions. That is, the molecules in our atmosphere are ionized by this invisible, short-wavelength radiation, but not by the sunlight we see with our eyes. When the solar X-rays reach the upper atmosphere, they rip electrons off the nitrogen and oxygen molecules, producing free electrons and atomic and molecular ions.

The energy of radiation is inversely proportional to its wavelength; so invisible waves with short wavelengths have more energy than visible light with longer ones (Focus 2.5). This can be understood by looking at your hand in different ways. In visible light, you see the skin of your hand, but shorter ultraviolet rays enter the skin to give you sunburn. The even shorter, more energetic X-rays penetrate through the skin and muscles of your hand to delineate its bones. Also, any heated gas emits its most intense radiation at a wavelength that decreases with increasing temperature, and the very hot corona, at a temperature of a million Kelvin, radiates most of its energy at X-ray wavelengths (Focus 2.5). Although this emission waxes and wanes in step with the 11-year activity cycle, it is always there at some level.

Focus 2.5 ▬▬▬▬▬▬▬▬▬▬▬▬▬▬▬▬▬▬▬▬▬▬▬▬▬▬▬▬▬▬

The Energy of Light

The Sun continuously radiates energy that spreads throughout space. This radiation is called "electromagnetic" because it propagates by the interplay of oscillating electrical and magnetic fields in space. Electromagnetic waves all travel through empty space at the same constant speed – the velocity of light. It is usually denoted by the lower case letter c, and has the value of $c = 2.9979 \times 10^8 \, \mathrm{m \, s^{-1}}$ or about $300{,}000 \, \mathrm{km \, s^{-1}}$. No energy can be transported more swiftly than the speed of light.

Astronomers describe radiation in terms of its wavelength, frequency, or energy. When light propagates from one place to another, it often seems to behave like waves or ripples on a pond. The light waves have a characteristic wavelength, often denoted by the symbol λ, the separation between wave crests. Sometimes radiation is described by its frequency, denoted by the symbol ν, instead of its wavelength. Radio stations are, for example, denoted by their call letters and the frequency of their broadcasts. The product of the wavelength and frequency is equal to the velocity of light, or

$$\lambda \nu = c.$$

When light is absorbed or emitted by atoms, it behaves like packages of energy, called photons. The photons are created whenever a material object emits electromagnetic radiation, and they are consumed when matter absorbs radiation. Moreover, each elemental atom can only absorb and radiate a very specific set of photon energies. The photon energy is given by:

$$\text{Photon energy} = h\nu = \frac{hc}{\lambda} ,$$

where Planck's constant $h = 6.6261 \times 10^{-34}$ J s. From this expression we see that photons of radiation at shorter wavelengths have greater energy. This is the reason that short, energetic X-rays can penetrate inside your body, while longer, less energetic visible light just warms your face.

A hot gas at temperature, T, will emit radiation at all wavelengths, but the peak intensity of that radiation is emitted at a maximum wavelength, λ_{max}, that varies inversely with the temperature. This can be seen by equating the thermal energy of the gas to the photon energy of its radiation, or by:

$$\text{Gas thermal energy} = \frac{3}{2} k T = \frac{h c}{\lambda_{max}} = \text{Radiation photon energy}$$

where Boltzmann's constant $k = 1.38066 \times 10^{-23}$ J K^{-1}. Collecting terms and inserting the values for the constants, this expression gives $\lambda_{max} = 2hc/(3kT) \approx 0.001/T$ m. The exact relationship, known as the Wien displacement law, is given by

$$\lambda_{max} = \frac{0.0028978}{T} \quad \text{m}$$

or $\lambda_{max} = 2.8978 \times 10^6/T$ nm =, where $1 \text{ nm} = 10^{-9}$ m. This tells us that the Sun's corona, with a temperature of about 2×10^6 K, will emit most of its radiation at X-ray wavelengths of $10^{-9} - 10^{-10}$ m, with photon energies of 1–10 keV, where $1 \text{ keV} = 1.602 \times 10^{-16}$ J. In contrast, the solar photosphere with a temperature of 5,780 K will radiate most of its energy at a visible wavelength of about 5×10^{-7} m.

Solar ultraviolet radiation is partially blocked by the Earth's atmosphere, and the Sun's X-rays are completely absorbed in our air. These components of solar radiation must therefore be observed with telescopes and other instruments in space. Scientists at the United States Naval Research Laboratory, or NRL for short, pioneered such investigations. Richard Tousey and his colleagues at NRL used V-2 rockets, captured from the Germans at the end of World War II, to obtain the first photographs of the solar ultraviolet spectrum in 1946, and to first detect the Sun's X-rays in 1948.

By the late 1940s, scientists knew that the corona had a temperature of about a million Kelvin, and should therefore produce X-rays, but it appeared so dilute that its radiant flux might not have a major effect on the Earth's atmosphere. But then in 1951, Herbert Friedman's team at NRL used instruments aboard a V-2 rocket to show that there are enough solar X-rays coming in from outside our atmosphere to create the ionosphere. Subsequent measurements made by this group using Aerobee sounding rockets over the course of a full sunspot cycle indicated that the intensity of the Sun's X-rays rises and falls with the number of sunspots, as does the temperature in the ionosphere.

The first X-ray image of the Sun was obtained by the NRL group in 1960 during a brief 5-min rocket flight. This primitive picture set the stage for the detailed X-ray images of the Sun taken from NASA sounding rockets in the early 1970s. During this time, solar X-ray instruments were also developed using NASA's series of small satellites known as the *Orbiting Solar Observatories*, or *OSOs*, launched from 1962 (*OSO 1*) to 1975 (*OSO 8*).

Our perspective of the X-ray Sun was forever changed with NASA's orbiting *Skylab*, launched on 14 May 1973, and manned by three-person crews until 8 February 1974. *Skylab's* X-ray telescope, and those to follow, could be used to look back at the Sun's hot corona, viewing it across the Sun's face for hours, days, and months at a time. This is because very hot material – such as that within the million-degree corona – emits most of its energy at these wavelengths. Also, the photosphere is too cool to emit intense X-ray radiation, so it appears dark under the hot gas.

Numerous *Skylab* X-ray photographs, first reported by Giuseppe Vaiana and his co-workers (1973a, b) showed that the X-ray corona is highly structured, containing coronal holes, coronal loops, and X-ray bright points (Fig. 2.11). Magnetized coronal loops mold, shape, and constrain the high-temperature, electrified gas (Fig. 2.12), and since the hot coronal gases are almost completely ionized, they cannot readily cross the intense, closed magnetic field lines. Typically, the electrons contained in coronal loops have temperatures of a few million Kelvin and a density of up to 100 million billion (10^{17}) electrons per cubic meter.

Although *Skylab* demonstrated the ubiquitous nature of the X-ray-emitting coronal loops, its X-ray telescope provided blurred images without fine detail. It lacked the resolution, sensitivity, and rapid, uniform imaging required to fully understand them. As described in Chap. 4, this has now largely been accomplished, first by the X-ray telescope aboard the *Yohkoh* spacecraft, then at extreme-ultraviolet wavelengths with telescopes aboard the *SOlar and Heliospheric Observatory*, or *SOHO* for short, and the *Transition Region And Coronal Explorer*, abbreviated *TRACE*, and most recently at X-ray wavelengths with the X-Ray Telescope aboard *Hinode*.

Skylab's inquisitive X-ray eyes also recorded black, seemingly empty places, called coronal holes. Max Waldmeier first observed them from the ground as

Fig. 2.11 X-ray corona. An X-ray photograph of the solar corona taken at wavelengths of 10–600 nm from a satellite of the American Science and Engineering Company. It shows active regions with coronal loops (large bright features distributed along the solar equator), small bright points nearly uniformly distributed over the Sun, and a coronal hole seen as a large dark area extending from the north pole of the Sun down through the middle of the disk. The coronal holes are regions where plasma might easily escape, giving rise to high-velocity streams in the solar wind. (Courtesy of Giuseppe Vaiana and the Solar Physics Group, American Science and Engineering, Inc.; presented by Vaiana and his colleagues at the 8 April 1974 meeting of the American Geophysical Union.)

long-lived regions of negligible intensity in coronal images of the green iron line – (Fe XIV) at 530.3 nm – naming them holes in 1957. Waldmeier used the green line to study polar coronal holes for nearly 40 years, from 1940 to 1978, reporting in 1981 that they exist permanently for about 7 years during each 11-year activity cycle, including activity minimum, but that the large polar coronal holes seem to disappear for a few years near sunspot maximum.

Dark, coronal holes were noted in the X-ray images obtained on NASA sounding rockets in 1970 and extensively described by *Skylab*; instruments aboard the *Yohkoh*, *SOHO*, and *Hinode* spacecraft routinely detect them. The large polar coronal holes are always present near the minimum of the solar activity cycle, but at activity maximum they are replaced by smaller coronal holes scattered all over the Sun. The large

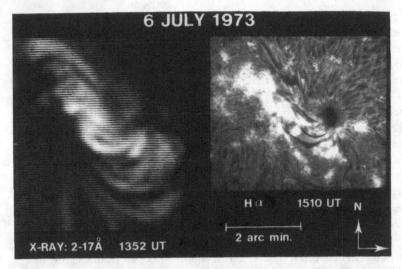

Fig. 2.12 X-ray coronal loops. A comparison of a *Skylab* X-ray image and a visible light (Hα) picture of a solar active region taken at the same time on 6 July 1973. The bright coronal loops shown in the X-ray image contain hot (million degree), dense (10 million billion, or 10^{16}, electrons per cubic meter) electrons trapped within magnetic loops that connect underlying regions of opposite magnetic polarity. The coronal loops are about 100,000 km in extent, or about ten times the Earth's diameter. (Courtesy of the Solar Physics Group, American Science and Engineering, Inc.)

polar coronal holes reappear soon after activity maximum, with opposite magnetic polarity.

Coronal holes are not empty. They are just more rarefied and cooler than other places in the corona, so their X-ray and extreme-ultraviolet emissions are faint. The electrons in coronal holes have densities a factor of ten smaller than the typical value in coronal loops. This is because most of the magnetic field lines in coronal holes do not form locally closed loops. Coronal holes are instead characterized by open magnetic fields that do not return to another place on the Sun, allowing the charged coronal particles to escape the Sun's magnetic grasp and flow outward into the surrounding space.

Allen S. Krieger, Adrienne F. Timothy, and Edmond C. Roelof concluded in 1973 that coronal holes are the sources of at least some of the recurrent high-speed streams observed in the solar wind near the ecliptic. They did this by comparing an X-ray photograph of the Sun, taken with a 5-min rocket flight in 1970, with in situ measurements of the solar wind velocity and flux from the *Vela* and *Pioneer VI* spacecraft. The recurrent high-speed streams were associated with equator-ward extensions of the polar coronal holes, rotating with a 27-day period. This result was substantiated in 1974 by detailed comparisons of the coronal holes mapped by *Skylab* with satellite measurements of the solar wind. Also in 1973, Giancarlo Noci showed that a strong solar wind is likely to originate in coronal holes as the result of their physical properties – open magnetic field lines, exceptionally low coronal density and temperature, outward thermal conduction and pressure.

Sounding rocket observations of extreme ultraviolet lines in coronal holes provided additional early evidence that they are the unknown font of the high-speed solar wind. As shown by Gary J. Rottman, Frank Q. Orrall, and James A. Klimchuk in 1982–1983, these spectral features had shorter wavelengths than their counterparts on the rest of the solar disk, indicating that material was moving out from the coronal holes at a velocity of $7–8\,\mathrm{km\ s^{-1}}$. As discussed in Chap. 5, observations with instruments aboard the *Ulysses* and *Yohkoh* spacecraft conclusively demonstrated that the polar coronal holes provide a fast lane for the high-speed solar wind near a minimum in the 11-year solar activity cycle.

As also shown in Chap. 5, it remained for instruments aboard *SOHO, TRACE*, and *Hinode* to delineate possible escape sites within holes, and to suggest how the liberated gases might be subsequently accelerated to create winds that move at speeds of up to $800\,\mathrm{km\ s^{-1}}$.

First discovered during rocket flights, but extensively studied during *Skylab's* coverage, the X-ray bright points appear all over the Sun, in active regions, coronal holes, and the places in between (Fig. 2.13). The ubiquitous small features glow like tiny, blinking lights in X-ray images of the Sun. The *Skylab* observations, described by Leon Golub and his colleagues (1974, 1976a, b), indicate that hundreds and even thousands of X-ray bright points appear each day, coming and going and usually lasting about 8 h; but some flare up and fade in just a few minutes while others last for days.

Fig. 2.13 X-ray bright points and jets in a coronal hole. This X-ray image, taken from the *Skylab* spacecraft in July 1973, shows several small, bright X-ray sources that project radially outward in a dark coronal hole at the Sun's north pole. The features are less than 10,000 km in width and have temperatures in excess of a million Kelvin. (Courtesy of the Solar Physics Group, American Science and Engineering, Inc.)

The X-ray bright points are concentrated regions of intense magnetism, some almost as large as the Earth. Many of them overlie magnetic regions that have both positive and negative polarity, and they fluctuate in brightness and emit X-ray jets that resemble small flares. The quantity of magnetic flux emerging in the form of X-ray bright points at high latitudes alone is estimated to be as large as the contribution from all active regions. *Skylab* showed that ubiquitous bright points dot the ultraviolet Sun as well. The varying ultraviolet and X-ray features have a temperature of about a million Kelvin. Observations from *Hinode* are describing how X-ray bright points, X-ray jets, and magnetic waves may help heat the corona and power the solar wind.

To complete our introductory survey of the discovery of space, we now turn to explosive solar outbursts that send intense radiation and high-speed charged particles into interplanetary space.

2.6 Solar Outbursts Send Energetic Radiation and Particles into Space

2.6.1 Solar Flares

Solar flares are sudden, rapid outbursts of awesome power and violence, on a scale unknown on Earth, the biggest explosions in the solar system. In minutes, the disturbance spreads along concentrated magnetic fields, releasing energy equivalent to millions of 100-megaton hydrogen bombs exploding at the same time, and raising the temperature of Earth-sized regions to tens of millions of Kelvin. Solar flares are therefore hotter than the corona. Sometimes they temporarily go out of control and lose equilibrium, becoming hotter than the center of the Sun for a short period of time.

The short-lived solar flares, which usually last only 10 min or less, unleash their vast power from a relatively small volume within solar active regions, the magnetized atmosphere in, around, and above sunspots. The largest solar flares cover but a few tenths of a percent of the solar disk, and are intimately related to the magnetized coronal loops that link sunspots and other regions of opposite magnetic polarity in the underlying photosphere. These incredible explosive outbursts become more frequent and violent when the number of sunspots is greatest; several solar flares can be observed on a busy day near the maximum of the sunspot cycle. They are not caused by sunspots, but are instead powered by the magnetic energy stored higher in the corona.

Despite the powerful cataclysm, most solar flares are not detected on the visible solar disk. Although emitting awesome amounts of energy, a solar flare releases less than one-tenth the total energy radiated by the Sun every second. And since most of this radiation is in visible sunlight, solar flares are only minor perturbations in the combined colors, or white light, of the Sun. The first record of a solar flare, observed in the white light of the photosphere, therefore did not occur until the mid

19th century – on 1 September 1859 when the English astronomers, Richard C. Carrington and Richard Hodgson, independently noticed one.

A new perspective, which demonstrated the frequent occurrence of solar flares, was made possible when they were observed in the red Balmer alpha transition of hydrogen at 656.3 nm, originating in the chromosphere. This was not possible until the early 20th century, after the invention of the spectroheliograph, which isolated the red hydrogen-alpha emission from the Sun's intense visible light at other wavelengths. Then, for more than half a century, astronomers throughout the world carried out a vigilant flare patrol, like hunters waiting for the sudden flash of game birds. They showed that at the chromospheric level in the solar atmosphere a solar flare consists in simplest form as two extended, parallel flare ribbons. But a fundamental understanding of the physical processes responsible for solar flares had to wait until they were detected at invisible wavelengths from both the ground and the space.

During World War II, it was discovered that sudden, intense radio outbursts from the Sun, associated with solar flares, could interfere with radio communications and radar systems. And in 1946, for example, the English physicists, Edward V. Appleton and James Stanley Hey, demonstrated that meter-wavelength solar radio noise originates in the vicinity of sunspot-associated active regions, and that sudden large increases in the Sun's radio output are associated with solar flares.

Soon after the War, in the 1950s, J. Paul Wild's group of Australian radio astronomers used swept frequency receivers to track the outward motions of the impulsive radio bursts, designated Type III, showing that solar flares are sending beams of electrons out into space at nearly the velocity of light. And at about the same time, Scott Forbush and his colleagues demonstrated that charged particles with low cosmic ray energies could be detected at the Earth during solar flares. Another slower kind of solar radio burst, denoted Type II, was also discovered by Wild's group, and attributed to shock waves being hurled out into space. These radio bursts were subsequently found to be associated with coronal mass ejections. Both types of radio bursts are discussed in greater detail in Chap. 6.

Although the radio emission from solar outbursts helps specify their magnetic and temperature structure, and identifies the expulsion of high-speed electrons or shock waves, the bulk of the radiation from high-temperature solar flares is emitted at much shorter, invisible wavelengths. These catastrophic explosions can briefly outshine the entire Sun at X-ray and extreme-ultraviolet wavelengths. This radiation is absorbed in the Earth's atmosphere, so astronomers have observed it with telescopes in outer space, beginning with primitive instruments aboard balloons or sounding rockets and continuing with increasingly sophisticated telescopes to the present day.

Since the ionosphere is created by the Sun's X-rays, it is perhaps not so surprising that X-rays from solar flares can alter the ionosphere and disrupt its ability to mirror radio waves. During moderately intense solar flares, long-wavelength radio communication can be silenced over the Earth's entire sunlit hemisphere, and it does not return to normal until the flaring activity stops. This was a major concern during

World War II, when scientists would use the number of sunspots to forecast such interruptions.

Milestones in the early history of space observations of solar flares began with a balloon flight in 1958, when Laurence E. Peterson and John Randolph Winckler observed a flare-associated burst of high-energy radiation, followed by pioneering rocket observations of X-rays emitted by solar flares in the late 1950s and early 1960s by Herbert Friedman and his colleagues, showing that the X-ray radiation emitted during solar flares can outshine the entire X-ray Sun. Then an instrument aboard *Mariner 2* observed interplanetary shocks associated with solar activity, reported by Charles P. Sonett and his co-workers in 1964.

NASA's *Orbiting Solar Observatory*, or *OSO*, series of satellites, next improved our knowledge of solar X-ray flares. In 1972, for example, Edward L. Chupp and his colleagues detected gamma ray lines from solar flares using an instrument aboard *OSO 7*, including the 0.511-MeV line caused by the annihilation of electrons with their anti-matter counterparts, the positrons. And the *Skylab* mission obtained high-resolution images of solar flares at ultraviolet and X-ray wavelengths in 1973–1974.

Satellites designed to study high-energy radiation from solar flares near the maximum in the 11-year solar activity cycle followed these pioneering investigations. They included NASA's *Solar Maximum Mission*, abbreviated *SMM*, launched in 1980, the Japanese *Hinotori* and *Yohkoh* spacecraft, respectively launched in 1981 and 1991, and NASA's *Ramaty High-Energy Solar Spectroscopic Imager*, launched in 1998. Instruments aboard other contemporary solar spacecraft, including *SOHO*, *TRACE*, *Hinode*, and *STEREO*, contribute to our understanding of coronal mass ejections or solar flares, and all of these results, from *SMM* to *Hinode*, are presented in Chap. 6.

2.6.2 Erupting Prominences

Arches of incandescent gas can be briefly detected during a total eclipse of the Sun, as illustrated by this account given in 1842 by Francis Bailey, a stockbroker and enthusiastic amateur astronomer in England:

> I had anticipated a luminous circle round the Moon during the time of total obscurity…, but the most remarkable circumstance attending this phenomenon was the appearance of three large protuberances apparently emanating from the circumference of the Moon, but evidently forming a portion of the corona.

The protuberances that Bailey observed were arches of incandescent gas that loop up from the chromosphere into the solar corona. Astronomers now call the looping features *prominences*, the French word for "protuberances." Early observations, by Richard A. Proctor in 1871, also indicated that prominences are often associated with eruptive phenomena.

Magnetic fields suspend and insulate the elongated prominences, which are filled with material at a temperature of about 10,000 K and are hundreds of times cooler and denser than the surrounding corona. Prominences can remain suspended and almost motionless above the photosphere for weeks or months, but there comes a time when they cannot bear the strain. Without warning, the supporting magnetism becomes unhinged, most likely because it got so twisted out of shape that it snapped. And a surprising thing happens! Instead of falling down under gravity, the stately, self-contained structures erupt, often rising as though propelled outward through the corona by a loaded spring (Fig. 2.14).

It is as if the lid had been taken off the caged material. Cool gas is then flung outward in slingshot fashion, tearing apart the overlying corona and ejecting large quantities of matter into space.

These eruptive prominences, as they are called when viewed at the Sun's apparent edge, are hurled outward at speeds of several hundreds of kilometers per second,

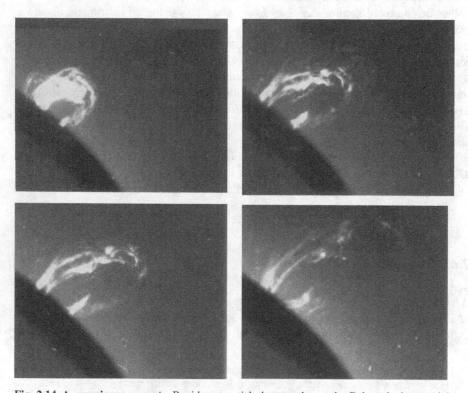

Fig. 2.14 A prominence erupts. Rapid, sequential photography, at the Balmer hydrogen alpha transition of 656.3 nm, catches an erupting prominence, which had not been detected as a filament during previous days. It suddenly rose from an active region and expanded at an apparent velocity of 375 km s^{-1}, hurling material far away from the Sun. Here the magnetic loops rise to a maximum visible extent of 360,000 km in just 16 min. This sequence of hydrogen-alpha images was taken on the west edge, or limb, of the Sun using the automatic flare patrol heliograph at the Meudon Station of the Observatoire de Paris; the solar disk has been occulted to give a better view of the event. (Courtesy of Madame Marie-Josephe Martres, and observer Michel Bernot, Observatoire de Paris, Meudon, DASOP.)

releasing a mass equivalent to that of a small mountain in just a few hours. Eruptive prominences can be larger, longer lasting, and more massive than solar flares.

Low-lying magnetic loops apparently support the long structures, which reside beneath magnetic aches that cross above them. This arcade of closed loops is anchored in the Sun, but is opened up by the rising prominence. It is as if some minor irritation builds up beyond the limit of tolerance, and the magnetic structure tosses off the pent-up frustration, like a dog shaking off the rain. The prominence rises and disappears, replaced by an elongated arcade of bright extreme-ultraviolet and X-ray loops that reform near their initial positions, and a new prominence often re-forms in the same shape and place after the explosive convulsion.

The disappearing prominences are strongly correlated with another form of energetic solar activity, the coronal mass ejections, which play an important role in solar–terrestrial interactions.

2.6.3 Coronal Mass Ejections

Powerful solar eruptions, originally termed coronal transients and now known as coronal mass ejections, were discovered using space-borne, white-light coronagraphs in the 1970s, first with the *Orbiting Solar Observatory* 7, or *OSO 7* for short, launched on 29 September 1971, then in greater profusion by *Skylab*, whose initial observations were reported by John T. Gosling and his colleagues in 1974. Since then, thousands of coronal mass ejections have been identified using coronagraph data from *Skylab* in 1973–1974, the U.S. Air Force's *P78-1* satellite from 1979 to 1985, NASA's *Solar Maximum Mission* in 1980 and from 1984 to 1989, the *SOlar and Heliospheric Observatory* from 1996 on, and *STEREO* from 2007.

The *P78-1* data, for example, resulted in the discovery of halo coronal mass ejections by Russell A. Howard and his colleagues in 1982. These mass ejections originate near the center of the solar disk, as viewed from Earth, and appear in coronagraph images as a ring or halo around the occulting disk. They may be headed toward the Earth, with threatening consequences, or away from it. More might be known about coronal mass ejections, at an earlier time, if the Air Force had not decided to shoot down the *P78-1* satellite during a test of an anti-satellite missile in 1985.

Coronal mass ejections are gigantic magnetic bubbles that can rush away from the Sun at supersonic speeds of more than $1,000 \, \text{km s}^{-1}$, expanding to become larger than the Sun in a few hours. They carry billions of tons of material out into space, produce intense shock waves, and accelerate vast quantities of energetic particles in interplanetary space. Arthur J. Hundhausen and his colleagues, for example, used the coronagraph aboard the *Solar Maximum Mission* satellite to specify the mass, velocity, energy, shape, and form of a large number of coronal mass ejections, fully reported in the literature in the 1990s and discussed in Chap. 6.

The physical size of the mass ejections dwarfs that of solar flares and even the active regions in which flares occur. Solar flares are nevertheless often associated with coronal mass ejections, and similar processes probably cause both of them. Like solar flares, the rate of occurrence of coronal mass ejections varies with the 11-year cycle of solar magnetic activity, ballooning out of the corona several times a day during activity maximum.

All kinds of solar activity therefore seem to be related to the sudden release of stored magnetic energy, but the exact relation between them is unclear. Large coronal mass ejections can occur together with an eruptive prominence, with a solar flare, or without either one of them. And although the frequency of solar eruptions increases with the number of sunspots, their strength does not necessarily increase. The most intense solar flares and coronal mass ejections are spread throughout the solar cycle.

Even though we are 150×10^6 km away, we still notice the disrupting effects of solar eruptions. Intense radiation from powerful solar flares can travel to the Earth in just 499 s, or about 8 min, altering our atmosphere, disrupting long-distance radio communications, and affecting satellite orbits. Very energetic particles, accelerated during the flare process, traveling almost as fast as light, can endanger unprotected astronauts or destroy satellite electronics. The coronal mass ejections arrive at the Earth 1–4 days after a major eruption on the Sun. They can result in strong geomagnetic storms with accompanying auroras and the threat of electrical power blackouts. All of these space weather effects are considered in Chap. 7, and they are of such vital importance that national centers now employ space weather forecasters to predict or warn us about storms in space.

In summary, there is a lot more going on at the Sun than meets the eye. Hot gases are caught within looping magnetic cages. Long-lasting magnetized holes, found at the Sun's poles, are locked open to expel a high-speed wind. The corona's magnetic energy is abruptly released, to power violent flares or mass ejections with threatening effects for the Earth. The entire solar atmosphere seethes and writhes in tune with the Sun's ever-changing magnetism.

Yet, even a decade ago, we did not understand how the Sun's magnetism originates and is regenerated in an 11-year cycle. We did not even have a firm grasp on the origins of sunspots. As we shall next see, the ancient mysteries of the sunspots and the Sun's cyclic magnetic variations have at least been partly solved by pinpointing the elusive dynamo that generates the solar magnetic fields. This has been accomplished by looking deep inside the Sun where internal motions amplify, compress, and slowly transform the Sun's spotty, ever-changing magnetism.

2.7 Summary Highlights: Discovery of Space

- The aurora borealis and aurora australis, or the northern and southern lights, were attributed to electrically charged, material corpuscles from the Sun, channeled to the Earth's polar regions by its magnetic fields and providing early evidence that space is not empty.

- Powerful geomagnetic storms are caused by explosive outbursts from the Sun, which travel through interplanetary space and can compress the Earth's magnetic field.
- The most abundant element in the Sun is hydrogen, and helium is the second most abundant solar element.
- The sharp outer edge of the visible solar disk, the photosphere, is somewhat illusory, for a hot, normally invisible, outer atmosphere, called the corona, envelops it.
- Coronal emission lines, observed during total eclipses of the Sun, were eventually attributed to iron, nickel, and calcium ionized at a temperature of a million Kelvin.
- The million-degree corona consists mainly of electrons and protons torn out of the Sun's abundant hydrogen atoms during frequent collisions, with lesser amounts of heavier ions and magnetic fields as well.
- The corona is always expanding away from the Sun, filling space with a relentless solar wind that envelops the Earth and all the other planets.
- The solar wind was inferred from comet ion tails, suggested by theoretical considerations, and fully confirmed by direct measurements with the Soviet *Lunik 2* (*Luna 2*) and American *Mariner 2* spacecraft.
- The ever-flowing solar wind blows at two supersonic speeds – a slow wind moving at about $400 \, \text{km s}^{-1}$ and a fast wind moving at about twice that speed.
- Numerous spacecraft have measured the physical properties of the solar wind for more than 45 years.
- Interplanetary space also contains very energetic atomic nuclei, the cosmic rays, which come from nearby interstellar space.
- The most abundant cosmic rays are protons, which are about a million times more energetic than solar wind protons, but with a flux at the Earth's orbit of 10×10^9 times less than their solar wind counterparts.
- Magnetic fields penetrate the visible solar disk, or photosphere, forming Earth-sized sunspots whose strength, measured by the Zeeman effect, is thousands of times greater than the strength of the Earth's magnetic field.
- The photosphere's magnetism is continuously varying as bipolar magnetic fields appear and disappear all over the Sun, creating ever-varying magnetic loops in the overlying chromosphere and corona.
- There is an 11-year cyclic variation in the number and positions of sunspots, the shape of the corona, the level of explosive solar activity, and even the amount of cosmic rays reaching the Earth.
- In the photosphere and below, gas pressure dominates the magnetic pressure, so the moving gas carries the magnetic fields around, but it is the other way around in the low overlying corona, where the magnetic fields shape, mold, and constrain the hot gas.

- Interplanetary magnetic fields are anchored at one end in the Sun, carried and stretched out into space by the solar wind, and twisted into a spiral shape by the Sun's rotation.
- The interplanetary magnetic field is divided into magnetic sectors that point in or out of the Sun and are separated by a magnetically neutral current sheet.
- The Sun's million-degree corona emits powerful X-rays that can only be detected from above the Earth's absorbing atmosphere.
- The Sun's X-rays and extreme-ultraviolet radiation create the Earth's ionosphere by tearing atmospheric molecules into their atomic components and ionizing the atoms.
- The corona's X-rays were first detected with instruments aboard 5-min rocket flights, confirmed from the *Orbiting Solar Observatory*, abbreviated *OSO*, satellites, and imaged from *Skylab*, which demonstrated the ubiquitous presence of coronal loops, X-ray bright points, and coronal holes.
- Comparisons of an X-ray image of a coronal hole, taken during a rocket flight, and solar wind velocity and flux measurements from the *Vela* and *Pioneer VI* satellites suggested that polar coronal holes are the source of the fast solar winds detected near the ecliptic, at least near a minimum in the 11-year cycle of solar magnetic activity. This was confirmed when *Skylab* X-ray photographs were compared to the solar wind measurements from the *Interplanetary Monitoring Platforms 6, 7,* and *8.*
- Although normally unseen in the Sun's visible light, solar flares can outshine the entire Sun at radio and X-ray wavelengths. Radio bursts observed with telescopes on the ground provide evidence for the ejection of electron beams or shock waves into interplanetary space during solar outbursts. The most intense radiation from solar flares is emitted at X-ray wavelengths, detected with increasing sophistication from balloons, rockets, and satellites.
- Looping arches of magnetism suspend elongated prominences seen in the red light of the Balmer alpha transition of hydrogen at 656.3 nm; they can suddenly and unpredictably open up and expel their contents into interplanetary space, defying the Sun's enormous gravity.
- Coronal mass ejections throw billions of tons of material into interplanetary space, and their associated shocks accelerate and propel vast quantities of high-speed particles and intense magnetic fields ahead of them. Coronal mass ejections were discovered using space-borne, white-light coronagraphs, first with the *Orbiting Solar Observatory 7*, and then in greater numbers from *Skylab*, the Air Force's *P78-1* satellite, and NASA's *Solar Maximum Mission*.
- The rates of occurrence of solar flares, erupting prominences, and coronal mass ejections all vary in tandem with the 11-year cycle of solar magnetic activity.
- X-rays from solar flares alter our atmosphere, disrupt long-distance radio communications, and affect satellite orbits; protons accelerated in solar flares or by coronal mass ejection shocks endanger unprotected astronauts and satellites; coronal mass ejections can produce strong geomagnetic storms and intense auroras, and they may cause electrical power blackouts.

2.8 Key Events in the Discovery of Space*

Date	Event
1600	William Gilbert, physician to Queen Elizabeth I of England, publishes a small treatise demonstrating that "the terrestrial globe is itself a great magnet."
1610–1613	Galileo Galilei first systematically studies sunspots through a telescope.
1716	Edmund Halley suggests that aurora rays delineate terrestrial magnetic field lines, and that the auroras are due to a magnetized fluid that circulates poleward along the Earth's dipole magnetic field lines.
1724	George Graham discovers large, irregular fluctuations in compass needles, later called magnetic storms; Anders Celsius saw them at about the same time.
1733	Jean Jacques d'Ortous de Mairan argues for a connection between the occurrence of auroras and sunspots.
1799–1804	Alexander von Humboldt makes regular, precise measurements of the strength, dip, and inclination of the Earth's magnetic field during his voyage to South America. The term "magnetic storm" came into common usage as the result of Humboldt's scientific analyses and reports.
1802	William Hyde Wollaston discovers that the solar spectrum is cut by several dark gaps, now called absorption lines.
1814	Joseph von Fraunhofer allows light to pass through a narrow slit and then a prism, detecting and cataloging more than 300 absorption lines in the spectrum of visible sunlight, assigning Roman letters to the most prominent; these absorption lines are now sometimes called Fraunhofer lines.
1838	Carl Friedrich Gauss publishes a mathematical description of the Earth's dipolar magnetic field, using it with observations to show that the magnetism must originate deep down inside the Earth's core.
1844	Samuel Heinrich Schwabe demonstrates that the number of sunspots varies from a maximum to a minimum and back to a maximum again in a period of about 11 years.
1852	Edward Sabine shows that global magnetic disturbances of the Earth, now called geomagnetic storms, vary in tandem with the 11-year sunspot cycle.
1858	Richard C. Carrington demonstrates that sunspots move from high latitudes to near the solar equator during the 11-year sunspot cycle.
1859–1860	In 1859, Richard C. Carrington and Richard Hodgson independently observe a solar flare in the white light of the photosphere, and in 1860 publish the first account of such a flare. Seventeen hours after the flare a large magnetic storm begins on the Earth.
1859–1861	By comparing the Sun's absorption lines with emission lines of elements vaporized in the terrestrial laboratory, Gustav Kirchhoff and Robert Bunsen identify in the solar atmosphere several elements known on Earth, including sodium, calcium, and iron.
1862	Anders Jonas Ångström announces his discovery of hydrogen in the spectrum of the Sun.

* See the References at the end of this book for complete references to these seminal papers. Also see the Key Event tables at the end of subsequent chapters for greater detail on the history of specific unsolved problems, including contributions after 1980 and the main results of the *Yohkoh, Ulysses, Wind, SOHO, ACE, TRACE, RHESSI*, and *Hinode* missions.

Date	Event
1868	Anders Jonas Ångström provides measurements of the wavelengths of more than 100 dark lines in the Sun's visible spectrum, in units that become known as the Ångström, abbreviated Å, where $1\,\text{Å} = 10^{-10}\,\text{m}$.
1868	During a solar eclipse on 18 August 1868, several observers noted a chromospheric line in the yellow. Most of them identified it as the sodium D line in emission, but Norman Lockyer pointed out that there was a discrepancy in the wavelength. Pierre Jules César (P. J. C.) Janssen and Lockyer independently realized how to observe the chromosphere outside the eclipse, resulting in a more leisurely method of measuring the wavelengths. It was probably not until the following year that Lockyer convinced himself that the yellow line could not be matched with any known terrestrial element, and coined the name "helium" for the new element. Helium was not identified on the Earth until 1895 when William Ramsay detected the line in gases given off by a uranium mineral called clevite.
1869	At the solar eclipse of 7 August 1869, Charles A. Young, and, independently, William Harkness, discovers a single, bright, green emission line in the spectrum of the solar corona. This conspicuous feature remained unidentified with any known terrestrial element for more than half a century, but it was eventually associated with highly ionized iron atoms missing 13 electrons, denoted Fe XIV, indicating that the corona has a million-degree temperature (see **1939–1941**).
1873	Elias Loomis demonstrates a correlation between intense auroras and the number of sunspots, both recurring with an 11-year periodicity and with similar times of maximum.
1882	Henry Rowland invents the concave diffraction grating, in which up to 20,000 lines to the inch are engraved on a spherical concave mirrored surface. The grating revolutionized spectrometry by widely dispersing light and permitting numerous spectral lines to be investigated with great accuracy and efficiency. Rowland used his grating to investigate the visible-light spectrum of the Sun, publishing the wavelengths for 14,000 lines.
1885	Johann Balmer publishes an equation that describes the regular spacing of the wavelengths of the four lines of hydrogen detected in the spectrum of visible sunlight, and predicts the wavelengths of other solar hydrogen lines at invisible ultraviolet, infrared, and radio wavelengths. In 1913, Niels Bohr developed an atomic model that explains Balmer's equation in terms of specific electron orbits with quantized values of energy.
1889–1890	Frank H. Bigelow argues that the structure of the corona detected during solar eclipses provides strong evidence for large-scale solar magnetic or electric fields. He correctly speculated that polar rays delineate open magnetic field lines along which material escapes from the Sun, and that equatorial elongations of the corona mark closed magnetic field lines.
1891	Henri Deslandres and George Ellery Hale independently invent the spectroheliograph used to image the Sun in the light of one particular wavelength only.
1895	William Ramsay discovers that a uranium mineral named cleveite produces the chemically inert gas helium when heated. Helium was known from spectrographic evidence to be present on the Sun, but had not yet been found on Earth.
1896–1913	Kristian Birkeland argues that polar auroras and geomagnetic storms are due to beams of electrons.
1901	On 12 December 1901, Guglielmo Marconi sends a long-wavelength radio signal in Morse code between Poldhu, Cornwall, and St John's, Newfoundland, across the Atlantic Ocean and over a distance of about 3,500 km.

Date	Event
1908	George Ellery Hale (1908b, c) uses the Zeeman splitting of spectral lines to measure intense magnetic fields in sunspots. They are thousands of times stronger than the Earth's magnetism.
1911	Arthur Schuster shows that a beam of electrons from the Sun cannot hold itself together against the mutual electrostatic repulsion of the electrons.
1912	During balloon flights, Victor F. Hess discovers an increase of atmospheric ionization at high altitudes, attributing the increase to energetic "radiation" that comes from outer space. His balloon flights at night and during the solar eclipse of 12 April 1912 additionally proved that the ionization was not caused by the Sun.
1919	George Ellery Hale and colleagues show that sunspots occur in bipolar pairs with a polarity orientation that varies over 22 years.
1919	Frederick Alexander Lindemann (later Lord Cherwell) suggests that an electrically neutral plasma ejection from the Sun is responsible for intense, non-recurrent geomagnetic storms.
1926	Robert A. Millikan confirms that penetrating radiation comes from beyond the Earth, and gives it the name cosmic rays.
1927	Jacob Clay shows that cosmic rays are less intense near the equator than at higher latitudes, suggesting that they are charged particles deflected toward the Earth's magnetic poles.
1928–1932	In 1928, Albrecht Unsöld uses solar absorption lines to show that hydrogen is at least a million times more abundant than any other element in the solar photosphere. In the following year, William H. Mc Crea similarly demonstrates the overwhelming abundance of hydrogen in the chromosphere, and by 1932 Bengt Strömgren shows that hydrogen has to be the most abundant element in the entire star.
1930–1960	Hydrogen alpha, or Hα, observations of chromosphere brightening, at the red Balmer alpha transition of hydrogen at 656.3 nm, by V. Bumba, Helen W. Dobson, Mervyn Archdall Ellison, Ronald G. Giovanelli, Harold W. Newton, Robert S. Richardson, A. B. Severny, and Max Waldmeier, show that chromosphere flares occur close to sunspots, usually between the two main spots of a bipolar group, that magnetically complex sunspot groups are most likely to emit flares, and that hydrogen alpha flare ribbons lie adjacent and parallel to the magnetic neutral line.
1930	Bernard Lyot invents the coronagraph, a telescope with an occulting disk that blocks out the intense light of the solar disk, permitting the observation of the solar corona.
1931–1940	Sydney Chapman and Vincent C. A. Ferraro propose that magnetic storms are caused when an electrically neutral plasma cloud ejected from the Sun envelops the Earth.
1932–1933	Arthur H. Compton demonstrates that the Earth's dipolar magnetic field is deflecting cosmic rays incident from outer space, resulting in an increase of their intensity with terrestrial latitude and showing that cosmic rays are very energetic charged particles.
1932	Carl D. Anderson (1932b) discovers the positron, or anti-electron, in cloud chamber observations of secondary particles produced by cosmic rays in our atmosphere
1934	Walter Baade and Fritz Zwicky propose that cosmic rays are accelerated and propelled into interstellar space during the explosion of massive dying stars called supernovae.
1935–1937	J. Howard Dellinger suggests that the sudden ionosphere disturbances that interfere with radio signals have a solar origin.

Date	Event
1938	Thomas H. Johnson uses the east–west effect to show that the most abundant cosmic rays are positively charged.
1939–1941	Walter Grotrian and Bengt Edlén identify coronal emission lines with highly ionized elements, indicating that the Sun's outer atmosphere has a temperature of about a million Kelvin. The conspicuous green emission line was identified with an iron atom missing 13 electrons, designated (Fe XIV).
1944	Robert S. Richardson proposes the term solar flare for sudden, bright, rapid, and localized variations detected in the hydrogen alpha, or Hα, light of the chromosphere.
1946	Edward V. Appleton and J. Stanley Hey demonstrate that meter-wavelength solar radio noise originates in sunspot-associated active regions, and that sudden large increases in the Sun's radio output are associated with chromosphere brightenings, also known as solar flares.
1946	Vitalii L. Ginzburg, David F. Martyn, and Joseph L. Pawsey independently confirm the existence of a very hot solar corona, with a temperature of about a million Kelvin, from the observations of the Sun's radio radiation.
1946	Richard Tousey and his colleagues at the U.S. Naval Research Laboratory use a V-2 rocket to obtain the first extreme-ultraviolet spectrum of the Sun on 10 October 1946.
1946, 1950	Scott E. Forbush and his colleagues describe flare-associated transient increases in the cosmic ray intensity at the Earth's surface, and attribute them to very energetic charged particles from the Sun.
1947	Ruby Payne-Scott, D. E. Yabsley, and John G. Bolton discover that meter-wavelength solar radio bursts often arrive later at lower frequencies and longer wavelengths. They attributed the delays to disturbances moving outward at velocities of 500–$750 \mathrm{km\ s^{-1}}$, exciting radio emission at the local plasma frequency.
1948–1949	Soft X-rays from the Sun are first detected on 5 August 1948 with a V-2 rocket experiment performed by the U.S. Naval Research Laboratory, abbreviated NRL, reported by T. Robert Burnight in 1949. Subsequent sounding rocket observations by the NRL scientists revealed that the Sun is a significant emitter of X-rays and that the X-ray emission is related to solar activity.
1948–1950	High-altitude balloon measurements by Helmut L. Bradt and Bernard Peters, and independently by Phyllis Freier and colleagues (1948a, b), show that cosmic rays consist of hydrogen nuclei (protons), with lesser amounts of helium nuclei (alpha particles) and heavier nuclei.
1950–1954	Scott F. Forbush demonstrates the inverse correlation between the intensity of cosmic rays arriving at Earth and the number of sunspots over two 11-year solar activity cycles.
1950–1959	John Paul Wild and his colleagues use a swept frequency receiver to delineate Type II radio bursts, attributed to shock waves moving out during a solar outburst at about $1,000 \mathrm{km\ s^{-1}}$, and Type III radio bursts, due to outward streams of high-energy electrons. The electrons are accelerated at the onset of a solar flare and move at nearly the velocity of light, or close to $300,000 \mathrm{km\ s^{-1}}$.
1951–1952	Herbert Friedman and his colleagues at the U.S. Naval Research Laboratory use instruments aboard sounding rockets to show that the Sun emits enough X-ray and extreme-ultraviolet radiation to create the ionosphere.

Date	Event
1951–1957	Ludwig F. Biermann argues that a continuous flow of solar corpuscles is required to push comet ion tails into straight paths away from the Sun, correctly inferring solar wind speeds as high as 500–$1,000$ km s^{-1}.
1951–1963	Herbert Friedman and his colleagues at the U.S. Naval Research Laboratory use rocket and satellite observations to show that intense X-rays are emitted from the Sun, that the X-ray emission is related to solar activity, and that X-rays emitted during solar flares are the cause of sudden ionosphere disturbances.
1955	Leverett Davis Jr. argues that solar corpuscular emission will carve out a cavity in the interstellar medium, now known as the heliosphere, accounting for some observed properties of low-energy cosmic rays.
1955	Horace W. Babcock and Harold D. Babcock use magnetograms, taken over a 2-year period from 1952 to 1954, to show that the Sun has a general dipolar magnetic field of about 10^{-4} T, or 1 G, in strength, usually limited to solar latitudes greater than $\pm 55°$. Bipolar magnetic regions are found at lower solar latitudes. Occasional extended unipolar areas, of only one outstanding magnetic polarity, are also found; they speculated that these unipolar regions might be related to 27-day recurrent terrestrial magnetic storms.
1956	Peter Meyer, Eugene N. Parker, and John A. Simpson argue that enhanced interplanetary magnetism at the peak of the solar activity cycle deflects cosmic rays from their Earth-bound paths.
1957	Max Waldmeier introduces the name coronal holes for seemingly vacant places with no detectable coronal radiation in the 530.3 nm emission line of ionized iron (Fe XIV). In 1981, Waldmeier reported his studies of these polar coronal holes from 1940 to 1978, noting that they are permanently present for about 7 years, including the minimum in the 11-year solar activity cycle, but that they seem to disappear for about 3 years near activity maximum.
1957	The first artificial satellite, *Prosteyshiy Sputnik*, was launched by the Soviet Union on 4 October 1957.
1957–1959	Sydney Chapman (1957, 1958, 1959a, b) shows that a very hot, static corona should extend to the Earth's orbit and beyond.
1958	Eugene N. Parker (1958d) suggests that a perpetual supersonic flow of electric corpuscles, which he called the solar wind, naturally results from the expansion of a very hot corona. He also demonstrates that the solar magnetic field will be pulled into interplanetary space, attaining a spiral shape in the plane of the Sun's equator due to the combined effects of the radial solar wind flow and the Sun's rotation.
1958–1959	The first American satellite, *Explorer 1*, was launched into orbit on 1 February 1958, followed by *Explorer 3* on 26 March 1958. James A. Van Allen and colleagues used instruments aboard these spacecraft to discover the belts of charged particles that girdle the Earth's equator.
1958–1959	During a balloon flight on 20 March 1958, Laurence E. Peterson and John Randolph Winckler observed a burst of high-energy, gamma-ray radiation (200–500 keV) coincident with a solar flare, suggesting non-thermal particle acceleration during such outbursts on the Sun.
1960–1961	Konstantin I. Gringauz reports that the Soviet spacecraft, *Lunik 2* (*Luna 2*), launched on 12 September 1959, has measured high-speed ions in interplanetary space outside the Earth's magnetic field with a flux of 200 million (2×10^8) ions (protons) per square centimeter per second.

Date	Event
1960–1961	Herbert Friedman and colleagues at the U.S. Naval Research Laboratory obtain the first, primitive X-ray picture of the Sun during a brief rocket flight.
1961	Cosmic ray electrons are discovered, but in much lower flux than protons, in cloud chamber observations by James A. Earl and by the balloon measurements of Peter Meyer and Rochus Vogt.
1961–1962	Herbert A. Bridge and co-workers use a plasma probe aboard the American space-craft *Explorer 10*, launched 25 March 1961, to provide rough measurements of the density, speed, and direction of the solar wind, but the spacecraft never reached the undisturbed interplanetary medium.
1962–1967	*Mariner 2* was launched on 7 August 1962. Using the data obtained during *Mariner's* voyage to Venus, Marcia Neugebauer and Conway W. Snyder demonstrate that a low-speed solar wind plasma is continuously emitted by the Sun, and discover high-speed wind streams that recur with a 27-day period within the orbital plane of the planets. Interplanetary shocks associated with solar activity are detected using instruments aboard the *Mariner 2* spacecraft in 1962, reported by Charles P. Sonett and colleagues in 1964.
1963	NASA launches *Explorer 18*, the first *Interplanetary Monitoring Platform* and abbreviated *IMP 1*, which first mapped the Earth's magnetotail.
1964–1966	Norman F. Ness and John M. Wilcox use magnetometers aboard NASA's *Interplanetary Monitoring Platform 1*, launched on 27 November 1963, to measure the strength and direction of the interplanetary magnetic field, to show that it has a spiral shape, and to discover large-scale magnetic sectors in interplanetary space that point toward or away from the Sun.
1968	John M. Wilcox suggests that the fast solar wind might originate in magnetically open regions near the poles of the Sun.
1968	Giuseppe Vaiana and his colleagues show that the soft X-ray emission of a solar flare corresponds spatially with the intense visible-light emission at the Balmer alpha transition of hydrogen, with roughly the same size, suggesting a close link between the two phenomena.
1968	Werner M. Neupert uses soft X-ray flare data, obtained with the third *Orbiting Solar Observatory*, abbreviated *OSO 3*, to confirm that soft X-rays slowly build up in strengths, and to show that the rise to maximum intensity resembles the time integral of the rapid, impulsive radio burst.
1969	Kenneth Schatten, John M. Wilcox, and Norman F. Ness develop a simple potential field model to predict the polarity of the interplanetary magnetic field in the ecliptic from the magnetic field measurements in the photosphere.
1971–1973	The first good, space-based observation of a coronal transient, now called a coronal mass ejection, was obtained on 14 December 1971 using the coronagraph aboard NASA's seventh *Orbiting Solar Observatory*, abbreviated *OSO 7*, reported by Richard Tousey in 1973.
1972–1973	Edward L. Chupp and his colleagues detect solar gamma-ray lines for the first time using a monitor aboard NASA's seventh *Orbiting Solar Observatory*, or *OSO 7*. They observed the neutron capture (2.223 MeV) and positron annihilation (0.511 MeV) lines associated with solar flares. The 2.223 MeV line had been anticipated theoretically by Philip Morrison.
1973	Allen S. Krieger, Adrienne F. Timothy, and Edmond C. Roelof compare an X-ray photograph of the Sun, obtained during a rocket flight on 24 November 1970, with *Vela* and *Pioneer VI* satellite measurements of the solar wind flux and velocity to show that coronal holes are the probable source of recurrent high-speed streams in the solar wind.

Date	Event
1973	Giuseppe S. Vaiana and his colleagues (1973a, b) use solar X-ray observations taken from rockets during the preceding decade to identify the three-fold magnetic structure of the solar corona – coronal holes, coronal loops, and X-ray bright points.
1973–1974	The manned, orbiting solar observatory, *Skylab*, is launched on 14 May 1973, and manned by three person crews until 8 February 1974. *Skylab's* Apollo Telescope Mount contained 12 tons of solar observing instruments that spatially resolved solar flares at soft X-ray and ultraviolet wavelengths, fully confirmed coronal holes, coronal loops, and X-ray bright points, and recorded many coronal mass ejections that sometimes moved fast enough to produce interplanetary shocks.
1973–1977	X-ray photographs of the Sun taken using the Apollo Telescope Mount on the manned, orbiting *Skylab* satellite, launched on 14 May 1973, fully confirm coronal holes, the ubiquitous coronal loops, and X-ray bright points. Detailed comparisons of *Skylab* X-ray photographs and measurements of the solar wind, made from *Interplanetary Monitoring Platforms*, abbreviated *IMPs*, 6, 7, and 8, confirm that solar coronal holes are the source of the high-velocity solar wind streams detected in the ecliptic.
1974	John Thomas Gosling and colleagues report observations of coronal mass ejections, then called coronal disturbances or coronal transients, with the coronagraph aboard *Skylab*, noting that some of them have the high outward speed of up to $1,000 \mathrm{km\ s^{-1}}$ needed to produce interplanetary shocks.
1974, 1976	Leon Golub and his colleagues (1974, 1976a, b) describe the physical properties of X-ray bright points observed with *Skylab*.
1974–1986	The *Helios 1* and *2* spacecraft, respectively launched in December 1974 and in January 1976, measured the solar wind parameters as close as 0.3 AU from the Sun for an entire 11-year solar cycle.
1978	Edward J. Smith, Bruce T. Tsurutani, and Ronald L Rosenberg use observations from *Pioneer 11* to show that the solar wind becomes unipolar, or obtains a single magnetic polarity, at high heliographic latitudes near 16°.
1979–1982	John L. Kohl and co-workers used 5-min rocket flights of an ultraviolet coronagraph spectrometer to pioneer spectroscopy of the extended corona in the absence of a solar eclipse. Observations over 40-h periods were obtained in the 1990s from *Spartan 201s* deployed and retrieved from the *Space Shuttle*, preparing the way for the UltraViolet Coronagraph Spectrometer, abbreviated UVCS, aboard the *SOlar and Heliospheric Observatory*, or *SOHO* for short.
1980, 1984–1989	NASA launches the *Solar Maximum Mission*, abbreviated *SMM*, satellite on 14 February 1980, with an in-orbit repair from the *Space Shuttle Challenger* on 6 April 1984 and a mission end on 17 November 1989. It excelled in X-ray and gamma ray spectroscopy of solar flares, and in coronagraph observations of coronal mass ejections, during a maximum in the 11-year solar activity cycle. Radiometric data taken from *SMM* during the decade after launch were used to show that the total solar irradiance of Earth, known as the solar constant, varies in step with the 11-year cycle of solar activity. The so-called solar constant has a total decline and rise of about 0.1%.
1981–1982	The Japanese spacecraft *Hinotori*, meaning *firebird*, was launched on 21 February 1981 and operated until 11 October 1982. It created images of solar-flare X-rays with an energy of around 20 keV, and measured solar flare temperatures of between 10 and 40×10^6 K using soft X-ray spectroscopy.

Date	Event
1980–1989	George Doschek, Ester Antonucci, and their colleagues use *P78-1* and *Solar Maximum Mission* observations of soft X-ray spectral lines to show that the impulsive phase of solar flares is associated with upward flows of heated chromosphere material with velocities of several hundred kilometers per second. Such an upflow is called chromospheric evaporation. In 1982, Katsuo Tanaka and colleagues used a spectrometer aboard the *Hinotori* spacecraft to independently confirm this up-flow associated with solar flares.
1982	Russell A. Howard and co-workers report the discovery of a halo coronal transient, or coronal mass ejection, using white-light coronagraph observations with the *P78-1* satellite; such halo events can be directed toward Earth.
1982–1983	Gary J. Rottman, Frank Q. Orrall, and James A. Klimchuk obtain rocket observations of extreme-ultraviolet resonance lines formed in the low corona and transition region, showing that the lines are systematically shifted to shorter wavelengths in large polar coronal holes. Outflow velocities of $7–8\,\mathrm{km\,s^{-1}}$ are inferred from these Doppler shifts.
1980–2007	See the key events at the end of Chaps. 3–7.

Chapter 3
Exploring Unseen Depths of the Sun

3.1 What Makes the Sun Shine?

3.1.1 The Sun's Size, Mass, and Temperature

To understand what makes the Sun shine, we must first describe the star's physical characteristics. The entire Sun is nothing but a big luminous ball of gas, hot and concentrated at the center and cooler and more tenuous further out. It is the most massive and largest object in our solar system. The Sun's mass, which is 333,000 times Earth's mass, can be inferred from Kepler's third law and the Earth's orbital period and mean distance from the Sun. And the Sun's size, at 109 times the diameter of the Earth, can be inferred from the Sun's distance and angular extent.

From the Sun's size and luminous output, we can infer the temperature of its visible disk, 5780 K (Focus 3.1). The gas we see in this photosphere is extremely rarefied, about 10,000 times less dense than the air we breathe. The pressure of the tenuous gas in the photosphere is less than that beneath the foot of a spider.

Focus 3.1

The Luminosity, Effective Temperature, and Central Temperature of the Sun

Satellites have been used to accurately measure the Sun's total irradiance just outside the Earth's atmosphere, establishing the value of the solar constant:

$$f_\odot = 1361 \, \mathrm{J \, s^{-1} \, m^{-2}},$$

where $1 \, \mathrm{J \, s^{-1}}$ is equivalent to 1 W. The solar constant is defined as the total amount of radiant solar energy per unit time per unit area reaching the top of the Earth's atmosphere at the Earth's mean distance from the Sun. We can use it to determine the Sun's absolute luminosity, $L_\odot$, from:

$$L_\odot = 4\pi f_\odot (\mathrm{AU})^2 = 3.854 \times 10^{26} \, \mathrm{J \, s^{-1}},$$

where the mean distance between the Earth and the Sun is $1 \, \mathrm{AU} = 1.496 \times 10^{11} \, \mathrm{m}$.

K.R. Lang, *The Sun from Space*, Astronomy and Astrophysics Library,
© Springer-Verlag Berlin Heidelberg 2009

The effective temperature, $T_\odot$, of the visible solar disk, called the photosphere, can be determined using the Stefan-Boltzmann law:

$$L_\odot = 4\pi\sigma\, R^2 T_\odot{}^4,$$

where the Stefan-Boltzmann constant $\sigma = 5.670 \times 10^{-8}$ J m^{-2} K^{-1} s^{-1}, and the Sun's radius is $R_\odot = 6.955 \times 10^8$ m. Solving for the temperature:

$$T_\odot = [L_\odot/(4\pi R_\odot{}^2)]^{1/4} = 5780 \,\text{K}.$$

Incidentally, the Stefan-Boltzmann law applies to other stars, indicating that at a given temperature bigger, giant stars have greater luminosity.

The temperature, T_C, at the center of the Sun can be estimated by assuming that a proton must be hot enough and move fast enough to counteract the gravitational compression it experiences from all the rest of the star. That is:

$$\text{Thermal energy} = \frac{3}{2}\,k\,T_C = \frac{G\,m_P\,M_\odot}{R_\odot} = \text{Gravitational potential energy},$$

where Boltzmann's constant $k = 1.38066 \times 10^{-23}$ J K^{-1}, the gravitational constant $G = 6.6726 \times 10^{-11}$ Nm^2 kg^{-2}, the Sun's mass $M_\odot = 1.989 \times 10^{30}$ kg, and the mass of the proton is $m_P = 1.6726 \times 10^{-27}$ kg. Solving for the central temperature we obtain:

$$T_C = \frac{2G\,m_P\,M_\odot}{3\,k\,R_\odot} = 1.56 \times 10^7 \,\text{K},$$

so the temperature at the center of the Sun is 15.6×10^6 K.

The material deep down inside the Sun must become hotter and more densely concentrated to support the overlying weight and to keep our star from collapsing. Calculations show that the temperature reaches 15.6×10^6 (1.5×10^7) K at the center of the Sun (also see Focus 3.1). The outward pressure generated by this hot gas is 233 billion (2.33×10^{11}) times Earth's air pressure at sea level. The center is also extremely compacted with a density of $1,500$ kg m^{-3}, which is about 10 times the density of gold or lead – but it still behaves like a gas. All of the essential physical properties of the Sun are given in Table 3.1.

3.1.2 Nuclear Reactions in the Sun's Central Regions

The lightest known element, hydrogen, is the most abundant ingredient of the Sun, about 75% by mass and 92.1% by number of atoms. Each hydrogen atom contains one proton at its center and one electron outside this nucleus, and they are both liberated from their atomic bonds in the hot solar gas. Helium accounts for almost all the rest of the solar material; the heavier elements only amount to 0.1% by number.

Table 3.1 Physical parameters of the Sun[a]

Mass	1.989×10^{30} kg (332,946 Earth masses)
Radius	6.955×10^{8} m (109 Earth radii)
Volume	1.412×10^{27} m^3 (1.3 million Earths)
Density (center)	$151,300$ kg m^{-3}
Density (mean)	$1,409$ kg m^{-3}
Pressure (center)	2.334×10^{11} bars
Pressure (photosphere)	0.0001 bar
Temperature (center)	15.6×10^{6} K
Temperature (photosphere)	5,780 K
Temperature (corona)	2–3×10^{6} K
Luminosity	3.854×10^{26} J s^{-1}
Solar constant	$1,361$ J s^{-1} m$^{-2} = 1,361$ W m^{-2}
Mean distance	1.4959787×10^{11} m $= 1.0$ AU
Age	4.55 billion years
Principal chemical constituents (by number of atoms)	
Hydrogen	92.1%
Helium	7.8%
All others	0.1%

[a]Mass density is given in kilograms per cubic meter, or kg m^{-3}; the density of water is 1,000 kg m^{-3}. The unit of pressure is bars, where 1.013 bars is the pressure of the Earth's atmosphere at sea level. The unit of luminosity is J s^{-1}, power is often expressed in watts, where 1.0 W $= 1.0$ J s^{-1}

As demonstrated by Francis W. Aston in 1917, the helium atom is slightly less massive, by just 0.7%, than the sum of the four hydrogen atoms that enter into it.

In a sequence of nuclear reactions at the center of the Sun, four protons are successively fused together into one heavier helium nucleus. The helium is slightly less massive (by a mere 0.7%) than the four protons that combine to make it. This mass difference, Δm, is converted into energy, ΔE, according to the famous equation $\Delta E = \Delta m c^2$, derived by Einstein in 1905–1906; since the velocity of light, c, is a large number, about $300,000$ km s^{-1}, a small mass difference, Δm, can account for the Sun's awesome energy.

The details of just how helium nuclei are synthesized from protons had to await the discovery of fundamental particles, such as the neutron and positron that were both discovered in 1932, by James Chadwick and Carl Anderson, respectively. (The positron, denoted as e$^+$, is the positive electron, or the anti-matter version of the electron, denoted by e$^-$, with the same mass and an opposite charge.) It was not until 1938 that Hans A. Bethe and Charles L. Critchfield demonstrated how a sequence of nuclear reactions, called the proton–proton chain, makes the Sun shine (Focus 3.2). Bethe was awarded the Nobel Prize in Physics in 1967 for this and other discoveries concerning energy production in stars.

Focus 3.2 ▬▬▬▬▬▬▬▬▬▬▬▬▬▬▬▬▬▬▬▬▬▬▬▬▬▬

The Proton–Proton Chain

The hydrogen-burning reactions that fuel the Sun are collectively called the proton–proton chain. It begins when two of the fastest moving protons, designated by the symbol p, collide head on. They move into each other and fuse together to make a deuteron, D^2, the nucleus of a heavy form of hydrogen known as deuterium. Since a deuteron consists of one proton and one neutron, one of the protons entering into the reaction must be transformed into a neutron, emitting a positron, e^+, to carry away the proton's charge, together with an electron neutrino, v_e, to balance the energy in the reaction. This first step in the proton–proton chain is written:

$$p + p \rightarrow D^2 + e^+ + v_e.$$

The deuteron next collides with another proton to form a nucleus of light helium, He^3, together with energetic gamma-ray radiation, denoted by the symbol γ. In the final part of the proton–proton chain, two such light helium nuclei meet and fuse together to form a nucleus of normal heavy helium, He^4, returning two protons to the solar gas. These reactions are written:

$$D^2 + p \rightarrow He^3 + \gamma$$
$$He^3 + He^3 \rightarrow He^4 + 2p.$$

Additional gamma rays are in the meantime produced when the positrons, e^+, combine with electrons, e^-, in the pair annihilation reaction:

$$e^+ + e^- \rightarrow 2\gamma.$$

The net result is that four protons have been fused together to make one helium nucleus, gamma rays, and electron neutrinos, or that:

$$4p \rightarrow He^4 + 6\gamma + 2v_e$$

with the creation of an amount of energy, ΔE, given by:

$$\Delta E = \Delta m \, c^2 = (4m_p - m_{He})c^2 = 0.007(4m_p) \, c^2 = 0.428 \times 10^{-11} \, J,$$

where $m_p = 1.6726 \times 10^{-27}$ kg and $m_{He} = 6.6465 \times 10^{-27}$ kg, respectively, denote the mass of the proton and the helium nucleus, and $c = 2.9979 \times 10^8 \, m \, s^{-1}$ is the velocity of light.

The number, N, of helium nuclei that are formed every second is $N = L_\odot / \Delta E \approx 10^{38}$, where the solar luminosity is $L_\odot = 3.854 \times 10^{26} \, J \, s^{-1}$. That is, a hundred trillion trillion trillion helium nuclei are formed each second to produce the Sun's energy at its present rate. In the process, the amount of mass, ΔM, consumed every second is $\Delta M = 10^{38} \, \Delta m = 4.76 \times 10^9 \, kg \approx 5 \times 10^6$ tons, where 1 ton is equivalent to 1,000 kg.

Each second, 2×10^{38} electron neutrinos are released from the Sun, moving out in all directions in space at the velocity of light. The Earth will intercept a small fraction of these, but it is still a large number, about $4 \times 10^{29} = 2 \times 10^{38} \, (R_E/AU)^2$, where the Earth's radius is $R_E = 6.371 \times 10^6$ m and its mean distance from the Sun is $AU = 1.496 \times 10^{11}$ m. This means that each second the number of solar electron neutrinos that pass through every square meter of the Earth is $4 \times 10^{29}/(\pi R_E{}^2) = 3$ million billion (3×10^{15}) electron neutrinos per square meter.

The Sun is consuming itself at a prodigious rate. Its central nuclear reactions turn about 700×10^6 tons of hydrogen into helium every second. In doing so, 5×10^6 tons (0.7%) of this matter disappears as pure radiation energy, and every second the Sun becomes that much less massive (also see Focus 3.2). Yet, the loss of material is insignificant compared to the Sun's total mass; it has lost only 1% of its original mass since the Sun began to shine about 4.6 billion years ago.

The center of the Sun is so densely packed that a proton can only move about 1 cm before encountering another proton. Moreover, at a central temperature of 15.6×10^6 K, the protons are darting about so fast that each one of them collides with other protons about 100 million times every second. Yet, the protons nearly always bounce off each other without triggering a nuclear reaction during a collision.

Since protons have the same electrical charge, they repel each other, and this repulsion must be overcome for protons to fuse together. Only a tiny fraction of the protons are moving fast enough to break through this barrier. Even then, they have to be helped along by a quantum mechanical process that operates in the realm of the very small.

George Gamow introduced the basic idea in 1928, when explaining why a radioactive element occasionally hurls an energetic particle out of its tightly bound nucleus. According to quantum theory, such a particle acts like a spread-out thing with a set of probabilities for being in a range of places. As a result of this location uncertainty, a sub-atomic particle's sphere of influence is larger than was previously thought. It might be anywhere, although with decreasing probability at regions far from the most likely location. This explains the escape of fast-moving, energetic particles from the nuclei of radioactive atoms like uranium; these particles usually lack the energy to overcome the nuclear barrier, but some of them have a small probability of escaping to the outside world.

A similar tunneling process, or barrier penetration, occurs the other way around in the Sun. It means that a proton has a very small but finite chance of occasionally moving close enough to another proton to overcome the barrier of repulsion and tunnel through it. Protons therefore sometimes fuse together, even though their average energy is well below that required to overcome their electrical repulsion.

Even with tunneling at work, it is still not hot enough at the center of the Sun for the vast majority of protons to fuse together. Most of them do not move fast enough, electrical repulsion between even the fastest protons keeps them from joining together, and tunneling does not happen very often. As a result, nuclear reactions proceed at a slow, stately pace inside the Sun. And that is a very good thing! Frequent fusion could make the Sun blow up like a giant hydrogen bomb.

All the Sun's nuclear energy is released in its dense, high-temperature core. Outside of the core, where the overlying weight and compression are less, the gas is cooler and thinner, and nuclear fusion cannot occur. It requires a high-speed collision that can only be obtained with both a very high temperature in excess of 5×10^6 K and the help of quantum mechanical tunneling. The energy-generating core therefore only extends to about one quarter the distance from the center of the Sun to the visible photosphere or about 175,000 km from the center, accounting for only 1.6% of the Sun's volume but about half its mass.

Astronomers have built mathematical models to deduce the Sun's internal structure. These models use the laws of physics to describe a self-gravitating sphere of protons, helium nuclei, electrons, and small quantities of heavier elements. At every point inside the Sun, the force of gravity must be balanced precisely by the gas pressure, which itself increases with the temperature. The energy released by nuclear fusion in the Sun's core heats the gas, keeps it ionized, and creates the pressure. This energy must also make its way out to the visible solar disk to keep it shining at the presently observed rate.

Detailed models of the interiors of the Sun and other stars were developed in the 1950s and 1960s, and then continued to be refined for decades using extensive computer programs. Such models must include and explain the evolution of the Sun, from the time it began to shine by nuclear reactions about 4.6 billion years ago. The reason for tracking the evolving Sun is that its structure changes as more and more hydrogen is converted into helium within the core, which slowly and inexorably heats up to make the Sun shine a bit brighter as time goes on. The evolutionary calculations are reworked until the simulated Sun ends up with its known mass, radius, energy output, and photosphere temperature. According to the models, the Sun has now used up about half the available hydrogen in its central parts since nuclear reactions began, and should continue to shine by hydrogen fusion for another 7 billion years.

3.1.3 Solving the Solar Neutrino Problem

Neutrinos, or little neutral ones, have no electric charge and almost no mass. They are so insubstantial, and interact so weakly with matter, that they move nearly unnoticed through any amount of material, including the Sun and Earth.

Every time a helium nucleus is synthesized within the Sun's energy-generating core, two electron neutrinos are created. And the helium is being produced at such a rapid rate that many trillions of electron neutrinos are produced faster than you can blink your eye. Each second, about 3 million billion (3×10^{15}) of them are passing through every square meter of the side of the Earth facing the Sun, and out the opposite side unimpeded. And a very small fraction of these electron neutrinos have been captured using massive detectors that are buried deep underground so that only neutrinos can reach them.

The first solar neutrino detection experiment, constructed by Raymond Davis Jr. in 1967, is a huge tank containing 378,000 liters of a cleaning fluid (C_2Cl_4 or perchloroethylene), which is located 1.5 km underground in the Homestake Gold Mine near Lead, South Dakota. Almost all of the solar electron neutrinos pass right through the tank, but when one of them very occasionally strikes a chlorine atom in the cleaning fluid head on, it produces a radioactive argon atom. Every 2 months, on average, Davis and his colleagues extracted just 130 argon atoms from their tank. Over a period of 30 years they succeeded in capturing a total of 2,000 argon atoms, thereby detecting 2,000 electron neutrinos and proving that nuclear fusion provides the Sun's energy – a feat that resulted in Davis sharing the 2002 Nobel Prize in Physics.

But there was a small, unexpected problem with these results. The Homestake instrument always detected about one-third the number of electron neutrinos expected from theoretical calculations – mainly by John N. Bahcall and co-workers. In 1998, for example, Bruce T. Cleveland and co-workers reported that the neutrino flux detected during three decades of Homestake observations was 2.56 ± 0.25 solar neutrino units, where the $\pm$ value denotes an uncertainty of one standard deviation. In contrast, the most recent theoretical result predicted that they should have observed a flux 8.5 ± 1.8 solar neutrino units. This discrepancy between the observed and the predicted number of electron neutrinos is known as the solar neutrino problem.

Another giant, underground neutrino detector, called Kamiokande, had begun to monitor solar neutrinos in 1987. This second experiment, located a kilometer underground in a zinc mine near Kamioka, Japan, initially consisted of a 4.5×10^6-liter tank of pure water. Nearly a thousand light detectors were placed in the tank's walls to measure signals emitted by electrons knocked free from water molecules by rare head-on collisions with neutrinos. After 1,000 days of observation, Yoji Totsuka, speaking for the Kamiokande collaboration led by Masatoshi Koshiba, reported in 1991 that the directional information obtained from the observed light signals indicated that the neutrinos are indeed coming from the Sun, and the instrument also confirmed the neutrino deficit observed by Davis. Moreover, the Japanese experiment had already detected, on 23 February 1987, a brief burst of just 11 neutrinos from a distant supernova explosion known as SN 1987A. Koshiba shared the 2002 Nobel Prize in Physics, for confirming Davis's results and detecting neutrinos from a supernova.

The solar neutrino problem was additionally substantiated in the 1990s by the Soviet-American Gallium Experiment, abbreviated SAGE, located some 2 km below the summit of Mount Andyrchi in the northern Caucasus, and a second multinational experiment, dubbed GALLEX for gallium experiment, placed about 1.4 km below a peak in the Appenine Mountains of Italy.

There are two possible explanations to the solar neutrino problem. Either we do not really know exactly how the Sun creates its energy, or we have an incomplete knowledge of neutrinos. But astronomers became convinced that they knew precisely how the Sun is generating energy, with the predicted quantities of solar electron neutrinos, and they therefore suspected that there is something wrong with our understanding of the neutrino. Already in 1990, John N. Bahcall and Hans A. Bethe, for example, announced that a more comprehensive knowledge of neutrinos was required if the Homestake and Kamiokande experiments were both correct.

And the astronomers became even more convinced of the veracity of their solar models when sound waves were used as a thermometer to measure the temperatures inside the Sun. If the Sun's central temperature is lower than their models assumed, then the number of electron neutrinos produced by nuclear reactions would be less. In 1996–1997, several investigators used observations with the Michelson Doppler Imager instrument aboard the *SOlar and Heliospheric Observatory* with the techniques of helioseismology, discussed a bit later in this chapter, to determine the internal temperature of the Sun. That is, they measured the sound wave velocities, showing that the sound speeds and temperatures are in agreement with predictions throughout almost the entire Sun, confirming the predicted central temperature of 15.6×10^6 K with a high degree of accuracy and strongly disfavoring any astrophysical explanation of the solar neutrino problem. By 2001, Sylvaine Turck-Chièze and her colleagues at Saclay, France, had refined these helioseismology temperature measurements to show that the neutrino deficit measured on Earth cannot be explained by adjustments to the solar model calculations.

The Japanese experiment had in the meantime been upgraded to Super-Kamiokande status, with a tank containing 50×10^6 liters of highly purified water surrounded by 13,000 light sensors, so sensitive that they could each detect a single photon of light generated by a neutrino interacting with the water. Super-Kamiokande can observe both solar electron neutrinos and atmospheric muon neutrinos, distinguishing between them by the tightness of the cone of light that is produced. (There are three types, or flavors, of neutrinos, each named after the fundamental, sub-atomic particle with which it is most likely to interact. All of the neutrinos generated inside the Sun are electron neutrinos, the kind that interacts with electrons. The other two flavors, the muon neutrino and the tau neutrino, interact with muons and tau particles, respectively.)

Super-Kamiokande continued to confirm the solar neutrino deficit, at about half the predicted amount using 3 years of data. But an even more startling result had already been obtained after monitoring the neutrino light patterns for about 500 days. As reported by Y. Fukuda and co-workers in 1998, the muon neutrinos produced by cosmic rays interacting with the Earth's atmosphere change type in mid-flight. Since roughly twice as many muon neutrinos were coming from the atmosphere directly over the detector than those coming up from the other side of the Earth, some of the ghost-like muon neutrinos were disappearing, transforming themselves into an invisible form during their journey through the Earth to arrive at the detector from below.

This suggested that some of the electron neutrinos created by nuclear reactions inside the Sun transform themselves during their journey to the Earth, escaping detection by changing character. In fact, Vladimir Gribov and Bruno Pontecorvo proposed such an idea in 1969, soon after Davis began his experiment. They reasoned that the solar electron neutrinos could switch from electron neutrinos to another type as they travel in the near vacuum of space between the Sun and the Earth. Almost a decade later, Lincoln Wolfenstein showed that the neutrinos could oscillate, or change type, more vigorously by interacting with matter, rather than in a vacuum, and in 1985 Stanislav P. Mikheyev and Alexei Y. Smirnov explained how the matter oscillations might explain the solar neutrino problem.

The chameleon-like change in identity is not one way, for a neutrino of one type can change into another kind of neutrino and back again as it moves along. The effect is called neutrino oscillation since the probability of metamorphosis between neutrino types has a sinusoidal, in and out, oscillating dependence on path length.

In order to change from one form to another, neutrinos must have some substance in the first place. They can only pull off their vanishing act if the neutrino, long thought to have no mass, possesses a very small one. And the Super-Kamiokande results proved that the atmospheric muon neutrinos had to have some mass. Nevertheless, the terrestrial muon neutrinos did not come from the Sun, and they are not directly related to nuclear fusion reactions in the solar core, which produce a different kind of neutrinos, the electron type. So the solution to the solar neutrino problem was not definitely known until 2001, when the Sudbury Neutrino Observatory demonstrated that solar electron neutrinos are changing type when traveling to the Earth.

The Sudbury Neutrino Observatory, abbreviated SNO and pronounced "snow", is a collaboration of Canadian, American, and British scientists. The detector is located 2 km underground in the Creighton nickel mine near Sudbury, Ontario. Unlike Kamiokande and Super-Kamiokande, which contain a purified form of the normal water you drink, the SNO detector contains heavy water, which does not appear any different but is toxic to humans in large quantities.

But the hydrogen in heavy water has a nucleus, called a deuteron, which consists of a proton and a neutron, rather than just a proton as in the nucleus of the hydrogen in ordinary water. And it is the deuteron that makes the Sudbury Neutrino Observatory sensitive to not just one type of neutrino but to all three known varieties of neutrinos. One million liters of heavy water has been placed in a central spherical cistern with transparent acrylic walls (Fig. 3.1), and a geodesic array of about 10,000 photo-multiplier tubes surrounds the vessel to detect the flash of light given off by heavy water when it is hit by a neutrino. Both the light sensors and the central tank are enveloped by a 7.8×10^6-liter jacket of ordinary water (Fig. 3.2), to shield the heavy water from weak natural radiation, gamma rays, and neutrons coming from the underground rocks. As with the other neutrino detectors, the overlying rock blocks energetic particles generated by cosmic rays interacting with the Earth's atmosphere.

The Sudbury Neutrino Observatory can be operated in two modes, one sensitive only to electron neutrinos and the other equally sensitive to all three types of neutrinos. When the experiment was operated in the first mode, measuring electron neutrinos from the Sun, it made history. In June 2001, the SNO Project Director, Arthur B. McDonald announced that:

We now have high confidence that the [solar neutrino problem] discrepancy is not caused by models of the Sun but by changes in the neutrinos themselves as they travel from the core of the Sun to the Earth. Earlier measurements had been unable to provide definitive results showing that this transformation from solar electron neutrinos to other types occurs. The new results from SNO, combined with previous work, now reveal this transformation clearly, and show that the total number of electron neutrinos produced in the Sun is just as predicted by detailed solar models.

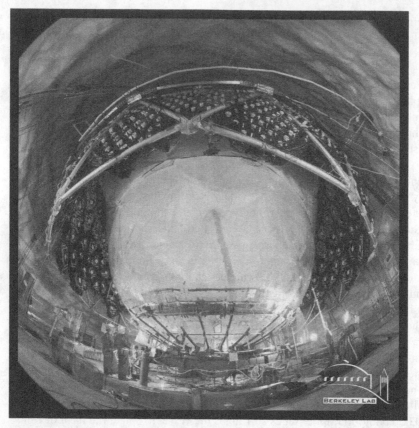

Fig. 3.1 Sudbury neutrino observatory. The central spherical flask of this neutrino observatory is 12 m in diameter, and is surrounded by a geodesic array of thousands of light sensors to detect the flash of light from the interaction of a neutrino with the heavy water. (Courtesy of Kevin Lesko, Lawrence Berkeley National Laboratory.)

By catching about 10 neutrinos a day, the SNO scientists detected about one-third of the solar electron neutrinos predicted by the standard solar model. The Super-Kamiokande detected about half the predicted number, but it had a small sensitivity to other neutrino types. The difference between the two measurements provided evidence that solar neutrinos oscillate, or change type, when traveling from the Sun, and showed that the total number of neutrinos emitted by the Sun agrees with predictions. After 33 years, the solar neutrino problem had been solved.

The epochal SNO results continued in succeeding years, when 2 tons of table salt was added to the 1,000 tons of heavy water. In 2003, the group announced that the salt experiment indicates that the total number of neutrinos of all types reaching the Earth from the Sun is precisely equal to the number of electron neutrinos produced by nuclear reactions in the core of the Sun, but that two-thirds of the electron-type neutrinos were observed to change to muon- or tau-type neutrinos before reaching the Earth.

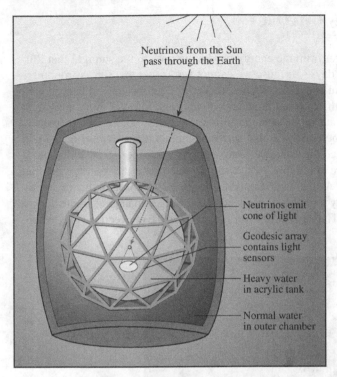

Fig. 3.2 How Sudbury works. Neutrinos from the Sun travel through more than 2 km of rock, entering an acrylic tank containing 1,000 tons (1×10^6 liters) of heavy water. When one of these neutrinos interacts with a water molecule, it produces a flash of light that is detected by a geodesic array of photo-multiplier tubes. Some 7,800 tons (7.8×10^6 liters) of ordinary water surrounding the acrylic tank absorbs radiation from the rock, while the overlying rock blocks energetic particles generated by cosmic rays in our atmosphere. The heavy water is sensitive to all three types of neutrinos

American and Japanese scientists next teamed up to construct the Kamioka Liquid scintillator Anti-Neutrino Detector, abbreviated KamLAND, at the site of the older Kamiokande solar neutrino detector. KamLAND detects electron anti-neutrinos that have traveled through the Earth from 51 nuclear reactors in Japan plus 18 reactors in South Korea. The measurements, reported in 2003–2005, show that some of the electron anti-neutrinos are disappearing when traveling to the detector. By 2005, the SNO salt-phase results for solar electron neutrinos had been combined with the KamLAND experiment to provide estimates for the tiny neutrino mass and for other properties of neutrino oscillation, while also indicating that most of the solar neutrino metamorphosis occurs inside the Sun.

So, the solar neutrino problem has been resolved without affecting our models of the solar interior. They indicate that nuclear reactions consume hydrogen to form helium in the Sun's central core, and that these nuclear reactions cannot continue beyond the outer edge of the core, located at about a quarter of the distance from the center of the Sun to the visible disk.

3.2 How the Energy Gets Out

So how does radiation get from the energy-producing core to the sunlight that illu-
minates and warms our days? Because we cannot see inside the Sun, astronomers
combine basic theoretical equations, such as those for equilibrium and energy generation or transport, with observed boundary conditions, like the Sun's mass and
luminous output, to create models of the Sun's internal structure. These models
consist of two nested spherical shells that surround the hot, dense core (Fig. 3.3). In
the innermost shell, called the radiative zone, energy is transported by radiation; it
reaches out from the core to roughly 70% of the distance from the center of the Sun
to the visible solar disk. As the name implies, energy moves through this region by
radiation. It is a relatively tranquil, serene, and placid place, analogous to the deeper
parts of Earth's oceans. The outermost layer is known as the convective zone, where

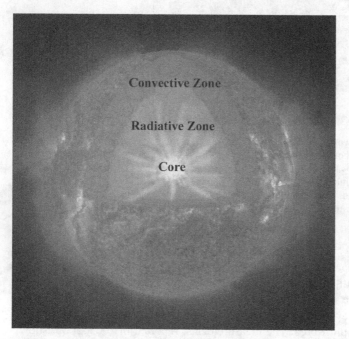

Fig. 3.3 Energy works its way out. The Sun is powered by the nuclear fusion of hydrogen in its
core at a central temperature of 15.6×10^6 K. Energy produced by these fusion reactions is transported outward, first by countless absorptions and emissions within the radiative zone, and then
by turbulent motion in the outer convective zone. The visible disk of the Sun, called the photosphere, has a temperature of 5,780 K. Just above the photosphere are the thin chromosphere and a
transition region to the rarefied, million-degree outer atmosphere of the Sun. They are represented
by an image at the extreme ultraviolet wavelength of 30.4 nm wavelength, emitted by singly ionized helium, denoted He II, at a temperature of about 60,000 K, taken from the Extreme-ultraviolet
Imaging Telescope, abbreviated EIT, aboard the *SOlar and Heliospheric Observatory,* or *SOHO*
for short. (Courtesy of the *SOHO* EIT consortium and NASA. *SOHO* is a project of international
cooperation between ESA and NASA.)

energy is transported in a churning, wheeling motion called convection; it is a lively, turbulent, and vigorous place, encompassing the radiative zone.

The Sun's interior cools with increasing distance from the core, as heat and radiation spread out into an ever-larger volume. The temperature always decreases outward, so the Sun is hottest at the core and coolest at the visible disk, the photosphere. Because the flux of radiation depends on temperature, there is slightly more outbound radiation than there is inbound. The energy is therefore always flowing outwards, at a rate controlled by the gas' opacity, its ability to absorb the radiation.

Radiation does not move quickly through the radiative zone, but instead diffuses slowly outward in a haphazard zigzag pattern. An energetic gamma ray produced by nuclear fusion in the solar core almost instantly collides with a particle in the radiative zone, and is re-radiated at a lower energy and longer wavelength. As the radiation travels outward, the process continues over and over again countless times, with continued ricocheting, innumerable collisions, and the steady loss of radiation photon energy.

Computations performed by R. Mitalas and K. R. Sills in 1992 indicate that it takes about 170,000 years, on the average, for radiation to work its way out from the Sun's core to the bottom of the convective zone. By this time, the radiation has shed so much energy that it emerges on the other side of the convective zone as visible light.

At the bottom of the convective zone, the temperature has become cool enough, at about 2×10^6 K, to allow some heavy nuclei to capture electrons; because of their lighter weight and greater speed, the abundant hydrogen and helium remain fully ionized. The less abundant heavy particles absorb light and block the flow of heat, obstructing the outward flow of radiation like dirt on a window. This causes the gas to become hotter than it would otherwise be, and the pent-up-energy is transported by convection, which occurs whenever a gas or fluid is sufficiently heated from below. The convective motions carry heat through the outer part of the Sun, from a depth of about 200,000 km up to the photosphere in about 10 days.

In response to heating, gases in the bottom layer of the convective zone expand and thereby become less dense than the gas in the overlying layers. The heated material, due to its low density, rises to the visible solar disk and cools by radiation. The cooled gas then sinks, because it is denser than the hotter one, only to be reheated and rise again. Such wheeling convective motions occur in a kettle of boiling water or a simmering pot of oatmeal, with hot rising bubbles and cooler sinking material.

The convective zone is capped by the photosphere, the place where the gaseous material changes from being completely opaque to radiation to being transparent. It is the visible layer of the Sun, where radiation last interacts with atoms before escaping from the Sun. This sunlight takes just 8.3 min to travel from the photosphere to the Earth.

As first noticed by William Herschel in 1801, the photosphere contains a fine granular pattern. These closely packed granulation cells can now be examined using high-resolution images taken from ground-based telescopes under conditions of

Fig. 3.4 Double, double toil and trouble. When optical, or visible light, telescopes zoom in and take a detailed look at the visible Sun, they resolve a strongly textured granular pattern. Hot granules, each about 1,400 km across, rise at speeds of 500 m s^{-1}, like supersonic bubbles in an immense, boiling cauldron. The rising granules burst apart, liberating their energy, and cool material then sinks downwards along the dark, inter-granular lanes. This photograph was taken at 430.8 nm with a 1-nm interference filter. It has an exceptional angular resolution of 0.2 s of arc, or 150 km at the Sun. (Courtesy of Richard Muller and Thierry Roudier, Observatoire du Pic-du-Midi et de Toulouse.)

excellent seeing (Fig. 3.4), or from spacecraft located outside the Earth's obscuring atmosphere. The images reveal a host of granules with bright centers surrounded by dark lanes, exhibiting a non-stationary, overturning motion caused by the underlying convection.

The bright center of each granule, or convection cell, is the highest point of a rising column of hot gas. The dark edges of each granule are the cooled gas, which sinks because it is denser than the hotter gas. Each individual granule lasts only about 15 min before it is replaced by another one, never reappearing in precisely the same location.

The mean angular distance between the bright centers of adjacent granules is about 2.0 s of arc, corresponding to about 1,500 km at the Sun. That seems very large, but an individual granule is about the smallest thing you can see on the Sun when peering through our turbulent atmosphere.

There are at least a million granules on the visible solar disk at any moment. They are constantly evolving and changing, producing a honeycomb pattern of rising and falling gas that is in constant turmoil, completely changing on time scales of minutes and never exactly repeating themselves.

The granules are superimposed on a larger cellular pattern, called the supergranulation, studied by Robert B. Leighton and his collaborators in the early 1960s. They modified a spectroheliograph, invented 65 years earlier by George Ellery Hale for photographing the Sun in a single wavelength, to image the Sun using the Doppler effect and to thereby study motions in the photosphere. The Doppler effect occurs whenever a source of light or sound moves with respect to the observer, and results in a change in the length of the waves (Fig. 3.5). When part of the Sun approaches us, the wavelength of light emitted from that region becomes shorter, the wave fronts or crests appear closer together, and the light therefore becomes bluer; when the light source moves away, the wavelength becomes longer and the light redder. The magnitude of the wavelength change, in either direction, establishes the velocity of motion along the line of sight, either approaching or receding from the observer (Focus 3.3).

Focus 3.3

The Doppler Effect

Just as a source of sound can vary in pitch or wavelength, depending on its motion, the wavelength of electromagnetic radiation changes when the emitting source moves with respect to the observer. This Doppler shift is named after Christian Doppler who discovered it in 1842. If the motion is toward the observer, the shift is to shorter wavelengths, and when the motion is away the wavelength becomes longer. You notice the sound-wave effect when listening to the changing pitch of a passing ambulance siren. The tone of the siren is higher while the ambulance approaches you and lower when it moves away.

If radiation is emitted at a specific wavelength, $\lambda_{emitted}$, by a source at rest, the wavelength, $\lambda_{observed}$, observed from a moving source is given by the relation:

$$z = \frac{\lambda_{observed} - \lambda_{emitted}}{\lambda_{emittted}} = \frac{V_r}{c},$$

where V_r is the radial velocity of the source along the line of sight away from the observer. The parameter z is called the redshift since the Doppler shift is toward the longer, redder wavelengths in the visible part of the electromagnetic spectrum. When the motion is toward the observer, V_r is negative and there is a blueshift to shorter, bluer wavelengths.

The Doppler effect applies to all kinds of electromagnetic radiation, including X-rays, visible light, and radio waves. Because everything in the Universe moves, the Doppler effect is a very important tool for astronomers, determining the radial velocity of all kinds of cosmic objects.

Our equation is strictly valid for radial velocities that are smaller than the velocity of light. For objects that move at speeds comparable to that of light, a somewhat more complicated expression describes the Doppler effect.

Leighton and his colleagues subtracted a long-wavelength spectroheliograph image from a short-wavelength one, creating Dopplergrams of motions in the photosphere. The Dopplergrams revealed a supergranulation pattern of horizontal flow, each an estimated 16,000 km across, or a bit larger than the Earth, with lifetimes of roughly 24 h. Because the motion is predominantly horizontal, the supergranules were not detected when looking directly at the center of the solar disk, but further

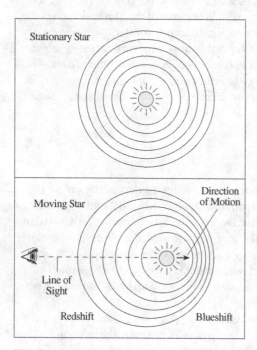

Fig. 3.5 Doppler effect. A stationary star (*top*) emits regularly spaced light waves that get stretched out or scrunched up if the star moves (*bottom*). Here we show a star moving away (*bottom right*) from the observer (*bottom left*). The stretching of light waves that occurs when the star moves away from an observer along the line of sight is called a redshift, because red light waves are relatively long visible light waves; the compression of light waves that occurs when the star moves along the line of sight toward an observer is called a blueshift, because blue light waves are relatively short. The wavelength change, from the stationary to moving condition, is called the Doppler shift, and its size provides a measurement of radial velocity, or the velocity of the component of motion along the line of sight. The Doppler effect is named after the Austrian physicist Christian Doppler, who first considered it in 1842

out toward the sides of the round solar disk, where the horizontal motion is partially directed along the line of sight.

Roughly 3,000 supergranules are seen on the visible solar disk at any moment. And like the ordinary granulation, the changing pattern of supergranulation is caused by convection. But unlike the granules, whose gases move up and down, the material in each supergranule cell rises in the center, and exhibits a sideways motion as it moves away from the center with a typical velocity of about $0.4\,\mathrm{km\,s^{-1}}$. Only after this prolonged horizontal motion does the material eventually sinks down again at the cell boundary. The supergranular flow carries the magnetic field across the photosphere, sweeping the magnetism to the edges of the supergranulation cells where it collects, strengthens, and forms a network of concentrated magnetic field.

Modern instruments, such as the Michelson Doppler Imager, or MDI for short, aboard the *SOlar and Heliospheric Observatory*, abbreviated *SOHO*, use a very

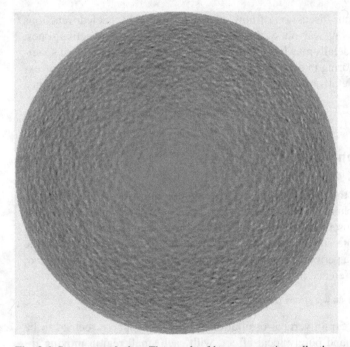

Fig. 3.6 Supergranulation. Thousands of large convection cells, the supergranules, are detected in this Dopplergram, which shows motion toward and away from the observer, along the line of sight, as light and dark patches, respectively. It was obtained using the Doppler effect of a single spectral line with the Michelson Doppler Imager, abbreviated MDI, instrument on the *SOlar and Heliospheric Observatory*, or *SOHO* for short. The image contains about 3,000 supergranules, each usually between 12,000 and 20,000 km across and typically lasting between 16 and 23 h. Near the disk center the supergranules do not contribute to the signal; their motion is predominantly horizontal and MDI measures only the component of motion directed towards or away from *SOHO*. Material in supergranules flows out and sideways from their centers, with a typical velocity of $400\,\mathrm{m\,s^{-1}}$. (Courtesy of the *SOHO* MDI consortium. *SOHO* is a project of international cooperation between ESA and NASA.)

similar Doppler cancellation technique, with full-disk, line-of-sight Doppler velocity measurements, to investigate the supergranules (Fig. 3.6). As reported by Marc
L. De Rosa and Juri Toomre in 2004 using MDI data, the supergranules have an
average diameter that lies between 12,000 and 20,000 km and average lifetimes in
the 16–23 h range.

In 2003, Laurent Gizon, Thomas L Duvall Jr. and Jesper Schou used *SOHO's*
MDI to discover possible wavelike properties of the supergranulation with periods
from 6 to 9 days. Since the waves move in the same direction as the Sun rotates,
the supergranulation appears to rotate faster than the other visible solar gas and
magnetic features like sunspots. This apparent motion could resemble the "wave"
of spectators at a baseball, football, or soccer game. When people in a stadium do
the wave, nobody actually moves in the direction of the wave – they just jump up
and sit down. As suggested by David H. Hathaway, Peter E. Williams, and Manfred
Cuntz in 2006, the supergranular superrotation may alternatively be just a projection
effect.

So that concludes our discussion of how energy generated by nuclear reactions
in the Sun's core works its way out of the overlying radiative and convective zones.
However, some of the details may have been overlooked by the assumption of spherical symmetry and ignoring internal movements or magnetic fields. We can now examine some of these details by using sound waves to open a window into the Sun's
interior.

3.3 Taking the Sun's Pulse

The unseen depths of the Sun are explored by watching the widespread throbbing
motions of the photosphere. Robert B. Leighton and his co-workers discovered these
vertical up and down oscillations in the early 1960s with the same instrument that
they used to investigate supergranules, finding that the subtracted images in the
Dopplergrams exhibit a periodic oscillation of about 5 min. As Leighton announced,
at an international conference in 1960:

> These vertical motions show a strong oscillatory character, with a period of 296 ± 3 s.

When discovered, the 5-min vertical oscillations of the photosphere seemed to be
a chaotic, short-lived, and purely local effect, with each small region moving independent of nearby ones. However, subsequent research demonstrated that all the
local motions are driven by sound waves that echo and resonate through the solar
interior, and that the entire Sun is vibrating with ponderous rhythms that extend to
its very core. The resulting global oscillations of the Sun have been identified as
steady, pressure-driven, or p-mode, sound waves generated in the convective zone.

The 5-min oscillations are not directly related to individual granules moving up
and down, but they are associated with turbulent convection. In 1952–1954, Michael
James Lighthill demonstrated how sound can be generated by turbulence, and in 1977
Peter Goldreich and Douglas A. Kelley showed how vigorous turbulent motion in the

convective zone produces intense, random noise, somewhat like the deafening roar of a jet aircraft or the hissing noises made by a pot of boiling water. When these sound waves strike the photosphere and rebound back down, they disturb the gases there, causing them to rise and fall, slowly and rhythmically at 5-min intervals, like the tides in a bay or a beating heart (Figs. 3.7 and 3.8). Each 5-min period is the time it takes for the localized motion to change from moving outward to moving inward and back outward again. Such 5-min oscillations are imperceptible to the unaided eye, for the photosphere moves a mere hundred-thousandth (0.00001) of the solar radius.

These sounds are trapped inside the Sun, for they cannot propagate through the near vacuum of interplanetary space. As suggested in 1970 by Roger Ulrich, and independently in 1971 by John Leibacher and Robert F. Stein, the sounds move within resonant cavities, or spherical shells, inside the Sun, like the shouts of a child traveling far across a cloud-covered lake on a summer night. The Sun's sound waves move around and around inside this circular waveguide, bouncing repeatedly against the photosphere, reverberating between the cavity boundaries and driving oscillations in the overlying material.

A real breakthrough came in 1975, when Franz-Ludwig Deubner showed that the Sun's oscillating power is concentrated into narrow ridges in a spatial-temporal display, which meant that they are due to the standing acoustic waves moving in the predicted cavity. Instead of meaningless, random fluctuations, orderly resonant motions produce enhanced oscillation power with specific combinations of size and frequency (Fig. 3.9). The frequencies of these resonant, pressure-driven p-modes correspond to periods of about 5 min.

Starting near the photosphere, a sound wave moves into the Sun toward its center. Since the speed of sound increases with temperature, which in the Sun increases with depth, the wavefront's deeper, inner edge travels faster than its shallower outer edge, so the inner edge pulls ahead. Gradually, an advancing wavefront is refracted, or bent, until the wave is once again headed toward the Sun's photosphere.

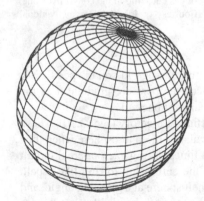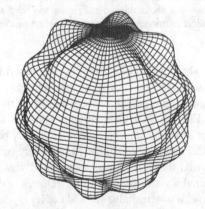

Fig. 3.7 Internal contortions. The Sun exhibits over a million shapes produced by its oscillations. Two of these shapes are illustrated here with exaggerated amplitude. (Courtesy of Arthur N. Cox and Randall J. Bos, Los Alamos National Laboratory.)

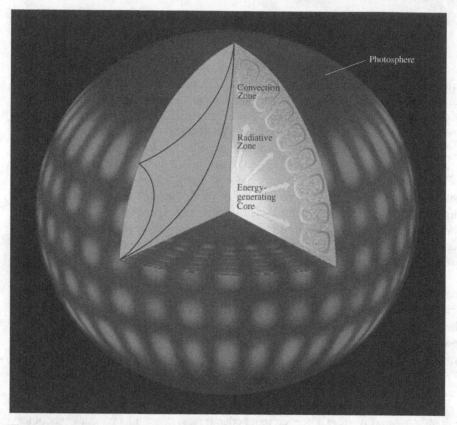

Fig. 3.8 The pulsating Sun. Sound waves inside the Sun cause the visible solar disk to move in and out. This heaving motion can be described as the superposition of literally millions of oscillations, including the one shown here for regions pulsing inward (*red regions*) and outward (*blue regions*). The sound waves, whose paths are represented here by black lines inside the cutaway section, resonate through the Sun. They are produced by hot gas churning in the convective zone, which lies above the radiative zone and the Sun's core. (Courtesy of John W. Harvey, National Optical Astronomy Observatories, except cutaway.)

The outer cavity boundary is caused by an abrupt change in sound speed associated with the enormous drop in density and steep temperature gradient near the photosphere, which reflects the waves traveling outward back in.

Each of these resonating sounds has a well-defined trajectory (Fig. 3.10). Those that travel far into the Sun move nearly perpendicular to the photosphere when they hit it, and they touch the photosphere only a few times during each internal circuit around the Sun. These deep sound waves cause the entire Sun to ring like a bell. Sound waves with shorter trajectories strike the photosphere at a glancing angle, and travel through shallower and cooler layers. They bounce off the visible photosphere more frequently, and describe motions and structure in the outer shell of the Sun.

Astronomers look inside the Sun by recording the 5-min oscillations in the photosphere and using them to examine sound waves that travel within the Sun on differing paths. They use the term helioseismology to describe such investigations of

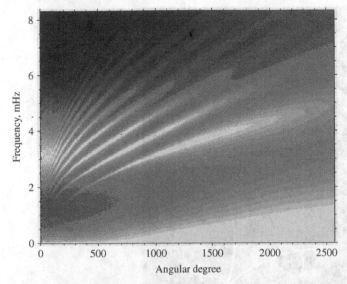

Fig. 3.9 Oscillation frequency and angular degree. Sound waves resonate deep within the Sun, producing photosphere oscillations with a frequency near 3 mHz, or 0.003 cycles per second, which corresponds to a wave period of 5 min. Here the frequency is plotted as a function of the spherical harmonic degree, l, for just 8 h of high-resolution data taken with the *SOHO* MDI instrument. The degree, l, is the inverse of the spatial wavelength, or surface size; an l of 400 corresponds to waves on the order of 10,000 km in size. The oscillation power is contained within specific combinations of frequency and degree, demonstrating that the photosphere oscillations are due to the standing waves confined within resonant cavities. The depth of the resonating cavity depends on the oscillation frequency and the degree, l. What is happening near the core of the Sun is described by oscillations in the lower left corner of the diagram. Moving up in frequency or degree tells more about what is happening near the photosphere. (Courtesy of the *SOHO* MDI consortium. *SOHO* is a project of international cooperation between ESA and NASA.)

the solar interior. It is a hybrid name combining the word *Helios* for the Greek god of the Sun, the Greek word *seismos* meaning quake or tremor, and *logos* for reasoning or discourse. So literally translated helioseismology is the logical study of solar tremors. Geophysicists similarly unravel the internal structure of the Earth by recording earthquakes, or seismic waves, that travel to different depths; this type of investigation is called seismology.

The helioseismology observations have been used to refine computer models that show, among other things, how the pressure, temperature, mass density, sound speed, and material content vary as a function of radius inside the Sun. In addition, the helioseismologists have demonstrated how gases rotate and flow inside the Sun, and pinpointed where the Sun's magnetism is probably generated. They have also discovered internal weather patterns that swirl around active regions, converging flows beneath sunspots that help hold them together, and hints of gravity waves within the Sun's core.

Many of these recent advances have been made from the *SOlar and Heliospheric Observatory*, abbreviated *SOHO*. Out in quiet, peaceful, and tranquil space,

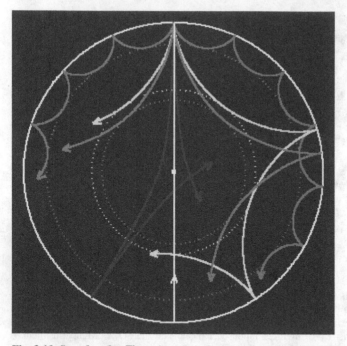

Fig. 3.10 Sound paths. The trajectories of sound waves are shown in a cross section of the solar interior. The rays are bent inside the Sun, like light within the lens of an eye, and circle the solar interior in spherical shells or resonant cavities. Each shell is bounded at the top by a large, rapid density drop near the photosphere and bounded at the bottom at the inner turning point (*dotted circles*), where the bending rays undergo total internal refraction, owing to the increase in sound speed with depth inside the Sun. How deep a wave penetrates and how far around the Sun it goes before it hits the photosphere depends on the harmonic degree, l. The white curve is for $l = 0$, the blue one for $l = 2$, green for $l = 20$, yellow for $l = 25$, and red for $l = 75$. (Courtesy of Jørgen Christensen-Dalsgaard and Philip H. Scherrer.)

unperturbed by terrestrial interference, it has a long, clear undistorted view of the Sun's pulsations, obtaining recordings of both large-scale global oscillations and small-scale ones with unprecedented quality. Two of its instruments, the Michelson Doppler Imager, or MDI for short, and the Global Oscillations at Low Frequency, abbreviated GOLF, record small periodic changes in the wavelength of light, measuring the oscillation speeds using the Doppler effect. A single, very narrow spectral feature is observed, and photosphere motions are inferred by taking solar images at different wavelengths on both the long-wavelength and the short-wavelength sides of it, determining the Doppler shift. Sequences of such measurements, taken at regular, successive intervals of time, are used to determine the periodicity of the motions.

When a lot of MDI's measurements are averaged over adjacent places or long times, the velocity of the photosphere oscillations can be determined with a remarkable precision of better than $1 \, \text{mm s}^{-1}$ – when the Sun's surface is moving in and out about a million times faster. Moreover, the MDI telescope detects these motions at a million points across the Sun every minute. Because the instrument is positioned well above Earth's obscuring atmosphere, MDI continuously resolves fine detail on small oscillation scales that cannot always be seen from the ground due to the blurring effects of our atmosphere, cloudy weather and the day–night cycle. The fantastic results and images from MDI can be found at http://soi.stanford.edu, as well as the *SOHO* home page at http://sohowww.nascom.nasa.gov/.

SOHO's GOLF instrument observes the entire disk of the Sun, without resolving tiny surface details, and thereby garners long, undisturbed views of global, whole-Sun oscillations that probe the deep solar interior. It gathers and integrates light over the visible solar disk and tends to average out the peaks and troughs of the smaller localized undulations that occur randomly at many different places and times within the field of view. But GOLF has the clearest view of either sound or gravity waves that penetrate deep within the Sun, even to its core. This instrument and its results are described at the GOLF site http://golfwww.medoc-ias.u-psud.fr and the *SOHO* home page.

Spacecraft have by no means rendered ground-based helioseismology obsolete. A worldwide, six-station network of observatories, known by the acronym GONG for the Global Oscillations Network Group, also observes the Sun around the clock. They form an unbroken chain that follows the Sun as the Earth rotates, providing additional studies of the internal structure and dynamics of the Sun using helioseismology. Their web site is located at http://gong.nso.edu.

The major accomplishments of *SOHO's* MDI and GOLF and the ground-based GONG are presented in the rest of this chapter. They will provide the foundation for future investigations with these instruments, as well as those from the Helioseismic and Magnetic Imager aboard NASA's *Solar Dynamics Observatory*, currently scheduled for launch in December 2008.

The serious student or curious reader may also want to consult reviews of these subjects prepared by professionals in the field (Focus 3.4).

Focus 3.4

Expert Reviews About Helioseismology

Professional solar astronomers and astrophysicists have reviewed important recent developments in helioseismology, often in technical terms. In alphabetical order, they are: Jørgen Christensen-Dalsgaard's (2002) review of helioseismology, the special issue of *Science* on GONG helioseismology in 1996, beginning with Douglas O. Gough, John W. Leibacher, Philip H. Scherrer, and Juri Toomre's perspectives on helioseismology, and including Christensen-Dalsgaard and co-workers account of the then-current state of solar modeling, Sami K. Solanki's (2006) review of the solar magnetic field, and Michael J. Thompson, Jørgen Christensen-Dalsgaard, Mark S. Miesch, and Juri Toomre's (2003) review of the internal rotation of the Sun.

Very thorough *Living Reviews* are also available on the web at the portal http://solarphysics.livingreviews.org/. In alphabetical order, the ones relevant to helioseismology include Paul Charbonneau's (2005) review of *Dynamo Models of the Solar Cycle*, Laurent Gizon and Aaron C. Birch's (2005) account of *Local Helioseismol-*

ogy, Yuhong Fan's (2004) discussion of *Magnetic Fields in the Solar Convection Zone*, Mark S. Miesch's (2005) review of *Large-Scale Dynamics of the Convection Zone and Tachocline*, and Neil R. Sheeley Jr.'s (2005) discussion of *Surface Evolution of the Sun's Magnetic Field: A Historical Review of the Flux Transport Mechanism*.

The contributions of scientists working with specific spacecraft data are often presented at workshops that are subsequently published in book form. The proceedings of *SOHO* or *SOHO*/GONG workshops that deal with helioseismology include, in chronological order: *SOHO-4: Helioseismology*, edited by J. Todd Hoeksema, Vicente Domingo, Bernhard Fleck, and Bruce Battric (1995); *SOHO-6/GONG 98: Structure and Dynamics of the Interior of the Sun and Sun-like Stars*, edited by Sylvain Korzennik and Andrew Wilson (1998); *SOHO-9: Helioseismic Diagnostics of Solar Convection and Activity*, edited by Thomas L. Duvall Jr., John W. Harvey, Alexander G. Kosovichev, and Zdenek Svetka (2000); *SOHO-10/GONG 2000: Helio- and Asteroseismology at the Dawn of the Millenium*, edited by Andrew Wilson and Pere L. Pallé (2001); *SOHO-12/GONG 2002: Local and Global Helioseismology: The Present and Future*, edited by Huguette Sawaya-Lacoste (2003); *SOHO-14/GONG 2004: Helio- and Asteroseismology: Towards a Golden Future*, edited by Dorothea Danesy (2004); *SOHO-18/GONG 2006: Beyond the Spherical Sun*, edited by Karen Fletcher and Michael Thompson (2006); and *SOHO-19/GONG 2007: Seismology of Magnetic Activity*, which will be published in a double Topical Issue of *Solar Physics* on *Heliosiesmology, Astroseismology and MHD Connections*, edited by John Leibacher and Lidia van Driel-Gesztelyi, joined by guest editors Laurent Gizon and Paul Cally.

3.4 Looking Within the Sun

The combined sound of all the notes reverberating inside the Sun has been compared to a gong in a sandstorm, being repeatedly struck with tiny particles and randomly ringing with an incredible din. The Sun produces order out of this chaos by reinforcing certain notes that resonate within it, like the rhythmic beat of a drum. This resonance effect is somewhat analogous to repeated pushes on a swing. If the pushes occur at the same point in each swing, they can increase the energy of the motion. In the absence of such a resonance, the perturbations would be haphazard and the effect would eventually fade away. When you regularly move water in a bathtub, the waves similarly grow in size, but when you swish it randomly, the water develops a choppy confusion of small waves. To put it another way, destructive interference filters out all but the resonant frequencies that combine and reinforce each other, transforming the random convective noise into a rich spectrum of resonant notes in the 5-min range.

Individual pulsations of the photosphere have velocity amplitudes of no more than $0.1 \, \text{m s}^{-1}$, but when millions of them are superimposed, they produce oscillations that move with thousands of times this speed. The low-amplitude components reinforce each other, producing the strong 5-min oscillations that grow and decay as numerous vibrations go in and out of phase to combine and disperse and then combine again, somewhat like groups of birds, schools of fish, or cars on a highway that gather together, move apart, and congregate again.

The photosphere oscillations are the combined effect of about 10 million separate notes resonating in the Sun – each of which has a unique path of propagation and samples a well-defined section inside the Sun. So, to determine the Sun's internal constitution all the way through – from its churning convective zone into its radiative zone and down to the core – we must determine the exact pitch of all the notes and measure the precise frequency of every one of them. Prolonged observations with high spatial resolution and detailed computer analyses are required to sort them all out.

The main goal of helioseismology is to infer the internal properties of the Sun from the oscillation frequencies of many different sound waves that travel along different paths within the Sun. The precise frequency depends on the propagation speed of the sound wave and the thickness of its resonant cavity. So the oscillation frequencies obtained for many different waves, which penetrate to different depths, can be inverted to determine the radial density and temperature profile inside the Sun, as well as the internal rotation and other motions.

To be precise, the observed frequencies are integral measures of the speed along the path of the sound wave, and helioseismologists have to invert these measured data to get the sound speed, which depends on the temperature and chemical composition of the material it is traveling in (Focus 3.5).

Focus 3.5

The Speed of Sound

Sound waves are produced by perturbations in an otherwise undisturbed gas. They can be described as a propagating change in the gas mass density, ρ, which is itself related to the pressure, P, and temperature, T, by the ideal gas law:

$$P = n\,k\,T = \frac{\rho\,k\,T}{m_u\,\mu},$$

where n is the number of particles per unit volume, Boltzmann's constant $k = 1.38066 \times 10^{-23}\,\mathrm{J\,K^{-1}}$, and the atomic mass unit $m_u = 1.66054 \times 10^{-27}\,\mathrm{kg}$. The mean molecular weight, μ, is given by $\mu = 2/(1 + 3X + 0.5Y)$ where X, Y, and Z represent the concentration by mass of hydrogen, helium, and heavier elements and $X + Y + Z = 1$. For a fully ionized hydrogen gas, $\mu = 1/2$, and for a fully ionized helium gas $\mu = 4/3$.

The wavelength, λ, and frequency, ν, of the sound waves are related by:

$$\lambda\,\nu = c_s,$$

and the velocity of sound, c_s, is given by:

$$c_s = \left(\frac{\partial P}{\partial \rho}\right)^{1/2} = \left(\frac{\gamma\,k\,T}{m_u\,\mu}\right)^{1/2},$$

where the symbol ∂ denotes a differentiation under conditions of constant entropy. For an adiabatic process, the pressure, P, and mass density, ρ, are related by $P = K\rho^{\gamma}$, where K is a constant. For adiabatic perturbations of a monatomic gas the index $\gamma = 5/3$ and for isothermal perturbations $\gamma = 1$. For both adiabatic and isothermal perturbations the sound speed is on the order of the mean thermal speed of the ions of the gas, or numerically:

$$c_s \approx 10(T/10^4\,\mathrm{K})^{1/2}\,\mathrm{km\ s}^{-1}.$$

By considering a sequence of resonant p-modes, which can travel to different depths, it is possible to peel away progressively deeper layers of the Sun and establish the radial profile of the sound speed (Fig. 3.11). When determining this profile in 1991, Jørgen Christensen-Dalsgaard, Douglas O. Gough, and Michael J. Thompson detected a small but definite change in sound speed, which marks the lower boundary of the convective zone, indicating that it has a radius of 0.713 ± 0.003 of the Sun's radius.

Since the speed of sound depends on both the composition and the temperature of the material it is moving in, the precise depth of the convective zone depends on the relative abundances of the elements within the Sun. A group led by Martin Asplund therefore caused quite a stir when they announced in 2004 that the solar photosphere appears to contain about 30% less oxygen than previously believed, and that this suggested lower abundances of other heavy elements such as neon, which does not show up in the photosphere's spectrum. These elements provide an opacity that impedes the outward flow of radiation, and when these opacities are used in detailed models of the Sun, they predict slightly different values for the speed of sound than those previously inferred from helioseismology at the same locations. This suggested a slight adjustment of the depth of the convective zone, from 0.287 to 0.274 solar radii, and perhaps more importantly, implied a disagreement between abundance determinations from spectroscopy and those inferred from solar models and helioseismology observations.

However, in 2007 Enrico Landi, Uri Feldman, and George A. Doschek used the Solar Ultraviolet Measurements of Emitted Radiation, abbreviated SUMER, instrument aboard *SOHO* to determine a substantially higher absolute neon abundance in a solar flare, in agreement with previous estimates, but higher than the one inferred from the low oxygen photosphere abundance and the relative amounts of neon and oxygen. This implied that the low oxygen abundance should be revised back upward by a factor of almost two, bringing it into closer accord with the earlier helioseismology results.

Future studies may resolve the differences and uncertainties in the photosphere abundances, but in the meantime we do not have to worry about the neutrino calculations. Even the worst abundance discrepancies occur at a temperature range that is too cool and distant from the Sun center to significantly affect the helioseismic measurements of the Sun's hotter, deeper temperatures. So the model calculations of the amount of neutrinos emitted by the Sun remain the same.

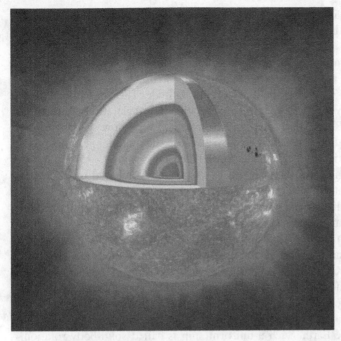

Fig. 3.11 Radial variations of sound speed. Just as scientists can use measurements of seismic waves, produced by earthquakes, to determine conditions under the Earth's surface, measurements of the Sun's oscillations, and the sound waves that produce them, can be used to determine the internal structure of the Sun. This composite image shows the extreme ultraviolet radiation of the solar disk (*orange*) and internal measurements of the speed of sound (cutaway). In the red-colored layers in the solar interior, sound waves travel faster than predicted by the standard solar model (*yellow*), implying that the temperature is higher than expected. Blue corresponds to slower sound waves and temperatures that are colder than expected. The conspicuous red layer, about a third of the way down, shows unexpected high temperatures at the boundary between the turbulent outer region (convective zone) and the more stable region inside it (radiative zone). The disk measurements were made at a wavelength of 30.4 nm, emitted by singly-ionized helium, denoted He II, at a temperature of about 60,000 K using the Extreme-ultraviolet Imaging Telescope, abbreviated EIT, aboard the *SOlar and Heliospheric Observatory*, or *SOHO* for short, and the MDI/SOI and VIRGO instruments on *SOHO* made the sound speed measurements over a period of 12 months beginning in May 1996. (Composite image courtesy of Steele Hill, *SOHO* is a project of international cooperation between ESA and NASA.)

SOHO helioseismologists have also used measurements of sound wave frequencies to infer rotational and other motions inside the Sun. The moving material produces slight changes in the frequency of a sound wave that passes through it, as expected from the well-known Doppler effect.

3.5 How the Sun Rotates Inside

It has been known ever since telescopes were used to carefully monitor sunspots that the visible solar disk rotates differently at different solar latitudes, at a faster rate near the solar equator than nearer the solar poles with a smooth variation in between. As on Earth, the solar latitude is the angular distance north or south of the equator.

The sidereal rotation period of the photosphere, from east to west against the stars, is about 25 days at the equator where the solar latitude is zero degrees; the rotation period reaches 34 days at about $\pm 75°$ solar latitude. These rotation periods can be converted into velocities – just divide the circumference at the latitude by its period (Table 3.2). Regions near the poles rotate with very slow speeds, in part because the rate of rotation is smaller, but also because the material near the poles is closer to the Sun's axis and the distance around the Sun is shorter.

The varying rotation of the photosphere is known as differential rotation, because it differs with latitude. The differential rotation pattern is remarkably smooth and steady, systematically varying with solar latitude and not changing by more than 5% since the first systematic measurements were made by Richard C. Carrington in 1863 – through careful tracking of sunspots at different latitudes. H. W. Newton and M. L. Nunn in 1951 and Herschel B. Snodgrass in 1983 have more recently provided differential rotation parameters for the magnetic features in the photosphere. Doppler shift measurements of photosphere rotation at different latitudes also reveal the effect, as documented by P. A. Gilman in 1974 and E. H. Schröter in 1985.

Every point on the surface of the Earth rotates at the same speed, so a day is 24 h long everywhere on the Earth. Such a uniform spin is called solid-body rotation. Only a gaseous or liquid body can undergo differential rotation; it would tear a solid body into pieces.

Until recently, we had no knowledge of how fast the Sun spins inside. This has now been determined using the photosphere oscillations driven by internal sound waves. The waves propagating in the direction of rotation will tend to be carried along by the moving gas, and will travel faster than they would in a static, non-moving medium. A bird or a jet airplane similarly moves faster when traveling with the wind and takes a shorter time to complete a trip. The resonating sound wave crests moving with the rotation will therefore appear, to a fixed observer, to travel faster and their measured periods will be shorter. Waves propagating against the rotation will be slowed down, with longer periods.

Thus, rotation imparts a clear signature to the oscillation periods, lengthening them in one direction and shortening them in the other. These opposite effects make

Table 3.2 Differential rotation of the Sun[a]

Solar latitude (degrees)	Rotation period (days)	Rotation speed $(km\,h^{-1})$	Rotation speed $(m\,s^{-1})$	Angular velocity (nHz)
0 (Equator)	25.67	7097	1971	451
15	25.88	6807	1891	447
30	26.64	5922	1645	434
45	28.26	4544	1262	410
60	30.76	2961	823	376
75	33.40	1416	393	347

[a]Data from the MDI instrument aboard the *SOHO* spacecraft

an oscillation period divide, splitting the frequencies of the p-modes. Such rotational splitting depends on both depth and latitude within the Sun. The solar oscillations have a period of about 5 min, so the rotational splitting is roughly 5 min divided by 25 days, or about one part in 7,000. The oscillations have to be measured 10 or a 100 times more accurately than this to determine subtle variations in the Sun's rotation, or as accurately as one part in a million.

In 1986, Thomas L. Duvall Jr., John W. Harvey, and Martin A. Pomerantz showed that the general pattern of differential rotation persists through the convective zone, and Timothy M. Brown and his co-workers confirmed this in greater detail in 1989. Years of *SOHO* oscillation data have more recently enabled researchers to determine the radial and latitude dependence of the Sun's internal rotation all the way down to the core (Fig. 3.12). In 1998, Jesper Schou and coworkers used data from the MDI instrument on *SOHO* to show that there is little variation of rotation with depth within the convective zone, where differential rotation is preserved. So at any specific solar latitude, the inside of the Sun does not rotate any faster than the outside. At greater depths, within the radiative zone, the rotation rate becomes independent of latitude, acting as if the Sun were a solid body. Though gaseous, the radiative interior of the Sun rotates at a nearly uniform rate intermediate between the equatorial and high-latitude rates in the overlying convective zone.

Thus the rotation velocity changes sharply at the top of the radiative zone, located nearly one-third of the way down to the center. At this place, the outer parts of the radiative interior, which rotates at roughly the same speed regardless of solar latitude, meets the overlying convective zone, which spins faster in its equatorial middle. This narrow interface region, known as the tachocline, probably marks the location of the solar dynamo that generates the Sun's magnetic field (see Sect. 3.9).

There is also a sharp rotational gradient just beneath the photosphere, for the fastest rotation rate at a given solar latitude occurs at about 0.95 solar radii instead of exactly 1.00 where the photosphere is located. This was suspected from differences between gas and magnetic rotation rates detected in the photosphere, and it is a pronounced feature in the radial rotation profiles from GONG and MDI. Thierry Corbard and Michael J. Thompson described the global MDI measurements of this rotational increase in 2002, while Rachel Howe and co-workers (2006b) discussed its local helioseismology analysis.

In 2003, Sebastien Couvidat and his co-workers combined observations with *SOHO's* MDI and GOLF instruments to infer the radial rotation profile down to the core. After removing the effects of differential rotation in the convective zone, a uniform rotation is obtained throughout the radiative zone and into the outer parts of the core, down to two tenths of the Sun's radius. To this depth, there is no indication of either a rapidly or slowly rotating center.

The central parts of the Sun's energy-generating core have not been resolved, so its rotation and structure remain a mystery, but there are hints that the core spins at a slightly faster rate than the radiative zone. Nevertheless, no asymmetry or oblateness in the shape of the Sun has been detected, as would be expected from exceptionally

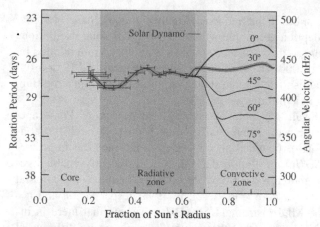

Fig. 3.12 Internal rotation of the Sun. The rotation rate inside the Sun, determined from helioseismology. The outer parts of the Sun exhibit differential rotation, with high solar latitudes rotating more slowly than equatorial ones. This differential rotation persists to the bottom of the convective zone at 28.7% of the way down. The rotation period in days is given at the left axis, and the corresponding angular velocity scale is on the right axis in units of nanoHertz, abbreviated nHz, where $1\,nHz = 10^{-9}\,Hz$, or a billionth, of a cycle per second. A rotation rate of $320\,nHz$ corresponds to a period of about 36 days (solar poles), and a rate of $460\,nHz$ to a period of about 25 days (solar equator). The rotation in the outer parts of the Sun, at latitudes of zero (solar equator), $30°$, $45°$, $60°$, and $75°$, has been inferred from 144 days of data using the Michelson Doppler Imager, abbreviated MDI, aboard the *SOlar and Heliospheric Observatory*, or *SOHO* for short. Just below the convective zone, the rotational speed changes markedly, and shearing motions along this interface may be the dynamo source of the Sun's magnetism. By examining more than 5 years of low-order acoustic modes, obtained using the GOLF and MDI instruments aboard *SOHO*, the rotation rate has been inferred for the deep solar layers (*error bars*), mainly along the solar equator. There is uniform rotation in the radiative zone, from the base of the convective zone at 0.713 solar radii to about 0.25 solar radii. The acoustic modes (sound waves) do not reach the central part of the energy-generating core. (Courtesy of Alexander G. Kosovichev for the MDI data showing differential rotation in the convective zone, and Sebastien Couvidat, Rafael García, and Sylvaine Turck-Chièze for the GOLF/MDI data in the radiative zone. *SOHO* is a project of international cooperation between ESA and NASA.)

rapid rotation inside the Sun, and this has important implications for theories of gravity (Focus 3.6).

Focus 3.6

Confirming Einstein's Theory

An interesting implication of the helioseismology rotation results involves tests of Einstein's theory of gravity. According to his *General Theory of Relativity*, space is distorted or curved in the neighborhood of matter, and this distortion is the cause of gravity. The result is a gravitational effect that departs slightly from Newton's expression, producing planetary orbits that are not exactly elliptical. Instead of returning to its starting point to form a closed ellipse in one orbital period, a planet moves slightly ahead in a winding path that can be described as a rotating ellipse.

Einstein first used this theory to describe such a previously unexplained twisting of Mercury's orbit.

Mercury's anomalous orbital shift, of only 43 s of arc per century, is in almost exact agreement with Einstein's prediction, but this accord depends on the assumption that the Sun is a nearly perfect sphere. If the interior of the Sun is rotating very fast, it will push the equator out further than the poles, so its shape ought to be somewhat oblate rather than perfectly spherical. The gravitational influence of the outward bulge, called a quadrupole moment, will provide an added twist to Mercury's orbital motion, shifting its orbit around the Sun by an additional amount and lessening the agreement with Einstein's theory of gravity.

Fortunately, helioseismology data indicate that most of the inside of the Sun does not rotate significantly faster than the outside, at least down to the energy-generating core. Even if the core of the Sun is spinning rapidly, a substantial asymmetry cannot be produced in the shape of the Sun. So, we may safely conclude that measurements of Mercury's orbit confirm the predictions of *General Relativity*. In fact, the small quadrupole moment inferred from the oscillation data by Thomas L. Duvall Jr. and colleagues in 1984, which is about one ten millionth rather than exactly zero as Einstein assumed, is consistent with a very small improvement in Mercury's orbit measured in recent times. So, the Sun does have an extremely small, middle-aged bulge after all.

3.6 Waves in the Sun's Core

In 2007, Rafael A. García and co-workers on *SOHO's* Global Oscillations at Low Frequency, abbreviated GOLF, team reported their detection of gravity waves. These long-period resonant oscillations are generated in the deeper parts of the Sun and have their largest amplitudes there (Fig. 3.13). And they are known as gravity waves, or g-modes, because it is the force of gravity that determines how quickly they rise and fall. (These are totally unrelated to the gravitational radiation predicted by Einstein's *General Theory of Relativity*.)

Gravity waves, or g-modes, occur in regions where there is a stable density difference, or stratification in density. They are produced when a parcel of gas oscillates above and below an equilibrium position, like waves in the deep sea. When a high-density parcel moves up into a lower-density region, it is pulled back into place by gravity, and then moves back due to the restoring force of buoyancy. In contrast, sound waves are restored by pressure, and are therefore designated p-modes with p for pressure.

Gravity waves become evanescent, or non-propagating, in regions where the gas is not stably stratified, such as the turbulent convective zone. As a result, they are largely confined to the Sun's deep interior where they are the strongest. Gravity waves that manage to reach the photosphere are expected to raise and lower an entire

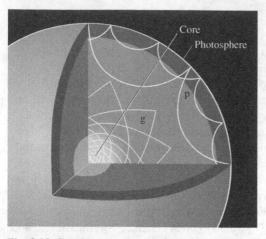

Fig. 3.13 Gravity waves. A cross section of the solar interior showing the ray paths of sound waves (p modes) and gravity waves (g modes). Sound waves produce oscillations detected in the photosphere, and can be used to determine the internal properties of the Sun down to about 0.2 of the solar radius. Gravity waves never reach the photosphere and are instead turned around inside the Sun, probing its central depths and reaching the core

hemisphere every few hours or so, but with a severe attenuation to low amplitude due to propagation through the convective zone. They are therefore very difficult to observe.

The detection of g-modes is nevertheless favored by the fact that they maintain phase coherence, or stay in sync, for years, so they could produce an observable signal when averaged over a very long time. The GOLF team therefore examined a 10-year sequence of observations, and looked for the distinctive, even spacing of the g-mode periods, which are unlike the sound-wave p-modes that have evenly spaced frequencies. And an evenly spaced, periodic structure was found in the GOLF data that is in agreement with the period separation predicted by the theory for gravity waves.

In addition to showing that g-modes may indeed exist, García and colleagues were able to make a rough estimate of the rotation rate of the Sun's core. By comparing their observations with theoretical models of the g-modes, they found some evidence that the core is spinning three to five times faster than the overlying radiative zone, but this may not be consistent with p-mode estimates of a uniform rotation in the radiative zone down to 0.2 solar radii. If there is an extra spin deep down in the solar core, it could be left over from the Sun's formation; the outer parts of the Sun would have been slowed over the eons by the braking action of the solar wind.

Since the g-mode detection has not been confirmed by any other measurements, there is still a possibility that it may not survive the test of time. If confirmed, however, these exciting reports of gravity waves and hints of a rapidly spinning core will become one of the key discoveries of helioseismology.

3.7 Internal Flows

3.7.1 Early Work and New Methods

Large-scale flows were first reported in 1979–1980 from the Doppler shift velocity measurements of spectral lines formed in the photosphere. More than a decade of full-disk velocity measurements from Mount Wilson were used by Robert Howard in 1979 to infer a poleward-directed motion of about $20\,\mathrm{m\,s^{-1}}$. Thomas L. Duvall Jr. endorsed this result in the same year using Doppler shift measurements with the Stanford magnetograph. This poleward motion in the photosphere is also called meridional flow, since the gas circulates north or south along meridians that pass through the Sun's poles.

The meridional flow was also inferred from numerical simulations of the observed movements of the magnetic remnants of former active regions during the 11-year sunspot cycle, conclusively demonstrated by Yi-Ming Wang, Ana G. Nash, and Neil Sheeley Jr. in 1989.

And well before that, in 1980, Robert Howard and Barry Labonte used the Mount Wilson velocity data to discover alternating latitude zones of slow and fast rotation – after subtracting the much larger differential rotation. The wave-like zones of anomalous rotations keep up with and flank the sunspot zones of magnetic activity, and drift from high latitudes, where they originate, to the equator in about 11 years. The east–west zonal flow bands are sometimes called torsional oscillations, since they were initially attributed to such an oscillation in which the solar rotation is periodically sped up or slowed down in certain zones of latitude.

Both the poleward motions and the zonal rotation bands move at a slow pace, with speeds that reach no more than $20\,\mathrm{m\,s^{-1}}$. They can only be detected when much faster and stronger motions are removed from the data. The Sun's equatorial rotation speed, for example, is about $2,000\,\mathrm{m\,s^{-1}}$, or hundreds of times greater than the flow speeds, and there are also granulation and supergranulation motions and the 5-min photosphere oscillations that might contaminate the Doppler-shift measurements in the photosphere.

Perhaps because of these uncertainties, the large-scale photosphere motions did not generate enthusiastic, widespread interest until accurate helioseismology measurements were used to confirm them and to show that the flows extend deeply into the Sun. And because it is not possible to use global helioseismology to detect flows in meridional planes, new techniques of local helioseismology were developed to investigate the structure and flows just below the photosphere. As with local anesthesia, it applies to localized areas on the Sun rather than its global, full-body properties.

One of the methods of local helioseismology, known as helioseismic tomography or time-distance helioseismology, is an adaptation of the tomographic techniques introduced by Ronald N. Bracewell in 1956 to infer the structure of cosmic radio sources from interferometer scans, and by Allan M. Cormack in 1963 to determine the internal properties of an object using beams of X-rays. This led to the development of Computer Assisted Tomography, or CAT, scans to derive clear views of the insides of living bodies from the numerous readings of X-rays criss-crossing the body from different directions, and incidentally led to Allan sharing the 1979 Nobel Prize in Medicine for his contribution. By recording sounds that have passed through the Sun at different angles, one can use triangulation to obtain similar tomographical information.

The time–distance, or helioseismic tomography, method measures and interprets the travel times of sound waves between any two locations on the photosphere. This travel time tells of both the temperature and the gas flows along the internal path connecting the two points. If the local temperature is high, sound waves move more quickly – as they do if they travel with the flow of gas. The travel time is obtained for a great many sets of points, and then inverted in a computer to chart the three-dimensional internal structure and dynamics of the Sun, including the sound speed, flow speed, and direction of motion.

Since the MDI instrument on *SOHO* can continuously obtain clear, sharp images with fine detail, it has proved particularly useful in helioseismic tomography, directly measuring travel times and distances for sound waves and enabling three-dimensional images of the structure and flows below the photosphere. It is one of several methods of local helioseismology described in Focus 3.7.

Focus 3.7

Local Helioseismology

Laurent Gizon and Aaron C. Birch reviewed the main mathematical techniques and results of local helioseismology in 2005, describing a number of different approaches that complement each other. Listed in chronological order of their first use, they are Fourier-Hankel decomposition, ring-diagram analysis, time–distance helioseismology or helioseismic tomography, helioseismic holography, and direct modeling.

Douglas C. Braun, Thomas L. Duvall Jr., and Barry J. Labonte introduced Hankel spectral analysis in 1987 in order to study the relationship between inward and outward traveling waves around sunspots, thereby demonstrating that sound waves are absorbed in and around sunspots.

The ring-diagram analysis can be used to infer the speed and direction of horizontal flows below the photosphere by obtaining the Doppler shifts of sound waves from a spectral analysis of solar oscillations. Frank Hill introduced this method in 1988, when he used it to demonstrate internal meridional flow directed from the equator toward the south pole.

Time–distance helioseismology, which is also known as helioseismic tomography, was introduced in 1993 by Thomas L. Duvall Jr., Stuart Mark Jefferies, John W. Harvey, and Martin A. Pomerantz, noting that it can be used to detect inhomogeneities and motions below sunspots and other flows beneath the photosphere. In 1996, Alexander G. Kosovichev used the method to detect converging downflows and an increase in sound speed below active regions. (Active regions are the areas in, around, above, and below sunspots, the seat of powerful explosions called solar flares.) And in 2000, Kosovichev, Duvall, and Philip H. Scherrer reviewed the time–distance inversion methods and results for active regions, sunspots, flows. and supergranulation.

Charles Lindsey and Douglas C. Braun described the basic principles of helioseismic holography in 1997, using it in 2000 to demonstrate how one can image active regions on the far side of the Sun. The basic holography concept, which was first suggested by Francois Roddier in 1975, is that the line-of-sight Doppler velocity observed at the solar photosphere can be used to estimate the sound wave field at any location in the solar interior at any instant of time.

And in 2002, Martin F. Woodard introduced the idea of estimating internal flows from direct inversion of the correlations seen in the wave field in the Fourier spectral domain, using it to determine the supergranular flow below the photosphere.

3.7.2 Poleward Flows

In 1996, David Hathaway and co-workers inferred a meridional circulation with a poleward flow of about $20\,\mathrm{m\ s}^{-1}$ from Doppler velocity observations obtained with the Global Oscillation Network Group, abbreviated GONG, which was consistent with the earlier photosphere observations of Robert F. Howard and independently Thomas L. Duvall in 1979. Then in the 1997, Peter M. Giles and his colleagues used MDI observations to demonstrate that the meridional flow penetrates deeply (Fig. 3.14). They found that the entire outer layer of the Sun, to a depth of at least 25,000 km, is slowly but steadily flowing from the equator to the poles with a speed of about $20\,\mathrm{m\ s}^{-1}$. At this rate, an object would be transported from the equator to a pole in a little more than 1 year. And by applying the time–distance helioseismology measurements to 6 years of MDI data, from 1996 to 2002, Junwei Zhao and Alexander G. Kosovichev found that meridional flows of order of $20\,\mathrm{m\ s}^{-1}$ remained poleward during the whole period of observation, and that extra meridional circulation cells converge toward the sunspot belts in both hemispheres.

3.7.3 Zonal Flow Bands

In the meantime, Kosovichev and Jesper Schou used rotational splitting of the solar oscillation frequencies observed with MDI to show in 1997 that alternating flow bands of faster and slower rotation, the so-called torsional oscillations, also extend to a considerable depth below the photosphere (Fig. 3.14). These broad zonal bands of gas lie parallel to the Sun's equator and sweep along at different speeds relative to each other, reminding us of the trade winds and jet streams in the Earth's atmosphere and the Gulf Stream in our oceans. The solar banding looks symmetric about the solar equator because the analysis method using global p-modes cannot distinguish between the

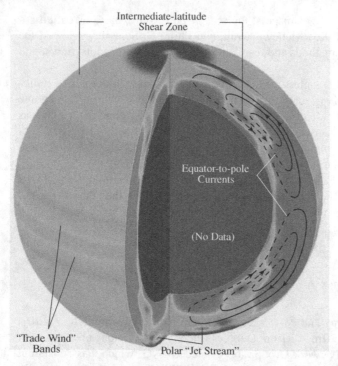

Fig. 3.14 Interior flows. Global helioseismology of internal flows in the Sun with rotation removed. Red corresponds to faster-than-average flows, yellow to slower than average, and blue to slower yet. On the left side, deeply rooted zonal flows (*yellow bands*), analogous to the Earth's trade winds, travel slightly faster than their surroundings (*blue regions*). The streamlines in the right-hand cutaway reveal a slow meridional flow toward the solar poles from the equator; the return flow below it is inferred. This image is the result of computations using 1 year of continuous observation, from May 1996 to May 1997, with the Michelson Doppler Imager, abbreviated MDI, instrument aboard the *SOlar and Heliospheric Observatory*, or *SOHO* for short. (Courtesy of Philip H. Scherrer and the *SOHO* SOI/MDI consortium. *SOHO* is a project of international cooperation between ESA and NASA.)

northern and the southern hemispheres of the Sun. There seem to be at least two zones of faster rotation and two zones of slower rotation in each hemisphere of the Sun. And a single zonal flow band is broad and deep, more than 65,000 km wide and penetrating as far as 20,000 km below the surface. The velocity of the faster zonal flows is about $5\,\mathrm{m\,s^{-1}}$ higher than gases to either side. This is substantially smaller than the mean velocity of rotation, which is about $2{,}000\,\mathrm{m\,s^{-1}}$, so the fast zones glide along in the spinning gas, like wide, lazy rivers of fire.

In 2000, Rachel Howe and co-workers used over 4 years of helioseismology data from GONG and *SOHO's* MDI to show that the bands of slower- and faster-than average rotation extend down to at least 60,000 km, and then in 2002 Sergey V. Vorontsov and co-workers used 6 years of *SOHO* MDI data to show that the zonal flow bands are coherent over the full depth of the convective zone, so the entire convection envelope appears to be involved.

3.7.4 Internal Changes Over the 11-year Cycle
of Magnetic Activity

The ways in which internal flows change with time are of particular interest since they may provide clues to the origin of the Sun's varying magnetism. And helio-seismologists have indeed shown that both the zonal flow bands and the poleward flows vary over the 11-year cycle of solar magnetic activity. Pioneering observations of these effects, by Jesper Schou in 1999, H. M. Anita and Sarbani Basu in 2000, and Dean-Yi Chou and De-Chang Dai in 2001 quickly evolved into more detailed investigations by Sergey V. Vorontsov and colleagues in 2002, John G. Beck, Laurent Gizon and Duvall in 2002, and Junwei Zhao and Alexander G. Kosovichev in 2004. More recently, Rachel Howe and her co-workers (2006b) used MDI and GONG data to obtain both global and local helioseismology of the zonal flows beneath the photosphere, confirming that zones of faster rotation approach the solar equator from mid-latitudes during the 11-year solar activity cycle.

Sunspots and the active regions that envelop them are located in solar latitude zones, or activity belts, which move from mid-latitudes to the solar equator during the sunspot cycle. The zonal bands of slower and faster rotation migrate in solar latitude together with these activity belts over the 11-year cycle. The zonal flow bands are most likely a side effect of the magnetic activity cycle and the internal flows around active regions, but this does not altogether explain why the bands persist at solar activity minimum.

The meridional flow also changes with time over the 11-year cycle, and is strongly correlated with the dominant latitude of magnetic activity. Time–distance helioseismology of *SOHO* MDI data indicates that the poleward meridional flow may be diverging out of the activity belts, and that extra meridional circulation cells converge toward the belts. These converging flow cells also migrate toward the solar equator, together with the activity belts as the solar cycle evolves.

Moreover, the internal structure of the Sun is not strictly spherically symmetric, with a latitude variation in sound speed. Sarbani Basu, H.M. Antia, and Richard S. Bogart showed in 2007 that this asphericity changes with solar activity, at least in the outer layers of the Sun.

After subtracting out the contributions from differential rotation, zonal rotation bands, and poleward meridional circulation, changing residual flows are found over a range of spatial scales. Deborah A. Haber, Bradley W. Hindman, and Rudi Komm and their co-workers, have investigated them, in 2002, 2004, and 2007, respectively. They reveal a pattern of solar subsurface weather that resembles the high- and low-pressure regions and swirling winds in the Earth's atmosphere, but on a much larger scale with hotter temperatures and no rain in sight. On the Sun, the meteorology and weather forecasts are linked to the varying solar magnetism, increasing in complexity as the 11-year solar cycle evolves towards activity maximum. And this brings us to a study of the structure and flows beneath active regions and sunspots.

3.8 Three-Dimensional Views of Sunspots and Active Regions

3.8.1 Looking Beneath Sunspots

For centuries, people have wondered about those strange dark spots on the Sun. What holds them together, so they last for weeks without breaking apart? The outward pressure of their strong magnetic fields ought to make sunspots expand at their edges and disperse into the surrounding photosphere, just as magnets with like polarity repel each other. And how far do the sunspots extend below the photosphere? Are they magnetic islands floating on the top of the convective zone, or are they anchored deep within it? These questions were finally answered when helioseismic tomography, or time–distance helioseismology, was used to zero in and see what lies beneath sunspots.

In 1996, Thomas L. Duvall Jr. and co-workers used the technique of helioseismic tomography with the data they obtained at the South Pole in 1991 to detect strong downflows beneath sunspots, with velocities of about $2\,\mathrm{km\,s^{-1}}$. The powerful flows bind the intense magnetic fields together, and stabilize the structure of sunspots.

Five years later, Junwei Zhao, working with his colleagues Alexander G. Kosovichev and Duvall, used the method with data from MDI to trace out the motions of hot flowing gas in, around, and below a sunspot (Fig. 3.15). They detected strong converging flows around the sunspot and downward directed flows in it. The adjacent streams of gas strengthen and converge towards the sunspots, pushing and concentrating the magnetic fields into them. Cool down-flowing material beneath the sunspots may also draw the surrounding gas and magnetic fields inward.

Further down, at depths of about 10,000 km, the flows seem to rip through a sunspot, apparently causing them to merge into deeper heated layers. This suggests that the sunspots are relatively shallow phenomena, extending about 5,000 km below the photosphere, so sunspots are about as deep as they are broad. It is as if the buoyant, concentrated magnetic fields rose up to the photosphere, gathering together and spreading out like a lotus flower on a lake, but perhaps connected to the depths with slender, thread-like roots.

Much of current helioseismology is devoted to the study of the influence of magnetic fields on sound waves and validation of the implications of their observations on the internal structure of sunspots. Detailed observations over several years, during various parts of the activity cycle, should tell us more about how the motions of hot solar gas interact with concentrations of the Sun's magnetism and give rise to its explosive activity. Current helioseismology suggests that swirling currents of gas beneath solar active regions may account for powerful outbursts known as solar flares.

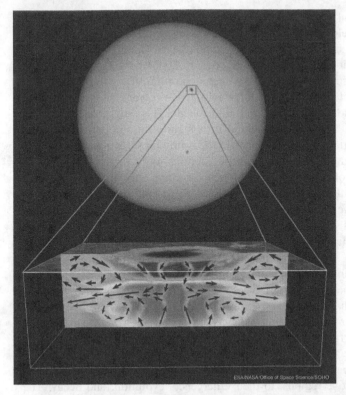

Fig. 3.15 What lies beneath a sunspot. The temperature and flow structure under a sunspot have been probed using local helioseismology techniques with data obtained by the Michelson Doppler Imager instrument aboard the *SOlar and Heliospheric Observatory*, or *SOHO* for short. The intense magnetic fields in and below a sunspot act as a plug that prevents the up-flow of energy from the hot solar interior. As a result, the sunspot is cooler and darker than its surroundings (*dark blue region in the bottom cross section*). Heat builds up below the magnetic plug, so the material underneath the sunspot's magnetic fields becomes hotter (*red area in cross section*). The converging flows of surrounding cooler material, denoted by the arrows, help hold a sunspot together. (Courtesy of the *SOHO* MDI consortium. *SOHO* is a project of international cooperation between ESA and NASA.)

3.8.2 Swirling Flows Beneath Active Regions

Active regions are locations in the solar atmosphere where intense magnetic fields, generated in the solar interior, have emerged through the photosphere and looped through the corona. The total numbers of active regions on the visible solar disk wax and wane with the 11-year cycle of solar activity, in tandem with the sunspots that they contain. Active-region coronal loops exhibit enhanced X-ray radiation, owing to the dense, hot gas within them, and the intensity of the X-ray emission varies in step with the activity cycle. Active regions are also the sites of intense outbursts called solar flares, which are more frequent at maximum solar activity.

The methods of local helioseismology are used to study the structure and dynamics of active regions below the photosphere. The helioseismic tomography that has been used to look beneath sunspots has, for example, been applied to the more extended and deeper regions beneath active regions. In 2003 and 2004, Junwei Zhao and Alexander G. Kosovichev applied this technique to *SOHO* MDI data to find deep, strong horizontal flows that converge into active regions and swirl around them, resembling cyclones in the Earth's atmosphere. They suggested that these sub-photospheric vortical flows might build up a significant amount of magnetic helicity and energy to power solar outbursts in the overlying solar atmosphere.

Twisting and shearing of the X-ray emitting coronal loops in active regions, as well as the interaction of existing loops with new emerging ones, may indeed cause solar flares and coronal mass ejections (see Sect. 6.10), and the helioseismic tomography results suggest that these coronal effects may be related to the internal vortical, swirling flows.

In 2006, Douglas Mason and co-workers used the ring-diagram technique of local helioseismology to survey flows beneath active regions using both GONG and *SOHO* MDI data. Their examination of more than 500 active regions indicated that the maximum sub-photospheric, horizontal vorticity, in both zonal east–west and meridional north–south components, is correlated with the X-ray flare intensity. In other words, the greater the circulating, sideways twist down below, the stronger the flare up above. The vertical vorticity component showed no clear relation to flare activity.

The methods of local helioseismology are also being applied to investigations of the life cycle of active regions, to their emergence, evolution, and decay. The initial results, obtained by Kosovichev and Duvall, show that some larger active regions are formed by repeated magnetic flux emergence from the deep interior, rooted at least 50,000 km below the photosphere. There is no initial evidence of a single large magnetic loop emerging from the interior to form an active region, but instead many thinner strands may be forming them. Large-scale, loop-like structures have nevertheless been detected below at least one active region.

Soon after formation, the active regions change the temperature structure and flow dynamics of the upper convective zone, forming large circulation cells of converging flows.

3.8.3 Detecting Active Regions on the Far Side of the Sun

Scientists are also now using sound waves to see right through the Sun to its hidden, normally invisible, backside, describing active regions on the far side of the Sun days before they rotate onto the side facing Earth. They use the technique of helioseismic holography with observations of the Sun's oscillations to create a sort of mathematical lens that focuses to different depths. A wide ring of sound waves is examined, which emanates from a region on the side of the Sun facing away from the Earth, the far side, and reaches the near side that faces the Earth (Fig. 3.16).

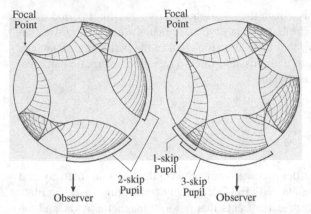

Fig. 3.16 Looking through the Sun. The arcing trajectories of sound waves from the far side of the Sun are reflected internally before reaching the front side, where they are observed with the Michelson Doppler Imager, abbreviated MDI, aboard the *SOlar and Heliospheric Observatory*, or *SOHO* for short, or the Global Oscillation Network Group, abbreviated GONG. Here we show a two-skip correlation scheme (*left*) and a one-skip/three-skip correlation scheme (*right*) of seismic holography used to image active regions (*focal point*) on the otherwise unseen far side of the Sun. These methods permit complete seismic imaging of the entire far hemisphere of the Sun, provided daily at http://soi.stanford.edu/data/full_farside/or http://gong.nso.edu/. This permits scientists to detect potentially threatening active regions on the far side of the Sun before the Sun's rotation brings them around to the front side that faces the Earth. [Adapted from Douglas C. Braun and Charles Lindsey, *Astrophysical Journal (Letters)* **560**, L189 (2001).]

When a large solar active region is present on the backside of the Sun, its intense magnetic fields cool the gas there, thus changing the level at which sound waves are reflected. A sound wave that would ordinarily take about 6 h to travel from the near side to the far side of the Sun and back again takes approximately 12 s less when it bounces off an active region on the far side. When nearside photosphere oscillations are examined, the quick return of these sound waves can be detected.

This remarkable result was introduced by Douglas C. Braun and Charles Lindsey in 2000, and first used by them in 2001 with *SOHO* MDI data to demonstrate how active regions can be detected on the side of the Sun facing away from Earth. These techniques have subsequently been refined to map the entire backside of the Sun, including the polar regions. In 2007, Junwei Zhao and Alexander G. Kosovichev, for example, provided new developments based on the Braun and Lindsey approach. Phase-sensitive helioseismic holography using MDI data is used on a routine basis to give daily images of magnetic activity on the far side of the Sun, which we cannot directly see from Earth. Images and movies are available at http://soi.stanford.edu/data/full_farside/ Similar work with GONG data also yields daily images of active regions on the far side of the Sun, available at http://gong.nso.edu/.

Solar astronomers are using this technique to monitor the structure and evolution of large regions of magnetic activity as they cross the back side of the Sun, thereby revealing the regions that are growing in magnetic complexity or strength and seem primed for explosive outbursts. Since the solar equator rotates with a period of 27 days, when viewed from the Earth, this can give more than a week's extra warning of potential solar flares or coronal mass ejections before the active region swings into view, threatening the Earth with possible intense radiation, energetic particles, or mass ejections from these outbursts.

3.9 The Solar Dynamo

Where do the Sun's magnetic fields come from, why do they vary in intensity and location, and how are they maintained? Hot, electrically conducting gases circulate within the Sun, generating electrical currents that produce magnetic fields, and the changing magnetic fields produce electric currents and sustain the generation of electricity, just as in a power-plant dynamo. The mechanical energy of the motion of the charged gas particles is thereby converted into the energy of magnetic fields. The Earth's magnetic field is supposed to be generated by such a dynamo, operating on a much smaller scale within its molten core.

The magnetic fields that have been spawned by the Sun's internal dynamo are entrained and "frozen" into the conducting gas that carries the magnetic fields along. As they move with the gas, the embedded magnetic fields are deformed, folded, stretched, twisted, and amplified, and eventually thread their way out of the Sun to form bipolar sunspots in the photosphere. This dynamo mechanism does not explain how the magnetic fields originated, but rather how they are amplified and maintained. The process of field amplification is nevertheless cumulative, so a dynamo can generate a strong magnetic field from an initially weak one.

A conceptually simple model of the solar dynamo, devised in 1961 by Horace W. Babcock, begins at sunspot minimum with a global, dipolar magnetic field that runs inside the Sun from south to north, or from pole to pole. Uneven, or differential, rotation shears the electrically conducting gases of the interior, so the entrained magnetic fields get stretched out and squeezed together. The magnetism is coiled, bunched, and amplified as it is wrapped around the inside of the solar globe, eventually becoming strong enough to rise to the surface and break through it in active-region belts with their bipolar sunspot pairs (Fig. 3.17), like a stitch of yarn pulled from a woolen sweater. The surrounding gas buoys up the concentrated magnetism, just as a piece of wood is subject to buoyant forces when it is immersed in water.

As Babcock expressed it:

> Shallow submerged lines of force of an initial, axisymmetric dipolar field are drawn out in longitude by the differential rotation. ... Twisting of the irregular flux strands by the faster shallow layers in low latitudes forms "ropes" with local concentrations that are brought to the surface by magnetic buoyancy to produce *bipolar magnetic regions* (BMRs) with associated sunspots and related activity.

Building upon a previous speculation with his father, Harold, in 1955, Babcock supposed that the poleward migration of the trailing parts of bipolar regions is responsible for the cancellation and reversal of the polar field, and that the leading

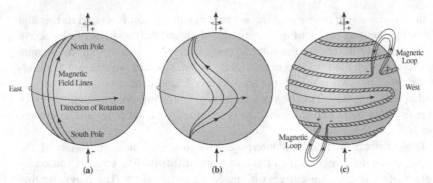

Fig. 3.17 Winding up the field. A simple model for generating the changing location, orientation, and polarity of the sunspot magnetic fields. At the beginning of the 11-year cycle of magnetic activity, when the number of sunspots is at a minimum, the magnetic field is the dipolar or poloidal field seen at the poles of the Sun (*left*). The internal magnetic fields then run below the photosphere from the south to the north poles. As time proceeds, the highly conductive, rotating material inside the Sun carries the magnetic field along and winds it up. Because the equatorial regions rotate at a faster rate than the polar ones, the internal magnetic fields become stretched out and wrapped around the Sun's center, becoming deformed into a partly toroidal field (*middle and right*). The fields are then concentrated and twisted together like ropes. With increasing strength, the submerged magnetism becomes buoyant, rises and penetrates the visible solar disk, or photosphere, creating magnetic loops and bipolar sunspots that are formed in two belts, one each in the northern and southern hemisphere (*right*). [Adapted from Horace W. Babcock, *Astrophysical Journal* **133**, 572–587 (1961).]

parts of the bipolar regions at low latitudes must migrate toward the equator where they are cancelled. Or in his own words:

> "Preceding" parts of BMRs expand toward the equator as they age, to be neutralized by merging; "following" parts expand or migrate poleward so that their lines of force neutralize then replace the initial dipolar field [with reversed polarity].

The sunspots appear in bipolar pairs of opposite magnetic polarity, and the leading spots (which are leading in the sense of rotation) in the northern belt of sunspots all have the same magnetic polarity, while the following spots have the opposite magnetic polarity. In the southern hemisphere, the leading and trailing sunspots of any sunspot pair also exhibit opposite polarities, but the magnetic direction of the bipolar sunspots in the southern belt is the reverse of that in the northern one.

In Babcock's model, the initial global dipole, or poloidal, magnetic field is sheared and stretched by differential rotation into a submerged toroidal, or ring-shaped, field running parallel to the equator, or east to west. Bipolar sunspot pairs are supposed to be produced by a lifting and twisting process related to the rising toroidal magnetic fields. Apparently, the dynamo generates two toroidal magnetic fields, one in the northern hemisphere and one in the southern hemisphere, but oppositely directed, which bubble up at mid- to low-latitudes to spawn the two belts of active regions, symmetrically placed on each side of the equator. Thus, according to Babcock's scenario, we may view the solar cycle as an engine in which differential rotation drives an oscillation between global poloidal and toroidal magnetic field geometries.

As the 11-year cycle progresses, the internal magnetic field is wound tighter and tighter by the shearing action of differential rotation, and the two belts of new active regions slowly migrate toward the solar equator. Because the active regions emerge, on the average, with their leading ends slightly twisted toward the equator, the leading sunspots in the two hemispheres tend to merge and cancel out, or neutralize, each other at the equator. This leaves a surplus of following-polarity magnetism in each hemisphere, north and south, which is eventually carried poleward at sunspot minimum.

In 1964, Robert B. Leighton interpreted the dispersal and migration of the photosphere magnetic regions as a random walk, diffusion-like process caused by supergranulation convection currents in the Sun's outer layers. The magnetic flux originates in sunspot groups and spreads outward in the photosphere via supergranular diffusion. And in 1969, he proposed a kinematical model of the solar cycle based on magnetic field amplification by solar differential rotation.

The resultant Babcock–Leighton model of the solar dynamo has served as the basis for decades of numerical simulations of photosphere magnetic observations, beginning with Kenneth H. Schatten and co-workers' 1972 computation of large-scale solar magnetic fields. They allowed the observed active-region fields to diffuse and to be sheared by differential rotation in accordance with Leighton's model.

More recent revisions of magnetic flux transport in the Babcock and Leighton models, reviewed by Neil R. Sheeley Jr. in 2005, have simulated the Sun's observed photosphere magnetic field over many sunspot cycles. They indicate that although most of the magnetic flux on the Sun originates in low-latitude sunspot belts, this flux is dispersed over a much wider range of solar latitudes by both supergranular convective motions and poleward meridional flow. As demonstrated by Yi-Ming Wang, Ana G. Nash, and Sheeley in 1989, the numerical simulations show that supergranular diffusion and poleward flows apparently sweep remnants of former active regions into streams, each dominated by a single magnetic polarity, which slowly wind their way from the low- and mid-latitude active-region belts to the Sun's poles. Deep meridional flow, or north–south circulation, is supposed to thereby help regenerate the global dipole field, also accounting for its reversal, and explain why sunspots do not form at high latitudes. By sunspot minimum, when the active regions have largely disintegrated, submerged, or annihilated each other, the continued poleward transport of their debris may form a global dipole. Because the Sun's polar field is created from the following polarity of decaying active regions, they reverse the overall dipole polarity at sunspot minimum, so the north and south pole switch magnetic direction or polarity. The internal magnetism has then readjusted to a poloidal form, and the magnetic cycle begins again. When the Sun's global dipolar magnetic flip is taken into account, we see that it takes two activity cycles, or about 22 years, for the overall magnetic polarity to get back where it started.

Thus, the revised Babcock–Leighton model seems to explain all of the repetitive aspects of the Sun's magnetism, including the 11-year periodic variation in the number of sunspots, their cyclic migration toward the equator, the roughly east–west orientation, location and polarity of bipolar sunspot pairs, and the periodic reversal of the overall global dipole.

The exact position of the solar dynamo has remained something of a mystery for several decades, leading some to doubt the dynamo theory. In the 1970s it was believed that the dynamo was located somewhere in the convective zone, but in the early 1980s the solar internal rotation profile inferred from helioseismology sent the dynamo theorists back to the drawing boards. That is, the persistence of differential rotation throughout the convective zone was not consistent with dynamo theories at the time. Theorists temporarily avoided this problem by placing the dynamo down near to the Sun's center, where no one could observe it and thereby contradict or confirm its hypothetical properties.

Many scientists now think that the solar dynamo may operate in both the convective zone and in a thin layer, called the tachocline, located near the bottom of the convective zone. Below the tachocline the Sun rotates like a solid object, with too little variation in spin to drive a solar dynamo, but there is a great deal of rotational shear at the tachocline, which is located at the boundary between uniform and differential rotation. In the vast majority of dynamo models, a poloidal magnetic field is generated in the convective zone and a toroidal field in the tachocline, and the two work together to give rise to the solar magnetic cycle.

Helioseismology results indicate that the tachocline is centered at a radius of 0.693 ± 0.003 solar radii near the solar equator. This is just below the bottom of the convective zone at 0.713 ± 0.003 of the Sun's radius. At higher solar latitudes, the location of the tachocline shifts upward, reaching 0.717 ± 0.003 at solar latitude of $60°$. The base of the convective zone does not exhibit any significant latitude variation.

So according to one modern theory, magnetic fields of about $100,000\,G$ in strength are generated by dynamo action at the tachocline. Buoyant magnetic flux tubes rise through the convective zone and emerge at the photosphere in active regions, where they form sunspots with magnetic field strengths of roughly $1,000\,G$ and overlying coronal loops with field strengths of 10–$100\,G$. Differential rotation winds up the photosphere magnetic field, which fragments under the stress and circulates meridionally to the poles. And a return meridional counterflow deep within the Sun most likely drives the migration of sunspots toward the equator.

The theory for dynamo models and the large-scale dynamics of the turbulent convective zone is quite complicated, and largely a subject for experts in fluid dynamics and magnetohydrodynamics. Reviews have been given by Peter A. Gilman in 2000, Yuhong Fan in 2004 updated in 2007, Paul Charbonneau in 2005, and Mark S. Miesch in 2005. Current large-scale solar dynamo models are of the flux-transport type, which involves three basic processes: generation of toroidal fields by shearing of pre-existing poloidal fields by differential rotation, called the Ω-effect; regeneration of poloidal fields by lifting and twisting the toroidal flux tubes, known as the α-effect; and meridional circulation.

As demonstrated by Mausumi Dikpati and co-workers in 2006, the flux-transport model can be used with helioseismology constraints to predict the strength and timing of future solar activity cycles. They forecast that cycle 24 will have a 30–50% higher peak than the previous one, and that it will only start in late 2007 or early 2008.

Nevertheless, despite all the mathematical complexity, or perhaps because of it, there is no solar dynamo model that explains the different aspects of the Sun's magnetic activity cycle in detail. And no one has yet observed the magnetic fields deep inside the Sun, so it is safe to say that we do not now have a complete, detailed understanding of just how the solar dynamo works.

3.10 Summary Highlights: Exploring the Inside of the Sun

- The Sun is energized by the fusion of hydrogen nuclei, or protons, within its hot, dense core, where mass is converted into energy in a sequence of nuclear reactions called the proton–proton chain.
- Two protons tunnel through the electrical barrier produced by the repulsion of their charges. This tunneling is a quantum mechanical process, made possible because protons do not have exact positions.
- When two protons fuse together, they produce a positron, the anti-particle of the electron, and an electron neutrino. The anti-matter positrons immediately collide with the material electrons, disappearing in a puff of energetic gamma-ray radiation, but the neutrinos pass effortlessly through both the Sun and the Earth.
- Electron neutrinos are produced in vast quantities by the central nuclear reactions in the Sun, and massive subterranean neutrino detectors have captured a small number of them. They detected only one-third to one-half the expected amounts, a discrepancy known as the solar neutrino problem.
- By measuring the internal velocity of sound waves, scientists have taken the temperature of the Sun's energy-generating core, showing that it agrees with model predictions, at 15.6×10^6 K, and apparently ruling out any astrophysical solution to the solar neutrino problem.
- The solar neutrino problem has now been resolved by observations at the Sudbury Neutrino Observatory, which indicate that solar neutrinos switch between types, or flavors, on their way to Earth. Some of the Sun's electron neutrinos transform into other types of neutrinos that were undetectable by the pioneering neutrino experiments.
- Only the core of the Sun is hot enough and dense enough to generate energy by nuclear fusion reactions. The energy produced in the core of the Sun slowly works its way out to the visible solar disk by radiation and convection.
- At any moment, about a million granules can be seen in the white light of the visible solar disk, the photosphere. They mark the tops of gases rising out of the Sun by convection, each granule lasting for about 15 min before another one replaces it. The granules are superposed on a larger cellular pattern called the supergranulation.
- Sound waves generated in the convective zone remain trapped inside the Sun and push the photosphere in and out, producing vertical oscillations with a period of about 5 min. They are detected by the Doppler effect of a spectral line seen in the visible sunlight of the photosphere.

- Observations of the 5-min solar oscillations have been used to detect low-pitched sound waves that travel to different depths within the Sun, enabling determination of the inner structure and dynamics of the Sun by the techniques of helioseismology.
- Significant helioseismology results have been obtained with instruments aboard the *SOlar and Heliospheric Observatory*, abbreviated *SOHO*, which has had a continued, uninterrupted view of the Sun for more than 10 years. The Sun has also been followed 24 h a day for years by a worldwide network of ground-based observatories known as the Global Oscillations Network Group, or GONG for short.
- A small but definite change in sound speed has pinpointed the bottom of the convective zone. It is located at a radius of 0.713 ± 0.003 of the Sun's radius.
- The visible solar disk rotates at different speeds at different solar latitudes, with a faster rate at the solar equator than the solar poles and a smooth variation in between. This differential rotation persists through the convective zone to about a third of the way inside the Sun, where the rotation becomes uniform from pole to pole.
- A 10-year sequence of photosphere observations with *SOHO's* Global Oscillations at Low Frequencies, abbreviated GOLF, instrument exhibits a periodic structure that could be due to gravity waves, or g-modes, generated in the deeper parts of the Sun. If confirmed, this could also suggest that the solar core could be rotating three to five times faster than the radiative zone.
- Helioseismic tomography, or time–distance helioseismology, has been used with *SOHO* Michelson Doppler Imager, abbreviated MDI, data to show that poleward motions, also called meridional flows, extend deep beneath the photosphere. They move with speeds of about $20 \, \mathrm{m \, s^{-1}}$.
- Helioseismology techniques have been used to show that after average rotation is removed from both MDI and GONG data, zonal bands of slower- and faster-rotation, sometimes called torsional oscillations, are detected, which extend throughout the convective zone. The zonal rotation bands migrate in solar latitude, together with belts of sunspots and active regions over the 11-year solar activity cycle.
- After subtracting the contributions of differential rotation, zonal rotation bands, and poleward meridional circulation, a swirling pattern of residual flows is found; this subsurface solar weather increases in complexity with the 11-year cycle of solar magnetic activity.
- The Sun is not strictly spherical, and its asphericity changes with solar activity, at least in the outer layers of the Sun.
- Helioseismic tomography has been used with *SOHO* MDI data to look beneath sunspots, and to discover that strong converging flows push and force the sunspot magnetic fields together, keeping sunspots from expanding at their edges and dispersing into the surrounding photosphere.

- GONG and *SOHO* MDI data have been used with the techniques of local helio-seismology to detect deep swirling flows that circulate horizontally around solar active regions, and to show that the strength of this twisting flow is correlated with the intensity of X-ray flares emitted from active regions.
- Helioseismic holography is used with *SOHO* MDI or GONG data to determine the location and magnitude of active regions anywhere on the backside of the Sun days before they rotate into view on the visible solar disk. Daily images of these unseen active regions are available on the web, and can be used to give advance warning of possible solar flares and coronal mass ejections when the active regions rotate to face the Earth.
- The Sun's magnetism is probably sustained by dynamo action in the convective zone and at the tachocline interface between the deep interior, which rotates at one speed, and the overlying gas that spins faster in the equatorial middle.

3.11 Key Events in the Understanding of the Internal Constitution of the Sun[*]

Date	Event
1905–1906	Albert Einstein uses the postulates of special relativity to show that energy radiated is equivalent to mass loss. Fifteen years later, Arthur Stanley Eddington and Jean Baptiste Perrin independently suggest that the transformation of mass into radiated energy might be the process that makes the Sun shine.
1917	Francis W. Aston demonstrates that the helium atom is slightly less massive, by a mere 0.7%, than the sum of the masses of the four hydrogen atoms that enter into it.
1919	Joseph Larmor proposes that the magnetic fields of the Earth and the Sun can be generated by the internal motions of conducting material through the action of a self-sustaining dynamo.
1920	In an article entitled *The Internal Constitution of the Stars*, Arthur Stanley Eddington suggests that nuclear processes in the hot cores of stars might provide the energy that makes them shine, and specifically describes the possibility that hydrogen could be transformed to helium, with the resultant mass difference released as energy to power a star. Jean Baptiste Perrin proposed a similar idea at about the same time.
1920	Ernest Rutherford announces that the massive nuclei of all atoms are composed of hydrogen nuclei, which he named protons. He also postulates the existence of an uncharged nuclear particle, later called the neutron, which was required to keep the positively charged protons from repelling each other.

[*] See the References at the end of this book for complete references to these seminal papers.

Date	Event
1928	Paul A. M. Dirac formulates the relativistic theory of the electron from which he predicted the existence of the positron, or positive electron, the anti-matter particle of the electron.
1928	George Gamow uses quantum mechanics to show how energetic charged particles can escape the nuclei of radioactive atoms, tunneling through the nuclear electrical barrier that constrains them. The tunneling process involves a quantum waviness, uncertainty, or spread-out character, which makes it possible for a sub-atomic particle to occasionally penetrate electrical barriers. It makes nuclear fusion possible at temperatures that exist inside the Sun and other stars.
1929	Robert d'Escourt Atkinson and Fritz Georg Houtermans provide the first attempt at a theory of nuclear energy generation in the hot central portions of stars, showing that the most likely nuclear reactions will involve light nuclei with low electrical charge. They also show that only a few, rare, high-velocity nuclei will be able to penetrate the electrical barrier between them, explaining why nuclear reactions proceed slowly inside stars.
1930	Wolfgang Pauli proposes that an invisible particle, with no charge and little or no mass, is removing energy during a kind of radioactivity called beta decay. In this process, the nucleus of a radioactive atom emits an energetic electron, or beta particle, whose energy is less than that lost by the nucleus, and the unseen particle was postulated to carry away the remaining energy. Four years later, Enrico Fermi provided further evidence for Pauli's conjecture and called the particle the *neutrino*, Italian for "little neutral one."
1931	In an article entitled *Atomic Synthesis and Stellar Energy*, Robert d'Escourt Atkinson argues that the most effective nuclear reactions in stars involve protons, the nuclei of hydrogen atoms, and proposes that the observed relative abundances of the elements might be explained by the synthesis of heavy elements from hydrogen and helium by successive proton captures.
1932	Carl Anderson (1932b) reports his discovery of the positrons, or positive electrons, the first anti-matter particle to be found, which are created when cosmic rays from outer space interact with particles in the Earth's atmosphere. At the time of his discovery, Anderson was unaware of Dirac's theoretical prediction of the positron in 1928.
1932	James Chadwick discovers the neutron when bombarding atoms with energetic particles.
1934	Enrico Fermi provides the correct theory for beta decay, incorporating neutrinos, which he named.
1938	Hans Bethe and Charles L. Critchfield demonstrate that the fusion of hydrogen nuclei, or protons, into helium nuclei by the proton–proton chain provides the energy that makes the Sun shine.
1939	Hans A. Bethe publishes a seminal paper on energy production in stars.
1952–1954	Michael James Lighthill demonstrates how sound is generated aerodynamically, and shows that turbulent motions generate acoustic power.
1954	A. B. Hart describes evidence for velocity variations from point to point in the photosphere, and presents some evidence for bands of constant velocity occurring perpendicular to the solar equator.
1956	Frederick Reines and Clyde L. Cowan detect electron anti-neutrinos emitted from the Savannah River nuclear reactor in South Carolina.
1960–1962	Robert B. Leighton and co-workers Robert W. Noyes and George W. Simon discover the supergranulation pattern, with predominantly horizontal motion, and the 5-min vertical velocity oscillations in the solar atmosphere.
1961	Horace W. Babcock proposes a model of the Sun's magnetic field and the 22-year magnetic cycle involving shallow, submerged magnetic fields of an initial dipolar field drawn out in solar longitude by differential rotation.

Date	Event
1964	Raymond Davis Jr. and John N. Bahcall, respectively, present the experimental and theoretical aspects of a proposal to study solar neutrinos with a chlorine detector.
1964	Robert B. Leighton interprets the dispersal and migration of solar magnetic regions as a random walk, diffusion-like process caused by supergranulation convection currents in the Sun's outer layers.
1967	Raymond Davis Jr. and co-workers detect neutrinos emitted from the Sun using a massive container of cleaning fluid buried deep underground in the Homestake Gold Mine near Lead, South Dakota. This pioneering experiment, reported by Davis et al. (1968), was continued for more than a quarter-century, always detecting about one third of the expected amount of solar neutrinos.
1967	Bruno M. Pontecorvo first discusses the possibility of solar neutrino oscillations between different types of neutrinos.
1968	Edward N. Frazier demonstrates that the power of the 5-min oscillations in the photosphere exists over a broad range of frequencies (or periods) and horizontal wave numbers (or sizes).
1969	Vladimir Gribov and Bruno Pontecorvo propose that the discrepancy between the observed flux of solar neutrinos and theoretical expectations could be explained if some solar neutrinos switch from electron neutrinos to another type of neutrino as they travel in the near vacuum of space from Sun to Earth, thereby escaping detection.
1970–1972	The resonant-cavity, spherical shell model for internal sound waves is further developed and used to make predictions for the solar 5-min oscillations – by Roger K. Ulrich in 1970, John Leibacher and Robert F. Stein in 1971, and Charles L. Wolff in 1972.
1972	Kenneth H. Schatten, Robert B. Leighton, Robert Howard, and John M. Wilcox present the first numerical simulation of the large-scale photosphere magnetic field with the diffusion of active-regions fields.
1975	Franz-Ludwig Deubner confirms the acoustic cavity hypothesis, showing that the oscillating power of the observed 5-min oscillations at low degrees is concentrated into narrow ridges in a spatial-temporal display.
1975	Francois Roddier describes a procedure for creating an acoustical hologram of the photosphere.
1975	Hiroyasu Ando and Yoji Osaki develop theoretical models of sound waves trapped in the Sun.
1976	Douglas O. Gough points out the diagnostic capabilities of helioseismology for the internal constitution of the Sun.
1977	Peter Goldreich and Douglas A. Keeley describe how sound waves are excited by turbulent convection in the Sun's convective zone.
1977–1978	Lincoln Wolfenstein discusses the oscillations of neutrinos traveling through matter.
1979	Franz-Ludwig Deubner, Roger K. Ulrich, and Edward J. Rhodes Jr. publish the first study of internal differential rotation using solar oscillations.
1979	Robert Howard and Thomas L. Duvall Jr. independently use observations of Doppler line shifts of material in the photosphere to infer a mean polewards flow, or meridional circulation, with a velocity of about $20\,\mathrm{m\,s^{-1}}$.
1979	First observational studies of the global, low-degree modes ($l \approx 1, 2, 3$) of the 5-min oscillations that penetrate deeply in the Sun by Andre Claverie, George R. Isaak, Clive P. McLeod, H. Bob van der Raay, and Teodoro Roca Cortes.

Date	Event
1980	Robert Howard and Barry J. Labonte use 12 years of Mount Wilson velocity measurements of the photosphere to discover alternating latitude zones of fast and slow rotation after subtracting a differential rotating frame. The residual zonal flow bands, also called torsional oscillations, have speeds of about $3\,\mathrm{m\ s^{-1}}$ and move from high solar latitudes, where they originate, to the equator with a period of 11 years.
1980–1983	High-quality, low-degree ($l \approx 0, 1, 2, 3$) spectra of solar oscillations obtained by Gérard Grec, Eric Fossat, and Martin A. Pomerantz from continuous observations (120 h) at the South Pole.
1982	Thomas L. Duvall Jr. demonstrates a dispersion law for solar oscillations.
1982	John H. Thomas, Lawrence E. Cram, and Alan H. Nye describe how the 5-min oscillations can be used as a subsurface probe of sunspot structure.
1983	Solar oscillations observed in the Sun's total irradiance of the Earth by Martin F. Woodard and Hugh S. Hudson (1983a, b) using the Active Cavity Radiometer Irradiance Monitor, abbreviated ACRIM, aboard the Solar Maximum Mission satellite.
1983	Thomas L. Duvall Jr. and John W. Harvey obtain the first observations connecting solar oscillations of low and high degree ($l = 1\text{–}139$).
1983	Eugene N. Parker describes how a loss of magnetic flux through the free surface of a star into the surrounding space has important implications for the generation of the field within the star.
1984	Thomas L. Duvall Jr., Wojciech A. Dziembowski, Philip R. Goode, Douglas O. Gough, John W. Harvey, and John W. Leibacher use solar oscillations to determine the Sun's internal rotation near its equatorial plane, and show that the rotation rate through the solar convective zone is very close to that observed at the photosphere. They infer a slow rotation for much of the solar interior, with very little solar oblateness (low quadrupole moment), increasing the accuracy of the confirmation of Einstein's *General Theory of Relativity* using Mercury's orbital motion.
1985	First determination of the speed of sound in the solar interior from inversions of the frequencies of the Sun's 5-min oscillations by Jørgen Christensen-Dalsgaard, Thomas L. Duvall Jr., Douglas O. Gough, John W. Harvey, and Edward J. Rhodes Jr.
1985	Stanislav P. Mikheyev and Alexei Y. Smirnov show that the idea of neutrino oscillations can provide an explanation for the solar neutrino problem.
1986–1989	Thomas L. Duvall Jr., John W. Harvey, and Martin A. Pomerantz use solar oscillation data to show that the entire convective zone mimics the observed surface differential rotation, with slower rotation at higher solar latitudes. Timothy M. Brown, Jørgen Christensen-Dalsgaard, Wojciech A. Dziembowski, Philip Goode, Douglas O. Gough, and Cherilynn A. Morrow confirmed this in greater detail.
1987–1988	Discovery of the absorption of sound waves in and around sunspots by Douglas C. Braun, Thomas L. Duvall Jr., and Barry J. Labonte.
1988	Frank Hill describes a ring method of determining local structure under the photosphere and presents evidence for a flow of $100\,\mathrm{m\ s^{-1}}$ directed from the equator toward the south pole.
1988–1989	Discovery that the frequencies of sound waves change during the 11-year solar activity cycle for both low-degree oscillations and intermediate or high-degree oscillations. The relevant observations and data analysis are reported by Pere L. Pallé, Clara Régulo, and Teodoro Roca-Cortés in 1989; Yvonne P. Elsworth, Rachel Howe, George R. Isaak, Clive P. McLeod, and Roger New in 1990; Kenneth G. Libbrecht in 1989; and Kenneth G. Libbrecht and Martin F. Woodard in 1990.

Date	Event
1989	Yi-Ming Wang, Ana G. Nash, and Neil R. Sheeley Jr. use numerical simulations to show that magnetic flux is dispersed poleward by supergranular convective motions and meridional circulation.
1990	John N. Bahcall and Hans A. Bethe argue that new particle physics is required if the Homestake and Kamiokande solar neutrino experiments are both correct; or in other words, the solar neutrino problem is due to an incomplete knowledge of neutrinos.
1990	Peter Goldreich and Pawan Kumar use theoretical arguments and oscillation data obtained by Kenneth Libbrecht to demonstrate that sound waves are excited by turbulent convection in the upper part of the solar convective zone.
1990	Charles Lindsey and Douglas C. Braun propose that helioseismic imaging could be used to produce seismic maps of magnetic regions on the far side of the Sun.
1991	Accurate determination of the radius of the bottom of the solar convective zone at 0.713 ± 0.003 solar radii by Jørgen Christensen-Dalsgaard, Douglas O. Gough, and Michael J. Thompson from the observed frequencies of solar oscillations.
1991	Yoji Totsuka reports that the Kamiokande solar neutrino detector confirms the neutrino deficit first observed by Raymond Davis Jr. with the Homestake detector, while additionally using directional information to show that the neutrinos were coming from the Sun.
1991–1992	The Soviet-American Gallium Experiment, abbreviated SAGE, in Russia and the multi-national GALLEX experiment in Italy confirm the solar neutrino deficit in radiochemical experiments involving gallium.
1993	Thomas L. Duvall Jr., Stuart Mark Jefferies, John W. Harvey, and Martin A. Pomerantz introduce helioseismic tomography, or time–distance helioseismology.
1995	The *SOlar and Heliospheric Observatory*, abbreviated *SOHO*, is launched on 2 December 1995. It is a project of international collaboration between ESA and NASA.
1995–1997	Observations of solar oscillations from the ground and space indicate that the deep solar interior, from the base of the convective zone down to about 0.2 solar radii, rotates like a rigid body with a uniform latitude-independent rate that is somewhat slower than the surface equatorial rate. The relevant observations and data analysis are reported by Yvonne P. Elsworth and colleagues in 1995, Steven Tomczyk, Jesper Schou, and Michael J. Thompson in 1995; Michael J. Thompson and colleagues in 1996; and Alexander G. Kosovichev and colleagues in 1997.
1996	Thomas L. Duvall Jr., Sydney D'Silva, Stuart M. Jefferies, John W. Harvey, and Jesper Schou detect downflows under sunspots using helioseismic tomography with the data they obtained at the South Pole in 1991. The downflows hold the magnetic fields together to form a sunspot.
1996	David H. Hathaway and co-workers use Doppler velocity data obtained with the Global Oscillation Network Group instruments to measure the Sun's meridional circulation with a poleward flow moving at about $20 \mathrm{m\ s^{-1}}$.
1996	Alexander G. Kosovichev uses helioseismic tomography to detect strong converging downflows and an increase in sound speed below an active region.
1996–1997	Helioseismological measurements of sound waves with low and intermediate frequencies are inverted to obtain sound speeds that are in agreement with predictions of numerical (standard) solar models to within 0.2% throughout almost the entire Sun. This agreement indicates that any nonstandard solar model or related astrophysical solutions cannot resolve the solar neutrino problem. The relevant observations and data analysis are reported by Sarbani Basu and colleagues in 1996; Alexander G. Kosovichev and colleagues in 1997; John N. Bahcall and colleagues in 1997; and D. B. Guenther and Pierre Demarque in 1997.

Date	Event
1997	Charles Lindsey and Douglas C. Braun describe the basic principles of helio-seismic holography of local regions beneath the photosphere.
1997	Alexander G. Kosovichev and Jesper Schou use the Michelson Doppler Imager on *SOHO* to demonstrate that zonal flow bands extend deep below the photosphere.
1997	Peter M. Giles, Thomas L. Duvall Jr., Philip H. Scherrer, and Richard S. Bogart use the Michelson Doppler Imager on *SOHO* to detect a subsurface flow of material from the Sun's equator to its poles.
1997	Martin F. Woodard describes the implications of localized acoustic absorption when using acoustic tomography to map the three-dimensional structure of the Sun.
1998	Bruce T. Cleveland and colleagues summarize three decades of solar neutrino flux measurements with the Homestake solar neutrino detector; the observed flux is 2.56 ± 0.25 solar neutrino units and less than the predicted flux of 8.5 ± 1.8.
1998	Y. Fukuda and co-workers report Kamiokande observations indicating that muon neutrinos, generated by cosmic rays in the terrestrial atmosphere, may be undergoing oscillations and changing type, and that some neutrinos may therefore possess a very small mass.
1998	Jesper Schou and co-workers present comprehensive helioseismic studies of differential rotation, meridional flow, and torsional oscillations.
1998	Alexander G. Kosovichev and Valentina V. Zharkova use the Michelson Doppler Imager aboard *SOHO* to detect seismic waves generated when a flare impacts the lower solar atmosphere.
1999	Jesper Schou uses the Michelson Doppler Imager aboard *SOHO* to detect the migration of zonal flows toward the equator.
2000	Rachel Howe and colleagues use helioseismic observations over 4 years from the Global Oscillation Network Group and the Michelson Doppler Imager on board *SOHO* to show that the torsional oscillations, the bands of slower- and faster-than-average rotation, extend downward at least 60,000 km.
2000	Rachel Howe and co-workers detect changes in the rotation of the Sun near the base of its convective zone, at the presumed site of the solar dynamo.
2000	Charles Lindsey and Douglas C. Braun (2000a, b) demonstrate how helioseismic holography can be used to image active regions on the far side of the Sun.
2001	Sylvaine Turck-Chièze and her colleagues use solar oscillation data obtained with instruments aboard the *SOHO* spacecraft to place constraints on the neutrino flux emitted by nuclear reactions in the Sun's core, showing that the neutrino deficit measured on Earth cannot be explained by adjustments to solar model calculations.
2001	Arthur B. McDonald announces that observations with the Solar Neutrino Observatory, when combined with previous work, indicate that solar electron neutrinos change type as they travel from the core of the Sun to Earth, explaining the solar neutrino problem, and that the total number of electron neutrinos produced in the Sun is just as predicted by detailed solar models.
2001	Junwei Zhao, Alexander G. Kosovichev, and Thomas L. Duvall Jr. use helioseismic tomography, or time–distance helioseismology, with *SOHO* Michelson Doppler Imager data to examine the structure and flows along, in, and under a large sunspot.
2001–2002	Q. R. Ahmad and colleagues (2001, 2002a) report evidence for solar neutrino flavor transformations using observations at the Sudbury Neutrino Observatory, indicating that some solar electron neutrinos change to another type during propagation from the Sun's core to Earth.

Date	Event
2002	John G. Beck, Laurent Gizon, and Thomas L. Duvall Jr. use time–distance helioseismology to provide maps of torsional oscillations and meridional flows in the Sun.
2002	Sergey V. Vorontsov, Jørgen Christensen-Dalsgaard, Jesper Schou, V.N. Strakhov, and Michael J. Thompson use 6 years of *SOHO* Michelson Doppler Imager data to show that the bands of slower and faster rotation, the so-called torsional oscillations, propagate poleward and equatorward from mid-latitudes at all depths throughout the convective zone.
2002	Martin F. Woodard uses *SOHO* Michelson Doppler Imager data to determine the subsurface, supergranular flow, showing that it is in general agreement with the flow detected in the photosphere.
2002–2007	Deborah A. Haber, Bradley W. Hindman, and Rudi Komm and their co-workers describe residual sub-photosphere flows, detected when differential rotation, zonal rotation bands, and meridional circulation are removed from helioseismology data.
2003	Sebastien Couvidat and co-workers (2003a) use helioseismology observations with the Global Oscillations at Low-Frequency and Michelson Doppler Imager instruments on *SOHO* and LOWL data to infer the radial rotation profile down to 0.2 solar radii.
2003	Sebastien Couvidat, Sylvaine Turck-Chièze, and Alexander G. Kosovichev (2003b) use helioseismology observations of the internal sound speed, obtained from both the Global Oscillations at Low-Frequency and Michelson Doppler Imager instruments on *SOHO*, to derive neutrino fluxes and oscillation properties, all compatible with the Sudbury Neutrino Observatory results.
2003	K. Equchi and colleagues report evidence for the disappearance of anti-electron neutrinos, generated by terrestrial nuclear reactors, when propagating through the Earth.
2003	Junwei Zhao and Alexander G. Kosovichev use *SOHO* Michelson Doppler Imager data with time–distance helioseismology to infer the structure and dynamics of a rotating sunspot beneath the solar photosphere.
2004	Junwei Zhao and Alexander G. Kosovichev use *SOHO* Michelson Doppler Imager data with time–distance helioseismology to determine torsional oscillation, meridional flows, and vorticity in the upper convective zone of the Sun.
2005	B. Aharmin and co-workers report salt phase observations from the Sudbury Neutrino Observatory, providing estimates for the mass and oscillation parameters for electron neutrinos.
2005	T. Araki and co-workers present KamLAND observations of neutrino oscillations based on anti-electron neutrinos generated by terrestrial nuclear reactors and sent through the Earth.
2005	Charles Lindsey and Douglas C. Braun (2005a, b) develop the acoustic showerglass technique for seismic diagnostics of photosphere magnetic fields and imaging active regions below the photosphere.
2005	Arthur B. McDonald summarizes the results of the Sudbury Neutrino Observatory. When combined with other measurements, they demonstrate and define the oscillation parameters and flavor change of neutrinos.
2006	J. Hosaka and co-workers report solar neutrino measurements in Super-Kamiokande-I with neutrino oscillation results.
2006	Douglas Mason and co-workers survey flows beneath more than 500 solar active regions, showing that horizontal, swirling motions are correlated with the intensity of X-ray flares emitted by the active regions.

Date	Event
2007	Sarbani Basu, H.M. Antia and Richard S. Bogart use *SOHO's* Michelson Doppler Imager data from 1996 to 2003 to show that the internal structure of the Sun is not strictly spherically symmetric, and that its asphericity changes with solar activity, at least in the outer layers.
2007	Rafael A. García and co-workers use data from the Global Oscillations at Low-Frequency instrument on *SOHO* to detect a periodic structure attributed to solar gravity modes in the Sun's core. If confirmed, this will be a significant discovery, and it may indicate that the core rotates at a faster rate than the radiative zone

Chapter 4
Solving the Sun's Heating Crisis

4.1 Mysterious Heat

More than half a century ago, astronomers discovered that the Sun's corona has an unexpectedly high temperature of a few million Kelvin (Sect. 2.1). The visible solar disk, the photosphere, is closer to the Sun's center than the corona, but the photosphere is several hundred times cooler, and this comes as a big surprise. The essential paradox is that energy should not flow from the cooler photosphere to the hotter corona anymore than water should flow uphill. When you sit far away from a fire, for example, it warms you less.

The temperature of the corona is just not supposed to be so much higher than that of the atmosphere immediately below it. It violates common sense, as well as the second law of thermodynamics, which holds that heat cannot be continuously transferred from a cooler body to a warmer one without doing work. This unexpected aspect of the corona has baffled scientists for decades, and they are still trying to explain where all the heat is coming from.

We know that visible sunlight emitted by the photosphere cannot heat the corona. There is so little material in the corona that it is transparent to almost all of the photosphere's radiation. Sunlight therefore passes right through the corona without depositing substantial quantities of energy in it, traveling out to warm the Earth and to also keep the photosphere cool.

So radiation cannot resolve the heating paradox, and we must look for alternate sources of energy. They are related to the motions and magnetic fields in the photosphere and below, which supply either the kinetic energy of moving material or the magnetic energy stored in magnetic fields. Unlike radiation, both of these forms of energy can flow from cold to hot regions, working to keep the corona hot.

Everything is in motion within the seething photosphere and the turbulent convective zone beneath it, and magnetic fields thread their way through the entire solar atmosphere. And somehow kinetic energy and/or magnetic energy are being transmitted into the vibrant, dynamic chromosphere and low corona and dissipated as heat within just a few hundred thousand kilometers above the photosphere, or in less than 1% of the solar radius.

K.R. Lang, *The Sun from Space*, Astronomy and Astrophysics Library,
© Springer-Verlag Berlin Heidelberg 2009

George Withbroe and Robert Noyes estimated the flux of energy that must be provided to heat the corona more than three decades ago. They showed that the total corona energy losses per unit area and time vary from about 30,000 to almost $10 \times 10^6 \, \mathrm{erg \, cm^{-2} \, s^{-1}}$, depending on the location. By way of comparison, the total amount of energy radiated into space by the Sun is about 63 billion (6.3×10^{10}) in the same units – just divide the solar luminosity by the Sun's area. So comparatively small amounts of energy must be supplied to heat the chromosphere and corona, it is just that sunlight cannot supply it.

The heating requirements vary between coronal structures – the active regions, the quiet Sun, and coronal holes (Table 4.1). The active regions have the hottest temperatures and greatest density of coronal material, so they require the most heating. The magnetic loops in the so-called quiet Sun outside active regions are slightly cooler and less dense, and coronal holes contain the coolest and most rarefied coronal material, since it flows out along the open magnetic fields there rather than being constrained within coronal loops.

The heating processes are selective in both space and time. They are correlated with magnetic structure, occur over a wide range of spatial scales, change over both short and long intervals of time, and depend on how magnetically active the Sun is at the moment. And it is likely that different heating mechanisms dominate at different places or times.

The way in which motions in and below the photosphere provide heat depends on the timescale, or how rapidly their energy is coupled to the solar atmosphere. If the motions are relatively rapid, changing within a few minutes, they can generate waves that propagate into the chromosphere or corona. Dissipation of wave energy is referred to as alternating current heating, or AC for short. Slower motions stress the overlying magnetic fields, twisting, braiding, and shearing them. The dissipation of magnetic stress is known as direct current heating, abbreviated DC.

Magnetic fields seems to play a fundamental role in channeling, storing, and transforming the energy into heat, supplying it on different timescales and sending it to various structures. When the magnetic geometry does not change, the magnetism plays a passive role, guiding the flow of charged particles, heat, and waves along the field lines. And when the magnetic configuration changes, the magnetism can play an active role by triggering instabilities and releasing stored magnetic energy through merging and reconnection of closed magnetic field lines.

Table 4.1 Coronal structures and their energy losses[a]

Feature	Largest extent (km)	Coronal temperature ($10^6 \, \mathrm{K}$)	Coronal electron density ($10^8 \, \mathrm{cm^{-3}}$)	Coronal energy loss ($\mathrm{erg \, cm^{-2} \, s^{-1}}$)
Active regions	10,000	≥ 2.0	10–100	0.2–2×10^6
Quiet Sun	–	≈ 1.5	1.9 ± 0.8	0.1–2×10^5
Coronal holes	900,000	≈ 1.0	0.8–1.1	0.5–1×10^4

[a]Adapted from Aschwanden and Acton (2001) and Aschwanden (2001a, b). Energy losses are in units of $\mathrm{erg \, cm^{-2} \, s^{-1}}$, to convert to $\mathrm{J \, m^{-2} \, s^{-1}}$ divide by 10^3.

As pointed out by James Klimchuk in his 2006 review, finding out where the energy is coming from is just the first step in solving the heating problem. Once the source of energy has been identified, one must determine where and how that energy is transported and dissipated as heat to different structures in the solar atmosphere, specify how the local gases respond to this heating, and identify the observed signatures of the various processes in the Sun's radiation spectrum. So it is not a simple problem!

But major discoveries have been made using the Soft X-ray Telescope aboard *Yohkoh*, and the extreme-ultraviolet telescopes on board the *SOlar and Heliospheric Observatory*, abbreviated *SOHO*, and the *Transition Region And Coronal Explorer*, or *TRACE* for short. And important contributions are being made using the visible light, extreme-ultraviolet, and X-ray telescopes on board *Hinode*. The instruments on these spacecraft have detected the signatures of sound and magnetic waves in the dynamic, changing solar atmosphere, shown that warm coronal loops in active regions contain numerous narrow strands that are intermittently heated from their lower regions, discovered ubiquitous, ever-changing loops that interact to release magnetic energy in the transition region or low corona, and found the signatures of unexpected heating processes in coronal holes, where the temperature depends on both the direction and the mass of the particles.

All of these discoveries will be discussed in the rest of this chapter, showing how they are related to heating the 10,000-K chromosphere, the multi-million Kelvin coronal loops in active regions, the million-Kelvin corona in "quiet" regions outside active regions, and the heating of the extended corona in coronal holes.

A number of plausible theories have been proposed to explain many of the detailed aspects of the coronal heating problem. But they are often exceptionally mathematical and thus outside the scope of this book. The serious student or curious reader who wants to delve into the topic in greater detail will want to consult any of the numerous reviews of these subjects prepared by professionals in the field (Focus 4.1) or read Markus Aschwanden's (2004, 2006) comprehensive book *Physics of the Solar Corona: An Introduction*, which includes topics such as instabilities and heating of plasmas, MHD oscillations and waves, and magnetic reconnection and particle acceleration processes.

Focus 4.1

Expert Reviews about Heating the Solar Atmosphere

Professional solar astronomers and astrophysicists have reviewed important developments in our knowledge of chromospheric and coronal heating, often in technical terms. In alphabetical order, they include Markus J. Aschwanden, Arthur I. Poland, and Douglas M. Rabin's (2001) review of the new solar corona; Sir William I. Axford and colleagues (1999) review of heating in coronal holes; James A. Klimchuk's (2006) approach to solving the coronal heating problems; John L. Kohl, Giancarlo Noci, Steven R. Cranmer, and John C. Raymond's (2006) review of ultraviolet spectroscopy of the extended solar corona; Max Kuperus, James A. Ionson, and Daniel S. Spicer's (1981) review of the theory of coronal heating mechanisms; Dana W.

Longcope's (2005) review of topological methods for the analysis of solar magnetic fields; David J. McComas and colleagues' (2007) summary of unsolved problems in understanding coronal heating. Udit Narain and Peter Ulmschneider's (1990, 1996) reviews of chromospheric and coronal heating mechanisms; Leon Ofman's (2005) review of MHD waves and heating in coronal holes; Sami K. Solanki's (2006) review of the solar magnetic field; Robert William Walsh and Jack Ireland's (2003) review of the heating of the solar corona; George L. Withbroe and Robert W. Noyes (1977) review of mass and energy flow in the solar chromosphere and corona; and Jack Zirker's (1993) review of coronal heating.

The contributions of scientists working with specific spacecraft data are often presented at workshops that are subsequently published in book form or in special issues of a journal. An example is the proceedings of the *TRACE* workshop *Physics of the Solar Corona and Transition Region*, edited by Oddbjorn Engvold and John W. Harvey in collaboration with Carolus J. Schrijver and Neal E. Hulburt (2004). The proceedings of *SOHO* workshops that deal with the chromosphere, the corona, and the transition region between them include *SOHO-1: Coronal Streamers, Coronal Loops, and Coronal and Solar Wind Composition* edited by Clare Mattok (1992); *SOHO-2: Mass Supply and Flows in the Solar Corona* edited by Bernhard Fleck, Giancarlo Noci, and Giannina Poletto (1994); *SOHO-5: The Corona and Solar Wind Near Minimum Activity* edited by Olav Kjeldseth-Moe and Andrew Wilson (1997); *SOHO-8: Plasma Dynamics and Diagnostics in the Solar Transition Region and Corona* edited by Jean-Claude Vial and Brigitte Kaldeich-Schürmann (1999); *SOHO-11: From Solar Min to Max: Half a Solar Cycle with SOHO* edited by Andrew Wilson (2002); *SOHO-13: Waves, Oscillations and Small-Scale Transient Events in the Solar Atmosphere: A Joint View from SOHO and TRACE* edited by Hugette Lacoste (2004); *SOHO-15: Coronal Heating* edited by Robert William Walsh, Jack Ireland, Dorothea Danesy, and Bernard Fleck (2004); and *SOHO-17: 10 Years of SOHO and Beyond* edited by Huguette Lacoste (2006).

4.2 Wave Heating

Ludwig Biermann, Martin Schwarzschild, and Evry Schatzman first suggested heating of the solar atmosphere by sound waves in 1948–1949, independently, with further considerations by Peter Ulmschneider in 1971 and Max Kuperus, James A. Ionson and Daniel S. Spicer in 1981. We now know that turbulent motions in the convective zone generate sound waves that course through the Sun (Sect. 3.3), and in the late 1940s it was thought that the up and down motion of the piston-like convection cells, the granules, will generate a thundering sound in the overlying atmosphere, in much the same way that a throbbing high-fidelity speaker drives sound waves in the air.

The sound (acoustic) waves should accelerate and strengthen as they travel outward through the increasingly rarefied solar atmosphere, until supersonic shocks occur that resemble sonic booms of jet aircraft. These shocks would dissipate their energy rapidly, and perhaps generate enough heat to account for the high-temperature corona.

Acoustic waves propagate at the sound speed $c_s = (\gamma P/\rho)^{1/2}$, where P is the gas pressure, ρ is the mass density, and the index γ is 5/3 for a monatomic gas and one for an isothermal perturbation. And since the gas pressure is a linear function of the temperature, T, the speed of sound varies as the square root of the temperature, or $c_s \propto (T)^{1/2}$, and it has the approximate values of about $10\,\mathrm{km\ s^{-1}}$ in the photosphere and chromosphere and up to $150\,\mathrm{km\ s^{-1}}$ in the corona.

Although the majority of the sound waves are reflected back into the solar interior at the photosphere, and remain trapped inside the Sun, a small percentage of them manage to slip through the photosphere, dissipating their energy rapidly within the chromosphere and generating large amounts of heat there. In the 1990s, for example, Mats Carlsson and Robert F. Stein were able to use one-dimensional simulations to show how acoustic shocks can form bright grains in the chromosphere. And in 2001, Philip G. Judge, Theodore D. Tarbell, and Klaus Wilhelm used a rapid sequence of observations with instruments on board *SOHO* and *TRACE* to show that the 5-min photospheric oscillations drive similar oscillations in the overlying chromosphere. Then in 2005, Bart De Pontieu, Robert Erdélyi, and Ineke De Moortel used numerical models to show that photospheric oscillations with periods around 5 min can propagate up into the overlying solar atmosphere; magnetic flux tubes that are tilted away from the vertical focus and guide the oscillations up into the chromosphere.

So the low chromosphere may indeed be heated by sound waves that are generated in the convective zone and dissipated by shocks in the chromosphere. This method of chromosphere heating is generally consistent with the fact that other stars with outer convective zones have chromospheres, while stars that have no outer convective zones do not exhibit a detectable chromosphere. There is still however some controversy about the issue, for Astrid Fossum and Mats Carlsson (2005a) used observations and numerical simulations to assert that acoustic waves cannot constitute the dominant heating mechanism of the solar chromosphere, falling short by a factor of at least 10, while Sven Wedemeyer-Böhm, and colleagues used various comparisons between *TRACE* data, ground-based data, and three-dimensional simulations in 2007 to come to a contradicting conclusion that acoustic flux provides sufficient energy for heating the solar chromosphere in internetwork regions within supergranules. But in 2007, Mats Carlsson and colleagues combined *TRACE* and *Hinode* observations to conclude that the acoustic-wave energy flux in the internetwork chromosphere of the quiet Sun cannot provide sufficient heat, so the controversy continues, as described by Wolfgang Kalkofen in 2008.

The supergranules, discussed in Sect. 3.2, outline a network of large convection cells with magnetic fields concentrated at their boundaries. As first modeled by Roger Kopp and Max Kuperus in 1968 and by Alan H. Gabriel in 1976, the intense magnetism at the narrow edges of this magnetic network expands with height, opening up into the overlying solar chromosphere and forming a canopy, like the trees in a rain forest; the tree trunks correspond to the magnetic flux tubes in the photosphere that rise in the vertical direction and spread out like branch foliage in the chromosphere. In 2003, Carolus J. Schrijver and Alan M. Title presented a different picture, in which as much as half the magnetic flux over the very quiet photosphere resides inside the network edges, rather than the funnel-shaped magnetic canopy.

Exactly how do sound waves, which are normally trapped beneath the photo-sphere, manage to leak out into the chromosphere to heat it? The magnetic field can act like a trap door, opening or closing to the waves that are constantly pass-ing by, depending on the magnetic field inclination. In 2006, Scott W. McIntosh and Stuart M. Jefferies used observations of a sunspot with the *Transition Region And Coronal Explorer*, abbreviated *TRACE*, spacecraft to demonstrate that such a mechanism lets some of the sounds out, using the magnetic fields as a guide. And in the same year, Jefferies and co-workers showed that the inclined magnetic field lines at the crack-like boundaries of supergranule convection cells provide the por-tals through which magnetoacoustic waves can break out and propagate into the chromosphere.

Although the chromosphere is often described as a thin, uniform layer of gas, about 2,000-km thick, it contains a jagged, dynamic, ever-changing set of little ver-tical spikes, which were described as early as 1877 by Pietro Angelo Secchi, and named spicules by the Walter Orr Roberts in 1945. Each needle-shaped spicule shoots up to heights as tall as 15,000 km in 5 min, moving at speeds of up to $250 \mathrm{km} \mathrm{s}^{-1}$. The spicule then falls back down again, but new spicules continually arise as old ones fade away. If one includes these varying, jet-like spicules, the ex-tended, dynamic chromosphere might average about 5,000-km thick.

These days, the vertically oriented structures are called mottles on the disk, and spicules at the limb. The material in the jet-like spicules and mottles is relatively cool, no hotter than about 20,000 K.

When you observe the Sun in the red spectral line of hydrogen, the Balmer alpha transition at 656.3 nm, with high spatial resolution, hundreds of thousands of the evanescent, flame-like spicules are observed dancing in the chromosphere at any given moment, rising and falling like waves on the sea or a prairie fire of burning, wind-blown grass (Fig. 4.1). For more than a century, no one knew for certain just what causes the upward-moving spicules, but the mystery now seems to have been solved. There were two clues to the solution. First, the spicule lifetimes are compa-rable to the 5-min period of the photosphere oscillations, and second, the spicules consist largely of ionized material that will follow the direction of magnetic field lines. In 2004, Bart De Pontieu, Robert Erdélyi, and Stewart P. James used a combi-nation of modeling and observations with the Swedish 1-m Solar Telescope to show how the 5-min solar oscillations leak sufficient energy along inclined magnetic fields into the chromosphere to power shocks that drive upward flows and form spicules. There are still some residual uncertainties about these new results, at least for some members of the solar physics community, but they do provide a startling new ap-proach to the origin of spicules.

And even more recently, De Pontieu and co-workers report the detection of two kinds of spicules with the Solar Optical Telescope aboard *Hinode*, distinguished by their dynamics and timescales. One type moves up and down on time scales of 3–5 min, driven by shock waves when global oscillations leak into the chromosphere along magnetic field lines. A second type of jet-like spicules is more dynamic, with

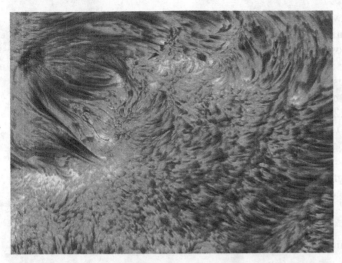

Fig. 4.1 Spicules. Thousands of dark, long, thin spicules, or little spikes, jet out of this high-resolution image of a solar active region, taken on 16 June 2003 on the blue-shifted wing of the Balmer hydrogen-alpha transition at 656.3-nm line with the Swedish 1-m Solar Telescope, abbreviated SST, on the Canary Island of La Palma, Spain. Layered, needle-shaped spicules (*right side*), each about a kilometer wide, shoot out to more than 15,000 km. The relatively narrow jets of gas are moving out of the solar chromosphere in magnetic channels, or flux tubes, at supersonic speeds of up to 250 km s^{-1}. Time-sequenced images have shown that these spicules rise and fall in about 5 min, driven by sound waves beneath them. The dimensions of the image are 65,000 × 45,000 km. (Courtesy of SST, Royal Swedish Academy of Sciences, and LMSAL.)

typical lifetimes of 10–60 s; they seem to be energized by magnetic reconnection and may be associated with rapid heating in the chromosphere and above.

In 1978, R. Grant Athay and Oran R. White, and independently by Elmo C. Bruner, used observations from the eighth *Orbiting Solar Observatory*, abbreviated *OSO 8*, to show that sound waves do not transport significant amounts of energy beyond the chromosphere. These measurements indicated that the sounds might have enough energy to heat the chromosphere, at a temperature of about 10,000 K, but that the inferred energy is far less than that required to heat the corona. The acoustic waves and shocks that are so effective in heating the chromosphere are severely damped by the time they reach the low corona. The sounds just cannot reach that far, being reflected inward by the steep temperature and density gradient in the transition region between the chromosphere and the corona.

For coronal heating, another kind of waves must be considered, the magnetic waves that can propagate into the corona and carry energy into it. The Sun's ever-changing coronal magnetic fields are always being jostled, twisted, and stirred around by motions deep down inside the Sun where the magnetism originates. A tension acts to resist the motions and pull the disturbed magnetism back, generating waves that propagate along magnetic fields, somewhat like a vibrating string. These waves do not form shocks, and once generated they can propagate large distances, directing their energy along open magnetic fields into the overlying corona.

Such waves are now called Alfvén waves, after Hannes Alfvén who first described them mathematically in 1942 and argued in 1947 that they might heat the corona. Ronald G. Giovanelli in 1949, Jack H. Piddington in 1956, and Donald E. Osterbrock in 1961 subsequently discussed the heating of the chromosphere and corona by Alfvén waves.

Alfvén pioneered a new field of study with the ponderous name of magnetohydrodynamics (Focus 4.2), and was awarded the Nobel Prize in Physics in 1970 for his discoveries in it. In technical jargon, the Alfvén waves are incompressible transverse oscillations that propagate along magnetic field lines with magnetic tension as the restoring force. They propagate at the Alfvén velocity $v_A = B/(4\pi\rho)^{1/2}$ for a magnetic field of strength B and a mass density ρ. This velocity has an approximate value of $10\,\mathrm{km\,s^{-1}}$ in the photosphere and up to $2,000\,\mathrm{km\,s^{-1}}$ in the corona.

Focus 4.2

Magnetohydrodynamics

The theory that deals with the interaction of a hot gas, or plasma, and a magnetic field is called magnetohydrodynamics, or MHD for short. As the name suggests, the equations are a combination of those of electromagnetism and fluid mechanics. The plasma is treated as an electrically conducting fluid of conductivity, σ, and magnetic diffusivity, $\eta = 1/(\mu_0\sigma)$, where μ_0 is the permeability of free space. The time dependence of the magnetic field, $\mathbf{B}$, is given by the induction equation:

$$\frac{\partial \mathbf{B}}{\partial t} = \nabla \times (\mathbf{v} \times \mathbf{B}) + \eta \nabla^2 \mathbf{B},$$

where the symbol ∂ denotes the partial derivative with respect to time, t, the ∇ indicates a three-dimensional spatial gradient, and $\mathbf{v}$ is the bulk speed of the fluid. This expression implies that the magnetic field changes in time due to transport of the magnetic field with the plasma (the first term on the right) and diffusion of the magnetic field through the plasma (the second term). For a fluid at rest, with velocity $\mathbf{v} = 0$, this relation describes magnetic diffusion in a time $\tau = L^2/\eta = \mu_0\sigma L^2$ over a linear scale L. The diffusion time for the Sun as a whole is larger than its age, so it takes a very long time for magnetic fields to leak out of a star by diffusion.

In the case of zero resistivity or "infinite conductivity," we have a perfectly conducting medium and the magnetic field satisfies the relation:

$$\frac{\partial \mathbf{B}}{\partial t} + \nabla \times (\mathbf{v} \times \mathbf{B}) = 0.$$

This equation expresses a condition in which the magnetic field is tied to, or frozen within, the plasma and moves with it.

For most of the Universe, the second term in our first equation is very much smaller than the first term, so the second equation is a good approximation. An important exception is in singularities called current sheets, where the magnetic gradient and electric current are extremely large. In such current sheets, the magnetic field lines can merge together, break and reconnect by slipping through the plasma and, in the process, magnetic energy is converted to heat, kinetic, and fast-particle energy. This process, called magnetic reconnection, is important in heating the solar corona and in energizing solar flares.

Another important equation of magnetohydrodynamics is the equation of motion:

$$\rho \frac{d\mathbf{v}}{dt} = -\nabla P + \mathbf{j} \times \mathbf{B},$$

where $\mathbf{j} \times \mathbf{B}$ is the force that the magnetic field exerts on a plasma of density, ρ, the pressure is denoted by P and the velocity by v, and the electric current is given by Ampere's law:

$$\mathbf{j} = \frac{1}{\mu_0} \nabla \times \mathbf{B}.$$

This equation of motion neglects the Sun's gravity, which is probably okay for the corona, but not for the photosphere or chromosphere.

The induction equation and the equation of motion can be combined to describe perturbations in density that act as waves, with a velocity, v, given by

$$v = (c_s^2 + v_A^2)^{1/2},$$

where c_s is the velocity of sound, and the Alfvén velocity, v_A, is given by

$$v_A = B/(\mu_0 \rho)^{1/2}.$$

The waves represent alternating compression and rarefaction of the gas and field. They are called fast magnetoacoustic waves since they are faster than both the sound and the Alfvén waves.

When the velocity of sound, c_s, is much smaller than the Alfvén velocity, v_A, or when $c_s \ll v_A$, we have a compressional Alfvén wave. These waves may propagate in a direction perpendicular to the magnetic field with gas particles oscillating in the direction of propagation.

A more general relation describes fast, slow, and Alfvén magnetohydrodynamic waves. When the direction of wave propagation is parallel to the magnetic field, one can have the slow wave and the Alfvén wave moving at the Alfvén velocity, where the particles oscillate in transverse motion to both the magnetic field and the direction of propagation.

Alfvén waves have been directly measured from spacecraft cruising through interplanetary space far from the Sun, detecting the waves that sweep by their instruments. In these locations, the solar magnetic field has been carried out into space, with one end tied to the Sun and the other end extending out to far beyond the planets, and the waves travel out along the open-ended magnetic fields. Plasma and magnetic field detectors aboard *Mariner 5* detected the Alfvén waves on its way to Venus in 1967, and they were subsequently detected from the two *Helios* spacecraft in the ecliptic and from *Ulysses* above the Sun's polar regions. But such waves have only recently been observed in the lower solar atmosphere where they are presumably generated.

The SUMER and UVCS instruments aboard *SOHO* have provided evidence for Alfvén waves close to the Sun; these results have been included in Steven R. Cranmer and Adriaan A. van Ballegooijen's 2005 summary of Alfvén wave measurements. Observations from *TRACE* have additionally demonstrated the widespread presence of magneto-hydrodynamic, or MHD, waves, both longitudinal and transverse, many of which travel through the chromosphere with unanticipated efficiency. And in 2007, several research teams reported evidence of Alfvén waves using instruments aboard *Hinode*, in articles by Jonathan Cirtain and colleagues, Takenori J. Okamoto and co-workers, and Bart De Pontieu and colleagues. These waves could potentially heat the corona to extreme temperatures by releasing energy as they travel outward from the Sun along magnetic field lines. Both types of spicules detected in the chromosphere by De Pontieu and colleagues, for example, are observed to carry Alfvén waves with significant amplitudes of about 20km s^{-1}.

The important question is whether or not Alfvén waves propagating through the corona dissipate sufficient energy to heat it. Steven Tomczyk and co-workers reported in 2007 the detection of numerous Alfvén waves in the corona itself, using an instrument at the National Solar Observatory in New Mexico. Their estimate of the energy carried by the Alfvén waves that they spatially resolved indicated that they are too weak to heat the solar corona; however, unresolved ones might carry enough energy. In contrast, De Pontieu's team showed in 2007 that the energy associated with the Alfvén waves they detected in the chromosphere may be sufficient to heat the corona and accelerate the solar wind.

And in 2005, Cranmer and van Ballegooijen teamed up with Richard J. Edgar to describe coronal heating from Alfvén waves along open magnetic field lines that reach from the photosphere into interplanetary space. As with earlier work by Yi-Ming Wang in 1993, the coronal base temperature increases with decreasing magnetic flux-tube divergence rate. They were continuing a long tradition of using Alfvén waves to produce coronal heating at places where open magnetic fields can channel the waves into the distant corona, beginning with John W. Belcher in 1971 and Belcher and Stanislaw Olbert in 1975. These investigations will be further discussed in Sect. 4.5 on the heating of polar coronal holes, but in the meantime let us consider the closed magnetic loops, which can contain the hottest coronal material requiring the greatest heat.

4.3 Heating Coronal Loops in Active Regions

An active region in the solar atmosphere develops when strong magnetic fields emerge from the inside of the Sun, and break through the photosphere in large, adjacent patches of positive and negative magnetic polarities, often marked by sunspots. Magnetic loops join these regions of opposite magnetic polarity, rising into the corona from the footpoint of one magnetic polarity and turning back into the other one. Hot, ionized gas is confined within the coronal loops, and when filled, these active-region coronal loops therefore shine brightly in extreme-ultraviolet and X-ray radiation.

The Soft X-ray Telescope aboard *Yohkoh* sharpened our understanding of the ubiquitous coronal loops, showing that they provide the woven fabric of the active corona (Fig. 4.2). The intense X-ray emission, at temperatures of $2-4 \times 10^6$ K, outlines the magnetic shape and structure of the loops, and demonstrates that the dynamic, ever-changing corona has no permanent features.

The Extreme-ultraviolet Imaging Telescope, or EIT, aboard the *SOlar and Heliospheric Observatory*, abbreviated *SOHO*, confirmed that the magnetically confined gas in solar active regions is extremely dynamic and time-variable, changing on time scales from hours to days. So the apparently steady corona that is frozen into a single image or photograph is an illusion. Using such a single picture to represent the corona provides no information about its varying spatial and temporal complexity.

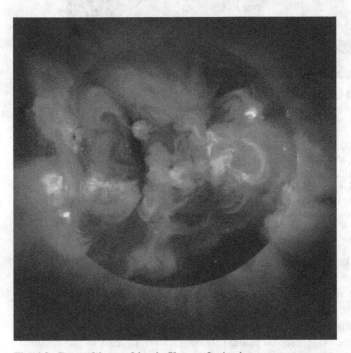

Fig. 4.2 Coronal loops shine in X-rays. Ionized gases at a temperature of a few million Kelvin produce the bright glow seen in this X-ray image of the Sun. It shows magnetic coronal loops that thread the corona and hold the hot gases in place. The brightest features are called active regions and correspond to the sites of intense magnetic fields. The Soft X-ray Telescope (SXT) aboard the Japanese *Yohkoh* satellite recorded this image of the Sun's corona on 1 February 1992, near a maximum of the 11-year cycle of solar magnetic activity. Subsequent SXT images, taken about 5 years later near activity minimum, show a remarkable dimming of the corona when the active regions associated with sunspots have almost disappeared, and the Sun's magnetic field has changed from a complex structure to a simpler configuration. (Courtesy of Gregory L. Slater, Gary A, Linford, and Lawrence Shing, NASA, ISAS, the Lockheed-Martin Solar and Astrophysics Laboratory, the National Astronomical Observatory of Japan, and the University of Tokyo.)

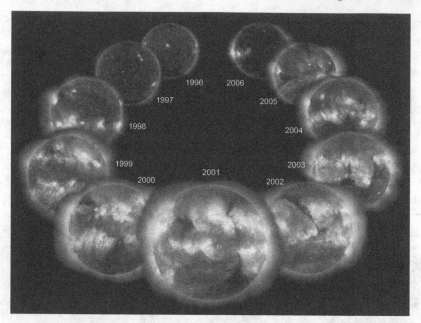

Fig. 4.3 Loops during the solar cycle. A complete 11-year solar cycle of magnetic activity observed from the Extreme-ultraviolet Imaging Telescope, abbreviated EIT, on the *SOlar and Heliospheric Observatory*, or *SOHO* for short. These images were taken in the light of 14 times ionized
iron, designated Fe XV, at 28.4 nm, formed at a temperature of about 2.5×10^6 K. In late 1996,
shortly after its launch, *SOHO* was able to observe the last activity minimum, when there were
hardly any active regions. The minimum was followed by a rapid rise in solar activity, peaking at
maximum in 2001 and 2002, with intense extreme-ultraviolet radiation from million-degree gas
constrained in active-region coronal loops. Activity levels slowly declined since then. These images were picked by Steele Hill to illustrate the relative activity of the Sun. (Courtesy of the *SOHO*
EIT consortium. *SOHO* is a project of international cooperation between ESA and NASA.)

SOHO's EIT has kept the low corona and underlying transition region under
careful watch for more than a decade, demonstrating that the abundance of active
regions, and the coronal loops they contain, rises and falls in step with the 11-year
cycle of solar magnetic activity, the sunspot cycle, becoming more intense and numerous near the peak of the cycle (Fig. 4.3). At activity maximum, coronal loops
in active regions are the brightest and hottest things around, at least in the corona,
dominating the Sun's extreme-ultraviolet and X-ray emission and making up about
80% of the total coronal heating energy.

The EIT takes full-disk images at three lines of ionized iron, Fe IX, Fe XII, and
Fe XV, and one line of ionized helium, He II; these are the permitted lines emitted
by ionized atoms, sensitive to temperatures from 60,000 to 2.1×10^6 K (Table 4.2),
and not the forbidden lines detected at visible wavelengths.

Because a given stage of ionization occurs within a narrow range of temperature,
the different spectral lines can be used to tune in coronal loops at particular temperatures. They have revealed that cool and hot loops are found side by side and
under and over each other, all over the Sun, each with its own unique temperature

Table 4.2 Some prominent solar emission lines in the transition region and low corona observed with the EIT instrument aboard *SOHO*[a]

Wavelength (nm)	Emitting ion	Formation temperature (Kelvin)
17.11	Iron, Fe IX, and Fe X	1,000,000
19.51	Iron, Fe XII	1,400,000
28.42	Iron, Fe XV	2,100,000
30.38	Helium, He II	60,000

[a]Subtract 1 from the Roman numeral to get the number of missing electrons. The wavelengths are in nanometers, abbreviated nm, where $1\,nm = 10^{-9}$ m. Astronomers sometimes use the Ångström unit of wavelength, abbreviated Å, where 1 Å $= 10^{-10}$ m $= 10\,nm$.

and location. In 1997, for example, Andre Fludra and colleagues used the Coronal Diagnostic Spectrometer, or CDS for short, on *SOHO* to show that coronal loops with different temperatures can co-exist in active region, sometimes very close to each other; the magnetic fields of individual loops apparently insulate them from adjacent ones and contain gas heated to a specific temperature for that particular loop.

So what makes these coronal loops so hot, and where does their heat come from? An early clue was provided in 1992 when Toshifumi Shimizu and co-workers showed that transient X-ray brightening is common in active regions, suggesting that these coronal loops are far from static. Six years later, Olav Kjeldseth-Moe and Paal Brekke used *SOHO's* CDS to detect large Doppler shifts, indicating fast motions and rapid time variation in the coronal loops, and suggesting that plasma flows play an important role in heating them. And James Klimchuk supplied another hint in 2000, demonstrating that the coronal loops observed by *Yohkoh* are approximately constant in cross-section. *Hinode* observations of transient active-region heating, reported in 2007 by Hui Li, Harry P. Warren, and their colleagues, indicate that emerging magnetic flux in the photosphere may produce the X-ray brightening, contributing heat by magnetic reconnection. Other *Hinode* results, reported by George Doschek et al. (2007a, b) and David H. Brooks et al. (2007), demonstrate the existence of strong non-thermal motions in active regions and approximately uniform loop cross sections that are difficult to reproduce with steady-state heating models, for both low- and high-temperature loops.

The widespread, dynamic evolution of the coronal magnetic field and the intermittent heating of the solar corona were established when the *Transition Region And Coronal Explorer*, abbreviated *TRACE*, was used to zero in on active regions. Like the EIT on *SOHO*, the *TRACE* telescope observes the Sun in the extreme-ultraviolet radiation of specific spectral lines sensitive to a wide range of temperatures (Table 4.3); but unlike EIT, which images the entire solar disk, *TRACE* observes specific regions on the Sun with higher spatial and temporal resolution, detecting fine details that could not be seen from previous spacecraft whose instruments integrated or smeared them out.

With its angular resolution of just 1.25 s of arc, or about 900 km on the Sun, *TRACE* demonstrated that the corona in active regions contains long, thin loops that

Table 4.3 *TRACE* spectral regions and temperature range[a]

Central wavelength (nm)	Ion	Region of solar atmosphere	Temperature (Kelvin)	Temperature range (Kelvin)
250	Continuum	Photosphere	5,000	4,000–6,300
170	Continuum	T_{min}/Chromosphere	6,300	4,000–10,000
160	CI, FeII, CIV	T_{min}/Chromosphere	6,300	4,000–250,000
155	CIV	Transition Region	126,000	63,000–250,000
121.6	Hydrogen Ly α	Chromosphere	16,000	10,000–32,000
28.4	Fe XV	Corona	2,000,000	1,000,000–5,000,000
19.5	Fe XII	Corona	1,259,000	100,000–2,500,000
17.1	Fe IX/X	Corona	795,000	200,000–2,000,000

[a]Adapted from the *TRACE* 2003 Senior Review Proposal. Subtract 1 from the Roman numeral to get the number of missing electrons. The wavelengths are in nanometers, abbreviated nm, where $1 \, nm = 10^{-9} \, ms$. Astronomers sometimes use the Ångström unit of wavelength, abbreviated Å, where $1 \, Å = 10^{-10} \, m = 10 \, nm$.

Fig. 4.4 Loops heated from below. Tall, thin coronal loops observed from the *Transition Region And Coronal Explorer*, abbreviated *TRACE*, satellite, taken on 9 August 1999, in the 17.1-nm passband sensitive to a temperature of about 1 million degrees. High arching loops stand out, to a height of approximately 120,000 km. The loops are visible along their entire length, and comparisons with the 19.5-nm observations indicate the temperature varies little along them. The fact that the temperature is so nearly constant along the length requires that most of the heating is concentrated low down, in the bottom 15,000 km or so. (Courtesy of the *TRACE* consortium and NASA. *TRACE* is a mission of the Stanford-Lockheed Institute for Space Research, and part of the NASA Small Explorer program.)

Fig. 4.5 Multiple narrow strands. The *Transition Region And Coronal Explorer*, abbreviated *TRACE*, demonstrated that the corona in active regions contains long thin strands. The multiple narrow loops stand alone without braiding, twisting, or otherwise interacting with neighboring strands. This *TRACE* image was taken in the 17.1-nm passband sensitive to a temperature of about 1×10^6 K. (Courtesy of the *TRACE* consortium and NASA. *TRACE* is a mission of the Stanford-Lockheed Institute for Space Research, and part of the NASA Small Explorer Program.)

break through the photosphere in multiple narrow strands or threads (Figs. 4.4, 4.5 and 4.6). As reported by Carolus Schrijver and his colleagues in 1999, the numerous thin, bright strands arch through the active-region corona, each connecting two photospheric regions of opposite magnetic polarity. The widths of the strands are often at or just above the instrumental resolution, but when resolved the loop cross-sections show no significant variation with height.

Fig. 4.6 Loops at the edge. The highly structured corona observed at the edge of the Sun using the *Transition Region and Coronal Explorer*, abbreviated *TRACE*, satellite on 21 June 2001. This image was taken in the 17.1-nm passband sensitive to a temperature of about 1×10^6 K. (Courtesy of the *TRACE* consortium and NASA: *TRACE* is a mission of the Stanford-Lockheed Institute for Space Research, and part of the NASA Small Explorer program.)

Narrow loops with significantly different temperatures are found within active regions, each standing alone without braiding, twisting, or otherwise interacting with their neighboring loops. High, cool loops, with temperatures of $1–2 \times 10^6$ K arch over hotter ones at $3–5 \times 10^6$ K. And each strand appears to have its own unique temperature.

The thin coronal loops seen in *TRACE* images of active regions continuously change in shape, with the hot, ionized gas moving back and forth within its magnetic cage. Each narrow strand is typically detected for only a few hours, while new ones constantly appear, as density and temperature evolve in response to the changing heating.

More than three decades ago, in 1978, Robert Rosner, Wallace H. Tucker, and Giuseppe Vaiana introduced one of the first analytical models for heating coronal loops in active regions. They assumed that the loops seen in *Skylab* X-ray photographs are in hydrostatic equilibrium with spatially uniform heating along an unchanging loop, deriving a scaling law between the loop temperature, pressure, and size that fit the observations. This then seemed like a reasonable approach, since the observed coronal loops appeared to be steady or unchanging with respect to the relevant physical time scales, such as the radiative and cooling times of roughly 10 min. Similar scaling relationships were derived and tested over the subsequent two decades using statistical samples of hot soft X-ray loops, with temperatures of a few million Kelvin, observed with the Soft X-ray Telescope aboard *Yohkoh*.

But now, on the basis of subsequent *SOHO*, *TRACE*, and *Hinode* results, the underlying assumption of hydrostatic equilibrium has been challenged, at least for warm loops with temperatures of about 1×10^6 K. Many of these loops are found to be far from equilibrium and are not in a steady state, thus requiring dynamic models. In 1999, for example, Dawn D. Lenz and her colleagues showed that long-lived loops observed with *TRACE* are incompatible with the traditional steady, uniformly heated loops model. And Markus J. Aschwanden (2001a, b) examined soft X-ray and extreme-ultraviolet observations from *Yohkoh, SOHO*, and *TRACE*, concluding that the observed coronal loops in active regions are overdense, by about an order of magnitude, when compared to the equilibrium, steady-state solutions for loop density and temperature. One way of explaining this apparent excess is to abandon the assumption of uniform heating, and supplying the extra density by upflows of heated material from the footpoints of the loops in the chromosphere.

The heating of these active-region loops may therefore be intermittent in both space and time, rather than steady. Upward pulses of hot material are apparently pumping up the loops from below, until they are crammed so full that they are on the verge of bursting apart at the seams. But then in minutes or tens of minutes the heating stops, and essentially shuts down over the entire loop volume for hours at a time.

The *TRACE* observations, described by Carolus J. Schrijver in 2001, indicate that whenever the heating is interrupted – which happens frequently – the loops cool down and the gas filling the overdense loops, no longer supported by its own pressure, rains back down into the transition region and chromosphere. In some of the new models, gas is heated to coronal temperatures in the chromosphere or transition region, with subsequent impulsive upflow into multiple narrow strands. The bright active-region loops are in the process of filling by heated upflows; and when they disappear from view, the material rains back down to the place it came from.

In 2008, Markus J. Aschwanden and David Tsiklauri provided another approach to the overdensity, or overpressure, of warm coronal loops observed at extreme-ultraviolet wavelengths. They conclude that steady-state equilibrium solutions cannot explain all of the observed loop physical parameters, but they can be explained by a non-stationary and non-equilibrium cooling process after the heating has ceased. This solution avoids discussion of how the loops are heated, by upflow from loop footpoints, by waves, or by some other mechanism.

The heating source, which is still not understood in detail, is presumed to be magnetic in origin, and is thus related to the larger questions of what is energizing the ubiquitous coronal loops found outside of active regions, which we will now consider.

4.4 Heating the "Quiet" Corona

Historically, the areas outside active regions were called quiet regions, but we now know that the term "quiet Sun" is only justified in relative terms. And for a working definition, we will describe the quiet Sun in terms of all the closed magnetic structures in the corona or below, except those in active regions and excluding the magnetically open coronal holes. These ubiquitous loops are the dominant structural element in the quiet solar atmosphere, popping out all over the Sun.

First results from the Extreme-ultraviolet Imaging Spectrometer aboard *Hinode*, for example, indicate that quiet-corona regions can exhibit high outflow velocities, on the order of $100 \mathrm{km \, s^{-1}}$, and that the quiet solar atmosphere is composed of both small, cool loops, with temperatures of about $0.4 \times 10^6 \mathrm{K}$, and larger, high-temperature ones at $1–2 \times 10^6 \mathrm{K}$. Kenneth P. Dere and colleagues and Keiichi Matsuzaki and co-workers, respectively, reported these results in 2007.

And although active regions are the sites of the most intense outbursts on the Sun, known as solar flares, the quiet Sun exhibits all sorts of small-scale explosive events observed at soft X-ray or extreme-ultraviolet wavelengths. There are X-ray jets and bright points, for example, and bright extreme-ultraviolet events that flash on and off. Millions of them have been observed from instruments aboard *Yohkoh*, *SOHO*, *TRACE*, and *Hinode*. And they have received all sorts of designations, such as jets, blinkers, explosive events, microflares, and nanoflares.

4.4.1 Heat from Jets, Bright Points, Blinkers, Explosive Events, and Interacting, Non-Flaring Loops

The soft X-ray telescope aboard *Yohkoh* showed that the gas and magnetic fields in the corona are forever changing in shape, intensity, and location throughout the

quiet Sun outside active regions. The hot gas outlines magnetic loops that continuously shift around, become twisted up or deformed, and break up and form new
connections, responding to internal motions. Coronal loops can suddenly appear out
of nowhere, filling up with superheated material. Hot, X-ray emitting gas can be propelled to remote locations within well-collimated jets, even across the solar equator
and to or from the polar coronal holes; these magnetic conduits are sometimes as
long as the Sun is wide.

These *Yohkoh* observations demonstrated that coronal magnetic energy could be
released, suddenly and catastrophically, when opposing magnetic fields merge and
cancel each other. The magnetic fields move into each other, but never end. They
reform or reconnect in new magnetic orientations and in so doing release energy into
the corona. Such magnetic reconnections may occur when newly emerging magnetic
fields rise through the photosphere to encounter pre-existing ones in the corona, or
when the existing coronal loops are forced together. In either case, the moving coils
are charged with pent-up energy and ready to erupt.

As an example, Kazunari Shibata and his colleagues demonstrated, in the 1990s,
that many collimated X-ray jets display the morphology and physical characteristics
of magnetic reconnection. They are sometimes propelled by magnetic connections
involving new magnetic flux, coming up through the photosphere, and pre-existing
open or closed magnetic fields (Fig. 4.7). Roughly a decade later, in 2007, Shibata
and his colleagues reported *Hinode* observations of ubiquitous jets in the chromosphere, detected in the ionized calcium line, Ca II H, at 396.85 nm. They often exhibit inverted Y-shapes implying magnetic reconnection of an emerging magnetic
dipole with a pre-existing unipolar, or open, magnetic field region.

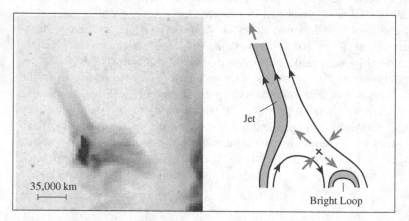

Fig. 4.7 X-ray jet. This figure shows how an X-ray jet (*left*) is produced by an emerging magnetic loop that connects with pre-existing magnetism (*right*). This magnetic configuration has been
called the anemone jet, since the active region at the footpoint of the jet looks like a sea anemone.
The collimated X-ray jets can shoot across large distances in the solar corona, sometimes nearly all
the way across the visible solar disk. The Soft X-ray Telescope on *Yohkoh* took the X-ray image.
[Courtesy of Kazunari Shibata, adapted from T. Yokoyama and K. Shibata (1995).]

The ubiquitous X-ray bright points, first observed extensively by Leon Golub and his colleagues during the *Skylab* mission (Sect. 2.5), also apparently result from magnetic reconnection in the low corona. Unlike sunspots and active regions, X-ray bright points are uniformly distributed over the Sun, even appearing at the poles and in coronal holes. However, like coronal loops in active regions, the X-ray bright points occur above regions of opposite magnetic polarity in the photosphere. The X-Ray Telescope aboard *Hinode* has resolved the so-called points into small-scale magnetic loops. About one of the loop-like bright points appears per hour averaged over the whole Sun.

The theory of magnetic reconnection was initiated in the 1950s and 1960s to explain the awesome release of energy during solar flares (Sect. 6.10). Harry E. Petschek, Peter A. Sweet, and Eugene Parker each made important contributions during that time, describing fast or slow reconnection. In both scenarios, oppositely directed field lines merge together and are effectively cut at the place where they touch, rejoining into a lower-energy configuration. The new, reconnected field lines are sharply bent, and so experience a strong magnetic tension force that snaps them apart. This accelerates and hurls material in opposite directions, like squeezing a tube of toothpaste open at both ends.

Davina Innes and colleagues observed such bi-directional collimated jets in 1997 (Fig. 4.8). Their *SOHO* ultraviolet spectra of explosive events in the chromosphere showed emission that was Doppler shifted to both longer wavelengths

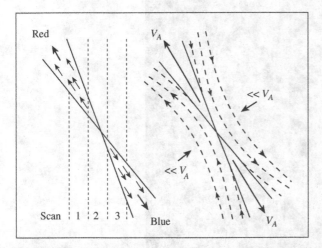

Fig. 4.8 Connecting magnetism. Ultraviolet spectral line observations with the *SOHO* SUMER instrument reveal Doppler shifts that change as the spectrometer scans across a jet structure (*left*). A magnetic reconnection model (*right*) explains the observations. It involves magnetic field lines (*dashed lines with arrows for magnetic direction*) and plasma flow (*solid arrows*). Material flowing inwards from each side, at speeds much less than the Alfvén velocity, V_A, carries anti-parallel magnetic fields together. At the center X, magnetic fields that point in opposite directions meet and join together. This catapults jets that move in both directions away from the point of magnetic contact at about the Alfvén velocity. [Adapted from Davina E. Innes, et al. (1997).]

and shorter ones. This indicated material emanating in opposite directions from a common point, presumably driven by the magnetic reconnection process. The structure of these jets evolved in the manner predicted by theoretical models of magnetic reconnection, suggesting that it is a fundamental process for heating material on the Sun. Oppositely directed "exhaust" jets attributed to such reconnection events have incidentally been detected in the solar wind just outside the Earth (Sect. 5.6).

When *SOHO* looks at the Sun in extreme-ultraviolet radiation, it is mottled all over with a granular appearance, like an orange, a stone beach, or a festering rash (Fig. 4.9). Each stone is a continent-sized bubble of hot gas, which flashes on and off in about 10 min and reaches temperatures of several hundred thousand to a million Kelvin. About 3,000 of these brightenings, known as blinkers, are seen erupting all over the Sun, including the darkest and quietest places at the solar poles. But even though there are a lot of these blinkers, their total thermal energy is still significantly less than that required to fully heat the corona.

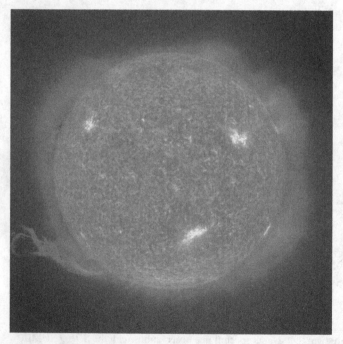

Fig. 4.9 Bright spots. Extreme-ultraviolet image of the low transition region at temperatures of 60,000–80,000 K, taken with the Extreme-ultraviolet Imaging Telescope, or EIT for short, aboard the *SOlar and Heliospheric Observatory*, abbreviated *SOHO*. It shows bright spots all over the Sun that can contribute to coronal heating. Two intense active regions with numerous magnetic loops are also seen, as well as a huge eruptive prominence at the solar limb (*left*). The image was taken on 14 September 1997, in the extreme ultraviolet resonance line of singly ionized helium, denoted He II, at 30.4 nm. (Courtesy of the *SOHO* EIT consortium. *SOHO* is a project of international cooperation between ESA and NASA.)

In another perspective, observations of extreme-ultraviolet lines, originating in the transition region at temperatures of about 100,000 K, have revealed the presence of localized hot spots that explode and hurl material outward at speeds of hundreds of kilometers per second. Guenter E. Brueckner and John-David F. Bartoe discovered such high-speed jets of matter in 1983 during brief rocket observations of the transition region. During the subsequent decade, Kenneth P. Dere and his colleagues at the Naval Research Laboratory carried out additional observations of the explosive transition-region jets, from rockets and the *Space Shuttle*, showing that they often occur in the magnetic network lanes. Jason G. Porter, David A. Falconer, Ronald L. Moore, and colleagues reasoned in the mid 1980s that ultraviolet flares might be frequent and powerful enough to heat the corona.

SOHO's ultraviolet and extreme-ultraviolet eyes have now shown that the tiny explosive events are going off all over the Sun and all the time – as many as 20,000 of them per minute, like an endless string of firecrackers. Richard A. Harrison and his colleagues have used *SOHO* instruments to show that the inner solar atmosphere remains a vigorous and violent place, even during the apparent lull at activity minimum when the whole Sun seems to sparkle in the ultraviolet light of thousands of localized bright spots formed at temperatures of up to a million Kelvin. Since these hot spots seem to be anchored within magnetic loops that move about and interact, magnetic reconnection is a prime candidate for powering the myriad of explosions.

Contorted and sheared coronal loops are commonly found in high-resolution *Yohkoh* soft X-ray images. Motions down below may imprint their twisted patterns up above, and the complex, oddly shaped magnetic geometry may be a signature of current flows in the corona. The free electrons in the coronal gas can move along the magnetic fields lines in thin sheets where oppositely directed magnetism is pressed together, creating very large electric currents that can generate their own magnetism and alter the overall magnetic shape. These electrical currents might heat the resistive gas in much the same way that currents heat the filaments of a light bulb.

New loops can rise up to interact with existing ones, releasing energy to heat the corona in non-flaring reconnection involving such electrical currents. In 1996, for example, Saku Tsuneta used *Yohkoh* soft X-ray observations to show how active regions in opposite hemispheres of the Sun can reconnect to form new transequatorial coronal loops with non-explosive coronal heating, and in 2000, Alexei Pevtsov showed that approximately one-third of all active regions on the Sun exhibit such soft X-ray, transequatorial loops that develop between existing active regions or between mature regions and newly emerging ones. In 2005, Dana W. Longcope and co-workers described the energy released when active regions, observed at extreme-ultraviolet wavelengths, emerge and reconnect in the corona, and in 2007, Loraine Lundquist and colleagues used observations with the X-ray telescope aboard *Hinode* to describe the interaction between emerging flux and ubiquitous large-scale loop systems that connect solar active regions to relatively distant reaches of the corona, both quiet and active. Such inter-connectivity, which commonly occurs over as much as one third of the solar disk area, may play a role in direct heating of the quiet solar corona by the interactions of coronal loops within it.

4.4.2 Nanoflares

Energy released by flare-like magnetic interactions might provide the searing heat of the corona. Already in 1960, Thomas Gold and Fred Hoyle argued that explosive solar flares are powered by stressed magnetic loops that interact, dissipating their energy in the corona. Then, in 1964, Gold suggested that the twisted magnetic field configurations might heat the corona. During the ensuing decades, Eugene N. Parker developed related heating models involving the winding and wrapping of coronal loops caused by the random, continuous motion of their footprints in the photosphere.

In 1988, Parker proposed, for example, that the solar X-ray corona is produced by the interaction of coronal loops that produce frequent numerous, small-scale explosions, which he dubbed nanoflares, which occur at random locations. In this picture, the coronal magnetic fields braid, twist and writhe in a perpetual dance, becoming entangled by the random motion of their footpoints caused by turbulent convection. Thus, energized from below, the ubiquitous coronal loops rise, fall, intertwine, and couple together to cumulatively produce the nanoflares that heat the quiescent corona. If nanoflares were true to their name, a billion of them would be required to release the same amount of energy as a normal flare.

But the fields cannot go on getting all wrapped up and entwined forever, entangling indefinitely and removing all signs of the coronal loops. As Parker proposed, the tangled fields eventually generate currents and reconnect into new configurations, releasing stored magnetic energy to power the nanoflares and disentangle the fields.

The nanoflare-heating concept has been a subject of "heated" controversy for about two decades, and it just would not go away. On the theoretical front, Peter J. Cargill examined its implications in 1994, proposing a related picture of multi-stranded, impulsively heated coronal loops. Cargill teamed up with James Klimchuk in 1997 to model coronal loops observed from *Yohkoh*, heated randomly and impulsively by nanoflares. And in 2005, Russell B. Dahlburg, Klimchuk, and Spiro K. Antiochos proposed a "switch-on" mechanism that can enable the magnetic energy release.

In the 1990s and early 2000s, the debate over the nanoflare heating scenario focused on extrapolations from observations of strong flares, to determine if undetected weaker ones might provide sufficient energy to heat and power the quiet corona. The Sun exhibits flares of all intensity, the frequency of occurrence increasing rapidly with decreasing strength, suggesting that the combined total energy released by the more numerous, but unseen, weaker flares might provide the necessary heat.

But as Hugh Hudson pointed out in 1991, the hypothetical heating flares have to be an entirely different breed from normal solar flares. Extrapolations from existing observations of solar flares to unobserved ones of smaller energy suggest that the low-energy variety does not occur often enough to collectively supply the corona's heat. Since that time various groups have characterized the distribution of energetic events observed in the corona or transition region, extrapolating the data to lower

energies and inferring their total contribution. In scientific parlance, it all depends
on the power law index of the flare frequency distribution. When there is a negative
index larger than 2.0 most of the energy is released by the sum of the smallest flares,
and when that index is less than 2.0 there is not sufficient combined power for the
small ones to do the job.

Arnold O. Benz and Säm Krucker have used *Yohkoh* and *SOHO* observations
to conclude, in 1998 and 1999, that microflares have a steep power law exponent
of −2.5. This suggested that the weak flaring multitudes could raise up to rule the
heating process. Other observers have disputed some of these conclusions. In 2000,
Clare E. Parnell and Peter E. Jupp used observations from *TRACE* to conclude
that there is insufficient energy input from nanoflares to explain the total energy
losses of the quiet corona, but they did not rule out the possibility that yet weaker,
and unobserved, events could heat it. And in 2002, Markus Aschwanden and Paul
Charbonneau gathered statistics of solar flares, microflares, and nanoflares over a
broad energy range, concluding that the power law index falls below the threshold
required to sustain appreciable coronal heating by nanoflares.

The nanoflare model predicts that coronal loops are collections of unresolved
strands, impulsively heated to high temperatures, but according to Aschwanden's
account in 2008 (a, b), this is in disagreement with *TRACE* observations of resolved
loop widths. He proposes instead that the hypothetical nanoflare events be relocated
from the low corona down to the chromosphere and transition region, where high-
resolution magnetograms reveal many low-lying magnetic loops.

4.4.3 The Magnetic Carpet

At any one time, the Michelson Doppler Imager, abbreviated MDI, instrument
aboard *SOHO* reveals a salt-and-pepper sprinkling of about 50,000 small magnetic
bipolar regions in the photosphere. Each one of them has a north and south pole connected by a rising magnetic arch. They together form a complex web of low-lying
coronal loops known as the magnetic carpet (Fig. 4.10). So in addition to the large
active-region coronal loops, there are numerous smaller loops being formed all the
time. It seems that there ought to be many more of the smaller loops than larger
ones since it would take less energy to thrust them up above the photosphere, just as
ocean waves toss many more smaller shells or pebbles on the beach than larger ones.

Uninterrupted by atmospheric turbulence, bad weather, and night, *SOHO's* MDI
magnetograms track the evolution of the small magnetic loops, showing that they
are continuously emerging at seemingly random locations in the photosphere, fragmenting in response to sheared flows, merging together or retracting inside the Sun.
The observations, reported by Carolus J. Schrijver, Alan M. Title and co-workers
from 1997 to 2002, indicate that the entire magnetic flux in the quiet solar photosphere disappears and is replenished every 15–40 h.

Tens of thousands of the small magnetic loops are constantly being generated,
rising up and out of the photosphere, interacting, fragmenting, and disappearing

Fig. 4.10 Magnetic carpet. Magnetic loops of all sizes rise up into the solar transition region from regions of opposite magnetic polarity (*black* and *white*) in the photosphere (*green*), forming a veritable carpet of magnetism in the low corona. Energy released when oppositely directed magnetic fields meet in the corona, to reconnect and form new magnetic configurations, is one likely cause for making the solar corona so hot. (Courtesy of the *SOHO* EIT and MDI consortia. *SOHO* is a project of international cooperation between ESA and NASA.)

within hours or days. And the numerous magnetic loops in this "magnetic carpet" are always being replaced, in a sort of self-cleaning action initiated from below. Motions in the photosphere and underlying turbulent convection push the carpet loops around, and when the magnetic fields of adjacent loops meet, they can break and reconnect with each other into simpler magnetic configurations, releasing energy that might be used to continuously heat the overlying corona.

How big are the numerous, smaller loops, and how far up do they rise above the photosphere? In 2003, Richard M. Close, Clare E. Parnell, Duncan H. Mackay, and Eric R. Priest examined the statistical properties of the quiet Sun magnetic fields displayed in the MDI data, discovering that the magnetic elements of one polarity connect preferentially to their nearest neighbors of opposite polarity, which seems logical since less energy would have to be expended to move over the shorter distance. And this means that most of the loops in the magnetic carpet are about as low as they can be. On average, 50% of them extend no higher than 2,500 km above the photosphere, and most of them never penetrate the corona (Fig. 4.11). They instead lie within the chromosphere and the transition region to the corona. In contrast, the coronal loops associated with active regions and their sunspots have sizes of more than 100,000 km and form the brightest part of the low corona.

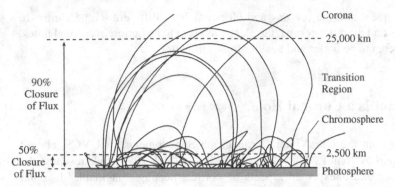

Fig. 4.11 Carpet loop heights. Side view of typical three-dimensional magnetic carpet fields. Fifty percent of the magnetic flux closes into loops below 2,500 km, while only 5–10% of the flux extends above 25,000 km. By way of comparison, the jet-like spicules in the chromosphere can rise to heights of 10,000 km, and the coronal loops in active regions can achieve heights of 100,000 km. [Adapted from Richard M. Close, Clare E. Parnell, Duncan H. Mackay and Eric Priest (2003).]

The loops in the magnetic carpet can produce heat by coming together and releasing stored magnetic energy when they make contact. Motions down inside the convective zone twist and stretch the overlying magnetic fields, slowly building up their energy. When these magnetic fields are pressed together, they can merge and join at the place where they touch, releasing their pent-up energy to heat the gas. The magnetic fields reform or reconnect in new magnetic orientations, so this method of heating is also termed magnetic reconnection.

But how is the energy propelled up to heat the corona high above? According to one explanation, proposed by Margarita Ryutova and Theodore Tarbell in 2003, merging carpet loops can form shock waves that are launched by the bent magnetic fields in slingshot fashion, flinging heated gas through the overlying transition region.

So to sum up, there is abundant evidence that the quiet corona outside active regions is linked to the interaction of magnetic loops, either lying within the corona or residing beneath it. Exactly which structures predominate is still unknown, a mystery to be solved by future observations with instruments aboard *Hinode*, in collaboration with those on *SOHO* and *TRACE*. After all, that is one of the main objectives of *Hinode*, to measure the emergence, structure, and motions of the Sun's magnetic field with continuous, fine detail, and to investigate how the high-temperature, ionized atmosphere responds to the changing magnetism. Similar investigations are being planned for the *Solar Dynamics Observatory* Mission, scheduled for launch in December 2008.

And even then, all of the heating problems will not be completely solved, for the proposed magnetic reconnection processes operate in or beneath the low corona where the ubiquitous magnetic loops rise and disappear. The outer corona must also be heated, in the extended corona above active regions, in coronal streamers, and above the coronal holes. At distances greater than a few tenths of a solar radius above the photosphere, the density of the corona is so low that collisions between particles become infrequent and the protons and electrons have different temperatures. In active regions and the quiet corona, the electrons seem to be hotter; in coronal holes, the protons seem to be hotter and heavier ions still hotter.

4.5 How Hot Is a Coronal Hole?

Observations from the UltraViolet Coronagraph Spectrometer, or UVCS, aboard *SOHO*, reported by John L. Kohl and co-workers in 1997–1999, have revealed the most surprising and unexpected aspects of coronal heating imaginable. In polar coronal holes at activity minimum, the heating is preferential and it has a direction. The heavy oxygen ions, designated O VI, in coronal holes are hotter than the protons there, which are in turn hotter than the relatively cool electrons. And the ion temperatures are anisotropic, with a higher temperature in the direction perpendicular to the radial direction than parallel to it. Moreover, the heating in coronal holes is extreme. While the protons attain plausible temperatures of several million Kelvin, the oxygen ions have searing kinetic temperatures, inferred from the Doppler broadening of their spectral lines, approaching 200×10^6 K. That is more than 10 times hotter than the center of the Sun! In 2003, Richard A. Frazin and co-workers used the UVCS to demonstrate that similar temperature differences exist at the edges of coronal streamers, but not in their dense central parts.

Unlike the magnetic carpet or active regions, the heating in coronal holes extends far away from the Sun, and that is because the magnetic fields in coronal holes are open instead of arching back down into the Sun to make loops. And the particles in coronal holes are not in thermal equilibrium, since they are so rarefied and spread out that they rarely collide with each other. As a result, the preferential and directional heating extends far out in the expanding solar corona (Fig. 4.12).

The unusual heating process in coronal holes occurs in nearly collisionless plasma where every ion species has its own unique temperature, and one plausible explanation is ion cyclotron heating by Alfvén waves, at least for the more massive ions. The magnetic waves might give them an extra boost, pushing them to higher speeds and kinetic temperatures, while also whirling them about in the direction perpendicular to the open, radial magnetic fields. Such Alfvén waves have been observed from the Sun to past the orbit of Earth (Fig. 4.13).

In 2000, Steven R. Cranmer demonstrated that the Alfvén wave heating must occur throughout the extended corona, and in 2005 Cranmer and his colleague Adriaan A. van Ballegooijen presented a comprehensive model of how Alfvén waves are generated in the photosphere, propagate up funnel-like, open magnetic flux tubes, and seed turbulence that determines the general properties of heating in the extended corona. In technical terms, the energy input comes from kink-mode magnetic field motion, generated by transverse shaking in intergranular lanes and transformed into Alfvén waves in the canopies of the transition region.

The ubiquitous X-ray jets formed in polar coronal holes may also play a role in heating them. As reported by Antonia Savcheva and colleagues in 2007, the

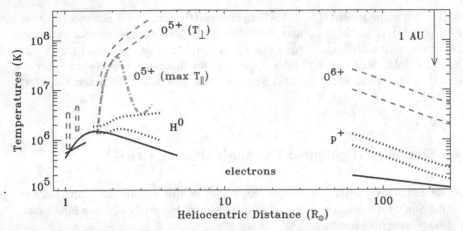

Fig. 4.12 Different temperatures. Summary of the radial dependence of temperature in coronal holes and the high-speed solar wind at the minimum of the 11-year solar activity cycle. The distance is given as height above the photosphere in units of the Sun's radius. Note the "gap" between telescopic observations near the Sun and direct particle detection by spacecraft far from the Sun. The temperatures of the electrons (*solid lines*), protons (*dotted lines*), and ionized oxygen (*dashed and dot-dashed lines*) are all different from one another. Paired sets of curves give representative ranges of observational uncertainty. [Courtesy of Steven R. Cranmer, adapted from John L. Kohl, Giancarlo Noci, Steven R. Cranmer and John C. Raymond (2006).]

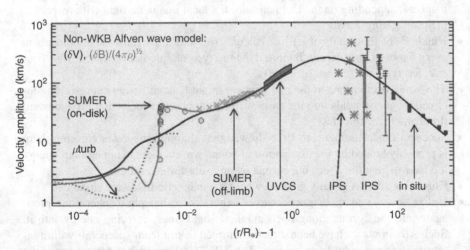

Fig. 4.13 Alfvén waves. The amplitudes of Alfvén waves, expressed as transverse velocities of the oscillating magnetic field lines, versus height above the photosphere, given in units of the solar radius. The solid curves denote a model constructed by Steven R. Cranmer and Adriaan A. Ballegooijen (2005); the other data points and curves denote various kinds of observational inferences of the wave amplitudes. The left-most two sets of data (*pink dotted curve* and *yellow points*) represent radial motions and may not correspond directly to the transversely oscillating Alfvén waves. (Courtesy of Steven R. Cranmer.)

X-Ray Telescope aboard *Hinode* has demonstrated the frequent occurrence, at about 60 jets per day, and high outflow velocities of the jets, suggesting that they may contribute to the high-speed wind that flows from these regions. And this brings us to the solar wind heating problem, or just how the expanding corona is accelerated to supersonic speed near the Sun, continuing to cruise out to the Earth and beyond.

4.6 Summary Highlights: The Sun's Heating Crisis

- Instead of growing colder at higher regions of its atmosphere, the temperature of the Sun's corona soars to several million Kelvin, hundreds of times hotter than the photosphere just below.
- Coronal heating processes are selective in both space and time, depending on the magnetic structure and how magnetically active the Sun is at the time. There are hot, dense coronal loops in active regions, slightly cooler and less dense coronal loops in the quiet Sun outside active regions, and even cooler and more rarefied coronal holes.
- One of the earliest explanations of coronal heating involved sound waves generated in the photosphere and turbulent convective zone. The sound (acoustic) waves produce shocks as they travel out and dissipate their energy to heat the chromosphere.
- Magnetic fields act like a trap door letting some sound waves through the photosphere, depending on the inclination of the local magnetic field with respect to the propagating waves.
- Hundreds of thousands of jet-like spicules rise and fall within the chomosphere every 5 min or so. A second, more dynamic type of spicule, observed from *Hinode*, has typical lifetimes of 10–60 s.
- The 5-min oscillations of the photosphere might leak sufficient energy along inclined magnetic fields into the chromosphere to power shocks that drive upward flows and form spicules.
- Spacecraft measurements in 1978 showed that although the lower chromosphere is probably heated by the dissipation of sound waves, there is not enough energy left over to heat the overlying corona by any substantial amount.
- Coronal heating is usually greatest where the magnetic fields are strongest.
- Unlike sound waves, magnetic waves generated in the photosphere can propagate into the corona along open magnetic field lines, carrying energy into it. Such Alfvén waves have been directly sampled in situ from spacecraft within interplanetary space far from the Sun. The *TRACE* telescope has demonstrated the widespread presence of magneto-hydrodynamic waves in the transition region, both longitudinal and transverse, many of which travel through the chromosphere with unanticipated efficiency. Several research teams have reported evidence of Alfvén waves in the chromosphere, transition region, or low corona using

instruments aboard *Hinode*, noting that the energy associated with these waves may be sufficient to heat the corona and accelerate the solar wind.

- Near the maximum in the 11-year cycle of solar magnetic activity, coronal loops in active regions make up about 80% of the total coronal heating energy.
- The *Skylab, Yohkoh*, and *SOHO* spacecraft have demonstrated that the hottest, densest material in the low corona, with the most intense X-ray and extreme-ultraviolet emission, is concentrated within strongly magnetized loops located in solar active regions, and that the active-region coronal loops are constantly varying on all detectable spatial and temporal scales.
- Observations from *TRACE* have demonstrated that the corona in active regions is structured on the smallest observable scale, and that it is comprised of numerous long, thin coronal loops with little significant cross-sectional variation or observable twists or braids.
- Coronal loops of different temperatures co-exist in active regions, with cool loops arching over hot ones.
- *SOHO* and *TRACE* extreme-ultraviolet observations of some coronal loops in active regions are inconsistent with steady, uniformly heated loop models. *TRACE* measurements of overdense coronal loops can be explained if the loops are intermittently heated by upflows of heated material from the loop footpoints or legs. They can also be explained by a non-stationary and non-equilibrium cooling process after the heating has ceased, regardless of the heating mechanism by upflows, waves, or some other mechanism.
- Bright active region loops are being filled with material heated to coronal temperatures in the chromosphere or transition region; and when the observed loops disappear from view, the hot gas is cooling and raining back down to the place it came from.
- Continued dynamic activity and forced magnetic connections are ubiquitous features throughout the low solar corona. Magnetic concentrations merge together and cancel all the time and all over the Sun, providing a plausible explanation for heating the low corona outside active regions.
- The so-called quiet corona outside active regions, and excluding coronal holes, could be at least partially heated when coronal loops merge and join together into new magnetic configurations. Observations from *Yohkoh's* soft X-ray telescope indicate that such magnetic merging and reconnection may occur when newly emerging magnetic fields rise through the photosphere to encounter pre-existing ones in the corona or when existing coronal loops are forced together.
- X-ray jets, X-ray bright points, ultraviolet explosive events, and large-scale, non-flaring, interacting X-ray loops can result from magnetic reconnection, contributing to the heating of the quiet solar corona. The large number of jets observed from *Yohkoh* and *Hinode* in polar coronal holes, for example, combined with their high outflow velocity, suggests that they may contribute to coronal heating and the high-speed solar wind.
- The combined total energy released by numerous, frequent weak flares, dubbed nanoflares, could contribute to the heating of the low corona or the underlying chromosphere and transition region.

- Magnetogram observations from *SOHO's* Michelson Doppler Imager indicate that a magnetic carpet of tens of thousands of small, low-lying loops continuously emerges from seemingly random locations in the photosphere, and that the entire magnetic flux in the quiet solar photosphere disappears and is replenished every 15–40 h.
- Most of the loops in the magnetic carpet lie within the chromosphere and transition region, never penetrating the corona.
- Magnetic energy generated when loops in the magnetic carpet interact can be used to heat the material in their vicinity, and the interacting loops can form shock waves that fling the heated gas into higher places.
- At distances of more than a few tenths of a solar radius above the solar photosphere, the density of the coronal plasma drops to a point where particle-particle collisions become infrequent. In such an environment, the temperature of the two most numerous kinds of particles, the protons and electrons, can become different from one another. In active regions and the quiet corona, the electrons seem to be hotter; in coronal holes, the protons seem to be hotter.
- Observations from *SOHO's* UltraViolet Coronagraph Spectrometer near activity minimum indicate that heating in polar coronal holes is preferential, extreme, and directional. The particles are not in thermal equilibrium, the massive ions are hotter than the less massive ions, and the massive ions have temperatures approaching 200×10^6 K with a higher temperature in the direction perpendicular to the radial direction than parallel to it.
- A plausible explanation for heating the extended, open-field coronal holes is ion cyclotron heating by Alfvén waves, at least for the heavy ions.

4.7 Key Events in Coronal Heating*

Date	Event
1939–1941	Walter Grotrian and Bengt Edlén identify coronal emission lines with highly ionized elements, indicating that the Sun's outer atmosphere has a temperature of millions of Kelvin. The conspicuous coronal green line was identified with (Fe XIV), an iron atom missing 13 electrons.
1946	Vitalii Ginzburg, David F. Martyn, and Joseph L. Pawsey independently confirm a coronal temperature of about a million Kelvin from observations of the Sun's radio emission.
1947	Hannes Alfvén argues that convective granulation can generate magnetohydrodynamic waves that can heat the inner corona.

*See the References at the end of this book for complete references to these seminal papers.

Date	Event
1947–1949	Ronald G. Giovanelli develops a theory of solar flares involving the magnetic fields in the solar atmosphere above sunspots, including electric currents at magnetic neutral points. He also describes how Alfvén waves might heat the solar corona.
1948–1949	Ludwig F. Biermann, Martin Schwarzschild, and Evry Schatzman independently reason that sound waves, produced by convective granulation, might transport mechanical energy into the chromosphere and corona. These sound waves would be quickly deformed into shock waves as they pass into regions of decreasing density, dissipating energy and heating the gas.
1958	Peter A. Sweet (1958a, b) develops a theory of slow magnetic reconnection.
1960	Thomas Gold and Fred Hoyle show that magnetic energy must power solar flares, and argue that these explosions are triggered when two magnetic loops of opposite sense or direction interact, merge, and suddenly dissipate their stored magnetic energy in the corona.
1963	Eugene N. Parker (1963a) derives detailed theory of slow magnetic reconnection.
1964	Thomas Gold suggests that the corona is heated by the relaxation of twisted or stressed coronal magnetic fields, driven by turbulent motion in the convective zone, and compares the energy dissipation mechanism to the magnetic interaction theory of solar flares developed by Gold and Fred Hoyle in 1960.
1964	Harry E. Petschek clarifies the process of fast magnetic field reconnection and shows that it can occur rapidly even in highly conducting plasma.
1968	Roger A. Kopp and Max Kuperus discuss the temperature structure and magnetic field of the transition region between the chromosphere and the corona.
1969–1971	Magnetic fluctuations are observed in the solar wind from *Mariner 5* on its way to Venus, and attributed to large-amplitude Alfvén waves by John W. Belcher, Leverett Davis Jr., and Edward J. Smith.
1970–1977	X-ray photographs from rockets, and then from the *Skylab* mission, are used to show that magnetic fields create a threefold structure in the inner corona, with its coronal holes, coronal loops, and X-ray bright points.
1974	Leon Golub and co-workers provide the first detailed studies of X-ray bright points using *Skylab* data, providing values for their densities, temperatures, sizes, lifetimes, and rate of occurrence.
1976	Alan H. Gabriel introduces a two-dimensional model of the chromosphere and corona, and the transition region between them, in which magnetic flux is concentrated at the boundaries of supergranular convection cells that produce a magnetic network. These flux tubes flare out as they rise into the overlying solar atmosphere to produce a magnetic canopy.
1976	George A. Doschek, Uri Feldman, and J. David Bohlin use Doppler wavelength shifts of ultraviolet lines observed from *Skylab* to suggest that most observable material in the transition region is falling down into the Sun rather than moving away from it.
1978	John T. Mariska, Uri Feldman, and George A. Doschek show that extreme ultraviolet lines observed from *Skylab* are wider than would be expected from thermal Doppler broadening alone.
1978	Robert Rosner, Wallace H. Tucker, and Giuseppe S. Vaiana use *Skylab* X-ray data and a theoretical model to derive a scaling law for coronal loops that are in hydrostatic equilibrium and uniformly heated, connecting their length, temperature, and density.

Date	Event
1978–1981	R. Grant Athay and Oran R. White, and independently Elmo C. Bruner Jr., observe chromosphere oscillations in the ultraviolet lines of ionized carbon from the eighth *Orbiting Solar Observatory*, abbreviated *OSO 8*, spacecraft. These observations indicate that the chromosphere might be heated by sound waves, but that there is not enough power in the sound waves to heat the overlying corona.
1983	Eugene N. Parker (1983b) argues that the main source of coronal heating is the dynamical dissipation of energy from coronal magnetic fields that have been twisted by sub-photosphere convection into neutral point reconnection.
1983	Rocket observations of the ultraviolet solar spectrum by Guenter E. Brueckner and John-David F. Bartoe reveal high-energy jets in the transition region outside active regions. Kenneth P. Dere and colleagues subsequently provide detailed observations of these compact, short-lived explosive events from rockets, the *Space Shuttle*, and satellites, showing that they occur in the solar magnetic network lanes at the boundaries of supergranular convective cells and might play a role in coronal heating.
1984–1998	Jason G. Porter, David A. Falconer, Ronald L. Moore, and colleagues observe low-lying, ultraviolet microflares in the transition region and low corona, using the *Solar Maximum Mission* satellite and *SOHO*. These events are located on magnetic neutral lines in active regions and in the magnetic network, and estimates suggest that they are frequent enough and powerful enough to heat the corona.
1986	Eric R. Priest and Terry G. Forbes put the idea of fast magnetic reconnection on a firm foundation, and discover a new family of fast regimes.
1988	Eugene N. Parker interprets the solar X-ray corona in terms of undetected nanoflares, occurring more often and with less intensity than observed solar flares. The nanoflares are related to bipolar magnetic fields driven by underlying convective motion.
1991	Kenneth P. Dere and colleagues report the detection of explosive events seen in extreme-ultraviolet light of the transition region, associating them with magnetic reconnection there.
1991	Hugh S. Hudson uses extrapolations from existing flare observations to show that similar, but less energetic, microflares cannot heat the corona.
1991	The *Yohkoh* Mission is launched on 31 August 1991.
1991–2000	The Soft X-ray Telescope on the *Yohkoh* satellite demonstrates the structured, dynamic nature of the inner corona more clearly than ever before.
1992	Toshifumi Shimizu and co-workers report observations of frequent transient brightening in active-region coronal loops observed with the *Yohkoh* Soft X-ray Telescope, concluding that the loops are far from static and must be dynamic.
1992, 1995	Mats Carlsson and Robert F. Stein show how acoustic shocks can dissipate energy in the chromosphere, producing bright grains there.
1992–1996	Kazunari Shibata, Toshifumi Shimizu, Saku Tsuneta, and their colleagues use data from *Yohkoh's* Soft X-ray Telescope to show that magnetic reconnection is rather common in the low corona, particularly inside active regions.
1993	Yi-Ming Wang uses model calculations based on the coronal energy balance to show how the magnetic flux-tube divergence rate can control the coronal temperature.
1994	Peter J. Cargill examines some implications of the nanoflare concept, in which the active-region corona is comprised of many hundreds of small elemental magnetic flux tubes randomly and impulsively heated by nanoflares.

Date	Event
1994	Eric R. Priest, Clare E. Parnell, Sara F. Martin, and Leon Golub (Parnell et al. 1994, Priest et al. 1994) give a model for heating X-ray bright points by magnetic reconnection, and show that this model explains the observed bright points.
1995	The *SOlar and Heliospheric Observatory*, abbreviated *SOHO*, is launched on 2 December 1995.
1996	Saku Tsuneta (1996a, b) uses *Yohkoh* soft X-ray observations to describe how solar active regions in opposite hemispheres can reconnect to form new transequatorial coronal loops and heat the corona in a less explosive way than in solar flares.
1997	Peter J. Cargill and James A. Klimchuk use a nanoflare model to explain the heating of active-region loops observed by the *Yohkoh* Soft X-ray Telescope.
1997	Andre Fludra and colleagues use the Coronal Diagnostic Spectrometer on *SOHO* to identify coronal loops with different temperatures that can co-exist in active regions.
1997	Davina Innes, Bernd Inhester, Sir William Ian Axford, and Klaus Wilhelm obtain *SOHO* observations that exhibit the bi-directional jets expected from Petschek-type magnetic reconnection.
1997–1998	Richard A. Harrison (1997b) reports *SOHO* observations of thousands of small-scale brightening events in the extreme-ultraviolet, inactive Sun. These blinkers flash on and off in about 10 min and reach temperatures of up to a million Kelvin.
1997–1998	Carolus J. Schrijver, et al. (1997a, 1998) use magnetograms from the Michelson Doppler Imager instrument aboard *SOHO* to show that the solar photosphere outside active regions contains numerous bipolar magnetic elements that mark the footpoints of low-lying loops, and that all the magnetic flux in this magnetic carpet is cancelled out and replenished every 15–40 h. Tens of thousands of small magnetic loops are being generated and renewed from inside the Sun, and magnetic reconnection associated with the interaction of these loops will heat the solar atmosphere.
1998	Olav Kjeldseth-Moe and Paal Brekke use the Coronal Diagnostic Spectrometer on *SOHO* to detect rapid flows in active-region coronal loops, suggesting models that combine very fine structure and episodic heating with disturbances propagating in the loop legs.
1998	Werner M. Neupert and co-workers use the Extreme-ultraviolet Imaging Telescope on *SOHO* to measure iosthermal coronal loops and conclude that the data are not consistent with a heating source high in the corona.
1998	The *Transition Region and Coronal Explorer*, abbreviated *TRACE*, is launched on 1 April 1998.
1998	Eric R. Priest and his colleagues show that the heating of the large-scale corona is uniform, and is likely to be due to turbulent reconnection of many small current sheets.
1998–1999	Arnold O. Benz and Säm Krucker use *SOHO* data to demonstrate that numerous low-level, unexpectedly frequent microflares could heat the quiet corona outside active regions. Subsequent observations by Clare E. Parnell and Peter E. Jupp (1999) and by Markus J. Aschwanden and colleagues (1999d) using *TRACE* observations suggest that there may not be enough energy in microflares or nanoflares to heat the entire quiet corona.

Date	Event
1998–2001	John L. Kohl, Steven R. Cranmer, and Mari Paz Miralles and their co-workers publish a series of papers reporting preferential and directional temperatures in polar coronal holes measured with the UltraViolet Coronagraph Spectrometer aboard *SOHO* near the minimum in the 11-year solar activity cycle. The oxygen ions are hotter than protons, which are in turn hotter than electrons, and the oxygen ions have greater temperatures perpendicular to the radial direction than along it. In 2003, Richard A. Frazin and colleagues reported similar oxygen ion measurements at the edges of coronal streamers but not in them.
1999	Dana W. Longcope and Charles C. Kankelborg present models of coronal heating by collision and cancellation of magnetic elements.
1999	Carolus J. Schrijver, Alan M. Title, and co-workers present *TRACE* observations that show that coronal loops in active regions are comprised of numerous long, thin strands, that loops of significantly different temperatures exist side by side, and that relatively cool loops tend to arch over hotter ones within active regions.
1999	Dawn D. Lenz and co-workers report that long-lived, active-region coronal loops observed from *TRACE* have no significant temperature stratification, and that they are denser than the classic steady, uniformly heated loop model predicts.
2000	Alexei A. Pevtsov uses *Yohkoh* soft X-ray data and magnetograms to show that approximately one-third of all active regions on the Sun exhibit transequatorial loops that may develop between existing active regions or between mature regions and newly emerging ones.
2000	Julie Anne Watko and James A. Klimchuk report that active-region coronal loops observed from *TRACE* have cross-sections that lie near the instrumental resolution, and that there is no significant change in width with height for the loops that are resolved. This is consistent with Klimchuk's report of unvarying cross-sections along X-ray coronal loops observed from *Yohkoh*, but for much larger widths.
2000–2001	Markus J. Aschwanden and colleagues demonstrate that coronal loops observed from *TRACE* are overdense, when compared with uniform heating models, and suggest that the loops are heated and filled from beneath the corona, at 10,000–20,000 km above the photosphere.
2001	Carolus J. Schrijver reports *TRACE* observations of catastrophic cooling and high-speed downflow in active-region coronal loops, indicating that they are heated intermittently and that when the heating stops the hot material rains down to the place it came from.
2002	Markus J. Aschwanden and Paul Charbonneau examine the statistics of solar flares, microflares, and nanoflares observed over a wide range of energy using the extreme-ultraviolet telescopes aboard the *SOHO* and *TRACE* satellites, concluding that these observations pose a serious challenge to coronal heating by nanoflares.
2002–2003	Amy R. Winebarger, Harry P. Warren, and their co-workers report the detection of upward flows in extreme-ultraviolet, active-region loops observed from *SOHO* and *TRACE*, which are inconsistent with previous hydrostatic loop models, and instead develop non-static, dynamic models with intermittent, impulsive heating from below
2003	Richard M. Close, Clare E. Parnell, Duncan H. Mackay, and Eric R. Priest examine the statistical properties of the magnetic carpet observed with *SOHO's* Michelson Doppler Imager, finding that the magnetic flux fragments connect preferentially to their nearest neighbors of opposite magnetic polarity, that half of the magnetic loops in the quiet Sun extend no higher than 2,500 km above the photosphere, and that most of them do not reach the corona

Date	Event
2003	Margarita Ryutova and Theodore Tarbell use simultaneous observations from *SOHO* and *TRACE* to support their proposal that shock waves produced by the interaction of low-lying magnetic elements can propel heated material and energy up into the overlying transition region.
2003	Carolus J. Schrijver and Alan M. Title estimate that as much as half of the coronal magnetic field over the very quiet Sun may be rooted in mixed-polarity internetwork fields within the supergranules rather than in the network flux concentrations at the supergranule edges.
2004	Bart De Pontieu, Robert Erdélyi, and Stewart P. James use observations with *TRACE* to demonstrate that photospheric oscillations leak into the chromosphere where they can power shocks, drive upward flows, and form spicules.
2005	Steven R. Cranmer and Adriaan A. van Ballegooijen model the properties of Alfvén waves from the photosphere to a distance past the Earth's orbit, with wave periods ranging from 3 s to 3 days.
2005	Markus J. Aschwanden and Richard W. Nightingale analyze the multithread structure of coronal loops observed from *TRACE*, concluding that the vast majority are isothermal with a unique temperature.
2005	Russell B. Dahlburg, James A. Klimchuk, and Spiro K. Antiochos propose a mechanism that "switches on" energy release via magnetic reconnection of coronal loops that merge together due to the random mixing of the magnetic loop footpoints by photospheric motions.
2006	The *Hinode* spacecraft is launched on 23 September 2006.
2006	Stuart M. Jefferies and co-workers use telescopes on board *SOHO* and *TRACE* and on the ground to show that inclined magnetic field lines at the boundaries of supergranule convection cells provide portals through which magnetoacoustic waves can propagate into the solar chromosphere.
2007	Steven R. Cranmer, Adriaan A. van Ballegooijen, and Richard J. Edgar present models for coronal heating by Alfvén waves, which propagate along open magnetic field lines into interplanetary space.
2007	Bart De Pontieu and co-workers (2007b) use data from the Solar Optical Telescope on board *Hinode* to show that the chromosphere is permeated with Alfvén waves with velocity amplitudes of 10–25 km s^{-1} and periods of 100–500 s, and note that spicules carry such waves.
2007	Loraine Lundquist and co-workers use observations from the X-Ray Telescope on *Hinode* to describe the interaction between emerging flux and large-scale loop systems that connect active regions with the distant reaches of the corona, both quiet and active, and commonly cover as much as one third of the solar disk.
2007	Steven Tomczyk and co-workers report the detection of Alfvén waves in the solar corona. The spatially resolved waves are too weak to heat the solar corona; however, unresolved Alfvén wave may carry sufficient energy.
2007	Jonathan Cirtain et al. (2007b), Takenorie J. Okamoto et al. (2007), and Bart De Pontieu et al. (2007b) present evidence of Alfvén waves using instruments aboard *Hinode*. The energy associated with these weaves is apparently sufficient to heat the corona and accelerate the solar wind.

Date	Event
2008	Markus J. Aschwanden (2008a) argues that the quasi-isothermal coronal loop structures whose widths have been resolved with the *TRACE* telescope contradict the nanoflare heating hypothesis, which predicts that they should contain unresolved strands, and proposes that the hypothetical nanoflare events be relocated from the corona down to the chromosphere or transition region.
2008	Markus J. Aschwanden and David Tsiklauri conclude that steady-state equilibrium solutions cannot explain all the physical parameters of coronal loops observed at extreme-ultraviolet wavelengths, but that they can be explained by a non-stationary and non-equilibrium cooling process after the heating has ceased, regardless of how the loops are heated.

Chapter 5
Winds Across the Void

5.1 The Fullness of Space

With a temperature of a few million Kelvin, the outer solar atmosphere, or corona, is so hot it cannot remain still. It has to expand out from the Sun, like heat escaping from a chimney on a cold day. The high pressure at the bottom of the corona naturally pushes the high-temperature gas away from the Sun in all directions (Focus 5.1).

Focus 5.1

Why a Solar Wind has to Exist

The theoretical concept of a solar wind grew out of consideration of a static, non-expanding, isothermal corona. At a distance, r, in the corona, the balance between the pressure gradient, dp/dr, and the gravitational force can be expressed by:

$$\frac{dp}{dr} = -\frac{GM_\odot \rho}{r^2},$$

where G is the gravitational constant, $M_\odot$ is the Sun's mass, and ρ is the corona's mass density. If we take the coronal protons and electrons to have the same temperature, T, the ideal gas law for the pressure, p, becomes

$$p = nk(T_e + T_p) = 2nkT,$$

where n is the number density of particles per unit volume, k is Boltzmann's constant, the subscripts e and p denote electrons and protons, and the coronal mass density is given by

$$\rho = n(m_e + m_p) = nm = \frac{mp}{2kT}$$

and m is the sum of the electron and proton mass, or essentially the proton mass since the electron has a relatively low mass.

By substitution, we obtain the differential equation:

$$\frac{1}{p}\frac{dp}{dr} = -\frac{GM_\odot m}{2kT}\frac{1}{r^2}.$$

K.R. Lang, *The Sun from Space*, Astronomy and Astrophysics Library,
© Springer-Verlag Berlin Heidelberg 2009

The solution of this equation for the pressure $p(r)$ at distance, r, is:

$$p(r) = p_0 \exp\left\{\frac{GM_\odot m}{2kT}\left(\frac{1}{r} - \frac{1}{R_\odot}\right)\right\},$$

where p_0 is the pressure at the base of the corona and $R_\odot$ is the radius of the Sun. As one would expect, the pressure decreases with increasing distance, r, but the difficulty is that it does not decrease fast enough. If we let the distance go to infinity, or $r \rightarrow \infty$, the pressure approaches the value:

$$p(\text{infinity}) = p_0 \exp[-GM_\odot m/(2kTR_\odot)],$$

where the coronal temperature can be taken to be $T = 10^6\,\text{K}$, since this temperature falls off very slowly with distance owing to the high thermal conductivity of the corona. The other constants are $G = 6.6726 \times 10^{-11}\,\text{N m}^2\,\text{kg}^{-2}$, $M_\odot = 1.989 \times 10^{30}\,\text{kg}$, $m = m_p = 1.6726 \times 10^{-27}\,\text{kg}$, $k = 1.38066 \times 10^{-23}\,\text{J K}^{-1}$, and $R_\odot = 6.955 \times 10^8\,\text{m}$.

So, the pressure at infinity is a small fraction of the high pressure at the base of the corona, but still many orders of magnitude higher than the pressure thought to exist in the interstellar medium. This means that the static, isothermal corona cannot be in equilibrium with the distant interstellar spaces, which is what led Eugene N. Parker to introduce in 1958 an expanding corona with nonzero flow speeds, v, obtaining the differential equation:

$$\rho v\frac{dv}{dr} = -2kT\frac{dn}{dr} - \rho\frac{GM_\odot}{r^2}.$$

Although Parker only had to wait 4 years until the *Mariner 2* spacecraft confirmed his prediction, they were a turbulent 4 years since some well-known scientists, including Sydney Chapman, publicly clashed with him about this (see Chap. 2).

So the wide-open spaces between the planets, once thought to be a tranquil, empty void, is swarming with hot, charged invisible pieces of the Sun. They expand and flow away from the Sun at supersonic speeds, faster than $300\,\text{km s}^{-1}$, forming a perpetual solar wind. Unlike any wind on Earth, the solar wind is an exceedingly rarefied mixture of electrons and protons set free from the Sun's abundant hydrogen atoms; but like terrestrial winds, the solar gale has strong gusts and periods of relative calm.

The Sun constantly blows itself away, sending a billion (10^9) kilograms, or a million tons, of electrons and protons into space every second. All of this material must be replaced from below, but this is a small amount compared with the enormous total mass of the Sun. At the present rate, it would take 10 billion years for the Sun to lose only 0.01% of its mass by the solar wind, and the Sun will evolve into a giant star long before it blows away completely.

The solar wind streams out in all directions from the Sun, and the planets move through it as if they were ships at sea. So the Earth is immersed within the seemingly eternal flow, and as a result we are living in the outer part of the Sun, encased within its relentless stormy wind.

The reason that space looks so empty is that the Sun's wind is very tenuous. Even at its origin near the visible Sun, the hot corona is so rarified that its pressure is no more than the pressure under the foot of a spider. By the time that it reaches the Earth's orbit, the solar wind has been further diluted by expanding into the increasing volume of space. There are about five million electrons and five million protons per cubic meter in the solar wind near the Earth. By way of comparison, there are 25 million billion billion (2.5×10^{25}) molecules in every cubic meter of our transparent air at sea level. The density of the solar wind is so low that if we could go out into space and put our hands in it, we would not be able to feel it.

The tenuous solar gale brushes past the planets and engulfs them, carrying the Sun's magnetic fields and outer atmosphere into the space between the stars. It thereby creates a teardrop-shaped bubble in interstellar space known as the heliosphere, from *Helios*, the "God of the Sun" in Greek mythology. In 1955, Leverett Davis Jr. postulated the existence of such a cavity in the interstellar medium to account for some observed properties of low-energy cosmic rays. The heliosphere is inflated by the solar wind, and threaded by open magnetic fields that have one end attached to the Sun (Fig. 5.1).

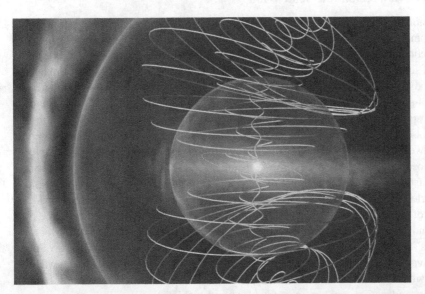

Fig. 5.1 The heliosphere. With its solar wind going out in all directions, the Sun blows a huge bubble within interstellar space called the heliosphere, with the Sun at its center and the planets inside. It is threaded by open magnetic fields that are connected to the Sun at one end, carried into space by the relentless solar wind on the other end, and twisted by the rotating Sun. Interstellar winds mold the heliosphere into a non-spherical teardrop shape, creating a bow shock (*left*) where interstellar forces encounter the solar wind. The heliosphere extends to about 100 times the distance between the Earth and the Sun. (Courtesy of Thomas H. Zurbuchen.)

Within the heliosphere, conditions are regulated by the Sun. Its domain extends out to about 100 times the distance between the Earth and the Sun, or about 100 astronomical units, marking the outer boundary or edge of the solar system. Out there, the solar wind has become so weakened by expansion that it can no longer repel interstellar forces.

As shown in Chap. 2, the relentless solar wind was inferred from comet tails, suggested by theoretical considerations, and fully confirmed by direct in situ measurements from spacecraft in the early 1960s. Instruments aboard numerous spacecraft have subsequently measured the detailed properties of the solar wind, finding gales that move at both slow and fast speeds. Scientists are just beginning to understand where these winds come from and how they might be accelerated to supersonic velocities.

Once released, the solar wind remains linked to the Sun by its magnetic field and explosive outbursts, so the winds vary with the Sun's tempestuous behavior. The solar wind also acquires its own independent characteristics, like children who have left home to make their own life. And it is additionally impacted from the outside by the local interstellar medium. All of these topics are discussed in greater detail within the rest of this chapter. The serious student or curious reader may want to consult reviews of these subjects prepared by professionals in the field (Focus 5.2).

Focus 5.2

Expert Reviews about the Solar Wind

Professional solar astronomers and astrophysicists have reviewed important developments in our knowledge of the Sun's winds, often in technical terms. In alphabetical order, they include Roberto Bruno and Vincenzo Carbone's (2005) review of solar wind turbulence, Benjamin D. G. Chandran's (2004) review of magnetohydrodyamic turbulence, Steven R. Cranmer's (2002) review of coronal holes and the high-speed solar wind, Johannes Geiss and George Gloeckler's (2001) review of heliospheric and interstellar phenomena deduced from pick-up ion observations, John T. Gosling's (1996) discussion of co-rotation and transient wind flow, Joseph V. Hollweg's (2006) overview of our changing perspectives of solar wind acceleration, Joseph V. Hollweg and Philip A. Isenberg's (2002) review of the generation of the fast solar wind emphasizing the resonant cyclotron interaction, John L. Kohl, Giancarlo Noci, Steven R. Cranmer, and John C. Raymond's (2006) comprehensive review of coronal ultraviolet spectroscopy that places constraints on the source regions and acceleration mechanisms of the solar wind, Serge Koutchmy and Moissei Livshits' (1992) review of coronal streamers, Eckart Marsch's (2006) review of our theoretical understanding of in situ measurements of solar wind particles and waves, and David J. Mc Comas and his colleagues' (2007) summary of unsolved problems in understanding coronal heating and solar wind acceleration.

The contributions of scientists working with specific spacecraft data are often presented at workshops that are subsequently published in book form. The *Ulysses* perspective of the heliosphere at solar minimum is given in the volume edited by André Balogh, Richard G. Marsden, and Edward J. Smith (2001). Specific issues of *Space Science Reviews* contain the results of other workshops, and have been republished in book form by Kluwer Academic Publishers or Springer Verlag. They include *Solar Dynamics and Its Effects on the Heliosphere*, edited by Daniel N. Baker and his colleagues (2006), *The Three-Dimensional Heliosphere at Solar Maximum*, edited by Richard G. Marsden (2001), *Co-rotating Interaction Regions*, edited by André Balogh and his colleagues (1999), and *The Heliosphere in the Local Interstellar Medium*, edited by Rudolf Von Steiger and colleagues (1996). The proceedings of *SOHO* workshops that deal with the solar wind include *SOHO-1, Coronal Streamers, Coronal Loops and Coronal and Solar Wind Composition*, edited by Clare Mattok (1992), *SOHO-2. Mass Supply and Flows in the Solar Corona*, edited by Bernhard Fleck, Giancarlo Noci, and Giannina Poletto (1994), *SOHO-5. The Corona and Solar Wind Near Minimum Activity*, edited by Olav Kjeldseth-Moe and Andrew Wilson (1997), *SOHO-7. Coronal Holes and Solar Wind Acceleration*, edited by John L. Kohl and Steven R. Cranmer (1999), *SOHO-16. Connecting Sun and Heliosphere*, edited by Bernhard Fleck and Thomas H. Zurbuchen (2005), and *SOHO-20. Transient Events on the Sun and in the Heliosphere*, edited by Bernhard Fleck, Joseph B. Gurman, Jean-Francois Hochedez, and Eva Robbrecht (2008).

5.2 The Two Solar Winds

The earliest in situ measurements of the solar wind, in the 1960s by the *Mariner 2* spacecraft on its way to Venus (Sect. 2.2), revealed that the solar wind travels with two main velocities, like an automobile with one high gear and one low gear. The fast wind moves at about $750 \mathrm{km\,s}^{-1}$ and the slow wind blows at about half that speed. Both the winds are always present, never disappearing during the more than four decades that the solar wind has been observed with spacecraft.

Helios 1 and *Helios 2* provided in situ analyses of the solar wind that tightened constraints on the two components. These twin spacecraft, respectively launched on 10 December 1974 and 15 January 1976, repeatedly looped as close as 0.3 AU from the Sun for years – until March 1986 for *Helios 1* and March 1980 for *Helios 2*. (The mean distance between the Earth and the Sun is 1 AU, or about 150×10^6 km.) Instruments aboard these spacecraft confirmed, in greater detail, that there are two kinds of solar wind flow, the fast and slow ones, with different physical properties (Table 5.1).

As the result of measurements from *Helios, Ulysses*, and other spacecraft, we now know that the high-speed wind is steady, uniform, and of relatively low proton density, while the slow-speed wind is variable, gusty, and of high proton density. When compared with the fast wind, a higher electron temperature, lower proton temperature, reduced helium abundance, and greater enrichment in easily ionized elements with low first ionization potentials also characterize the slow wind.

The massive protons dominate the energy transported by the solar wind. *Helios* instruments showed that the proton density is high whenever the wind is slow, and

Table 5.1 Average solar wind parameters measured from *Helios 1* and 2 between December 1974 and December 1976 normalized to the distance of the Earth's orbit at 1 AU[a]

Parameter	Fast wind	Slow wind
Source	Coronal holes	Equatorial steamers
Composition, temperature and density	Uniform	Highly variable
Proton density, N_p	$3 \times 10^6 \, \mathrm{m}^{-3}$	$10.7 \times 10^6 \, \mathrm{m}^{-3}$
Proton speed, V_p	$667 \, \mathrm{km \, s}^{-1}$ $(750 \, \mathrm{km \, s}^{-1})$[b]	$348 \, \mathrm{km \, s}^{-1}$
Proton flux, $F_p = N_p V_p$	$1.99 \times 10^{12} \, \mathrm{m}^{-2} \, \mathrm{s}^{-1}$	$3.66 \times 10^{12} \, \mathrm{m}^{-2} \, \mathrm{s}^{-1}$
Proton temperature, T_p	280,000 K	55,000 K
Electron temperature, T_e	130,000 K	190,000 K
Helium temperature, T_α	730,000 K	170,000 K
Helium to proton abundance, A^a	0.036 (constant)	0.025 (very variable)

[a]Adapted from Rainer Schwenn (1990). Measurements are referred to a distance of 1 AU = 1.496×10^8 km. The helium ion to proton abundance $A = N_\alpha/N_p$, where N is the number density and the subscripts α and p, respectively, denote the helium ions, or alpha particles, and the protons
[b]The *Helios 1* and 2 spacecraft traveled near the ecliptic where the slow solar wind dominates the flow, and this led to an underestimate of the velocity of the high-speed component. It has a speed of about $750 \, \mathrm{km \, s}^{-1}$

that the proton density is low when the wind speed is high. The product of the proton density and velocity, or the proton flux, is about the same in the fast and slow winds, with a value of between 1.5 and 4 million million $[(1.5–4) \times 10^{12}]$ protons per square meter per second at the Earth's distance from the Sun.

Instruments aboard the two *Helios* spacecraft measured temperatures of the charged particles blowing in the wind, showing significant and unexpected differences. In the high-speed wind, heavier particles have higher temperatures; but it is the other way around in the slow wind, where lighter particles are hotter. In the high-speed wind, the protons are a few times hotter than the electrons, and the helium nuclei, or alpha particles, are even hotter than the protons.

The existence of two solar winds, the fast and slow ones, was also indicated by observing remote radio sources that fluctuate, or scintillate, when observed through the rarefied solar wind, in much the same way that stars twinkle when seen through the Earth's wind-blown atmosphere. The radio waves are perturbed when they pass through the solar wind, producing a hazy, blinking, and distorted image. It is something like looking at a light from the bottom of a swimming pool.

The fluctuating radio signals sweep past the Earth like the beacon of a searchlight, and the velocity of the solar wind can be inferred from the time it takes the sweeping signal to move between two antennas. You could similarly watch the shadow of a wind-blown cloud and determine the cloud's speed by seeing how long it takes the shadow to move from place to place. Barney J. Rickett and William A. Coles in the United States and Takakiyo Kakinuma and Masayoshi Kojima in Japan used this interplanetary scintillation technique to study the solar wind throughout the 1970s and 1980s.

Their radio scintillation data indicated that the average wind velocity increases from the solar equator to higher solar latitudes toward the solar pole near the minimum in the 11-year activity cycle, suggesting that the fast wind then originates near the poles. And because the slow wind was confined to low solar latitudes near the solar equator, it most likely originated there, at least near activity minimum. The blinking radio signals also indicated that the slow-speed wind does not blow steadily; it exhibits squalls and calms. In contrast, a uniform, fast wind seemed to spill out of higher solar latitudes.

The bimodal, fast–slow nature of the solar wind was fully confirmed during *Ulysses'* pioneering polar orbit at the 1994–1995 minimum in the 11-year solar activity cycle. It provided the first measurements of the wind speed all around the Sun, over the full range of solar latitudes, or angular distances from the solar equator. This velocity data, reported by David J. McComas and his colleagues, conclusively proved the existence of two basic types of solar wind at activity minimum. The fast gale blows at a smooth, uniform, and steady clip at high latitudes, filling over 60%

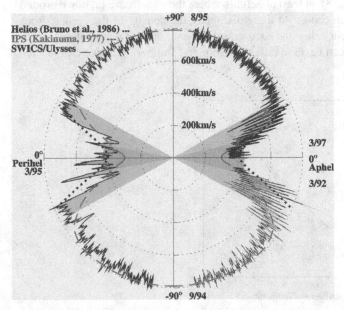

Fig. 5.2 Solar wind speeds at activity minimum. A polar diagram of the solar wind speed variation with solar latitude, from 0° at the solar equator to ±90° at the Sun's north and south poles (*top* and *bottom* respectively). These 1-day averages of the proton speed, derived from *Ulysses* SWICS observations from March 1992 to March 1997, show that the fast wind dominates the outflow at most latitudes near the minimum in the 11-year cycle of solar activity, when the slow wind is localized near the equatorial regions. The equatorial slow-speed wind exhibits strong fluctuations; the polar high-speed one also fluctuates, but at much lower levels. There are fewer fluctuations in the slow wind at the *left-center* because *Ulysses* was then moving rapidly over the equatorial regions. No shading represents fast coronal hole streams, while the *reddish color* denotes the slow solar wind and the streamer belt. These data are compared with previous measurements at lower latitudes by the *Helios* spacecraft [*dotted line*, Bruno et al. (1986)] and by interplanetary scintillations of extragalactic radio sources [*dashed red line*, Kakinuma (1977)]. [Courtesy of Joachim Wach, and adapted from Joachim Woch et al. (1997).]

of the heliosphere near the 1994–1995 solar minimum, while the slower, more gusty, and variable breeze was confined to low latitudes near the solar equator (Fig. 5.2).

The radio scintillation and *Ulysses* discoveries completely reversed some earlier interpretations, based mainly on spacecraft that had stayed close to the ecliptic. This previous view, which was now found to be wrong, interpreted the high-speed streams as an occasional perturbation of the slow wind, typically called the "ambient solar wind." But a uniform fast wind was now found to be the dominant component, at least during the minimum in the solar activity cycle, with the slow wind contributing a varying input at low solar latitudes.

Instruments aboard *Ulysses* also showed that the two solar wind types have differing ion compositions. As ions move out into the increasingly rarefied corona, collisions become infrequent and the ions decouple from other particles, retaining, or "freezing-in", an unchanging temperature. Johannes Geiss, George Gloeckler, and Rudolf Von Steiger used *Ulysses* measurements of these thermometers to show that the fast wind originates from regions where the coronal electron temperature is relatively low, and that the slow wind comes from regions of higher coronal electron temperature (Fig. 5.3), at least at heights where the ions freeze in, down around 1.1 solar radii from Sun center. At these distances, the density of the corona drops to a point where particle collisions become infrequent, and the temperatures of the protons and electrons can become different from one another. In active regions and

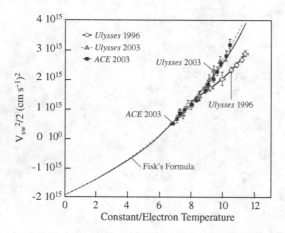

Fig. 5.3 Solar wind velocity and coronal electron temperature. Solar wind velocity, V_{SW}, plotted as a function of inverse coronal electron temperature, $1/T_e$, from the SWICS instruments on *Ulysses* (*filled triangles* and *open circles*) and *ACE* (*filled circles*). The temperatures are inferred from the oxygen ion abundance ratio, O^{7+}/O^{6+}, or oxygen ionized seven and six times. This establishes the freezing-in temperature at which the ions were formed. The temperature when the oxygen ions were created is high in the slow-speed wind, at about 1.6×10^6 K, while the formation temperature for the ions found in the high-speed wind is relatively low, about 1.2×10^6 K. The *dashed* and *solid curves* are fits to the *Ulysses* and *ACE* observations, respectively, using Len Fisk's (2003) model in which coronal loops reconnect with open magnetic field lines to feed the solar winds. The 1996 *Ulysses* high-latitude curve (*open red circles*) has a flatter shape from which small loop heights might be inferred. (Adapted from both the *ACE* and *Ulysses* 2005 Senior Review Proposals.)

the quiet corona, the electrons seem to be hotter than the protons; in coronal holes, the protons seem to be hotter than the electrons.

Some theoretical proposals attribute this result to the radiation energy losses of hot coronal structures. This radiated energy is no longer available to accelerate the solar wind and hence the solar wind originating in hot regions is slower. The temperature trend could reverse at higher altitudes, say between 2 and 10 solar radii, with high coronal electron temperatures that might be associated with the fast wind and vice versa, but there are still no direct measurements of the electron temperature at those heights.

In 2007, Steven R. Cranmer and his co-workers proposed an alternative explanation, incorporating models in which both the slow and the fast winds arise from open magnetic flux tubes rooted in coronal holes, streamers, and active regions, while also tackling the issue of a possible reversal of electron temperature above the freezing-in height.

5.3 Where Do the Sun's Winds Come From?

The two solar winds do not blow uniformly from all points of the Sun, but instead depend on solar latitude. The spatial distribution of the two types of winds also depends upon the Sun's magnetic field configuration, which varies dramatically with the 11-year solar activity cycle.

5.3.1 Source Regions of the Solar Winds Near Solar Activity Minimum

Near activity minimum, the Sun can be described as a simple magnet with north and south poles where large coronal holes are located. The northern hole is of one magnetic polarity, or direction, and the southern one of opposite magnetic polarity (Fig. 5.4). According to a model proposed by Gerald W. Pneuman and Roger A. Kopp in 1971, the negative and positive field lines loop outside the Sun and meet in between the poles, near the solar equator, where a magnetically neutral layer, or current sheet, is dragged out into space by the radially outflowing solar wind (Fig. 5.5). Near the Sun, the current sheet is rooted in a belt of streamers that seems to meander across the star like the seam of a baseball.

The wind's magnetic sectors, detected by spacecraft in the ecliptic (Sect. 2.4), have been described by a "ballerina model", which was advocated in the late 1970s by Hannes Alfvén. Since the Sun's magnetic dipole axis is tilted with respect to its rotation axis, the current sheet is warped. And as the Sun rotates, the current sheet wobbles up and down, like the edge of a spinning ballerina's skirt, sweeping sectors of opposite magnetic polarity past the Earth (Fig. 5.6).

Comparisons of solar X-ray images, obtained from rockets and the *Skylab* mission, with satellite wind measurements taken at about the same time in the 1970s

Fig. 5.4 Coronal structures near sunspot minimum. The Sun's corona becomes visible to the unaided eye during a total solar eclipse, such as this one observed from Oranjestad, Aruba on 26 February 1998, close to a minimum in the Sun's 11-year cycle of magnetic activity. Several individual photographs, made with different exposure times, were combined and processed electronically in a computer to produce this composite image, which shows the solar corona approximately as it appears to the human eye during totality. Note the fine rays and helmet streamers that extend far from the Sun along the equatorial regions (*left* and *right*), and the polar rays (*top* and *bottom*) that suggest a global, dipolar magnetic field. (Courtesy of Fred Espenak.)

established that at least some of the high-speed solar wind has its origin in the extended, low-density coronal holes (Sect. 2.5).

As suggested by Sir William Ian Axford in the 1980s and 1990s, the steady, uniform, high-speed wind emanates from magnetically open configurations in the corona. The open magnetic fields in coronal holes provide a conduit for the fast wind, like the express lane of a divided highway. In contrast, the slow wind, which is filamentary and transient, involves the intermittent release of material from previously closed magnetic regions, like cars leaving a traffic jam, so the slow wind may not be treated as an equilibrium flow in a steady state.

The intrepid explorer, *Ulysses*, has clarified our knowledge of the two solar wind sources by venturing into the polar regions never before visited by any spacecraft. However, *Ulysses* always kept its distance, never passing closer to the Sun than the Earth does. Scientists therefore had to rely on other instruments to tell exactly where the winds come from. Fortunately, the first *Ulysses* measurements, which occurred during its polar orbit in 1994–1995, were obtained near activity minimum with a particularly simple corona characterized by marked symmetry and stability. There were pronounced coronal holes at the Sun's north and south poles, and coronal streamers encircled the solar equator.

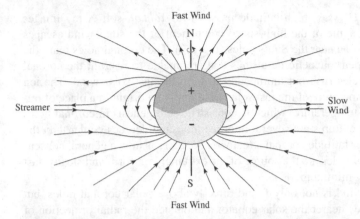

Fig. 5.5 Magnetism at minimum. Theoretical cross section of the magnetic field lines expected in the Sun's corona with a dipole-type geometry and equatorial current sheet near the minimum in the 11-year cycle of solar magnetic activity. The high-speed wind escapes along the open magnetic field lines in the polar regions. At the equator, where the slow wind originates, the magnetic field lines have been pulled outward by the solar wind into oppositely directed, parallel magnetic fields separated by a neutral current sheet. Here, the transition from closed to open field lines at the equator occurs at two and a half times the Sun's radius. [Adapted from Pneuman and Kopp (1971).]

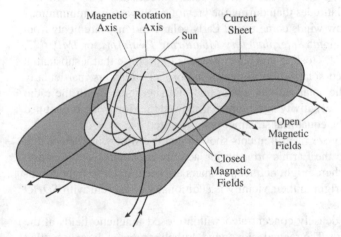

Fig. 5.6 Ballerina model of current sheet. Near the Sun, closed magnetic loops begin and end on the Sun. Further out, magnetic field lines have only one end in the Sun, with the other "end" being carried off by the solar wind. The shaded surface is the heliospheric current sheet, which is bounded by magnetic fields that are directed in and out of the Sun (*arrows*). It is shown as warped, because the equivalent magnetic symmetry axis does not coincide with the Sun's rotation axis. The tilt angle between the two axes causes the current sheet to take the form of a wavy ballerina skirt as it rotates past the Earth

Comparisons of *Ulysses'* high-latitude passes with *Yohkoh* soft X-ray images showed that at least some of the high-speed component of the solar wind escapes from holes in the corona near the Sun's poles (Fig. 5.7). The coronal holes mark the base of extended, open magnetic field lines that are unable to restrain the coronal gas, so the hot particles rush out unimpeded to form the fast solar wind. The ion composition and temperature of the fast solar wind imply an origin in a place of relatively low electron temperature, which is consistent with genesis in coronal holes whose electron temperatures are several hundred thousand degrees cooler than the temperatures in lower latitude coronal streamers. So, there is now a general consensus that polar coronal holes are the source of the extremely uniform and steady, fast solar wind at activity minimum.

The high-speed wind is not only found directly above polar coronal holes, but also at lower latitudes nearer the solar equator and outside the radial projection of the coronal hole edges. One explanation is that the magnetic field lines emerging from coronal holes splay and bend outward, like the petals of an opening flower, allowing the fast wind to spread out as it escapes.

At the next activity minimum, in 2007–2008, *Ulysses* found steady, fast solar wind flows at high solar latitudes, which were virtually indistinguishable from the polar coronal hole winds observed during the spacecraft's first polar orbit, indicating a return to a simple bimodal, fast–slow wind structure associated with large, persistent polar coronal holes and quiescent equatorial streamers. During this minimum, the polar high-speed wind was confined to solar latitudes above 45°, with less spillover to lower solar latitudes than during the previous solar activity minimum.

And where do the slow winds come from? Early solar wind measurements from the sixth, seventh, and eighth *Interplanetary Monitoring Platforms*, or *IMP, 6, 7, and 8*, in 1981 led John T. Gosling and his colleagues to propose that a substantial fraction of the low-speed solar wind originates in coronal streamers, particularly near the minimum in the solar activity cycle, and this site agreed with the radio scintillation observations at activity minimum that showed the slow wind is confined to low latitudes where the equatorial streamer belt is located.

A few years later, *Ulysses* measurements showed that the slow wind detected at distances comparable to the Earth's orbit during activity minimum is confined to the equatorial regions where bright helmet streamers are seen in white-light coronagraphs. A slow-wind origin in their vicinity is additionally consistent with *SOHO* abundance measurements.

Electrified matter is densely concentrated within closed magnetic fields at the base of coronal streamers. The ragged, slow wind probably remains largely bottled up in this magnetic cage. The part that manages to escape seems to be worn out and varies in strength as the result of the effort.

As shown by Neil R. Sheeley Jr., Yi-Ming Wang, and their colleagues (Sheeley et al., 1997; Wang et al., 2000), *SOHO*'s Large Angle and Spectrometric COronagraph, abbreviated LASCO, detects a gusty flow of bright, dense material concentrations, or blobs, from the cusp-like base of the equatorial streamers, which are therefore indeed streaming. The blobs accelerate away from the Sun and flow along the magnetically open streamer axes, or stalks, like leaves floating on a moving stream, tracing the outflow of the slow solar wind.

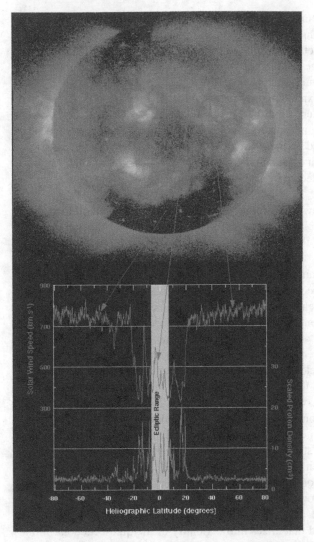

Fig. 5.7 Winds and holes. An X-ray image (*top*) taken with the Soft X-ray Telescope (SXT) on *Yohkoh* is compared with measurements of the solar wind proton flow speed and density obtained with the *Ulysses* SWOOPS instrument (*bottom*) near a minimum in the 11-year solar activity cycle. At least some of the high-speed solar wind originates in coronal holes (the *dark areas* shown at both poles of the X-ray image). At high latitudes the velocity is high and the density is low; near the equator the velocity is low and the density is high. [Adapted from John L. Phillips, et al. (1995), and courtesy of the *Ulysses* SWOOPS consortium. *Ulysses* is a project of international cooperation between ESA and NASA.]

In 1992, Yutaka Uchida and his colleagues reported observations with *Yohkoh's* Soft X-ray Telescope showing that the corona above active regions expands, sometimes continually. Interplanetary scintillation measurements by Paul Hick and his colleagues in 1995 also suggested that active regions might be the source of the slow winds near activity minimum, noticing that this would be consistent with *Yohkoh* soft X-ray observations of expanding active-region loops.

Taro Sakao and co-workers used the X-Ray Telescope aboard *Hinode* to identify, in 2007, X-ray-emitting plasma continually flowing into the upper corona. The continuous outflow occurred at the edge of a solar active region, adjacent to an equatorial coronal hole, with an apparent velocity of $140 \mathrm{km\ s^{-1}}$. Like the expanding active-region corona detected by Uchida, this is another possible source region of the low-speed solar wind.

At least some of the slow wind might also leak out of the bottled-up part of the streamers, like water working its way through a beaver dam or down a clogged sink, but it is not known if the spurt-like blobs or expanding loops dominate the low-speed wind.

In 2002, Leonard Strachan and co-workers used another *SOHO* instrument, the UltraViolet Coronagraph Spectrometer or UVCS, to derive the wind outflow speeds in the vicinity of coronal streamers. They found no detectable outflow in the apparently closed fields of the central core region of a streamer, but flows were found outside the bright streamer edges and along the streamer axis. Moreover, John Raymond had previously used the UVCS instrument to conclude, in 1999, that the abundances in the closed-field parts of the steamers are in complete disagreement with the abundances seen in the slow solar wind, while the open-field "legs" of the streamers exhibit slow-solar-wind abundances. And in 2005, Ester Antonucci and colleagues used UVCS to conclude that the slow wind during the 1996–1997 minimum came from the region external to and running along the streamer boundary and in the region above the streamer core beyond 2.7 solar radii, where the transition between closed and open field lines takes place. The UVCS results collectively suggest that the slow wind comes mainly from the vicinity of the streamer legs, and that the lower "core" never, or only very rarely, opens up.

Still, there is no general agreement about the exact genesis and acceleration mechanism of the slow wind that arises in the vicinity of the quiescent solar minimum streamers. One possible source for the slow solar wind is the magnetic fields in the coronal holes that border the streamers. If there were a magnetic connection between one leg of a streamer and an adjacent open field line, some of the streamer material would be released there. Or the slow solar wind could originate directly from the rapidly diverging, open flux tubes inside coronal hole boundaries.

In another explanation, the streamer stalks serve as the wellspring of the slow solar wind. The inward- and outward-directed magnetic fields on each side of the narrow stalks could be pressed together, joining at the place where they touch. The lower parts of a streamer would then close down and collapse, and the outer disconnected segment would be propelled out to form a gust in the slow solar wind.

In summary, the likely source regions of the two solar winds near activity minimum are coronal holes and coronal streamers. A steady torrent of high-speed wind rushes out of the open magnetic fields in the Sun's polar regions, emerging from the central regions of large coronal holes. A slow, gusty and variable wind moves away from regions near the Sun's equator, associated with streamer cusps, legs, stalks, or the streamer/coronal-hole boundary regions.

5.3.2 Source Regions of the Solar Winds Near Solar Activity Maximum

Ulysses data indicate that the simple, bimodal fast–slow wind structure disappears near the maximum in the 11-year solar activity cycle, when the wind speeds are much more variable and typically lower. A chaotic and complex mixture of varying solar wind flows is found at all solar latitudes, including near the poles (Fig. 5.8). And the winds arise from a variety of sources including coronal holes, coronal streamers, coronal mass ejections, and solar active regions.

These sources are related to the complex magnetic structure near solar activity maximum, which differs from the simpler one at minimum activity. When the Sun's activity cycle peaks, active regions, streamers and long, narrow rays emerge at solar latitudes of up to 45°, with strong magnetic fields that extend far out into the corona

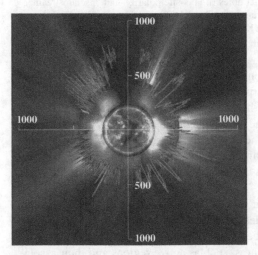

Fig. 5.8 Solar wind speeds at activity maximum. The distribution of solar wind speeds as a function of solar latitude, obtained with the SWOOPS instrument aboard *Ulysses* during its second polar orbit near a maximum in the 11-year solar activity cycle. The north and south poles of the Sun are at the *top* and *bottom* of the plot, the solar equator is located along the *middle*, and the velocities are in units of kilometers per second, abbreviated km s^{-1}. At solar maximum, small, low-latitude coronal holes give rise to fast winds, and a variety of slow-wind and fast-wind sources resulted in little average latitudinal variation of the solar wind speed. (Courtesy of David J. McComas and Richard G. Marsden. The *Ulysses* mission is a project of international collaboration between ESA and NASA. The central images are from the Extreme-ultraviolet Imaging Telescope, or EIT, and the Large Angle and Spectrometric Coronagraph, denoted LASCO, aboard the *Solar and Heliospheric Observatory*, abbreviated *SOHO*, as well as the Muana Loa K-coronameter.)

Fig. 5.9 Coronal structures following sunspot maximum. Miloslav Druckmüller and Peter Aniol took this image of the solar corona during the total eclipse of 29 March 2006 from Libya. At this time, the Sun was in a declining phase from a maximum in the solar activity cycle, in 2000, so polar coronal holes, with their radial magnetic fields, were well developed, but active regions were still present at a variety of solar latitudes. The bulk of the long fine rays arise from polar and low-latitude plumes that overlie small magnetic bipoles inside coronal holes, helmet streamer rays that overlie large loop arcades and separate coronal holes of opposite polarity, and "pseudostreamer" rays that overlie twin loop arcades and separate coronal holes of the same polarity. [Courtesy of Miloslav Druckmüller, Peter Aniol and Serge Koutchmy, and adapted from Wang et al. (2007).]

(Figs. 5.9 and 5.10). These active-region streamers tend to be narrower and brighter than their quiescent equatorial counterparts at solar minimum. And the remnants of these active regions are carried to higher latitudes by supergranular diffusion and meridional flow (see Sect. 3.9). The global magnetic structure is then changing radically, with a switch in overall magnetic polarity; the north magnetic pole eventually becomes south and vice versa, and the dipolar, warped-current-sheet model falls apart. The main current sheet then moves about and secondary sheets are often present.

At activity maximum, the large polar coronal holes shrink and disappear, smaller coronal holes appear at all solar latitudes, the slow and fast winds seem to emanate from all over the Sun, and the high-speed winds abate. The slow winds seem to be associated with closed magnetic structures, such as active regions and their associated streamers, or small coronal holes in their vicinity, while the fast winds rush out of the interiors of the largest of the smaller coronal holes (Fig. 5.11). Coronal mass ejections briefly provide a noticeable third flow as they pass through interplanetary space near the maximum of the activity cycle.

The speed of the solar wind emerging from the two main coronal sites may be related to the structure of the chromosphere underlying them. By combining measurements from two NASA spacecraft, Scott W. McIntosh and Robert J. Leamon were able to show in 2005 that the solar wind speed and composition near Earth

Fig. 5.10 Coronal rays in eclipse composite. Total eclipse of the Sun on 11 August 1999, near a maximum in the 11-year solar activity cycle. These simultaneous images are in white light, taken from the ground (*blue background*) and from a coronagraph in space (LASCO 2 on *SOHO, orange background*). The contrast in both images has been enhanced to reveal the large-scale coronal structures and their sources in the lower corona. Careful alignment shows numerous long, narrow rays in both images, extending several solar diameters into space from both low and high solar latitudes. [Courtesy of Serge Koutchmy and Steele Hill, description in Koutchmy and co-workers (2004).]

seem to be rooted deep within chromosphere structures of different magnetic topology. The *Advanced Composition Explorer*, abbreviated *ACE*, spacecraft measured the wind velocity near Earth, while the *Transition Region and Coronal Explorer*, or *TRACE* for short, was used to measure the time sound waves took to travel between two heights in the chromosphere. The comparison suggested that the speed of the solar wind emerging from a given area of the solar corona might be estimated from the properties of the underlying chromosphere and the magnetic fields threading it. A dense chromosphere might be correlated with closed magnetic structures and a slow, dense solar wind outflow. A less dense chromosphere may be associated with open magnetic fields and fast, tenuous solar winds.

ACE and *Ulysses* studies of solar wind source regions near solar maximum, reported by Paulett Liewer, Marcia Neugebauer, and Thomas Zurbuchen in 2004, indicated that some solar wind originates from open magnetic flux rooted in active regions with observational signatures of small coronal holes. Richard Woo and Shadia Habbal reasoned in 2005 that active regions can be a source of the slow solar wind when the Sun is in a more active phase, even when there is no streamer around. In fact, there are very convincing correlations of coronal composition and the properties of the neighborhoods near active regions, as found by Marcia Neugebauer and co-workers.

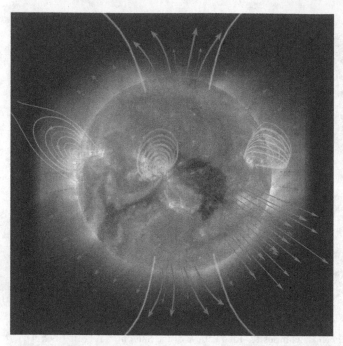

Fig. 5.11 Fast and slow winds, open and closed magnetism. The solar atmosphere, or corona, is threaded with magnetic fields (*yellow lines*). Regions with open magnetic fields, known as coronal holes, give rise to fast, low density, solar wind streams (*long, solid red arrows*). In addition to permanent coronal holes at the Sun's poles (*top* and *bottom*), coronal holes can sometimes occur closer to the Sun's equator (*center*). Areas with closed magnetic fields yield the slow, dense wind (*short, dashed red arrows*). Comparisons of *TRACE* images with solar wind *ACE* data indicate that the speed and composition of the solar wind emerging from a given area have deep roots in the chromosphere. There is a shallow dense chromosphere below the strong, closed magnetic regions with a slow, dense, solar wind outflow; deep, less dense chromosphere is found below the open magnetic regions with fast, tenuous, solar wind outflow. This image was taken on 11 September 2003, with the Extreme-ultraviolet Imaging Telescope, abbreviated EIT, aboard the *SOlar and Heliospheric Observatory*, or *SOHO* for short. (Courtesy of the *SOHO* EIT consortium. *SOHO* is a project of international collaboration between ESA and NASA.)

And in the meantime other spacecraft were beginning to zero in on the detailed source regions of the fast solar wind in polar coronal holes.

5.3.3 Fast Wind Sources in Coronal Holes

The open, nozzle-like polar plumes might be an important source of the high-speed wind. These dense, long, narrow features are the brightest ultraviolet things found in a dark coronal hole and something has to be energizing them. The tall, conspicuous features rise out of Earth-sized magnetic regions in the photosphere, and extend millions of kilometers into space like long ropes, lasting for about a day but recurring from the same footpoint.

In 1998, Klaus Wilhelm and his colleagues used the Solar Ultraviolet Measurements of Emitted Radiation, abbreviated SUMER, instrument on *SOHO* to determine the physical parameters above a polar coronal hole, stating that the polar plumes are not a major source of the fast solar wind streams. Since the high-speed wind was apparently coming out of the entire coronal hole, with no substantial difference in speed between plumes and adjacent places, and since the interplume regions occupy most of the polar-hole area, the interplume regions were probably the main source of the polar high-speed wind. More recently, in 2005, Alan H. Gabriel and his colleagues extended the observations to higher heights, with *SOHO's* UltraViolet Coronagraph Spectrometer, or UVCS for short, to conclude that the outflow velocities in the plumes exceed those in the interplume regions at low heights and fall below interplume velocities at heights greater than 1.6 solar radii, presumably through mass transfer to the interplume regions, so that approximately half of the fast solar wind arises from plumes. Thus, there is a controversial uncertainty about how much of the mass, momentum, and energy flux of the fast solar wind comes from the dense polar plumes or in the low-density interplume region.

And if the plumes are not the main source, there are other bright places that could provide most of the action in coronal holes, such as the X-ray bright points and jets. Leon Golub and his colleagues discovered the ubiquitous X-ray bright points in 1974 *Skylab* observations (Sect. 2.5), observing them at all latitudes from equator to poles. The X-ray jets, associated with flare-like bright points, were discovered by Kazunari Shibata and his colleagues in 1992 using observations with the soft X-ray telescope aboard *Yohkoh*; they proposed that the X-ray jets are caused by reconnection between emerging magnetic flux and the overlying coronal magnetic field. In 1998, Yi-Ming Wang and his co-workers reported correlated *SOHO* observations of white-light and extreme-ultraviolet jets from polar coronal holes, proposing that they are triggered by small-scale magnetic reconnection events between emerging bipolar loops and neighboring open, unipolar flux in coronal holes. And in 2007, Shibata and his co-workers used *Hinode* observations to discover large numbers of jets in the chromosphere, with inverted Y-shapes suggesting such magnetic reconnections.

Observations of X-ray jets in polar coronal holes, with the soft X-ray telescope aboard *Hinode* and reported by Jonathan W. Cirtain and colleagues in 2007, indicate that these jets may contribute significantly to the high-speed wind from coronal holes, but the relative amounts of the fast solar wind contributed by plumes, interplume regions, and jets from polar coronal holes are still uncertain.

By examining the ultraviolet lines emitted at the base of the corona in coronal holes, Donald M. Hassler and his colleagues have used the SUMER instrument on *SOHO* to show that the high-speed outflow emerges mainly from the funnel-like boundaries of the magnetic network formed by underlying supergranular convection cells (Fig. 5.12). They seem to gush out of the crack-like edges of the network, like grass or weeds growing in the dirt where paving stones meet. This implies that the high-speed winds could be expelled by explosive magnetic interactions along the boundaries or boundary intersections of the chromosphere's magnetic network inside coronal holes.

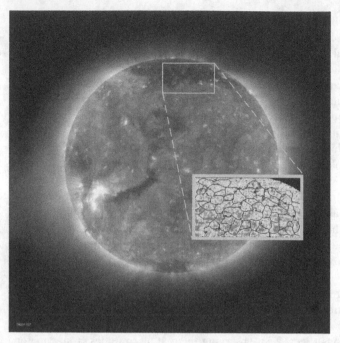

Fig. 5.12 Source of the high-speed solar wind. This full-disk image of the Sun shows coronal gas at 1.5×10^6 K, shining in the extreme-ultraviolet light of 11 times ionized iron, the Fe XII line at 19.5 nm. Bright regions indicate the hot, dense material confined within strong magnetic loops, while the dark polar regions (*top* and *bottom*) imply open magnetic fields. These polar coronal holes are the source of much of the high-speed solar wind. The inset provides a close-up, Doppler velocity map of the million-degree gas at the base of the corona where the fast solar wind originates, taken in the extreme-ultraviolet light of six times ionized neon, the Ne VIII line at 77 nm. *Dark blue* represents an outflow, or blueshift, at a velocity of 10 km s^{-1}; it marks the beginning of the high-speed solar wind. *Dark red* indicates a downflow at the same speed. Superposed are the edges of the "honey-comb" shaped pattern of the magnetic network, where the strongest outward flows (*dark blue*) are found. The relationship between the outflow velocities and the network suggests that the high-speed wind emanates from the boundaries and boundary intersections of the magnetic network. These observations were taken on 22 September 1996, with the Extreme-ultraviolet Imaging Telescopes, abbreviated EIT (*full disk*) and the Solar Ultraviolet Measurements of Emitted Radiation, or SUMER for short (*velocity inset*) spectrometer on the *SOlar and Heliospheric Observatory*, abbreviated *SOHO*. [Adapted from Hassler et al. (1999), courtesy of the *SOHO* EIT consortium and the *SOHO* SUMER consortium. *SOHO* is a project of international collaboration between ESA and NASA.]

The network boundaries are places where the magnetic fields are concentrated into funnel-shaped magnetic fields that open up and rapidly fan out to fill the overlying corona. Chuan-Yi Tu and his colleagues have used the *SOHO* SUMER instrument to show that the solar wind indeed escapes from coronal holes in funnels of open magnetic fields (Fig. 5.13). The winds start flowing out of the corona at between 5,000 and 20,000 km above the photosphere, with outflow speed of 10 km s^{-1} or less. As we shall next see, other mechanisms are required to accelerate the fast winds to their interplanetary speeds of about 750 km s^{-1}.

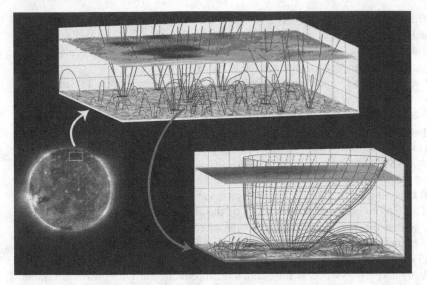

Fig. 5.13 Coronal funnels. The fast solar wind emerges from the coronal funnels located in polar coronal holes. The magnetic curves are open magnetic field lines; *dark gray arches* show closed ones. The *lower* diagram shows the magnetic vertical components measured by the *SOHO* MDI instrument (*blue to red*). The *upper* diagram shows Ne VIII Doppler shifts from the *SOHO* SUMER instrument (*hatched regions have large outflows*), together with the inclination of the magnetic field (*blue to red*), as extrapolated from MDI measurements. [Courtesy of the *SOHO* MDI and SUMER consortia, and adapted from Tu et al. (2005). *SOHO* is a project of international collaboration between ESA and NASA.]

5.4 Getting Up to Speed

What forces propel the solar wind to its supersonic velocities of hundreds of kilometers per second? Since the winds are due to the expansion of the hot corona, their acceleration is related to the heating of the solar corona, discussed in the preceding Chap. 4. But we will now separately discuss the acceleration of the slow and fast winds, describing measurements of the relevant velocities and proposed acceleration mechanisms.

The solar wind outflow begins slowly near the Sun, where the solar gravity is the strongest, and then continuously accelerates out into space, gaining speed with distance. But the wind cannot go on accelerating forever, for there is a limit to the amount of energy being pumped into it. The slow and fast winds therefore break away from the Sun, and eventually cruise along at a roughly constant speed, called the asymptotic or terminal velocity.

More than one mechanism is required to explain the acceleration of the two types of solar winds, the slow and fast ones, for they move in very different ways. The slow wind naturally reaches terminal velocities of a few hundred kilometers per second as the corona expands away from the Sun. Additional energy must be deposited in the low corona to give the fast wind an extra boost and double its speed. In technical terms, the fast wind has a velocity and mass flux density that are too high to be explained by heat transport and classical thermal conduction alone. And when the complicated magnetic field geometry of the slow wind source regions is taken into account, additional energy may be needed there as well.

And where are the winds accelerated? Near the minimum in the 11-year solar activity cycle, the slow winds that are detected by *Ulysses* at distances comparable to the Earth are found at low solar latitudes where coronal streamers reside. And time-lapse sequences of white-light coronagraph images, obtained during activity minimum conditions in 1996 with *SOHO's* Large Angle and Spectrometric Corona-graph, or LASCO for short, indicate that the coronal streamers are far more dynamic than was previously thought. The LASCO images show that blobs of high-density material are being ejected at the streamer cusps and accelerate outward along their stalks. These blobs are regarded as tracers of the ambient slow solar wind flow, and they show that the slow wind takes a while to get up to speed. In 1997, Neil R. Sheeley Jr. and coworkers showed that they accelerate at a leisurely pace and move out to 25 solar radii before reaching a terminal velocity of about 300 km s^{-1}, and Yi-Ming Wang and his co-workers substantiated this conclusion with greater amounts of data in 2000 (Fig. 5.14).

In 2002, Leonard Strachan and co-workers used another *SOHO* instrument, the UltraViolet Coronagraph Spectrometer or UVCS, to derive the slow wind outflow in coronal streamers, finding low-speed winds along the streamer axis at heights above about 3.5 solar radii from Sun center, with a gradual increase to only 90 km s^{-1} at about 5 solar radii in the streamer stalk.

Radio scintillation measurements indicate that the fast solar wind accelerates quickly, like a racehorse breaking away from a starting gate. The high-speed wind flowing from polar coronal holes reaches terminal speeds of 750 km s^{-1} within 10 solar radii (Fig. 5.15). And in 1998 John Kohl, Steven R. Cranmer, and co-workers used UVCS measurements of the velocities, derived from the dimming of spectral lines caused by Doppler shifts, to suggest that the high-speed solar wind emerging from coronal holes accelerates to supersonic velocity within just 2.5 solar radii from the Sun center. So the fast wind is accelerated relatively close to the Sun.

In 1977, Randolph H. Levine, Martin D. Altschuler, and John W. Harvey reported that the fastest solar-wind streams are correlated with those magnetic flux tubes that expand least in cross-sectional area over the distance between the photosphere and the coronal height where the solar wind begins. Then in 1990, Yi-Ming Wang and Neil R. Sheeley Jr. (1990a) showed that the solar-wind speed is inversely correlated with the rate of magnetic flux-tube expansion in the corona. If the field strength falls off slowly with height, the "heating" energy extends to greater heights, permitting acceleration of the winds to higher speeds. In other words, the faster winds are ac-celerated in the slowly diverging, open field lines. When the field diverges rapidly with height, less of the deposited energy goes into accelerating the wind. It is some-thing like the rapid jet of water from an open nozzle on a hose, and the slower, fountain-like spray from a partially closed nozzle.

The fast solar wind could be accelerated to its high speeds in open magnetic funnels located in coronal holes. Closed magnetic loops may be swept by underlying

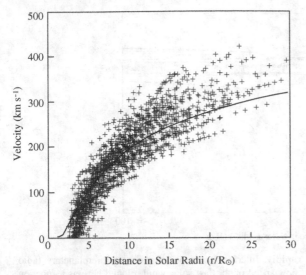

Fig. 5.14 Slow winds break out slowly. Time-lapse sequences of white-light images, obtained with the *SOHO* LASCO instrument, show prominent features, sometimes called blobs, that trace out the flow of the slow wind during a minimum in the 11-year solar activity cycle. The blobs move radially out from the Sun in its equatorial regions at the vicinity of helmet streamers. The velocity of about 80 blobs, observed along edge-on structures, is plotted versus distance from the Sun center in units of the solar radius. The speed does not reach an asymptotic speed of roughly 300 kilometers per second, denoted 300 km s^{-1}, until about 15 solar radii from the Sun center. The *solid curve* shows Eugene Parker's (1963) solution for a radially expanding, isothermal wind at a temperature of 1×10^6 K, whose sonic point is located at 5.8 solar radii. [Courtesy of Yi-Ming Wang, and adapted from Wang et al. (2000).]

convection into the funnel regions where they undergo magnetic reconnection with the existing open magnetic field lines, releasing energy to power the fast wind in polar coronal holes.

In 1999, for example, Len A. Fisk, Nathan A. Schwadron, and Thomas H. Zurbuchen proposed that the fast solar wind is accelerated by the continual emergence of small magnetic loops that reconnect with open field lines near their base. Fisk extended the idea in 2003, reasoning that both the fast and the slow winds arise from this process, and that the inverse dependence of solar wind speed on electron temperature is an intrinsic signature of the material in the coronal loops that are releasing mass into the open regions. An alternative model, proposed by Steven R. Cranmer and his co-workers in 2007, associates most of the solar wind, both slow and fast, with open, "unipolar" regions on the Sun where loops do not seem to be prevalent.

The *SOHO* UVCS instrument has demonstrated that the fast solar wind in coronal holes at solar minimum exhibits the most bizarre and unexpected motions imaginable. In 1998, John L. Kohl, Giancarlo Noci, Steven Cranmer, and their co-workers announced that the Doppler broadening of ultraviolet lines indicates that oxygen ions have velocities that depend on direction, with larger velocities in the direction perpendicular to the radial than along it. And since the open magnetic fields extend

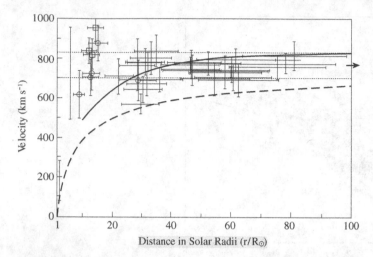

Fig. 5.15 Fast winds accelerate rapidly. Interferometric observations of interplanetary radio scintillations measure the apparent flow speed of the fast solar wind, plotted here as a function of distance from the Sun center in units of the solar radius. The results indicate that the fast solar wind accelerates to its high velocity very close to the Sun, within 10–20 solar radii. The *vertical bar* on each data point is the 90% confidence limit, and the *horizontal bar* indicates the distance range over which the scintillation estimate is averaged. Measurements with the Very Long Baseline Array, abbreviated VLBA, are marked with circles and squares. The upper and lower bounds of the *Ulysses* velocity measurements, at much larger distances from the Sun, are plotted as *horizontal dotted lines*, and the mean *Ulysses* fast-wind velocity is marked with an arrow at 100 solar radii. The flow speed from a wave-assisted acceleration model is plotted as a *dashed line*, and the apparent scintillation velocity calculated from this model is plotted as a *heavy solid line*. The point nearest the Sun is estimated from *Spartan-201* coronagraph measurements. [Adapted from Grall et al. (1996).]

out of coronal holes in the radial direction, the ion velocity measured perpendicular to the magnetic field is greater than that in the parallel direction. A similar velocity anisotropy is found further out, by in situ measurements of the protons in the solar wind.

Moreover, the same heavy ions, the less abundant ones, are preferentially accelerated, even though they experience larger gravitational forces due to their extra mass (Fig. 5.16). Above two solar radii from the Sun's center, oxygen ions within coronal holes move at about twice the speed of neutral hydrogen, which in the lower corona should be strongly coupled to the protons. The oxygen ions and the protons seem to live on their own, isolated from each other and cut off from their neighbors. They just do not have enough time to jostle together and smooth out their velocity differences.

So the heavier particles in coronal holes are moving faster, but even if they are separated from their lighter neighbors it is still very surprising. It would be something like heavier adults jogging around a racetrack much more rapidly than lighter, slimmer youngsters, or overweight people at a party dancing wildly about when all their slim companions are waltzing at a slow tempo.

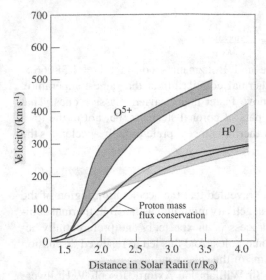

Fig. 5.16 Heavier ions move faster than lighter ones in coronal holes. Outflow speeds at different distances over the solar poles for hydrogen atoms, denoted by H^0 or H I, and ionized oxygen, designated O^{5+} or O VI. Here the distances are given in units of the solar radius, denoted $R_\odot$. These data were taken in late 1996 and early 1997, with the UltraViolet Coronagraph Spectrometer, abbreviated UVCS, aboard the *SOlar and Heliospheric Observatory*, or *SOHO* for short. They show that the heavier oxygen ions move out of coronal holes at faster speeds than the lighter hydrogen, and that the oxygen ions attain supersonic velocities within 2.5 solar radii from the Sun center. The proton mass flux conservation area denotes the hydrogen outflow speed derived from mass flux conservation; for a time-steady flow, the product of the density, speed, and flow-tube area should be constant. (Courtesy of the *SOHO* UVCS consortium. *SOHO* is a project of international collaboration between ESA and NASA.)

In contrast, within equatorial regions where the slow-speed wind begins, the lighter hydrogen moves faster than the oxygen, as one would expect for a gas with thermal equilibrium among different types of particles that remain in contact with each other (Focus 5.3). Frequent collisions within the dense central parts of coronal streamers, where the density is greater than in coronal holes, adjust particle temperatures to similar values, while also wiping out any memory of the initial acceleration mechanism and erasing signatures of it. The UVCS observations, reported by Richard Frazin and co-workers in 2003, nevertheless indicate particle velocity, or temperature, differences at the low-density edges of coronal streamers that resemble those in coronal holes.

Focus 5.3

Temperature, Mass, and Motion

The kinetic temperature, T, of a particle is obtained by equating the thermal energy of the particle to its kinetic energy of motion, or by:

$$\text{Thermal energy} = \frac{3}{2}kT = \frac{1}{2}mV_{\text{thermal}}^2 = \text{Kinetic energy}$$

or equivalently:

$$T = \frac{mV_{\text{thermal}}^2}{3k},$$

where m is the mass of the particle and Boltzmann's constant $k = 1.38066 \times 10^{-23}$ J K^{-1}. When particles are in thermal equilibrium at the same temperature, lighter particles with smaller mass move faster than heavier, massive ones. That is what happens in the dense, central pars of coronal steamers, but not in the low-density coronal holes where some non-thermal process preferentially accelerates the heavier particles.

So the UVCS instrument on *SOHO* revealed that the acceleration region of the fast solar wind, in polar coronal holes at activity minimum, is far from thermal equilibrium, and that some non-thermal process is unexpectedly and preferentially accelerating the heavier particles in coronal holes. And this led to renewed discussion of collisionless acceleration mechanisms of the solar wind.

In one explanation, developed by Sir William Ian Axford, Joseph V. Hollweg, Phillip A. Isenberg, Eckart Marsch, James F. McKenzie, and others, magnetic waves pump energy into the heavy ions by driving their gyrations around the magnetic field lines. The charged particles move around the magnetic fields in coronal holes in the same way that charged particles move in laboratory particle accelerators or cyclotrons. As the waves move along the magnetic fields, they will produce rapid gyrations in the direction perpendicular to the fields and little extra motion along them, something like a hula-hoop that moves in and out from your hips but not up and down them.

According to the cyclotron resonance theory, the heavier ions gyrate with lower frequencies where the magnetic waves are most intense. The ions resonate with the waves that match their natural vibration in a magnetic field, at the cyclotron frequency (Focus 5.4). These waves damp and dissipate energy, losing it when they resonate with the ions, and thereby accelerate them. And because the heavier ions consume more magnetic-wave energy, they are accelerated to higher speeds. This process might explain the rapid acceleration of the fast wind close to the Sun, in coronal holes at activity minimum.

Focus 5.4

Particles Gyrating about Magnetic Fields

If a magnetic field of strength, **B**, acts on a particle with charge eZ, for electron charge e, and velocity, **V**, the particle experiences a force, **F**, called the Lorentz force:

$$\mathbf{F} = eZ(\mathbf{V} \times \mathbf{B})$$

and from Newton's law for a particle of mass, m, and momentum, $m\mathbf{V}$, we have

$$m\frac{d\mathbf{V}}{dt} = \mathbf{F} = eZ(\mathbf{V} \times \mathbf{B}),$$

where we have assumed that gravitational forces are negligible.

The motion of the charged particle is a circle, and it does not change the particle's kinetic energy. If $V_\perp$ denotes the component of velocity perpendicular to the magnetic field, we can rewrite our equation as:

$$\frac{d^2 V_\perp}{dt^2} = \frac{eZB}{m} \frac{dV_\perp}{dt} = \left(\frac{eZB}{m}\right)^2 V_\perp,$$

which describes circular motion with a cyclotron frequency, v_c, given by:

$$v_c = eZB/m = 1.8 \times 10^{11} (Zm_e/m)B \text{ Hz},$$

where m_e denotes the electron mass, the magnetic field strength B is in units of Tesla, and 1 Hz is equivalent to one cycle per second.

The radius of gyration, R_C, is obtained from

$$R_C = \frac{V_\perp}{v_C} = \frac{mV_\perp}{eZB}.$$

In the context of the acceleration of particles by waves in coronal holes, there is more power in the lower frequencies, and heavier particles gyrate at these lower frequencies.

Observations, models, and theory have all been applied to the ion cyclotron resonance process in recent years. In 2000, for example, Steven R. Cranmer considered the wave-dissipation arising from more than 2000 low-abundance ion species, concluding that the waves that accelerate the high-speed solar wind must be generated throughout the extended corona. Five years later, Xing Li and Shadia Habbal (2005) focused on the acceleration process due to the resonant interaction between the ion cyclotron waves and both oxygen ions and protons.

But there is no agreement on where these waves come from. According to one hypothesis, proposed by Joseph V. Hollweg and W. Johnson in 1988, the Sun could launch low-frequency Alfvén waves that cascade up to high-frequency waves where they are dissipated by cyclotron resonance. Yet, Philip A. Isenberg, one of the pioneering experts in theoretical aspects of the process, cast a shadow of uncertainty on all these speculations, declaring in 2004 "the generation of fast solar wind is not caused by the collisionless dissipation of parallel propagating ion cyclotron waves." But perhaps, oblique waves dominate the solar wind, rather than parallel propagating ones, or there might be some other assumption that could be corrected to alleviate the model's failure.

One model attributes the rapid, close acceleration of the solar wind to protons that have been pumped up by cyclotron resonance. Others argue that the whole resonance idea is wrong for protons, the main ingredients of the solar wind, but still viable for heavy ions. So even if the ion cyclotron resonance process does preferentially accelerate the minor heavy ions, we still might not understand the physical processes that accelerate the other charged particles in the solar wind – the protons and electrons.

This brings us to the observation of Alfvén waves blowing further out in the winds.

5.5 Riding the Waves

The solar wind is described by the equations of magnetohydrodynamics, or MHD for short, which specify the dynamics of an electrically conducting fluid in the presence of a magnetic field. This is just what the solar corona and wind are all about – the motion of electrically charged particles in a magnetic field. And the general properties of the solar wind, derived by Eugene Parker in 1958, are specified by the MHD equations of the steady, radial expansion of the million-degree corona.

More than 40 years of in situ measurements have nevertheless demonstrated that the solar wind is by no means steady, spherically symmetric, or without structure. The wind's physical properties are instead variable in both space and time, and these variations are often attributed to waves that can be observed by spacecraft and specified by the MHD equations.

Hannes Alfvén described one such oscillation in 1942; he received the Nobel Prize in Physics in 1970 for his pioneering MHD work. Alfvén showed that when charged particles are perturbed in the presence of a magnetic field, they will move up, down, and forward, oscillating and propagating like a wave. These oscillations are now known as Alfvén waves, which move at the Alfvén velocity. Their speed increases with increasing magnetic field strength and decreasing particle density, with a value of roughly $1,000 \, \text{km s}^{-1}$ in the low solar corona and about $32 \, \text{km s}^{-1}$ in the solar wind at the Earth's orbit.

The presence of Alfvén waves in the solar wind has been confirmed by direct observations ever since the late 1960s. Paul J. Coleman Jr., for example, noticed in 1968 that the solar wind flow is turbulent, as evidenced by variations in the magnetic field strength and wind velocity observed from *Mariner 2*. He used these measurements to estimate the power spectra of the turbulent variations, and noticed that they would produce Alfvén waves, especially in the fast wind. And within a year, John W. Belcher, Leverett Davis Jr., and Edward J. Smith attributed the magnetic fluctuations observed from *Mariner 5* to large-amplitude Alfvén waves. And in 1971, both Belcher and independently G. Alazraki and P. Couterier described solar wind acceleration by the gradient in Alfvén wave pressure.

The ubiquitous Alfvén waves have been continuously reconfirmed for decades, usually from spacecraft moving in the ecliptic and near the Earth's orbit; the observed waves have periods of hours, and predominantly propagate away from the Sun.

Alvfén waves were detected far outside the ecliptic, above the Sun's polar regions, when magnetometers aboard *Ulysses* detected magnetic fluctuations in the fast, high-latitude wind, with periods of 10–20 h. They were changes in the direction of the magnetic field, and not variations in its strength. Andre Balogh, Edward J. Smith, Bruce T. Tsurutani, and their colleagues attributed the fluctuations to large-amplitude Alfvén waves, similar in properties to those seen in the fast solar wind near the ecliptic by *Helios 1* and 2 and previous spacecraft.

The magnetic waves rippling through the Sun's polar regions tend to block incoming high-energy cosmic rays, intruders from outer space. Scientists expected that the charged cosmic ray particles would be able to penetrate deep within the regions above the Sun's poles, where the magnetic fields stretch out radially and smoothly with little twist. Since solar rotation winds up the solar wind's magnetism in the equatorial plane of the Sun, cosmic rays might have more difficulty in penetrating these regions. After all, increased magnetism in the solar wind, associated with a maximum in the 11-year solar activity cycle, was known to cut off cosmic rays so that fewer of them reach the Earth at activity maximum (Sect. 2.4).

Ulysses' instruments surprised nearly everyone; they did not register substantially more cosmic rays over the poles than near the ecliptic. As suggested by J. Randy Jokipii and Joseph Kóta in 1989, strong magnetic waves in the polar regions can repel the high-energy ions, sending them back into space. The incoming cosmic rays meet an opposing force, like a swimmer encountering powerful currents or pounding surf. To put it in more scientific language, the Alfvén waves are very long, with wavelengths reaching one-third the distance between the Earth and the Sun, or 0.33 AU, so they can resonate with the energetic cosmic rays and oppose their entry into the polar regions.

The Alfvén waves exert a pressure that can provide an extra boost to the heat-driven solar wind, accelerating it to higher speed. This has led to a long series of models of wave-driven winds, ever since Alfvén waves were discovered in the solar wind and John W. Belcher proposed that they push the solar wind to its higher speed. In effect, the fast winds are pushed along by the pressure of the Alfvén waves, resembling the way the Sun's radiation pressure might propel a spacecraft with a large enough solar sail.

When reviewing the wind-acceleration mystery in 2006, Joseph V. Hollweg nevertheless pronounced that wave-driven models all ultimately fail to explain the rapid acceleration of the fast wind close to the Sun. But in 2005, Steven R. Cranmer and Adriaan van Ballegooijen described the generation, propagation, and reflection of Alfvén waves all the way from the solar photosphere to the distant heliosphere, concluding that their model of these waves gives the right amount of wave damping and associated heating of the extended solar corona. One unsettled question is whether there is enough power in the Alfvén waves at low frequencies close to the Sun to drive the fast solar wind.

Some evidence is provided from the observations of cosmic radio sources or radio signals from spacecraft as their radiation passes through the solar wind. Characteristics of coronal Alfvén waves inferred from such measurements of natural radio sources led Salvatore Mancuso and Steven R. Spangler to report in 1999 that there is not enough Alvén wave flux to heat and accelerate the solar wind. On the other hand, V. E. Andreev and colleagues reported in 1997 that observations of radio signals from the two *Helios* spacecraft demonstrate substantial transfer of wave energy to the solar wind. And John K. Harmon and William A. Coles' announced in 2005 the possible indirect evidence for Alfvén waves near the Sun in radio scattering and scintillation observations of the inner solar wind.

As mentioned in Sect. 4.5, instruments aboard *Hinode* have detected Alfvén waves lower down and closer to the Sun. Observations with its X-Ray Telescope reveal ubiquitous X-ray jets in polar coronal holes and provide evidence for Alfvén waves generated at the bottom of coronal holes, leading Jonathan W. Cirtain and his colleagues to speculate in 2007 that these waves may contribute to the high-speed solar wind. And observations with the Solar Optical Telescope on *Hinode*, reported by Bart De Pontieu and co-workers (2007b), indicate that the chromosphere is permeated with Alfvén waves that are energetic enough to accelerate the solar wind.

So the controversy continues, and they say old theories never die. They are just resurrected in a slightly different form. We can therefore count on future acceleration models that incorporate magnetic Alfvén waves, and new observations might yet substantiate one of them.

5.6 Magnetic Connections

5.6.1 Flowing Wide Open

Magnetic fields generated within the Sun can emerge into the outer solar atmosphere, the corona, forming magnetic loops that are often filled with material heated to temperatures of a few million Kelvin. But, owing to its high temperature, some of the electrified coronal material escapes the Sun's gravitational pull, and carries the solar magnetic field into the heliosphere (Fig. 5.17). These magnetic fields are open, pointing in just one direction – either in or out of the Sun, with one end rooted in the Sun and the other transported out into the heliosphere by the solar wind.

Whenever spacecraft have ventured into interplanetary space, either within the ecliptic or at higher solar latitudes, the large-scale magnetic field extending from the Sun is open, never closed into a loop-like structure. The only exception is the short-lived, loop-like magnetic structures hurled through space by coronal mass ejections, which can contribute a noticeable fraction of the magnetic fields in the heliosphere near a maximum in the 11-year solar activity cycle.

As *Ulysses* traveled along its orbit, the spacecraft's magnetometers always measured a similar component of the magnetic field, pointing in the radial direction away from the Sun and when normalized to the same distance from it. This indicated that the Sun's distant, open magnetic flux is uniform all around the Sun, or at all solar latitudes and longitudes.

This is not what you would expect for the distribution of magnetic fields near a dipole, which is more concentrated over the poles. The magnetic fields of a dipole come together at the poles and spread out between them. But by the time the solar wind has traveled out to *Ulysses'* distance, which is comparable to that of the

Fig. 5.17 Magnetic fields near and far. In the low solar corona, strong magnetic fields are tied to the Sun at both ends, trapping hot, dense electrified gas within magnetized loops. Far from the Sun, the magnetic fields are too weak to constrain the outward pressure of the hot gas, and the loops are opened up, allowing electrically charged particles to escape, forming the solar wind, and carrying magnetic fields away. (Courtesy of Newton Magazine, the Kyoikusha Company.)

Earth or more, the solar magnetism had been redistributed, producing a uniform radial field.

Moreover, the open magnetic flux remained approximately constant during the entire 11-year solar activity cycle. Near sunspot and activity minimum, most of the open flux resides in large polar coronal holes, whereas at sunspot and activity maximum it is rooted in relatively small, low-latitude holes located near active regions and characterized by strong footpoint fields. Since the decrease in total area occupied by the holes is offset by the increase in magnetic field strength, the open magnetic flux, or magnetic field strength per unit area, remains roughly constant from minimum to maximum.

Something is transforming a complex and varied magnetic structure near the Sun into a uniform, open magnetic flux further out. And this apparently happens both near activity minimum, when magnetic fields extending from coronal holes dominate much of the heliosphere, and near the maximum in solar activity when the open flux also originates in the vicinity of active regions.

So what keeps the magnetic flux open, uniform, and radial far from the Sun? Although the technical details differ, several groups have proposed that the Sun's open magnetic regions form as the result of the emergence, dispersal, and/or interaction of magnetic fields in the underlying photosphere, particularly in coronal holes where most of the open flux is concentrated. These holes rotate at the same speed at all solar latitudes, in sharp contrast with the differential rotation of the underlying photosphere, which spins faster at lower solar latitudes. In 1988, Ana G. Nash, Neil Sheeley Jr., and Yi-Ming Wang proposed, for example, that the uniform, rigid rotation of large, polar coronal holes is due to the ongoing field-line reconnections, as the result of differential rotation, supergranular diffusion, and meridional flows in the photosphere and below.

And Len Fisk and his colleagues have also attributed the Sun's open magnetic flux to a diffusive displacement process resulting from the reconnection of open field lines with closed loops. This reconnection between closed solar and open heliospheric field lines has been dubbed an interchange reconnection, since their identities are interchanged. A footpoint initially attached to a closed loop becomes open, and vice versa.

Wang and Sheeley addressed the problem again in 2004, arguing that footpoint exchanges between open and closed magnetic field lines, the interchange reconnections, cause open flux to jump from one location to another when active regions emerge. They act to untie the rotation of coronal holes from the underlying photosphere, counteracting differential rotational shear in the holes and maintaining their uniform rotation.

The Sheeley–Wang duo also noticed that some of the open magnetic flux eventually closes down again, producing inflowing material that helps to keep the total amount of open magnetic flux roughly constant over the solar cycle. The episodic downflows generally occur below 5.5 solar radii from the Sun center, the point of no return for the escaping solar gas and magnetic fields. The returning inflow most likely occurs after magnetic reconnection of adjacent open fields that point in and out of the Sun at magnetic sector boundaries.

5.6.2 Unexpected Behavior

The heliosphere's magnetic field does not just placidly stream away from the Sun, entrained in the steady flow of the solar wind. The magnetic field lines are instead going where they are not supposed to be, and making forbidden liaisons there. They reorganize, merge, and join together into new configurations, reconnecting with each other and even disconnecting from the Sun. Significant amounts of magnetic energy can be extracted to accelerate particles in the process.

Direct evidence that such magnetic reconnection occurs in the solar wind has been found in *Advanced Composition Explorer*, abbreviated *ACE*, observations near the Earth's orbit. As reported by John T. Gosling and his co-workers (2005a, b), the main evidence consists of an accelerated ion flow within magnetic field reversals in the solar wind. The high-speed ions are hurled out in a direction perpendicular to the X-line where oppositely directed magnetic field lines meet and reconnect.

These jetting flows have been called Petschek-type reconnection exhausts, as the result of Harry E. Petschek's 1963 theoretical discussion of the annihilation of oppositely directed fields and the related conversion of magnetic energy into particle energy during solar flares. The solar wind exhausts are often found in the compression regions of coronal mass ejections arriving near Earth.

As reported by Tai D. Phan and coworkers in 2006, and Gosling and his colleagues in 2006 and 2007, multi-spacecraft observations of the reconnection exhausts in the solar wind suggest that the accelerated flow associated with reconnection in the solar wind is a large-scale, prolonged, and steady process that can extend along enormous distances. In 2007, for example, Gosling and his co-workers reported observations of exhaust jets from a magnetic reconnection X-line extending more than 4×10^6 km, or hundreds of Earth radii, in the solar wind at the Earth's orbit, persisting for at least 5 h and 20 min. These oppositely directed plasma jets were observed from a flotilla of five well-separated spacecraft – *STEREO A* and *B*, *ACE*, *Wind*, and *Geotail*.

As the solar wind streams radially outward, pulling the Sun's magnetism with it, the rotating star twists the magnetic fields into a spiral shape within the plane of the Sun's equator. It is sometimes called the Parker spiral, after Eugene Parker who explained it in 1958. The shape is also that of an Archimedean spiral described in the third century BC by the Greek mathematician Archimedes. (An Archimedean spiral is the locus of points described by a point moving away from a fixed point with a constant speed along a line that rotates with constant angular velocity.) Measurements from spacecraft have demonstrated that the interplanetary magnetic field in the ecliptic is oriented in a simple spiral pattern, if averaged over a sufficiently long time.

Observations from *Ulysses* also showed that the large-scale, global magnetic field is in good agreement with Parker's model throughout the heliosphere, although the magnetic spiral is less tightly wound than predicted at high solar latitudes. The high-latitude distortions of the spiral shape may be due to the magnetic waves propagating out and above the Sun's polar regions.

Since charged particles cannot easily cross magnetic field lines, high-speed electrons that are hurled into space by solar flares must follow a curved spiral path rather than a straight line. And these electrons have trajectories, inferred from their radio emissions, which confirm this expectation (Fig. 5.18).

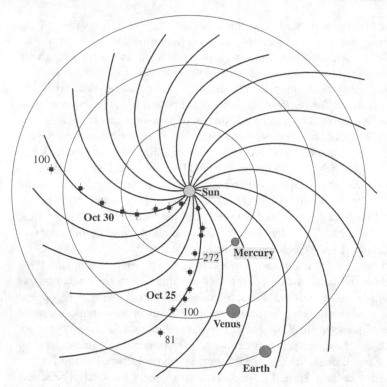

Fig. 5.18 Magnetic spiral. The trajectory of flare electrons in interplanetary space as viewed from above the polar regions using the *Ulysses* spacecraft. The squares and crosses show *Ulysses* radio measurements of type III radio bursts on 25 October 1994 and 30 October 1994. As the high-speed electrons move out from the Sun, they excite radiation at successively lower plasma frequencies. The numbers denote the observed frequency in kiloHertz, or kHz. Since the flaring electrons are forced to follow the interplanetary magnetic field, they do not move in a straight line out from the Sun, but instead move along the spiral pattern of the interplanetary magnetic field, shown by the *solid curved lines*. The magnetic fields are drawn out into space by the radial solar wind, and attached at one end to the rotating Sun. The approximate locations of the orbits of Mercury, Venus, and the Earth are shown as *circles*. [Courtesy of Michael J. Reiner, and adapted from Reiner, Fainberg and Stone (1995). *Ulysses* is a project of international collaboration between ESA and NASA.]

But the magnetic fields are also linked to their sources in the ever-varying photosphere. Random motions and systematic differential rotation of these magnetic anchoring points can produce unexpected behavior. They can, for example, carry open field lines across coronal hole boundaries, making them connect and creating "sub-Parker" spirals.

The movements of the magnetic fields down at the Sun can also enable magnetic reconnections between different solar latitudes out in the solar wind, enhancing the propagation of energetic particles from low latitudes to higher ones. Recurrent electron enhancements have, for example, been unexpectedly observed up to the highest solar latitudes observed by *Ulysses*.

Louis J. Lanzerotti, George M. Simnett, and their co-workers reported such enhancements during *Ulysses'* first polar orbit, at activity minimum when they were only expected at low solar latitudes. The electrons are accelerated in the low places by the shocks produced when the slow and fast winds meet. But once produced, the accelerated electrons follow the outward spiral of the magnetic field. So they were not expected at high solar latitudes, beyond the extent of the interaction regions that occur at low latitudes where the minimum slow wind is concentrated.

In 1996, Len Fisk attributed the unexpected, high-latitude appearance of recurrent electron enhancements to the motions of the footpoints of open magnetic field lines from low to high solar latitudes. An alternative theory, proposed by J. Randy Jokipii in 1966, proposed that energetic particles might propagate to higher solar latitudes as the result of perpendicular diffusion driven by random solar motions at the feet of magnetic field lines.

Motions of magnetic fields at the photosphere are coupled to the outer heliosphere, resulting in the braiding, mixing, and reconnection of interplanetary magnetic fields lines. Such mixing was inferred in 2000 by Joseph E. Mazur and co-workers from *ACE* flare observations of magnetic flux tubes that are alternately filled and devoid of flare ions even though they had a common flare source on the Sun.

Roberto Bruno and Vincenzo Carbone in 2005, Eckart Marsch in 2006, and B. R. Ragot (2006b, c) have described the theoretical aspects of interacting magnetic fields in the turbulent solar wind. In 2005, Benjamin D.G. Chandran discussed weakly turbulent Alfvén waves and fast magetosonic waves, including their effects on minor ions in the solar corona. On the observational front, in 2002 Steven R. Spangler and colleagues have reported Very Long Baseline Interferometry measurements of turbulence in the inner solar wind.

5.7 Ingredients of the Sun's Winds

The material input to the solar wind is by no means uniform, and this is reflected in the composition of the winds. The slow wind, the fast wind, and energetic gusts in the winds exhibit different ion abundances, and some of these differences are attributed to their varying circumstances of origin. Processes in the solar wind can subsequently accelerate the ions to higher energies, alter their composition, and affect their motion. Ions arriving at the Earth's orbit therefore exhibit a variety of compositions and energies, which are detected by instruments aboard the *ACE, Ulysses, STEREO A* and *B*, and *Wind* spacecraft. The ions in the solar wind, for example, usually have relatively low energies, but the winds occasionally contain a fewer number of ions that have been accelerated to higher energies by explosive outbursts near the Sun or by interplanetary shocks further out. And spacecraft can also detect relatively rare and exceptionally energetic cosmic ray ions that arrive from outside the solar system.

5.7.1 Different Sources, Varying Ionization

The electron that is most easily extracted from an atom, and least tightly bound to it, is the electron in the outermost, first orbit. And the first ionization potential, abbreviated FIP, is the energy required to remove that electron. Those elements with the lowest first ionization potentials, the lowest FIPs, are more abundant in the solar wind than in the underlying photosphere.

Ove Havnes discovered such an enhancement in the abundance of elements with low FIPs in the early 1970s for cosmic rays. In the late 1970s and early 1980s, instruments aboard the *Voyager 1* and *2* spacecraft recorded the FIP effect in the solar energetic particles emitted by solar flares, leading Walter R. Cook, Edward C. Stone, and Rochus E. Vogt to conclude in 1984 that "both the solar energetic particles and the solar wind composition are significantly and persistently different from that measured for the photosphere."

Over the ensuing decades, spacecraft observations have confirmed the systematic overabundance of low FIP elements, when compared to photosphere abundances, for particles hurled into interplanetary space during explosive solar outbursts. And they have established a similar FIP effect for the slow solar wind at times of quiescent solar activity – by Peter Bochsler and Johannes Geiss in 1989. Measurements from *Ulysses*, presented by Rudolf Von Steiger and co-workers in 2000, indicated that the low-FIP elements in the unperturbed slow wind are about three or four times more abundant than those in photosphere. The FIP effect is also present in the fast wind, but with a smaller enhancement.

These abundance differences most likely reflect the element abundances in the wind's place of origin. If the physical conditions, such as temperature, density, or magnetic field strength, are different in various locations within the chromosphere and low corona, where the winds originate, then it seems logical that the easily ionized elements will vary in abundance. In the 1990s, Uri Feldman and co-workers (1992, 1993) reported measurements with the Solar Ultraviolet Measurements of Emitted Radiation, or SUMER for short, instrument aboard *SOHO*, which indicate the high-temperature chromosphere and corona indeed exhibit element abundances similar to those seen further out in the solar wind and solar energetic particles, and different from the abundances of these elements in the photosphere, also noticing that the element abundances change from region to region. And in 1999, John Raymond reviewed the composition variations observed with *SOHO*'s UltraViolet Coronagraph Spectrometer, abbreviated UVCS, instrument, confirming that the FIP effect is found in both the low corona and the solar wind, and must originate in the chromosphere by some unknown physical process. One possibility is that the fast wind is related to small, short-lived loops interacting with open fields, while the slow wind may come from larger, longer-lived coronal structures.

5.7.2 Shocking Times

Some ions found in the solar wind move far too swiftly to have been accelerated at the wind's source near the Sun. They have been propelled to higher speeds within interplanetary space, at places where the fast solar wind overtakes the slower wind ahead of it, producing compressed regions that are bounded by forward and reverse shocks and co-rotate with the Sun (Fig. 5.19). The leading edge of the co-rotating interaction regions is a forward pressure wave that propagates into the sluggish flow ahead, while the trailing edge is a reverse pressure wave that propagates into the trailing high-speed flow. At large distances, the bounding pressure waves steepen into forward and reverse shocks that can accelerate particles.

Theoretical investigations of co-rotating interaction regions extend back to 1963 when Eugene Parker (1963b) noticed that solar wind flows of different speeds will interact in co-rotating regions of high pressure, roughly aligned with the Archimedean spiral pattern of the interplanetary magnetic field. In the early 1970s, Arthur J. Hundhausen (1972a, b, 1973) predicted that these compression regions would be bounded by forward–reverse shock pairs not too far beyond Earth's orbit, which was confirmed by the observations in the mid-1970s with the *Pioneer 10* and *11* deep space probes, reported by Edward J. Smith and John H. Wolfe in 1976. In the same year, Frank B. McDonald and his colleagues showed that the observed

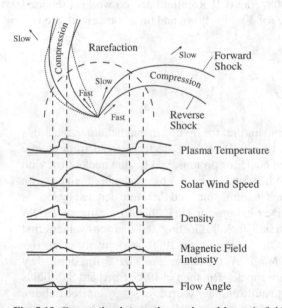

Fig. 5.19 Co-rotating interaction regions. Magnetic fields and solar wind compression and rarefaction regions when the fast winds overtake the slow ones are shown in the upper figure in a coordinate system rotating with the Sun. Variations in the velocity, density and magnetic field strength are shown in the lower part of the figure, for conditions near the Earth's orbit. At larger distances, the edges of the interaction/compression region steepen into shocks. [Adapted from Belcher and Davis (1971).]

co-rotating interaction regions also accelerate ions to high energies. Instruments aboard the *Helios 1* and 2 spacecraft subsequently detected the shocks in the inner heliosphere, within the Earth's orbit, and *Voyager 2* found that the forward–reverse shock pairs are a common feature of the solar wind at distances beyond the Earth's orbit.

We now know that co-rotating regions are the dominant large-scale structure in the heliosphere during both the declining and the minimum parts of the 11-year solar activity cycle, when they are the main source of accelerating particles to energies from several keV to several MeV. They are also responsible for 27-day recurrent geomagnetic storms observed at Earth during activity minimum.

Shocks formed by coronal mass ejections dominate interplanetary acceleration processes during the active phase of the solar cycle, and these ejections can trigger intense geomagnetic storms when arriving at the Earth with the right magnetic orientation.

The Solar Wind Ion Composition Spectrometer, abbreviated SWICS, instruments aboard the *Ulysses* and *ACE* spacecraft have confirmed and extended the role of co-rotating interaction regions in accelerating particles, especially near activity minimum and in the low solar latitude parts of the heliosphere where the fast–slow wind interaction occurs – see the review by Glenn M. Mason and colleagues in 1999 and Ian G. Richardson's account in 2004. As previously mentioned, magnetic connections provide an access for these particles to higher latitudes, where they have been observed from *Ulysses*. And in 2008, Alexis P. Rouillard and co-workers reported the first observations of co-rotating solar winds flows and their source with the two *STEREO* spacecraft.

5.7.3 Pick-Ups Near and Far

Owing to the corona's very high temperature, the hydrogen and helium in the Sun's expanding corona are fully ionized with all of their electrons freed from atomic bonds. The free electrons, hydrogen nuclei or protons, and helium nuclei flow with the solar wind unattached to each other. Such helium ions are doubly ionized, with both the electrons removed from each helium atom, and therefore denoted by He^{++}.

The observations of singly ionized helium, or He^+, therefore came as an unexpected surprise. Rainer Schwenn and Jack T. Gosling and their co-workers first found the singly ionized helium in the solar wind in 1980, respectively using *Helios 1* and seventh *Interplanetary Monitoring Platform*, or *IMP 7* for short; they attributed these ions to interplanetary shocks. But then in 1985, Eberhard S. Möbius and his colleagues called attention to quite another source for singly ionized helium, using instruments aboard the *Active Magnetospheric Particle Tracer Explorers*, abbreviated *AMPTE*, spacecraft. They showed that the observed solar wind He^+ has a velocity distribution that unambiguously indicates an interstellar origin.

Perfectly normal, unionized helium atoms in nearby interstellar space drift far inside the heliosphere with relatively slow velocities of about $26 km s^{-1}$, as the result of the Sun's motion through the local interstellar cloud. And because they are uncharged, the electrically neutral atoms are completely oblivious of the solar wind's magnetic field. As pointed out by Möbius, the Sun's gravity focuses the incoming helium atoms into a conical region in the direction opposite to the direction from which they came.

But just one electron is eventually removed from each incoming helium atom when it gets close to the Sun, as the result of the increasing intensity of ionizing ultraviolet sunlight and/or the greater density of solar wind material that can interact with the helium to alter its charge. And now that they are electrically charged, the newborn He^+ ions are picked up by the magnetic field in the out-flowing solar wind and taken for a ride. So Möbius and his colleagues called them He^+ pick-up ions.

The basic idea had been introduced by Len A. Fisk, Benzion Kozlovsky, and Reuven Ramaty in 1974, when they proposed that observed enhancements of cosmic ray oxygen and nitrogen, with energies of about 10 MeV, could result from neutral interstellar particles that are swept into the solar vicinity by the motion of the Sun through the interstellar medium, and which are subsequently ionized, carried along by the solar wind magnetic field, and taken back to the edge of the solar system. There they can rebound at higher velocities due to shock waves, something like a squash or tennis ball bouncing off a wall, and become accelerated to even higher energies. Because the abundances and compositions of these particles are unusual, when compared to other types of cosmic rays, scientists call them anomalous cosmic rays.

In 1996, Johannes Geiss, George Gloeckler, and Rudolf von Steiger announced that pick-up carbon, denoted C^+, had been detected from *Ulysses* close to the Earth's orbit, and that it originates in a different inner source located within a few times the Earth's distance from the Sun. The ionized carbon could not possibly originate from interstellar carbon atoms, because they would have been ionized by solar ultraviolet radiation much further from the Sun and carried out by the solar wind. An inner source close to the Sun also produces singly ionized oxygen or nitrogen, but it does not significantly contribute to their total amounts. The inner source pick-ups have a velocity distribution that peaks at or below the solar wind speed, and exhibits a composition similar to the solar wind. And since their amounts decrease roughly with increasing distance from the Sun, they most likely originate near the Sun, perhaps from abundant interplanetary dust grains.

The SWICS instruments on *Ulysses* and *ACE* have revealed a vast population of interstellar pick-up ions, many for the first time and each with just one electron missing. These investigations have been reported in several articles published by George Gloeckler and his co-workers ever since the launch of the two spacecraft. The *Ulysses* instrument, for example, discovered interstellar pick-up hydrogen, denoted H^+, in the solar system, identifying it by its distinct velocity distribution. And both spacecraft have carried out continuous and comprehensive observations of pick-up helium, He^+, nitrogen, N^+, oxygen, O^+, and neon, Ne^+.

The properties of the pick-up ions have been used to extrapolate back into the place they came from, permitting inferences about the abundance of elements and the chemical and physical conditions in the local interstellar cloud, as well as the extent of the heliosphere. The observations have also fully confirmed the hypothesis that pick-up ions derived from the interstellar gas are the dominant source of anomalous cosmic rays, which are pre-accelerated inside the heliosphere and re-accelerated at the terminal shock of the solar wind out at the edge of the heliosphere. Moreover, the pick-up ions already residing within the solar wind are preferentially boosted to higher energies by the interplanetary shocks associated with either co-rotating interaction regions or coronal mass ejections.

5.7.4 Super-Hot and Undemocratic

A Maxellian distribution, first derived by James Clerk Maxwell in 1860, describes the velocities of particles in any gas at thermal equilibrium at a given temperature. In this velocity distribution, most of the particles have an average speed governed by the gas temperature, and a very few are found in the low-velocity or high-velocity tails of the distribution. It is the same distribution that describes the grades of a large number of students taking a test, most with an average grade and a small number with the lowest or highest ones.

Instruments aboard *Ulysses* and *ACE* have unexpectedly shown that some ions in the solar wind move at exceptionally fast speeds, up to at least 50 times the velocity of other solar wind ions. As reported by George Gloeckler in 2003, these high-speed particles are observed in the velocity distributions of both the solar wind and the pick-up ions, for protons and all heavier ions that can be detected and extending to the highest energies that can be measured. And since there are more of the ubiquitous high-speed ions than expected from a thermal, Maxwellian velocity distribution of the corona, at a temperature of a few million Kelvin, they have been dubbed "suprathermal" particles. They are also super-hot and decidedly undemocratic, forming an exceptional group that attains the highest energies.

Moreover, the abundant pre-accelerated suprathermal solar wind and pick-up ions are further accelerated to higher energies by interplanetary shocks. In 2006, for example, Mihir I. Desai and his colleagues published a comprehensive survey of heavy-ion abundances observed by *ACE* during solar energetic particle events arriving at Earth's orbit. They concluded that shocks generated by coronal mass ejections accelerate a suprathermal seed population that is already enriched in high-speed, heavy ions, and that the solar energetic particles are not the result of the shocks accelerating ambient coronal or solar wind ions.

5.8 Edge of the Solar System

How far does the Sun's influence extend, and where does it all end? The relentless solar wind streams out in all directions, rushing past the planets and carving an immense heliosphere in interstellar space. But since the solar wind thins out as it expands into a greater volume, it eventually becomes too dispersed to repel interstellar forces. The winds are no longer dense or powerful enough to withstand the pressure of gas and magnetic fields coursing between the stars. The radius of this celestial standoff distance, in which the pressure of the solar wind falls to a value comparable to the interstellar pressure, has been estimated at about 100 AU, or 100 times the mean distance between the Earth and the Sun (Focus 5.5).

Focus 5.5

The Heliosphere's Outer Boundary

The solar wind carves out a cavity in the interstellar medium known as the heliosphere. The radius of the heliosphere can be estimated by determining the standoff distance, or stagnation point, in which the ram pressure, P_W, of the solar wind falls to a value comparable to the interstellar pressure, P_I. As the wind flows outward, its velocity remains nearly constant, while its density decreases as the inverse square of the distance. The dynamic pressure of the solar wind therefore also falls off as the square of the distance, and we can use the solar wind properties at the Earth's distance of 1 AU to infer the pressure, P_{WS}, at the stagnation-point distance, R_S. Equating this to the interstellar pressure we have:

$$P_{WS} = P_{1\,AU} \times \left(\frac{1\,AU}{R_S}\right)^2 = \left(N_{1\,AU}V_{1\,AU}^2\right) \times \left(\frac{1\,AU}{R_S}\right)^2 = P_I,$$

where the number density of the solar wind near the Earth is about $N_{1\,AU} = 5$ million particles per cubic meter and the velocity there is about $V_{1\,AU} = 500\,km\,s^{-1}$.

To determine the distance to the edge of the solar system, R_S, we also need to know the interstellar pressure, which is the sum of the thermal pressure, the dynamic pressure, and the magnetic pressure in the local interstellar medium. It is estimated at $P_I = (1.3 \pm 0.2) \times 10^{-12}\,dyn\,cm^{-2}$, resulting in $R_S = 100$ AU or more, far beyond the orbits of the known outer planets. However, the estimates by different authors give a broad range for the distance to the edge of the heliosphere, depending on the uncertain values of various components of the interstellar pressure.

Instruments aboard the twin *Voyager 1* and *2* spacecraft, launched in 1977 and now cruising far beyond the outermost planets, have approached this edge of the solar system from different directions, *Voyager 1* moving in the northern hemisphere of the heliosphere and *Voyager 2* in the southern hemisphere (Fig. 5.20). In 2005, three articles in the same issue of *Science*, by Leonard F. Burlaga, Richard B. Decker, and Edward C. Stone and their co-workers, announced that *Voyager 1* had crossed the termination shock of the supersonic flow of the solar wind on 16 December 2004, at a distance of 94 times the mean distance between the Earth and the Sun, or at 94 AU from the Sun. At this distance, the spacecraft's instruments recorded a sudden increase in the strength of the magnetic field carried by the solar wind, as expected when the solar wind slows down and its particles pile up at the termination shock.

Voyager 2 crossed the termination shock on 30 August 2007, at a distance of 84 AU from the Sun. The observations, published by Leonard F. Burlaga, Richard B.

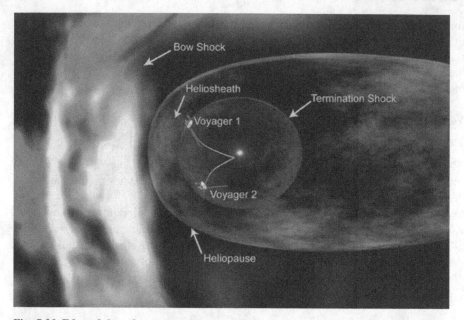

Fig. 5.20 Edge of the solar system. *Voyager 1* and *2* spacecraft, shown at respective distances of about 90 and 70 AU, approach the place where the solar system ends and interstellar space begins. One astronomical unit is the mean distance between the Earth and the Sun, and the edge of the solar system is located at roughly 100 times this distance. At the termination shock, the supersonic solar wind abruptly slows from an average speed of $400\,\mathrm{km\ s^{-1}}$ to less than one-quarter that speed. Beyond the termination shock is the heliosheath, a vast region where the turbulent and hot solar wind is compressed as it presses outward against the interstellar wind. The edge of the solar system is found at the heliopause, where the pressure of the solar wind balances that of the interstellar medium. A bow shock likely forms as the interstellar wind approaches and is deflected around the heliosphere, forcing it into a teardrop-shaped structure with a long, comet-like tail. (Courtesy of JPL and NASA.)

Decker, Donald A. Gurnett, John D. Richardson, and Edward C. Stone and their co-workers in 2008, indicated that there were multiple crossings, three of which occurred when data were being received from the spacecraft. It appears that there is a significant north/south asymmetry in the heliosphere, likely due to the direction of the local interstellar magnetic field.

Both *Voyager 1* and *2* have therefore now crossed into the vast, turbulent heliosheath, the region where the interstellar gas and solar wind interact, due to the reflection and deflection of the solar wind ions by the magnetized wind beyond the heliosheath. In technical terms, the solar wind ions in the heliosheath are deflected by magnetosonic waves reflecting off of the heliopause, causing the ions to flow parallel to the termination shock toward the heliotail.

But the Sun's winds are not alone in the dark, cold outer fringes of the solar system. About a million million, or 10^{12}, unseen comets have been hibernating out there ever since the solar system formed (Focus 5.6). These comets are so small and widely spaced that it is exceedingly unlikely that the *Voyager 1* or *2* spacecraft will encounter even one of them, just as the *Voyagers* passed through the asteroid belt unaffected by the billions of asteroids there. Before they come close to the Sun, comets are tiny balls of ice about 10 km across, and large asteroids are battered chunks of rock of about the same size.

Focus 5.6

The Oort Comet Cloud

In 1950, Jan Hendrik Oort postulated the existence of a remote spherical shell of comets occupying the outer parts of the solar system. His careful examination of the trajectories of observed long-period comets, which became visible when they entered the inner parts of the solar system, could only be explained if these comets came from a distant reservoir, which he located at between 50,000 and 150,000 AU, where 1 AU is the mean distance between the Erath and the Sun. At greater distances, the stars in the neighborhood of our solar system compete for gravitational control of the comets. And since the observed comets come from all possible direction in the sky, the reservoir had to be in the shape of a spherical shell.

Modern calculations suggest that there is an inner Oort cloud, with an inner edge at around 3,000 AU and a density falling off with greater distance. The outer Oort cloud is continuous with this, but is defined to be those objects at distances greater than 20,000 AU. The cloud fades away with increasing distance, and its tenuous outer edge is dynamically limited to about 200,000 AU by the galactic gravity field.

But how do comets fall from the Oort comet cloud to the heart of the solar system? The distant comets are only weakly bound to the solar system, and are easily perturbed by the gravitation of nearby, moving objects, which throw some of the comets back into the planetary system. The random gravitational jostling of individual stars passing nearby, for example, knocks just a few of the comets in the Oort cloud from their stable orbits, either injecting them into interstellar space or gradually deflecting their paths toward the Sun. Every one million years, about a dozen stars pass close enough to stir up the comet cloud, sending a steady trickle of comets into the inner solar system on very long elliptical orbits. A giant interstellar molecular cloud can also impart a gravitational tug when it moves past the Oort comet cloud, helping to jostle some of them out of their remote resting place. Tidal forces generated in the cloud by the disk of our galaxy, the Milky Way, also help to feed new long-period comets into the planetary region.

As time goes on, the accumulated effects of these tugs will send a few comets in toward the Sun – or outward to interstellar space. If the several hundred new comets observed during recorded history have been shuffled into view by the perturbing action of the nearby stars or molecular clouds, then there are at least one hundred billion, or 10^{11}, comets in the Oort cloud. There may be a trillion, 10^{12}, or even 10 trillion, 10^{13}, of them, each a frozen ball of ice no larger than New York City or Paris. This large population of unseen comets can sustain the visible long-period comets and persist without serious depletion for many billions of years, until long after the Sun expands to consume Mercury and boil the Earth's oceans away.

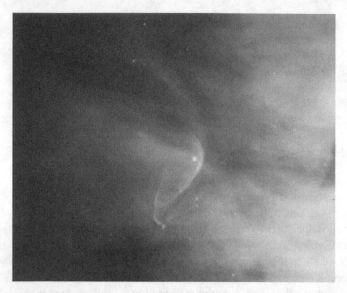

Fig. 5.21 Stellar bow shock. A crescent-shaped bow shock is formed when the material in the fast wind from the bright, very young star, LL Ori (*center*) collides with the slow-moving gas in its vicinity, coming from the lower right. The stellar wind is a stream of charged particles moving rapidly outward from the star. It is a less energetic version of the solar wind that flows from the Sun. This image was taken from the *Hubble Space Telescope*. This star is located in the Orion Nebula; an intense star-forming region located about 1,500 light-years from the Earth. (Courtesy of NASA, the Hubble Heritage Team, STScI, and AURA.)

Both *Voyager* spacecraft are equipped with plutonium power sources expected to last until at least 2020 and perhaps 2025. So they ought to eventually measure the heliopause, at the outer edge of the heliosheath. It is the place where interstellar space begins.

The motion of the interstellar gas, with its own wind, compresses the heliosphere on one side, producing a teardrop-like, non-spherical shape with an extended tail. A bow shock is formed when the interstellar wind first encounters the heliosphere; just as a bow shock is created when the solar wind strikes the Earth's magnetosphere. And the graceful arc of a bow shock, created by an interstellar wind, has been detected around the young star LL Orionis (Fig. 5.21).

Closer to home, space physicists are concerned about the impact of powerful solar eruptions on the Earth's environment in space.

5.9 Summary Highlights: The Sun's Expanding Corona

- The Sun's hot and stormy atmosphere is forever expanding in all directions, filling the solar system with a ceaseless flow – called the solar wind – that contains electrons, protons, and also heavier ions as well as magnetic fields.
- The solar wind creates a teardrop-shaped bubble in interstellar space, known as the heliosphere, which extends out to about 100 times the mean distance between the Earth and the Sun or to about 100 AU.

- The Earth is immersed in the solar wind, which engulfs all the planets, so we live inside the expanding outer atmosphere of the Sun.
- Early spacecraft measurements showed that there are two kinds of solar wind, a fast one moving at about $750 \, \text{km s}^{-1}$ and a slow one with about half that speed.
- The twinkling, or scintillation, of radio sources suggested that a fast wind is streaming out at high solar latitudes near a minimum in the 11-year solar activity cycle, and coordinated X-ray and interplanetary particle and magnetic field data showed that the fast winds originate from polar coronal holes near the activity minimum.
- Measurements from the *Helios 1* and *2* spacecraft indicated that the electrons, protons, and helium nuclei in the solar wind have different temperatures; in the high-speed wind, the more massive particles are hotter.
- The differing ion composition of the fast and slow solar winds indicates that the wind speed is inversely correlated with the electron temperature at the wind sources, and that the slow wind originates from very different sources than the fast wind.
- The *Ulysses* spacecraft has made measurements all around the Sun, at a distance comparable to that of the Earth. Its first polar orbit occurred near a minimum in the Sun's 11-year activity cycle, while the second polar orbit was near an activity maximum, and the third polar orbit was near the cycle minimum again, but with the Sun's magnetic poles reversed from the first time around.
- A comparison of *Ulysses* velocity data and X-ray and white-light coronagraph observations indicated that at activity minimum the uniform, tenuous fast wind rushed out along the open magnetic field lines of polar coronal holes, and that the variable, dense slow wind was confined to low solar latitudes in the vicinity of coronal streamers.
- Near the maximum in the 11-year solar activity cycle, the large polar coronal holes are replaced by a collection of smaller holes scattered over a wide range of solar latitudes, and the fast wind at this time comes from the interiors of the largest of these "smaller holes." The smallest coronal holes, including those associated with active regions with or without streamers, may give rise to a slow solar wind at all latitudes. Coronal mass ejections provide a transient contribution to the solar wind flow near the maximum in the solar cycle.
- The slow, dense solar wind outflow emerges from a dense chromosphere with closed magnetic structures; the fast, tenuous winds are associated with a less dense chromosphere and open magnetic fields.
- The fast-wind sources in coronal holes have been attributed to polar plumes, X-ray bright points and jets, the boundaries of the chromosphere's magnetic network, and/or coronal magnetic funnels.
- The high-speed wind is accelerated very close to the Sun, within just a few solar radii, and the slow component obtains full speed much further away.
- The fast and slow winds could be accelerated by different-size coronal loops that emerge from the underlying photosphere and reconnect with open magnetic fields lines near their base, releasing the bottled up solar material, or they might

be associated with open "unipolar" regions on the Sun where loops do not seem to be prevalent.

- *SOlar and Heliospheric Observatory*, abbreviated *SOHO*, observations of ultraviolet line broadening in polar coronal holes at solar activity minimum demonstrate that heavy ions have velocities that depend on direction, with faster speeds in a direction perpendicular to the open, radial magnetic field.
- Ultraviolet line observations from *SOHO* indicate that heavy ions in polar coronal holes move faster than light ions in polar coronal holes.
- The velocity anisotropy and preferential acceleration of heavier ions in coronal holes could be due to waves that resonate with ion cyclotron motion in a magnetic field.
- Alfvén waves have been observed in the solar wind for decades, leading to speculations that these waves accelerate the fast solar wind to its higher speed.
- Instruments aboard *Ulysses* have detected magnetic fluctuations, attributed to Alfvén waves, far above the Sun's poles; they may block cosmic rays trying to enter these regions.
- Magnetic field measurements from *Ulysses* indicate that the large-scale magnetic field in the solar wind at distances comparable to the Earth's orbit is always open, uniform, and radial, pointing away from the Sun and never closed into loop-like structures, and that the solar wind at these distances contains the same open magnetic flux at all solar latitudes and during the entire 11-year solar activity cycle.
- Instruments aboard the *Advanced Composition Explorer*, abbreviated *ACE*, have provided direct evidence that oppositely directed magnetic fields in the solar wind merge and join near the Earth's orbit, releasing magnetic energy to accelerate ions in prolonged, steady magnetic reconnection layers that extend for hundreds of Earth radii.
- The spiral shape of the open magnetic field in the solar wind is underwound at high solar latitudes.
- The heliosphere's magnetic field is anchored in the Sun's ever-changing photosphere, whose random motions and differential rotation can enhance the propagation of electrons to high solar latitudes and result in magnetic reconnections between solar latitudes.
- The composition of ions in the solar wind, as well as those produced during solar outbursts, are significantly and persistently different from those measured in the photosphere. The enhanced abundance of elements with low first ionization potential in solar outbursts and in the slow and fast solar wind, when compared with their abundance in the underlying photosphere, reflects the physical conditions in their place of origin.
- The interaction of fast and slow solar winds produces co-rotating interaction regions whose associated shocks accelerate particles within the solar wind.
- The *Wind* spacecraft measures magnetic fields and energetic particles arriving in the solar wind at the dayside of the Earth's magnetosphere, contributing to the investigations of magnetic field reconnection in the solar wind and the origin, acceleration, and propagation of energetic particles associated with solar flares

or coronal mass ejections. Instruments aboard *Wind* also track the radio signals generated by shocks that are driven by coronal mass ejections.

- The Solar Wind Ion Composition Spectrometer, abbreviated SWICS, instruments on *ACE* and *Ulysses* have made comprehensive and continuous measurements of singly ionized particles, including those of hydrogen, helium, carbon, nitrogen, and oxygen, and many of them for the first time. Some of these pick-up ions, with just one electron missing, have been attributed to interstellar atoms that once moved slowly into the heliosphere, became ionized by the Sun's ultraviolet radiation, electron impact, or charge exchange, and were picked up and carried away by the magnetic field in the outflowing solar wind; but pick-up carbon has been found too close to the Sun to be of interstellar origin and has been attributed to an inner source.

- *ACE* and *Ulysses* observations of the velocity distributions of both solar wind ions and pick-up ions include a ubiquitous suprathermal tail, indicating motions much faster than expected from a thermal distribution for the corona. The suprathermal ions can move at least 50 times the speed of other ions in the solar wind.

- The *Voyager 1* spacecraft crossed the termination shock of the supersonic flow of the solar wind on 16 December 2004, at a distance from the Sun of 94 times the mean distance between the Earth and the Sun, or 94 AU, becoming the first spacecraft to begin exploring the heliosheath, the outermost layer of the heliosphere.

- The *Voyager 2* spacecraft crossed the termination shock of the solar wind on 30 August 2007, at a distance of 84 AU from the Sun.

- *Hinode* telescopes detect Alfvén waves that are thought to be energetic enough to accelerate the solar wind.

- Observations from the *Hinode* spacecraft indicate that some of the X-ray corona can continuously flow from an active region next to an equatorial coronal hole, suggesting a source of the slow solar wind.

5.10 Key Events in Studies of the Solar Wind*

Date	Event
1869	At the solar eclipse of 7 August 1869, Charles A. Young, and, independently, William Harkness, discovered a single, bright, green emission line in the spectrum of the solar corona. This conspicuous feature remained unidentified with any known terrestrial element for more than half a century, but it was eventually associated with highly ionized iron atoms missing 13 electrons, designated [Fe XIV], indicating that the corona has a temperature of about a million Kelvin (see 1939–1941).

---

*See the References at the end of this book for complete references to these seminal papers. An AU is the mean distance of the Earth from the Sun, or about 146×10^6 km.

Date	Event
1889–1990	Frank H. Bigelow (1889, 1890a, b) argues that the structure of the corona detected during solar eclipses provides strong evidence for large-scale solar magnetic or electric fields. He correctly speculated that polar rays delineate open magnetic field lines along which material escapes from the Sun, and that equatorial elongations of the corona mark closed magnetic field lines.
1896–1913	Kristian Birkeland argues that polar auroras and geomagnetic storms are due to beams of electrons from the Sun.
1908	George Ellery Hale measures intense magnetic fields in sunspots, thousands of times stronger than the Earth's magnetism.
1919	Frederick Alexander Lindemann (later Lord Cherwell) suggests that an electrically neutral plasma ejection from the Sun is responsible for non-recurrent geomagnetic storms.
1939–1941	Walter Grotrian and Bengt Edlén identify coronal emission lines with highly ionized elements, indicating that the Sun's outer atmosphere has a temperature of about a million Kelvin. The conspicuous green emission line was identified with [Fe XIV], an iron atom missing 13 electrons.
1942	Hannes Alfvén uses theoretical equations to demonstrate the existence of electromagnetic–hydrodynamic waves, subsequently known as Alfvén waves.
1944–1956	Herman Bondi, Fred Hoyle, and William H. Mc Crea develop a theory for spherically symmetric accretion of interstellar matter by a star, including a critical solution for transonic accretion flow; it is applicable to the solar wind, with flow away from, instead of into, the star.
1946	Vitalii L. Ginzburg, David F. Martyn, and Joseph L. Pawsey independently confirm the existence of a very hot solar corona, with a temperature of about a million Kelvin, from the observations of the Sun's radio radiation.
1948–1949	Soft X-rays from the Sun are first detected on 5 August 1948, with a V-2 rocket experiment performed by the U.S. Naval Research Laboratory, reported by T. Robert Burnight in 1949. Subsequent sounding rocket observations by the NRL scientists revealed that the Sun is a significant emitter of X-rays and that the X-ray emission is related to solar activity.
1950–1954	Scott F. Forbush demonstrates the inverse correlation between the intensity of cosmic rays arriving at Earth and the number of sunspots over two 11-year solar activity cycles.
1951–1952	Herbert Friedman and his colleagues at the U.S. Naval Research Laboratory use instruments aboard sounding rockets to show that the Sun emits enough X-ray and ultraviolet radiation to create the ionosphere.
1951–1957	Ludwig F. Biermann argues that a continuous flow of solar corpuscles is required to push comet ion tails into straight paths away from the Sun, correctly inferring solar wind speeds of between 500 and $1,000 \text{km s}^{-1}$.
1955	Leverett Davis Jr. argues that solar corpuscular emission will carve out a cavity in the interstellar medium, now known as the heliosphere, accounting for some observed properties of low-energy cosmic rays.
1955	Horace W. Babcock and Harold D. Babcock use magnetograms taken over a 2-year period, 1952–1954, to show that the Sun has a general dipolar magnetic field of about 10^{-4}T, or 1 G, in strength usually limited to solar latitudes greater than $\pm 55°$. Bipolar regions are found at lower solar latitudes. They argued that occasional extended unipolar areas of only one outstanding magnetic polarity might be related to 27-day recurrent terrestrial magnetic storms.
1956	Peter Meyer, Eugene N. Parker, and John A. Simpson argue that enhanced interplanetary magnetism at the peak of the solar activity cycle deflects cosmic rays from their Earth-bound paths.

Date	Event
1957	Max Waldmeier introduces the name coronal holes for seemingly vacant places with no detectable coronal radiation in the 530.3 nm emission line of ionized iron [Fe XIV]. In 1981, Waldmeier reported his studies of these polar coronal holes from 1940 to 1978, noting that they are permanently present for about 7 years, including the minimum in the 11-year solar activity cycle, but that they seem to disappear for about 3 years near activity maximum.
1957–1959	Sydney Chapman (1959a, b) shows that a hot, static corona should extend to the Earth's orbit and beyond.
1957–1967	Rocket observations by Richard Tousey and colleagues at the U.S. Naval Research Laboratory indicate that the brightest line in the ultraviolet spectrum of the solar corona is the Lyman alpha transition of hydrogen atoms.
1958	Eugene N. Parker suggests that a perpetual supersonic flow of electrically charged corpuscles, which he called the solar wind, naturally results from the expansion of a very hot corona. He also demonstrates that the solar magnetic field will be pulled into interplanetary space, attaining a spiral shape in the plane of the Sun's equator due to the combined effects of the Sun's rotation and radial wind flow.
1960–1961	Konstantin I. Gringauz reports that the Soviet spacecraft, *Lunik 2* or *Luna 2*, launched on 12 September 1959, has measured high-speed ions in interplanetary space outside the Earth's magnetic field, with a flux of 2 million million (2×10^{12}) ions (protons) per square meter per second.
1962–1967	*Mariner 2* was launched on 7 August 1962. Using the data obtained during *Mariner's* voyage to Venus, Marcia Neugebauer and Conway W. Snyder demonstrated that a low-speed solar wind plasma is continuously emitted by the Sun, and discovered high-speed wind streams that recur with a 27-day period within the orbital plane of the planets. Interplanetary shocks associated with solar activity are detected using instruments aboard the *Mariner 2* spacecraft in 1962, reported by Charles P. Sonett and colleagues in 1964.
1964–1966	Norman F. Ness and John M. Wilcox use magnetometers aboard NASA's *Interplanetary Monitoring Platform 1*, launched on 27 November 1963, to measure the strength and direction of the interplanetary magnetic field. They confirm that the interplanetary magnetic field is pulled into a spiral shape by the combined effects of the Sun's rotation and radial wind flow. They also discover large-scale magnetic sectors in interplanetary space, which point toward or away from the Sun.
1966	Peter A. Sturrock and Richard E. Hartle propose a two-component (electrons and protons) model of the solar wind driven by electron heat conduction from a hot corona. However, the wind is too slow and the protons are too cool when compared with the observations of the fast solar wind, normalized to the distance of the Earth's orbit.
1967	Edmund J. Weber and Leverett Davis Jr. consider the effects of solar rotation and magnetic fields on a steady solar wind flow in the equatorial plane. They show that co-rotation of the wind and Sun exists out to the Alfvén critical point, or radial distance, where the Alfvén Mach number is one.
1968	Paul J. Coleman Jr. determines the power-law spectra of solar wind magnetic fields and velocities measured from the *Mariner 2* spacecraft, concluding that the solar wind flow is often turbulent in the region near the Earth's orbit at 1 AU and that Alfvén waves will be formed, particularly in the fast wind.
1968	Roger A. Kopp and Max Kuperus discuss the temperature structure and magnetic field of the transition region between the chromosphere and the corona.
1969–1971	Magnetic fluctuations are observed in the solar wind from *Mariner 5* on its way to Venus, and attributed to large-amplitude Alfvén waves by John W. Belcher, Leverett Davis Jr., and Edward J. Smith.

Date	Event
1970	Klaus Jockers and others demonstrate that totally collisionless (exospheric) models of the solar wind do not work.
1970	Thomas E. Holzer and William Ian Axford review the theory of steady, radial, spherically symmetric solar-wind flow, introducing heating of the corona to several million Kelvin (with thermal velocities above the Sun's escape velocity) and showing that rapid wind acceleration occurs close to the Sun. They additionally pointed out that ionization states could be used to determine coronal temperatures.
1970	D. E. Robbins, Arthur J. Hundhausen, and Samuel J. Bame describe differential flows of protons and alpha particles (helium nuclei) in the solar wind.
1971	John W. Belcher and Leverett Davis Jr. note the ubiquitous presence of large-amplitude Alfvén waves in the solar wind, and show how the interaction of slow and fast solar wind streams will lead to co-rotating interaction regions with the shocks that affect the internal properties of the solar wind.
1971	John W. Belcher, and independently G. Alazraki and P. Couterier, showed how solar wind acceleration might be caused by the gradient of Alfvén wave pressure.
1971	Alan H. Gabriel explains the coronal Lyman alpha line in terms of resonant scattering of ultraviolet light generated below the corona.
1971	Ove Havnes discovers systematic differences between the abundances of cosmic rays of low energy and universal abundances. These differences are correlated with the first ionization potentials of the corresponding elements, and it has become known as the FIP effect.
1971	Gerald W. Pneuman and Roger A. Kopp propose a dipolar magnetic model for the Sun, in which the solar wind drags the Sun's magnetic field into a neutral current sheet of oppositely directed magnetism near the plane of the solar equator.
1972	Johannes Geiss and Hubert Reeves publish measurements of the solar wind helium abundance, using the foil collectors left by American astronauts on the Moon.
1973	Allen S. Krieger, Adrienne F. Timothy, and Edmond C. Roelof compare an X-ray photograph, obtained during a rocket flight on 24 November 1970, with *Vela* and *Pioneer VI* satellite measurements of the solar wind to show that coronal holes are the source of recurrent high-speed streams in the solar wind.
1973	Giuseppe S. Vaiana and his colleagues use solar X-ray observations taken from rockets during the preceding decade to identify the threefold magnetic structure of the solar corona – coronal holes, coronal loops, and X-ray bright points.
1973–1977	X-ray photographs of the Sun taken using the Apollo Telescope Mount on the manned *Skylab* satellite, launched on 14 May 1973, fully confirm coronal holes, the ubiquitous coronal loops, and X-ray bright points. Detailed comparisons of *Skylab* X-ray photographs and measurements of the solar wind, made from the sixth, seventh, and eighth *Interplanetary Monitoring Platforms*, abbreviated *IMP* 6, 7, and 8, confirm that solar coronal holes are the source of the high-velocity solar wind streams. John Wilcox had previously suggested that the fast streams might originate in magnetically open, unipolar regions on the Sun, but the *Skylab* X-ray photographs definitely identified the place.
1974	Len A. Fisk, Benzion Kozlovsky, and Reuven Ramaty propose that anomalous enhancements in cosmic ray oxygen and nitrogen, observed at energies of about 10 MeV per nucleon, could result from neutral interstellar atoms that are swept into the heliosphere by the Sun's motion through the interstellar medium and subsequently ionized and accelerated.
1974	Leon Golub and his colleagues announce the discovery, and describe the observed properties, of ubiquitous X-ray bright points found at all solar latitudes from equator to the poles.

Date	Event
1974–1978	Hannes Alfvén, and independently Lief Svalgaard and John M. Wilcox, interpret the magnetic structure of the solar wind, at activity minimum, in terms of a warped neutral current sheet dividing the solar wind into two hemispheres of opposite magnetic polarity.
1974–1986	The *Helios 1* and *2* spacecraft, respectively launched on 10 December 1974 and on 15 January 1976, measure the solar wind parameters as close as 0.3 AU from the Sun for a whole 11-year solar cycle. They confirmed the existence of two kinds of solar wind flow. There is a steady, uniform high-speed wind and a varying, slow-speed wind.
1976	Alan H. Gabriel introduces a two-dimensional model of the chromosphere and corona, and the transition region between them, in which magnetic flux is concentrated at the boundaries of supergranular convection cells that produce a magnetic network. These flux tubes flare out as they rise into the overlying solar atmosphere to produce a magnetic canopy.
1976	Edward J. Smith and John H. Wolfe report *Pioneer 10* and *11* observations of the interaction regions and co-rotating shocks between 1 and 5 AU, and Frank B. McDonald and his colleagues show that these shocks accelerate ions to high energies.
1977	Randolph H. Levine, Martin D. Altschuler, and John W. Harvey report that the fastest solar wind streams are correlated with those magnetic flux tubes that expand least in cross-sectional area over the distance between the photosphere and the coronal height where the solar wind begins.
1977–1991	Barney J. Rickett and William A. Coles, and independently by Takakiyo Kakinuma and Mayoshi Kojima, use interplanetary scintillations of extragalactic radio sources to investigate the solar wind speed outside the plane of the ecliptic. Their data show that near the minimum in the 11-year solar activity cycle, the slow wind is confined to low solar latitudes, while the fast wind emanates from high solar latitudes. Near activity maximum, the slow wind is dominant over the whole range of observable latitudes.
1977–1980	Helmuth R. Rosenbauer, Wolfgang K. H. Schmidt, and their colleagues use measurements from *Helios 1* and *2* and the first *International Sun-Earth Explorer*, abbreviated *ISEE-1*, spacecraft to show that helium and other heavy ions move faster than protons in the high-speed wind. In addition, the electrons are cooler than the protons in this fast component of the solar wind.
1978	Edward J. Smith, Bruce T. Tsurutani, and Ronald L. Rosenberg use observations from *Pioneer 11* to show that the solar wind becomes unipolar, or obtains a single magnetic polarity, at solar latitudes near 16°.
1980	Rainer Schwenn and John T. Gosling and their colleagues respectively use instruments aboard *Helios 1* and the seventh *Interplanetary Monitoring Platform*, abbreviated *Imp 7*, to detect singly-ionized helium, He^+, produced in the solar wind by interplanetary shocks.
1981	John T. Gosling and his colleagues use solar wind data from the sixth, seventh, and eighth *Interplanetary Monitoring Platforms*, abbreviated *IMP 6, 7*, and *8*, to suggest that a substantial fraction of the low-speed solar wind originates in the vicinity of coronal streamers, particularly near the minimum in the solar activity cycle.
1982–1983	Gary J. Rottman, Frank Q. Orrall, and James A. Klimchuk obtain rocket observations of extreme-ultraviolet resonance lines formed in the low corona and transition region, showing that the lines are systematically shifted to shorter wavelengths in large polar coronal holes with well-developed, low-latitude extensions. Outflow velocities of between 7 and $8\,km\,s^{-1}$ are inferred from these Doppler shifts.

Date	Event
1983	Philip A. Isenberg reviews early attempts to explain the high ion speed and temperatures in the solar wind by ion-cyclotron resonance, which accelerates and heats the ions.
1984	Walter R. Cook, Edward C. Stone, and Rochus E. Vogt report measurements from the *Voyager 1* and *2* spacecraft of the FIP effect in solar energetic particles emitted by solar flares, and conclude that both the solar energetic particles and the solar wind composition are significantly different from that measured for the photosphere.
1985	Eberhard S. Möbius and his colleagues use instruments aboard the *Active Magnetospheric Particle Tracer Explorers*, abbreviated *AMPTE*, spacecraft to detect singly ionized helium, He^+, attributing it to interstellar helium atoms that have entered the solar system, become ionized there, and then picked-up and entrained in the solar wind.
1986	James F. Dowdy Jr., Douglas M. Rabin, and Ronald L. Moore show that narrow magnetic funnels open up into the base of the corona, emerging from only a fraction of the magnetic network.
1986–1988	Joseph V. Hollweg and colleagues describe ion–cyclotron resonance effects in the solar wind close to the Sun.
1989	J. Randy Jokipii and Joseph Kóta argue that Alfvén waves streaming out of the Sun's polar regions may block the incoming cosmic rays.
1989	George Gloeckler and Johannes Geiss describe the abundance enrichment of solar wind ions with low first ionization potential, the FIP effect.
1990	Yi-Ming Wang and Neil R. Sheeley Jr. (1990a, b) develop a model of the solar wind that includes the inverse correlation between solar wind speed at 1 AU and the rate of coronal magnetic flux-tube expansion.
1990	The *Ulysses* spacecraft is launched on 6 October 1990.
1991	The *Yohkoh* spacecraft is launched on 30 August 1991.
1992	Uri Feldman reports that element abundances in the upper solar atmosphere are similar in nature to those in the solar wind and solar energetic particles, but different from abundances in the underlying photosphere.
1992	Yutaka Uchida and his colleagues use data from the Soft X-ray Telescope aboard *Yohkoh* to show that the active-region corona is continuously expanding, perhaps as the source of the slow solar wind. This idea was subsequently supported by *SOHO* LASCO coronagraph images that show expanding coronal loops near the Sun's equatorial regions.
1993	George Gloeckler and co-workers use an instrument on *Ulysses* to detect interstellar pick-up hydrogen in the solar system.
1993	Kazunari Shibata and his colleagues discover X-ray jets in the solar corona using the Soft X-ray Telescope onboard the *Yohkoh* satellite.
1994	The *Ulysses* spacecraft begins its first passage over the Sun's polar regions.
1994	The *Wind* spacecraft is launched on 1 November 1994.
1995	David J. McComas and his coworkers use the Solar Wind Observations Over the Poles of the Sun, abbreviated SWOOPS, instrument aboard the *Ulysses* spacecraft to determine the latitudinal structure of the three-dimensional solar wind near a minimum in the Sun's 11-year activity cycle, finding fast wind over the Sun's polar regions and variable wind confined near the solar equator. John L. Phillips and his colleagues use correlations with other data to show that the slow wind is narrowly confined to low latitudes above an equatorial steamer belt. They also showed that at least some of the fast wind is emitted from polar coronal holes, and that the fast wind extends to lower latitudes than the radial extension of coronal holes during activity minimum.

Date	Event
1995	Andre Balogh and Edward J. Smith, and their colleagues, show that the radial component of the magnetic field detected, and normalized to 1 AU by *Ulysses* does not vary with solar latitude.
1995	Johannes Geiss, George Gloeckler, and Rudolf Von Steiger (1996a) use *Ulysses* ion composition measurements to suggest that the fast solar wind originates in a region of low electron temperature and that the slow solar wind originates in a region of high electron temperature.
1995	Louis J. Lanzerotti and George M. Simnett and their co-workers discover unexpected, recurrent enhancements of electrons at high solar latitudes using *Ulysses* data.
1995	The *SOlar and Heliospheric Observatory*, abbreviated *SOHO*, is launched on 2 December 1995
1995–1996	*Ulysses'* measurements of cosmic rays, obtained by John A. Simpson and colleagues (1995a, b), do not show substantially more cosmic rays above the Sun's poles than near the ecliptic. This may be explained by Alfvén waves observed in the polar fast wind by Edward J. Smith, Bruce T. Tsurutani, and colleagues with *Ulysses*; the magnetic waves repel the incoming cosmic rays above the polar regions.
1995–1997	Sir William Ian Axford, Joseph V. Hollweg, Phillip A. Isenberg, Eckart Marsch, James F. Mc Kenzie, and others develop fast solar wind models in which magnetic waves heat the corona and preferentially accelerate heavier ions.
1995–1997	Data from the *Ulysses, Yohkoh* and *SOHO* spacecraft independently show that the polar fast wind originates in a relatively low electron temperature region when compared with the electron temperature of the source of the slow wind.
1997	Eckart Marsch and Chuan-Yi Tu provide a theoretical model in which the solar wind is heated and accelerated by Alfvén waves in magnetic funnels opening into the corona from the chromosphere magnetic network.
1997	The *Advanced Composition Explorer*, or *ACE* for short, was launched on 25 August 1997.
1997–2000	Neil R. Sheeley Jr. et al. (1997) and Yi-Ming Wang et al. (2000) use time-lapse coronagraph images, taken from *SOHO's* Large Angle and Spectrometric COronagraph instrument, abbreviated LASCO, to show that one component of the slow wind may be emitted far out in coronal steamers, and that it does not accelerate to terminal velocity until 20 or 30 solar radii from the Sun center.
1997–1998	John L. Kohl, Giancarlo Noci, and their colleagues use *SOHO* UltraViolet Coronagraph Spectrometer, abbreviated UVCS, measurements to show that oxygen ions flowing out of coronal holes accelerate to supersonic outflow velocities within 2.5 solar radii of the Sun center. The outflow velocities of the oxygen ions are faster than protons in coronal holes, and electrons in these regions move at even slower speeds. The ion velocities in the direction perpendicular to the radial direction are greater than those parallel to the radial direction.
1998–2001	John L. Kohl, Steven R. Cranmer, and Mari Paz Miralles and their co-workers publish a series of papers reporting preferential and directional temperatures in polar coronal holes measured with the UltraViolet Coronagraph Spectrometer aboard *SOHO* near the minimum in the 11-year solar activity cycle. The oxygen ions are hotter than protons, which are in turn hotter than electrons, and the oxygen ions have greater temperatures perpendicular to the radial direction than along it. In 2003, Richard A. Frazin and colleagues reported similar oxygen ion measurements at the edges of coronal streamers, but not in them.

Date	Event
1998	George Gloeckler and Johannes Geiss describe pick-up ions observed in the solar wind with the Solar Wind Ion Composition Spectrometer, abbreviated SWICS, on *Ulysses*, including hydrogen, helium, nitrogen, oxygen, and neon ions of interstellar origin. They also report the discovery in 1993 of pick-up hydrogen from interstellar space, the discovery of singly ionized carbon in the inner heliosphere that cannot be of interstellar origin, and the discovery of a new extended source of carbon, oxygen, and nitrogen pick-up ions in the inner solar system.
1998	Martin Hilchenbach and his colleagues use the Charge, Element, and Isotope Analysis System, abbreviated CELIAS, instrument on *SOHO* to detect energetic hydrogen atoms assumed to be coming from the heliosheath.
1999	Donald M. Hassler and his colleagues use the *SOHO* Solar Ultraviolet Measurements of Emitted Radiation, abbreviated SUMER, spectrometer to show that the high-speed, solar wind outflow velocity, observed in the low corona in a polar coronal hole, is spatially correlated with the boundaries of the magnetic network seen in the underlying chromosphere.
1999	John Raymond reviews the composition variations observed with *SOHO's* Ultra-Violet Coronagraph Spectrometer, abbreviated UVCS, instrument to confirm that the FIP effect is found in both the low corona and the solar wind.
2000	Len A. Fisk, Nathan A. Schwadron, and Thomas H. Zurbuchen propose that the fast solar wind is accelerated by the continual emergence of small magnetic loops that reconnect with open magnetic field lines near their base, and attribute large-scale motions of the heliosphere magnetic field to differential rotation of the photosphere and non-radial expansion of the solar wind near the Sun
2000	Joseph E. Mazur and his co-workers use impulsive flare particles to demonstrate the mixing of interplanetary magnetic fields.
2000	Sheela Shodhan and her colleagues use the third *International Sun-Earth Explorer*, abbreviated *ISEE 3*, the eighth *Interplanetary Monitoring Platform*, abbreviated *IMP 8*, and *Wind* observations of counter-streaming electrons to infer the topology of interplanetary magnetic clouds associated with coronal mass ejections.
2000	Rudolf Von Steiger and colleagues use the Solar Wind Ion Composition Spectrometer, abbreviated SWICS, instrument on *Ulysses* to confirm the enhanced abundance of elements with low first ionization potential in the slow solar wind, and report its presence at lower amounts in the fast wind.
2002	Leonard Strachan and co-workers use the UltraViolet Cornonagraph Spectrometer, abbreviated UVCS, instrument on *SOHO* to detect slow solar winds along the axis of coronal streamers at heights above three solar radii from the Sun center, with velocities that gradually increase to about $90\,\text{km s}^{-1}$ at five solar radii from the Sun center.
2003	Len A. Fisk proposes that the reconnection of coronal loops with open magnetic fields can account for both the fast and the slow winds, and that the inverse dependence of solar wind speed on electron temperature is an intrinsic signature of the material in the coronal loops that release mass.
2003	George Gloeckler uses the Solar Wind Ion Composition Spectrometer, abbreviated SWICS, instruments on *Ulysses* and the *Advanced Composition Explorer*, abbreviated *ACE*, to detect ubiquitous high-speed suprathermal ions moving from twice to 50 times the solar wind speed.
2003	David J. McComas and his colleagues use the Solar Wind Observations Over the Poles of the Sun, abbreviated SWOOPS, instrument aboard *Ulysses* to determine the three-dimensional latitudinal structure of the solar wind during the spacecraft's second polar orbit, performed near solar maximum, revealing a mixture of slow and fast winds at all solar latitudes, including near the poles.

Date	Event
2003	John D. Richardson, Chi Wang, and Leonard F. Burlaga use *Voyager 2* observations to demonstrate correlated variations in speed, density, and magnetic fields within the outer heliosphere. The density and speed are generally anti-correlated in the inner heliosphere.
2004	Nancy U. Crooker and her colleagues use *Wind* observations to describe large-scale magnetic field inversions at solar wind sector boundaries.
2005	Three articles published in the same issue of *Science*, by Leonard F. Burlaga, Robert B. Decker, and Edward C. Stone, report that *Voyager 1* crossed the termination shock of the supersonic flow of the solar wind on 16 December 2004, at a distance of 94 AU from the Sun, becoming the first spacecraft to begin exploring the heliosheath, the outermost layer of the heliosphere.
2005	Jack T. Gosling and his colleagues (2005a, b) provide direct evidence for magnetic reconnection in the solar wind near the Earth's orbit at 1 AU.
2005	Chuan-Yi Tu and his colleagues use *SOHO* observations to suggest a model in which the solar wind in coronal holes originates from coronal funnels.
2006	The *Hinode* spacecraft is launched on 23 September 2006
2006	The *Solar TErrestrial RElations Observatory*, abbreviated *STEREO*, is launched on 25 October 2006.
2006–2007	In 2006, Tai D. Phan and his colleagues report three-spacecraft observations that show magnetic reconnection can occur over extended regions in the solar wind, at least 390 times the Earth in size. In 2007, John T. Gosling and co-workers (2007a, b) report five-spacecraft observations of oppositely directed exhaust jets from a magnetic reconnection X-line extending 4.26×10^6 km in the solar wind at 1 AU.
2007	*Voyager 2* crosses the termination shock of the solar wind on 30 August 2007, at a distance of 84 AU from the Sun, suggesting a north (*Voyager 1*, termination shock at 94 AU) – south (*Voyager 2* at 84 AU) asymmetry in the heliosphere, likely due to the direction of the local interstellar magnetic field. Leonard F. Burlaga, Richard B. Decker, Donald A. Gurnett, John D. Richardson, and Edward C. Stone, and their colleagues, report these observations in 2008.
2007	Bart De Pontieu and co-workers (2007b) report observations with the Solar Optical Telescope on *Hinode* that show the chromosphere is permeated with Alfvén waves that are energetic enough to accelerate the solar wind, and Jonathan W. Cirtain and colleagues describe evidence for Alfvén waves in the solar X-ray jets seen in polar coronal holes, which may contribute to the high-speed solar wind.
2007	Jonathan Cirtain, Takenorie J. Okamoto, and Bart De Pontieu and their colleagues present evidence of Alfvén waves using instruments aboard *Hinode*. The energy associated with these waves is apparently sufficient to heat the corona and accelerate the solar wind.
2007	Taro Sakao and colleagues use the X-ray Telescope aboard *Hinode* to detect the continuous outflow of X-ray emitting material at the edges of active regions adjacent to an equatorial coronal hole, suggesting a plausible source of the solar wind possibly related to the observations by Yutaka Uchida in 1992 with *Yohkoh*.
2007–2008	Instruments aboard *Ulysses* detect a steady, fast solar wind flow at high solar latitudes, originating from polar coronal holes and confined to solar latitudes above 45°.

Chapter 6
Our Violent Sun

6.1 Solar Outbursts of Awesome Power

Without warning, the relatively calm solar atmosphere can be torn asunder by brief and catastrophic outbursts of incredible energy, called solar flares, which occur in active regions, the magnetized atmosphere in and around sunspots. They are sudden and brief, usually lasting no more than 10 min, and rapidly increase the temperature of relatively small, Earth-sized region of the corona to temperatures as high as 20×10^6 K. As a result, solar flares can outshine the entire Sun in X-rays and extreme-ultraviolet radiation (Fig. 6.1). Solar flares are normally not detected in bright, visible sunlight, but they can be seen at visible wavelengths when focusing in on the red spectral line of hydrogen, the Balmer alpha transition at 656.3 nm.

Since the invisible extreme-ultraviolet and X-ray flare radiation is absorbed in the Earth's atmosphere, it is observed from telescopes in space, such as *Yohkoh*, the *SOlar and Heliospheric Observatory*, or *SOHO* for short, the *Transition Region And Coronal Explorer*, abbreviated *TRACE*, the *Ramaty High Energy Solar Spectroscopic Imager*, abbreviated *RHESSI*, and *Hinode*.

Solar flares accelerate particles to nearly the velocity of light, hurling them out into the solar system and down into the Sun. As the high-speed electrons move outward, they spiral around magnetic fields in the low corona, shining brightly in radio waves. High-speed electrons that are thrown down into the Sun emit hard X-rays when entering the lower solar atmosphere. Energetic flare protons create nuclear reactions when they are tossed down into the chromosphere or photosphere, emitting gamma rays in the process.

Solar flares are the biggest explosions in the solar system. The most powerful ones involve the explosive release of incredible amounts of energy, sometimes amounting to as much as a million, billion, billion (or 25 orders of magnitude and 10^{25}) Joule in a relatively short time between 100 and 1,000 s. This is comparable in strength to the simultaneous explosion of 20 million nuclear bombs, each blowing up with an energy of 100 megatons of TNT.

A substantial fraction of this energy goes into accelerating electrons and ions to high speeds at a substantial fraction of the velocity of light. These high-energy particles result in enhanced radio, soft X-ray, hard X-ray, and gamma-ray radiation from the Sun.

K.R. Lang, *The Sun from Space*, Astronomy and Astrophysics Library,
© Springer-Verlag Berlin Heidelberg 2009

Fig. 6.1 Extreme-ultraviolet flare. An X3.9 flare on 24 August 2002, observed with the *Transition Region And Coronal Explorer*, abbreviated *TRACE*, telescope in the 19.5-nm passband sensitive to gas temperatures of about 1.4×10^6 K. This image shows the cooling phase of the flare, in which part of the material involved in the flare has cooled from several million degrees to some $10,000$–$20,000°$ K. That cooler material is no longer transparent for the extreme ultraviolet radiation emitted by the hot gases behind it; it therefore appears as dark strands as it slides along different magnetic field lines. (Courtesy of Dick Shine, the *TRACE* consortium and NASA; *TRACE* is a mission of the Stanford-Lockheed Institute for Space Research, and part of the NASA Small Explorer Program.)

To classify the power of a solar flare, scientists use the maximum soft X-ray flux, in W m^{-2}, measured near the Earth with a *GOES* spacecraft in the wavelength range 0.1–0.8 nm. *GOES* is an acronym for a *Geostationary Operational Environmental Satellite*, which remains above a given location on Earth by orbiting the planet once every 24 h at a height of 35,790 km. Once a *GOES* spacecraft is launched successfully, it is given a number; *GOES 1* was launched in 1975 and we are up to *GOES 13*, with several others still in operation, imaging the Earth beneath them for short-term terrestrial weather forecasting and storm tracking and monitoring the space environment – including the temporal variation of the Sun's soft X-ray flux received at the spacecraft.

The *GOES* scheme classifies the soft X-ray radiation of solar flares as A, B, C, M, or X, from the weakest to the strongest, according to the peak detected flux (Table 6.1). Each class has a peak flux 10 times greater than the preceding one; with X class flares greater than 0.0001, or 10^{-4}, W m^{-2}. A given class of X-ray flares has nine linear subdivisions; numbered from 1 to 9 in increasing flux, so an X2

Table 6.1 Solar flare X-ray classification[a]

Class	Peak soft X-ray flux $(W\ m^{-2})$
A	Less than 10^{-7}
B	$10^{-7} - 10^{-6}$
C	$10^{-6} - 10^{-5}$
M	$10^{-5} - 10^{-4}$
X	Greater than 10^{-4}

[a]The peak flux is measured from the *GOES* spacecraft near Earth at soft X-ray wavelengths between 0.1 and 0.8 nm or 1 and 8 Å

flare is twice as powerful as an X1 flare. The largest X-ray flare on record occurred on 4 November 2003; it saturated the *GOES* X-ray detectors at X17.4 and had an estimated classification of X28, or $0.0028\,Wm^{-2}$, and perhaps even up to X45.

The X-class flares are major events that can trigger planet-wide radio blackouts and longlasting radiation storms; M-class flares produce brief radio blackouts and minor radiation storms, while C-, B- and A-class flares have few noticeable consequences here on Earth.

The rate of occurrence of solar flares varies with the 11-year cycle of solar magnetic activity, becoming more frequent near activity maximum. Truly outstanding flares are infrequent, occurring only a few times a year even at times of maximum solar activity; like rare vintages, they are often denoted by their date or given a special name. An example is the Bastille Day Flare that occurred on 14 July 2000, near the peak of a maximum in the solar activity cycle. Flares of lesser magnitude occur much more frequently; several tens of such events may be observed from spacecraft on a busy day near the cycle maximum.

There are other types of solar outbursts such as the erupting prominences. A prominence consists of relatively cool material, with a temperature of about 10,000 K, suspended above the solar photosphere. It can stretch up to half way across the Sun, and remain there for weeks or months at a time. Prominences can be detected by focusing in on the red spectral line of hydrogen with telescopes on the ground, or by observing their extreme-ultraviolet spectral signatures from space.

These magnetically supported features are called prominences when detected at the apparent edge of the Sun, where they stand out against the dark background. They appear as dark snaking features, called filaments, when projected against the bright solar chromosphere. So, prominence and filament are essentially two words that describe different perspectives of the same thing. And sometimes, their magnetic support becomes unhinged, and a prominence erupts or a filament lifts off, ejecting large quantities of matter into space (Fig. 6.2).

Coronal mass ejections are another type of outburst from the Sun. They carry billions of tons of coronal material into interplanetary space, expanding as they balloon out from the corona and becoming bigger than the Sun in a few hours or less. They are often accompanied by solar flares and erupting prominences, and the shocks associated with coronal mass ejections accelerate and propel vast quantities of high-speed particles ahead of them.

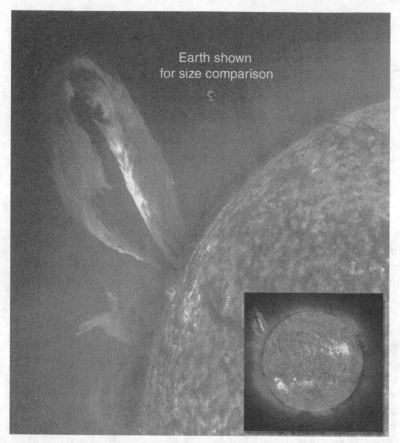

Earth shown
for size comparison

Fig. 6.2 Eruptive prominence. The Extreme-ultraviolet Imaging Telescope, abbreviated EIT,
aboard *SOHO* imaged this large erupting prominence in the extreme ultraviolet light of ionized
helium (He II at 30.4 nm) on 24 July 1999. The comparison image of the Earth shows that the
prominence extends over 35 Earths out from the Sun, while the inset full-disk solar image indicates that the eruption looped out for a distance almost equal to the Sun's radius. (Courtesy of the
SOHO EIT consortium. *SOHO* is a project of international collaboration between ESA and NASA.)

These mass ejections are detected in an entirely different manner from either
solar flares or erupting prominences, by using a space-borne coronagraph whose
occulting disk blocks out the bright light of the Sun's photosphere. Such a coronagraph is used to record sequential images of the corona's white light, the combined
colors of all the sunlight reflected by the rarefied coronal particles.

Coronal mass ejections are routinely imaged with the Large Angle and Spectrometric Coronagraph, abbreviated LASCO, aboard *SOHO* (Fig. 6.3). And the coronagraphs aboard the twin spacecraft of the *Solar TErrestrial RElations Obervatory*,
or *STEREO A* and *B* for short, are poised to help follow their trajectory from the Sun
to the orbit of the Earth.

Fig. 6.3 Ejection from the corona. The occulting disk of a coronagraph blocks the intense light from the photosphere, revealing the surrounding faint corona and a coronal mass ejection (*bottom left*) on 23 April 2000. In this composite image, created by Steele Hill, an extreme-ultraviolet image, taken on the same day, has been superposed at the location corresponding to the visible solar disk and masked by the coronagraph. The coronagraph image was taken with the Large Angle and Spectrometric COronagraph, abbreviated LASCO, on the *SOlar and Heliospheric Observatory*, or *SOHO* for short, and the superposed image was taken with the Extreme-ultraviolet Imaging Telescope, abbreviated EIT, on *SOHO*. (Courtesy of Steele Hill, the *SOHO* LASCO and EIT consortia and NASA. *SOHO* is a project of international collaboration between ESA and NASA.)

The appearance of a solar flare depends on how you look at it. That is, a flare produces copious radiation across the full electromagnetic spectrum, and one sees varied aspects of it when using different wavelengths. Both the spatial and the temporal behaviors of a solar flare depend on the perspective you choose, and each view focuses in on a different aspect of the flaring mechanism. In this chapter, we will therefore first discuss solar flares as seen from different viewpoints – as flare ribbons in the chromosphere, in soft and hard X-rays – with a model for the flaring X-ray emission, in white-light visible to the aided eye, as solar radio bursts, and in gamma rays.

Erupting prominences or filaments are the next topic, followed by coronal mass ejections. An account of the aftermath of solar flares and coronal mass ejections follows, including loop oscillations, global waves, and coronal dimming.

  Possible explanations of the magnetic triggering and energy release of these various kinds of energetic solar outbursts – the solar flares, erupting prominences, and coronal mass ejections – are then discussed.

  This chapter therefore introduces an enormous range of interesting and important topics. The serious student or curious reader may want to explore some of them in greater detail, and they should consult reviews of these subjects prepared by professionals in the field (Focus 6.1).

Focus 6.1

Expert Reviews about Energetic Solar Outbursts

Professional solar astronomers and astrophysicists have reviewed important developments in our knowledge of solar flares, erupting prominences, and coronal mass ejections, sometimes in technical terms. In alphabetical order, they include Markus J. Aschwanden's (2004, 2006) book entitled *Physics of the Solar Corona: An Introduction*, which includes discussions of magnetic reconnection and particle acceleration during flares and coronal mass ejections, Aschwanden's (2006) review of the localization of particle acceleration sites in solar flares and coronal mass ejections, Aschwanden and co-worker's (2006) review of theoretical modeling for the *STEREO* mission, Amitava Bhattacharjee's (2004) review of impulsive magnetic reconnection in the Earth's magnetotail and the solar corona, James Chen's (2001) review of the physics of coronal mass ejections, Edward L. Chupp's (1984, 1987) reviews of the observations and physics of gamma ray and neutron production during solar flares, Terry G. Forbes's (2000) review of the genesis of coronal mass ejections, Terry G. Forbes and co-worker's (2006) review of the theory and models for coronal mass ejections, Natchimuthuk Gopalswamy and co-worker's (2006a) review of the pre-eruptive Sun, Hugh S. Hudson, Jean-Louis Bougeret and Joan Burkepile's (2006a) review of observations of coronal mass ejections, Hugh S. Hudson, C. Jacob Wolfson, and Thomas R. Metcalf's (2006b) review of white-light flares, Stephen W. Kahler's (1992) review of solar flares and coronal mass ejections and Kahler's (2007) review of the solar sources of heliospheric energetic electron events, James A. Klimchuk's (2001) review of the theory of coronal mass ejections, Horst Kunow and co-worker's (2006) review of coronal mass ejections, Robert P. Lin's (1987) review of particle acceleration by the Sun, Jun Lin, Willie Soon, and Sallie L. Baliunas' (2003) review of theories of solar eruptions, David E. McKenzie's (2002) review of the signatures of reconnection in eruptive flares, Zoran Mikic and M. A. Lee's (2006) introduction to the theory and models of coronal mass ejections, shocks and solar energetic particles, Ronald L. Moore and Alphonse C. Sterling's (2006) review of the initiation of coronal mass ejections, Valery M. Nakariakov and Erwin Verwichte's (2005) review of coronal waves and oscillations, Donald F. Neidig's (1989) review of the photometric and spectrographic observations of white-light flares, Monique Pick and colleagues (2006) review of radio emission from the Sun and the interplanetary medium, Eric R. Priest and Terry G. Forbes's (2002) review of the magnetic nature of solar flares, Rainer Schwenn and co-worker's review of coronal observations of coronal mass ejections,

Sami K. Solanki's (2006) review of the solar magnetic field, and Mei Zhang and Boon Chye Low's (2005) review of the hydromagnetic nature of solar coronal mass ejections. Arnold O. Benz (2008) has provided a complete review of *Flare Observations* in a *Living Review*, available at http://solarphysics.livingreviews.org/.

The contributions of scientists working on specific topics are often presented at meetings or workshops and subsequently published in book form. They include *Coronal Mass Ejections*, edited by Nancy Crooker, Jo Ann Joselyn, and Joan Feynman (1997), *Multi-wavelength Observations of Coronal Structure and Dynamics*, edited by Petrus C. H. Martens and David P. Cauffman (2002), *SOHO-13: Waves, Oscillations and Small-Scale Transient Events in the Solar Atmosphere: A Joint View from SOHO and TRACE*, edited by Hugette Lacoste (2004), *Solar Eruptions and Energetic Particles*, edited by Natchimuthuk Gopalswamy, Richard Mewaldt, and Jarmo Torsti (2006b), and *SOHO-20: Transient Events on the Sun and in the Heliosphere*, edited by Bernhard Fleck, Joseph B. Gurman, Jean-Francois Hochedez, and Eva Robbrecht (2008).

6.2 Flare Ribbons

Routine and frequent visual observations of solar flares have been carried out from ground-based telescopes for at least 80 years, by monitoring the Sun's chromosphere in the red light of the Balmer alpha transition of hydrogen, designated Hα, at 656.3 nm wavelength. When viewed in this way, a solar flare appears as a sudden brightening, lasting from a few minutes to an hour, usually in strong, complex magnetic regions. They emit a sudden flash of red light followed by a slower decay, somewhat like igniting a fire in a pool of gasoline.

The Hα flares are nearly always located close to sunspots and comparable in area to them, often occupying less than one ten thousandth (0.01%) of the Sun's visible disk. They do not occur directly on top of sunspots, but are instead located between regions of opposite magnetic polarity, near the line or place of magnetic neutrality. They often appear on each side of the magnetic neutral line as two extended, parallel ribbons, like the double yellow line at the center of a highway (Fig. 6.4).

Much, of not most, of a flare's energy goes into accelerating large numbers of electrons and protons to a good fraction of the speed of light. These charged particles move away from the flare-initiation site in the low corona, down along magnetic fields into the upper layers of the chromosphere, where they form the bright ribbons visible in hydrogen alpha.

Such flare ribbons have also been detected at extreme-ultraviolet wavelengths using instruments aboard *SOHO* and *TRACE*. In 2001 and 2004, for example, Lyndsay Fletcher and colleagues used the high cadence and resolution of *TRACE* to obtain detailed measurements of the magnetic structure and generation of the flare ribbons, as well as the motion of the ribbons as the flare progresses. These results

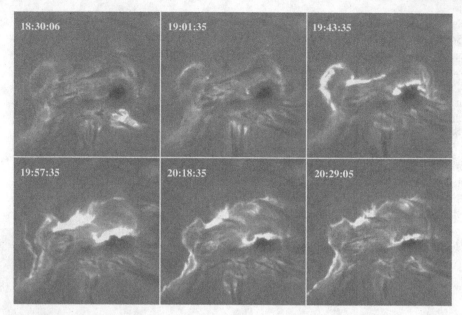

Fig. 6.4 Flare ribbons. A large solar flare observed in the red light of the Balmer alpha transition of hydrogen, called hydrogen alpha or Hα for short. It portrays two extended, parallel flare ribbons in the chromosphere. Each image is 200,000 km in width, subtending an angle of 300 s of arc, or about one-sixth the angular extent of the Sun. These photographs were taken at the Big Bear Solar Observatory on 29 April 1998. (Courtesy of Haimin Wang.)

support the hypothesis that the flare ribbons map out the chromospheric footpoints of reconnected, newly formed coronal loops, and indicate that the footpoints move apart as the flare progresses, at a speed of about 15 km s^{-1}.

The magnetic reconnection rate is measured by using *TRACE* to track the particle precipitation sites as they move across the photosphere magnetic field in *SOHO* Michelson Doppler Imager, abbreviated MDI, magnetograms. In 2003, Haimin Wang and colleaques inferred similar ribbon separation speeds and magnetic reconnection rates from Hα observations and magnetic fields obtained from MDI. As reported by Fletcher and Hugh S. Hudson in 2002, the hard X-ray footpoint ribbons also exhibit systematic motions.

When impacting the chromosphere, the kinetic energy of the flare-accelerated particles is transformed into heat. The heated chromospheric matter responds by moving into the corona, where it can increase the gas density within a flaring loop a 1000-fold. As proposed by Tadashi Hirayama in 1974, the dense, heated material comes from the chromosphere "by a process like evaporation while the flare is in progress," giving rise to the soft X-ray radiation of the flare. The heat is then quickly lost into space in the form of bright X-ray radiation.

Hydrogen-alpha or extreme-ultraviolet images of the Sun's flaring atmosphere provide a two-dimensional, flatland picture of the chromosphere or transition region without information about what is happening above them. Observations at X-ray wavelengths from the *Yohkoh* and *Hinode* spacecraft provide a three-dimensional perspective, leading to a more complete understanding of the physical processes responsible for solar flares. As we shall next see, they have confirmed Hirayama's hypothesis of chromospheric evaporation, and identified the particle acceleration site in the low solar corona.

6.3 X-Ray Flares

6.3.1 Soft X-Rays and Hard X-Rays from Solar Flares

Much of the energy radiated during a solar flare is emitted as X-rays. This radiation provides detailed information about the flare process including why and where flares occur. The wavelength of an X-ray is on the order of, or smaller than, the size of an atom, or between 10^{-11} and 10^{-9} m.

Researchers also describe X-rays by the energy of the radiation photons. There are soft X-rays with relatively low energy and modest penetrating power. The hard X-rays have higher energy and greater penetrating power. As a metaphor, one thinks of the large, pliant softballs and the compact, firm hardballs, used in the two kinds of American baseball games.

The energy of the X-ray radiation is a measure of the energy of the electrons that produce it. The high-energy, hard X-ray radiation of solar flares is produced by non-thermal electrons accelerated to nearly the velocity of light. They tell us about the acceleration, propagation, and confinement of very energetic electrons. The soft X-rays describe the thermal radiation of hot electrons of lower energy.

Like energetic charged particles, the energy of flaring X-rays is often specified in kilo-electron volts, denoted keV. Soft X-rays have energies between 1 and 10 keV, and hard X-rays lie between 10 and 100 keV. Gamma rays are even more energetic than X-rays, exceeding 100 keV in energy. The wavelength of radiation is inversely proportional to its energy, so hard X-rays are shorter than soft X-rays, and the wavelengths of gamma rays are still smaller. Electrons with a given amount of energy produce X-rays or gamma rays with about the same energy.

Although X-rays can penetrate small amounts of material substances, including your skin and muscles, solar X-rays are totally absorbed in the atmosphere. This radiation is therefore now observed from satellites orbiting the Earth above our air. Pioneering observations of the X-ray emission from solar flares using rockets, the *Orbiting Solar Observatories* and *Skylab* were discussed in Sect. 2.5, and are included in the key events given at the end of this chapter.

The radiation at different X-ray wavelengths, the hard and soft varieties, describes different parts of the flare time profile and is attributed to different physical mechanisms. They are known as the impulsive hard-X-ray phase and gradual soft X-ray decay phases of a solar flare (Fig. 6.5). At the impulsive stage of a solar flare, electrons are accelerated rapidly, in a second or less, to energies that can exceed

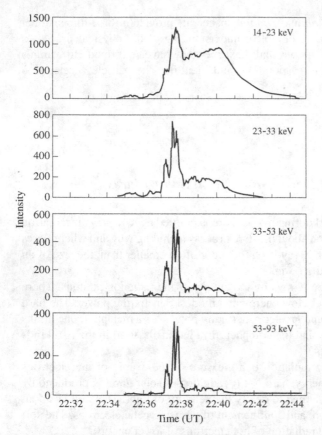

Fig. 6.5 Impulsive and gradual phases of a flare. This time profile of a solar flare observed at hard X-ray energies, above 30 keV (*bottom*), is characterized by an impulsive feature that lasts for about 1 min. This impulsive phase coincides with the acceleration of high-speed electrons that emit non-thermal bremsstrahlung at hard X-ray wavelengths and non-thermal synchrotron radiation at centimeter radio wavelengths. The less-energetic emission, below 30 keV (*top*), can be composed of two components, an impulsive component followed by a gradual one. The latter component builds up slowly and becomes most intense during the decay phase of solar flares when thermal radiation dominates. At even lower soft X-ray energies (about 10 keV), the gradual phase dominates the flare emission. These data were taken on 15 November 1991 with the Hard X-ray Telescope (HXT) aboard *Yohkoh*. (Courtesy of NASA, ISAS, the Lockheed-Martin Solar and Astrophysics Laboratory, the National Astronomical Observatory of Japan, and the University of Tokyo.)

1 MeV. The high-energy electrons emit hard X-rays and gamma rays that mark the flare onset. The soft X-rays emitted during solar flares gradually build up in strength and peak a few minutes after the impulsive emission, so the soft X-rays are a delayed effect of the main flare outburst.

The soft X-rays emitted during solar flares are thermal radiation, released by virtue of their intense heat and dependent upon the random thermal motions of very hot electrons. At such high temperatures, the electrons are set free from atoms

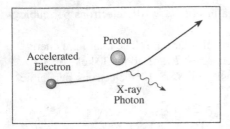

Fig. 6.6 Bremsstrahlung. When an electron moves rapidly and freely outside an atom, it inevitably moves near a proton in the ambient gas. There is an electrical attraction between the electron and the proton because they have equal and opposite charge, and this pulls the electron toward the proton, bending the electron's trajectory and altering its speed. The electron emits electromagnetic radiation in the process. This radiation is known as *bremsstrahlung* from the German word for "braking radiation"

and move off at high speed, leaving the ions (primarily protons) behind. When an electron moves through the surrounding material, it is attracted to the oppositely charged protons and emits a kind of thermal radiation called bremsstrahlung (Fig. 6.6, Focus 6.2).

Focus 6.2

Bremsstrahlung

When a hot electron moves rapidly and freely outside an atom, it inevitably moves near a proton in the ambient gas. The free electron is deflected from a straight path and changes its speed during its encounter with the proton, emitting electromagnetic radiation in the process. This radiation is known as *bremsstrahlung* from the German word for "braking radiation."

Because of their greater mass, the protons move more slowly than the electrons with the same temperature or kinetic energy. So, one can picture a proton at rest with the electron moving by. There is an electrical attraction between the electron and the proton because they have equal and opposite charge, and this pulls the electron toward the proton, bending the electron's trajectory and altering its speed (Fig. 6.6). The bremsstrahlung is produced at the point of the electron's closest approach to the proton. Bremsstrahlung can be emitted at all wavelengths, from long radio waves to short X-rays, but during solar flares it becomes very intense at X-ray wavelengths.

The greater the number of electrons, the stronger the thermal bremsstrahlung is. To be precise, the thermal bremsstrahlung power, P, increases with the square of the electron density, N_e, and the volume of the radiating source, V. It also depends upon the temperature of the electrons, T_e. A formula for the bremsstrahlung power is:

$$P = \text{constant} \times N_e^2 V\, T_e^{1/2},$$

where the constant is also a function of the electron temperature. Scientists can use this expression with measurements of the X-ray power during a flare to determine the density of electrons participating in the radiation, assuming that they completely fill the observed volume. Electron density values of $N_e \approx 10^{17}$ electrons per cubic meter are often obtained.

How much energy is released during a typical solar flare? The total flare energy, E_T, expended in producing electrons with energies, E_e, of about 30 keV, or 5.8 × 10^{-15} J, in a radius R with an electron density of $N_e \approx 10^{17}$ electrons per cubic meter is:

$$E_T = (4 \pi R^3 E_e N_e)/3 \approx 2 \times 10^{24} \, \text{J},$$

where the radius of a compact, impulsive flare is $R = 10^7$ m, subtending an angular radius of 14 s of arc when viewed from the Earth.

What fuels these catastrophic eruptions on the Sun, and where does such a large amount of energy come from? The almost universal consensus is that solar flares are powered by magnetic energy. The magnetic energy, E_M, for a magnetic field of strength, B, in a volume with radius, R, is:

$$E_M = [4\pi/(6 \, \mu_0)]B^2 R^3 = 0.166 \times 10^6 \, B^2 R^3 \, \text{J},$$

where the permeability of free space is $\mu_0 = 4\pi \times 10^{-7}$ H m^{-1}, the radius is in meters and the magnetic field strength in Tesla.

Magnetic reconnection in the low solar corona serves as an efficient method for converting magnetic energy into plasma kinetic energy and thermal energy. Let us suppose that the reconnection site is just above a coronal loop of radius $R = 10^7$ m, with a comparable size. To provide the flare energy, $E_T = 2 \times 10^{24}$ J in a volume of radius $R = 10^7$ m, a magnetic field of about 0.03 T is required. Solar astronomers often use the c.g.s. unit of Gauss, where $1 \, \text{T} = 10,000 \, \text{G}$, so the required magnetic field change in the corona is roughly 300 G.

There are two kinds of bremsstrahlung emitted at X-ray wavelengths during solar flares. They are called thermal bremsstrahlung and non-thermal bremsstrahlung, which distinguishes both the method of electron acceleration and the energy of the X-rays. The thermal, soft X-ray bremsstrahlung predominates during the decay phase of a solar flare. This radiation is produced when electrons are heated to high temperatures of about 10×10^6 K, moving at speeds of about 0.05 times the velocity of light and emitting soft X-rays when they encounter protons.

The electrons accelerated to high velocities during the impulsive phase of a flare radiate hard X-rays by non-thermal bremsstrahlung, which is the same process as the thermal one except the electrons are moving at non-thermal speeds near the velocity of light, faster than possible in a hot gas, and the photon energy of the radiation is much greater.

The intensity of thermal bremsstrahlung falls off rapidly with increasing photon energy, while that of the non-thermal hard X-ray flare radiation follows a less steep, power-law drop at increasing energy (Fig. 6.7). For the power law situation, the number of non-thermal electrons with energy, E, varies as E^{-P}, where the power-law index P is a small positive number. Observation of non-thermal bremsstrahlung provides a way to study the accelerated electrons and specify this index.

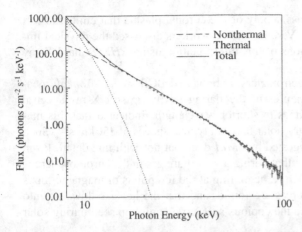

Fig. 6.7 Energy spectrum of flare electrons. The spatially integrated energy spectrum of the radiation photons during a 14-s time interval at the peak of a solar flare on 20 February 2002. The observed low-energy spectrum is described by a thermal spectrum at a temperature of 15×10^6 K. The high-energy emission is attributed to non-thermal emission of energetic electrons with a power-law spectral index of 4.4. These data were obtained from NASA's *Ramaty High Energy Solar Spectroscopic Imager*, abbreviated *RHESSI*. (Courtesy of Brian R. Dennis, NASA.)

The physical processes that give rise to the soft X-ray and hard X-ray flare emission were first suggested by observations from the *Solar Maximum Mission*, abbreviated *SMM*, in the 1980s, and amplified, confirmed, and extended with instruments aboard *Yohkoh*, with improved angular and temporal resolution and a wider energy range. Images from *Yohkoh's* Hard X-ray Telescope, or HXT, clarified, for example, the double-source, loop-footpoint structure of impulsive hard X-ray flares with unprecedented clarity. It established a double-source structure for the hard X-ray emission of roughly half the flares observed in the purely non-thermal energy range above 30 keV. The other half of the flares detected with *Yohkoh* were either single sources, which could be double ones that are too small to be resolved, or multiple sources that could be an ensemble of double sources.

When the two hard X-ray sources (above 30 keV) are seen from *Yohkoh*, they are located on opposite sides of the line that separates regions of different magnetic polarity (the magnetic neutral line), strongly suggesting that the hard X-rays are emitted from the footpoints of a flaring magnetic loop. The double hard X-ray sources occur and vary nearly simultaneously in time, within 0.1 s or less, a result that excludes transport mechanisms other than high-energy electrons in coronal loops.

In 1968, Werner M. Neupert reported that the slow, smooth rise of the soft X-ray flares, observed from the third *Orbiting Solar Observatory*, resemble the time integration of the rapid, impulsive radio bursts at centimeter wavelengths. After *SMM* was launched, the temporal correlation, which is now known as the Neupert effect, was extended and confirmed for soft and hard X-rays, showing that the impulsive hard X-ray flux corresponds to the time derivative of the soft X-ray flux. This suggests that the energetic electron beams that gives rise to the hard X-ray flare is the main source of heating and mass supply of the coronal plasma that emits the soft X-ray flare. In 2005, Astrid M. Veronig and co-workers discussed the physical implications and possible limitations of this interpretation using *RHESSI* hard X-ray and *GOES* soft X-ray data.

The four Bragg Crystal Spectrometers, abbreviated BCS, aboard *Yohkoh* measured the motions and temperatures of the flaring gas using soft X-ray spectral lines. Doppler shifts of these lines to shorter wavelengths indicated that this material was moving upwards during solar flares at typical speeds of $350\,\mathrm{km\ s^{-1}}$. Such upward flare motions were suggested by *SMM* data, but not with fine detail. From the *Yohkoh* results, it was shown that during a solar flare gas in the chromosphere is heated from $10,000\,\mathrm{K}$ to $20 \times 10^6\,\mathrm{K}$, beginning at the footpoints of magnetic loops where the hard X-ray flare sources are found, and that the hot gas flows up into flaring coronal loops to produce the copious soft X-ray emission seen during solar flares.

This rise in heated material during solar flares is described by the theory of chromospheric evaporation, first suggested by Tadashi Hirayama in his 1974 interpretation of the soft X-ray flare, but it has nothing to do with the evaporation of any liquid. Initially cool chromospheric material, heated by down flowing, or precipitating, flare electrons, expands upwards into the low-density corona along magnetic loops that shine brightly in soft X-rays after filling. The *Yohkoh* BCS data indicate that when replenished in full, the post-impulsive flare loops contain gas heated to a maximum temperature of about $23 \times 10^6\,\mathrm{K}$, even hotter than the center of the Sun at $15.6 \times 10^6\,\mathrm{K}$.

Fig. 6.8 Post-flare loops. This image was taken from the *Transition Region And Coronal Explorer*, abbreviated *TRACE*, telescope in the 17.1-nm passband sensitive to gas temperatures of about $1 \times 10^{6°}$. It shows some post-flare loops, cooling after an M2 flare that started 2 h prior to this image. (Courtesy of Dawn Myers, the *TRACE* consortium and NASA; *TRACE* is a mission of the Stanford-Lockheed Institute for Space Research, and part of the NASA Small Explorer program.)

The *SMM* observations of chromospheric evaporation include those of Loren Acton and co-workers in 1982, Ester Antonucci and colleagues in 1983 and 1984, and Andre Fludra and co-workers in 1989. The *Yohkoh* results were reported by George A. Doschek in 1990, 1991, and John T. Mariska in 1994, and revisited in the context of numerical models by Doschek and Harry P. Warren in 2005. In the meantime, Anja Czaykowska and colleagues in 1999 and by Jeffrey W. Brosius and Kenneth J. H. Phillips in 2004 had used the Coronal Diagnostic Spectrometer aboard *SOHO* to describe the material flowing out from loop footpoints.

While the *Yohkoh* observations reveal magnetic structures that become filled by hot flare plasma, at temperatures of $10–20 \times 10^6$ K, the *Transition Region And Coronal Explorer*, abbreviated *TRACE*, shows the locations of the post-flare plasma, after it has cooled down to 1 or 2×10^6 K, often tracing out post-flare loops (Fig. 6.8).

6.3.2 A Model of Flaring X-Ray Emission

A well-developed model describes the hard X-ray and soft X-ray radiation of solar flares (Fig. 6.9). According to this picture, solar flare energy release occurs mainly

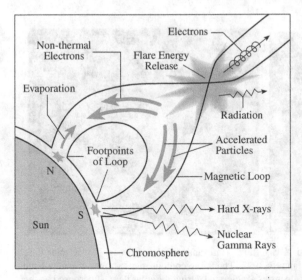

Fig. 6.9 Solar X-ray flare model. A solar flare is powered by magnetic energy released from a magnetic interaction site above the top of the loop shown schematically here. Electrons are accelerated to high speed, generating a burst of radio energy as well as impulsive loop-top hard X-ray emission. Some of these non-thermal electrons are channeled down the loop and strike the chromosphere at nearly the speed of light, emitting hard X-rays by electron-ion bremsstrahlung at the loop footpoints. When beams of accelerated protons enter the dense, lower atmosphere, they cause nuclear reactions that result in gamma-ray spectral lines and energetic neutrons. Material in the chromosphere is heated very quickly and rises into the loop, accompanied by a slow, gradual increase in soft X-ray radiation. This upwelling of heated material is called chromospheric evaporation, and occurs in the decay phase of the flare

during the rapid, impulsive phase, when charged particles are accelerated and non-thermal hard X-rays are emitted. The thermal decay phase, detected by the gradual build up of soft X-rays, is viewed as an atmospheric response to the energetic particles generated during the impulsive hard X-ray phase. The historical development of this model is deferred until Sect. 6.10, where we introduce early proposals for magnetic reconnection at the flare particle acceleration site, with refinements of the X-ray model developed by Kazunari Shibata and colleagues in 1995 and by Saku Tsuneta in 1996 and 1997, as well as the subsequent inclusion of coronal mass ejections. But for now, we will focus on the implications for the X-ray emission of solar flares.

It is generally believed that a solar flare is triggered by an instability or rearrangement in the magnetic configuration in the low corona. This results in the rapid release of stored, non-potential magnetic energy and the acceleration of non-thermal particles by processes that are still not well understood. In 1994–1995, Satoshi Masuda and co-workers used *Yohkoh* hard X-ray and soft X-ray observations of compact impulsive flares near the solar limb to infer the site of primary flare energy release. In addition to double-footpoint sources, a hard X-ray source was found well above the corresponding soft X-ray loop structure at around the peak time of the impulsive phase (Fig. 6.10). The energetic particles that give rise to the loop-top hard X-rays were accelerated in the corona and above the bright soft X-ray flare loops,

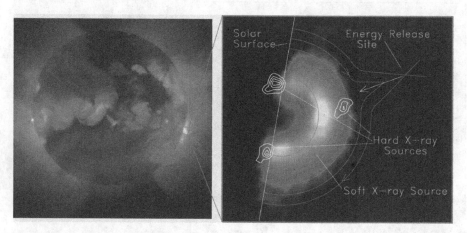

Fig. 6.10 Loop-top impulsive hard X-ray flare. Hard and soft X-ray images of a solar flare occurring near the solar limb on 13 January 1992. The white contour maps show three impulsive hard X-ray sources from high-energy electrons accelerated during the solar flare, superposed on the loop-like configuration of soft X-rays emitted by plasma heated during the flare gradual or decay phase to temperatures of 10–20×10^6 K. In addition to the double-footpoint sources, a hard X-ray source exists above the corresponding soft X-ray magnetic loop structure, with an intensity variation similar to those of the other two hard X-ray sources. This indicates that the flare is energized from a site near the magnetic cusp. These simultaneous images were taken with the Hard X-ray Telescope (HXT) and the Soft X-ray Telescope (SXT) aboard the *Yohkoh* satellite. (Adapted from Satoshi Masuda, 1994; Courtesy of NASA, ISAS, the Lockheed-Martin Solar and Astrophysics Laboratory, the National Astronomical Observatory of Japan, and the University of Tokyo.)

which are formed during the flare process and filled with evaporated plasma from the chromosphere.

This particle acceleration site was confirmed in 1996 by Markus S. Aschwanden and colleagues from observations of fast (subsecond) pulses of hard X-ray flare emission, which exhibit electron time-of-flight delays as the result of propagation between the coronal acceleration site and the hard X-ray footpoints in the chromosphere. The energy-dependent time delays were measured with an instrument aboard the *Compton Gamma Ray Observatory*, abbreviated *CGRO*. And the corresponding time-of-flight distances indicated that the flare acceleration site is located about 50% higher than the soft X-ray flare loop height inferred from *Yohkoh* observations. Such an acceleration site was additionally confirmed in 1997 when Aschwanden and Arnold O. Benz used radio observations to infer the location between upward-moving and downward-moving electron beams.

In less than a few seconds, electrons and protons are accelerated in the low corona, and beamed into the lower, denser reaches of the solar chromosphere, along newly linked coronal loops, or hurled out into space along open magnetic field lines. As the non-thermal electrons move either out or down along magnetic channels, they generate intense radio emission. Further down, at the footpoints of the arching magnetic loops, the high-speed electrons emit non-thermal hard X-ray bremsstrahlung via interactions with the ambient protons.

The chromosphere at the loop footpoints is heated very rapidly (in seconds) by the accelerated particles that slam into it. The high-temperature material in the chromosphere is driven upward by the large pressure gradients and "evaporates" along the guiding magnetic field to get rid of the excess energy. Relatively long-lived (tens of minutes) soft X-ray radiation is then emitted by thermal bremsstrahlung as the flaring loop is filled with the hot, rising material, and the coronal loop relaxes into a more stable configuration during the cooling, decay phase of a solar flare. And in the meantime, the rarely seen, white-light flares can be produced by the downward impact of the impulsive, non-thermal electrons when they strike the chromosphere or photosphere.

6.4 White-Light Flares

Exceptionally powerful solar flares can be detected in the combined colors of sunlight, or in white light. These white-light flares, as they are now called, create only a minor perturbation in the steady luminous output of the photosphere, so they are rarely seen. Richard Carrington and Richard Hodgson independently published the first account of one in 1860. But during the ensuing centuries, no more than about 100 white-light flares have been reported from ground-based observatories.

In recent decades, it has been shown that the white-light emission correlates well with the hard X-ray emission from solar flares in both time and space (Fig. 6.11). That is, the white-light flare usually occurs during the impulsive phase of a solar flare and is found at the footpoints of coronal loops, suggesting that the visible

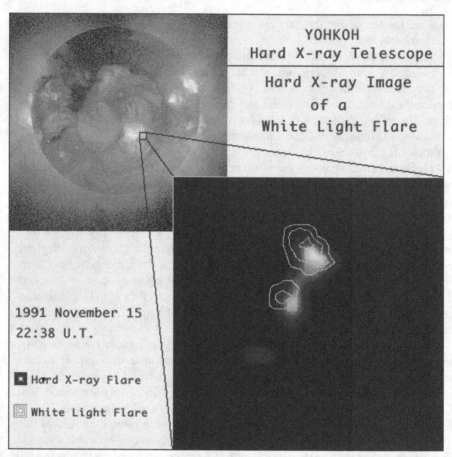

Fig. 6.11 Double hard X-ray and white light flare. Hard X-ray sources in solar flares often occur in simultaneous pairs that are aligned with the photosphere footpoints of a flaring magnetic loop detected at soft X-ray wavelengths. These footpoints can also be the sites of white-light flare emission. The time profiles of this flare, detected on 15 November 1991, show that the increase of white-light emission matches almost exactly that of the hard X-ray flux. This and the simultaneity of hard X-ray emission from the two footpoints establish that non-thermal electrons transport the impulsive-phase energy along the flare loops. These soft X-ray, hard X-ray, and white-light images of the solar flare were taken with telescopes aboard the *Yohkoh* mission. (Courtesy of NASA, ISAS, the Lockheed-Martin Solar and Astrophysics Laboratory, the National Astronomical Observatory of Japan, and the University of Tokyo.)

radiation may be related to non-thermal electrons accelerated in the low corona and beamed down into the chromosphere and upper photosphere. An acoustic pulse associated with white-light flares marks the location where the flare impulse hits the photosphere.

White-light flares have been observed with *TRACE*, and compared to simultaneous hard X-ray observations with *RHESSI*. These results, reported by Hugh S. Hudson, C. Jacob Wolfson. and Thomas R. Metcalf (2006b), confirmed the strong association of white-light emission with hard X-ray sources. The *TRACE* imaging capability was also used to show that the white-light emission is highly localized and intermittent, and commonly contains unresolved features less than 1 s of arc in size.

In 2007, Lyndsay Fletcher and colleagues compared co-spatial *TRACE* white-light continuum and *RHESSI* hard X-ray observations of loop footpoints, showing that the energy required to power the white-light emission was comparable to the total energy carried by the electron beam giving rise to the hard X-rays, and that the white-light flares are energized by the more numerous low-energy electrons which carry most of the total energy. Such low-energy electrons cannot penetrate into the photosphere or lower chromosphere, so they deposit energy in the upper chromosphere where flare ribbons are detected.

It was not until 2004 that Thomas N. Woods and colleagues reported the first detection of a flare in the Sun's total irradiance, which is dominated by visible sunlight, using NASA's *SOlar Radiation and Climate Experiment*, abbreviated *SORCE*. In 2006, Woods, Greg Kopp, and Phillip C. Chamberlin demonstrated that the flare energy inferred from the total solar irradiance of large flares is about 100 times greater than the soft X-ray measurements from the *GOES* spacecraft at 0.1–0.8 nm.

6.5 Gamma Rays from Solar Flares

Normally you cannot observe gamma rays from the Sun, even from outer space, for their intensity is too low to be detected with the available instruments. But when protons and heavier ions are accelerated to high speed during solar flares, and beamed down into the chromosphere, they produce nuclear reactions and generate gamma rays that have been detected. They are the most energetic kind of radiation detected from solar flares.

The gamma rays have energies of about 1 MeV, equivalent to a 1,000 keV, so the gamma rays are 10–100 times more energetic than the hard X-rays and soft X-rays detected during solar flares. Like X-rays, the gamma rays are totally absorbed in our atmosphere and must be observed from space.

Some of the protons accelerated during a solar flare slam into the dense, lower atmosphere, like a bullet hitting a concrete wall, shattering abundant heavy nuclei in a process called spallation. The lighter nuclear fragments are left in an excited state, but promptly calm down and get rid of their excess energy by emitting gamma rays. Other abundant nuclei are directly excited by collision with the flare-accelerated protons, and radiate the extra energy in gamma rays, thereby relaxing to their former unexcited state. Thus, nuclear reactions can occur on the visible disk of the Sun, as well as deep down in its energy-generating core (Table 6.2).

During bombardment by flare-accelerated ions, energetic neutrons can be torn out of the nuclei of atoms. Many of these neutrons are eventually captured by ambient, or non-flaring, hydrogen nuclei, the protons, in the photosphere, making deuterons, the nuclei of deuterium atoms, and emitting one of the Sun's strongest gamma-ray lines, at 2.223 MeV. The neutrons must slow down and lose some energy

Table 6.2 Some important gamma-ray lines from solar flares[a]

Process	Particles or elements	Energy (MeV)
Pair annihilation	$e^+ + e^-$	0.511*
Neutron capture	$^2H (=^2 D)$	2.223*
Spallation	^{12}C	4.438*
Spallation	^{16}O	6.129*
		6.917*
		7.117*
		2.741
Alpha excitation	^7Be	0.431*
Alpha excitation	^7Li	0.478*
Proton excitation	^{14}N	5.105*
		2.313
Proton excitation	^{20}Ne	1.634*
		2.613
		3.34
Proton excitation	^{24}Mg	1.369*
		2.754
Proton excitation	^{28}Si	1.779*
		6.878
Proton excitation	^{56}Fe	0.847*
		1.238*
		1.811

[a] The most prominent lines are marked with an asterisk * and have been detected in the flare of 27 April 1981 (Murphy et al., 1985). The energy is given in MeV, where $1\,MeV = 1,000\,keV = 1.602 \times 10^{-13}\,J$ or $1.602 \times 10^{-6}\,erg$.

by collisions before the protons can capture them, so the gamma-ray line is delayed by a minute or two from the onset of impulsive flare emission.

Neutrons produced by accelerated particle interactions during solar flares can also escape from the Sun, avoiding capture there. Neutrons with energies above 1,000 MeV have even been directly measured in space near Earth, in the 1980s from the *SMM*, and in the 1990s from the *CGRO*. In the most energetic flares, the associated neutrons can reach the Earth, and produce a signal in ground-level neutron monitors.

Another strong gamma-ray line emitted during solar flares is the 0.511 MeV positron annihilation line. Positrons, the anti-matter counterpart of electrons, are released during the decay of radioactive nuclei produced when flare-accelerated protons and heavier nuclei interact with the lower solar atmosphere. The positrons annihilate with the electrons, producing radiation at 0.511 MeV, which is the energy contained in the entire mass of a non-moving electron. Before the positrons can interact with the ambient thermal electrons, they must also slow down by collisions until they have similar velocities. The positron can even combine with an electron to briefly produce a positronium "atom", before self-destructing with the production of the 0.511 MeV spectral feature.

The detection of gamma-ray lines during solar flares is not a recent thing. Edward L. Chupp observed the positron-annihilation, neutron capture, and carbon and oxygen de-excitation lines in 1971, using an instrument aboard NASA's seventh *Orbiting Solar Observatory*, abbreviated *OSO 7*. But our understanding of the high-energy processes that produce solar-flare gamma rays and neutrons has been significantly improved as the result of observations in the 1980s, using the Gamma Ray Spectrometer aboard the *SMM*, reviewed by Chupp in 1984 and 1987, in the 1990s with the *CGRO*, and after 2002 from the *Ramaty High Energy Solar Spectroscopic Imager*, abbreviated *RHESSI*. Instruments aboard *RHESSI* have detected the delayed neutron capture line, the pair-annihilation line, and the gamma-ray lines from excited nuclei of carbon, oxygen, nitrogen, neon, magnesium, silicon, and iron during solar flares (Focus 6.3).

Focus 6.3

Nuclear Reactions on the Sun

Nuclear reactions during solar flares produce gamma-ray lines, emitted at energies between 0.4 and 7.1 MeV. They result from the interaction of flare-accelerated protons and helium nuclei, having energies between 1 and 100 MeV, with nuclei in the dense atmosphere below the acceleration site.

When protons with energies above 300 MeV interact with the abundant hydrogen in the solar atmosphere, they can produce short-lived fundamental particles, called mesons, whose decay leads to gamma-ray emission. The decay of neutral mesons produces a broad gamma-ray peak at 70 MeV, and the decay of charged mesons leads to bremsstrahlung giving a continuum of gamma rays with energies extending to several MeV. Neutrons with energies above 1,000 MeV are also produced.

Narrow gamma-ray lines (≤ 100 keV in width) have been observed during solar flares from deuterium formation, electron–positron annihilation, and excited carbon, nitrogen, oxygen and heavier nuclei (also see Table 6.2). These reactions are often written using letters to denote the nuclei, a Greek letter γ to denote gamma-ray radiation, and an arrow $\rightarrow$ to specify the reaction; nuclei on the left side of the arrow react to form products given on the right side of the arrow. The letter p is used to denote a proton, the nucleus of a hydrogen atom; the Greek letter α signifies an alpha particle, which is the nucleus of the helium atom.

As an example, the collision of a flare-associated proton, p, or alpha particle, α, with a heavy nucleus in the dense solar atmosphere may result in a spallation reaction that causes the nucleus to break up into lighter fragments which are left in excited states denoted by an asterisk (*). They subsequently de-excite to emit gamma-ray lines. Important examples are the production of excited carbon, $^{12}C^*$, and oxygen, $^{16}O^*$, by flaring protons, p, that break up oxygen, ^{16}O, or neon, ^{20}Ne, nuclei by the reactions:

$$p + {}^{16}O \rightarrow {}^{12}C^* + \alpha$$

and

$$p + {}^{20}Ne \rightarrow {}^{16}O^* + \alpha,$$

with de-excitation and emission of a gamma-ray line, γ, of energy, $h\nu$, by:

$$^{12}C^* \rightarrow ^{12}C + \gamma \quad (h\nu = 4.438 \text{ MeV})$$

and

$$^{16}O^* \rightarrow ^{16}O + \gamma \quad (h\nu = 6.129 \text{ MeV}).$$

Flare-accelerated protons, p, can also interact with abundant nuclei, N, in the solar atmosphere, exciting nuclei that emit gamma-ray lines on de-excitation. The inelastic scattering of the nucleus, N, with a proton, p, can be written:

$$N + p \rightarrow N^* + p',$$

where the unprimed and primed sides, respectively, denote the incident and scattered proton, p, and the excited nucleus N^* reverts to its former state emitting a gamma-ray line, γ, by:

$$N^* \rightarrow N + \gamma.$$

Elements and energies of prominent gamma-ray lines resulting from proton excitation of abundant elements were given in Table 6.2.

Heavy excited nuclei can also be generated by the fusion of flare-associated particles with lighter nuclei in the solar atmosphere. Examples include beryllium, $^7Be^*$, and lithium, $^7Li^*$, produced by flaring alpha particles, α, and ambient helium, ^4He.

$$^4He + \alpha \rightarrow ^7Be^* + n$$

and

$$^4He + \alpha \rightarrow ^7Li^* + p,$$

with the excited nuclei reverting to their former unexcited state accompanied by the emission of a gamma-ray line, γ, of energy, $h\nu$, by:

$$^7Be^* \rightarrow ^7Be + \gamma \quad (h\nu = 0.431 \text{ MeV})$$

and

$$^7Li^* \rightarrow ^7Li + \gamma \quad (h\nu = 0.478 \text{ MeV}).$$

Additional gamma-ray lines that are observed during solar flares include those associated with electron–positron annihilation at 0.511 MeV, denoted by:

$$e^- + e^+ \rightarrow \gamma + \gamma \quad (h\nu = 0.511 \text{ MeV}),$$

where e^- and e^+, respectively, denote an electron and a positron. The positron, or positive electron, is the anti-matter particle of the electron. Another common gamma-ray line is the feature at 2.223 MeV associated with neutron, n, capture by hydrogen nuclei, ^1H or p, to produce excited deuterium, $^2H^*$, that relaxes with emission of gamma rays, γ.

$$^1H + n \rightarrow ^2H^*$$

with

$$^2H^* \rightarrow {}^2H + \gamma \quad (h\nu = 2.223 \text{ MeV}).$$

The neutrons are produced when nuclei in the solar atmosphere are shattered during collisions with ions that have been accelerated in the flare. The line emission is usually delayed from the time of flare acceleration in order to allow the neutrons to thermalize, or slow down, before they are captured by the abundant hydrogen nuclei in the photosphere.

As an interesting aside, instruments aboard *CGRO* and *RHESSI* have also detected gamma ray flashes associated with lightning high in the Earth's atmosphere (Focus 6.4).

Focus 6.4

Terrestrial Gamma-Ray Flashes

In 1994, G. Jerry Fishman and colleagues reported observations from the *CGRO* of intense, millisecond gamma-ray flashes in the Earth's atmosphere. And in 2005, David M. Smith and co-workers announced *RHESSI* observations of more frequent terrestrial gamma-ray flashes of even greater energy. These flashes are generated by high-speed electrons, with energies of up to 20 MeV, occur every day or two, and appear to be due to high-altitude lightning discharge above thunderstorms.

High-resolution spectroscopy from *RHESSI* has resulted in spectrally resolved electron–positron annihilation and nuclear de-excitation lines for the first time, respectively reported in 2003 by Gerald H. Share and David M. Smith and their colleagues. The 0.511 MeV electron–positron annihilation line exhibited variable and unexpectedly broad widths, reflecting conditions where the annihilation positrons reside. The nuclear de-excitation lines of neon, magnesium, silicon, iron, carbon, and oxygen showed mass-dependent Doppler redshifts of 0.1–0.8%, attributed to tilted magnetic loops or downward motion of the accelerated protons.

The first gamma-ray images of solar flares have also been obtained from an instrument aboard *RHESSI*, and reported by Gordon J. Hurford and his colleagues in 2003 and 2006. The images, obtained in the 2.223 MeV neutron-capture line, showed gamma ray sources that were spatially separated from the hard X-ray flare-loop footpoints. This unexpected displacement indicates that the accelerated protons, which result in the neutron-capture line, are not co-spatial with the accelerated electrons, which produce the hard X-rays. Perhaps, the protons and electrons are accelerated in different places, or once accelerated they may propagate and lose their energy differently. One of the flares, occurring on 28 October 2003, yielded the first image of double-footpoint, gamma-ray line sources. The gamma-ray images from solar flares observed so far are located in the flare active region, demonstrating that the responsible protons and other ions are accelerated by a flare process and not by shocks such as those generated during fast coronal mass ejections.

RHESSI has additionally been used to image the flaring regions at hard X-ray energies. In 2003, for example, Linhui Sui and Gordon Holman discovered a compact coronal X-ray source whose motions, initially downward and subsequently upward, reveal the site of flare energy release, located between the coronal source and the underlying flare loops. In 2005 and 2006. Sui and Holman teamed up with Brian R. Dennis to present additional evidence for impulsive flare energy release in the corona, between the tops of the flare loops and a separate, higher coronal source.

The high-speed electrons that are accelerated during the impulsive phase of a solar flare are not all beamed down into the Sun along magnetic loops. Some of them break free of the arching magnetic structures, and are tossed into surrounding interplanetary space along open magnetic fields where they emit radio radiation. So this takes us from the shortest waves emitted during a solar flare, the gamma rays, to the longest ones, the radio waves.

6.6 Solar Radio Bursts

The radio emission of a solar flare is often called a radio burst to emphasize its brief, energetic, and eruptive characteristics. During such bursts, the Sun's radio emission can increase up to a million times normal intensity in just a few seconds, so they can outshine the entire Sun at radio wavelengths.

Solar radio bursts are very effective probes of the physical state of the flaring solar atmosphere, providing an important diagnostic tool for magnetic and temperature structures and displaying signatures of electrons accelerated to very high speeds, approaching that of light. The expulsion of these energetic electrons has been confirmed by direct in situ measurements in interplanetary space (Sect. 7.3).

As first shown by Ruby Payne-Scott, Donald E. Yabsley, and John G. Bolton in 1947, the bursts do not occur simultaneously at different radio frequencies or wavelengths, but instead drift to later arrival times at lower frequencies and longer wavelengths. This is explained by a disturbance that travels out through the progressively more rarefied layers of the solar corona, making the local electrons vibrate at their natural frequency of oscillation, called the plasma frequency (Focus 6.5).

Focus 6.5

Exciting Plasma Oscillations in the Corona

At the high temperature of the solar corona, of about a million Kelvin, electrons are stripped from the gaseous atoms by innumerable collisions, leaving electrons and ions that are free to move about. The electrons have a negative charge, and since ions are atoms that are missing one or more electrons, they are positively charged. An un-ionized atom is electrically neutral without charge. In the solar corona, the negative charge of the electrons equals the positive charge of the protons, so the mixture of electrons and protons, called plasma, has no net charge. The entire Sun, including its outer atmosphere, is nothing but a giant, hot ball of plasma.

When a flare-associated disturbance, such as an electron beam or a shock wave, moves though the coronal plasma, the local electrons are displaced with respect to the protons, which are more massive than the electrons. The electrical attraction between the electrons and protons pulls the electrons back in the opposite direction, but they overshoot the equilibrium position. The light, free electrons therefore oscillate back and forth when a moving disturbance passes through the corona.

The natural frequency of oscillation, called the plasma frequency, depends on the local electron concentration, with a higher plasma frequency at greater coronal electron densities. The exact expression for the square of the plasma frequency, $v_p{}^2$, is:

$$v_p^2 = e^2 N_e / (4\pi^2 \varepsilon_0 m_e) = 81 N_e \text{ Hz}^2,$$

where the electron density, denoted by N_e, is in units of electrons per cubic meter, the electron charge $e = 1.60 \times 10^{-19}$ coulomb, the electron mass $m_e = 9.11 \times 10^{-31}$ kg, and the permittivity of free space $\varepsilon_0 = 10^{-9}/(36\pi)$ F m^{-1}. Low in the solar corona, where $N_e \approx 10^{14}$ m^{-3}, the plasma frequency is $v_p \approx 9 \times 10^7$ Hz $= 90$ MHz, where 1 MHz is a million Hz or a million cycles per second. Thus, a solar flare can set the corona oscillating at radio frequencies, and this plasma frequency decreases with diminishing electron density at greater distances from the Sun.

As the explosive disturbance moves out through the progressively more rarefied layers of the corona, it excites radiation at lower and lower radio frequencies. Ground-based radio telescopes can identify these radio signals by changing the frequency to which their receiver is tuned. As an example, the radiation at a frequency of 200 MHz, or at 1.5 m wavelength, might arrive about a minute before the 20 MHz (15 m) outburst. The product of frequency and wavelength is equal to the velocity of light $c = 2.9979 \times 10^8$ m s^{-1}.

With an electron density model of the solar atmosphere, the emission frequency can be related to height, and combined with the time delays between frequencies to obtain the outward velocity of the moving disturbance. One can alternatively use the relationship between the drift rate, dv/dt, near frequency v, and the disturbance velocity, V:

$$dv/dt = (N^{-1} \, dN/ds) \, v \, V/2,$$

where the gradient of density, N, with height, s, is $(N^{-1} \, dN/ds) = 10^{-8}$ m^{-1} for the solar corona at a temperature of 2×10^6 K.

For type III radio bursts, a velocity between 0.2 and 0.8 c is determined, with an average velocity of 0.4 c, where c is the velocity of light. Beams of high-speed electrons, thrown out from solar flares, excite progressively lower plasma frequencies. The kinetic energy of the flare-associated electrons is 10–100 keV. One kiloelectron volt is equivalent to an energy of 1.6×10^{-16} J, and the rest mass energy of the electron, $m_e c^2 = 511$ keV, where m_e is the mass of the electron and c is the velocity of light.

As the hot corona expands into space, creating the solar wind, it fills an ever-greater volume and thins out, becoming more and more rarefied at greater distances from the Sun. There are only about 10 million electrons per cubic meter in interplanetary space near the Earth's orbit, corresponding to a plasma frequency of about

28,000 Hz, or 28 kHz. Radio experiments aboard spacecraft have followed the high-speed electrons of type III bursts as they have moved out into these more rarefied regions, mapping out the spiral structure of the interplanetary magnetic field that guides the electron motion.

The ionosphere is also plasma, with a plasma frequency of up to 10 MHz corresponding to wavelengths longer than 30 m. The Sun's X-ray and extreme-ultraviolet radiation create this electrically charged layer, located between 50 and 1,000 km above the Earth's surface. It reaches a maximum density of almost one million million-lion (10^{12}) electrons per cubic meter at an altitude of a few hundred kilometers.

The ionosphere reflects radiation with frequencies lower than the ionosphere's plasma frequency. Terrestrial long-wavelength radio communication utilizes this mirroring capability of the ionosphere to get around the curvature of the Earth.

The ionosphere similarly reflects incoming solar radio waves back into space if their frequency is lower than about 10 MHz. Spacecraft lofted above the ionosphere must therefore be used to track high-speed electrons or shock waves at remote distances from the Sun, where the density and plasma frequency are lower. They have monitored the corona's plasma radiation at frequencies from 0.01 to 10 MHz for more than three decades, beginning in 1963 from the *Alouette-1* satellite followed by the first *Radio Astronomy Explorer*, abbreviated *RAE-1*, launched in 1968.

Ground-based radio telescopes observe the dynamic spectra, or frequency drift, for solar radio bursts between 10 and 8,000 MHz. In the 1950s, John Paul Wild's group in Australia, for example, used a swept-frequency receiver to distinguish at least two types of meter-wavelength radio bursts (Fig. 6.12, Table 6.3). Designated as type II and type III bursts, they both show a drift from higher to lower frequencies, but at different rates. Solar radio astronomers usually measure this frequency in units of MHz, where 1 MHz is equal to a million, or 10^6, Hz. A frequency of 300 MHz corresponds to a wavelength of 1 m.

In 1963, Wild, Stefan F. Smerd and A. A. Weiss reviewed their pioneering investigations of radio bursts from the solar corona. Monique Pick and her colleagues gave a recent review of solar radio emissions from the Sun and the interplanetary medium in 2006, and in 2008 Arnold Benz provided a comprehensive review of flare observations as a *Living Review in Solar Physics*.

Modern instruments that detect the changing frequency of solar radio bursts include those aboard the *Wind, Ulysses*, and the *STEREO A* and *B* spacecraft. They have been used to triangulate the trajectories of type II bursts, which are produced by the shocks driven by coronal mass ejections. These results permit a three-dimensional determination of the shock's motion through space from the low corona near the Sun to the Earth's orbit.

The most common bursts detected at meter wavelengths are the fast-drift type III bursts, which provide evidence for the ejection of very fast electrons from the Sun. These radio bursts last for only a few minutes at the very onset of solar flares and extend over a wide range of radio frequencies (also see Fig. 6.12). The rapid

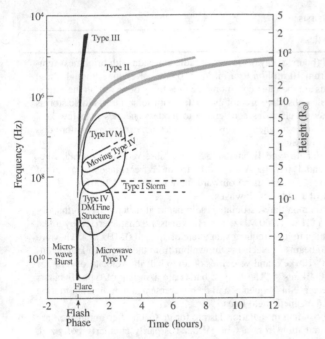

Fig. 6.12 Solar radio bursts. Schematic representation of the radio spectrum during and after a large solar flare. It can be associated with several different kinds of intense radio emission, depending on the frequency or wavelength (*left vertical axis*) and time after the explosion (*bottom axis*). In these plots, the impulsive, or flash, phase of the solar flare is indicated at 0 h; it normally lasts about 10 min and is associated with a powerful microwave burst. Dynamic spectra at frequencies of about 10^8 Hz, or 100 MHz, show bursts that drift from high to low frequencies as time goes on, but at different rates depending on the type of burst. Types II and III radio bursts have been respectively attributed to shock waves and electron beams moving outward into the solar atmosphere, or corona, exciting plasma oscillations. The height scale (*right vertical axis*) corresponds to the height, in units of the Sun's radius of 696×10^6 m, at which the coronal electron density yields a plasma frequency corresponding to the frequency on the left-hand side. At frequencies less than about 10^7 Hz, or 10 MHz, the bursts must be observed from spacecraft, above the ionosphere that deflects incoming radio waves at these low frequencies and long wavelengths. [Adapted from H. Rosenberg (1976).]

frequency drift corresponds to an outward velocity of about half the speed of light (also see Focus 6.5).

Electrons have to be accelerated to very high energies to move this fast. An electron's energy is often specified in units of kilo-electron-volts, or keV for short. One kiloelectron volt is equivalent to 1,000 eV, and to an energy of 1.6×10^{-16} J. An electron volt is the energy acquired by an electron when it is accelerated through a potential difference of 1 V. The electrons responsible for type III bursts have been accelerated to energies of about 100 keV.

A type III radio burst emits radiation generated by non-thermal processes, and cannot be due to the thermal radiation of a hot gas. Thermal radiation is emitted by a collection of particles that collide with each other and exchange energy frequently,

Table 6.3 Types of solar radio bursts[a]

Burst	Characteristics
Type I	Long-lived (hours to days) sources of radio emission with brightness temperatures from 10 million to a billion (10^7–10^9) Kelvin. Although these noise storms are the most common type of activity observed on the Sun at meter wavelengths, they are not associated with solar flares. Noise storms are attributed to electrons accelerated to modest energies of a few keV within large-scale magnetic loops that connect active regions to more distant areas of the Sun.
Type II	Meter-wavelength type II bursts have been observed at frequencies between 0.1 and 100 MHz. A slow drift to lower frequencies of a rate of about 1 MHz s^{-1} suggests an outward motion at about 1,000 km s^{-1} and has been attributed to shock waves.
Type III	The most common flare-associated radio bursts at meter wavelengths, observed from 0.1 to 1,000 MHz. Type III bursts are characterized by a fast drift from high to low frequency, at a rate of up to 100 MHz s^{-1}. They are attributed to beams of electrons thrown out from the Sun with kinetic energies of 10–100 keV, and velocities of up to half the velocity of light, or up to 150,000 km s^{-1}. The U-type bursts are a variant of type III bursts that first decrease in frequency and then increase again, indicating motion along closed magnetic field lines.
Type IV	Broadband continuum radiation lasting for up to 1 h after impulsive flare onset. The radiation from a type IV burst is partly circularly polarized, and has been attributed to synchrotron emission from energetic electrons trapped within magnetic clouds that travel out into space with velocities from several hundred to a 1,000 km s^{-1}.
Centimeter	Impulsive continuum radiation at centimeter wavelengths that lasts just a few minutes at flare onset. These microwave bursts are attributed to the gyrosynchrotron radiation of high-speed electrons accelerated to energies of 100–1,000 keV. The site of acceleration is located above the tops of coronal loops.
Millisecond	Radio flares observed at 200–1,400 MHz can include literally thousands of spikes, each lasting a few milliseconds, suggesting sizes less than a 1,000 km across and brightness temperatures of up to a million billion (10^{15}) Kelvin, requiring a coherent radiation mechanism.

[a]A frequency of 1 MHz corresponds to a million Hz, or 10^6 Hz. An energy of 1 keV corresponds to 1.6×10^{-16} J or 1.6×10^{-9} erg

giving a distribution of particle energy that can be characterized by a single temperature. Impulsive solar flares do not have enough time to achieve this equilibrium, and the flaring electrons are far too energetic for a thermal process to work. To make electrons travel at half the speed of light, a thermal gas would have to be heated to implausible temperatures of 1.5×10^9 K, or about 100 times hotter than the center of the Sun.

Electron beams excite the resonance frequency at which electrons in the coronal plasma oscillate versus protons. This oscillation, in turn, emits radio waves at the same frequency or its harmonic.

The radio emission from the slow-drift type II bursts consists of two bands that drift to lower frequencies at a leisurely pace, corresponding to an outward velocity of about $1,000 \, \text{km s}^{-1}$. They are excited by shock waves set up at the time of a solar outburst and moving out into space. Spatially resolved radio interferometry observations in the 1950s confirmed the outward motion of type II bursts at these speeds, and also indicated that they can be very large, with angular extents that can become comparable to that of the Sun. As it turned out, at least some of the type II bursts were the radio counterparts of the coronal mass ejections that were subsequently discovered using space-borne coronagraphs.

We now know that radio bursts at meter wavelengths occur at altitudes far above the site where the flare energy is released and particles are accelerated, but it took solar physicists surprisingly long to realize that the bursts originate much lower in the corona. In the 1980s and 1990s, Arnold O. Benz and collaborators systematically studied the decimeter bursts ($1 \, \text{dm} = 0.1 \, \text{m} = 10 \, \text{cm}$). They discovered low-altitude signatures of electron beams at frequencies up to 8,000 MHz, or at wavelengths as short as 3.75 cm, moving downward from the acceleration regions into the low corona. In 1997, Markus Aschwanden and Benz used swept frequency measurements of solar bursts to specify the acceleration site, located in the low corona between upward and downward electron beams. A giant array of radio telescopes, located near Socorro, New Mexico, and called the Very Large Array, can zoom in at the very moment of a solar flare, taking snapshot images with just a few seconds exposure. It has pinpointed the location of the impulsive decimeter radiation and the electrons that produce it. These radio bursts are triggered low in the Sun's atmosphere, unleashing their vast power in a relatively small area within solar active regions, often accelerating the radio-emitting electrons just above the apex of coronal loops.

In 1998, Timothy Bastian, Arnold Benz, and Dale Gary reviewed the available knowledge of the radio emission from solar flares, including incoherent gyrosynchrotron emission at millimeter and centimeter wavelengths and coherent plasma radiation at decimeter and meter wavelengths. The high-speed electrons that emit centimeter bursts spiral around the magnetic field lines, moving rapidly at velocities near that of light, and sending out radio waves called gyrosynchrotron radiation after the man-made synchrotron particle accelerator where a similar kind of radiation was first observed (Fig. 6.13). Synchrotron radiation was used to explain the radio emission from our Galaxy and other cosmic radio sources in the 1950s, even before being applied to the radio emission of solar flares.

Some interesting subsequent work includes radio measurements of the height and strength of magnetic fields above sunspots, by Jeffrey Brosius and Stephen White in 2006, the triangulation of type II bursts from solar spacecraft, and the still controversial suggestion by Gregory D. Fleishman, Gelu M. Nita, and Dale Gary in 2005 that some decimeter solar bursts may be produced by resonant transition radiation.

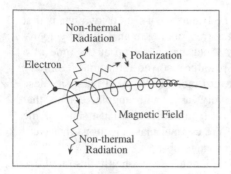

Fig. 6.13 Synchrotron radiation. High-speed electrons moving at velocities near that of light emit a narrow beam of synchrotron radiation as they spiral around a magnetic field. This emission is sometimes called non-thermal radiation because the electron speeds are much greater than those of thermal motion at any plausible temperature. The name "synchrotron" refers to the man-made, ring-shaped synchrotron particle accelerator where this type of radiation was first observed; a synchronous mechanism keeps the particles in step with the acceleration as they circulate in the ring

6.7 Filaments Lift Off, Prominences Erupt

A filament, or prominence, is a region of relatively cool gas embedded within the much hotter corona, and suspended above the photosphere by a low-lying magnetic loop. It can stretch horizontally for large distances across the Sun, above and along the magnetic neutral line dividing opposite polarity in the photosphere, and like all magnetic loops, it is connected at both ends to photosphere magnetic fields that point in or out of the Sun. But there are differences that make filaments unique. The core magnetic loop of a filament is long and low-lying, with oppositely curved elbows on each end, giving the filament a twisted, sinuous form. And the filament magnetic loop dips down in its long middle, forming a reservoir where the cool gas collects like water in a valley.

The filament material has temperatures characteristic of the chromosphere, at roughly 10,000 K. On the solar disk, these cool, dense features appear dark in hydrogen-alpha or extreme-ultraviolet images, in absorption against the bright background. And a prominence is just a filament viewed at the edge or limb of the Sun, in emission against the less-intense background (Figs. 6.14 and 6.15).

A further distinction is made between filaments, or prominences, in active regions and those in the quiet Sun outside active regions. The active-region filaments are relatively compact, with typical lengths of about 30,000 km, and often connected to opposite polarity sunspots. The quiescent filaments can stretch across up to half a solar radius, with lengths up to 300,000 km, and are relatively long-lived, lasting for up to several months. Both types of filaments are much lower than they are long, as if pressed down by the weight of all the material the magnetic loop is holding up. The active region filaments lie only a few thousand kilometers above the photosphere, while the large quiescent filaments have typical heights of 50,000 km.

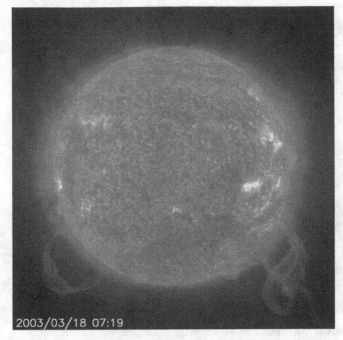

2003/03/18 07:19

Fig. 6.14 Twin prominences. Two solar prominences arise from the Sun's southern *(lower)* hemisphere in this image taken on 21 March 2003 with the Extreme-ultraviolet Imaging Telescope, or EIT for short, aboard the *SOlar and Heliospheric Observatory*, abbreviated *SOHO*. The prominences are composed of relatively cool, dense plasma, appearing as dark filaments against the brilliant solar disk, but as bright arcs when seen against the thin, million-degree corona at the edge of the visible Sun. This image was taken at a wavelength of 30.4 nm, emitted by singly ionized helium, denoted He II, at a temperature of about 60,000 K. (Courtesy of the *SOHO* EIT consortium. *SOHO* is a project of international collaboration between ESA and NASA.)

An arcade of magnetic loops arches above the filament's core magnetic loop. In contrast to the core magnetism, which holds the filament up, the anchoring fields are directed across the magnetic neutral line in the photosphere, rather than along it, and they are rooted in regions of opposite magnetic polarity whose separation is considerably greater than the height of the low-lying filament core. The higher, arching fields help restrain the filament, and keep it from expanding into space.

At first glance, filaments look grand and stable, as if they could stay there forever. But the loop material that is held within their magnetic confines is in a state of continued agitation. As indicated in the 1990 review by Bodgan Rompolt, motions in small-scale filamentary structures of prominences have been studied for more than a century. In 1998, Sara F. Martin, for example, reviewed the conditions for their formation and maintenance, and in the same year, Jack B. Zirker, Oddbjorn Engvold and Martin described counter-streaming gas flows in solar prominences, as evidence for vertical magnetic fields.

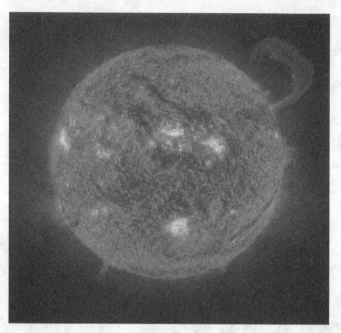

Fig. 6.15 Erupting prominence. An extreme ultraviolet image of the Sun with a huge, handle-shaped prominence (*top right*). It was taken on 14 September 1999 at a wavelength of 30.4 nm, emitted by singly ionized helium, designated He II, with the Extreme-ultraviolet Imaging Tele-scope, abbreviated EIT, aboard the *SOlar and Heliospheric Observatory*, or *SOHO* for short. Prominences are huge clouds of relatively cool, dense plasma suspended in the Sun's hot, thin corona. At times, they can erupt, escaping from the Sun's gravitational embrace. Emission in this helium spectral line shows the upper chromosphere at a temperature of about 60,000 K. Every feature in the image traces magnetic field structure. The hottest areas appear almost white, while the darker red areas indicate cooler material. (Courtesy of the *SOHO* EIT consortium. *SOHO* is a project of international collaboration between ESA and NASA.)

Observations from *TRACE* also reveal strands of continually moving cool gas, appearing to slosh back and forth within the nearly horizontal, core magnetic field. As reported by Carolus Schrijver and co-workers in 1999 and Therese Kucera and colleagues in 2003, 2006, the *TRACE* observations indicate that the material in low, elongated filaments is extremely dynamic, exhibiting counter-streaming on nearby field lines. Such flows suggest that cool gas is fed into filaments from their magnetic ends near the chromosphere.

Both active-region and quiescent filaments or prominences can erupt, rising out away from the Sun. The very largest filament eruptions are the quiescent fila-ments that reside outside active regions and have relatively weak magnetic fields. Although small in extent, active-region filament eruptions are more powerful, ex-pending 10–100 times as much magnetic energy as their quiescent counterparts. Many solar flares are often accompanied by the disruption of an active-region

Fig. 6.16 Filament lifts off. A filament is caught at the moment of erupting from the Sun. The dark matter is relatively cool, around 20,000 K, while the bright material is at a temperature of about a million Kelvin. The structure extends 120,000 km from top to bottom. This image was taken from the *Transition Region And Coronal Explorer*, abbreviated *TRACE*, on 19 July 2000 at a wavelength of 17.11 nm, emitted by eight and nine times ionized iron, denoted Fe IX and Fe X, at a temperature of about 1×10^6 K. (Courtesy of the *TRACE* consortium and NASA; *TRACE* is a mission of the Stanford-Lockheed Institute for Space Research, and part of the NASA Small Explorer program.)

filament, whose closed magnetic fields can be blown open during the flare's impulsive phase.

And what makes the filament or prominence erupt? As pointed out by Ronald L. Moore in 1988, the chromosphere material in both the active-region and the quiescent filaments resides in, and traces out, sheared magnetic fields lying above magnetic neutral or inversion lines, and the eruption is related to the expansion and untwisting of the flux tube in which the core erupting filament is embedded. The highly sheared magnetic fields, which suspend and insulate the filament, can become so twisted at their roots that they lose their equilibrium, and the filament lifts off or the prominence erupts (Fig. 6.16). The erupting fields can expand out, forming either a twisted, arched magnetic flux rope (Fig. 6.17) or a uniform, untwisted one. Or the core field can untwist and restructure without forming a rising flux rope, resulting in a failed filament or prominence eruption.

The rising magnetic fields of an erupting prominence can break through the arches straddling it, but after the prominence lifts off the magnetic arcade regroups beneath it and closes up again, forming bright, glowing magnetic loops detected at extreme-ultraviolet or X-ray wavelengths, aligned like the bones in your rib cage or the arched trestle in a rose garden (Fig. 6.18). First observed from *Skylab* and subsequently investigated with both the Soft X-ray Telescope on *Yohkoh* and from *TRACE*, the arcade of loops retains a memory of its former stability,

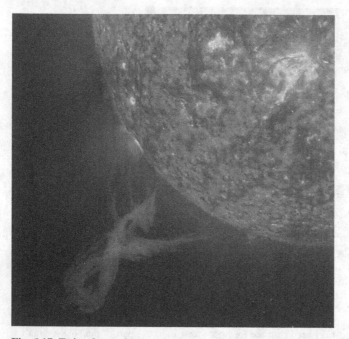

Fig. 6.17 Twisted prominence. An erupting prominence twists out into space on 12 January 2000, caught in this image taken with the Extreme-ultraviolet Imaging Telescope, or EIT for short, aboard the *SOlar and Heliospheric Observatory*, abbreviated *SOHO*. A prominence is composed of relatively cool, dense plasma immersed in the hot, thin, corona and seen as bright material at the edge of the visible Sun. This image was taken at a wavelength of 30.4 nm, emitted by singly ionized helium, denoted He II, at a temperature of about 60,000 K. (Courtesy of the *SOHO* EIT consortium. *SOHO* is a project of international collaboration between ESA and NASA.)

stitching together and healing the wound inflicted by emptying that part of the corona. The original low-lying magnetic channel above and along the magnetic neutral line is also restored, so the filament or prominence can reform at the same place.

The disappearing filaments and prominences are strongly correlated with another sort of solar outburst, the coronal mass ejections. In 1979, for example, Richard H. Munro and co-workers found that more than 70% of coronal mass ejections are associated with eruptive filaments, while G. P. Zhou and colleagues showed in 2006 that ejected filaments accompany 94% of Earth-directed "halo" coronal mass ejections. So, we will next discuss these fantastic coronal mass ejections, which can form an expanding bubble around and above the filament, before considering, in Sect. 6.10, the various theories for the formation of solar flares, erupting prominences, and coronal mass ejections.

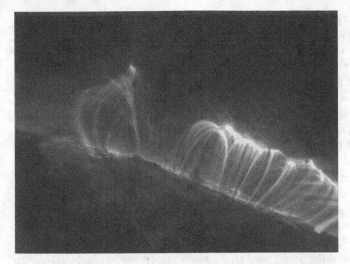

Fig. 6.18 Magnetic arcade. The *Transition Region And Coronal Explorer*, abbreviated *TRACE*, observed this arcade of cooling loops almost 3 h after the peak of an X28 flare and a very fast coronal mass ejection on 4 November 2003. It is taken in the 19.5-nm passband, sensitive to gas at a temperature of about 1.4×10^6 K. The arcade of magnetic loops was part of the disrupted magnetic field during the early eruptive phase of the flare, and the heat deposited by the flare caused the loops to fill with hot gases. Now, these are cooling and raining back down toward the solar photosphere. (Courtesy of the *TRACE* consortium and NASA; *TRACE* is a mission of the Stanford-Lockheed Institute for Space Research, and part of the NASA Small Explorer program.)

6.8 Coronal Mass Ejections

6.8.1 Coronagraph Observations of Coronal Mass Ejections

The most spectacular solar eruptions are gigantic magnetic bubbles, called coronal mass ejections, which expand outward from the Sun. A typical coronal mass ejection carries about 10 billion tons, or 10 million million kilograms, of coronal material as it lifts off into space, removing about a tenth of the total coronal mass. By way of comparison, the weight of water in a large lake, measuring 10 km wide and long and 100 m deep, is also about 10 billion tons.

The outward-moving coronal mass ejections can sometimes stretch the magnetic field until it snaps, leaving behind only bright rays rooted in the Sun. And at other times the expanding mass ejection develops in a magnetic cloud stretching all the way to the Earth, but rooted at both ends in the Sun.

Nearly everything we know about coronal mass ejections has been learned in just a few decades. They could not be clearly identified until special telescopes known as coronagraphs were flown in space beginning in the early 1970s (See Sect. 2.6). These instruments have a small occulting disk to mask the Sun's face and block out the photosphere's direct sunlight. The bright solar glare previously hid the corona from view, except during rare and brief total eclipses of the Sun. Space-borne coronagraphs provide nearly continuous edge-on views of the corona extending far out into space around their miniature-occulting moon.

You might wonder why these spectacular mass ejections have not been observed during total eclipses of the Sun, when the Moon passes in front of the photosphere and the corona becomes briefly visible. A thorough search of eclipse records, after the discovery of coronal mass ejections, did result in two or three possible coronal transient events during the past two centuries – including a possible one during the total solar eclipse of 18 July 1860 (Eddy, 1974). When combined with the rarity and short duration of a total solar eclipse, and the known occurrence rate and speeds of the ejections, there is in fact about one chance per century of seeing one during a total eclipse of the Sun.

But since 1970, a multitude of coronal mass ejections has been observed using the space-borne coronagraphs listed in Table 6.4. More than 10,000 of them have been observed with the Large Angle and Spectrometric COronagraph, abbreviated LASCO, aboard the *SOlar and Heliospheric Observatory*, or *SOHO* for short, since its launch in 1996. They are listed and described in an online catalog at http://cdaw.gsfc.nasa.gov/CME/list. And now we have the inner and outer whitelight coronagraphs, designated COR1 and COR2, aboard the twin *Solar TErrestrial RElations Observatory*, or *STEREO A* and *B*, spacecraft launched 25 October 2006. The inner coronagraph has a field of view from 1.3 to 4.0 solar radii, with some overlap with the outer one, from 2 to 15 solar radii.

Russell A. Howard gave an overview of the *P 78-1* Solwind observations of 1979–1980 in 1985. The *Solar Maximum Mission* results, in 1980 and from 1984 to 1989, are included in the 1998 book edited by Keith T. Strong and colleagues. Natchimuthuk Gopalswamy and colleagues in 2003 and Seiji Yashiro and coworkers in 2004 have described the statistical properties of nearly 7,000 events observed from LASCO on *SOHO* from 1996 to 2002.

Table 6.4 Space-borne orbiting coronagraphs[a]

Satellite	*OSO 7*	*Skylab*	*P78-1*	*SMM*	*SOHO*
Coronagraph		ATM	Solwind	C/P	LASCO/C3[b]
Observation Time	Oct. 71 to June 74	May 73 to Feb. 74	March 79 to Sept. 85	Feb.–Sept. 80 Apr. 84 to Nov. 89	Feb. 96 on
Field of view	$3–10R_\odot$	$1.6–6R_\odot$	$2.6–10R_\odot$	$1.6–8R_\odot$	$3.7–30R_\odot$
Resolution	75 arc s	8 arc s	1.25 arc m	6.4 arc s	56 arc s

[a]Here $R_\odot$ denotes the solar radius, or 696 million (6.96×10^8) meters, and an angular resolution of 1 arc s corresponds to 725,000 (7.25×10^5) m at the Sun. Not included are the inner and outer coronagraphs, COR1 and COR2, on both the *STEREO A* and *B* spacecraft launched on 25 October 2006, with respective fields of view of 1.3–4.0 and 2–15 solar radii
[b]LASCO consists of three coronagraphs, designated C1, C2, and C3, with respective overlapping fields of view of 1.1–3.0, 1.7–6.0, and 3.7–32.0 solar radii and respective pixel angular resolutions of 5.6, 11.2, and 56.0 arc s. C1 has not been operational since the temporary loss of *SOHO* in June 1998

Ground-based coronagraphs are limited by the brightness and variation of the sky, but they can complement space-based coronagraphs by providing higher temporal resolution. Routine observations have been carried out at the Mauna Loa Solar Observatory for many years. The Mark IV coronagraph has been in use since 1998, recording polarized brightness images of the inner corona, from 1.08 to 2.85 solar radii, every 3 min, from 17 to 22 h universal time, weather permitting.

The book *Coronal Mass Ejections*, edited by Horst Kunow and colleagues and published in 2006, provides a fine overview of our current knowledge of these outbursts, including summaries of their observations by Rainer Schwenn and his coworkers and by Hugh S. Hudson, Jean-Louis Bougeret, and Joan Burkepile (2006a). Ultraviolet observations of coronal mass ejections, from the UltraViolet Coronagraph Spectrometer, abbreviated UVCS, on *SOHO*, are included in the review by John L. Kohl and colleagues in 2006.

6.8.2 Physical Properties of Coronal Mass Ejections

So what exactly is a coronal mass ejection? All of the space-borne, white-light coronagraphs show what appear to be bright magnetic loops accelerating off the solar

Fig. 6.19 Three-part coronal mass ejection. A coronal mass ejection (*top bright loop*) rises above a dark cavity, followed by a rising prominence (*central bright oval*). An occulting disk of a coronagraph has blocked the intense sunlight from the photosphere, revealing the surrounding faint corona, and the white circle denotes the edge of the photosphere. This image was taken with the Large Angle and Spectrometric COronagraph, abbreviated LASCO, on the *SOlar and Heliospheric Observatory*, or *SOHO* for short. (Courtesy of the *SOHO* LASCO consortium and NASA. *SOHO* is a project of international collaboration between ESA and NASA.)

limb into space, moving outward across the coronagraph field of view in a few minutes to several hours. They can expand to become larger than the visible solar disk, streaming outward past the planets and dwarfing everything in their path. Such events work only in one direction, always moving away from the Sun into interplanetary space and almost never falling back in the reverse direction. As announced by Yi-Ming Wang and Neil R. Sheeley Jr. in 2002, significant amounts of material can nevertheless fall back into the Sun during the outward motion of some coronal mass ejections.

Coronal mass ejections often exhibit a three-part structure – a bright outer front, followed by a darker, underlying cavity, surrounding a brighter core (Fig. 6.19). The outer leading edge may be a region where an expanding magnetic bubble-like shell has compressed the overlying gas, piling the corona up and shoving it out like a snowplow. The bright outer edge has also been pictured as an expanding magnetic loop filled with dense, shining gas. The cavity or void is an expanding, low-density region whose high magnetic pressure and strong magnetic field might push coronal material aside. The core is the brightest part of the coronal mass ejection, because of its high density. It is often identified as an erupting prominence (Fig. 6.20), on the

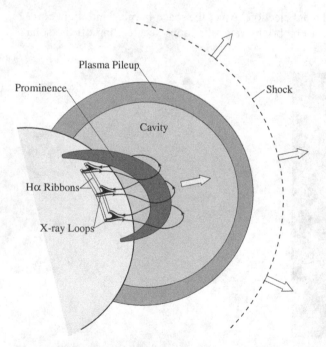

Fig. 6.20 Model of three-part coronal mass ejection. In this model of a three-part coronal mass ejection, portrayed by Terry Forbes (2000), swept-up, compressed mass and a bow shock have been added to the eruptive-flare portrayal of Tadashi Hirayama (1974). The combined representation includes compressed material at the leading edge of a low-density, magnetic bubble or cavity, and dense prominence gas. The prominence and its surrounding cavity rise through the lower corona, followed by sequential magnetic reconnection and the formation of flare ribbons at the footpoints of a loop arcade. [Adapted from Hugh S. Hudson, Jean-Louis Bougeret and Joan Burkepile (2006a).]

basis of its visibility in chromospheric emission lines. These three parts are usually well maintained as the ejection expands, almost with a nearly perfect circular cross section.

Not all coronal mass ejections display the classical three-part structure; some exhibit splayed and twisted forms (Fig. 6.21). Coronal mass ejections can have voids with no prominence in them, and others have no detectable leading edge, perhaps because it is too faint to be seen. Narrow ejections with no resemblance to the usual images have also been observed.

But the observed shapes are two-dimensional projections of three-dimensional structures, and it is often difficult to disentangle the overlapping features that can produce distortions and confuse the interpretation. The situation should be improved with the *STEREO* mission, since its two spacecraft are poised to gather information from two perspectives, or angles, outside the Sun–Earth line, and to combine them into a three-dimensional view that removes some ambiguities and uncertainties in the two-dimensional image of a single coronagraph.

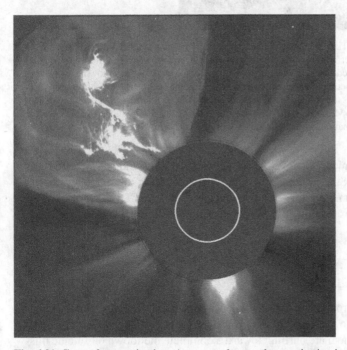

Fig. 6.21 Coronal mass ejection. A contorted coronal mass ejection is seen in this coronagraph image, taken on 4 January 2002. The white circle denotes the edge of the photosphere, so the length of the ejected material is about twice the size of the visible disk of the Sun. The dark area corresponds to the occulting disk that blocked the intense sunlight from the photosphere, revealing the surrounding faint corona. This image was taken with the Large Angle and Spectrometric COronagraph, abbreviated LASCO, on the *SOlar and Heliospheric Observatory*, or *SOHO* for short. (Courtesy of the *SOHO* LASCO consortium and NASA. *SOHO* is a project of international collaboration between ESA and NASA.)

Fig. 6.22 Fatal impact. This composite image records a comet plunging into the Sun on 23 December 1996. The innermost image (*center*) records the bottom of the million-degree solar atmosphere, known as the corona. The electrically charged coronal gas is seen blowing away from the Sun just outside the inner dark circle, which marks the edge of one instrumental occulting disk. Another instrument records the comet (*lower left*), as well as the coronal streamers at more distant regions and the stars of the Milky Way. (Courtesy of the *SOHO* EIT, UVCS and LASCO consortia. *SOHO* is a project of international collaboration between ESA and NASA.)

The LASCO instrument aboard *SOHO* has watched thousands of coronal mass ejections leaving the Sun, in beautiful images available at the *SOHO* web site http://sohowww.nascom.nasa.gov/. And in the process LASCO unexpectedly located more than a thousand comets passing near the Sun or even into it. (Fig. 6.22, Focus 6.6).

Focus 6.6

More than a Thousand Comets

SOHO's LASCO has provided unexpected measurements of over a thousand comets. Most of them belong to the Kreutz family of sungrazers, which pass very close to the Sun, and some of them are hurtling toward a complete meltdown on suicide missions into the Sun. The sungrazers are named after the German astronomer Heinrich Kreutz, who found that many of the comets that came closest to the Sun in the 19th century seemed to have a common origin with similar orbits.

Nearly half of all the comets for which orbital elements have been determined, since 1761, were discovered in LASCO images, and over two-thirds of those by amateur astronomers accessing LASCO data via the Web. Each day, numerous people from all over the world download the near-real-time images to search for new comets.

Spacecraft observations have demonstrated that coronal mass ejections are big, massive, fast, and energetic. They are events of surprising proportion, blowing away huge pieces of the corona. Each time a mass ejection rises out of the corona, it carries away $1-50 \times 10^9$ tons $(1 \times 10^{12} - 5 \times 10^{13} \, kg)$.

The frequency of occurrence, location on the Sun, and average speed of coronal mass ejections all depend on the 11-year cycle of solar activity (Table 6.5). Data obtained from *SOHO's* LASCO between 1996 and 2002 indicate an order of magnitude increase in the detection rate, from 0.5 per day at activity minimum to 6 per day at activity maximum. The average apparent speed of coronal mass ejections increases from 300 to $500 \, km \, s^{-1}$ from activity minimum to maximum. Coronal mass ejections were detected around the Sun's equatorial regions during activity minimum, while during activity maximum they appeared at all solar latitudes, even at the high ones near the solar poles. The high-latitude events are prominence-associated, while the ones at lower latitudes are associated with active regions, and there appears to be a north–south asymmetry to the high-latitude ones.

As the coronal mass ejection rises, it probably starts out slowly and then accelerates to higher speed in order to overcome the Sun's gravity and move into space. The observed apparent speeds in the coronagraph field of view range from 50 to $3,400 \, km \, s^{-1}$, but the escape velocity of the Sun is about $614 \, km \, s^{-1}$. The average mass ejection speed, of between 300 and $500 \, km \, s^{-1}$, is comparable to that of the slow solar wind and less than the fast solar wind, at about $800 \, km \, s^{-1}$, as if the mass

Table 6.5 Physical properties of coronal mass ejections near the Sun[a]

Characteristic	Value
Mass ejected	$1 \times 10^{12} - 5 \times 10^{13} \, kg$
	$(1-50 \times 10^9 \, tons)$
Angular width (Heliocentric)	$20-120°$
Average angular width	$47°$ (activity minimum)
	$61°$ (activity maximum)
Solar latitudes	Equatorial region (activity minimum)
	All latitudes (activity maximum)
Frequency of occurrence	0.5 events per day (activity minimum)
	6 events per day (activity maximum)
Apparent speed	$100-3,400 \, km \, s^{-1}$
Average apparent speed	$300 \, km \, s^{-1}$ (activity minimum)
	$500 \, km \, s^{-1}$ (activity maximum)
Mass-flow rate	About $2 \times 10^8 \, kg \, s^{-1}$
Average time to reach Earth	About $100 \, h$
Average kinetic energy	Approximately $10^{23}-10^{24} \, J$

[a]Adapted from Seiji Yashiro and colleagues (2004), Natchimuthuk Gopalswamy and co-workers (2003), John Gosling (1993), and Arthur Hundhausen (1997). One kilogram $= 1,000 \, g$, 1 ton $= 1,000 \, kg$, $1 \, m \, s^{-1} = 0.001 \, km \, s^{-1}$, and $1 \, J = 10 \times 10^6 \, erg$

ejections were riding along with the solar wind. Those with higher speeds, greater than $1,000 \mathrm{km} \mathrm{s}^{-1}$, move faster than the characteristic velocity, sound speed, and Alfvén speed of the fast or slow solar wind.

In 1999, Neil Sheeley Jr. and co-workers proposed two classes of coronal mass ejections – the extremely impulsive events with high speeds and strong, rapid acceleration of more than $1,000 \mathrm{m} \mathrm{s}^{-2}$, and the gradual ones, characterized by slow speeds and weak, persistent acceleration of less than $20 \mathrm{m} \mathrm{s}^{-2}$. The impulsive class has observed apparent speeds in excess of $750–1,000 \mathrm{km} \mathrm{s}^{-1}$, and prompt acceleration followed by either constant speed or deceleration at distances of 2 solar radii or more. The gradual class, with slow apparent speeds of $400–600 \mathrm{km} \mathrm{s}^{-1}$, is gradually accelerated for several hours throughout the inner and outer corona and to distances of about 30 solar radii. Since the gradual coronal mass ejections move at speeds comparable to the average speed, most coronal mass ejections belong to this category.

Nevertheless, the statistical studies by Jie Zhang and Kenneth P. Dere in 2006 indicate that both the magnitude and duration of the acceleration do not demonstrate a strong either-or, bimodal distribution, and that most coronal mass ejections are characterized by a moderate acceleration of hundreds of meters per square second.

Perhaps the more important point is that the major acceleration of the fast coronal mass ejections occurs low in the corona, below 2 solar radii, with subsequent evolution that may even include deceleration, and they usually carry the most kinetic energy. The slower mass ejections take their time getting up to speed, and are usually less energetic.

The faster, more impulsive coronal mass ejections also tend to be associated with active regions and solar flares, while the gradual class is often associated with erupting prominences, but these are somewhat murky distinctions. Flares can be associated with both fast and slow coronal mass ejections. Both fast and slow mass ejections can be accompanied by an eruptive prominence or a solar flare, or they might not be accompanied by either one of them. And no associated coronal mass ejection has been detected for nearly half of the big flares, of *GOES* class M and X.

By the time that they are a few solar radii above the Sun's edge, most coronal mass ejections have reached a cruising velocity of about $400 \mathrm{km} \mathrm{s}^{-1}$. At that speed, the expelled mass can reach the Earth in about 100 h, carrying with it an average kinetic energy of $10^{23}–10^{24} \mathrm{J}$ or $10^{30}–10^{31}$ erg (Focus 6.7). The amount of energy that a coronal mass ejection liberates in producing the motions of the expelled mass and in lifting it against the Sun's powerful gravity is roughly comparable to the energy of a typical solar flare. However, most of the energy of a mass ejection goes into the expelled material, while a flare's energy is mainly transferred into accelerated particles that subsequently emit intense X-ray and radio radiation and travel into interplanetary space or move down into the Sun. Some very fast and wide coronal mass ejections can have kinetic energies exceeding 10^{33} erg; they generally originate from large active regions and are accompanied by powerful flares.

Focus 6.7

Mass, Mass Flux, Energy, and Time Delay of Coronal Mass Ejections

Coronal mass ejections are detected as localized brightness increases in white-light coronagraph images. Integration of the brightness increase, which depends only on the electron density, N_e, permits evaluation of the total mass, M, of the ejection. For a sphere of radius, R, we have:

$$M = 4 \pi R^3 N_e m_p / 3,$$

where the proton mass $m_p = 1.67 \times 10^{-27}$ kg. The corona is a fully ionized, predominantly (90%) hydrogen plasma, so the number density of protons and electrons are equal; but since the protons are 1,836 times more massive than the electrons, the protons dominate the mass. For a mass ejection with an electron, or proton, density of $N_e = 10^{13}$ electrons per cubic meter, that has grown as large as the Sun, with $R = 6.96 \times 10^8$ m, this expression gives

$$M = 10^{13} \text{ kg} = 10 \times 10^9 \text{ tons.}$$

At the rate of one ejection per day, and 10^{13} kg per ejection, this amounts to a mass flow rate of about 10^8 kg s^{-1}, since there are 86,400 s per day, and a mass loss rate of about 2×10^{-11} kg m^{-2} s^{-1}; just divide the mass by the time of 1 day and by $4\pi R^2$, where the Sun's radius $R = 6.955 \times 10^8$ m.

By way of comparison, the solar wind flux observed in the ecliptic at the orbit of the Earth is about 5×10^{12} protons per square meter per second, or 8.3×10^{-15} kg m^{-2} s^{-1}. If this flux is typical of that over the entire Sun-centered sphere, with an average Sun–Earth distance of $D = 1.5 \times 10^{11}$ m, we can multiply by the sphere's surface area, $4\pi D^2$, to obtain a solar wind mass flow rate of about 2×10^9 kg s^{-1}. That is 20 times the mass flow rate from coronal mass ejections, or in other words the coronal mass ejection rate is about 5% that of the steady, perpetual solar wind. Other measurements indicate that the mass outflow of the solar wind in coronal holes amounts to about 2×10^{-9} kg m^{-2} s^{-1}, or about 100 times that of coronal mass ejections. Thus, coronal mass ejections contribute roughly 1–5% of the solar wind mass flux.

The kinetic energy, KE, of a coronal mass ejection with a speed of $V = 400$ km s^{-1} and a mass $M = 10^{13}$ kg is:

$$\text{KE} = MV^2/2 \approx 10^{24} J = 10^{31} \text{ erg.}$$

This is comparable to the energies of large solar flares that lie between 10^{21} and 10^{25} J.

At a speed of $V = 400$ km s^{-1}, the time, T, to travel from the Sun to the Earth, at an average distance, D, is:

$$T = D/V = 3.75 \times 10^5 \text{ s} = 104 \text{ h} = 4.34 \text{ days.}$$

Although most coronal mass ejections are observed as magnetic bubbles ejected from one side of the Sun, these events are not likely to collide with the Earth. The outward rush of an Earth-directed mass ejection appears in coronagraph images as a gradually expanding, Sun-centered ring or halo around the occulting disk. Russell A. Howard and colleagues first observed such a halo mass ejection in 1982, using the Solwind coronagraph on the *P78-1* satellite. Since that time, about 3% of the coronal mass ejections detected in the LASCO images have been halos, expanding into a circumsolar ring in the coronagraph images.

When ejected from the front side of the Sun facing Earth, a halo ejection can cause intense geomagnetic storms, provided it has the right magnetic orientation when striking the Earth. In contrast, coronal mass ejections that are expelled from near the visible edge of the Sun will not impact Earth, but threaten other parts of space.

Images from a single white-light coronagraph are normally unable to determine if the halo-like ejection is traveling toward or away from the observer. In 2004, however, Thomas G. Moran and Joseph M. Davilla demonstrated how polarization techniques could be used with LASCO data to obtain three-dimensional images of coronal mass ejections, determining their direction and computing their true speeds, rather than apparent ones.

The twin *STEREO* spacecraft can potentially remove all the ambiguity, determining the trajectories of coronal mass ejections from their onset at the Sun to the orbit of the Earth. The earliest results from the *STEREO A* and *B* coronagraphs demonstrate their potential; in 2008, Richard A. Harrison and colleagues reported the first images of a coronal mass ejection viewed from outside the Sun–Earth line. Also in 2008, J. A. Davies and co-workers published the first observations of coronal mass ejections from the Sun to distances greater than that of the Earth's orbit, using the Heliospheric Imagers aboard the *STEREO* spacecraft.

The Earth-directed coronal mass ejections may be identified by associated coronal activity at extreme ultraviolet and X-ray wavelengths near the center of the solar disk, as viewed from the Earth. These coronal waves and dimming are discussed with other outburst consequences in the next Sect. 6.9.

6.9 After the Blast

6.9.1 Solar Flares Excite Coronal Loop Oscillations

The explosive solar flares can set coronal loops quivering in an oscillatory motion, like a lone tree leaf vibrating in a gust of wind. These loop oscillations were discovered using rapid, sequential images from *TRACE* and *SOHO*. Markus J. Aschwanden and co-workers and Valery M. Nakariakov and colleagues reported the *TRACE* results in 1999, noting that the loops are displaced up and down, in the transverse direction, with a period of between 2 and 7 min, and that they damp very

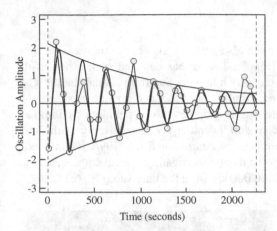

Fig. 6.23 Loop oscillation. An example of a transverse loop oscillation observed from the *Transition Region And Coronal Explorer*, abbreviated *TRACE*, in the 17.1-nm passband. The observed data points are denoted by diamonds and fitted with a damped sinusoidal oscillation (*dark, wide solid line*). Transverse magnetohydrodynamic, or MHD, kink-mode oscillation amplitudes (*light solid line*) have been fitted to the data. [Adapted from Markus J. Aschwanden and colleagues (2002).]

quickly, within just a few oscillations (Fig. 6.23). A review of the early *TRACE* observations of transverse loop oscillations, by Carolus J. Schrijver and co-workers and Aschwanden and colleagues in 2002, indicates that they usually occur in loops that lie within an active region, or in loops that connect an active region to a neighboring site, and that the loops do not resonate; they are just a brief response excited by the explosive power of a large flare in their vicinity.

In 2002, Tongjiang J. Wang and colleagues reported oscillations of very hot loops in active regions, revealed by Doppler shifts of spectral lines detected with the SUMER instrument aboard *SOHO*. A review of these hot coronal loop oscillations, provided by Wang in 2003, shows that their periods lie between 7 and 31 min, with a comparable decay time, that they are only found in hot flare spectral lines at temperatures higher than 6×10^6 K, and that they are excited by small flares.

The coronal loop oscillations detected from both *TRACE* and *SOHO* are attributed to magenetohydrodynamic, or MHD, waves that are strongly damped. Aschwanden has reviewed the theoretical aspects of these MHD waves in his 2004, 2006 book *The Solar Corona: An Introduction*. Nakariakov and Erwin Verwichte have additionally described them in a 2005 Living Review on *Coronal Waves and Oscillations*. Comparisons of the observations with theoretical expectations have given rise to the field of coronal seismology, in which observed waves or oscillations are used to infer properties of the coronal loops.

6.9.2 Sunquakes

An exceptionally powerful solar flare can also generate major seismic waves in the Sun's interior. *SOHO's* Michelson Doppler Imager has detected such a powerful sunquake caused by electron beams sent down into the Sun from a solar flare originating in the low corona, and observed circular flare-generated seismic waves moving out in all directions from the flare site, like ripples from a pebble tossed into a pond. In 1998, Alexander G. Kosovichev and Valentina V. Zharkova reported the helioseismic response to an X-class flare, which occurred on 8 July 1996. It consisted of a series of concentric, expanding wave fronts, propagating at an average speed of about $50\,\mathrm{km\,s^{-1}}$ out to a distance of $120,000\,\mathrm{km}$ from the flare site (Fig. 6.24).

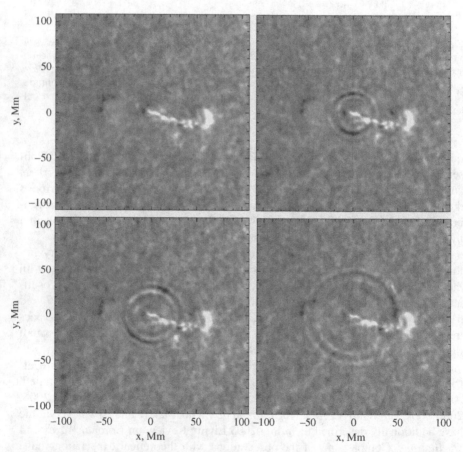

Fig. 6.24 Sunquake. A powerful flare in the solar atmosphere sent shock waves into the underlying gases, causing a massive sunquake that resembles a seismic earthquake on our planet. The explosion, on 9 July 1996, produced concentric rings of a sunquake that spread away from the flare, somewhat like ripples from a rock dropped into a pool. Over the course of an hour, the solar waves traveled for a distance equal to 10 Earth diameters. The Sun was quaking with about 40,000 times the energy released in the great earthquake that devastated San Francisco in 1906. (Courtesy of Alexander G. Kosovichev and Valentina V. Zharkova, and the *SOHO* SOI/MDI consortium. *SOHO* is a project of international cooperation between ESA and NASA.)

Most flares nevertheless do not generate detectable sunquakes. Alina-Catalina Donea and Charles Lindsey published the next report of seismic waves emitted by powerful solar flares in 2005, for X-class flares that occurred on 28 and 29 October 2003. They used helioseismic holography to image multiple acoustic sources of the waves associated with the footpoints of a coronal loop where the flare hard X-ray emission was located in *RHESSI* images. The high-speed electrons generated during the flare's impulsive phase were sent down into the Sun with such force that they generated powerful shocks in the dense solar atmosphere, which then produced the seismic waves.

6.9.3 Coronal Dimming After Coronal Mass Ejections

Another repercussion of solar outbursts accompanies the launch of a coronal mass ejection. They rip huge pieces out of the corona, noticeably reducing the intensity of its radiation at extreme-ultraviolet and X-ray wavelengths. Richard T. Hansen and his colleagues first recorded the scars of such abrupt depletions of the corona in 1974 using a ground-based, white-light coronagraph, but with limited spatial and temporal resolution. David M. Rust and Ernest Hildner noted the corresponding X-ray effects in *Skylab* data in 1976.

Yohkoh's Soft X-ray Telescope, abbreviated SXT, was subsequently used to observe large-scale dimming of the soft X-ray corona near the time and location of coronal mass ejections. In 1996, for example, Hugh S. Hudson, Loren W. Acton, and Sam L. Freeland noticed that large regions of the X-ray corona appeared empty in SXT images during a long-duration flare, identifying the missing material with a coronal mass ejection in the process of launching an estimated mass of 4×10^{11} kg of material. Then in 1997, Alphonse C. Sterling and Hugh S. Hudson reported a sudden intensity decrease of X-rays just prior to a halo coronal mass ejection, suggesting that the dimming might be used to predict coronal mass ejections headed toward the Earth.

SOHO's EIT has also been used to monitor the dimming associated with halo coronal mass ejections, for instance, by Dominic M. Zarro and co-workers in 1999. Spectral data, obtained by Richard A. Harrison and colleagues in 2003, indicated that the dimming is due to mass loss, rather than a temperature change, and that the extracted mass is comparable to that estimated for the coronal mass ejection. This means that the dimming can be used to describe the volume, structure, and source of the ejection. In 2007, Louise K. Harra and her colleagues combined observations from *EIT* and *Hinode* to show that the dimming regions marks the locations of strong outflows at velocities of up to 60 km s^{-1}, apparently at the footpoints of extended loops.

6.9.4 Large Flares Excite Global Waves

Large flares can generate concentric waves that propagate across the Sun in the low corona, away from the ejection site at speeds of 200–500 km s^{-1}. The powerful disturbances, observed with *SOHO's* EIT and first reported by Barbara J. Thompson and her colleagues in 1998–1999, move away from the erupting region like tidal waves or tsunami going across the ocean (Fig. 6.25), usually affecting most of the visible solar disk and traversing a solar diameter in less than an hour. Curiously, the waves do not cross a coronal hole, either because the hole stops them or there is too little material to sustain the waves.

They have been identified as the coronal manifestation of Moreton waves, first observed in the chromosphere by Gail E. Moreton in 1960–1961. He used rapid,

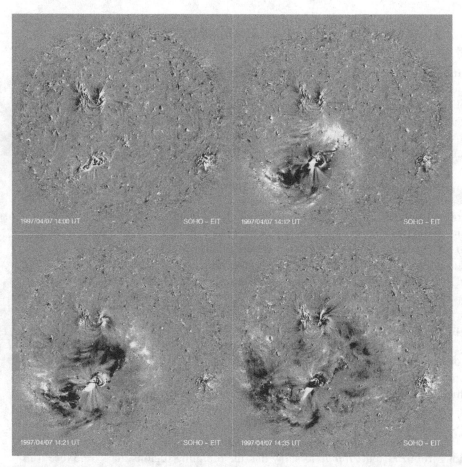

Fig. 6.25 Waves across the Sun. This sequence of images, taken with the Extreme-ultraviolet Imaging Telescope, or EIT for short, instrument on the *SOlar and Heliospheric Observatory*, abbreviated *SOHO*, shows a wave running across the solar disk in the corona at supersonic velocities of about 1.5×10^9 m h^{-1}. It originated in the vicinity of a solar explosion, or flare, on 7 April 1997. The images were taken at 19.5 nm in the emission lines of iron ions (Fe XII) formed at $1.5 \times 10^{6°}$ K. Each image is the difference between the one taken at the time shown and the previous one. (Courtesy of the *SOHO* EIT consortium. *SOHO* is a project of international collaboration between ESA and NASA.)

time-lapse photographs taken in the light of the Balmer alpha transition of hydrogen to record wave-like chromosphere disturbances initiated by solar flares; these Moreton waves move across the Sun from the flare site for distances of hundreds of thousands of kilometers with velocities of around $1,000$ km s^{-1}. So the EIT phenomena have been variously called EIT, Thompson, or coronal-Moreton waves. Although the vast majority of solar flares do not produce these waves, many, but not all, coronal mass ejections do.

The EIT waves and Moreton waves are not seismic waves, but instead magnetohydrodynamic, or MHD, waves set up in the low corona. In 1968, and again in 1973, Yutaka Uchida interpreted the chromospheric Moreton waves in terms of a MHD shock, predicting that it would have a strong coronal counterpart. The shock, generated by a powerful impulse, propagates through the low corona, and the Moreton waves track the intersection of the expanding coronal disturbance and the chromosphere. In the case of the EIT waves, the initiating impulse is a large solar flare associated with a halo coronal mass ejection, and the waves record the shock-generated coronal disturbance. This shock wave may also generate a type II radio burst.

So this completes our overview of the aftermath of powerful solar flares or coronal mass ejections, detected at the Sun, and we now turn to a discussion of why these solar outbursts occur, as a prelude to the following space-weather chapter, which includes possible methods of predicting them as well as their consequences in interplanetary space and at planet Earth.

6.10 Explaining Why Solar Outbursts Happen

6.10.1 What We Do Not Know About Solar Outbursts

Why does the Sun suddenly release vast quantities of energy in solar flares, erupting prominences, or coronal mass ejections? All three types of explosive outbursts include a reconfiguration of the Sun's magnetic field; they are all most likely triggered by a magnetic instability or interaction; and they all derive their power from magnetic energy. Nevertheless, we do not yet understand in detail how that energy is converted into the thermal energy of intense flare radiation or into the expulsive mechanical energy of erupting prominences or coronal mass ejections. And we are still unable to understand exactly how particles are accelerated to such high energies during solar flares.

The possible relationship between solar flares and coronal mass ejections continues to be a controversial topic. It was once thought that solar flares acted like colossal bombs which blasted the coronal mass ejections out of the Sun. But now scientists are not so sure. Many strong flares have no associated coronal mass ejection, and some coronal mass ejections arise from erupting prominences without detectable flares.

The strongest, fastest coronal mass ejections are nevertheless almost always accompanied by solar flares in active regions, as noticed by Murray Dryer, Hugh Hudson, and others over the years. Impulsive coronal mass ejections, with rapid acceleration close to the Sun, are, for example, usually correlated with soft X-ray flares detected from *GOES*.

The notion that there is a cause–effect relation between the two phenomena, with flares causing the coronal mass ejections or the other way around, is not now in vogue. There instead seems to be a growing consensus that coronal mass ejections and flares are a different manifestation of the same magnetic energy-release process in the corona, possibly via magnetic reconnection. That is, both kinds of outbursts may involve a process in which oppositely directed magnetic fields merge together and reconnect into new configurations, releasing free magnetic energy in the process.

Still, there is no consensus regarding the magnetic field configuration that causes a coronal mass ejection, and we are unable to determine with certainty when and where the hypothetical magnetic reconnection occurs. It might initiate the outburst or result from it. And if reconnection drives a coronal mass ejection, its relation to other features, such as prominences or solar flares, is uncertain. Magnetic reconnection might not even account for the loss of equilibrium that triggers solar outbursts; alternative candidates include the kink and tearing-mode instabilities.

Part of the difficulty is observational. Owing to the coronagraph occulting disks, coronal mass ejections are first detected at some distance from the Sun, away from their place of origin, and they are most clearly viewed when erupting from the Sun's edge. In contrast, the details of solar flares and active regions are best detected on the solar disk.

The spatial and temporal resolution required to study pre-eruptive magnetic configurations on the Sun have only become available in relatively recent times, with spacecraft such as *SOHO, TRACE*, and *Hinode*. And solar astronomers are currently using instruments aboard these satellites to study just how the magnetic fields and plasmas in the low corona, chromosphere, and photosphere shear, twist, writhe, interact, and otherwise change prior to and during solar outbursts, on both a local and global scale, seeking to unravel their detailed mechanisms. Future observations from *STEREO A* and *B*, as well as the *Solar Dynamics Observatory*, are expected to additionally help constrain or test the theories.

Current research into the origins of coronal mass ejections, and what triggers and drives them, involves the interplay between new observational results, numerical simulations, and theoretical considerations, with comparisons of observations with models for specific events. There have been several excellent reviews of this work, including the difficulties involved when the models confront the observations. In 2000, for example, Terry G. Forbes reviewed the possible instabilities involved in the genesis of coronal mass ejections, concluding that none of the proposed explanations provide an accurate prediction of their occurrence. A year later, James A. Klimchuk described the physical attributes of the various theories. He favors a storage and release model that involves the slow buildup of magnetic energy by gradual stressing of the coronal magnetic field related to the slow motion of photospheric footpoints, with energy release triggered by magnetic reconnection. In 2006, Ronald L. Moore and Alphonse C. Sterling reviewed magnetic reconnection processes that might result in a coronal mass ejection. But back in 2001, James Chen advocated an alternative scenario, involving twisted magnetic flux ropes that exist prior to the eruption, suspended in the corona and extending deep inside the Sun, with eruption driven and energized from below and magnetic reconnection as a consequence of the eruption. And in 2006, Ilia I. Roussev and Igor V. Sokolov described the theory and simulations of both the storage release and the erupting flux rope scenarios, concluding that the observations do not convincingly prove or disprove either explanation.

So there are all kinds of theories. Perhaps none of them are correct, perhaps several are, or maybe different models are applicable in different situations. The best we can do at this stage is to understand the possibilities. And since the fastest and most powerful coronal mass ejections are practically always associated with flares, and the notion of magnetic reconnection on the Sun originated with explanations of solar flares, we will begin there. Because solar flares and coronal mass ejections could be a byproduct of a common magnetic instability or reconfiguration, this introduction will help us understand some of the subsequent proposed explanations for coronal mass ejections.

6.10.2 *Magnetic Reconnection During Solar Flares*

To explain how solar flares happen, we must know where their colossal energy comes from; why that energy is suddenly and rapidly released; and how it is transferred to accelerated particles and intense radiation. As stressed by Thomas Gold and Fred Hoyle in 1960, the Sun's magnetic field provides the only plausible source of energy for solar flares, and we now think it is most likely released during magnetic reconnection.

We can see how the required magnetic energy might be produced in the corona by considering a single magnetic loop that connects regions of opposite magnetic polarity in the photosphere, divided by a magnetic neutral line. If the magnetic fields connect the two underlying magnetic elements in the shortest, most direct path, and therefore run perpendicular to the magnetic neutral line, scientists say that the magnetic field has a potential configuration. It can be distorted into a non-potential shape when flows at or below the photosphere shear and twist the looping fields. These non-potential magnetic fields have more magnetic energy than the potential ones, and this extra energy is called free magnetic energy, perhaps because it is free to power solar outbursts.

Strong coronal magnetic fields are tied to the photosphere at both ends, trapping hot, dense electrified gas within low-lying coronal loops. Higher up, the relatively weak, outermost coronal loops cannot constrain the outward pressure of the hot gas. Their magnetic fields get caught up by the solar wind and are pulled into interplanetary space. The overall result consists of closed loops down low, and two extended stalks of oppositely directed magnetic field further out. The ensemble resembles the shape of an inverted Y.

Such a magnetic geometry is inferred from the observations of coronal streamers above active regions, and it was used in one of the first magnetic theories for the origin of solar flares, proposed by Ronald G. Giovanelli in 1946–1948. According to Giovanelli's model, varying magnetic fields generate currents at a special place in the corona, called the neutral point. It is located at the interface between the closed and open magnetic structures. The currents were supposed to build up excess energy at the neutral point until an electrical discharge releases it as a solar flare.

During the 1950s, James W. Dungey further considered the growth of currents and subsequent electrical discharge at neutral points of the coronal magnetic field. He noticed that the effect of the discharge would be to quickly "reconnect" the initially open, oppositely directed magnetic lines of force at the place where they touch. According to Dungey's model, beams of accelerated particles are shot out from the neutral-point discharge in two directions, up into space and down toward the Sun. Such a bi-directional flow, driven by magnetic reconnection at the neutral point, is part of the CSHKP model of solar flares and erupting prominences developed between 1964 and 1976 (Focus 6.8).

Focus 6.8

Energizing Solar Flares and Erupting Prominences by Magnetic Reconnection – the CSHKP Model

The theory of magnetic reconnection at the neutral-point interface between closed and open magnetic structures took several decades to develop. Following the pioneering work by Ronald G. Giovanelli in the 1940s and James W. Dungey in the 1950s, Peter A. Sweet considered magnetic fields that are brought into contact at a neutral point boundary. This magnetic reconnection was attributed to twists in complex sunspot magnetic configurations in active regions, and it was supposed to release energy to accelerate high-energy particles in solar flares. In 1958, he showed that the electric current formed between merging, oppositely directed magnetic fields would flow in a flat, two-dimensional plane. It is shaped like a sheet on a well-made bed, and is hence called a current sheet. And since the magnetic direction cancels into neutrality at the place where oppositely directed fields meet, the term neutral current sheet is also used.

By 1963, Eugene Parker had evaluated this "slow" magnetic reconnection qualitatively and derived its detailed mathematics. And in the following year, Harry E. Petschek described how a faster rate of magnetic field annihilation might be achieved; showing that fast reconnection in the current sheet could permit the rapid conversion of magnetic energy within the observed minute-length duration of solar flares. The two models of magnetic reconnection are now known as the slow Sweet-Parker reconnection and fast Petschek reconnection.

In 1964, Hugh Carmichael proposed that magnetic field lines high above the photosphere could be forced open by the solar wind, and discussed how they might be brought together above coronal loops to accelerate high-energy particles. The magnetic geometry would resemble that of a coronal helmet streamer, with closed magnetic loops beneath open magnetic stalks, forming an upside-down Y shape. In the 1966–1968 model of Peter Sturrock, a tearing-mode instability near the inverted Y neutral point might trigger a solar flare, as the result of footpoint motions of sheared coronal loops. The stretched-out, open magnetic fields would become unstable to tearing, like caramel candy pulled until it snaps apart, thereby releasing stored magnetic energy in ways that might account for particle acceleration in a downward direction and shock waves and plasma ejection in the outward direction.

Tadashi Hirayama added an erupting prominence to the magnetic-reconnection solar flare model in 1974, proposing that the prominence rises above an X-type neutral point between oppositely directed open magnetic fields. The X marks the spot where magnetic reconnection and energy dissipation occur, with electric currents flowing parallel to the neutral line in the underlying photosphere. Some of the non-thermal particles accelerated at the reconnection current sheet move down into the chromosphere, forming flare ribbons and resulting in chromospheric evaporation that fills the newly linked magnetic fields with hot X-ray emitting material.

By 1976, *Skylab* had obtained results showing that pre-existing, closed magnetic loops in the corona can be torn open by an erupting prominence, leading Roger A. Kopp and Gerald W. Pneuman to refine Hirayama's comprehensive model, predicting a continuous rise of the reconnection point, with newly connected loops that increase in height and have wider footpoints, resulting in the separation of flare ribbons.

This theoretical model for both solar flares and erupting prominences, involving magnetic reconnection at a Y or X neutral point between closed and open magnetic fields in the low corona, became known as the CSHKP model after the first letters of the last names of the five people who developed it, in a chronological listing by publication date between 1964 and 1976.

For several decades, theoretical models involving free magnetic energy, associated with non-potential magnetic fields, primarily dictated our knowledge of solar flares. The coiled magnetic fields were supposed to hold the excess energy in place, remaining without substantial change for days, weeks, and even months at a time. Then, suddenly and unpredictably, they might go out of control, ripping the magnetic cage open and breaking its grip apart. The free magnetic energy was supposed to be released to accelerate particles and expel them away from the site of energy release by mechanisms that were not clearly understood.

Then, *Yohkoh* came along to demonstrate that magnetic interactions could indeed strike the match that ignites solar explosions. Its Soft X-ray Telescope, abbreviated SXT, observed the sharp, peaked cusp-like structure attributed to magnetic reconnection along a neutral line, or current sheet above a coronal loop (Fig. 6.26). First reported by Saku Tsuneta and co-workers in 1992, the peaked cusp shape marks the place where oppositely directed magnetic field lines stretch out nearly parallel to each other and are brought into close proximity. Here the magnetism comes

Fig. 6.26 Cusp geometry. A large helmet-type structure is seen in the south west (*lower right*) of this negative soft X-ray image obtained on 25 January 1992 following a coronal mass ejection. The cusp, seen edge-on at the top of the arch, is the place where the oppositely directed magnetic fields, threading the two legs of the arch, are stretched out and brought together. Several similar images have been taken with the Soft X-ray Telescope (SXT) aboard *Yohkoh*, showing that magnetic reconnection is a common method of energizing solar explosions. (Courtesy of Loren W. Acton, NASA, ISAS, the Lockheed-Martin Solar and Astrophysics Laboratory, the National Astronomical Observatory of Japan, and the University of Tokyo.)

together, merges and annihilates part of itself, releasing the energy needed to power a solar flare.

An X-point neutral sheet, located just above the detectable cusp, marks the site of magnetic-reconnection energy release and particle acceleration, while the rounded soft X-ray shape seen just below the cusp tip outlines magnetic fields newly linked by the reconnection. This scenario is consistent with other *Yohkoh* flare observations showing an impulsive hard X-ray source above the top of the soft X-ray flare loops, discovered by Satoshi Masuda and colleagues in 1994 and presumed to mark the site of magnetic reconnection. It is also supported by *Yohkoh* observations of chromospheric evaporation.

The combined *Yohkoh* observations resulted in a more detailed version of the CSHKP model of magnetic reconnection in solar flares and erupting prominences. These elaborate extensions, provided by Kazunari Shibata and co-workers in 1995 and by Saku Tsuneta in 1996 and 1997, included reconnection inflows at the current sheet, plasma ejections, shocks, outflows, and other inflows.

As shown by Terry Forbes and Loren Acton in 1996, some initially open magnetic fields lines will reconnect to form cusp-shaped, closed loops that will then relax and shrink inward into round ones. And by 2002, Eric Priest and Forbes could review the situation, declaring that the flare eruption stretches out the magnetic field lines, which reconnect to form a rising arcade of soft X-ray loops and separating flare ribbons.

That is all in accord with the extended CSHKP interpretation, in which the magnetic reconnection starts low and progressively rises with time. As the outburst progresses, newly formed, post-flare loops will become progressively higher and wider with an increase in footpoint and flare-ribbon separation, brightening as they are filled by chromospheric evaporation and fading away quickly as their material cools and rains back down.

Observations from the *Ramaty High Energy Solar Spectroscopic Imager*, abbreviated *RHESSI*, have apparently confirmed that large-scale magnetic reconnection in the low corona is the most likely explanation for how solar flares suddenly release so much energy. It has detected hot post-flare loops that move upward with time, in agreement with the classical CSHKP model as extended by Shibata and Tsuneta. In some flares, instruments aboard *RHESSI* have also detected a much weaker, compact X-ray source above the bright X-ray loops during the impulsive phase of intense solar flares.

In 2003, Linhui Sui and Gordon D. Holman interpreted this extra source in terms of a neutral current sheet that formed above the nested coronal loops. Other magnetic loops, formed by magnetic reconnection in the current sheet, moved downward onto the existing ones, which then emitted bright X-ray radiation, and other reconnected magnetic fields and plasma flowed and twisted upward, again consistent with the classical flare model. Then in 2004 and 2006, Sui, Holman and Brian R. Dennis reported that the coronal source detected from *RHESSI* above flare loops moved downward for a few minutes during the impulsive phase of a flare, before the flare loops moved upward. They associated the initial, contracting feature, which was not predicted by the existing flare models, with the formation and development of a current sheet between the loop tops and the coronal source.

But not all the features of the magnetic-reconnection, solar flare model have been fully confirmed or explained. It requires inflow from the sides of the X-point current sheet to push the oppositely directed magnetic fields into a new connection. Observations of such inflows during solar flares are scarce, with just one reported by Takaaki Yokoyama and colleagues in 2001. There are nevertheless different kinds of solar outbursts, known as coronal jets, in which the outflow of material expected from magnetic reconnection has quite definitely been detected (Focus 6.9, Fig. 6.27).

Focus 6.9

Coronal Jets, Emerging Flux and Magnetic Reconnection

Observations of numerous X-ray jets with the Soft X-ray Telescope aboard *Yohkoh* inspired a reconnection model between open and closed magnetic structures,

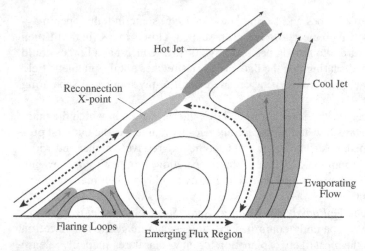

Hot Jet

Reconnection
X-point

Cool Jet

Evaporating
Flow

Flaring Loops

Emerging Flux Region

Fig. 6.27 Reconnection model of coronal jet. In this model of X-ray jets, developed by Kazunari Shibata and co-workers in the 1990s, a closed magnetic loop emerges from beneath the photosphere and encounters open magnetic field lines in the low corona, shown here at an oblique angle – also see Fig. 4.7. The closed and open magnetic structures merge together and reconnect at the X-point, releasing magnetic energy that powers the collimated, jet-like ejections. Hot jets can emerge from the reconnection region, accelerated by fast shocks, and cool jets can arise as the result of evaporation occurring when high-energy particles are sent down into the chromosphere from the reconnection site. As shown here, small footpoint flares often accompany X-ray jets. The arrows denote the direction of the magnetic field lines

developed by Kazunari Shibata and colleagues in the 1990s. It followed an emerging flux model for solar flares proposed by Jean Heyvaerts, Eric R. Priest, and David M. Rust in 1977, which included a rising magnetic loop that reconnects with overlying, open coronal magnetic fields, with ejection of material along the open fields (Fig. 6.27). This scenario is fundamentally different from magnetic reconnection models for solar flares, erupting prominences, and coronal mass ejections, which involve a sudden disruption of a closed magnetic field configuration.

Kazunari Shibata and his colleagues summarized the physical parameters of solar X-ray jets in 1992, and Masumi Shimojo and Shibata further specified them in 2000. They are bursts of hot plasma, with temperatures of $3–8 \times 10^6$K, ejected in a collimated, radial motion, outward from the Sun along presumably open magnetic fields with apparent velocities that are usually near or below the sound speed of a few hundred kilometers per second. A large percentage of these jets are associated with small footpoint flares, and they have been observed in polar coronal holes, the quiet Sun, and active regions. Shibata distinguishes between high-temperature jets from active regions, which might emerge from the reconnection point, and cooler jets in coronal holes, which could result from chromospheric evaporation along open magnetic field lines (Fig. 6.27).

In 2007, Shibata and his co-workers used observations with an instrument aboard *Hinode* to discover large numbers of jets in the chromosphere, with inverted Y-shapes that can be attributed to magnetic reconnection. Another type of coronal jet has been detected in white-light, from the LASCO instrument aboard *SOHO*. In 1998, Yi-Ming Wang and colleagues combined LASCO observations of these jets, which arise from polar coronal holes, with data from the EIT on *SOHO*, showing that the white-light coronal jets are also due to closed magnetic loops reconnecting with overlying open ones. And in 2006, Wang, Monique Pick, and Glenn Mason showed that impulsive solar energetic particles, observed near the Earth from 1997 to 2003, are associated with jets arising from magnetic reconnection between closed and open fields.

Recent observations with the X-Ray Telescope aboard *Hinode*, reported by Jonathan W. Cirtain and colleagues in 2007, indicate that X-ray jets from polar coronal holes have two distinct velocity components. One component moves at near the sound speed of about $200 \, \mathrm{km \, s^{-1}}$, but the other one travels near the Alfvén speed of about $800 \, \mathrm{km \, s^{-1}}$. Both components could be generated as the result of magnetic reconnection, and the faster one might contribute to the fast solar wind arising from polar coronal holes. So there is abundant evidence that the process is at work in jets.

It still is not clear, from the theoretical point of view, how the vertical current sheet can be formed, since it may break apart as the result of the tearing-mode instability. And all the expected outflows, down flows, shocks, and waves are just beginning to be scrutinized, partly as the result of their possible association with coronal mass ejections as well as solar flares.

6.10.3 Extension of Magnetic Reconnection to Coronal Mass Ejections

What forces expel the erupting prominences and coronal mass ejections? And what causes the Sun's magnetism to suddenly erupt with enough force to drive a large part of the corona out against the restraining forces of closed magnetic loops and solar gravity?

There are two competing magnetic forces, magnetic pressure and tension. Regions with strong magnetic fields have greater magnetic pressure, and tend to expand into places of weaker magnetic field. That pressure would push an isolated flux rope or coronal loop outward, but magnetic tension holds them in place. An outburst might occur when an instability or magnetic reconfiguration tips the balance in favor of the outward pressure gradient. Coronal mass ejections, for example, erupt from the Sun as self-contained structures of hot material and magnetic fields, with a magnetic imbalance possibly triggered by reconnection events similar to those that ignite solar flares.

Once magnetic reconnection had been successfully applied to solar flares and erupting prominences; it was indeed natural and inevitable that the models would be extended to include coronal mass ejections. The close proximity of all three types of solar outbursts in space and time suggests that they are consequences of a similar magnetic restructuring, with the mass ejections strongly coupled to the processes that cause solar flares and/or erupting prominences.

One possibility, suggested by Ulrich Anzer and Gerald W. Pneuman in 1982, is that the impulsive prominence stretches the erupting magnetic fields, which couple together in a current sheet below the prominence, pinching off the rising material like a hot-air balloon with its tethers cut (Fig. 6.28). That would be something like blowing soap bubbles and watching them pinch off at the bottom as they waft into space, except in this case it is magnetic bubbles blasting off at supersonic speeds and growing larger than the Sun.

In this type of model, a coronal mass ejection, with an accompanying prominence, rises up, blowing open the previously closed, overlying magnetic fields. But the newly opened magnetism promptly couples together again to reform closed loops below the rising material. So the magnetism opens and closes like a sea anemone. Post-eruptive arcades, frequently detected at extreme-ultraviolet and soft X-ray wavelengths in the aftermath of a coronal mass ejection, are identified with these newly linked, post-flare loops, which rise to successively higher altitudes as the reconnection site is dragged outward along the outstretched neutral line by the rising coronal mass ejection, like closing up a zipper.

This theoretical model for coronal mass ejections has been widely developed in recent years. In 1994, Zoran Mikic and Jon A. Linker, for example, examined

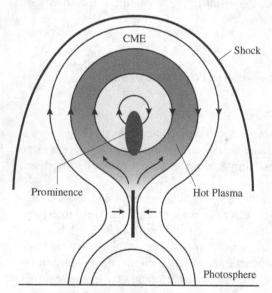

Fig. 6.28 Model of solar eruption. A magnetic reconnection takes place at a current sheet (*dark vertical line*) beneath a prominence and above closed magnetic field lines. The coronal mass ejection, abbreviated CME, traps hot plasma below it (*dark region*). The solid curve at the top is the bow shock driven by the CME. The closed field region above the prominence (*center*) is supposed to become a flux rope in the interplanetary medium. [Adapted from Petrus C. Martens and N. Paul Kuin (1989).]

how coronal magnetic-field arches, subject to shearing photospheric flows, can be disrupted, opening up to form current sheets and magnetic reconnection, while in 2000 Jun Lin and Terry Forbes examined the effects of magnetic reconnection in permitting a magnetic flux rope to escape.

A contemporary version of one model for the coronal mass ejection and solar flare system is shown in Fig. 6.29. It employs some of the detailed events described in Saku Tsuneta's 1997 description of solar flares, and extends this interpretation to include a larger, overlying, and pre-existing magnetic flux rope that is severed by reconnection, as portrayed in the 2000 model of Jun Lin and Terry G. Forbes. In such an interpretation, a rising flux rope can be identified with an erupting prominence at the dense core of a three-component coronal mass ejection.

Then in 2001, Ronald L. Moore and co-workers presented observations of the evolving, photospheric magnetic field recorded in Kitt Peak magnetograms during several eruptive solar flares observed with *Yohkoh's* Soft X-ray Telescope, concluding that each magnetic explosion was unleashed by runaway tether-cutting via reconnection in the middle of a sheared and twisted bipolar loop. And in 2007, Sujin Kim and colleagues combined *Hinode* X-Ray Telescope observations with *TRACE* and *SOHO* MDI images to delineate an evolving flare structure consistent with a tether-cutting model involving a single-bipolar explosion. The tethers are the magnetic field lines that provide the tension which was holding the magnetic loop in place, and the reconnection occurs underneath the erupting structure.

An alternative breakout model, introduced by Spiro K. Antiochos in 1998, and by Antiochos, Carl Richard DeVore, and James A. Klimchuk in 1999, involves a highly sheared magnetic field held down by an overlying, unsheared fields. The new aspects of this model are that it involves at least two magnetic loops instead of one, in quadrupolar geometry, and that magnetic reconnection occurs above the erupting structure. When the overlying, restraining magnetic fields, which represent the tethers, are weakened or partially removed by merging with neighboring magnetism, the low-lying, sheared loops can "break on through to the other side" as the *Doors* song puts it. Observational support for the breakout model has been provided by Guillaume Aulanier and co-workers in 2000, for the renown Bastille day flare on 14 July 2000; by G. Allen Gary and Ronald L. Moore in 2004; and by B. J. Lynch and colleagues, also in 2004. In the same year, Alphonse C. Sterling and Moore reported that observations of some other outbursts were consistent with both runaway tether-cutting and breakout reconnection.

In 2007, however, Ignacio Ugarte-Urra, Harry P. Warren, and Amy R. Winebarger found that the magnetic topology of only 7 of 23 coronal mass ejections could be interpreted in terms of the breakout model, and Moore and Sterling discussed a new subclass of ejections with an explosion from one foot of a loop that blows its top off.

In another interpretation, the coronal mass ejection is driven and energized from below, resulting in the eruption of a pre-existing magnetic flux rope. James Chen and Jonathan Krall and their colleagues have described observational evidence for such a catastrophic event in 1997 and 2006, respectively and Yi-Ming Wang and Neil R. Sheeley Jr. provided other observational evidence for erupting flux ropes in 2006. In the catastrophe model, magnetic reconnection occurs as a byproduct

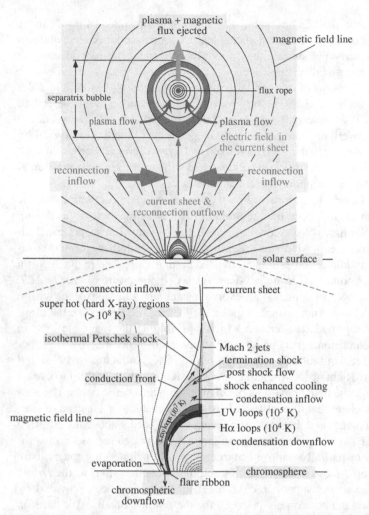

Fig. 6.29 Detailed model of coronal mass ejection and solar flare. Schematic magnetic field configuration and flow pattern for a coronal mass ejection and flare system. The upper part of the diagram portrays the flux-rope model of coronal mass ejections advocated by Jun Lin and Terry Forbes (2000), showing the eruption of the flux rope, the current sheet formed behind it, and the post-flare loops below, as well as the inflows and outflows associated with the magnetic reconnection at the current sheet. The lower part of the diagram is an enlarged view of the post-flare loops, adapted from Terry Forbes and Loren Acton (1996). The upper tip of the reconnection cusp rises as reconnection proceeds. Ultraviolet spectroscopy of coronal mass ejections has contributed to this cross-sectional model of magnetic reconnection inflow that cuts off a magnetic bubble and flux rope with underlying coronal loops and chromosphere flare ribbons. [The combined diagram is courtesy of John N. Kohl and Steven R. Cranmer. It is adapted from Jun Lin, John C. Raymond, and Adriaan A. van Ballegooijen (2004) as well as John L. Kohl, Giancarlo Noci, Steven R. Cranmer, and John C. Raymond (2006).]

of the eruption; while in both the tether-cutting and breakout models the magnetic reconnection is responsible for the onset and growth of the coronal mass ejection, not a consequence of it.

Observational evidence has also been obtained for the magnetic-reconnection current sheet developed when outstretched and oppositely directed magnetic field lines are pushed together. In 2003, David F. Webb and colleagues suggested that bright rays formed in the wake of many coronal mass ejections mark trailing current sheets that can last for several hours and extend more than five solar radii into the outer corona; these ejections were observed from the *Solar Maximum Mission* satellite. The UltraViolet Coronagraph Spectrometer, abbreviated UVCS, aboard *SOHO* has also been used to describe the current sheets, which appear as geometrically long and narrow features in high-temperature emission lines – see the reports by Angela Ciaravella, Yuan-Kuen Ko, and Jun Lin and their colleagues in 2002, 2003, and 2005, respectively.

Other astronomers have focused on the later evolution of coronal mass ejections, when a rising locus of newly filled post-flare loops has been observed from *TRACE*. In 2004, for example, Neil Sheeley Jr., Harry P. Warren, and Yi-Ming Wang used *TRACE* observations to describe the post-flare loops as "the end result of the formation, filling, deceleration and cooling of magnetic loops produced by the reconnection of field lines blown open by the flare."

Already in 2001, Markus Aschwanden and David Alexander had described *TRACE* observations of the Bastille-day flare, including a curved arcade of about 100 post-flare loops that brightened in a sequential manner from highly-sheared, low-lying bipolar loops to higher-lying less sheared ones. As described by Lyndsay Fletcher and Hugh Hudson, also in 2001, there was a systematic increase in the flare ribbon separation observed in both extreme-ultraviolet and hard X-ray radiation during this outburst. Mei Zhang and Leon Golub demonstrated in 2003 that flares associated with fast coronal mass ejections usually show footpoint-separating, two-ribbon brightening observed with *TRACE*. And in 2006, Lidia Contarina and colleagues combined *TRACE* and *RHESSI* observations of separating, extreme-ultraviolet footpoints and rising X-ray loop tops as evidence for the formation of an X-point current sheet and consequent reconnection.

Observations have nevertheless been found that are compatible with every one of the models, and it appears that no single model is consistent with all of the observed coronal mass ejections. Moreover, no one knows for sure just what triggers the sudden and apparently unpredictable solar flares and/or coronal mass ejections? A comparison of solar flares with avalanches and earthquakes is instructive.

6.10.4 Flares, Avalanches, and Earthquakes

The occurrence rate, or frequency distribution, of solar flares exhibits a well-ordered dependence on the flare energy, with systematically more frequent flares at lower energies, from giant flares to nanoflares. In technical terms, the frequency distribution $N(E)$ of flares of energy, E, varies as $E^{-\alpha}$ over eight orders of magnitude or energies from 10^{17} to 10^{25} J and 10^{24} to 10^{32} erg. The power law exponent α has a value of about 1.55.

As proposed by Edward T. Lu and Russell J. Hamilton in 1991, such a distribution can be explained if the coronal magnetic field in solar active regions is in a self-organized critical state analogous to avalanches in a pile of sand. The sand can be added to the pile until a critical state is reached. After that, the addition of more material causes avalanches, keeping the system in the same critical state. In this analogy, the electric currents and stored magnetic energy in coronal loops slowly build up until they are on the brink of instability, in a critical condition where further perturbation results in avalanche-like disruptions. And the time profile of the solar flare might be pictured as avalanches of very small magnetic reconnection events. However, Marina Battaglia, Paolo C. Grigis, and Arnold O. Benz showed in 2006 that small flares have different hard X-ray characteristics than large flares. Thus the latter cannot be just the sum of many small flares.

A solar flare has also been likened to the loss of equilibrium during an earthquake. According to this comparison, the moving footpoints of a sheared magnetic loop are analogous to two tectonic plates. As the plates move in opposite directions along a fault line, they build up stress and energy. When the stress is pushed to the limit, the two plates cannot slide further and the accumulated energy is released as an earthquake. After the earthquake, that part of the fault line then lurches back to its original, equilibrium position, waiting for the next big one.

The sheared magnetic loop may be sent into an increasingly stressed situation like the tectonic plates, until the loop can no longer bear the strain and releases its pent-up energy as a solar flare. In this analogy, the flare time profile is interpreted as a sequence of earthquake aftershocks. And when the Sun's explosive convulsion dies down, the magnetic fields regain their composure, fusing together and becoming primed for the next outburst.

Solar flares do, in fact, occur in regions of strong magnetic shear in the photosphere, and the coronal loops are constantly being reconfigured as they twist and writhe in response to internal differential rotation and convection motions. And an essential part of both the tether-cutting and breakout models for coronal mass ejections is strongly sheared magnetic fields. Both the solar flares and the mass ejections, in these models, are powered by magnetic free energy that is slowly built up in stressed conditions.

6.10.5 All Twisted Up

So it is now widely believed that many solar flares and coronal mass ejections are magnetic explosions, powered by free magnetic energy. They usually happen in active regions with initially closed coronal loops located above one or more neutral lines separating regions of opposite magnetic polarity in the photosphere. And the free magnetic energy is generated when one or more of these closed magnetic bipoles is sheared or twisted into a non-potential configuration.

The hot coronal plasma that is constrained within the contorted loops is illuminated by its soft X-ray emission, which is twisted into a large sigmoid with an S or reversed-S shape (Fig. 6.30). Such sinuous X-ray emission, within active regions and not between them, received their designation by David M. Rust and Ashok Kumar who in 1996 noticed that many of the sigmoid features detected in *Yohkoh* Soft X-ray Telescope images evolved into arcades of bright loops which are often associated with coronal mass ejections. In 1997, Alphonse C. Sterling and Hugh S. Hudson used the *Yohkoh* instrument to study a halo coronal mass ejection, and found that the pre-eruption sigmoid disappeared, leaving a soft X-ray arcade and two "transient coronal holes" behind. Then 2 years later, in 1999, Richard C. Canfield, Hudson, and David E. McKenzie examined 2 years of full-Sun *Yohkoh* images, demonstrating a correlation between the appearance of large twisted sigmoid features in solar active regions and the probable occurrence of eruptive activity there, identified by arcades and cusped loops. When the bright X-ray emitting, active-region coronal

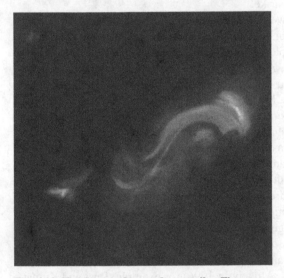

Fig. 6.30 The Sun getting ready to strike. The strong magnetic fields that constrain hot, X-ray emitting gas in this active region have been contorted into an S, or sigmoid, shape that has been resolved at the moment of eruption on 12 February 2007 with the X-Ray Telescope, abbreviated XRT, aboard *Hinode*. Statistical studies of such features, detected from the *Yohkoh* satellite with less detail, indicate that the appearance of such a large S or inverted S shape in soft X-rays is likely to be followed by a solar flare or coronal mass ejection in just a few days. The twisted coronal loops then release their pent-up magnetic energy in a solar explosion also detected in X-rays. The fine details observed from *Hinode's* XRT are helping decide between competing models for the formation of these structures and their subsequent eruption (McKenzie and Canfield, 2008). The sigmoid portrayed here, for example, consists of many fine strands gathered into two back-to-back J shapes, which merge together into the familiar S when observed with poorer angular resolution. [Courtesy of Monica Bobra, Leon Golub, Katharine Reeves, the XRT consortium, SAO, NASA, JAXA and NAOJ. *Hinode* is a Japanese mission developed and launched by ISAS/JAXA, with NAOJ as domestic partner and NASA and STFC (UK) as international partners. It is operated by these agencies in co-operation with ESA and NSC (Norway).]

loops were twisted into a sigmoid shape, they were 68% more likely to be eruptive than non-sigmoid regions. In other words, the magnetism tends to be distorted into a sigmoid before it explodes, and then relaxes to a simpler, less stressful situation.

In their synopsis in 2006, Moore and Sterling concluded that the basic process that triggers and drives the explosion is a core magnetic loop structure, which is sheared and twisted into the shape of a sigmoid with an oppositely curved elbow at each end. The sigmoid may or may not be embedded in strong magnetic fields, and it may or may not contain a flux rope before it starts to explode.

And in 2007, Canfield and his colleagues examined 107 sigmoid active regions over the full decade-long span of the *Yohkoh* mission, showing that they are frequently accompanied by subsequent coronal mass ejections. Altogether 163 eruptions were associated with these sigmoids, and 83% of these were also detected as mass ejections by the LASCO instrument on *SOHO*. So, when the coronal magnetism in active regions gets stirred up into a complex, stressed, and twisted S or reverse-S shape, it is likely to be a prelude to an explosive outburst, something like a coiled-up rattlesnake waiting to strike.

But unlike the snake's rattle, we do not have a signal that will predict exactly when the solar outburst might occur. Scientists have therefore trained the *Hinode* X-Ray Telescope onto coronal sigmoids, hoping that the high-resolution observations might reveal their inner workings. All of the sigmoids detected with *Yohkoh*, with poorer resolution, were composed of multiple loops, without exception, and the initial *Hinode* results, reported by McKenzie and Canfield in 2008, indicated that during the pre-eruptive phase the loops or strands, which extend along the sigmoid's length, consist of two separate J shapes whose straight sections lie anti-parallel to each other in the middle of the S, on opposite sides of the magnetic polarity inversion line (also see Fig. 6.30). Images during the eruptive phase reveal that the sigmoid lifts off in its middle, even before the soft X-ray flare begins.

Similar observations in the future may be important for understanding, and perhaps predicting, coronal mass ejections. But there is more than one way to describe and detect an active region that is getting all twisted up.

Soon after the discovery of the soft X-ray sigmoids, astronomers tried to explain how they are formed and evolve, and to describe the conditions that drive them to eruption as coronal mass ejections. Already in 1996, for example, Rust and Kumar suggested that the observed sigmoids exhibit a twisted and bent, or kinked, magnetic flux rope topology, and a measure of this twisted structure is known as helicity (Focus 6.10). Theoreticians then described the mathematics of twisted flux ropes, and performed numerical simulations of their evolution, culminating in an eruption.

Focus 6.10

Helicity

Magnetic helicity quantifies how the magnetic field is twisted, and the direction or sign of the twist, known as chirality or handedness, tends to change at the solar equator. Magnetic structures in the Sun's northern hemisphere most often exhibit a negative, left-handed helicity, while those in the southern hemisphere usually display a positive, right-handed one. So a magnetic screw or vortex would turn in one direction in the north and in the opposite one in the south, like water running out of a bathtub in the northern or southern hemispheres of the Earth – but for a different reason.

In 1990, Norbert Seehafer first attracted attention to the use of active region vector magnetograms to study the hemispheric helicity trend, but with a small number of active regions. Alexei A. Pevtsov, Richard C. Canfield, and Thomas R. Metcalf described it in greater detail in 1995, and in 1999 Pevtsov and Canfield were careful to note that the handedness of the magnetic twist is a dominant tendency rather than a definite rule. Then in 2008, Pevtsov and colleagues examined the solar active-region hemispheric helicity behavior over a 19-year period, concluding that it is a weak tendency with significant scatter, and that almost 100 active regions are needed to get a plausible result. This indicates that annual subsets used to study temporal variations of the helicity rule are not likely to be reliable.

Filaments, or prominences, also seem to be helically twisted in different directions in the two solar hemispheres. And in 1994, David Rust noticed that interplanetary magnetic clouds usually have the same direction of magnetic helicity as the erupting prominences that presumably spawned them; at about the same time, Sara F. Martin and her colleagues investigated the handedness of the twist and shear of the solar magnetic field in prominences, or filaments. *SOHO* LASCO observations also indicate that a significant fraction of coronal mass ejections exhibit a helical structure in their cores. The soft X-ray sigmoids that often appear before coronal mass ejections also tend to exhibit this magnetic helicity segregation.

So it appears that magnetic helicity is conserved in the emergence of magnetic flux into the solar corona and its expulsion in solar outbursts, which has been stressed by John W. Bieber and Rust in 1995 and in the review by Mei Zhang and Boon Chye Low in 2005. If so, it would be a good thing, for magnetic helicity is continuously being accumulated inside the Sun, as the result of local shearing motions, differential rotation, and the emergence of twisted flux systems. If there was not some mechanism to shed the magnetic helicity and stop the internal build up, it might even stop the solar dynamo that generates the Sun's magnetic fields.

It turned out that the observed sigmoids tend to be twisted one way in the Sun's southern hemisphere, say into a forward-S shape, and in the opposite reverse or reverse-S direction in the other hemisphere. To be consistent with this hemisphere rule, the sheared magnetic flux ropes should dip in the middle, and be heated there to shine in soft X-rays. And in 1999, Viacheslav S. Titov and Pascal Démoulin proposed a way that a twisted coronal loop, which is tied to the photosphere at both ends, can generate electric currents and heat up in its middle, sort of like a filament in a light bulb.

Moreover, as this twisted flux tube emerges from below the photosphere to a certain height in the corona, it can become unstable and form an erupting prominence and a two-ribbon flare. This theory, known as the bald-patch separatrix model, receives its ponderous name from the bald place where the dipped field touches the photosphere and the separate surfaces that are found there during flux rope emergence.

In 2004 and 2006, Yuhong Fan and Sarah E. Gibson proposed an alternative kinked flux rope model. Their numerical simulations indicated that current sheets in the twisted rope can form a sigmoid shape, with the observed orientation, and that the flux rope will erupt as it crosses a threshold of magnetic twist. When the kinked flux rope is twisted beyond a few turns, it becomes unstable to the kink-mode instability and an outburst occurs, which can be likened to a rubber band that snaps when wound too tightly.

The application of the kink instability to solar outbursts has a long history. In 1976, Takashi Sakurai was the first to interpret an erupting prominence as an unstable kinked flux rope. Three years later, Alan William Hood and Eric R. Priest suggested that solar flares are caused by the kink instability, which sets in when the amount of magnetic twist in a flux tube exceeds a critical value. In their 1996 discussion of sigmoids, Rust and Kumar attributed both coronal mass ejections and erupting prominences to the helical kink instability, and in 2000 Tahar Amari and co-workers presented a twisted flux rope interpretation of the soft X-ray sigmoids, which cannot stay in equilibrium.

In 2005, Yuhong Fan and Tibor Török and Bernhard Kliem independently presented numerical simulations that showed how coronal magnetic field lines could become twisted and lead to an erupting prominence or a coronal mass ejection from a kink-unstable flux rope. Török and Kliem considered both confined and ejective eruptions, and it turns out that their model provided a good description of a failed filament eruption observed from *TRACE* (Fig. 6.31). And by 2006, Gibson, Fan, Török, and Kliem had teamed up to describe the evolving sigmoid as evidence for magnetic flux ropes in the corona before, during, and after coronal mass ejections.

But there is still controversy over the way in which sigmoids erupt. The 2008 *Hinode* observations of McKenzie and Canfield favor the Titov and Démoulin approach rather than the kink instability. And there are contradictory reports about whether or not the observed level of twist is sufficient for the kink instability to trigger solar flares and coronal mass ejections – see the descriptions by Robert J. Leamon and colleagues in 2003, Kimberly D. Leka and co-workers in 2005, and Rust and Barry J. LaBonte, also, in 2005.

And although the magnetic energy that fuels solar outbursts seems to be released in the low corona, the eruptions might well be triggered from below, even deep down inside the Sun. That is where all of the action originates – with convection motions that shear and twist coronal loops or emerging magnetic loops which interact with those already there. So scientists have been keeping a careful watch on magnetic activity in the photosphere and below, using magnetograms and helioseismology.

The photospheric magnetic structures before, during, and after solar flares are being scrutinized with the hope of discovering the magnetic activity that might trigger solar flares. These outbursts have indeed long been known to often occur within complex delta configurations of sunspots. Rapid and permanent changes in the photosphere magnetic field occur during some strong solar flares, as reported by Jeffrey Sudal and John W. Harvey in 2005 and by Haimin Wang and colleagues in 2002

Fig. 6.31 Kinky flux. A confined, or failed, filament eruption observed at 18.5 nm wavelength from the *Transition Region And Coronal Explorer*, or *TRACE* for short, on 27 May 2002. Tibor Török and Bernard Kliem (2005) have presented numerical simulations of twisted magnetic field lines, projected from the photosphere magnetograms, to show how the kink instability can produce the observed evolution of this event, rising to a maximum altitude of 80,000 km and outlining the twisted core of a kink-unstable flux rope. (Courtesy of the *TRACE* consortium and NASA; *TRACE* is a mission of the Stanford-Lockheed Institute for Space Research, a joint program of the Lockheed-Martin Solar and Astrophysics Laboratory, or LMSAL for short, and Stanford's Solar Observatories Group.)

and 2005. And in 2005 Valentina Abramenko used active-region magnetograms to show that a steep magnetic power spectrum is a measure of flare productivity.

All the motions and twists in the solar atmosphere are ultimately driven by internal flows. Comparisons of *Yohkoh* soft X-ray sigmoids with photosphere vector magnetograms indicated to Alexei Pevtsov, Richard Canfield, and Alexander McClymont in 1997 that currents responsible for the coronal structures originate below the photosphere; their twist, or magnetic helicity, could be injected from their footpoints. And in 2005, Carolus J. Schrijver and colleagues used *TRACE* and *SOHO* MDI observations of 95 active regions to demonstrate that significant non-potentiality, or free magnetic energy, in their coronal configurations is driven by flux emergence from below, with a characteristic growth and decay of 10–30 h.

Motions beneath the photosphere can be examined with the techniques of helioseismology (Sects. 3.7 and 3.8). As an example, Douglas Mason and his colleagues used the method of local helioseismic tomography to examine the swirling flows beneath hundreds of active regions, showing that the intensity of X-ray flares from these regions is greater when there is a stronger, sideways circulation beneath them.

So, to sum up, there are several possible explanations for the initiation and expulsion of solar outbursts, and many of them are related to twisted, sheared magnetic configurations. Solar flares appear to be energized by magnetic reconnection of sheared coronal loops. Coronal mass ejections seem to be related to sheared and twisted magnetic loops – by tether-cutting reconnection in the low corona below the erupting structure, by breakout reconnection above it, by the expulsion of helically twisted flux ropes driven from beneath the photosphere, and perhaps by the tearing mode or kink instabilities of sheared, twisted magnetic configurations that do not intimately involve reconnection.

So what is all the fuss about anyway? A coronal mass ejection could be triggered from down below, from up above, or from in between. And it might happen anytime, and no one knows precisely how. Hundreds of scientists have been nibbling away at the problem for decades, and several costly spacecraft are dedicated to finding some clues to the unsolved problem.

The sense of urgency, and one of the dominant underlying objectives, is to someday forecast threatening solar outbursts that can cause potential harm in outer space or at the Earth. As we shall next see, these hazardous events are related to strong solar flares, powerful coronal mass ejections, and to shocks and other processes in interplanetary space.

6.11 Summary Highlights: Solar Flares, Erupting Prominences, and Coronal Mass Ejections

- The relatively calm solar atmosphere can be torn asunder by sudden, brief outbursts called solar flares, the most powerful explosions in the solar system. In minutes, they release energy equivalent to millions of 100-megaton hydrogen bombs exploding at the same time, and raise the temperature of Earth-sized regions in the low corona up to 20×10^6 K.
- Large looping arches of magnetism, containing relatively cool material with a temperature of about 10,000 K, can suddenly expand out into space; such erupting prominences or filaments are often associated with coronal mass ejections.
- Coronal mass ejections rip out billions of tons of material from the corona, hurling it into interplanetary space in expanding magnetic bubbles that rapidly rival the Sun in size.
- The soft X-ray emission arriving at the Earth from solar flares is monitored by the *Geostationary Operations Environmental Satellites*, abbreviated *GOES*, classifying them by their peak flux; those of X-class or M-class are the most powerful ones and have noticeable consequences on Earth.
- Flares are detected in the chromosphere by tuning into the red spectral line of hydrogen, the Balmer alpha transition at 656.3 nm. These chromospheric hydrogen-alpha flares often exhibit two parallel ribbons of light, attributed to

energetic particles beamed down into the chromosphere along newly linked coronal loops.

- Flare ribbons can be detected at extreme-ultraviolet wavelengths from *TRACE*, with high spatial and temporal resolution. It has been used to measure the separation speed of loop-footpoint ribbons and the rate of magnetic reconnection above them.

- Hard X-rays are detected during the impulsive phase of solar flares, when electrons are accelerated to high velocities and hurled down to the footpoints of coronal loops.

- Soft X-rays are emitted during the decay phase of solar flares, when material from the chromosphere rises up to fill newly linked coronal loops in a process called chromospheric evaporation.

- Solar-flare X-rays are emitted by bremsstrahlung, or braking radiation, when electrons encounter protons; thermal electrons give rise to soft X-ray bremsstrahlung and more energetic, non-thermal electrons produce hard X-ray bremsstrahlung.

- The slow, smooth rise of the flare emission at soft X-rays resembles the time integration of a flare's impulsive hard X-ray radiation. This Neupert effect suggests that the electron beams that give rise to the hard X-rays produce chromospheric evaporation that fills coronal loops and gives rise to the soft X-ray emission.

- Observations with the Hard X-ray Telescope aboard *Yohkoh* have confirmed that the hard X-rays of many flares are emitted from the footpoints of a flaring coronal loop.

- The Bragg Crystal Spectrometer aboard *Yohkoh* has been used to show that flaring gas is heated in the chromosphere from 10,000 to 20×10^6 K, flowing up into flaring coronal loops that produce the flare soft X-ray emission.

- *Yohkoh's* Hard X-ray Telescope has been used to discover non-thermal, loop-top hard X-ray sources located just above flaring loops detected by the Soft X-ray Telescope, suggesting that very energetic electrons are accelerated above the flare loops.

- A flare particle acceleration site that is located about 50% higher than the soft X-ray flare loop heights has been inferred from electron time-of-flight measurements with the *Compton Gamma Ray Observatory* and from ground-based observations of upward and downward moving radio bursts.

- Rare white-light flares correlate well with the flare hard X-ray emission, in both space and time, suggesting a similar origin.

- The total irradiance of solar flares, at all wavelengths including the visible ones, has been measured, indicating a total energy about 100 times that of the soft X-ray flare emission.

- Flare-associated protons and heavier ions can be beamed down into the lower solar atmosphere, producing nuclear reactions with the emission of gamma-ray spectral lines, meson decay gamma rays, and neutrons that move nearly at the speed of light.

- High-resolution, gamma-ray spectroscopy from *RHESSI* has spectrally resolved the electron-positron annihilation and nuclear de-excitation lines for the first time, with unexpectedly wide annihilation lines and mass-dependent Doppler shifts of the de-excitation lines.

- The first gamma-ray images of solar flares, in the 2.223 MeV neutron-capture line, have been obtained from an instrument aboard *RHESSI*, showing that the flaring gamma-ray sources are spatially displaced from the hard X-ray ones.

- The *RHESSI* imaging instrument has located the region of flare energy release in the corona between the tops of flare loops and a higher coronal source of X-rays.

- A solar flare can outshine the entire Sun at radio wavelengths. These radio bursts are used to probe the flare acceleration site.

- Type II and type III radio bursts are respectively attributed to shock waves and high-speed electron beams that move outward through the corona, exciting plasma oscillations at successively lower plasma frequencies.

- A filament is a region of relatively cool gas embedded in the corona, seen in absorption on the solar disk and suspended by a long, low-lying core magnetic loop.

- A prominence is a filament seen in emission at the edge, or limb, of the Sun.

- An erupting prominence, or filament, can be caused by excessive shear of its low-lying, core magnetic loop, which can rise to open the overlying magnetic arches that straddle it.

- Many thousands of coronal mass ejections have been observed since the 1970s from space-borne coronagraphs. The LASCO instrument aboard *SOHO* has recorded images of more than 10,000 coronal mass ejections, and incidentally resulted in the discovery of more than a thousand comets.

- Coronal mass ejections can exhibit a three-part structure – a bright outer front, followed by a darker, underlying cavity, surrounding a brighter core; but not all coronal mass ejections display these nearly circular shapes.

- Coronal mass ejections hurl between 1 and 50 billion tons, or $10^{12} - 5 \times 10^{13}$ kg, of material out from the Sun at apparent speeds near the Sun of up to $3,400 \, \text{km s}^{-1}$. But most coronal mass ejections exhibit apparent speeds of between 300 and $500 \, \text{km s}^{-1}$.

- At the minimum in the Sun's 11-year activity cycle, coronal mass ejections appear near equatorial latitudes with an average rate of 0.5 events detected per day; near cycle maximum they appear at all latitudes, including the poles, with an average rate of six events per day.

- Coronal loops are sent into quivering oscillations in the aftermath of solar flares. Such oscillations can be used to infer the physical properties of the loops, in a technique known as coronal seismology.

- Coronal mass ejections can generate waves that propagate across the entire Sun, which have been detected with the EIT instrument on *SOHO*.

- Coronal mass ejections are associated with depleted regions of the corona, which have been detected as reductions in the soft X-ray emission observed by instruments aboard the *Yohkoh* and *Hinode* spacecraft; a related dimming of the Sun's extreme-ultraviolet radiation has been detected from *SOHO*.

- The rate of occurrence of solar flares, erupting prominences, and coronal mass ejections varies in step with the 11-year cycle of solar magnetic activity, becoming more frequent near the cycle maximum and suggesting an origin related to strong magnetic fields.

- Coronal mass ejections and solar flares may be a manifestation of a similar energy-release process in the solar corona. Exceptionally fast and energetic coronal mass ejections are commonly accompanied by solar flares, and vice versa, but each type of solar outburst can occur without the other one and there appears to be no general cause–effect relation between the two.

- Solar flares can be ignited when oppositely directed and current-carrying magnetic loops come together and coalesce in a process called magnetic reconnection. During this coronal merging process, the stressed magnetic fields partially annihilate each other, release energy stored in them, and reconnect into less-energetic, more stable configurations.

- The Soft X-ray Telescope aboard *Yohkoh* has been used to discover a soft X-ray, cusp-type geometry in the low corona, seen edge-on at the apex of long-lived, gradual (hours) flaring loops that can be associated with coronal mass ejections. This morphology was predicted by magnetic reconnection theory.

- Coronal jets have been attributed to magnetic reconnection between rising magnetic loops and overlying, open coronal magnetic fields.

- Despite more than 30 years of study, there is no known, universal mechanism that can account for all the observations of coronal mass ejections, but different models are supported by observations of different specific events.

- There are several plausible models for coronal mass ejections in which magnetic reconnection is responsible for the onset and growth of the ejection. They include the breakout model involving at least two magnetic loops and reconnection above the erupting structure, and the tether-cutting model with just one magnetic loop and reconnection below the erupting structure. An alternative catastrophe model involves a pre-existing magnetic flux rope anchored and driven from below, with magnetic reconnection as a byproduct of the eruption.

- The frequency distribution of solar flares exhibits a well-ordered, power-law dependence on energy over a broad energy range, with systematically more flares of lower energy, suggesting that the coronal magnetic field in solar active regions is in a self-organized critical state similar to avalanches in a sand pile or to earthquakes.

- Solar flares and coronal mass ejections most likely arise from coronal loops that have been sheared and twisted into non-potential configurations with free magnetic energy available to power the outbursts. This twist can be detected as active-region sigmoid structures in soft X-rays, and in images of both eruptive prominences and coronal mass ejections. The twisted structures may be related to the helicity seen in photospheric magnetograms of active regions, which tend to display an opposite helicity sign or direction in the two solar hemispheres, and/or to swirling flows beneath active regions.

- A twisted magnetic flux tube that emerges into the corona can become unstable to form an erupting prominence or coronal mass ejections. There are at least two models for the outbursts, and one of them includes the kink instability that sets in when there is too much twist.

6.12 Key Events in Understanding Solar Outbursts*

Date	Event
1852	Edward Sabine demonstrates that global magnetic disturbances of the Earth, now called geomagnetic storms, vary in tandem with the 11-year sunspot cycle.
1859–1960	In 1859, Richard C. Carrington and Richard Hodgson independently observe a solar flare in the white light of the photosphere, and in 1860 publish the first account of such a flare. There was no perceptible change in the sunspots after the "sudden conflagration", leading Carrington to conclude that it occurred above and over the sunspots. Seventeen hours after the flare a large magnetic storm began on the Earth.
1908	George Ellery Hale uses the Zeeman splitting of spectral lines to measure intense magnetic fields in sunspots. They are thousands of times stronger than the Earth's magnetism.
1919	Frederick Alexander Lindemann (later Lord Cherwell) suggests that an electrically neutral plasma ejection from the Sun is responsible for powerful non-recurrent geomagnetic storms.
1919	George Ellery Hale and his colleagues show that sunspots occur in bipolar pairs with an orientation that varies with a 22-year period.
1930–1960	Observations of solar flares in the chromosphere, at the Balmer alpha transition of hydrogen, by Václav Bumba, Helen W. Dobson, Mervyn Archdall Ellison, Ronald G. Giovanelli, Harold W. Newton, Robert S. Richardson, Andrei B. Severny, and Max Waldmeier, show that chromosphere flares occur close to sunspots, usually between the two main spots of a bi-polar group, that magnetically complex sunspot groups are most likely to emit flares, and that hydrogen-alpha flare ribbons lie adjacent and parallel to the magnetic neutral line in the photosphere.
1935–1937	J. Howard Dellinger suggests that the sudden ionosphere disturbances that interfere with short-wave radio signals have a solar origin.
1944	Robert S. Richardson proposes the term solar flare for sudden, bright, rapid, and localized variations detected in the chromosphere at the Balmer alpha transition of hydrogen.
1946	Edward V. Appleton and J. Stanley Hey demonstrate that meter-wavelength solar radio noise originates in sunspot-associated active regions, and that sudden large increases in the Sun's radio output are associated with chromosphere brightening, also known as solar flares.
1946–1948	Ronald G. Giovanelli develops a theory of solar flares involving the magnetic fields in the solar atmosphere above sunspots, including electric currents at magnetic neutral points.
1946, 1950	Scott E. Forbush and his colleagues describe brief, flare-associated increases in the intensity of cosmic rays arriving at the Earth's surface, and attribute the transient increases to very energetic charged particles from the Sun. They were originally designated solar cosmic rays, but are more recently known as solar energetic particles.

*See the References at the end of this book for complete references to these seminal papers

Date	Event
1947	Ruby Payne-Scott, Donald E. Yabsley, and John G. Bolton discover that meter-wavelength solar radio bursts often arrive later at lower frequencies and longer wavelengths. They attributed the delays to disturbances moving outward at velocities of 500–750 km s^{-1}, exciting radio emission at the local plasma frequency.
1948–1949	Soft X-rays from the Sun were first detected on 6 August 1948, with a V-2 rocket experiment performed by the U.S. Naval Research Laboratory, reported by T. Robert Burnight in 1949.
1949	Alfred H. Joy and Milton L. Humason show that main sequence (dwarf M) stars other than the Sun emit flares.
1950–1959	John Paul Wild and his colleagues use a swept frequency receiver to delineate type II radio bursts, attributed to shock waves moving out during a solar outburst at about a 1,000 km s^{-1}, and type III radio bursts, due to outward streams of high-energy electrons, accelerated at the onset of a solar flare and moving at nearly the velocity of light, or at almost 300,000 km s^{-1}.
1950	Hannes Alfvén and Nicolai Herlofson argue that synchrotron radiation of high-speed electrons spiraling about magnetic fields might generate the observed radio emission from discrete cosmic sources, and Karl Otto Kiepenheuer reasons that the synchrotron radiation mechanism can account for the radio emission of our Galaxy.
1950–1954	Scott E. Forbush demonstrates the inverse correlation between the long-term intensity of cosmic rays arriving at Earth and the number of sunspots over two 11-year solar activity cycles.
1951–1963	Herbert Friedman and his colleagues at the U.S. Naval Research Laboratory use rocket and satellite observations to show that intense X-rays are emitted from the Sun, that the X-ray emission is related to solar activity, and that X-rays emitted during solar flares are the cause of sudden ionosphere disturbances.
1953	James W. Dungey proposes a magnetic neutral point discharge theory for solar flares.
1954	Philip Morrison proposes that magnetized clouds of gas, emitted by the active Sun, account for worldwide decreases in the cosmic ray intensity observed at Earth, lasting for days and correlated roughly with geomagnetic storms.
1957	André Boischot discovers moving type IV radio bursts, and Boischot and Jean-Francoise Denisse explain them in terms of magnetic clouds of high-energy electrons propelled into interplanetary space.
1958	Alan Maxwell and Govind Swarup call attention to U-type radio bursts, a spectral variation of type III fast-drift bursts that first decrease and then increase in frequency, suggesting motions away from and into the Sun along closed magnetic field lines.
1958	Peter A. Sweet develops the neutral point theory of solar flares, including magnetic reconnection in a current sheet resulting from the twist and shear of photosphere magnetic fields.
1958–1959	During a balloon flight on 20 March 1958, Laurence E. Peterson and John Randolph Winckler observed a burst of high-energy, gamma ray radiation (200–500 keV) coincident in time with a solar flare, suggesting non-thermal particle acceleration during such outbursts on the Sun.

Date	Event
1959	Thomas Gold argues that solar flares will eject material within magnetic clouds, which remain magnetically connected to the Sun, and suggests that an associated shock front can produce sudden geomagnetic storms. He also coined the term magnetosphere for the region in the vicinity of the Earth in which the Earth's magnetic field dominates all dynamical processes involving charged particles.
1960	Thomas Gold and Fred Hoyle show that magnetic energy must power solar flares, and argue that flares are triggered when two magnetic loops of opposite sense or direction interact, merge, and suddenly dissipate their stored magnetic energy.
1960–1961	Gail E. Moreton uses rapid, time-lapse photography of the chromosphere in the red light of the Balmer alpha transition of hydrogen to discover wave-like disturbances initiated by solar flares. These Moreton waves move away from the site of impulsive flares, across the visible solar disk with velocities of around a thousand kilometers per second.
1961–1963	Mukul R. Kundu demonstrates the similar time profiles of centimeter wavelength, impulsive radio bursts and hard X-ray radiation from solar flares. In 1963, Kees de Jager and Kundu explained the similarity in the profiles in terms of the same energetic electrons producing the radio and hard X-ray emission.
1962–1967	Interplanetary shocks associated with solar activity are detected using instruments aboard the *Mariner 2* spacecraft in 1962, reported by Charles P. Sonett and colleagues in 1964 and by Marcia Neugebauer and Conway W. Snyder in 1967.
1963	Eugene Parker develops the details of Sweet's 1958 method of magnetic reconnection, subsequently known as slow Sweet-Parker reconnection.
1964	Hugh Carmichael proposes that the magnetic field lines above low-lying coronal loops could be forced open by the solar wind, and discusses how magnetic fields might be brought together at the interface of the closed and open magnetic structures to accelerate high-energy particles.
1964	T. R. Hartz obtains the first spacecraft observations of solar type III bursts using a swept frequency receiver from 1.5 to 10 MHz.
1964	Harry E. Petschek clarifies the process of magnetic field reconnection, showing how magnetic energy might be quickly released within the short duration of solar flares, in a process of magnetic field annihilation now known as fast Petschek reconnection.
1966–1968	Peter A. Sturrock develops a model for solar flares in which sheared coronal loops trigger a tearing mode instability near the Y neutral point between closed and open magnetic fields in the low corona, accounting for particle acceleration in a downward direction and shock waves and plasma ejection in the outward direction.
1967–1968	Roger L. Arnoldy, Sharad R. Kane, and John Randolph Winckler demonstrate a flux correlation for centimeter-wavelength impulsive radio bursts and hard X-ray solar flares.
1968	Giuseppe Vaiana and his colleagues show that the soft X-ray emission of a solar flare corresponds spatially with the chromospheric brightening, with roughly the same size, indicating a close association between the two phenomena.
1968	Werner M. Neupert uses soft X-ray flare data, obtained with the third *Orbiting Solar Observatory*, abbreviated *OSO 3*, to confirm that soft X-rays slowly build up in strength, and to show that the rise to maximum intensity resembles the time integral of the rapid, impulsive radio burst.

Date	Event
1968	Yutaka Uchida interprets Moreton waves by a shock-initiated, expanding coronal disturbance intersecting the chromosphere.
1970–1973	Ke Chiang Hsieh, John A. Simpson, Joan Hirshberg, William F. Dietrich, and their colleagues demonstrate that impulsive solar flares can produce a large relative abundance of helium in the solar wind with an enrichment of the helium-3 isotope.
1971–1973	The first good, space-based observation of a coronal disturbance or transient, now called a coronal mass ejection, was obtained on 14 December 1971, using the coronagraph aboard NASA's seventh *Orbiting Solar Observatory*, or *OSO 7* for short, reported by Richard Tousey in 1973.
1972–1973	Edward L. Chupp and his colleagues detect solar gamma-ray lines for the first time using a monitor aboard NASA's seventh *Orbiting Solar Observatory*, abbreviated *OSO 7*. They observed the neutron capture (2.223 MeV) and electron–positron annihilation (0.511 MeV) lines associated with solar flares. The 2.223 MeV line had been anticipated theoretically by Philip Morrison.
1973–1974	Yutaka Uchida and colleagues explain chromosphere Moreton disturbances in terms of magnetohydrodynamic waves responsible for type II bursts.
1973–1974	The manned, orbiting solar observatory, *Skylab*, is launched on 14 May 1973, and manned by three-person crews until 8 February 1974. *Skylab's* Apollo Telescope Mount contained 12 tons of solar observing instruments that spatially resolved solar flares at soft X-ray and ultraviolet wavelengths.
1974	John Thomas Gosling and colleagues report observations of coronal mass ejections, then called coronal disturbances or coronal transients, with the coronagraph aboard *Skylab*, noting that some of them have a high outward speed of up to a $1,000 \, \text{km s}^{-1}$ needed to produce interplanetary shocks. They also found that some coronal mass ejections are not associated with solar flares, but instead with the eruption of filaments or prominences.
1974	Richard T. Hansen and colleagues observe sudden depletions of background regions of the inner solar corona using the Mark I coronagraph at Mauna Loa.
1974	Tadashi Hirayama develops a theoretical model of two-ribbon solar flares and erupting prominences or filaments in which magnetic reconnection occurs at an X-type neutral line under a prominence and rises up as the prominence moves out into space. The oppositely directed magnetic fields are brought together at the X-point to release free magnetic energy, accelerating particles that move down into the chromosphere to form flare ribbons and undergo chromospheric evaporation of material heated to about $10 \times 10^6 \, \text{K}$, producing the soft X-ray flare emission.
1976	Roger A. Kopp and Gerald W. Pneuman develop a theory for two-ribbon flares in which coronal magnetic fields above active regions come together during the outward expulsion of a prominence, predicting a continuous rise of the reconnection point, the increase in height of the newly connected magnetic loops, and separating loop footpoints and flare ribbons.
1976	Takashi Sakurai interprets solar erupting prominences, or filament eruptions, as unstable, kinked flux ropes.
1977	Jean Heyvaerts, Eric R. Priest, and David M. Rust propose an emerging flux model of solar flares, in which a magnetic loop rises into the corona, merging and reconnecting with pre-existing magnetic structures and forming a current-sheet that heats up and becomes unstable. Although this mechanism is apparently only applicable to small flares, it is related to Kazunari Shibata's 1992–1996 interpretations of coronal X-ray jets.

Date	Event
1977	Roberto Pallavicini, Salvatore Serio, and Giuseppe S. Vaiana use *Skylab* observations to define two classes of soft X-ray flares: compact, brief (minutes) events, and extensive, long-enduring (hours) ones associated with soft X-ray arcades, filament eruptions, and coronal mass ejections.
1977–1978	Franz Dröge and Cornelius Slottje independently discover microwave spikes that are a few milliseconds in duration, estimating sizes of less than a thousand kilometers across and brightness temperatures of a million billion (10^{15}) Kelvin.
1979	Alan William Hood and Eric R. Priest suggest that the kink instability of coronal loops, or magnetic flux tubes, is the cause of solar flares; the instability sets in when the amount of magnetic twist in the flux tube exceeds a critical value.
1980	The *Solar Maximum Mission*, abbreviated *SMM*, satellite is launched on 14 February 1980, to study the physics of solar flares during a period of maximum solar activity. It excelled in X-ray and gamma ray spectroscopy of solar flares, as well as observing the white-light emission of coronal mass ejections in 1980 and from 1984 to 1989.
1980–1989	George Doschek, Ester Antonucci, and their colleagues use *P78-1* and *Solar Maximum Mission* observations of soft X-ray spectral lines to show that the impulsive phase of solar flares is associated with the upward flow of heated chromosphere material with velocities of several hundred kilometers per second. Such an upflow is called chromospheric evaporation, predicted by Tadashi Hirayama in 1974. Katsuo Tanaka and colleagues use a spectrometer aboard the *Hinotori* spacecraft to independently confirm chromospheric evaporation in 1982.
1980 1984–1989	Arthur J. Hundhausen and colleagues use the coronagraph aboard the *Solar Maximum Mission* satellite to specify the mass, velocity, energy, shape, and form of a large number of coronal mass ejections, fully reported in the literature in the 1990s (Hundhausen, 1994).
1980–1982	Edward L. Chupp and his colleagues use the Gamma Ray Spectrometer, or GRS, on the *Solar Maximum Mission* satellite to detect energetic solar neutrons near the Earth following a solar flare, which occurred on 21 June 1980 (Chupp, 1984).
1980–1984	Kenneth A. Marsh and Gordon J. Hurford use the Very Large Array, or VLA, in 1980 to resolve a two-centimeter burst source and locate it near the top of a flaring coronal loop. This implies that the initial flare energy release occurs near or above the loop apex, and may indicate trapping of the energetic electrons there. In 1984, Robert F. Willson and Kenneth R. Lang used the VLA to show that flaring emission at 20 cm wavelength also originates near the apex of coronal loops, marking the site of flare energy release above the loop tops in the low corona.
1981	Robert P. Lin and colleagues find that solar flares can produce thermal sources with high enough temperatures to be detectable as hard X-rays.
1981–1982	The Japanese spacecraft *Hinotori*, meaning *firebird*, was launched on 21 February 1981 and operated until 11 October 1982. It created images of solar flare X-rays with an energy of around 20 keV, and measured solar flare temperatures of between 10 and 40 million $(1–4 \times 10^{7})$ Kelvin using soft X-ray spectroscopy.
1981–1982	Peter Hoyng, André Duijveman, and their colleagues use instruments aboard the *Solar Maximum Mission* satellite to resolve hard X-ray solar flares into double sources found at the two footpoints of coronal loops. Tatsuo Takakura and his colleagues found similar structures using the *Hinotori* spacecraft, but most of the detected flares were apparently single rather than double.

Date	Event
1981–1983	In 1981–1983, Arnold O. Benz and colleagues discover decimetric type II bursts that drift rapidly from low to high frequencies, indicating downward-directed electron beams. This suggests flare energy release and electron acceleration in the low corona above the downward beams. Markus Aschwanden and Benz subsequently pinpoint the acceleration site at the demarcation between the downward-moving and upward-moving electron beams.
1982	Ulrich Anzer and Gerald W. Pneuman introduce a theoretical model that combines a two-ribbon solar flare and a coronal mass ejection, or a coronal transient as it was then called. The rising flare loop system undergoes magnetic reconnection, and employs this reconnection as the coronal transient's driving force. As in previous models of solar flares and erupting prominences, a lower loop is created that is rooted in the photosphere and an upper, disconnected loop is produced, which rises and drives the coronal transient.
1982	Russell A. Howard and colleagues report the first detection of an Earth-directed halo coronal mass ejection, and its associated interplanetary shock observed near the Earth, traveling at a speed of nearly $2{,}000$ km s^{-1}. The coronal mass ejection, then known as a coronal transient, was detected from the Solwind coronagraph aboard the *P78-1* satellite; the shock wave was detected with instruments aboard the third *International Sun-Earth Explorer*, abbreviated *ISEE 3*, spacecraft.
1982	Zdenek Svestka and colleagues use instruments aboard the *Solar Maximum Mission* satellite to discover giant X-ray post-flare arches above eruptive flares.
1982–1990	Edward L. Chupp, Hermann Debrunner, and colleagues report the observation of neutron emission at the Earth from the 3 June 1982 flare, giving signals in both the *Solar Maximum Mission* detector and the neutron monitor on Jungfraujoch in Switzerland.
1983	David J. Forrest and Edward L. Chupp use *Solar Maximum Mission* satellite observations to demonstrate the simultaneous acceleration of relativistic electrons (hard X-rays) and energetic ions (gamma rays) to within a few seconds. Masato Yoshimori and colleagues confirmed this simultaneity using observations from the *Hinotori* spacecraft.
1985	David J. Forrest and colleagues report the observation of gamma rays from the decay of neutral mesons during the 3 June 1982 solar flare.
1985, 95	Raghunath K. Shevgaonkar and Mukul R. Kundu use the Very Large Array to resolve a double, loop-footpoint source at 2-cm wavelength during an impulsive solar flare in 1985. A decade later, Kundu and his colleagues use the Nobeyama radioheliograph and the *Yohkoh* spacecraft to show that the two-centimeter and hard X-ray sources coincide spatially. The two non-thermal radio sources were circularly polarized with opposite polarities, indicating oppositely directed magnetic fields, consistent with the footpoints of a single coronal loop.
1987	Hilary V. Cane, Neil R. Sheeley Jr., and Russell A. Howard show that strong interplanetary shocks are associated with fast coronal mass ejection moving at speeds greater than 500 km s^{-1}. They used the third *International Sun-Earth Explorer*, abbreviated *ISEE 3*, low-frequency (< 1 MHz) radio data and Solwind coronagraph observations from the *P78-1* satellite.
1988	Marcos E. Machado and colleagues use X-ray images, taken with the *Solar Maximum Mission* spacecraft, to show that the interaction of magnetic loops in the corona is an essential ingredient in triggering flare energy release.

Date	Event
1988	Donald V. Reames proposes that the abundances of solar energetic particles arriving at Earth, observed from the *International Sun-Earth Explorer* spacecraft over an 8.5-year period, imply two distinct population of particles of separate origin. The ^{3}He and electron-rich events are attributed to impulsive solar flares; the other population, which has lower helium and electron abundances and is responsible for most large proton events seen at Earth, is supposed to be accelerated by coronal mass ejections or interplanetary shocks.
1989	Petrus C. Martens and N. Paul Kuin derive a circuit model for solar filament eruptions and two-ribbon flares, involving the filament and a current sheet at the top of post-flare loops.
1989	Peter A. Sturrock argues that the loss of equilibrium concept cannot be applied to prominence eruptions or coronal mass ejections, and that they are probably caused by instability of a plasma configuration.
1990	Eric R. Priest and Terry G. Forbes give a model for prominence eruptions from twisted magnetic fields, including magnetic reconnection and the creation of currents below the erupting prominence.
1990	Norbert Seehafer demonstrates that magnetic helicity in a small number of active regions at the photosphere level is predominantly negative in the Sun's northern hemisphere and positive in its southern hemisphere.
1991	Edward T. Lu and Russell J. Hamilton propose that the solar coronal magnetic field is in a self-organized critical state, thus explaining the observed power-law dependence of the frequency distribution of solar flares on their energy, with systematically more frequent flares at lower energy. They picture solar flares as avalanches of many small reconnection events, in analogy to avalanches of sand.
1991	The Japanese *Yohkoh*, meaning *sunbeam*, satellite is launched on 30 August 1991, to study the soft and hard X-ray emission from solar flares and soft X-rays from non-flaring structures.
1991–2001	The Soft X-ray Telescope, or SXT, on the *Yohkoh* satellite reveals the magnetically structured, dynamic nature of the inner corona more clearly than ever before.
1991–2001	The Bragg Crystal Spectrometers, or BCS, aboard *Yohkoh* are used to show that gas heated in the chromosphere from 10,000 K up to about 20×10^6 K flows into flaring coronal loops at typical speeds of 350 km s^{-1}, producing the copious soft X-ray emission seen during flares. These motions were studied and compared with the prediction of theoretical models of flares. The BCS instruments were able to show that the bulk of the gas in solar flares reaches a maximum temperature of about 23×10^6 K, in contrast to some stellar flares that can be much hotter. The reason for the solar limit is still unknown.
1991–1993	Gottfried Kanbach and colleagues report observations with the *Compton Gamma Ray Observatory* of meson decay, gamma-ray flare on 11 June 1991, lasting 8 h.
1991–1998	Philip A. Isenberg, Terry G. Forbes, Pascal Démoulin, and Jun Lin develop the catastrophe model of eruptive flares and coronal mass ejections involving the loss of mechanical equilibrium of a force-free magnetic flux rope.
1992	Saku Tsuneta and his colleagues discover a cusp geometry in *Yohkoh* Soft X-ray Telescope images of a longlasting (hours) soft X-ray flare at the solar limb on 21 February 1992. The observations included a soft X-ray arch with a rising cusp structure and separating footpoints, at speeds of 10–30 km s^{-1}, and were explained by magnetic reconnection at a neutral sheet above the loop created by pre-flare magnetic restructuring.
1992–1996	Kazunari Shibata and co-workers describe the physical parameters of coronal X-ray jets, and develop a model for them involving reconnection between closed and open magnetic field lines.

Date	Event
1992–1996	Satoshi Masuda and his colleagues use the Hard X-ray Telescope aboard *Yohkoh* to discover loop-top hard X-ray sources in compact two-ribbon and impulsive solar flares, suggesting that flare particle acceleration and magnetic reconnection occur in the cusp area above the closed soft X-ray flaring loop. The two expected hard X-ray sources at the footpoints of that loop were also detected.
1994	Zoran Mikic and Jon A. Linker introduce a theoretical model in which solar flares and coronal mass ejections are initiated by sudden disruptions of coronal magnetic field arcades when photospheric flows drive the arcades beyond a critical shear. The magnetic fields expand outward, opening the field lines, and magnetic reconnection releases magnetic energy with the ejection of plasma.
1994	David M. Rust proposes that helical magnetic fields spawned inside the Sun, and detected in active regions and filaments, are expelled during solar outbursts to give rise to observed helical magnetic clouds.
1994	Taro Sakao uses *Yohkoh* data to show that hard X-ray bursts occur simultaneously at the footpoints of coronal loops, to within 0.2 s. This confirms that high-speed electrons are accelerated in closed magnetic loops, with the emission of hard X-rays as the electrons stream downwards along the magnetic field lines and enter the dense lower layers of the solar atmosphere.
1994	The *Wind* spacecraft is launched on 1 November 1994. It provides nearly continuous, direct, in-situ measurements of the solar wind, magnetic fields, and energetic particles arriving at the Earth's magnetosphere. *Wind* has also investigated the shocks generated by coronal mass ejections, examined the characteristics of magnetic clouds, and with other spacecraft measured long, steady reconnection layers in the solar wind near Earth.
1994–1997	Yoichiro Hanaoka and Kazunari Shibata use data from *Yohkoh's* Soft X-ray Telescope to show that solar flares are associated with the magnetic interaction of coronal loops, confirming *Solar Maximum Mission* observations by Marcos E. Machado and colleagues in 1988.
1995	John W. Bieber and David M. Rust propose that emerging toroidal magnetic flux, generated by dynamo action inside the Sun, is ejected into space by coronal mass ejections, filament eruptions, and active region flares or loop expansion, collectively removing some magnetic helicity from the Sun.
1995	Alexei A. Pevtsov, Richard C. Canfield, and Thomas R. Metcalf study the average magnetic helicity in magnetographs of solar active regions over an 8-year period, confirming the predominant hemisphere sign difference reported in 1990 by Norbert Seehafer. Pevtsov and Canfield additionally confirmed such a hemisphere helicity rule in 1999, but at a low level, and in 2008 Pevtsov and co-workers showed that the solar active-region hemispheric helicity rule is a weak tendency with significant scatter and with reliable results only for a large number of active regions.
1995	The *SOlar and Heliospheric Observatory*, abbreviated *SOHO*, is launched on 2 December 1995. The Large Angle and Spectometric Coronagraph, or LASCO for short, aboard *SOHO* carried out more than a decade of investigations of coronal mass ejections. *SOHO's* Extreme-ultraviolet Imaging Telescope, abbreviated EIT, observed solar flares and the aftermath of coronal mass ejections during the same period. The UltraViolet Coronagraph Spectrometer, or UVCS for short, aboard *SOHO* has also been used to study coronal mass ejections, as well as the solar wind.
1995–1997	Kazunari Shibata and co-workers and Saku Tsuneta develop models of solar flares, in which the main energy release and magnetic reconnection occur at an X-point above the soft-X-ray flare loops. These loops are interpreted as newly linked, closed magnetic fields, subsequently heated by slow shocks and filled by chromospheric evaporation. These are detailed and elaborate extensions of the CSHKP X-point magnetic-reconnection model of solar flares and erupting prominences.

Date	Event .
1996	Terry G. Forbes and Loren Acton use images of flare loops taken with the Soft X-ray Telescope aboard *Yohkoh* to describe the decrease in height, or shrinkage, of open field lines after they reconnect to form closed loops during solar flares.
1996	Alexei A. Pevtsov, Richard C. Canfield, and Harold Zirin present vector magnetograms and X-ray images of reconnection and helicity during a solar flare with an observed twist that exceeds the threshold for the kink instability.
1996	David M. Rust and Ashok Kumar demonstrate that many *Yohkoh* soft X-ray images of the Sun exhibit transient brightening, known to be associated with erupting prominences and coronal mass ejections, which are sigmoid, or S-shaped, with ratios of length to width consistent with an eruption or ejection caused by the helical kink instability of a kinked and twisted magnetic flux rope.
1996, 98	Markus J. Aschwanden and colleagues use observations with the *Compton Gamma Ray Observatory*, abbreviated *CGRO*, to model electron time-of-flight distances of fast, sub-second hard X-ray flare pulses, from the coronal acceleration site to the chromosphere, showing that the acceleration occurs 50% higher than the soft X-ray flare loop heights observed from *Yohkoh*.
1997	The *Advanced Composition Explorer*, or *ACE* for short, was launched on 25 August 1997. From a vantage point just outside the Earth, *ACE* monitors the solar wind, with its charged particles and magnetic fields, and observes high-energy particles accelerated at the Sun, within the solar wind, or in galactic regions beyond the heliosphere.
1997	Markus J. Aschwanden and Arnold O Benz compare *Yohkoh* observations of soft X-ray flare loops with electron densities inferred from the plasma frequencies of decimetric and type III radio bursts, showing that the flare acceleration site is located above the bright soft X-ray flare loops.
1997	J. Daniel Moses and colleagues describe the first observations with the Extreme-ultraviolet Imaging Telescope, abbreviated EIT, aboard the *SOlar and Heliospheric Observatory*, or *SOHO* for short. New observations of coronal structures with temperatures of 1–2×10^6 K are presented, including jets in polar coronal holes.
1997	Alphonse C. Sterling and Hugh S. Hudson report *Yohkoh* Soft X-ray Telescope observations of X-ray dimming associated with a halo coronal mass ejection, and two years later Dominic M. Zarro and co-workers described *SOHO* EIT observations of extreme-ultraviolet dimming just before such a halo event.
1997–1999	Barbara J. Thompson and her colleagues report the detection of large-scale transient waves in the low corona, initiated by large solar flares that are also associated with coronal mass ejections. The coronal waves were detected using the EIT instrument aboard *SOHO*. They were identified as the coronal counterpart of Moreton waves discovered in the chromosphere by Gail Moreton in 1960–1961 and explained by Yutaka Uchida in 1968 and 1973–1974, and have been variously called EIT waves, Thompson waves, and coronal-Moreton waves.
1998	Sarah E. Gibson and Boon Chye Low present a theoretical model for the expulsion of a coronal mass ejection involving a twisted magnetic flux rope.
1998	Sara F, Martin reviews the conditions for the formation and maintenance of filaments, or prominences, and Jack B. Zirker, Oddbjorn Engvold, and Martin describe counter-streaming flows in them. Bogdan Rompolt previously reviewed the small-scale filamentary structure and dynamics of prominences in 1990.
1998	The *Transition Region And Coronal Explorer*, abbreviated *TRACE*, is launched on 1 April 1998; the high spatial resolution and rapid sequential images of the *TRACE* telescope have, when combined with other simultaneous observations, provided evidence for the mechanisms that trigger solar flares or coronal mass ejections, as well as detailed information about flare ribbons and post-flare loops.

Date	Event
1998	Alexander G. Kosovichev and Valentina V. Zharkova use the Michelson Doppler Imager aboard *SOHO* to detect seismic waves generated when a powerful solar flare sent beams of electrons into the Sun. Although most flares do not generate detectable seismic waves, they were generated by flares in October 2003, and reported by Alina-Catalina Donea and Charles Lindsey in 2005.
1998–1999	Spiro K. Antiochos, and his colleagues Carl Richard DeVore and James A. Klimchuk, present a magnetic breakout model for coronal mass ejections involving at least two sets of closed magnetic loops, with a quadrupolar or multi-polar topology, in which magnetic reconnection occurs between the overlying unsheared loops and highly-sheared, low-lying ones, removing the unsheared field and allowing the core sheared flux to burst open. In 2000, Guillaume Aulanier and colleagues presented observational support for this magnetic breakout model, as did G. Allen Gary and Ronald L. Moore in 2004, but observations of other coronal mass ejections are not consistent with it.
1999	Markus J. Aschwanden and colleagues, and Valery M. Nakariakov and co-workers report *TRACE* observations of coronal loop oscillations excited by large flares. They have periods of 2–7 min and are rapidly damped in comparable times. Carolus J. Schrijver and co-workers gave an overview of these transverse loop oscillations in 2002, and Aschwanden and colleagues described the relevant geometrical and physical parameters in the same year. In 2005, Nakariakov and Erwin Verwiche prepared a *Living Review* about coronal waves and oscillations.
1999	Kenneth P. Dere and co-workers report observations of helical structure in coronal mass ejections observed with the LASCO and EIT instruments aboard *SOHO*.
1999	Viacheslav S. Titov and Pascal Démoulin describe the topology of a force-free, twisted magnetic flux rope that emerges from below the photosphere and becomes unstable and erupts in the low corona.
2000	Tahar Amari and co-workers carry out numerical simulations of a twisted flux rope model for coronal mass ejections, erupting prominences, and two-ribbon flares. The twisted flux rope, which represents a prominence and exhibits an S-shaped structure as observed in soft X-ray sigmoids, cannot stay in equilibrium and releases considerable magnetic energy on disruption.
2000	Jun Lin and Terry G. Forbes investigate how magnetic reconnection affects the acceleration of coronal mass ejections and vice versa. They describe a model of a pre-existing magnetic flux rope that drives the ejection by means of a catastrophic loss of mechanical equilibrium. The erupting field creates reconnection in a current sheet, allowing the flux rope to overcome magnetic tension and escape from the Sun.
2000	Angelos Vourlidas and colleagues use observations with *SOHO's* LASCO to determine the potential and kinetic energies of coronal mass ejections that exhibit flux-rope morphologies, showing that they are magnetically driven.
2001	Lyndsay Fletcher and Hugh Hudson use *SOHO, TRACE*, and *Yohkoh* images of flare ribbons at extreme-ultraviolet and X-ray wavelengths, to support the hypothesis that the flare ribbons mark out the chromospheric footpoints of magnetic loops newly linked by reconnection.
2001	Petrus C. Martens and Cornelis Zwaan present a model for the formation, evolution, and eruption of solar filaments or prominences.

Date	Event
2001	Ronald L. Moore and colleagues provide observational evidence in support of a tether-cutting model of coronal mass ejections. It is a single-loop, bipolar model for eruptive solar flares, and their associated coronal mass ejections, in which a magnetic explosion is unleashed by runaway tether-cutting via internal magnetic connection in the middle of a single sheared and twisted core bipole with a sigmoid shape.
2001	Jie Zhang and co-workers use instruments aboard *SOHO* to study the temporal relationship between coronal mass ejections and solar flares, describing a three-phase speed profile for coronal mass ejections – the initiation, impulsive acceleration, and propagation phases. The initiation phase is characterized by a slow rise with a speed of about 80 kilometers per second for a period of tens of minutes. The subsequent acceleration phase, which coincides with the flare impulsive rise and lasts a few to tens of minutes, ceases near the peak of the soft X-ray emission and is followed by the propagation phase.
2002	The *Ramaty High Energy Solar Spectroscopic Imager*, abbreviated *RHESSI*, is launched on 5 February 2002 to investigate high-energy acceleration processes close to the Sun by observing flare radiation from soft X-rays to gamma rays.
2002–2003	Tongjiang J. Wang and co-workers discover and describe hot coronal loop oscillations, with temperatures greater than 6×10^6 K and oscillation and decay periods between 7 and 31 min. They were detected as Doppler shifts of spectral lines using the SUMER instrument aboard *SOHO*.
2003	Gordon J. Hurford and co-workers obtain the first image of a solar flare at gamma-ray wavelengths, in the 2.223 MeV neutron capture line using an imaging instrument aboard *RHESSI*. The gamma-ray sources were displaced from the hard X-ray ones, which was not expected. The first image of a double-footpoint, gamma-ray line, flare source was also obtained from *RHESSI*.
2003	Ilia I. Roussev and colleagues (2003a, b) present numerical simulations of the eruption of a three-dimensional magnetic flux rope balanced between magnetic compression and tension forces, and investigate the loss of equilibrium that might initiate a coronal mass ejection, as suggested by Titov and Démoulin in 1999.
2003	Gerald H. Share and David M. Smith, and their colleagues, report the spectral resolution of the positron-annihilation and nuclear de-excitation lines at gamma-ray wavelengths during solar flares using an instrument aboard *RHESSI*. The annihilation lines were unexpectedly broad and the de-excitation lines exhibited mass dependent Doppler shifts.
2003	Haiman Wang and colleagues use observations of the separation of two flare ribbons, associated with a quiescent filament eruption and fast coronal mass ejection, to provide evidence that the impulsive flare energy release is governed by fast magnetic reconnection in the corona.
2003	David F. Webb and co-workers show that about half the coronal mass ejections observed from 1984 to 1989 with the coronagraph aboard the *Solar Maximum Mission* spacecraft are followed by coaxial, bright rays, suggesting trailing current sheets and magnetic reconnection in an X-type neutral point that rises in the wake of the ejection.
2003	Mei Zhang and Leon Golub investigate *TRACE* observations of flares associated with fast and slow coronal mass ejections detected with the LASCO instrument on *SOHO*, showing that footpoint-separating, two-ribbon brightenings are often present during the fast mass ejections while appearing less often in flares associated with slow coronal mass ejections.

Date	Event
2003–2005	Linhui Sui and Gordon Holman use an imaging instrument aboard *RHESSI* in 2003 to discover a compact coronal X-ray source with a temperature gradient increasing downward, opposite to that at the top of flare X-ray loops below the compact source, demonstrating that energy release occurred in the corona between the top of the loops and the coronal source. In 2005, they teamed up with Brian Dennis to show that the weak coronal X-ray sources propagate outward following a coronal mass ejection, supporting a model in which free magnetic energy is released in a coronal current sheet where magnetic reconnection occurs.
2003–2007	Angela Ciaravella, Yuan-Kuen Ko, and Jun Lin, and their colleagues, report *SOHO* UVCS observations of the dynamical and physical properties of long, narrow current sheets formed in the wake of coronal mass ejections detected with LASCO on *SOHO*.
2004	Yuhong Fan and Sarah E. Gibson provide numerical simulations of three-dimensional coronal magnetic fields resulting from the emergence of twisted flux tubes.
2004	Lyndsay Fletcher and co-workers use *TRACE* to measure the motions of a solar flare's foopoint ribbons, obtaining an average speed of $15 \, \mathrm{km \, s^{-1}}$ and measuring the magnetic reconnection rate.
2004	Jun Lin, John C. Raymond, and Adriaan A. Van Ballegooijen investigate the role of magnetic reconnection on the observable features of a coronal mass ejection, tentatively identifying the outer shell, the expanded bubble and a pre-existing flux rope with the leading edge, void, and core of a three-component mass ejection, and show that the observed features and physical processes are similar to those of a model in which a flux rope is made during the eruption by magnetic reconnection.
2004	Thomas G. Moran and Joseph M. Davila present the first three-dimensional polarimetric imaging of a coronal mass ejection using data taken with the LASCO instrument aboard *SOHO*.
2004	Thomas N. Woods and colleagues report the first detection of a solar flare in the total solar irradiance, on 28 October 2003, using NASA's *SOlar Radiation and Climate Experiment*, abbreviated *SORCE*.
2004	Jie Zhang and co-workers report correlations between the observed features of solar flares and coronal mass ejections, which indicate that when the two kinds of outbursts occur together they are strongly coupled, with the same processes involved in the mass ejection and flare energy release.
2005	Marina Battaglia, Paolo C. Grigis, and Arnold O. Benz find that small flares have on the average a softer hard X-ray spectrum than large flares. Thus large flares cannot be composed of many small flares.
2005	George A. Doschek and Harry P. Warren revisit chromospheric evaporation in flares using soft X-ray spectra obtained by the Bragg Crystal Spectrometer experiment on *Yohkoh*, with comparisons to different models.
2005	Gregory D. Fleishman, Gelu M. Nita, and Dale E. Gary present controversial evidence that some decimetric continuum radio bursts, in the 10–30 cm wavelength range, are produced by an incoherent emission mechanism, most likely resonant transition radiation resulting from the interaction of fast electrons with small-scale inhomogeneities in the background plasma.
2005	Carolus J. Schrijver and colleagues use *TRACE* and *SOHO* MDI observations to show that non-potential coronal magnetic fields in active regions emerge from below.

Date	Event
2005	Vasyl Yurchyshyn and co-workers study the statistical distribution of the observed speeds of 4,315 coronal mass ejections in the plane of the sky, obtaining a log normal distribution. It suggests that the same non-linear driving mechanism is acting in both slow and fast coronal mass ejections.
2006	Sarah E. Gibson and Yuhong Fan further develop the twisted magnetic flux rope interpretation of soft X-ray sigmoids and prominence eruptions, and with Tibor Török and Bernhard Kliem review the evolution of evolving sigmoids as evidence for magnetic flux ropes before, during, and after coronal mass ejections.
2006	The Japanese *Hinode*, meaning *sunrise*, spacecraft is launched on 23 September 2006 to investigate how magnetic interactions and related processes generate the solar atmosphere and solar activity, including how magnetic energy is converted into intense ultraviolet and X-ray radiation and how magnetic interactions cause solar flares and coronal mass ejections.
2006	The twin spacecraft of the *Solar TErrestrial RElations Observatory*, abbreviated *STEREO A* and *B*, are launched on 25 October 2006, to obtain a three-dimensional, stereoscopic view of coronal mass ejections from their onset at the Sun to the orbit of the Earth, and to thereby investigate the origin, evolution, mechanisms, and interplanetary propagation of coronal mass ejections.
2006	Thomas N. Woods, Greg Kopp, and Philip C. Chamberlin report the detection of very energetic flares in total solar irradiance measurements from NASA's *SOlar Radiation and Climate Experiment*, abbreviated *SORCE*, and demonstrate that the total flare energy is about 100 times the energy observed in soft X-rays from *GOES* in the 0.1–0.8 nm range.
2007	Richard C. Canfield and colleagues study the structure, formation, and evolution of 107 bright sigmoids viewed over a decade of observations with the Soft X-ray Telescope aboard *Yohkoh*, showing that they are all composed of multiple loops, and compare their X-ray configurations to magnetic field extrapolations from photospheric magnetograms. They identified 163 eruptions associated with the sigmoids, 83% of which were detected as coronal mass ejections by the LASCO instrument aboard *SOHO*.
2007	Sujin Kim and colleagues combine *Hinode* X-Ray Telescope images of an impulsive flare with those from *TRACE* and *SOHO's* MDI to support a flare tether-cutting model involving a single-bipole explosion with a sigmoidal structure and two-step, pre-flare and main phase, magnetic reconnection.
2008	David E. McKenzie and Richard C. Canfield present high-resolution *Hinode* observations of a longlasting coronal sigmoid that erupted as a solar flare, and show that the sigmoid is composed of two separate J-shaped, anti-parallel bundles of loops.

Chapter 7
Space Weather

7.1 The Space Weather Concept

The Sun is the ultimate power source. It warms the ground we walk on, lights our days, sustains life, and provides directly or indirectly most of the energy on our planet. And it is solar heat that powers the winds and cycles water from sea to rain, the source of our weather and arbiter of our climate. Nowadays, and in all former times, it is the Sun-driven seasons that dominate weather on Earth.

Once it was realized that the space between the Sun and the Earth is not empty, and just more rarefied than our transparent atmosphere, it was natural to suppose that the Sun also powered weather in space. In 1959, for example, Thomas Gold proposed that space vehicles should measure the permanent and variable features of the interplanetary medium, as the "counterpart of meteorology on Earth," determining the space equivalents of "meteorological measurements of temperature, pressure, and wind." And that is just what happened, beginning just a few years later when the Soviet and American spacecraft, *Luna 2* and *Mariner 2*, reached out to sample the density and speed of the solar wind.

In 2006, R.P. Kane traced the term *space weather* to a 1967 technical report by T.M. Georges and to the preface of a 1970 book edited by Patrick McIntosh and Murray Dryer. So Gold apparently did not introduce the designation, but in his 1959 discussion he did coin the word *magnetosphere*, as the region in space near the Earth where the terrestrial magnetic field dominates the motions of charged particles.

The modern definition of the term space weather, adopted by the U.S. National Space Weather Plan, refers to conditions on the Sun and in the solar wind, magnetosphere, ionosphere, and thermosphere, which can influence the performance and reliability of space-borne and ground-based technological systems and which can affect human life and health. The ionosphere and thermosphere refer to the upper layers of the Earth's atmosphere, which are heated and ionized by the Sun's variable extreme-ultraviolet and X-ray radiation.

So space weather includes effects near the Earth, driven by processes on the Sun, and two of these effects, intense auroras and geomagnetic storms, have long been attributed to solar activity. Already in 1716, Edmond Halley, for example, had suggested that a spectacular aurora display was due to particles, or "magnetical

K.R. Lang, *The Sun from Space*, Astronomy and Astrophysics Library,
© Springer-Verlag Berlin Heidelberg 2009

effluvia" moving along the Earth's magnetic field lines, and in 1773 Jean Jacques d'Ortous de Mairan had asserted that the aurora are due to a mingling of the solar and terrestrial atmospheres.

In 1722, George Graham developed a sensitive compass and used it to show that the Earth's magnetic field undergoes large and rapid variations, which are now called geomagnetic storms. Anders Celsius, who independently discovered the storms around the same time, collaborated with Graham to compare the magnetic variations in London and Uppsala, showing that the disturbances are planetary rather than local in scale. And in 1747, Celsius' student, Olaf Hioter, noted that auroras are accompanied by magnetic deflections.

About a century later, in 1843, Samuel Heinrich Schwabe announced his discovery of the decade-long variation in the number of sunspots, and at about this time Edward Sabine was analyzing the magnetic measurements obtained at army stations in Britain and its colonies. In 1852, the astronomer, John Herschel, wrote to Sabine about the 11-year sunspot period and its possible relationship to the magnetic field measurements. In reply, Colonel Sabine reported that the frequency of global magnetic storms rose and fell with the number of sunspots.

Sunspots do not cause geomagnetic storms, but for quite a long time it was supposed that solar flares do produce them. The first recorded white-light flare, independently discovered by Richard Carrington and Richard Hodgson in 1859, was followed some 17.5 h later by one of the largest magnetic storms on record, measured at the Kew Observatory in London. And George Ellery Hale presented systematic evidence for a relation between solar flares and geomagnetic storms in 1931. He used systematic, Balmer hydrogen-alpha flare observations and terrestrial magnetic data to show that the great magnetic storms tend to follow solar flares in about 26 h, the time it might take flare particles to reach the Earth. Harold W. Newton further bolstered the case for a flare connection in 1943, when he examined about a half-century of observations and found a significant correlation between very intense solar flares and great magnetic storms that occurred about a day later.

So for many decades, it was widely supposed that the great, non-recurrent geomagnetic storms that shake the Earth's magnetic field to its very foundations are indeed caused by powerful solar flares. After all, these storms occur most often when the Sun is more spotted and solar flares are also more frequent.

The solar flare theory nevertheless had its problems. Since electrons repel each other, a focused stream of electrons sent into space from solar flares will disperse before it reaches Earth. Frederick Lindemann (later Lord Cherwell) temporarily saved the explanation in 1919 when he suggested that an electrically neutral cloud or stream, containing equal numbers of electrons and protons, would hold itself together when traveling from the Sun to the Earth. About a decade later, Sydney Chapman and Vincent C.A. Ferraro suggested that electric currents are induced in such a plasma cloud when it encounters the Earth's magnetic field, resulting in geomagnetic storms.

The prophetic Thomas Gold noticed in 1955 that interplanetary shocks generated by ejections from the Sun could cause the abrupt, sudden onsets of geomagnetic storms when striking the Earth. Of course, coronal mass ejections were not even discovered until the 1970s, but it was eventually realized that they push strong shocks ahead of them. If directed toward the Earth, these shocks can ram into the terrestrial magnetic fields and trigger the initial phase, or sudden commencement, of a large geomagnetic storm. The most intense, non-recurrent geomagnetic storms occur when fast coronal mass ejections hit the Earth's magnetosphere with the right magnetic alignment, and they also generate power surges on transmission lines that could cause electrical power blackouts of entire cities. And since solar flares often accompany the fast, intense coronal mass ejections, the observed correlations of flares and great geomagnetic storms would be explained.

But solar flares do have important space-weather consequences. During World War II, for example, it was realized that strong flares could disrupt radio communications on Earth. Intense X-ray and extreme-ultraviolet radiation emitted during the flares heats the Earth's atmosphere, disrupting communications and also altering satellite orbits.

Sudden signals in cosmic ray detectors at Earth were associated with solar flares, by Scott Forbush in 1946 and Peter Meyer and colleagues in 1956. This suggested that the flares could accelerate charged particles to extraordinarily high, cosmic ray energies in excess of 5 GeV.

We now realize that very energetic particles can also be accelerated by the interplanetary shocks associated with coronal mass ejections, and not just at the Sun during solar flares. Charles P. Sonett and colleagues and Marcia Neugebauer and Conway W. Snyder first observed interplanetary shock waves, in 1964 and 1967, respectively, using instruments aboard the *Mariner 2* spacecraft. Early theoretical insights were summarized by Eugene Parker in his concise 1963 book entitled *Interplanetary Dynamical Processes*, while pioneering observations were reviewed by David S. Colburn and Sonett in 1966, Leonard F. Burlaga in 1971, and Murray Dryer in 1974 and 1982.

As our civilization deploys ever more sophisticated, space-borne technological systems, it becomes increasingly at the mercy of storms in space. Its gusts and squalls, the cosmic equivalent of terrestrial blizzards or hurricanes, are related to explosive outbursts on the Sun, and to dynamic processes in interplanetary space, in near-Earth space, and in the magnetosphere.

Down here on the ground, we are shielded from much of this space weather by the Earth's atmosphere and magnetic fields, keeping us from bodily harm. But out in deep space there is no place to hide, and both humans and satellites are vulnerable. Energetic protons accelerated by solar flares or coronal mass ejections can cripple spacecraft and seriously endanger unprotected astronauts that venture into outer space. Sun storms can also disrupt global radio communications and disable satellites used for navigation, military reconnaissance or surveillance, and communication, from cell phones to pagers, with considerable economic, safety, and security consequences. This technology has become part of our everyday lives, enhancing our vulnerability to space weather and increasing the importance of understanding and predicting it.

Exceptionally powerful solar outbursts, producing the greatest damage at Earth, are legendary. In March 1989, for example, a geomagnetic storm produced by a powerful coronal mass ejection disrupted the electrical power system in Quebec, Canada, plunging the entire province into complete darkness. In March and April 2001, the space weather spawned by solar flares and coronal mass ejections cut off radio communications, and disrupted or damaged several military and commercial satellites. And damage from the Bastille Day flare (Fig. 7.1), on 14 July 2000, was mitigated by alerts and warnings to industry, the military, and space agencies.

There were the record-breaking Halloween storms of 18 October and 5 November 2003, when scores of coronal mass ejections traveled fast and wide through interplanetary space. During these Halloween storms, there were 11 flares of the most powerful X-class, including the most intense one ever recorded from the *GOES* series of spacecraft, 90 coronal mass ejections detected by *SOHO*, at least 16 shocks observed near the Sun, with 8 of them intercepted by spacecraft along the Sun–Earth line. Powerful solar energetic particle events arrived at Earth, where they were observed from the *ACE* and *GOES* spacecraft, including some of the highest proton fluxes on record. And intense geomagnetic storms were also measured, including 2 of the 12 biggest ones since records began in 1932.

The Halloween storms damaged 28 satellites, knocking two out of commission, and caused power blackouts in Sweden. Astronauts in the *International Space Station* were ordered into the aft portion of the station five times during this time interval. Power-grid operators on Earth modified their system operations to avoid damage and outages, and nuclear power stations reduced their power output or delayed power switching. Global positional system and satellite communication sys-

Fig. 7.1 Solar flare produces energetic particle storm. A powerful solar flare (*left*), occurring at 10 h 24 min Universal Time on Bastille day 14 July 2000, unleashed high-energy protons that began striking the *SOlar and Heliospheric Observatory*, abbreviated *SOHO*, spacecraft near Earth about 8 min later, continuing for many hours, as shown in the image taken on 22 h 43 min Universal Time on the same day (*right*). Both images were taken at a wavelength of 19.5 nm, emitted at the Sun by 11 times ionized iron, denoted Fe XII, at a temperature of about 1.5×10^6 K, using the Extreme-ultraviolet Imaging Telescope, abbreviated EIT, on the *SOlar and Heliospheric Observatory*, or *SOHO* for short. (Courtesy of the *SOHO* EIT consortium. *SOHO* is a project of international cooperation between ESA and NASA)

tems experienced difficulty, and airplanes were diverted to avoid dangerous radiation and communication outages.

The field of space weather is growing rapidly. A space-weather monograph, edited by Paul Song, Howard J. Singer, and George L. Siscoe, was published in 2001, and a well written book entitled *Storms from the Sun: The Emerging Science of Space Weather* was published by Michael Carlowicz and Ramon Lopez in 2002. Volker Bothmer and Ionannis A. Daglis published *Space Weather – Physics and Effects* in 2006; it is written by a team of international experts on topics that include space-weather impacts on the terrestrial atmosphere and magnetosphere, communications, power grids, spacecraft hardware and operations, and satellite navigation, as well as space-weather forecasting. An international journal devoted to the topic, and appropriately named *Space Weather*, was launched by the American Geophysical Union in 2003, with the help of the U.S. National Science Foundation. The serious student or curious reader may want to explore some of these topics in greater detail, and they should consult reviews prepared by professionals in the field (Focus 7.1).

Focus 7.1

Expert Reviews about Space Weather Topics

Professional solar astronomers and astrophysicists have reviewed important developments in our knowledge of space weather. In alphabetical order, they include Markus Aschwanden and co-worker's (2006) review of theoretical modeling for the *STEREO* mission, Edward L. Chupp's (1984, 1987) reviews of the observations and physics of gamma ray and neutron production during solar flares, Walter D. Gonzalez, Bruce T. Tsurutani, and Alicia D. Clau De Gonzales's (1999) review of the interplanetary origin of geomagnetic storms, Natchimuthuk Gopalswamy and co-worker's (2006b) review of the pre-eruptive Sun, John T. Gosling's (1996) review of co-rotating and transient solar wind flows in three dimensions, Stephen W. Kahler's (1992) review of solar flares and coronal mass ejections and Kahler's (2007) review of the solar sources of heliospheric energetic electron events, Berndt Klecker and co-workers (2006) review of energetic particle observations, Horst Kunow and co-worker's (2006) review of coronal mass ejections, Robert P. Lin's (1987) review of particle acceleration by the Sun, Glenn Mason and Trevor R. Sanderson's (1999) review of co-rotating interaction region associated energetic particles in the inner and middle heliosphere, Zoran Mikic and M. A. Lee's (2006) introduction to the theory and models of coronal mass ejections, shocks, and solar energetic particles, Donald V. Reames's (1999) review of particle acceleration at the Sun and in the heliosphere, and Rainer Schwenn and co-worker's (2006) review of coronal observations of coronal mass ejections.

The contributions of scientists working on specific topics are often presented at meetings or workshops and subsequently published in book form. They include *Coronal Mass Ejections*, edited by Nancy Crooker, Jo Ann Joselyn, and Joan Feynman (1997), a book of the same title edited by Horst Kunow in 2006, *Solar Eruptions and Energetic Particles*, edited by Natchimuthuk Gopalswamy, Richard Mewaldt, and Jarmo Torsti (2006a), and *SOHO-20: Transient Events on the Sun and in the Heliosphere*, edited by Bernhard Fleck, Joseph B. Gurman, Jean-Francois Hochedez, and Eva Robbrecht (2008).

And there are no less than five recent living reviews on related topics, at http://www.livingreviews.org. They are: Rainer Schwenn's (2006) review of *Space Weather: The Solar Perspective*; Tuija Pulkkinen's (2007) *Space Weather: Terrestrial Perspective*; on longer time scales, Joanna D. Haigh's (2007) *The Sun and the Earth's Climate* and Manuel Güdel's (2007) *The Sun in Time: Activity and Environment*, and Arnold O Benz's (2008) *Flare Observations*. Mark Moldwin has created a fascinating timeline of solar-terrestrial physics at http://measure.igpp.ucla.edu/solar-terrestrial-luminaries/timeline.html.

To fully understand space weather, we must first learn how a storm develops on the Sun, and our beginning investigations of the origin of solar flares and coronal mass ejections were discussed in the previous Sect. 6.10. We would also like to anticipate when a solar storm is likely to occur, and the next Sect. 7.2 describes some of the forecast possibilities. And like everyday weather forecasts here on Earth, we want to know if a storm is headed our way, how it changes while traveling toward us, when it will arrive, and how dangerous it is likely to be. Section 7.3 describes the hazardous solar energetic particles that are accelerated to high energy on the Sun or in interplanetary space and travel to the Earth, while their effects when impacting our planet are presented in Sect. 7.4. The concluding Sect. 7.5 considers longer-term effects of the Sun on Earth, on time scales of decades, centuries, and millennia, which we term Sun – climate. This concluding section also discusses human-induced global warming and the next astronomically induced ice age.

7.2 Forecasting Space Weather

What everyone wants to know is when a Sun storm is going to occur. Most of the fastest coronal mass ejections, with the largest amount of energy, come from magnetic explosions in active regions with sunspots, producing a flare in tandem with the ejections and often beginning with an erupting prominence or filament. So a good place to begin our space weather forecasts is to know when a threatening active region, with its sunspots and strong magnetic fields, is on the Sun.

Active regions appear more frequently near the maximum of the 11-year sunspot cycle, as do solar flares and coronal mass ejections. So long-term solar activity can be forecast in a general way using this cycle.

On a shorter time scale of weeks, we can use helioseismology to detect large solar active regions on the hidden backside of the Sun, with techniques introduced by Douglas C. Braun and Charles Lindsey in 2001 (also see Sect. 3.8). Since the solar equator rotates with a period of 27 days, when viewed from Earth, the detection of a magnetically complex and strong active region on the far side of the Sun can give more than a week's warning before it swings into view to threaten the Earth. Daily images of the unseen, far side of the Sun are available on the web at http://soi.stanford.edu/data/full_farside and http://gong.nso.edu.

Once a strong active region rotates into view, we know it's there, perhaps primed for an outburst like a dark storm cloud looming on the horizon. Its strong, complex magnetic fields can be detected in the photosphere using magnetograms, from the ground or space, and its intense extreme-ultraviolet and X-ray radiation can be monitored from spacecraft such as *SOHO, TRACE, Hinode*, and in the near future from the *Solar Dynamics Observatory*.

So we know when there is a threatening cloud on the horizon, so to speak, but we still do not know, with certainty, if and when the storm might occur, with an active region emitting a powerful solar flare and/or coronal mass ejection. But timely space weather forecasts can be linked to observations of the active-region magnetic fields, determining when they are sheared and twisted away from the potential, current-free state (also see Sect. 6.10). Moreover, there are also the occasional erupting filaments or prominences that are not associated with sunspots or active regions; they can also affect the Earth and additionally trace the magnetic field topology involved in the associated coronal mass ejection.

One indication of such a stressed magnetic configuration is the presence of a soft X-ray, active-region sigmoid. As demonstrated by Richard Canfield, Alphonse Sterling, and their colleagues in 1999 and 2000, such sigmoid regions are more likely to produce coronal mass ejections than non-sigmoid active regions. Pre-eruptive sigmoid patterns are apparently present in over half the coronal mass ejections, but there is not a one-to-one correspondence. Scientists are now using *Hinode* to study the detailed evolution of sheared magnetic fields in a sigmoid, hoping to understand the mechanisms that lead to solar outbursts from them – see for example the studies by David McKenzie and Richard Canfield and by Yingna Sun and colleagues in 2008.

Another approach uses magnetograms to assess the non-potentiality of active-region magnetic fields. In 2002, for example, David A. Falconer, Ronald L. Moore, and Gilmer Allen Gary demonstrated that the coronal mass ejection productivity of solar active regions is correlated with their overall non-potentiality, or their overall twist and shear, detected in vector magnetograms.

The signature of an immanent explosion might be found deeper down, under the photosphere. In 2005, Carolus J. Schrijver and colleagues used *TRACE* and *SOHO's* MDI to show that the non-potentiality, or magnetic free energy, in the coronal configuration of active regions is driven by flux emergence from below. And in the following year, Douglas Mason and co-workers used the techniques of local helioseismology to demonstrate that the strength of flares from active regions is correlated with the amount of circulating, sideways flows beneath them. Helioseismic inversions using GONG data also show the development of strong vorticity flows below sunspots prior to flare activity.

However, some regions that exhibit magnetic shear and twist never erupt, so contorted magnetism may be a necessary, but not sufficient condition, for solar flares or coronal mass ejections. And the Sun's sudden and unexpected outbursts often remain as unpredictable as most human passions. They just keep on happening, and even seem to be necessary to purge the Sun of pent-up frustration and to relieve it of twisted, contorted magnetism.

And to be honest, scientists have not solved the question of what exactly initiates a solar flare or coronal mass ejection, igniting the explosion from stressed coronal magnetic fields? They think the storms might be triggered when magnetized coronal loops are pressed together, driven by motions beneath them, meeting to touch each other, merging to break open the magnetic fields and release free magnetic energy. But no one has identified a signature that allows the prediction of exactly when such an outburst might occur. So far, we only have signs of a possible solar storm; it is something like seeing that dark storm cloud but not knowing if it is going to rain.

But even after a solar outburst has occurred, there is still time to take cover from some of the most damaging results. The exact warning time depends on the type of solar hazard, since they travel with different velocities and on various trajectories in space. Powerful solar flares can be detected just 8.3 min after they happen, in the time it takes for their intense radiation to travel from the Sun to just outside the Earth where *SOHO* or the *GOES* satellites are located. But once you see the radiation, it is too late, for it has already arrived.

In comparison to radiation, the forecast times for coronal mass ejections can be quite comfortable. The fastest ones are usually accompanied by intense flares, which can signal the mass ejection. Then, at a speed of about $1,000 \, \mathrm{km \, s^{-1}}$, it takes 42 h or 1.7 days for the coronal mass ejection to reach Earth. Such coronal mass ejections have a direct impact on the Earth, including intense geomagnetic storms and aurora (Sect. 7.4).

Of course, coronal mass ejections are not always accompanied by solar flares, and vice versa; indeed, the large majority of flares occur without ejecting substantial mass. But there are other ways to know when an Earth-approaching ejection has occurred on the Sun.

If it is headed toward our planet, a coronal mass ejection must occur on the visible solar disk, and such front-side halo events are seen in white-light coronagraphs, such as *SOHO's* LASCO. The actual source regions are most likely hidden by the coronagraph-occulting disk, but there are extreme-ultraviolet and X-ray signatures that can be detected with other instruments. Hugh S. Hudson and Edward W. Cliver summarized these non-coronagraphic manifestations in 2001, including the coronal X-ray dimming discussed in Sect. 6.9. Natchimuthuk Gopalswamy and colleagues have also reviewed possible signatures of an impending ejection in 2006.

Solar energetic particles, accelerated by solar flares or coronal mass ejection and interplanetary shocks, pose a real threat to unprotected astronauts and equipment in space. Although the high-energy electrons move about as fast as radiation, the more dangerous protons apparently move at slower speeds. As demonstrated by Arik Posner in 2007, the measured arrivals of 50 MeV protons are delayed on average 63 min from the first-arriving high-speed electrons. Solar electron monitoring can therefore give about an hour's advance warning of the arrival of energetic protons and heavier ions, which carry a significantly higher long-term risk to astronaut health than the electrons. For robotic missions, forecasts by the direct, in situ electron measurements can be used to switch off sensitive equipment, protecting both the instruments and their data.

Spacecraft that are situated at the first Lagrangian point, at 1.4×10^6 km from the Earth, provide measurements of the incoming solar energetic particles and the solar wind an additional 40 min to an hour before they reach the outer edges of the magnetosphere, traveling at solar wind velocities, so they can be used for forecasting with up to an extra hour's warning time. As an example, the *Advanced Composition Explorer*, abbreviated *ACE*, uses real-time observations to provide short-term forecasts of shock-accelerated, high-energy protons at http://www.srl.caltech.edu/ACE/ASC/rtsw.html, reducing the risk to astronauts and space hardware located closer to the Earth.

But what exactly are these threats, and how do they move through space from the Sun to the Earth? These are our next topics.

7.3 Solar Energetic Particles

7.3.1 Energetic Particles Accelerated by Solar Flares or Coronal Mass Ejection Shocks

For more than half a century, we have known that extraordinarily energetic particles can be hurled into space during a solar flare, with some of them striking the Earth. In the 1940s, for example, Scott Forbush and his colleagues used ground-based cosmic ray detectors to rather unexpectedly record transient increases in the number of energetic charged particles arriving at Earth after solar flares. Balloon and rocket observations in the late 1950s and early 1960s indicated that the most energetic particles detected near the Earth following solar flares are mainly protons. And since they seemed to be created at or near the Sun, they were dubbed solar cosmic rays. The other "galactic" cosmic rays also include energetic protons and heavier ions that rain down on the Earth, coming in all directions from interstellar space. They travel with even greater energy than the solar protons, but with lower flux (Table 7.1). Today, we use the term solar energetic particles to denote high-energy electrons, protons, or other ions arriving at the Earth from the Sun.

Solar energetic particle events, which can be associated with either coronal mass ejections or solar flares, are a major element of space weather. They can severely

Table 7.1 Energy and flux of protons arriving at Earth[a]

Source	Energy (MeV)	Flux (protons $m^{-2}\,s^{-1}$)
Cosmic rays	1,000	6×10^2
Coronal mass ejections	10	3×10^8
Solar flares	10	1×10^7
Solar wind	0.001	5×10^{12}

[a] An energy of $1\,\mathrm{MeV} = 10^6\,\mathrm{eV} = 1.6 \times 10^{-6}\,\mathrm{erg} = 1.6 \times 10^{-13}\,\mathrm{J}$

affect the health of unprotected astronauts traveling outside the Earth's protective magnetosphere, and they are capable of penetrating spacecraft to damage or disrupt sensitive technical systems. The strongest events produce radiation doses that might be lethal to astronauts fixing a spacecraft in outer space or taking a walk on the Moon or Mars. So there is danger blowing in the gusts and squalls of the Sun's winds.

Energetic particles coming at the Earth from the Sun have the same main ingredients as the steady, ever-flowing solar wind, but with much faster speeds and vastly greater energy. During solar flares, protons and electrons can be accelerated to speeds of more than 100 times that of the solar wind, achieving energies as high as 20,000 million electron volts (20 GeV) and 100 million electron volts (100 MeV), respectively; this is at least 10,000 times more energetic than solar wind particles. Because protons are 1,836 times as massive as electrons, and move with comparable speed, protons are far more energetic than electrons in both flares and the wind.

Observations of flaring hard X-ray continuum and gamma ray spectral lines, which are respectively produced by energetic electrons and protons hurled down into the Sun, have shown that intense flares can accelerate electrons up to hundreds of MeV and protons up to many GeV in energy. This radiation has been observed for decades, and so have the flare-associated protons and electrons, observed in situ with particle detectors aboard spacecraft stationed near the Earth or traveling in interplanetary space. In 1984 and 1987, Edward L. Chupp reviewed some of the early results obtained with instruments aboard the *Solar Maximum Mission*.

Only exceptionally rare and energetic solar protons are detected at the Earth's surface. They need energies of about 1 GeV to spiral around the terrestrial magnetic fields with a large enough radius to reach the ground. Lower-energy protons move along the terrestrial magnetic fields in a tighter spiral, and are channeled into the polar regions where they enhance the ionosphere.

Energetic electrons produced during solar outbursts have been inferred from ground-based radio astronomical observations. In 1963, John Paul Wild, Stephan F. Smerd, and A.A. Weiss reviewed the pioneering radio investigations of the 1950s, inferring two separate phases of electron acceleration. The impulsive type III bursts were attributed to electrons moving at with energies of 10–100 keV. In large flares, electrons were supposed to be additionally accelerated to higher energies of up to 1 GeV, by shocks that propagated away from the Sun; and these shocks were also supposed to produce type II radio bursts. Radio astronomers spotted these electrons leaving the Sun, stimulating oscillations of lower and lower frequency as they passed through and jostled the progressively more rarefied coronal atmosphere, with a fast drift for the type III bursts and a slower drift for type II radio bursts (Sect. 6.6).

Energetic charged particles generated during solar flares will only threaten our planet if they occur at just the right place on the Sun, at one end of a spiral magnetic field line that connects the flaring region to the Earth. Given the right circumstances, with a flare near the west limb and the solar equator, the magnetic spiral acts like an interplanetary highway that connects the flaring electrons to the Earth. Spacecraft observations of type III radio bursts have been used to track the electrons as they move away from the Sun, confirming that open magnetic field lines connect flares directly into the interplanetary medium. And studies combining *Helios 1* and *2* and *IMP 8* observations have demonstrated that the spacecraft with a spiral magnetic connection closest to the flare nearly always detects the highest intensity of energetic particles.

Moving along its spiral magnetic conduit at about half the speed of light, a 100-keV electron generated during a type III radio burst travels from the Sun to the Earth in about 20 min. These interplanetary electrons have been directly sampled by in situ measurements using interplanetary probes and Earth-orbiting satellites. In 1965, James A. Van Allen and Stramatios M. Krimigis, for example, used a detector on NASA's Mars-bound spacecraft *Mariner 4* to observe impulsive bursts of high-speed electrons in deep space, with energies greater than 40 keV.

Direct particle detections from spacecraft in the 1970s indicated that the relative abundances of solar energetic particles arriving at the Earth depend on the initiating event at the Sun or in interplanetary space. In 1970, for example, Robert P. Lin presented evidence obtained from the *Interplanetary Magnetic Platform*, abbreviated *IMP*, satellites in the 1960s, to propose two separate electron acceleration and/or emission mechanisms. One of them produces electrons with about 40 keV in energy, or more, and is accompanied by the type III radio bursts; the other produces relativistic MeV electrons, and is associated with solar proton events and type II bursts.

And in the same year, Ke Chiang Hsieh and John A. Simpson used observations from the fourth *IMP* satellite to discover solar flare events with an enhanced abundance of the rare helium isotope ^{3}He, which was at least a hundred times more abundant in solar flares than in the solar corona. Gordon J. Hurford and co-workers then used instruments aboard *IMP 7* to show, in 1975, that ^{3}He-rich solar flares also exhibit enhancements of heavy ions like iron, by about an order of magnitude. This heavy-ion enhancement of ^{3}He-rich solar particle events was confirmed and extended in 1986 by Glenn M. Mason and colleagues using an instrument aboard the third *International Sun-Earth Explorer*, or *ISEE-3* for short. But there were other solar energetic particle events, not necessarily associated with flares, which did not exhibit the abundance enhancements, just as there were some other events that mainly consisted of abundant high-energy protons. And this suggested that although some of the particle events arriving at Earth were accelerated in solar flares, different processes might accelerate other ones, either at the Sun or within interplanetary space.

Interplanetary shocks do provide an alternative mechanism for particle acceleration, as anticipated by Thomas Gold in 1955 and 1959. His suggestions involved closed magnetic fields ejected from a region of the corona that did not previously contribute to the solar wind, and which generated shocks as they moved into interplanetary space. Such shock waves are not blast waves from flares, but are instead driven by magnetic structures ejected from the Sun, with shocks developing between the ejected material and the interplanetary medium already present. And according to Gold, the expanding magnetic loops might even remain attached to the Sun at both ends while moving all the way to the Earth.

When coronal mass ejections were discovered in the 1970s, it was realized that they might develop the interplanetary shocks proposed by Gold in the 1950s and detected from spacecraft in the 1960s. And comparisons of coronagraph data with type II radio bursts, detected from space at low frequencies, suggested that energetic interplanetary shocks were indeed being propelled by fast coronal mass ejections.

In coronagraph images, the CMEs look like big magnetic bubbles, with no exceptionally high-speed particles in sight. And observations of type II radio bursts associated with CMEs suggest relatively slow shock waves moving at speeds of $1,000 \mathrm{km\ s}^{-1}$ or less. So you do not "see" the acceleration of high-energy particles very near the Sun; the particles are instead accelerated and detected further out in space after the shocks have left the Sun.

And in contrast to the energetic particles hurled out from solar flares, which follow the interplanetary magnetic spiral, a coronal mass ejection can move right through the interplanetary magnetic field, hardly noticing it and continuously pumping up the energy of particles and accelerating them all the way from the Sun to the Earth.

By 1985, Neil R. Sheeley Jr. and co-workers were able to conclusively demonstrate that coronal mass ejections are producing interplanetary shocks. They combined *P78-1* coronagraph observations of coronal mass ejections with *Helios 1* observations of interplanetary shocks, showing that nearly all of the shocks were associated with fast mass ejections. And within a decade or two, sophisticated instruments aboard *SOHO* were used to directly detect the shock waves as they were being driven by the coronal mass ejections, by John C. Raymond and colleagues in 2000, Angelous Vourlidas and co-workers in 2003, and Angela Ciaravella and colleagues in 2005–2006.

Strong mass ejections plow into the slower-moving solar wind, like a car out of control, serving as pistons to drive huge shock waves millions of kilometers ahead of them (Fig. 7.2). The shock waves propelled by coronal mass ejections carry along electrons and ions in the interplanetary medium, crossing magnetic field lines and accelerating particles as they go, much as ocean waves propel surfers. When slow coronal mass ejections move outward into interplanetary space, with speeds below that of the ambient solar wind, no shock waves are generated and energetic solar particle events are not observed.

It is generally thought that metric type II radio bursts, observed from radio telescopes on the ground, are due to the electrons accelerated at an outward propagating coronal shock front near the Sun, within a few solar radii. Another kind of solar radio burst, called the kilometric type II radio burst and only observable from space, is attributed to the electrons accelerated at interplanetary shock waves between the Sun and the Earth. Some argue that both the metric and the kilometric bursts are produced by CME-driven shocks. Others, such as William J. Wagner and Robert M. MacQueen in 1983, have proposed that the type II radio bursts observed at meter wavelengths could stem from the coronal shocks driven by flares or blast waves near the Sun, but the interplanetary shocks signaled by the kilometric type II bursts, detected from spacecraft such as *Wind*, are driven by CMEs.

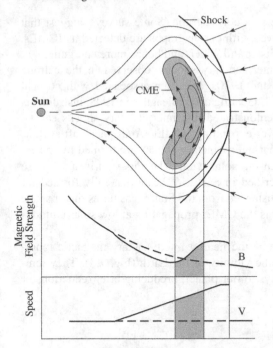

Fig. 7.2 Interplanetary CME shock. As it moves away from the Sun (*top left*), a fast coronal mass ejection (CME, *top right*) pushes an interplanetary shock wave before it, amplifying the solar wind speed, V, and magnetic field strength, B (*bottom*). The CME produces a speed increase all the way to the shock front, where the wind's motion then slows down precipitously to its steady, unperturbed speed. Compression, resulting from the relative motion between the fast CME and its surroundings, produces strong magnetic fields in a broad region extending sunward from the shock. The strong magnetic fields and high flow speeds commonly associated with interplanetary disturbances driven by fast CMEs are what make such events effective in stimulating geomagnetic activity

Whatever the detailed connection with type II radio bursts might be, the available evidence indicates that solar particle events that contain abundant, threatening high-energy protons arise from coronal mass ejections that are moving fast enough to generate interplanetary shocks, usually with speeds greater than $750\,\mathrm{km\,s^{-1}}$ – see for example Donald V. Reames, Stephen W. Kahler, and Chee K. Ng's article in 1997 and the review by Berndt Klecker and colleagues in 2006.

7.3.2 Interplanetary Coronal Mass Ejections, Interplanetary Shocks, and Magnetic Clouds

When a coronal mass ejection, or CME, travels out into interplanetary space, it is often called an interplanetary coronal mass ejection, abbreviated ICME, to distinguish it from the CME seen near the Sun in a coronagraph. Some surveys suggest that only about 10% of the CMEs observed with coronagraphs are detected as ICMEs. These mass ejections are generally faster and wider, and hence more energetic.

As we have previously noted, interplanetary shocks, observed in the ecliptic at low solar latitudes in the 1960s and 1970s, are often associated with coronal mass ejections. As reported by John T. Gosling and colleagues in 1998, instruments aboard *Ulysses* have additionally demonstrated that interplanetary shocks can be formed at high solar latitudes by "over-expanding" CMEs. And spacecraft such as *Voyager 1* and *2* have also shown that interplanetary shocks are formed by the interaction of fast and slow streams in the solar wind. The shocks driven by these co-rotating interaction regions, described in Sect. 5.7.2, are generally formed far beyond the Earth's orbit, at greater distances from the Sun. So as far as major space weather at the Earth is concerned, it is the CMEs propagating at low solar latitudes that are of dominant interest.

Spacecraft have been used to probe the nearby interplanetary magnetic topology at the time the ejecta arrive in the vicinity of the Earth (Fig. 7.3). They temporarily block the flow of cosmic rays to our planet, producing brief reductions or

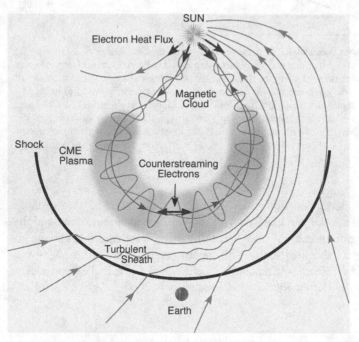

Fig. 7.3 Magnetic cloud. When a coronal mass ejection travels into interplanetary space, it can create a huge magnetic cloud containing bidirectional, or counter-streaming, beams of electrons that flow in opposite directions within the magnetic loops that are rooted at both ends in the Sun. The magnetic cloud also drives an upstream shock ahead of it. Magnetic clouds are only present in a subset of observed interplanetary coronal mass ejections. (Courtesy of Deborah Eddy and Thomas Zurbuchen)

depressions in the cosmic ray density of up to 70%. This suggests that the interplanetary ejecta are predominantly closed magnetic structures. As suggested by Philip Morrison in 1954, long before the discovery of coronal mass ejections, ionized magnetic clouds emitted by the active Sun will deflect cosmic rays from an Earth-bound path, thereby accounting for the observed worldwide decreases in cosmic ray intensity, lasting for days and correlated roughly with geomagnetic storms. We now realize that these clouds can originate at the Sun as coronal mass ejections.

Other evidence, reviewed by Ian G. Richardson in 1997, includes the bidirectional particle flows present in regions of some ejections, which is consistent with particle circulation and reflection, or mirroring, within closed magnetic structures. The observations of solar particle event onsets inside some ejections suggest that their magnetic field lines are rooted at the Sun. So, at least some of the mass ejections travel out into interplanetary space as looped magnetic fields that remain rooted in the Sun at both ends, producing a sort of magnetic bottle as they move into space, much as Gold had predicted in 1959.

As a coronal mass ejection expands into interplanetary space, it often takes the form of a magnetic cloud that contains a well-organized, twisted magnetic flux tube. Leonard F. Burlaga and colleagues discovered one in 1981, using the data from five spacecraft (*Voyager 1* and *2, Helios 1* and *2,* and the eighth *Interplanetary Magnetic Platform,* or *IMP 8*) to describe a magnetic cloud moving behind an interplanetary shock as it expanded away from the Sun. This cloud was characterized by magnetic loops that attained a radial extent of about half the distance between the Earth and the Sun. The following year, Burlaga and co-workers showed that an interplanetary magnetic cloud, observed from *Helios 1,* was associated with a coronal mass ejection observed with the Solwind coronagraph on the *P78-1* spacecraft. Also in 1982, Larry W. Klein and Burlaga described 45 interplanetary magnetic clouds arriving at Earth's orbit, some associated with CMEs, and by 1984 Robert M. Wilson and Ernest Hildner presented striking evidence linking magnetic clouds with coronal mass ejections. Four years later, Guowei Zhang and Burlaga described the basic characteristics of magnetic clouds, including their ability to block the flow of cosmic rays toward the Earth and to produce geomagnetic activity. A magnetic cloud observed at the Earth's orbit is defined by the following characteristics relative to the background solar wind: enhanced magnetic field strength, a smooth rotation of the field direction through a large angle, and a low proton temperature with a low ratio of particle pressure to magnetic pressure. The clouds can arrive behind interplanetary shocks and can extend in the radial direction from the Sun for more than half its distance from the Earth.

Such clouds can twist and spiral out, like a giant corkscrew, carrying intense, helical magnetic fields with them. As described by Volker Bothmer and Rainer Schwenn in 1994, and independently by David Rust in the same year, the twisted, helical fields in the expanding flux ropes of interplanetary magnetic clouds are comparable to those of the erupting prominences, or filaments, from which they arise. Moreover, in 2004, Jarmo Torsti, Esa Riihonen, and Leon Kocharov used an instrument aboard *SOHO* to detect energetic protons inside a magnetic cloud, associated with a previous coronal mass ejection, and demonstrated that the magnetic flux-tube structure of the CME provided a "highway" for the transport of solar energetic protons.

But single spacecraft observations suggest that as few as one-third of the interplanetary coronal mass ejections observed near Earth are characterized as magnetic clouds. The other two-thirds could exhibit magnetically complex ejecta, which could be attributed to the interaction of coronal mass ejections moving at different speeds. As described by Natchimuthuk Gopalswamy and colleagues in 2001–2002, radio signatures suggest that a fast CME can overtake a slow one, cannibalizing it and enhancing the acceleration of solar energetic particles from its material. Isolated magnetic clouds nevertheless appear to be more effective than complex ejecta in producing geomagnetic storms.

7.3.3 Observations of Solar Energetic Particles with Contemporary Spacecraft

In recent decades, instruments aboard several spacecraft have been used to examine the detailed properties of solar energetic particles. They have determined their intensity-time profiles, energy spectra, and elemental, isotopic, and charge composition. These observations help to determine the sources of the solar energetic particles, the physical processes that accelerate them, and how they escape and propagate through space, all crucial information for protecting spacecraft and astronauts.

NASA's *Advanced Composition Explorer*, abbreviated *ACE*, and *Wind* spacecraft have been keeping a careful watch on the vast and shifting web of subatomic particles as they bombard the Earth, inferring their origin and subsequent transformations both at the Sun and in interplanetary space. They also measure conditions in the solar wind at the Earth's orbit, including its magnetic fields and shocks.

The source of the threatening solar energetic particles arriving at Earth has also been investigated by combining *ACE* and *Wind* observations with those of spacecraft that look back at the Sun, such as the *SOlar and Heliospheric Observatory*, or *SOHO* for short, and the *Ramaty High Energy Solar Spectroscopic Imager*, abbreviated *RHESSI*, to investigate the solar origin of the particles. Comparisons of *ACE* observations of ^{3}He-rich impulsive solar energetic particles with solar images from *SOHO*, by Yi-Ming Wang, Monique Pick, and Glenn M. Mason in 2006, indicate that every one of the impulsive events observed over a 6-year period originated from a small flaring active region that lies next to a coronal hole containing Earth-directed open magnetic field lines. The impulsive particles are interpreted in terms of magnetic reconnection, or footpoint exchange, between closed and open field lines.

And when the *ACE* and *Wind* detections are compared to *RHESSI* observations of particles accelerated in large solar flares, it is found that the flares are not responsible for the most energetic solar particles arriving at Earth. As reported by Robert P. Lin in 2005, they are most likely accelerated by the interplanetary shocks driven by fast coronal mass ejections. The relative roles of flare acceleration and shock acceleration are nevertheless still controversial, and particles accelerated in modest flares still appear to be similar to weaker concentrations of energetic particles arriving near Earth, which are rich in electrons, ^{3}He, and heavy ions. And in some situations there can also be difficulty in distinguishing between direct flare acceleration and acceleration by CME-driven or interplanetary shocks.

Statistical timing comparisons of impulsive type III radio bursts and electron events observed at the Earth's orbit from the *Wind* spacecraft, by Säm Krucker and co-workers in 1999, indicated that the radio burst onset coincided with the electron release times, suggesting a common acceleration mechanism. However, additional comparisons with *Wind* and *ACE* detections, by Dennis K. Haggerty and Edmond C. Roelof in 2002, and George M. Simnett, Roelof, and Haggerty in 2002, indicated that some near-relativistic electron events, with energies of 40–300 keV, were released about 10 min after their associated radio or soft X-ray emission. This led to controversial proposals that these electrons arrived at Earth after being accelerated by shocks associated with coronal mass ejections. But as reviewed by Stephen W. Kahler in 2007, the acceleration of the near-relativistic electron events usually occurs in solar flares, while the CME-driven shocks are the dominant acceleration mechanism for the most energetic electron events, with relativistic MeV energies.

And an additional caveat resulting from *ACE* observations, as reviewed by Mihir I. Desai and colleagues in 2006, suggests that solar energetic particles arriving at the Earth are not accelerated by CMEs from the bulk solar wind, but instead consist of suprathermal ions that have been further accelerated to higher energies by the interplanetary shocks. That is, the shocks preferentially accelerate these exceptionally hot, suprathermal nuclei, discussed in Sect. 5.7.4, which have already been pre-accelerated by some currently unknown process.

7.3.4 Two Classes of Solar Energetic Particles

Summing up 8.5 years of energetic-particle observations from *ISEE-3* in 1988, Donald V. Reames argued for two classes of solar energetic particle events based on their probable acceleration mechanism, the particle composition when arriving at the Earth, and the duration of the associated radio or soft X-ray emission at the Sun. The two categories of solar energetic particle events are known as impulsive and gradual events, a designation that may have been derived from Roberto Pallavicini, Salvatore Serio, and Giuseppe Vaiana's use of *Skylab* data to define, in 1977, two classes of soft X-ray flares based on their duration – compact, brief (minutes) impulsive events, and gradual, long-enduring (hours) ones associated with soft X-ray arcades, filament eruptions, and coronal mass ejections. But the classification has been extended to include differences in composition and acceleration sites for the solar energetic particles detected near Earth (Table 7.2).

The impulsive solar particle events are associated with rapid, impulsive solar flares and accompanied by type III radio bursts. They are dominated by near-relativistic electrons and have relatively weak ion abundances, albeit greatly enriched

Table 7.2 Properties of impulsive and gradual solar energetic particle events[a]

	Impulsive events	Gradual events
Particles	Electron-rich	Proton-rich
$^3\mathrm{He}/^4\mathrm{He}$	≈ 1	≈ 0.0005
Fe/O	≈ 1	≈ 0.1
H/He	≈ 10	≈ 100
Duration of X-ray flare	Impulsive (minutes, hard X-rays)	Gradual (hours, soft X-rays)
Duration of particle event	Hours	Days
Radio bursts	Types III and V[b]	Types II and IV
Coronagraph	Nothing detected	Coronal mass ejections, 96%
Acceleration site	Solar flares	Interplanetary shocks
Solar wind	Energetic particles	Very energetic particles
Longitudinal extent	$< 30°$	$\approx 180°$
Events per year	$\approx 1,000$	≈ 100

[a]Adapted from Gosling (1993) and Reames (1997)
[b]Impulsive type III and V bursts can be followed by type II and IV

in ^{3}He and heavy ions, with charge states characteristic of the high temperatures of up to 10×10^6 K found in impulsively heated solar flares. The impulsive events are the most commonly observed solar events detected at the Earth's orbit, occurring roughly 100 times more frequently than gradual ones. They can last for hours, at the Earth's orbit, are not associated with coronal mass ejections, and extend over a limited region in solar longitude of less than 30°. The narrow excursion in solar longitude is attributed to acceleration of the charged particles in localized flares on the Sun, which then follow a well-defined magnetic pathway along the spiral interplanetary magnetic fields.

The gradual events are associated with long-lasting (hours) soft X-ray emission, and are accompanied by coronal mass ejections and by type II radio emission. They are responsible for most of the large proton events seen at Earth, and are probably accelerated from ambient material by coronal and interplanetary shocks. The smooth, extended time profile of the gradual solar particle events, which last for days in the vicinity of the Earth, comes from continuous acceleration by shocks moving away from the Sun to the Earth. These gradual events often spread over more than 180° of solar longitude, which is attributed to large-scale shock waves that can easily propagate across magnetic field lines. Although fast coronal mass ejections are usually accompanied by solar flares, any flare-accelerated particles would follow the nearest spiral magnetic field lines and could not be transported to the very distant magnetic field lines where the gradual solar particle events are observed.

Of course, for every rule there is an exception, and the separation of solar energetic particle events into two distinct classes may be an oversimplification. There are no doubt cases that blur the distinctions, but they remain useful for a first-order appraisal of the events.

So coronal mass ejections, or CMEs, now seem like the most dangerous culprit as far as the high-speed proton events go. And as we shall next see, the shock waves, ejected mass, and magnetic fields associated with coronal mass ejections can produce intense geomagnetic storms, create power surges in terrestrial transmission lines, and trigger brilliant auroras in the polar skies. And astronomers are now developing ways to follow their trajectories in space using multiple spacecraft, numerical simulations, and theoretical considerations, with the hope of predicting threatening dangers coming along their path (Focus 7.2).

Focus 7.2

Tracking Solar Storms in Space

The combination of data from space-based instruments and ground-based telescopes allows us to track solar outbursts from their beginning on the Sun, to their passage through the interplanetary medium, and in some cases on to their ending impact at Earth. Instruments aboard *SOHO* and *Hinode*, for example, monitor the varying magnetic fields in the photosphere, as well as the extreme-ultraviolet or X-ray emission from solar flares. Coronagraphs aboard *SOHO* and the twin *STEREO A* and *B* spacecraft routinely detect coronal mass ejections, abbreviated CME. And the progress of solar flare electrons and CME-driven shocks can be followed into space by type III and type II radio bursts, respectively – by ground-based radio telescopes near the Sun and from the ACE, *Wind, Ulysses,* and *STEREO* spacecraft further out in interplanetary space. The spacecraft observations of these radio bursts can be triangulated to specify their three-dimensional trajectory. Near the Earth's orbit, *GOES* monitors flaring X-ray radiation and energetic particles, while *ACE* and *Wind* measure the arrival of interplanetary shocks, magnetic clouds, and solar energetic particle events. Down on Earth, routine magnetic field measurements detect geomagnetic storms.

The combined set of measurements, taken at a multitude of points in space, specify how energy is generated on the Sun and transferred to the space near Earth. They provide the boundary conditions for theoretical models and numerical simulations that describe the initiation of solar flares or coronal mass ejections, and the evolution and propagation of the CMEs with their associated interplanetary shocks and particle acceleration. In 2006, Markus J. Aschwanden and colleagues, Terry G. Forbes and co-workers, and Zoran Mikic and M. A. Lee provided reviews of some of the relevant theories and models. Models of the CME initiation include magnetic shearing, kink instability, filament eruption, and magnetic reconnection in the flaring lower corona, while the modeling of CME propagation entails interplanetary shocks, interplanetary particle acceleration and beams, solar energetic particles, and geoeffective connections.

The magnetohydrodynamic, or MHD, modeling and numerical simulation of flares, coronal mass ejections, and interplanetary shocks were pioneered by Murray Dryer, who has reviewed early work in 1974 and provided subsequent accounts of later developments, including one in 1998 when real-time, continuous three-dimensional, numerical MHD simulations were being used to forecast solar-interplanetary space weather when solar flares or CMEs were observed – for U.S. Air Force Space Weather interests. Several other groups, including one at the Center for Space Environment Modeling at the University of Michigan, are currently developing three-dimensional models. They have developed an end-to-end Space Weather Modeling Framework, abbreviated SWMF, that includes everything from the solar corona, eruptive filaments, inner heliosphere, solar energetic particles, global magnetosphere, inner magnetosphere, radiation belts, ionosphere electron dynamics and the upper atmosphere in a high-performance coupled model – see http://csem.engin.umich.edu/swmf/. Integrated space weather modeling is also described at http://www.bu.edu/cism.

In 2007, Gábor Tóth and co-workers used the SWMF to perform the first Sun-to-thermosphere simulation of a real space storm, of 28–30 October 2003. And in 2004 Murray Dryer and colleagues described real-time shock arrival predictions during the Halloween storm period. When combined with observations from space and the ground, these models and numerical simulations provide the framework for tracking the movements of future space storms and forecasting their possible dangers at Earth or in outer space.

7.4 Impacting Planet Earth

7.4.1 Our Protective Magnetic Cocoon

Our planet is immersed within the hot, gusty, electrically charged solar wind that blows out from the Sun in all directions and never stops, carrying with it a magnetic field rooted in the Sun. Fortunately, we are protected from the full force of this relentless, stormy gale by the Earth's magnetic field.

As demonstrated by William Gilbert in 1600, the Earth is itself a great magnet. The magnetic fields emerge from the south geographic pole, loop through nearby space, and re-enter at the north geographic pole, so the north geographic pole corresponds to the south magnetic pole and vice versa. The Earth's magnetic field has an intensity of about 50,000 nT in the polar regions, where the magnetic fields bunch close together, and about 30,000 nT above the equator where the magnetic fields spread out. And because charged particles do not cross magnetic field lines, and instead move around and along them, most of the solar wind is diverted around our planet at a distance far above the atmosphere, like a stream flowing around a rock or air deflected around the windshield of a car. Magnetic clouds associated with coronal mass ejections have magnetic field strengths of just 15–30 nT, so the much stronger terrestrial magnetic field also provides a good protection from them.

Although the solar wind is exceedingly rarefied, far less substantial than a terrestrial breeze, it possesses the power to bend and move things in its path. Measurements from NASA's first *Interplanetary Monitoring Platform*, or *IMP 1*, in the mid-1960s showed that the never-ending flow from the Sun shapes the Earth's magnetic cocoon into the form of a comet or a teardrop (Fig. 7.4). It produces a

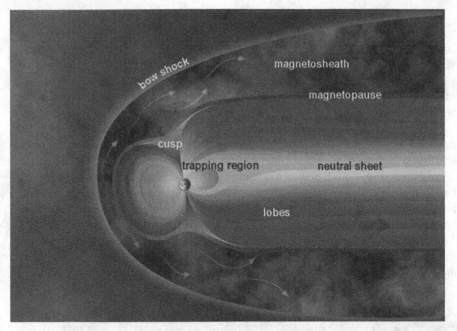

Fig. 7.4 Asymmetric magnetosphere. The Earth's magnetic field carves out a hollow in the solar wind, creating a protective cavity, called the magnetosphere (*blue*). It is sculpted into an asymmetric shape by the solar wind, with a bow shock that forms at about 10 Earth radii on the sunlit, day-side facing the Sun (*left*). The location of the bow shock is highly variable since it is pushed in and out by the gusty solar wind. The magnetopause marks the outer boundary of the magnetosphere, at the place where the solar wind takes control of the motions of charged particles. The solar wind is deflected around the Earth, pulling the terrestrial magnetic field into a long magnetotail on the night side (*right*). The red regions in the inner magnetosphere contain both the ring current and the outer Van Allen belt, where electrons, protons, and other ions are trapped on closed drift paths. (Courtesy of ESA.)

shock wave, called a bow shock, when first encountering the Earth's magnetism. Under typical conditions the bow shock is located at about 10 Earth radii upstream, or upwind, of the Earth, but very strong wind gusts can push the bow shock in to about half that distance. The relentless wind drags and stretches the terrestrial magnetic field out into a long magnetotail on the night side of Earth. The stretched-out, terrestrial magnetic field points roughly toward the Earth in the northern half of the tail and away in the southern half. And the field strength drops to nearly zero at the center of the tail, where the opposite magnetic orientations meet in a neutral current sheet.

Thus, the terrestrial magnetic field hollows out a cavity in the solar wind, called the magnetosphere, bounded at the manetopause where the plasma and magnetic fields of the solar wind and magnetosphere are in pressure balance. The Earth's magnetosphere is not precisely spherical, so the term magnetosphere does not refer to form or shape. It instead implies a sphere of influence. The magnetosphere of the Earth, or any other planet, is that region surrounding the planet in which its magnetic field dominates the motions of energetic charged particles such as electrons, protons, and other ions.

Particles in the solar wind transport only one ten-billionth the energy of that carried by sunlight, and Earth is protected from the full blast of even this dilute, varying solar wind by the terrestrial magnetosphere. The Earth's magnetic shield is so perfect that only 0.1% of the mass of the solar wind that hits it manages to penetrate inside. Yet, even that small fraction of the wind particles has a profound influence on the Earth's nearby environment in space. They create an invisible world of energetic particles and electric currents that flow, swirl, and encircle the Earth.

Some charged particles flowing from the Sun can enter the Earth's magnetic domain and become trapped within it. They can be stored along the stretched dipolar field lines, earthward of the tail magnetic connection site, in a region called the plasma sheet (Fig. 7.5). It acts as a holding tank of electrons and ions, suddenly releasing them when stimulated by the ever-changing Sun. Nearer the Earth particles are stored in the radiation belts, donut- or torus-shaped regions of unexpectedly high flux of high-energy electrons and protons that girdle the Earth at roughly 2 and 3 Earth radii in the equatorial regions. These places are sometimes called the inner and outer Van Allen radiation belts, named after James A. Van Allen, whose instruments aboard the *Explorer 1* and *3* satellites first observed them in 1958; they have been

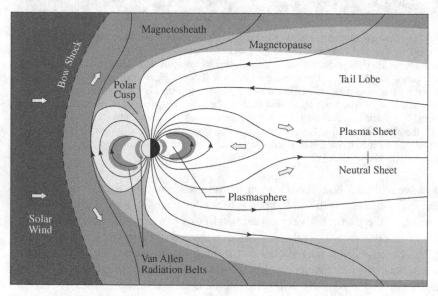

Fig. 7.5 Elements of the magnetosphere. The Earth's magnetic field carves out a hollow in the solar wind, creating a protective cavity, called the magnetosphere. A bow shock forms at about 10 Earth radii on the sunlit side of our planet. The location of the bow shock is highly variable since it is pushed in and out by the gusty solar wind. The magnetopause marks the outer boundary of the magnetosphere, at the place where the solar wind takes control of the motions of charged particles. The solar wind is deflected around the Earth, pulling the terrestrial magnetic field into a long magnetotail on the night side. Plasma in the solar wind is deflected at the bow shock (*left*), flows along the magnetopause into the magnetic tail (*right*), and is then injected back toward the Earth within the plasma sheet (*center*). The Earth, its auroras, atmosphere and ionosphere, and the two Van Allen radiation belts all lie within this magnetic cocoon

dubbed radiation belts since the charged particles that they contain were known as corpuscular radiation at the time of their discovery. The nomenclature is still used today, but it does not imply either electromagnetic radiation or radioactivity. The Van Allen radiation belts mainly consist of high-energy electrons and protons that bounce back and forth between polar mirror points in seconds to minutes and drift around the planet on time scales of hours. A ring current also circulates around the Earth just outside the Van Allan belts. Particles in the ring current and the Van Allen belts originate from both the external solar wind and the ionosphere down below.

Although the solar wind never actually reaches the Earth's surface, it can cause dramatic changes in the Earth's magnetic field. Under constant buffeting by the varying solar wind, the terrestrial magnetic fields are buckled, distorted, and reshaped, producing invisible magnetic storms far above any rain, sleet, or snow. These large and rapid variations in the Earth's magnetic field produce wide, irregular movements in the direction that compass needles point.

7.4.2 Earth's Magnetic Storms

As the solar wind and coronal mass ejections brush past the Earth, they carry some of the Sun's magnetic field with them. And since magnetic fields have a direction, their magnetism can point toward or away from the direction of the Earth's magnetic field. As postulated by James Dungey in 1961, a small portion of the solar wind can gain entry into the magnetosphere through the magnetopause when the magnetic field in the solar wind points in the opposite direction to the magnetosphere fields. With this orientation, they can join each other and become linked, just as the opposite poles of two toy magnets stick together. The magnetic field in the solar wind is then broken, and reconnects with the terrestrial magnetic field.

So geomagnetic activity is primarily driven by magnetic reconnection between the interplanetary magnetic field and the terrestrial magnetic field. That may happen if the northward pointing Earth field on the front of the magnetosphere is hit by solar wind carrying southward pointing interplanetary magnetic fields. They can connect and merge together at the place they touch, and as a result, the solar wind energy, momentum, and mass can interact with the magnetosphere.

In 1966, Donald H. Fairfield and Larry J. Cahill Jr. compared *Explorer 12* measurements of the magnetic field outside the magnetosphere to ground magnetograms from arctic observations to show that exterior fields with a southerly component tend to be associated with a magnetic disturbance detected on the ground. A northward exterior field was associated with magnetically quiet conditions. These results were confirmed and extended by comparisons of magnetic variations on Earth with interplanetary magnetic field measurements by *Explorer 33* and *35*, published by Joan Hirshberg and David S. Colburn in 1969, and by Rande K. Burton, Robert L. McPherron, and Christopher T. Russell in 1975.

Semi-annual variations in geomagnetic activity have been attributed to the change in the orientation of the Earth's dipole axis relative to the Sun–Earth line over the course of a year, which causes a change in its orientation with respect to the interplanetary magnetic field. As demonstrated by Christopher T. Russell and Robert L. McPherron in 1973, the Earth's magnetic activity is largest when its dipole is tilted to increase the projection of the southward component of the interplanetary magnetic field on the geomagnetic field. In the following year, Russell, McPherron and Rande K. Burton provided other evidence that geomagnetic storms are associated with strong southward interplanetary magnetic fields, confirming Dungey's magnetic reconnection theory for the storms.

Energy transfer is most efficient when the reconnection takes place at the dayside of the magnetopause, when the interplanetary field points southward and is thus antiparallel to the intrinsic geomagnetic field. But when the magnetic coupling occurs, it can be dragged downstream all along the length of the Earth's magnetotail. Since the immense magnetic tail forms the bulk of the magnetosphere, it provides the main location for breaching the Earth's magnetic defense.

Dynamic processes that occur in the magnetotail's plasma sheet, where the energy entering the magnetosphere from the solar wind is focused, largely drive magnetosphere space weather events. This focusing action leads to structural changes in the neutral current sheet separating the oppositely directed magnetic fields in the magnetotail, including reconnection which can provide a back door entry that funnels some of the gusty solar wind into the magnetosphere (Fig. 7.6). The passing solar wind is slowed down by the connected fields and decelerates in the vicinity of the tail. Energy is thereby extracted from the nearby solar wind and drives a large-scale circulation, or convection, of charged particles within the magnetosphere.

Thus, while creating and sustaining the magnetotail, the solar wind brings the oppositely directed tail lobes into close contact, where they can merge together. The magnetotail then snaps like a rubber band that has been stretched too far. The snap catapults part of the tail downstream into space, creating a gust-like eddy in the solar wind. The other part of the tail, propelled by energy released in the magnetic

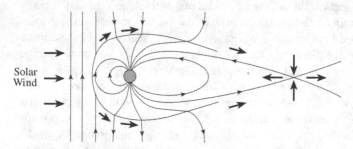

Fig. 7.6 Magnetic connection on the backside. The Sun's winds bring solar and terrestrial magnetic fields together on the night side of Earth's magnetosphere, in its magnetotail. Magnetic fields that point in opposite directions (*thin arrows*), or roughly toward and away from the Earth, are brought together and merge, reconnecting and pinching off the magnetotail close to Earth. Material in the plasma sheet is accelerated away from this disturbance (*thick arrows*). Some of the plasma is ejected down the magnetotail and away from the Earth, while other charged particles follow magnetic field lines back toward Earth

merging, rebounds back toward the Earth. The solar wind is then plugged into the Earth's electrical socket, and our planet becomes wired to the Sun. Electrons and ions hurtle along magnetic conduits that are connected to the Earth, linking the solar wind to both equatorial storage regions and down into the polar caps.

Events on the Sun leading to large perturbations of the Earth's magnetosphere are called geoeffective. And coronal mass ejections can be geoeffective, producing strong geomagnetic storms, primarily because they can bring to Earth strong southward magnetic fields of long duration. In fact, we now know that the most powerful geomagnetic storms, the great non-recurrent ones that shake the Earth's magnetic field to its very foundations, are caused by coronal mass ejections. But not every coronal mass ejection produces a great storm. The faster ones are more effective, and they also have to impact our planet with the right orientation. When the loop-like fields have strong components directed southward, they merge and reconnect with the Earth's magnetic fields that point to the north. With this alignment, the two fields become linked during a reconnection process, and this both triggers and energizes the most powerful geomagnetic storms.

Studies of Earth-directed, halo coronal mass ejections from *SOHO*, by Hilary V. Cane, Ian G. Richardson, and O. Christopher St. Cyr in 2000, indicate that their ability to produce geomagnetic storms depends on the southward magnetic field strength. Such coronal mass ejections are the dominant cause of major, intense, non-recurrent geomagnetic storms that occur most often near the maximum in the 11-year solar activity cycle.

But only about half of the front-side halo coronal mass ejections even encounter the Earth, and only about one-quarter of the ones that do manage to strike our planet result in noticeable geomagnetic activity. As we shall see, a CMEs effectiveness in disturbing terrestrial magnetism depends on the magnetic orientation of the CME; the solar wind speed and dynamic pressure are also factors.

The energy gained from the coronal mass ejection drives currents that create the intense magnetic storm. The Earth intercepts about 70 coronal mass ejections per year when solar activity is at its peak, and several of them will produce great magnetic storms and exceptionally intense auroras. And the cosmic flow of electric currents associated with the most intense, non-recurrent geomagnetic storms can even interfere with electrical power grids here on Earth, creating voltage surges on long-distance power lines (Focus 7.3). Strong aurora currents in the ionosphere can also induce currents in oil and gas pipelines, railway systems, and telecommunication cables, especially at high terrestrial latitudes where auroras are commonly seen.

Focus 7.3

Turning Off the Lights

The whole Earth has become wired together, first with telegraph wires, then by telephone lines and electrical-power grids. And when solar storms produce changes in the Earth's magnetism, disabling electrical currents and voltages can be produced in the wires. This threat is greatest in high-latitude regions where the currents are strongest, such as Canada, the northern United States, Scandinavia, and Russia.

Even back in the 1840s, when telegraph lines were first deployed, operators noticed extra current whenever overhead auroras signaled the presence of an intense geomagnetic storm. And about a century and a half later, on 13 March 1989, a particularly severe geomagnetic storm, produced by a coronal mass ejection, plunged virtually all of the Canadian province of Quebec into complete darkness, without warning and within a few seconds. The disturbed magnetic fields induced electric currents in the Earth's surface, which in turn created voltage surges on the long-distance power lines, blowing circuit breakers, overheating or melting the windings of transformers, and causing the massive electrical failure.

As demand for electricity increases, utility companies rely more and more on large, interconnected grids of power transmission lines that can span continents, providing rapid response to the diverse energy demands of users scattered throughout the globe. In the United States alone, nearly a million kilometers of electrical transmission lines connect more than 10,000 power stations. Such power distribution systems are threatened by severe geomagnetic storms, initiated when a coronal mass ejection with the right magnetic orientation makes contact with the magnetosphere. They can plunge major urban centers, like New York City or Montreal, into complete darkness, causing social chaos, threatening safety, and resulting in large losses in revenue. The threat does not occur very often, perhaps once a year, but the potential consequences are serious enough to employ early warning systems.

The clear-cut association between geomagnetic storms and some coronal mass ejections falls apart when moderate and low-level storms are considered, especially near the minimum in the 11-year solar activity cycle when the weaker storms are easier to detect. They can recur with a 27-day period corresponding to the apparent rotation period of the solar equator.

The moderate, recurrent geomagnetic storms have been associated with the high-speed solar wind emanating from coronal holes, but with an interplanetary origin and a geoeffectiveness that also depends on the southward component of the interplanetary magnetic field. In 1987, for example, Bruce T. Tsurutani and Walter Gonzalez attributed low-level geomagnetic substorms to interplanetary, co-rotating interaction regions, where the fast wind overtakes the slow one, but caused by the related compressed southward component of the interplanetary magnetic field striking Earth. Near cycle maximum, coronal mass ejections dominate the variable interplanetary medium, producing the most intense geomagnetic storms, and the low-level storms are less noticeable.

7.4.3 The Auroras – Cosmic Neon Signs

The great geomagnetic storms are also associated with intense aurora. At such times, energy is also expended in accelerating both the infiltrating solar wind particles and the local particles to make polar auroras. The accelerated electrons are guided along the Earth's magnetic fields into the upper atmosphere at the Earth's polar regions, generating spectacular northern and southern lights, named the *aurora borealis* and *aurora australis* in Latin. Residents in far northern locations can see the green and red lights shimmering far above the highest clouds every clear and dark winter night. Rare, brilliant auroras, associated with great magnetic storms, can even extend down toward the Earth's equator. Auroras have also been observed around the poles of Jupiter and Saturn, two other planets in our solar system that have extensive magnetospheres.

Energetic electrons bombarding the upper atmosphere principally cause the multi-colored aurora light show. As the electrons cascade down the polar magnetic field lines into the atmosphere, they are slowed down by collisions with the increasingly dense air, exciting oxygen and nitrogen atoms. The pumped-up atoms then give up the energy acquired from the electrons, emitting a burst of color and fluorescing like a cosmic neon sign.

Today, spacecraft look down on the aurora from high above the north polar region, showing the northern lights in their entirety as an immense aurora oval centered at the north magnetic pole (Fig. 7.7). Currents can be produced along the aurora ovals that are as strong as a million amperes. These currents flow down from the magnetosphere, through the ionosphere in the upper atmosphere, around the aurora oval, and back out and up to the magnetosphere.

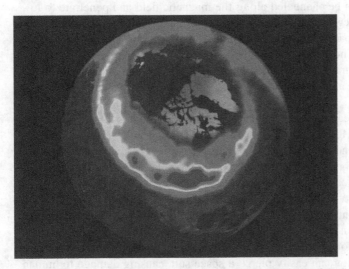

Fig. 7.7 The aurora oval. Instruments aboard the *POLAR* spacecraft look down on the aurora from high above the Earth's north polar region on 22 October 1999, showing the northern lights in their entirety. The glowing oval, imaged in ultraviolet light, is 4,500 km across. The most intense aurora activity appears in bright red or yellow, toward the night side of the Earth; it is typically produced by magnetic reconnection events in the Earth's magnetotail. The luminous aurora oval is constantly in motion, expanding toward the equator or contracting toward the pole, and always changing in brightness. Such ever-changing aurora ovals are created simultaneously in both hemispheres. (Courtesy of the Visible Imaging System, University of Iowa and NASA.)

Visual auroras normally occur in the ionosphere, at 100–250 km above the ground. This height is much smaller than either the average radius of the aurora oval, at 2,250 km, or the radius of the Earth, about 6,380 km. An observer on the ground therefore sees only a small, changing piece of the aurora oval, which can resemble a bright, thin, wind-blown curtain hanging vertically down from the Arctic sky.

The *Space Shuttle* has even flown right through the aurora. When inside the display, astronauts have closed their eyes and seen flashes of light by the beams of high-energy particles, which ripped through the satellite walls and passed through their eyeballs, making them glow inside. And as astronauts travel further out into space, they become immersed within a cosmic shooting gallery of solar energetic particles, which can threaten their health and even their lives.

7.4.4 High-Flying Humans at Risk

Energetic solar particles can pose a threat to people in airplanes flying along polar routes. The Earth's magnetic field deflects many of these particles, and the atmosphere usually absorbs all but the ones of greatest energy, which can gain access to its lower layers. However, the high-energy particles created by solar flares or coronal mass ejections can be channeled along the magnetic field and penetrate to low altitudes in the polar regions, exposing airline crews and passengers to elevated levels of particles from space. The higher the plane is flying and the closer to the poles, the greater the dose. The health risk is small, but most for frequent fliers, pilots and flight attendants who travel polar routes often. Pregnant women are advised to not take an airplane flying a polar route during a storm on the Sun, to avoid risk of birth defects.

There are even greater hazards aboard spacecraft at higher altitudes. Go far enough into space and the chemical bonds in your molecules will be broken apart by storms from the Sun, increasing the risk of cancer and errors in genetic information. Space agencies therefore set limits to the exposure to solar energetic particles and radiation an astronaut can have while traveling or working unprotected in space.

Solar energetic particle events can endanger the health and even the lives of astronauts when they venture into outer space, completely unprotected by the Earth's magnetic field or even a spacecraft (Fig. 7.8), to unload spacecraft cargo, construct a space station, or walk on the Moon or Mars. High-energy protons from a solar flare or coronal mass ejection can easily pierce a space suit, causing damage to human cells and tissues.

A future trip to Mars will involve considerable risks. Astronauts would spend 6 months or more in transit each way, and stay on the Martian surface for as long as a year-and-a-half, until the red planet again moved closest to the Earth. Some estimate that every third human cell would be damaged by solar energetic particles during the flight, and others worry about how to keep the astronauts from being irradiated to death. Long exposures to cosmic rays in space also increase the risk of getting

Fig. 7.8 Unprotected from space weather. The first untethered walk in space, on 7 February 1984, where there is no place to hide from Sun-driven storms. Astronaut Bruce McCandless II, a mission specialist, wears a 300-pound (136 kg) Manned Maneuvering Unit (MMU) with 24 nitrogen gas thrusters and a 35 mm camera. The MMU permits motion in space where the sensation of gravity has vanished, but it does not protect the astronaut from solar flares or coronal mass ejections. High-energy particles resulting from these explosions on the Sun could injure or even kill the unprotected astronaut. (Courtesy of NASA.)

cancer, apparently to a 40% lifetime chance after a voyage to Mars and far above acceptable thresholds of government agencies. A future return trip to the Moon, with an extended stay, or to explore Mars, must include methods of protection of the crew from the harmful effects of Sun-driven space weather and cosmic rays.

Solar astronomers, and employees of national space-weather forecast centers, therefore keep careful watch over the Sun during space missions, to warn of possible solar activity occurring at just the wrong place and time. Space flight controllers can then postpone space walks during solar storms, keeping astronauts within the heavily shielded recesses of a satellite or space station. The astronauts would also be told to curtail any strolls on the Moon or Mars, and to move inside underground storm shelters.

7.4.5 Failing to Communicate

Eight minutes after the outburst of an energetic flare on the Sun, a strong blast of X-rays and extreme-ultraviolet radiation reaches the Earth traveling at the speed of

light, and radically alters the structure of the planet's upper atmosphere, the ionosphere, by producing an increase in the amount of free electrons that are no longer
attached to atoms. And enhanced aurora currents at times of intense geomagnetic
storms also affect the ionosphere.

Changes in the ionosphere resulting from solar flares or coronal mass ejections
can attenuate or disrupt high-frequency radio wave communications that utilize reflection from the ionosphere to carry signals to distances beyond the local horizon.
Even during moderately intense flares, long-distance radio communications can be
temporarily silenced over the Earth's entire sunlit hemisphere. The radio blackouts
are particularly troublesome for the commercial airline industry, which uses radio
transmissions for weather, air traffic, and location information; the United States Air
Force and Navy are also concerned about this solar threat to radio communications.

The Air Force operates a global system of ground-based radio and optical telescopes and taps into the output of national, space-borne X-ray telescopes and particle detectors in order to continuously monitor the Sun for intense flares that might
severely disrupt military communications and satellite surveillance. The Air Force
has also recently tested a Solar Mass Ejection Imager, abbreviated SMEI, aboard its
Coriolus spacecraft, designed to help forecast the arrival of coronal mass ejections –
as reported by David F. Webb and colleagues in 2006 and 2008.

Space weather interference with radio communication can be avoided by using short-wavelength, ultra-high-frequency signals that pass right through the ionosphere to satellites that can relay the transmissions to other locations. Signals in this
frequency range are nevertheless also vulnerable to the aurora currents in the ionosphere, and can be degraded or completely lost during times of high geomagnetic
activity.

The telecommunications industry is also threatened by the loss of their satellites
due to disabling solar outbursts.

7.4.6 Satellites in Danger

Solar energetic particles arising from solar flares or coronal mass ejections can de-
grade, disrupt or destroy a satellite. And there are now roughly 1,000 of them in daily
use by governments, corporations, and ordinary citizens. Geosynchronous satellites,
which orbit the Earth at the same rate that the planet spins, stay above the same
place on Earth to relay and beam down signals used for aviation and marine navigation, cellular phones, global positioning systems, national defense, and internet
commerce and data transmission. Other satellites whip around the planet, scanning
air, land, and sea for environmental change, weather forecasting and military reconnaissance. All of these spacecraft can be temporarily or permanently disabled by
solar energetic particle events, causing engineers to design spacecraft with greater
shielding and increased redundancy in their components.

Geosynchronous satellites, for example, are endangered by the coronal mass
ejections that cause intense geomagnetic storms. These satellites orbit our planet

once every 24 h at about 6.6 Earth radii, and thus remain at constant longitude above the Earth. A powerful coronal mass ejection can compress the magnetosphere from its usual location at about 10 Earth radii to below the satellites' geostationary orbits, exposing them to the full brunt of the gusty solar wind and its charged, energized ingredients. Geosynchronous satellites are also affected by high-energy electrons associated with high-speed, solar-wind streams during the declining phase of the solar activity cycle.

When intense solar storms buffet the magnetosphere, they can greatly increase the amounts of high-speed electrons trapped in the Van Allen radiation belts. The high-energy electrons can move right through the thin metallic skin of a spacecraft, penetrating electronic equipment. The excess negative charge can give rise to potential differences and intense voltage discharges, severe current pulses, and damaging surges of electric energy. Metal shielding and radiation-hardened computer chips are used to guard against this recurrent hazard, and satellite orbits can be designed to minimize time in the radiation belts, or to avoid them altogether.

Nothing can be done to shield the solar cells used to power nearly all Earth-orbiting satellites; the photovoltaic cells convert sunlight to electricity and therefore have to be exposed to space. When satellites repeatedly pass through the radiation belts, its energetic particles slowly deteriorate and shorten the useful lives of their solar cells.

Infrequent, anomalously large eruptions on the Sun can hurl very energetic protons toward the Earth and elsewhere in space. The solar protons can easily enter a spacecraft to produce single event upsets in electronic components by ionizing a track along parts of their circuits. The ionized tracks can occur in transistors and memory devices, producing erroneous commands and crippling their microelectronics. Such single event upsets have already destroyed at least one weather satellite and disabled several communications satellites. Space weapons can also wipe out a satellite; so if you did not know the Sun was at fault, you might think someone was trying to shoot down our satellites. But error-correcting software has been developed to decrease damages by single event upsets to satellite operations.

To put the space-weather threat in perspective, just a few satellites have been lost to storms from the Sun out of thousands deployed. And the U.S. military is more concerned with disruption of communications, since they build satellites that can withstand the effects of a nuclear bomb exploded in space. Commercial satellites are less expensive and more vulnerable.

7.4.7 Varying Solar Radiation and Earth's Changing Atmosphere

Unlike the charged particles in the solar wind, the Sun's electromagnetic radiation passes right through the Earth's magnetic fields without noticing them. Short-wavelength radiation, that contributes only a tiny fraction of the Sun's total luminosity, is mainly absorbed high in the Earth's atmosphere, ionizing the air and transforming its physical condition. And although the temperature of the atmosphere at first drops with height, where the air expands and becomes colder in the lower density and pressure, the temperature then increases at higher altitudes where it becomes hotter than the ground – in the stratosphere where the Sun's ultraviolet rays produce the ozone layer and in the ionosphere, produced by solar X-ray radiation (Fig. 7.9).

The increased solar output at extreme-ultraviolet and X-ray wavelengths during the maximum in the Sun's 11-year activity cycle can cause the temperature of the upper atmosphere to soar to more than twice the values encountered at activity minimum (Fig. 7.10). The enhanced radiation caused by increasing solar activity heats the atmosphere and causes it to expand. This brings higher gas densities to a given altitude, increasing the friction and drag exerted on a satellite, pulling it to a lower altitude, and sometimes causing ground controllers to lose contact with them.

Increased atmospheric friction caused by rising solar activity has sent several satellites to a premature, uncontrollable, and fatal spiral toward the Earth, including *Skylab* and the *Solar Maximum Mission*. Both spacecraft were ungratefully destroyed by the very phenomenon they were designed to study – solar flares and/or coronal mass ejections. Space stations have to be periodically boosted in altitude to higher orbit to avoid a similar fate.

Precise monitoring of all orbiting objects depends on accurate knowledge of the atmospheric change caused by storms from the Sun. The U.S. Space Command, for

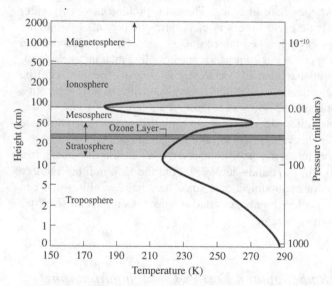

Fig. 7.9 Sun-layered atmosphere. The pressure of our atmosphere (*right scale*) decreases with height (*left scale*). This is because fewer particles are able to overcome the Earth's gravitational pull and reach greater heights. The temperature (*bottom scale*) also decreases steadily with height in the ground-hugging troposphere, but the temperature increases in two higher layers that are heated by the Sun. They are the stratosphere, with its critical ozone layer, and the ionosphere. The stratosphere is mainly heated by ultraviolet radiation from the Sun, and the ionosphere is created and modulated by the Sun's X-ray and extreme-ultraviolet radiation

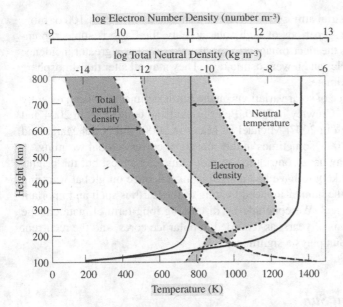

Fig. 7.10 Varying solar heating of the upper atmosphere. During the Sun's 11-year activity cycle, the upper-atmosphere temperatures fluctuate by factors of 2, and neutral (un-ionized atom) and electron densities by factors of 10. The bold lines (*right side*) register maximum values and the less bold (*left side*) the minimum values. Enhanced magnetic activity on the Sun produces increased ultraviolet and X-ray radiation that heats the Earth's upper atmosphere and causes it to expand, resulting in higher temperatures and greater densities at a given altitude in our atmosphere. (Courtesy of Judith Lean.)

example, often has to recompute the orbits of many hundreds of low-Earth-orbit objects affected by the Sun-induced increase in air friction.

But it is visible sunlight that penetrates our transparent atmosphere to warm the ground and produce winds, rain, and snow in the lowest level of the atmosphere, the troposphere. And the amount of this visible radiation also varies in tandem with the 11-year solar activity cycle.

7.5 Sun – Climate

Over the eons, the annual seasons and weather patterns on Earth have been driven by the sunlight that reaches the lower atmosphere and ground, and modulated by the winds, the ocean currents, and the shifting configurations of land and sea. So it is possible, even likely, that the terrestrial climate is also related to varying activity on the Sun.

In recent decades, it has indeed been shown that the total amount of sunlight received by the Earth varies with the 11-year cycle of solar activity, albeit by very small amounts. And the amount of the Sun's ultraviolet and X-ray radiation, which is absorbed in the terrestrial atmosphere, varies by enormous factors of 100 or more over the activity cycle. At periods of high solar activity, the Earth is subject to enhanced radiation across the electromagnetic spectrum, as well as a greater incidence of solar energetic particles and fewer cosmic rays. They may all alter the atmosphere and contribute to our climate.

The possible influence of solar variations on the Earth's climate has been recently discussed by Thomas J. Crowley in 2000, Peter Foukal and colleagues in 2004 and 2006, Johanna D. Haigh in 2007, Michael E. Mann in 2000, and Nigel Marsh and Hendrik Svensmark in 2003, sometimes in the context of current global warming.

The variable Sun may have contributed to past climate changes, but the brightening of the Sun is unlikely to have had a significant influence on global warming since 1950. And since the human-induced heating of the Earth is such an important topic, it is considered next. We conclude by discussing long-term climate change over millions and billions of years, as recorded in polar ice cores, and the recurrent ice ages, in which the Sun plays a significant role.

7.5.1 An Inconstant Sun

Day after day the Sun rises and sets in a seemingly endless cycle, illuminating our days and warming our world. And the total amount of its life-sustaining energy has been called the "solar constant," because no variations could be reliably detected from anywhere beneath the Earth's changing atmosphere. The solar constant is the average amount of radiant energy per second per unit area reaching the top of the Earth's atmosphere at a mean distance of 1 AU. Nowadays, it is also known as the total solar irradiance, abbreviated TSI, perhaps because it is not constant.

The discovery of variations in the total solar radiation was somewhat unexpected. Throughout most of the 20th century, most astronomers and climatologists insisted that the Sun shines steadily, and any fluctuations in the amount of sunlight reaching the ground were attributed to variable absorption and scattering in the atmosphere. After all, when a cloud passes by and casts its shadow on the ground, we do not attribute the temporary cooling to a varying Sun. Nevertheless, as reliable as it appears, the Sun is an inconstant companion, brightening and fading in step with the 11-year sunspot cycle.

The measurement precision required to detect the Sun's varying radiation output was not obtained until the early 1980s, with an exquisitely sensitive detector, created by Richard C. Willson and known as the Active Cavity Radiometer Irradiance Monitor, or ACRIM for short. It was aboard NASA's *Solar Maximum Mission*, placed just outside the Earth on February 1980, and detected the total solar radiation reaching the satellite with an incredible precision of 0.01% or one part in 10,000, when averaged over 1 day. At this level of accuracy, the Sun's total radiation output is almost always changing, at amounts of up to a few tenths of a percent and on all time scales from 1 s to 10 years. Similar variations were also detected by the Earth

Radiation Budget, or ERB, radiometer aboard the *Nimbus 7* satellite, launched in October 1978, with an even longer record, but with less precision.

Hugh S. Hudson teamed up with Willson to help him identify the irradiance variations with well-known features on the Sun. The largest, relatively brief, downward excursions correspond to, and are explained by, the rotation of a large group of sunspots across the face of the Sun. The concentrated magnetism in sunspots acts as a valve, blocking the energy outflow, which is why sunspots are dark and cool, producing reductions in the Sun's output of up to 0.3% that last a few days.

Bright patches on the visible solar surface cause the increased luminosity. They are called *faculae*, from the Latin for "little torches." On the short-term scale of minutes to days, sunspot blocking is greater than facular brightening; on the long-term scale of months and years, faculae are dominant because they have significantly longer lifetimes than sunspots and cover a larger fraction of the solar disc.

As demonstrated by Peter V. Foukal and Judith Lean in 1990, the decade-long variations in the Sun's total radiation output are the result of a competition between dark sunspots, in which radiation is depleted, and bright faculae, which are sources of enhanced radiation, with the faculae winning out. As magnetic activity increases, they both appear more often, but the excess radiance from bright faculae is greater than the loss from sunspots.

Five instruments on different spacecraft have measured the total solar irradiance for three decades, determining the average amount of radiant solar energy per unit time per unit area just outside the Earth (Fig. 7.11). Their combined results, described by Claus Fröhlich and Judith Lean in 2004 and Fröhlich in 2006, indicate an average minimum value over 30 years of $1,365.5 \pm 0.04 \, \text{W m}^{-2}$.

There are significant uncertainties between these instruments, which disagree among themselves by a few watts per square meter. They are also about $5 \, \text{W m}^{-2}$ higher, on average, than recent measurements from the *SOlar Radiation and Climate Experiment*, abbreviated *SORCE* and reported by Greg Kopp and colleagues in 2005, which suggest a somewhat lower average value of about $1361 \, \text{W m}^{-2}$ but with similar varying quantities and apparently no instrument or calibration errors.

The solar-cycle variation in the total solar irradiation is known with much greater accuracy, showing a change of about 0.07% from cycle minimum to maximum. Although the most intense part of the solar radiation, including that at visible wavelengths, exhibits this relatively modest variation, the entire spectrum of the Sun's radiation is modulated by solar activity.

Radiation at the short-wavelength, invisible parts of the solar spectrum change significantly during the solar cycle, though contributing only a tiny fraction of the Sun's total radiation. As we have discussed in the previous Sect. 7.4, the Earth's upper atmosphere acts as a sponge, soaking up the unseen ultraviolet and X-ray radiation. At times of high solar activity, the Sun pumps out much more of the invisible rays, the air absorbs more of them, and our upper atmosphere heats up; when solar activity diminishes, the high-altitude air absorbs less and cools down. Only the least variable, visible portion of the solar spectrum penetrates through to the relatively placid lower atmosphere, but this sunlight might still vary enough to warm and cool the air.

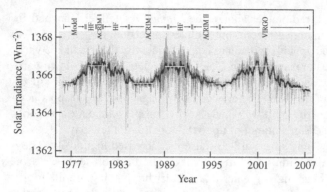

Fig. 7.11 Variations in the solar constant. Observations with very stable and precise detectors on several Earth-orbiting satellites from 1978 to 2007 show that the Sun's total radiation input to the Earth, termed the solar irradiance, is not a constant, but instead varies over time scales of days and years. Here the total irradiance just outside our atmosphere, also called the solar constant, is given in units of watts per square meter, abbreviated W m^{-2}, where 1 W is equivalent to $1\,\mathrm{J\,s^{-1}}$. The observations show that the Sun's output fluctuates during each 11-year cycle of solar magnetic activity, changing by about 0.1% between maximums (1979, 1990, and 2001) and minimums (1987, 1997, and 2008) in magnetic activity. The average minimum value is $1,365.509 \pm 0.426\,\mathrm{W\,m^{-2}}$, and the cycle amplitudes are 0.934 ± 0.019, 0.897 ± 0.020 and $0.829 \pm 0.017\,\mathrm{W\,m^{-2}}$ above the average minimum value. Measurements with the Total Irradiance Monitor aboard *the SOlar Radiaton and Climate Experiment*, abbreviated *SORCE*, indicate a lower value of about $1,361\,\mathrm{W\,m^{-2}}$. Temporary dips of up to 0.3% and a few days' duration are due to the presence of large sunspots on the visible hemisphere. The larger number of sunspots near the peak in the 11-year cycle is accompanied by a rise in magnetic activity that creates an increase in luminous output which exceeds the cooling effects of sunspots. The capital letters given at the top are acronyms for the different radiometers. They are the Hickey-Frieden (HF) radiometer of the Earth Radiation Budget (ERB) experiment on the *Nimbus-7* spacecraft, the two Active Cavity Radiometer Irradiance Monitors (ACRIM I and II) placed aboard the *Solar Maximum Mission* satellite and the *Upper Atmosphere Research Satellite* (UARS), respectively, and the Variability of solar IRradiance and Gravity Oscillations (VIRGO) radiometers flying on the *SOlar and Heliospheric Observatory*, abbreviated *SOHO*. [Courtesy of Claus Fröhlich, see Frölich and Judith Lean (2004) for a discussion.]

The precise lengths of the 11-year solar cycle might be indicators of the solar activity closely associated with climate. In 1991, Eigil Friss-Christensen and Knud Lassen published an interesting correlation between cycle length and temperature variations over the past 130 years. That is, the yearly mean air temperature over land in the Northern Hemisphere has moved higher or lower, by about 0.2° Centigrade, in close synchronism with the solar-cycle length during this period. Short cycles are characteristic of greater solar activity that seems to produce a temperature increase, while longer cycles signify decreased activity on the Sun and cooler times at the Earth's surface.

Peter Laut and Jesper Gundermann updated this work in 2000, using longer records and finding a weaker correlation between temperature variations and cycle length. Although there is still a correspondence in the 20th century, it is not detectable in more recent times, so the measurements cannot be used in support of any connection between the Sun and global warming since 1970.

Nevertheless, surface temperatures, cloud cover, drought, rainfall, tropical monsoons, and forest fires all show some correlation with solar activity, but the correlations are variable and uncertain, and climate experts usually cannot explain why they might occur. They just do not know how the climate responds to the solar variability. It is difficult to understand how the feeble changes in the solar brightness, that have been observed so far, can have a noticeable effect on the Earth's climate. But climate modelers find that the Sun's heating effect can be amplified, especially for specific regions due to subtle changes in the ocean circulation patterns. In 2001 and 2003 Drew T. Shindell and co-workers demonstrated, for example, that long-term regional changes in the climate during the pre-industrial era were dominated by the Sun's variable activity. Their models show that solar irradiance reductions in the late 17th and most of the 18th century produced significant regional temperature changes that were much larger than global average changes, with cooler temperatures in the Northern Hemisphere agreeing with historical records of cold surface temperatures during the prolonged Maunder Minimum of low solar activity.

Other scientists have reasoned that the Sun may also be influencing the Earth by powerful indirect routes geared to the solar activity cycle. As an example, one solar-cycle mechanism, suggested by Edward P. Ney in 1959, involves changes in the solar wind, galactic cosmic rays, and terrestrial clouds. In 1975, Robert E. Dickinson proposed related ideas in which solar variability modulates the cosmic ray input to Earth and therefore the cloudiness of its atmosphere. At times of enhanced activity on the Sun, the solar wind is pumped up with intense magnetic fields that extend far out into interplanetary space, blocking cosmic rays that would otherwise arrive at Earth. The resulting decrease in cosmic rays means that fewer energetic charged particles penetrate to the lower atmosphere where they may help produce clouds, particularly at higher latitudes where the shielding by Earth's magnetic field is less. The reduction in clouds, which reflect sunlight, would explain why the Earth's surface temperature gets hotter when the Sun is more active. In 1997, Henrik Svensmark and Eigil Friis-Christensen indeed found that during the previous 11-year activity cycle of the Sun, the cloudiness on a global scale decreased with increasing solar activity and a lower intensity of cosmic rays entering the atmosphere.

The mechanism proposed by Dickinson, in which changing cloud properties via cosmic ray ionization influences atmospheric transparency, was advocated by Nigel Marsh and Svensmark in their 2003 review of the solar influence on Earth's climate. But the idea still remains controversial. Also in 2003, Peter Laut challenged the statistical certainty of the proposed correlations between cosmic ray intensity and total or low cloud cover, and in 2007 Joanna D. Haigh questioned if the relatively small ions produced by cosmic rays could act directly as cloud condensation nuclei. And even if the cloud formation mechanism is feasible, its possible impact on the climate remains uncertain.

The direct effects of enhanced solar activity on the upper atmosphere are far less controversial. We know with certainty that the Sun's ultraviolet rays create ozone in the stratosphere, and the pioneering work of Karin Labitzke and Harold Van Loon in the 1990s demonstrated that the varying intensity of solar ultraviolet radiation produces a solar-cycle variation in the temperature of the stratosphere. These temperature changes drive high-altitude winds and global circulation patterns that move downward through the ground-hugging troposphere, where all our weather occurs. So the entire lower atmosphere might beat to the 11-year solar rhythm. Climate dynamists, such as Joanna D. Haigh and colleagues in 2005, are studying how the resultant solar-cycle temperature changes in the lower stratosphere might produce important changes in the underlying troposphere where all our weather and climate occur. One difficulty may nevertheless be that the solar ultraviolet variations during the 20th century do not correlate well with global temperatures in that century.

A related discovery, by Udo Schüle and colleagues in 2000, is that the quiet, inactive Sun at far-ultraviolet wavelengths also changes with the solar activity cycle. Moreover, the 11-year cycle that clocks the rise and fall of magnetic activity on the Sun may not repeat with the same strength at minimum. It instead seems to be modulated over century-long time intervals.

A substantial community of scientists is now seeking methods of amplifying small variations in the Sun's radiative output to produce significant variations in global temperatures. When, and if, that link is identified, we will have a more complete understanding of the Sun–climate interaction.

7.5.2 Solar Variability and Climate Change
Over the Past 1,000 Years

A crucial question is whether the Sun's total irradiance exhibits pronounced changes over time scales greater than the solar activity cycle. For the three decades that it has been measured, the total radiation from the Sun might show a systematic brightness increase or decrease smaller than the solar-cycle one. In 2003, Richard C. Willson and Alexander V. Mordvinov reported an increase of 0.05% from the minimum in 1986 to the minimum in 1996; however, this result is controversial owing in part to the difficulty in combining measurements from several different missions and instruments. The more recent and reliable comparison for the minimum in 2008 indicates a decrease of 0.02% from the minimum in 1996.

But past observations of sunspots and observations of other stars once suggested that more dramatic variations could occur, and may have happened during the past millennium. Despite diligent observations by European astronomers, for example, very few dark spots were found on the Sun between 1645 and 1715, a 70-year period that included the reign of France's "Sun King" Louis XIV. Gustav Spörer called attention to the 70-year absence in 1887. An indirect consequence of the missing sunspots was reported much earlier, in 1733, by Jean Jacques D'Ortous de Mairan, as a decrease in the number of auroras seen on Earth, but he was ridiculed for thinking that the northern lights could be related to increases in the number of sunspots.

E. Walter Maunder fully documented the dearth of sunspots using extensive historical records covering hundreds of years. His accounts, entitled *A Prolonged Sunspot Minimum*, were presented to the Royal Astronomical Society in 1890 and 1894, but they remained largely ignored until the 1970s when John A. "Jack" Eddy provided further evidence for the "Maunder Minimum", as he called it, using the growth rings of trees.

As the trees lay down their rings each year, they record the amount of atmospheric carbon dioxide captured in the process of photosynthesis. The carbon intake comes in two varieties, or isotopes, stable carbon 12 and radioactive carbon 14, and the radioactive type tells us how active the Sun was at the time. Cosmic rays from outer space produce radioactive carbon 14 when they strike atoms in the air. Because the cosmic rays are deflected away from the Earth by the Sun's magnetic fields during high solar-activity levels, there is less radioactive carbon in the air during episodes of high solar activity, and more of it at times of low activity on the Sun.

An analysis of the world's longest-lived trees, the bristle cone pines, suggests that the Sun's output has been turned low for several extended periods in the past millennia. Eddy used the technique to read the history of solar activity all the way back to the Bronze Age, and showed that the tree-ring data are supported by other evidence such as the ancient sightings of terrestrial auroras (Fig. 7.12). In 1976, Eddy concluded that the Sun has spent nearly a third of the past 2,000 years in a relatively inactive state. He pinpointed several periods of low activity with significantly more radioactive carbon 14, each about a century long, naming them the Maunder, Spörer, and Wolf minima.

Since the Sun's total radiative output has only been precisely measured from space since 1978, earlier brightness changes must be inferred from the historical records of solar activity and observations of Sun-like stars. This two-part reconstruction of the past involves the use of ancient sunspot observations to estimate

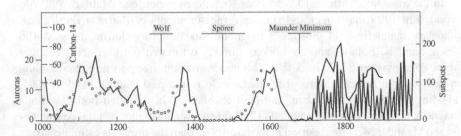

Fig. 7.12 Long periods of solar inactivity. Three independent indices demonstrate the existence of prolonged decreases in the level of solar activity. The observed annual mean sunspot numbers (*scale at right*) also follows the 11-year solar activity cycle after 1700. The curve extending from AD 1000 to 1900 is a proxy sunspot number index derived from measurements of carbon-14 in tree rings. Increased carbon-14 is plotted downward (*scale at left-inside*), so increased solar activity and larger proxy sunspot numbers correspond to reduced amounts of radiocarbon in the Earth's atmosphere. Open circles are an index of the occurrence of auroras in the Northern Hemisphere (*scale at left-outside*). The pronounced absence of sunspots from 1645 and 1715 is named for the English astronomer E. Walter Maunder, who fully documented it, and another noticeable lack of solar activity is named for the German astronomer Gustav Spörer who previously called attention to the prolonged absence between 1645 and 1715. The third prolonged absence of sunspots is named for the Swiss astronomer, Johann Rudolf Wolf, who investigated the connection of the 11-year sunspot cycle with geomagnetic activity and devised what is now known as the Wolf sunspot number. (Courtesy of John A. "Jack" Eddy.)

the variable radiation from solar active regions, and extrapolations from the brightness variations of other solar-type stars permit estimates for the possible total range in brightness variation of the Sun in past and future centuries, from the inactive, non-cycling lulls to the active, cycling luminosity highs.

Observations of Sun-like stars prompted Sallie L. Baliunas and Robert L. Jastrow to speculate in 1990 that the Sun might have undergone substantial luminosity variations on time scales of centuries, associated with dramatic changes in the Earth's climate. And in 1995, Judith Lean, Jürg Beer, and Raymond Bradley used such stellar comparisons in combination with proxies of solar activity to estimate the Sun's effect on the Earth's climate since 1610. They concluded that the Sun's low luminosity during the Maunder Minimum could account for a simultaneous long, cold spell on Earth, and that the Sun's subsequent brightness increase might account for much of the global warming since then. Already in 1994, however, Peter Foukal suggested that the levels of solar activity required for this to occur are not observed in the carbon-14 records over the past several millennia, which indicated to Foukal that such changes in the Sun's luminosity are now unlikely.

Then in 2002, Lean and co-workers retracted the stellar evidence, and the solar irradiance reconstructions based on it, with the admission that the long-term irradiance variations used in climate models for the previous decade may be a factor of five larger than justified. In other words, the variations during the past several centuries should be more in accord with the spacecraft measurements during the past three decades. So, Peter Foukal and his colleagues reviewed the situation, in 2004 and again in 2006, concluding that the brightening of the Sun is unlikely to have had a significant influence on global warming since the 17th century.

In the meantime, Michael Lockwood, Richard Stamper, and Matthew N. Wild reported in 1999 that direct and surrogate measurements of the near-Earth interplanetary magnetic field indicate a magnetic flux doubling during the past 100 years. Sami K. Solanki and co-workers quickly estimated the long-term changes of the solar magnetic field back to the Maunder Minimum, supporting the doubling of the Sun's large-scale, coronal magnetic field in the past century. And in 2005, Yi-Ming Wang, Judith L. Lean, and Neil R. Sheeley Jr. modeled both the Sun's magnetic field and irradiance since 1713, inferring much smaller irradiance changes for the Maunder Minimum period in the late 1600s than the previous estimates using stellar variability results. Interestingly, Nicola Scafetta and Bruce J. West showed in 2006 that both the new lower levels and previous higher levels of solar variations for the Maunder Minimum can empirically fit the reconstructed temperature trend from the 1600s up to about 1950, with a good correspondence between global surface temperature records and solar-induced temperature curves during the pre-industrial era from 1600 to 1800.

But removal of these effects from the climate records results in a very large temperature increase in the late 20th century, which cannot be explained by natural effects. Michael Mann and colleagues noted this dramatic temperature rise in 1999; it was independently noted by Mann and Thomas Crowley in 2000. And in 2007, Lockwood and Fröhlich reaffirmed that the observed rise in global mean temperatures over the previous 20 years could not be attributed to the Sun.

Although there is some controversy about some remaining solar influence on the climate during the past few decades, the temperature increase observed on Earth certainly coincides with the unprecedented release of carbon dioxide and other heat-trapping gases into the Earth's atmosphere, suggesting that human activity is taking control of the world's temperature, boosting it to unheard levels.

7.5.3 The Earth's Rising Fever

We have been turning up the global thermostat by dumping more and more carbon dioxide and other heat-trapping gases into the atmosphere. Because carbon dioxide is colorless, odorless, and disperses immediately in the atmosphere, few realized how much of the potentially dangerous gas enters into the air. Half a century ago, no one even knew if any of the carbon dioxide stays in the atmosphere or if it was all being absorbed in the forests and oceans. But in 1958, Charles Keeling began measurements of the atmospheric concentration of carbon dioxide (Keeling, 1960), which demonstrated its smooth exponential increase over the past half century (Fig. 7.13). Both the increase and its acceleration have been as inexorable as the expansion of the world's population, human industry, and pollution. This systematic rise in the atmospheric concentrations of carbon dioxide is the direct result of burning coal, oil, and natural gas.

Carbon dioxide and other gases, generated as the result of human activity, trap heat radiation near the Earth's surface and elevate its temperature by the greenhouse effect (Focus 7.4). Without remedial action, the levels of atmospheric carbon dioxide in this century will, for example, become twice those of the previous century and this will eventually raise the temperature of the Earth's surface by alarming amounts. So much of the world's population has, to put it mildly, become alarmed about continued global warming by this effect if the irreversible buildup of carbon dioxide and other heat-trapping gases continues unabated. The concern was demonstrated by the award of the Nobel Peace Prize for 2007 to an Intergovernmental Panel on Climate Change, and to Albert A. Gore Jr. "for their efforts to build up and disseminate greater knowledge about man-made climate change, and to lay the foundations for the measures that are needed to counteract such change."

Focus 7.4 ▬▬▬▬▬▬▬▬▬▬▬▬▬▬▬▬▬▬▬▬▬▬▬▬▬▬

The Greenhouse Effect

Our planet's surface is now comfortably warm because the atmosphere traps some of the Sun's heat and keeps it near the surface. The thin blanket of gas acts like a one-way filter, allowing sunlight through to warm the surface, but preventing the escape of some of the heat into the cold, unfillable sink of space. Much of the ground's heat is re-radiated out toward space in the form of longer infrared waves that are less energetic than visible ones and thus do not pass through the atmosphere's gas as easily as sunlight.

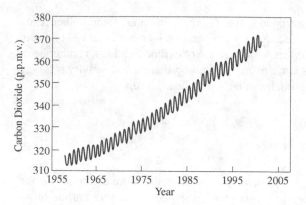

Fig. 7.13 Rise in atmospheric carbon dioxide. The average monthly concentration of atmospheric carbon dioxide, denoted CO_2, in parts per million by volume, abbreviated ppmv, of dry air plotted against time in years observed continuously since 1958 at the Mauna Loa Observatory, Hawaii. It shows that the atmospheric amounts of the principal waste gas of industrial societies, carbon dioxide, have risen steadily for nearly half a century. The up and down fluctuations, which are superimposed on the systematic increase, reflect a local seasonal rise and fall in the absorption of carbon dioxide by trees and other vegetation. Summertime lows are caused by the uptake of carbon dioxide by plants, and the winter highs occur when the plants' leaves fall and some of the gas is returned to the air. Carbon dioxide ice-core data indicate that the exponential increase in the amount of carbon dioxide has been continuing for the past two and a half centuries. (Courtesy of Dave Keeling and Tim Whorf, Scripps Institution of Oceanography.)

This is why cloudy nights tend to be warmer than clear nights; escaping infrared heat radiation is blocked by the water vapor in the clouds, keeping the ground warm at night. The glass windows of a greenhouse also retain heat and humidity inside of a greenhouse, so the atmospheric warming has also come to be called the greenhouse effect. But the designation is a misnomer, since the air inside a garden greenhouse is not heated by retaining infrared radiation; it is heated because it is enclosed, preventing the circulation of air currents that would carry away heat and cool the interior.

As first shown by John Tyndall in 1861, the main ingredients of the atmosphere, nitrogen (78%) and oxygen (21%), play no part in the warming of the Earth's surface, since these diatomic molecules do not absorb noticeable amounts of infrared radiation, but the less abundant, triatomic molecules, like water vapor and carbon dioxide, absorb the infrared heat, retaining it and raising the temperature of the planet.

The greenhouse effect is literally a matter of life and death. If the Earth had no atmosphere, it would be directly heated by the Sun's light to only 255 K, which is well below the freezing point of water at 273 K. Fortunately for life on Earth, the greenhouse gases in the air warm the planet to as much as 288 K, and this extra heat can keep the oceans, lakes and streams from turning into ice. Most of this "natural" greenhouse warming comes from water molecules, and carbon dioxide only provides about 10% of it.

Since the industrial revolution, humans have released heat-trapping gases into the atmosphere at an ever-increasing rate, creating an "unnatural" greenhouse effect. The amount of carbon dioxide in our air has, for example, been steadily accumulating as the result of the rapid growth of the world population and increased burning of coal, oil, and natural gas. During the past few decades, other heat-trapping gases, such as methane and nitrous oxide, have been accumulating noticeably in the atmosphere; methane, also known as natural gas, is released from swamps, coalmines, and even cows, and nitrous oxide, or laughing gas, originates from fertilizers.

There can be no doubt that the temperatures are already rising. The evidence comes from direct measurements of rising surface air temperatures and subsurface ocean temperatures, as well as retreating glaciers, increases in average global sea levels, and changes to many physical and biological systems.

Residents of northern countries may welcome the increasing heat. The growing season of crops will be lengthened, there will be less snow to shovel, and many days will be warmer, comparable to current more temperate climes in Mediterranean countries.

But some disturbing consequences of global warming are forecast for the future. The details cannot be precise, owing to the uncertain effects of clouds and oceans, but that should not distract us from the overall, worldwide problems. If current emissions of carbon dioxide and other greenhouse gases go unchecked, the increased heat and violent weather will drastically change the climate we are used to.

The associated rise in sea level will flood coastal cities, inundate island nations, and erase fertile deltas at the mouths of major rivers, like the Nile, Yangtze, Mekong, and Mississippi. More than a hundred million people live within a meter of mean sea level. And when many of the world's glaciers experience complete meltdown, in about 100 years or less at the present rate, some of these people will have their homes destroyed. But the Antarctica glaciers are not expected to melt, and some cities are likely to save homes by building protective dams. Some of the poorest regions on Earth will be hit hardest, such as densely populated Bangladesh. The resultant flooding will disrupt major cities such as Alexandria, Bangkok, Boston, New York City, Shanghai, Tokyo, and Venice.

And there are other catastrophes that might result from significant global warming in the future. Hurricanes will become stronger and wetter; water supplies will be reduced and forest fires will become more common; more species will become extinct; drought will be intensified within the interiors of many continents; the American Midwest might become a colossal dust bowl; power companies will be unable to air condition our sweltering cities; and extreme heat waves will cause great human stress and more deaths, particularly among the poor, elderly, and weak or those with cardiovascular and respiratory disease.

Although the most severe consequences of global warming are not likely to be noticed by you or your children, we've already initiated changes that will affect future generations. Once in the atmosphere, carbon dioxide stays there for centuries, so our grandchildren and their children will have to contend with the consequences of our present actions. The invisible waste gases that we have already dumped in the air will slowly change the climate of the Earth regardless of future actions, and sometime in the future a lot of people might be feeling like the world is melting down in a pool of sweat.

Most scientists therefore support prudent steps to curb the continued buildup of heat-trapping gases, even asserting that the evidence warrants a sense of urgency. In 2005, for example, the world's most influential scientific academies warned world leaders that they can no longer ignore the "clear and increasing" threat posed by global warming, and that "the scientific understanding of climate change is now sufficiently clear to justify nations taking prompt action." The unprecedented joint statement included the heads of the scientific academies of Brazil, Canada, China, France, Germany, India, Italy, Japan, Russia, the United Kingdom, and the United States.

Current international agreements to curtail the production of heat-trapping gases are not going to solve the problem. The *Kyoto Protocol*, which calls for mandatory reductions in the emissions of greenhouse gases such as carbon dioxide and methane, has not been signed by the United States, which contributes about 18% of the total emissions with just 4% of the world's population. And developing nations are not bound by the treaty restrictions even though their emissions are expected to surpass even the unrestrained emissions of the richer nations in a few decades. China, for example, is rapidly increasing its consumption of coal and oil and is expected to overtake the United States as the world's largest carbon dioxide emitter. And the combined increase in greenhouse gases contributed by these two countries will outstrip any reductions agreed to by other countries.

While waiting for the world's governments to get their act together, and effectively curtail the global emissions of heat trapping gases, state and national governments can curtail future global warming by setting carbon dioxide emission limits on cars, which account for 21% of the world's total carbon dioxide emission. They can adopt energy policies that shift from coal and oil to gas, and eventually to wind, water, solar, or nuclear power. Countries can avoid the clearing of their forests and plant a lot more trees, and farmers can use plants to pull carbon out of the air, while not plowing the soil that releases the carbon back into the air. By protecting existing forests and planting new ones, or by practicing no-till farming, countries might offset up to 20% of the expected carbon dioxide build up during this century.

Individuals can also take steps to help. They can stop buying gas-guzzling sports utility vehicles, and instead purchase compact, hybrid cars. Ordinary people can also reduce their consumption of coal, oil, or natural gas that electrify and heat their homes, offices, and schools, power their vehicles, and fuel their factories. And they can additionally use energy-efficient appliances, reduce their daily electricity use, drive their cars less, and insulate their buildings.

And if government and individual action do not solve the global-warming problem, nature will. In perhaps 100 years or even less, we will completely exhaust oil supplies, and the entire world will run out of gas. Once that happens, the Earth's climate should cool gradually, as the deep ocean waters slowly absorb the carbon dioxide pumped into the air during the recent frenetic pulse of activity. Then a long-overdue ice age might be on its way.

7.5.4 Climate Change Over Millions and Billions of Years

During the past two million years, huge ice sheets have advanced across the Northern Hemisphere and retreated again more than 20 times. The great, extended ice sheets last roughly 100,000 years, keeping the climate cold and the sea level low. The warm periods that punctuates the cold spells, called an interglacial, lasts roughly 10,000 years.

The most recent advance of the glaciers started about 120,000 years ago in Canada, Scandinavia, and Siberia. By the time the ice had spread to its maximum southern extent, most of northern Europe, New England, and the Midwestern United States were buried under ice a kilometer thick. The sea level had fallen to about 100 m lower than it is today, enlarging the size of continents above their surrounding waters and making it possible to walk from England to France, from Siberia to Alaska, and from New Guinea to Australia.

We now live during an interglacial, known as the Holocene period, which began about 10,000 years ago, when the world became warmer and wetter, about 5 K warmer on average, and human civilization flourished. The ice sheets melted and shrank back to their present-day configurations, leaving only the glacial ice in Greenland and parts of arctic Canada, as well as the massive ice sheets of Antarctica, and the sea level rose around the world.

The rhythmic ebb and flow of the great continental glaciers are affected by three astronomical rhythms that slowly alter the distances and angles at which sunlight strikes the Earth (Fig. 7.14). They are sometimes called the Milankovitch cycles, after the Milutin Milankovitch who described how variations in the planet's orbit, wobble, and tilt could influence the pattern of incoming solar radiation at different locations on the globe. Joseph Alphonse Adhémar previously suggested, in 1842, that the ice ages might be due to variations in the way the Earth moves around the Sun, and James Croll took up the idea in greater detail in 1876, showing how long periodic variations in the Earth's distance from the Sun might change the terrestrial climate. But the theory received its fullest mathematical development from 1920 to 1941 by Milankovitch.

The shortest astronomical rhythm is a periodic wobble in the Earth's rotation axis that is repeated in periods of 23,000 years. It determines whether the seasons in a given hemisphere are enhanced or weakened by orbital variations. A longer periodic variation, of the Earth's axial tilt from 21.5° to 24.5° and back again, occurs every 41,000 years. It is currently 23.5°, and accounts for our yearly seasons. The greater the tilt is, the more intense the seasons in both hemispheres, with hotter summers and colder winters.

The third and longest cycle is due to a slow periodic change in the shape of the Earth's orbit every 100,000 years. As the orbit becomes more elongated, the Earth's distance from the Sun varies more during the year, intensifying the seasons in one hemisphere and moderating them in the other.

The astronomical theory for the recurring ice ages was not strongly supported until 1976, when climate scientists James D. Hays, John Imbrie, and Nicholas J. "Nick" Shackleton demonstrated that variations in the Earth's orbit serve as a

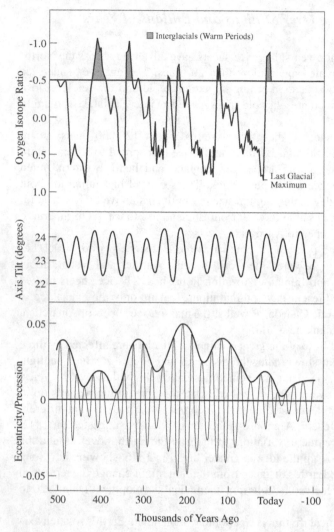

Fig. 7.14 Astronomical cycles cause the ice ages. The advance and retreat of glaciers are controlled by changes in the Earth's orbital shape or eccentricity, and variations in its axial tilt and wobble. They alter the angles and distances from which solar radiation reaches Earth, and therefore change the amount and distribution of sunlight on our planet. The global ebb and flow of ice is inferred from the presence of lighter and heavier forms of oxygen, called isotopes, in the fossilized shells of tiny marine animals found in deep-sea sediments. During glaciations, the shells are enriched with oxygen-18 because oxygen-16, a lighter form, is trapped in glacial ice. The relative abundance of oxygen-18 and oxygen-16 (*top*) is compared with periodic 41,000-year variations in the tilt of the Earth's axis (*middle*) and in the shape, or eccentricity – longer 100,000-year variation, and wobble, or precession, of the Earth's orbit – shorter 23,000-year variation (*bottom*).

pacemaker of the ice ages. They used an analysis of different types, or isotopes, of oxygen atoms in deep-sea sediments to infer the proportion of the world's water that was frozen within the glacial ice sheets at different times, revealing all three astronomical rhythms, with a dominant 100,000-year one. Analysis of cores extracted from the glacial ice in Antarctica and Greenland between 1985 and 2005 confirmed that the major ice ages are initiated every 100,000 years by orbital-induced changes in the intensity and distribution of sunlight arriving at Earth.

It was somewhat surprising that the glaciers have advanced and retreated in synchronism with this longer rhythmic stretching of the Earth's orbit. The shorter cycles have a greater, direct effect on the seasonal change in incident sunlight, but apparently produce smaller changes in ice volume than the longer one that has a weaker seasonal effect. By itself the 100,000-year cycle does not appear strong enough to bring about direct alterations of the terrestrial climate, so it must be leveraged by some other factor, and it has been found in ice cores from Greenland and Antarctica.

Microscopic air bubbles, which have been trapped in falling snowflakes and entombed in the glacial ice, record long-term climate changes over hundreds of thousands of years. Successive layers of the frozen snow build up on top of each other, like layers of sediment in geological strata, and the air bubbles entrapped in deep ice cores can be used to determine ancient variations in the temperature and atmospheric composition (Fig. 7.15). Analyses of the deep ice core taken from Vostok, Antarctica, reveal the roughly 100,000-year periodicity of the ice ages over the past 420,000 years, and indicate that transitions from glacial to warm epochs are accompanied by an increase in the atmospheric concentration of the three principal greenhouse gases – carbon dioxide, methane, and nitrous oxide. The temperatures go up whenever the levels of the heat-trapping gases increase, and they decrease together as well, rising and falling in tandem as the glaciers come and go and come again. These results were published by Jean-Robert Petit and colleagues in 1999 and reviewed by Bernhard Stauffer in 2000.

Scientists cannot yet agree whether the increase in greenhouse gases preceded or followed the rising temperatures, but the increase does answer the riddle of why the largest climate variations occur every 100,000 years. Changing orbital parameters initiate the end of a glacial epoch, through a relatively small increase in the intensity of incident solar radiation, and an increase in greenhouse gases amplifies the weak orbital signal. Melting of the large ice sheets in the Northern Hemisphere then further increases warming.

The ice cores can also be used to study the solar activity at times that pre-date historical records. In 1990, for example, Jürg Beer and co-workers demonstrated how the beryllium concentrations in polar ice can be used to study the 11-year solar activity cycle long before the first telescopic observations of sunspots, and in 1998 Gisela Dreschhoff and Edward J. Zeller described how high nitrate concentrations in ice cores signal solar proton events injected into the polar stratosphere during past centuries.

And the die is cast for the next advance of the glaciers, when the ice will come again. But because the current level of greenhouse gases, recently deposited in our

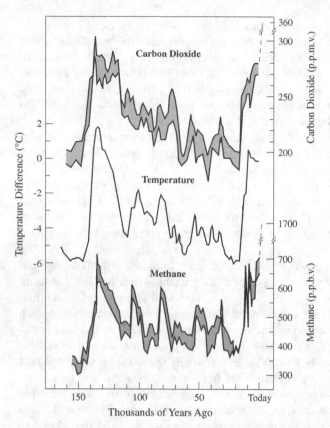

Fig. 7.15 Ice age temperatures and greenhouse gases. Ice-core data indicate that changes in the atmospheric temperature over Antarctica closely parallel variations in the atmospheric concentrations of two greenhouse gases, carbon dioxide and methane, for the past 160,000 years. When the temperature rises, so does the amount of these two greenhouse gases, and vice versa. A deeper Vostok ice core has extended the correlation to the past 420,000 years – see Jean Robert Petit and colleagues (1999) and Bernhard Stauffer (2000). The carbon dioxide (parts per million per volume) and methane (parts per billion per volume) increases may have contributed to the glacial–interglacial changes by amplifying orbital forcing of climate change. The ice-core data do not include the past 200 years, shown as dashed and broken lines at the right. They indicate that the present-day levels of carbon dioxide and methane are unprecedented during the past four 150,000-year glacial-interglacial cycles. (Adapted from Claude Lorius, *EOS*, **69**(26), 1988.)

atmosphere by humans, far surpasses any natural fluctuation of these substances recorded during past ice ages, we are not sure if the next ice age will dampen future global warming or whether recent global warming is delaying the coming of the ice. Nevertheless, when we completely exhaust oil supplies, in 100 years or less, and when our sources of coal and natural gas are also depleted, perhaps in a few hundred years, the ice should be on its way. And the next time it happens, the advancing glaciers could bury Copenhagen, Detroit, and Montreal under mountains of ice.

Climate changes over even longer periods, of about 100 million years, are very difficult to forecast. For instance, there was a time, some 65 million years ago, when there were no polar ice caps and dinosaurs roamed the Earth in a climate that was perhaps 15 K warmer than today. And many millions of years from now, continents will have collided, creating towering mountain ranges, or spilt open to make way for new oceans, altering the flow of our air and sea and strongly influencing the future climate in unforeseen ways.

And when we step back to view the Earth's probable climate on cosmic time scales of billions of years, we realize that the Sun's long, gradual evolution plays a role in both the remote past and the distant future. Our star began it life, for example, shining with only 70% of its present luminosity, slowing growing in luminous intensity as it aged. So the total solar irradiance has been steadily increasing at the rate of about 1% every 150 million years, with a total increase of roughly 30% since the Sun formed 4.6 billion years ago. This inexorable increase in the Sun's brightness is a consequence of increasing amounts of helium in the Sun's core; the greater mean density produces higher core temperatures, faster nuclear reactions, and a steady increase in luminosity.

If the Sun was 70% dimmer billions of years ago, the Earth should have been in a deep freeze, provided its atmosphere was not fundamentally different from today. That is, assuming an unchanging atmosphere, with the same composition and reflecting properties as today, the decreased solar luminosity would have caused the Earth's global surface temperature to drop below the freezing point of water during its first 2.5 billion years. Yet, there is clear geological evidence that the Earth was never this cold, and must have had a warm climate in its early history. Sedimentary rocks, which had to be deposited in liquid water, date from 3.8 billion years ago, when the Earth was less than one billion years old, and there is fossil evidence in these rocks for the emergence of life at least 3.5 billion years ago.

The discrepancy between the Earth's warm climatic record and an initially dimmer Sun has come to be known as the faint-young-Sun paradox, which has been reviewed by James F. Kasting and Owen B. Toon in 1989. It can be resolved if the Earth has a long-lasting climate control system that maintained relatively constant surface temperatures throughout the four billion years of recorded geological history. One possibility, proposed by Carl Sagan and George Mullen in 1972, Tobias C. Owen and colleagues in 1979, and Sagan and Christopher Chyba in 1997, is that there was a stronger atmospheric concentration of greenhouse gases, such as carbon dioxide, methane, or ammonia, in the Earth's early history. If the planet's primitive atmosphere contained hundreds of times more carbon dioxide than it does now, the greater heating of the enhanced greenhouse effect could have kept the oceans from freezing. The Earth could then only maintain a temperate climate by turning down its greenhouse effect as the Sun grew warmer and turned up the heat.

A different remedy of the faint-young-Sun problem, proposed by I.-Juliana Sackmann and Arnold Boothroyd in 2003, is that the young Sun was not faint. It might have began shining as a bigger, brighter, hotter, and more massive star, subsequently losing much of that mass in strong solar winds associated with its youth – see Manuel Güdel's 2007 review of evidence for an active young Sun. Or perhaps because of its faster rotation, the young Sun might have had stronger magnetic fields with enhanced extreme-ultraviolet and X-ray radiation and a greater output of high-energy particles. They could have played a role in modifying our planet's atmosphere back then. And as with humans, the Sun's bright or active youth might have evolved into a calmer, statelier old age, with lower luminosity, slower winds, and moderate magnetic activity.

An alternative explanation of the paradox involves the ancient oceans, during the first 1.5 billion years on Earth, and the regulatory effects of plants and animals thereafter. The very young Earth contained very little dry land, and the greater ocean surface would have absorbed more of the incoming solar radiation than it does today. Then, for the past three billion years, plants and animals could have developed the capability to control the environment, transforming the atmosphere and regulating the surface temperature. According to this *Gaia* hypothesis, developed by James E. Lovelock and Lynn Margulis in the 1970s, it is life that continues to control the environment, making it comfortable for living things in spite of adverse physical and chemical changes. For example, there was little or no oxygen in our atmosphere billions of years ago, but oxygen now makes up about a fifth of our air. If plants did not continuously replenish the oxygen, animals that breathe oxygen would use it all up.

But there is probably no escape in the end. Whatever life on Earth does, its remote future is not secure. As the Sun continues to brighten, the planet will eventually become a burned-out cinder, a dead and sterile place. From both astrophysical theory and observations of other stars, we know that the Sun will grow enormously in size and luminosity billions of years from now (Fig. 7.16). Astronomers calculate that the Sun will be hot enough in three billion years to boil the Earth's oceans away, and four billion years thereafter, our star will balloon into a giant star, engulfing the planet Mercury and becoming 2,000 times brighter than it is now. Its light will be intense enough to melt the Earth's surface. So, our long-term prospects are not all that great, and we might as well concentrate on protecting, improving, and experiencing the magnificent world that we are so privileged to inhabit.

7.6 Summary Highlights: Space Weather

- Space weather refers to conditions on the Sun and in the solar wind, magnetosphere, ionosphere, and thermosphere that can influence the performance and reliability of space-borne and ground-based technological systems and can affect human life and health.
- The Sun powers weather in space, with explosive outbursts that produce gusts and squalls in the tempestuous solar wind.
- Energetic protons accelerated by solar flares or coronal mass ejections can cripple spacecraft and seriously endanger unprotected astronauts that venture into outer space. Sun storms can also disrupt global radio communications and disable

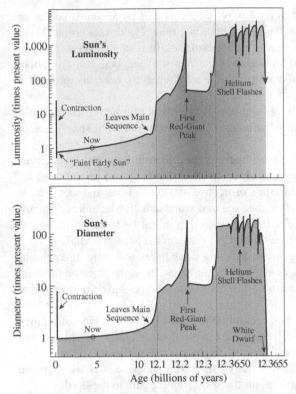

Fig. 7.16 The Sun's fate. In about eight billion years the Sun will become much brighter (*top*) and larger (*bottom*). The time scale has been expanded near the end of the Sun's life to show relatively rapid changes. (Courtesy of I-Juliana Sackmann and Arnold I. Boothroyd.)

satellites used for navigation, military reconnaissance or surveillance, and communication, from cell phones to pagers, with considerable economic, safety, and security consequences.

- Stormy weather in space occurs after the outburst of coronal mass ejections and/or solar flares, which erupt from solar active regions that appear more frequently near the maximum of the 11-year solar activity cycle.
- Helioseimology is used to detect magnetically complex and strong active regions on the hidden back side of the Sun, and to provide more than a week of warning before they rotate into view to threaten the Earth.
- Potential solar outbursts can be forecast by monitoring the twist and shear of active-region magnetic fields using vector magnetograms on the ground or in space, by the detection of twisted soft X-ray sigmoid shapes, or by observing emerging or swirling flows beneath the active regions.
- No one has yet identified a forecast signature that predicts exactly when a solar flare or coronal mass ejection will begin, but once they occur there is still some warning time – of about 1 to 2 days for a fast coronal mass ejection.

- When initiated, threatening coronal mass ejections can be accompanied at or near the photosphere by solar flares, halo coronal mass ejections, coronal X-ray or extreme-ultraviolet dimming and global extreme-ultraviolet waves.
- The direct in situ detection of high-speed electrons at a spacecraft can give about an hour's advance warning of the arrival of more dangerous, energetic protons at the same location.
- Solar energetic particle events, of high-energy electrons, protons, or heavier ions, arrive at the Earth within tens of minutes from the Sun following solar flares or coronal mass ejections.
- The solar proton events are the most energetic and therefore the most dangerous solar energetic particles. They can severely affect the health of unprotected astronauts traveling outside the Earth's protective magnetosphere, and they are capable of penetrating spacecraft to damage or disrupt sensitive technical systems. The strongest events produce radiation doses that might be lethal to astronauts fixing a spacecraft in outer space or taking a walk on the Moon or Mars.
- Energetic charged particles generated during solar flares will only threaten our planet or any other place in space if they occur at just the right place on the Sun, at one end of a spiral magnetic field line that connects the flaring region to that location.
- Solar energetic particles can be accelerated to very high energies in solar flares or by interplanetary shocks driven by fast coronal mass ejections, abbreviated CMEs. The flare-associated particles are produced at the Sun and follow the interplanetary magnetic fields, while the CME-driven shocks can cross magnetic fields lines and accelerate particles all the way from the Sun to the Earth.
- Type II radio bursts are due to the electrons accelerated at an outward propagating shock front. The meter-wavelength type II bursts are generated in the corona near the Sun and can be observed from the ground, while kilometer-wavelength ones originate further out in interplanetary space and can only be observed from spacecraft.
- Coronal mass ejections propagating in interplanetary space are known as interplanetary coronal mass ejections, or ICMEs for short. Their magnetic topology when arriving near Earth has been probed by spacecraft, indicating that many of them are closed, looping magnetic structures rooted in the Sun at both ends and extending radially from the Sun to the Earth.
- Interplanetary magnetic clouds often travel behind interplanetary shocks, which are driven by coronal mass ejections. Such a magnetic cloud contains a well-organized, twisted magnetic flux tube, which can provide a "highway" for the transport of solar energetic particles. When arriving near Earth, a magnetic cloud exhibits an enhanced magnetic field strength, a smooth rotation of the field direction through a large angle, and a low proton temperature with a low ratio of particle pressure to magnetic pressure.
- Only about a third of the ICMEs arriving near Earth can be characterized as magnetic clouds. The other, magnetically complex ejecta might be attributed to the interaction of coronal mass ejections moving at different speeds, with the fast ones even cannibalizing the slower ones.

- Impulsive solar energetic particle events are accelerated at the Sun during rapid solar flares, accompanied by type III radio bursts, rich in near-relativistic electrons, the rare helium isotope ^{3}He and heavy ions, and can last minutes at the Sun and hours at the Earth's orbit.
- Gradual solar particle events are accelerated by CME-driven interplanetary shocks, accompanied by type II radio bursts and long-lasting (hours) soft X-ray emission. They are responsible for most of the large proton events seen at Earth, and they can last days in the vicinity of the Earth.
- We are protected from the full blast of the Sun's violent activity by a dipolar magnetic field that diverts charged particles around the Earth and forms a cavity, called the magnetosphere, within the Sun's relentless winds.
- Doughnut-shaped belts of energetic electrons and protons girdle the Earth's equator.
- The Sun's winds bring the solar and terrestrial magnetic fields together on the night side of the Earth, where magnetic fields that point in the opposite direction can merge together. Electrons and protons can enter the magnetosphere at this point of magnetic reconnection, and these particles can become accelerated within the magnetosphere.
- Intense, non-recurrent geomagnetic storms, accompanied by exceptionally bright aurora, are caused when coronal mass ejections encounter the Earth's magnetosphere with the right magnetic alignment.
- Coronal mass ejections that strike the Earth can generate power surges on transmission lines that could cause electrical power blackouts of entire cities.
- Low-level, recurrent geomagnetic storms, with a 27-day repetition period, are produced by co-rotating interaction regions in the solar wind. When the fast-speed and slow-speed solar winds meet, they produce one of these co-rotating interaction regions.
- The auroras are caused by high-speed electrons that pump up oxygen and nitrogen molecules in the atmosphere, causing them to fluoresce like a cosmic neon sign.
- When viewed from space, the auroras form an oval centered on each magnetic pole, where magnetic fields guide energetic electrons down into the Earth's upper atmosphere.
- High-speed protons generated during explosive outbursts on the Sun can cripple satellites and endanger space-walking astronauts.
- Changes in the ionosphere resulting from solar flares or coronal mass ejections can attenuate or disrupt high-frequency radio wave communications that utilize reflection from the ionosphere to carry signals to distances beyond the local horizon. Even during moderately intense flares, long-distance radio communications can be temporarily silenced over the Earth's entire sunlit hemisphere.
- Space weather interference with radio communication can be avoided by using short-wavelength, ultra-high-frequency signals that pass right through the ionosphere to satellites that can relay the transmissions to other locations. Signals in this frequency range are nevertheless also vulnerable to the aurora currents in the ionosphere, and can be degraded or completely lost during times of high geomagnetic activity.

- When encountering the Earth, coronal mass ejections can compress the magnetosphere below the orbits of geosynchronous satellites that hover above one place on Earth, exposing the satellites to the full force of the solar wind.
- Solar X-rays and extreme-ultraviolet radiation both produce and significantly alter the Earth's ionosphere. The solar X-rays fluctuate in intensity by two orders of magnitude, or a factor of 100, during the Sun's 11-year magnetic activity cycle. Near activity maximum greater amounts of X-rays produce increased ionization, greater heat, and expansion of our upper atmosphere, altering satellite orbits and disrupting communications.
- The total solar irradiance of the Earth, the so-called solar constant, rises and falls in step with the 11-year magnetic activity cycle, but with a total recent change of only about 0.1%. When sunspots cross the visible solar disk, they produce, in themselves, a brief dimming of the Sun's radiative output, amounting to a few tenths of 1% for just a few days; a brightness increase caused by faculae and plage exceeds the overall sunspot decrease at times of high solar activity.
- The Earth is now hotter than any time during the previous 1,000 years. This global warming is attributed to carbon dioxide and other heat-trapping greenhouse gases released into the atmosphere by humans.
- The major ice ages, that repeat every 100,000 years, are caused by astronomical rhythms that alter the angles and distances from which sunlight strikes the Earth.
- Analysis of the deep ice core taken from Vostok, Antarctica, reveal the roughly 100,000-year periodicity of the ice ages over the past 420,000 years, and indicate that transitions from glacial to warm epochs are accompanied by an increase in the atmospheric concentration of the three principal greenhouse gases – carbon dioxide, methane, and nitrous oxide. A relativity small increase in the intensity of incident solar radiation, associated with a 100,000-year periodic change in the Earth's orbit, may be amplified by an increase in greenhouse gases to produce a warm, interglacial epoch.
- Our star began its life shining with only 70% of its present luminosity, slowly growing in luminous intensity as it aged. Assuming an unchanging atmosphere, with the same composition and reflecting properties as today, the past decreased solar luminosity would have caused the Earth's global surface temperature to drop below the freezing point of water during its first 2.5 billion years. Yet, there is clear geological evidence that the Earth was never this cold, and must have had a warm climate in its early history. The discrepancy between the Earth's warm climatic record and an initially dimmer Sun has come to be known as the faint-young-Sun paradox.
- The faint-young-Sun paradox can be resolved if there was a stronger atmospheric concentration of greenhouse gases, such as carbon dioxide, methane, or ammonia, in the Earth's early history; the greater heating of the enhanced greenhouse effect could have kept the oceans from freezing. Another solution is that the young Sun was a bigger, brighter, hotter, and more massive star than it is now, subsequently losing much of that mass in strong solar winds associated with its youth.

- The Sun will be hot enough in three billion years to boil the Earth's oceans away, and four billion years thereafter, our star will balloon into a giant star, engulfing the planet Mercury and becoming hot enough to melt the Earth's surface.

7.7 Key Events in the Discovery of Solar-Terrestrial Interactions*

Date	Event
1600	William Gilbert, physician to Queen Elizabeth I of England, publishes a small treatise demonstrating that "the terrestrial globe is itself a great magnet".
1610–1613	Galileo Galilei first systematically studies sunspots through a telescope.
1716	Edmund Halley suggests that aurora rays delineate terrestrial magnetic field lines, and that the auroras are due to a magnetized fluid that circulates poleward along the Earth's dipole magnetic field lines.
1724	George Graham discovers large, irregular fluctuations in compass needles, later called magnetic storms; Anders Celsius saw them at about the same time.
1733	Jean Jacques d'Ortous de Mairan argues for a connection between the occurrence of auroras and sunspots.
1799–1804	Alexander von Humboldt makes regular, precise measurements of the strength, dip, and inclination of the Earth's magnetic field during his voyage to South America. The term "magnetic storm" came into common usage as the result of Humboldt's scientific analyses and reports.
1838	Carl Friedrich Gauss publishes a mathematical description of the Earth's dipolar magnetic field, using it with observations to show that the magnetism must originate deep down inside the Earth's core.
1842	Joseph Alphose Adhémar suggests that the ice ages might be caused by variations in the way the Earth moves around the Sun, basing his theory on changes in the tilt of the Earth's rotation axis over long periods of time.
1844	Samuel Heinrich Schwabe demonstrates that the number of sunspots varies from a maximum to a minimum and back to a maximum again in a period of about 11 years.
1852	Edward Sabine demonstrates that global magnetic disturbances of the Earth, now called geomagnetic storms, vary in tandem with the 11-year sunspot cycle.
1859–1860	In 1859, Richard C. Carrington and Richard Hodgson independently observe a solar flare in the white light of the photosphere, and in 1860 publish the first account of such a flare. Seventeen hours after the flare a large magnetic storm begins on the Earth.
1876	James Croll publishes his theory of climate change based on long-term changes in the shape of the Earth's orbit and the wobble of its spin axis, arguing that ice ages develop when winters are colder than average.
1887–1889	Gustav Friedrich Wilhelm Spörer draws attention to a long, continued "damping-down" in solar activity between 1645 and 1715, when almost no sunspots were seen.

---

*See the References at the end of this book for complete references to these seminal papers.

Date	Event
1892	William Thomson (Baron Kelvin) shows that geomagnetic storms cannot be due to the direct magnetic action of the Sun, arguing that the "supposed connection between magnetic storms and sunspots is unreal."
1892, 1900	George Francis Fitzgerald and Oliver Lodge independently suggest that terrestrial magnetic disturbances might be due to electrified particles emitted by the Sun.
1896–1913	Kristian Birkeland argues that polar auroras and geomagnetic storms are due to beams of electrons from the Sun.
1899–1902	Guglielmo Marconi successfully sends radio signals across the Atlantic Ocean. Arthur E. Kennelly and Oliver Heaviside independently postulate the existence of an electrically conducting atmospheric layer, now called the ionosphere, to explain Marconi's radio transmission. The radio waves found their way around the curved Earth because they were reflected from the ionosphere.
1905	Edward Walter Maunder shows that geomagnetic storms tend to recur at 27-day intervals, the rotation period (relative to Earth) of low solar latitudes, and argues that the recurrent storms are due to narrow streams emanating from active areas on the Sun of limited extent.
1908	George Ellery Hale uses the Zeeman splitting of spectral lines to measure intense magnetic fields in sunspots, thousands of times stronger than the Earth's magnetism, but points out that the sunspot magnetic fields are still inadequate to account for geomagnetic storms by direct magnetic action.
1911	Arthur Schuster shows that a beam of electrons from the Sun cannot hold itself together against the mutual electrostatic repulsion of the electrons.
1919	Frederick Alexander Lindemann (later Lord Cherwell) suggests that an electrically neutral plasma ejection from the Sun is responsible for non-recurrent geomagnetic storms.
1920–1941	Milutin Milankovitch describes how the major ice ages might be produced by rhythmic fluctuations in the wobble and tilt of the Earth's rotational axis and the shape of the Earth's orbit with periods of 23,000, 41,000 and 100,000 years. They cause cold northern summers that prevent winter snow from melting and produce the ice ages.
1922	Edward Walter Maunder provides a full account of the 70-year dearth of sunspots, from 1645 to 1715, previously noticed by Gustav Friedrich Wilhelm Spörer in 1887–1889. This interruption in the normal sunspot cycle is now referred to as the Maunder Minimum.
1925	Edward Victor Appleton, and his research student M. A. F. Barnett, uses radio transmissions in the United Kingdom to verify the existence of the electrically conducting ionosphere, or the Kennelly-Heaviside layer as it was subsequently known. Gregory Breit and M. A. Tuve in America confirmed this in 1926. A height of about 100 km was inferred for the radio-reflecting layer by measuring the time delay between the transmission of the radio signal and the reception of its echo.
1929	William M.H. Greaves and Harold W. Newton distinguish between great, non-recurrent geomagnetic storms and smaller ones with a 27-day recurrence.
1931–1940	Sydney Chapman and Vincent C.A. Ferraro propose that a magnetic storm is caused when an electrically neutral plasma cloud ejected from the Sun envelops the Earth.
1931, 1943	George Ellery Hale and H.W. Newton present evidence that great geomagnetic storms are associated with solar flares observed with the spectrohelioscope.
1935–1937	J. Howard Dellinger suggests that the sudden ionosphere disturbances that interfere with short-wave radio signals have a solar origin.

Date	Event
1946	Edward V. Appleton and J. Stanley Hey demonstrate that meter-wavelength solar radio noise originates in sunspot-associated active regions, and that sudden large increases in the Sun's radio output are associated with chromosphere brightening, also known as solar flares.
1946, 1950	Scott E. Forbush and his colleagues describe flare-associated transient increases in the cosmic ray intensity at the Earth's surface, and attribute them to very energetic charged particles from the Sun. They were originally designated solar cosmic rays, but are more recently known as solar energetic particles.
1947	Ruby Payne-Scott, Donald E. Yabsley, and John G. Bolton discover that meter-wavelength solar radio bursts often arrive later at lower frequencies and longer wavelengths. They attributed the delays to disturbances moving outward at velocities of 500–750 km s^{-1}, exciting radio emission at the local plasma frequency.
1950–1954	Scott E. Forbush demonstrates the inverse correlation between the intensity of cosmic rays arriving at Earth and the number of sunspots over two 11-year solar activity cycles.
1951–1963	Herbert Friedman and his colleagues at the U.S. Naval Research Laboratory use sounding rocket and satellite observations to show that intense X-rays are emitted from the Sun, that the Sun emits enough X-ray and extreme ultraviolet radiation to create the ionosphere, that the X-ray emission is related to solar activity, and that X-rays emitted during solar flares are the cause of sudden ionosphere disturbances.
1950–1959	John Paul Wild and his colleagues use a swept frequency receiver to delineate type II radio bursts, attributed to shock waves moving out during a solar outburst at about a 1,000 km s^{-1}, and type III radio bursts, due to outward streams of high-energy electrons, accelerated at the onset of a solar flare and moving at nearly the velocity of light, or at almost 300,000 km s^{-1}.
1951–1957	Ludwig F. Biermann argues that a continuous flow of solar corpuscles is required to push comet ion tails into straight paths away from the Sun, correctly inferring solar wind speeds as high as 500–1,000 km s^{-1}.
1954	Philip Morrison proposes that magnetized clouds of gas, emitted by the active Sun, account for worldwide decreases in the cosmic ray intensity observed at Earth, lasting for days and correlated roughly with geomagnetic storms.
1955–1966	Cesare Emiliani shows that cyclic oxygen isotopic variations in deep-sea sediments can record the ebb and flow of the ice ages over the past 700,000 years. Although he initially thought the data recorded temperature variations, Nicholas J. Shakleton subsequently shows that it records changes in the global ice volume.
1956	Peter Meyer, Eugene N. Parker, and John A. Simpson argue that enhanced interplanetary magnetism at the peak of the solar activity cycle deflects cosmic rays from their Earth-bound paths.
1957	André Boischot discovers moving type IV radio bursts, and Boischot and Jean-Francoise Denisse explain them in terms of the magnetic clouds of high-energy electrons propelled into interplanetary space.
1958	Eugene N. Parker suggests that a perpetual supersonic flow of electric corpuscles, which he called the solar wind, naturally results from the expansion of a very hot corona. He also demonstrates that the solar magnetic field will be pulled into interplanetary space, attaining a spiral shape in the plane of the Sun's equator due to the combined effects of the radial solar wind flow and the Sun's rotation.

Date	Event
1958–1981	Stellar evolution theory is used by several different authors, including Douglas O. Gough, C. B. Haselgrove, Fred Hoyle, Martin Schwarzschild, and Roger K. Ulrich, to reliably predict that the solar luminosity has risen steadily by about 30% during the 4.5 billion-year life of the Sun. Schwarzschild speculates in 1958 that the changing solar brightness might have detectable geological and geophysical consequences.
1958–1959	The first American satellite, *Explorer 1*, was launched into orbit on 1 February 1958, followed by *Explorer 3* on 26 March 1958; instruments aboard these spacecraft, provided by James A. Van Allen and colleagues, discovered belts of charged particles that girdle the Earth's equator.
1958–1959	During a balloon flight on 20 March 1958, Laurence E. Peterson and John Randolph Winckler observed a burst of high energy, gamma ray radiation (200–500 keV) coincident in time with a solar flare, suggesting non-thermal particle acceleration during such outbursts on the Sun.
1958, 1965	Wolfgang Gleissberg finds an 80-year cycle in the record of the number of sunspots, and subsequently in the frequency of auroras.
1959	Thomas Gold suggests that geomagnetic storms are caused by a shock front associated with magnetic clouds ejected from the Sun, and coins the term magnetosphere for the region in the vicinity of the Earth in which the Earth's magnetic field dominates all dynamical processes involving charged particles.
1959	Edward P. Ney argues that the 11-year, solar-cycle modulation of cosmic rays may produce a climate effect in the Earth's lower atmosphere, the troposphere, by producing enhanced ionization, more stormy weather, greater cloud cover, and reduced ground temperatures at times of activity minimum and vice versa.
1961	James W. Dungey proposes a mechanism for transmitting solar wind energy to the magnetosphere by direct magnetic linkage or merging between the interplanetary and the terrestrial magnetic fields in a process now known as magnetic reconnection. In 1946–1948, Ronald G. Giovanelli developed a theory of solar flares involving magnetic neutral points, and in 1953 Dungey had shown how magnetic reconnection might energize solar flares.
1961	William Ian Axford and Colin O. Hines raise the possibility that the magnetosphere is energized by fluid friction at its boundary with the solar wind.
1961	Minze Stuiver makes a comparison of variations in sunspot activity and fluctuations in the radiocarbon, or carbon 14, concentration during the past 13 centuries, with suggestions of some correspondence between the two. The larger the number of sunspots, the greater the depression of cosmic ray intensity in the higher atmosphere with a corresponding decrease in radiocarbon.
1962–1967	*Mariner 2* was launched on 7 August 1962. Using the data obtained during the spacecraft's voyage to Venus, Marcia Neugebauer and Conway W. Snyder demonstrate that a low-speed solar wind plasma is continuously emitted by the Sun, and discover high-speed wind streams that recur with a 27-day period within the orbital plane of the planets.
1962–1964	An interplanetary shock associated with solar activity is detected using instruments aboard the *Mariner 2* spacecraft in 1962, reported by Charles P. Sonett, David S. Colburn, Leverett Davis Jr., Edward J. Smith, and Paul J. Coleman Jr. in 1964 and by Marcia Neugebauer and Conway W. Snyder in 1967.
1964	Syun-Ichi Akasofu develops the notion of a magnetosphere substorm.
1964	T. R. Hartz obtains the first spacecraft observations of solar type III bursts using a swept frequency receiver from 1.5 to 10 MHz.

Date	Event
1964–1968	Norman F. Ness and colleagues use instruments aboard NASA's first *Interplanetary Monitoring Platform*, or *IMP-1*, launched on 27 November 1963, to detect a large bow shock formed in the solar wind ahead of the magnetosphere and a long magnetic tail on the night side of the Earth. An extended magnetotail was previously measured by Charles P. Sonett in 1960 and Edward J. Smith in 1962, using instruments on *Explorer 6* and *10*. Ness and John M. Wilcox also use magnetometers aboard *IMP-1* to measure the strength and direction of the interplanetary magnetic field. They show that the interplanetary magnetic field is pulled into a spiral shape by the combined effects of the Sun's rotation and radial wind flow. They also discover large-scale magnetic sectors in interplanetary space that point toward or away from the Sun.
1965	Hans E. Suess discovers an approximate 200-year periodicity in the amount of radioactive carbon 14 within tree rings back to 5000 years BC.
1966	Donald H. Fairfield and Larry J. Cahill Jr. argue that geomagnetic activity is greatest when the interplanetary magnetic field is southward, leading to an enhanced merging rate between it and the geomagnetic fields. Also see the 1969 paper by Joan Hirshberg and David S. Colburn and the 1975 paper by Rande K. Burton, Robert L. McPherron, and Christopher T. Russell.
1966–1978	Olin C. Wilson measures the long-term magnetic activity of roughly 100 Sun-like stars by way of variations in their violent emission lines of singly ionized calcium – the H and K lines at 393.3 and 396.7 nm.
1969–1971	Magnetic fluctuations are observed in the solar wind from *Mariner 5* on its way to Venus, and attributed to large-amplitude Alfvén waves by John W. Belcher, Leverett Davis Jr., and Edward J. Smith.
1970	Ke Chiang Hsieh and John A. Simpson use the fourth *Interplanetary Monitoring Platform*, or *IMP-4*, to discover greatly enhanced interplanetary abundances of the rare helium isotope ^{3}He associated with solar flares.
1970	Robert P. Lin uses observations from *Interplanetary Monitoring Platform* satellites in 1964–1967 to propose two separate acceleration and/or emission mechanisms for solar energetic particles – one for solar proton events and the other for enhanced abundances of 40 keV electrons.
1971	John W. Belcher and Leverett Davis Jr. note the ubiquitous presence of large-amplitude Alfvén waves in the solar wind and show how the interaction of slow and fast solar wind streams will lead to co-rotating interaction regions with shocks that affect the internal properties of the solar wind.
1971	Ove Havnes discovers systematic differences between the abundances of cosmic rays of low energy and universal abundances. These differences are correlated with the first ionization potentials of the corresponding elements, and it has become known as the FIP effect.
1971–1973	The first good, space-based observation of a coronal disturbance or transient, now called a coronal mass ejection, was obtained on 14 December 1971 using the coronagraph aboard NASA's seventh *Orbiting Solar Observatory*, or *OSO 7* for short, reported by Richard Tousey in 1973.
1972	Carl Sagan and George Mullen show that the faint brightness of the young Sun is in conflict with the temperature history of the Earth, and suggest compensating terrestrial action by modification of the Earth's early atmosphere. With an unchanging atmosphere, the Earth's oceans would have frozen over about two billion years ago, which is conflict with geological evidence for liquid water on Earth more than three billion years ago.

Date	Event
1972–1973	Edward L. Chupp and his colleagues detect solar gamma ray lines for the first time using a monitor aboard NASA's seventh *Orbiting Solar Observatory*, abbreviated *OSO 7*. They observed the neutron capture (2.223 MeV) and electron–positron annihilation (0.511 MeV) lines associated with solar flares. The 2.223 MeV line had been anticipated theoretically by Philip Morrison.
1973	James E. Lovelock and Lynn Margulis suggest the *Gaia* hypothesis in which plants and animals have developed the capability to control their environment, keeping it comfortable for living things in spite of threatening changes.
1973–1974	In 1973, Christopher T. Russell and Robert L. McPherron propose that the semiannual variation in geomagnetic activity is caused by a change in the effective southward component of the interplanetary magnetic field, related to the changing tilt of the Earth's dipole axis over the course of a year. In 1974, Russell, McPherron and Rande K. Burton show that strong southward interplanetary magnetic fields can be associated with the development of geomagnetic storms.
1973–1974	The manned, orbiting solar observatory, *Skylab*, is launched on 14 May 1973, and manned by three person crews until 8 February 1974. *Skylab's* Apollo Telescope Mount contained 12 tons of solar observing instruments that spatially resolved solar flares at soft X-ray and ultraviolet wavelengths.
1974	John Thomas Gosling and colleagues report observations of coronal mass ejections, or CMEs, then called coronal disturbances or coronal transients, with the coronagraph aboard Skylab. They found that many CMEs are not associated with solar flares and that some CMEs have the high outward speed of up to a 1,000 km s^{-1} needed to produce interplanetary shocks.
1974–1978	Hannes Alfvén, and independently Lief Svalgaard and John M. Wilcox, interpret the magnetic structure of the solar wind, at activity minimum, in terms of a warped neutral current sheet dividing the solar wind into two hemispheres of opposite magnetic polarity.
1974–1986	The *Helios 1* and *2* spacecraft, respectively launched on 10 December 1974 and on 15 January 1976, measure the solar wind parameters as close as 0.3 AU from the Sun for a whole 11-year solar cycle. They confirmed the existence of two kinds of solar-wind flow. There is a steady, uniform high-speed wind and a varying, slow-speed wind.
1975	Robert E. Dickinson reviews mechanisms connecting solar activity to the meteorology of the Earth's lower atmosphere, showing that related variations in cloudiness, caused by cosmic rays, could be important.
1975	Gordon J. Hurford and colleagues use an instrument aboard the seventh *Interplanetary Monitoring Platform*, or *IMP-7*, to show that ^{3}He-rich solar flares exhibit an enrichment of heavy nuclei like iron. This enhanced abundance of interplanetary helium and iron associated with solar flares was confirmed and extended by Glenn M. Mason and co-workers in 1986 using the third *International Sun-Earth Explorer*, abbreviated *ISEE-3*.
1975–1977	A meeting held at the California Institute of Technology to discuss the solar constant and the Earth's atmosphere stimulates interest in spacecraft measurements of the solar irradiance of Earth. The results of these discussions were published in a book edited by Oran R. White in 1977, and contributed to the inclusion of the Active Cavity Radiometer Irradiance Monitor, abbreviated ACRIM, aboard the *Solar Maximum Mission*, or *SMM*, satellite.
1976	John A. Eddy shows that the prolonged sunspot minimum from 1645 to 1715 coincided with a decrease in solar activity, as characterized by a marked absence of terrestrial auroras, an abnormally high terrestrial carbon 14 abundance, and exceptionally low temperatures on Earth.

Date	Event
1976	Roger A. Kopp and Gerald W. Pneuman model eruptive solar flares at the base of reconnecting magnetic loops left behind an erupting prominence or coronal mass ejection.
1976	Edward J. Smith and John H. Wolfe use *Pioneer 10* and *11* data to investigate co-rotating interaction regions of the fast and slow solar wind between 1 and 5 AU, and Frank B. McDonald and his colleagues show that the shocks generated by these regions accelerate ions to high energies. Charles P. Sonett and colleagues first detected them during studies of the geomagnetic field about 1970. CIRs are the probable source of the 27-day repetition period of the so-called substorm, but there is still controversy about this conclusion.
1976	James D. Hays, John Imbrie, and Nicholas J. Shackleton use an analysis of oxygen isotopes in deep-sea sediments to show that the major ice ages during the past half million years recurred at intervals of 19,000, 23,000, 41,000, and 100,000 years, with a dominant 100,000-year recurrence. This supported the idea that the ice ages are caused by a variation in the intensity and distribution of solar energy arriving at the Earth. The double wobble period of 19,000 and 23,000 years was explained by refined astronomical calculations by André Berger in the following year.
1977	Stephen W. Kahler finds that X-ray long duration events are well associated with white light coronal mass ejections and are the X-ray analogs of the post-flare loop systems modeled by Roger A. Kopp and Gerald W. Pneuman.
1977	Michael J. Newman and Robert T. Rood suggest that the faint-young-Sun paradox could be resolved by a strong greenhouse effect in the past. Carl Sagan and Christopher Chyba also considered this in 1997.
1977	Roberto Pallavicini, Salvatore Serio, and Giuseppe S. Vaiana use *Skylab* observations to define two classes of soft X-ray flares: compact, brief (minutes) events, and extensive, long-enduring (hours) ones associated with soft X-ray arcades, filament eruptions, and coronal mass ejections.
1977–1980	Helmuth R. Rosenbauer, Wolfgang K.H. Schmidt, and their colleagues use measurements from *Helios 1* and *2* and the first *International Sun-Earth Explorer*, abbreviated *ISEE-1*, spacecraft to show that helium and other heavy ions move faster than protons in the high-speed wind. In addition, the electrons are cooler than the protons in this fast component of the solar wind.
1978	Edward J. Smith, Bruce T. Tsurutani, and Ronald L. Rosenberg use observations from *Pioneer 11* to show that the solar wind becomes unipolar, or obtains a single magnetic polarity, at solar latitudes near 16°.
1980	Rainer Schwenn and John T. Gosling and their colleagues respectively use instruments aboard *Helios 1* and the seventh *Interplanetary Monitoring Platform*, abbreviated *Imp 7*, to detect singly-ionized helium, He^+, produced in the solar wind by interplanetary shocks.
1980	The *Solar Maximum Mission*, abbreviated *SMM*, satellite is launched on 14 February 1980, to study the physics of solar flares during a period of maximum solar activity. It excelled in X-ray and gamma ray spectroscopy of solar flares, as well as observing the white-light emission of coronal mass ejections in 1980 and from 1984 to 1989.
1980–1982	Edward L. Chupp and his colleagues use the Gamma Ray Spectrometer, or GRS, on the *Solar Maximum Mission* satellite to detect energetic solar neutrons near the Earth following a solar flare, which occurred on 21 June 1980.
1980, 1984	Minze Stuiver and Paul D. Quay attribute the changing atmospheric carbon 14 to a variable Sun, and Charles P. Sonett provides further evidence for the 200-year periodicity in the radiocarbon data.

Date	Event
1980 **1984–1989**	Arthur J. Hundhausen and colleagues use the coronagraph aboard the *Solar Maximum Mission* satellite to specify the mass, velocity, energy, shape, and form of a large number of coronal mass ejections, fully reported in the literature in the 1990s.
1981	Robert P. Lin and colleagues find that solar flares can produce thermal sources with high enough temperatures to be detectable as hard X-rays.
1981	Jo Ann Joselyn and Patrick S. McIntosh make a convincing case that large magnetic storms can occasionally be associated with disappearing solar filaments. Filament disruptions were previously found to be associated with coronal mass ejections.
1981	Richard C. Willson, Samuel Gulkis, Michael Janssen, Hugh S. Hudson, and Gary A. Chapman report high-precision measurements of variations in the total solar irradiance of the Earth, or the solar constant, with an amplitude of up to 0.2%, made with the Active Cavity Radiometer Irradiance Monitor (ACRIM) on the *Solar Maximum Mission* satellite.
1981–1982	The Japanese spacecraft *Hinotori*, meaning *firebird*, was launched on 21 February 1981 and operated until 11 October 1982. It created images of solar flare X-rays with an energy of around 20 keV, and measured solar flare temperatures of between 10 and 40 million $(1–4 \times 10^7)$ Kelvin using soft X-ray spectroscopy.
1981–1984	In 1981, Leonard F. Burlaga and colleagues discover interplanetary magnetic clouds using measurements from five spacecraft in 1981, the *Voyager 1* and *2, Helios 1* and *2*, and the eighth *Interplanetary Magnetic Platform*, abbreviated *IMP-8*, and in 1982 they use *Helios 1* and the Solwind coronagraph on the *P78-1* spacecraft to demonstrate that the magnetic clouds can be associated with coronal mass ejections. In 1984, Robert M. Wilson and Ernest Hildner present strong evidence linking magnetic clouds with coronal mass ejections. In interplanetary space near the Earth, these magnetic clouds are associated with an increase in the magnetic field strength, a smooth rotation of the magnetic field direction over a large angle, and a low proton temperature. They arrive behind interplanetary shocks and can extend to more than half the distance toward the Sun.
1982	Russell A. Howard and colleagues report the first detection of an Earth-directed halo coronal mass ejection, and its associated interplanetary shock observed near the Earth, traveling at a speed of nearly $2,000 \text{km s}^{-1}$. The coronal mass ejection, then known as a coronal transient, was detected from the Solwind coronagraph aboard the *P78-1* satellite; the shock wave was detected with instruments aboard the third *International Sun-Earth Explorer*, abbreviated *ISEE 3*, spacecraft.
1982–1990	Edward L. Chupp, Hermann Debrunner, and colleagues report the observation of neutron emission at the Earth from the 3 June 1982 flare, giving signals in both the *Solar Maximum Mission* detector and the neutron monitor on Jungfraujoch in Switzerland.
1983	David J. Forrest and Edward L. Chupp use *Solar Maximum Mission* satellite observations to demonstrate the simultaneous acceleration of relativistic electrons (hard X-rays) and energetic ions (gamma rays) to within a few seconds. Masato Yoshimori and colleagues confirmed this simultaneity using observations from the *Hinotori* spacecraft.
1984	Walter R. Cook, Edward C. Stone, and Rochus E. Vogt report measurements from the *Voyager 1* and 2 spacecraft of the FIP effect in solar energetic particles emitted by solar flares, and conclude that both the solar energetic particles and the solar wind composition are significantly different from that measured for the photosphere.

Date	Event
1985	Eberhard S. Möbius and his colleagues use instruments aboard the *Active Magnetospheric Particle Tracer Explorers*, abbreviated *AMPTE*, spacecraft to detect singly ionized helium, He$^+$, attributing it to interstellar helium atoms that have entered the solar system, become ionized there, and then picked-up and entrained in the solar wind.
1985	Neil R. Sheeley Jr. and colleagues combine *P78-1* Solwind coronagraph observations of coronal mass ejections with *Helios 1* observations of interplanetary shocks to demonstrate that fast mass ejections drive the shocks.
1985–1999	Ice cores drilled in Greenland and Antarctica are used to determine the local temperature and atmospheric carbon dioxide and methane content in the polar regions for up to 420,000 years ago, including four glacial-interglacial cycles. This data confirmed the evidence from deep-sea sediments for initiation of the major ice ages every 100,000 years by changes in the intensity and distribution of sunlight arriving at Earth. The ice-core data also showed that the climate repeatedly warms and cools when the amounts of carbon dioxide and methane increase or decrease, respectively. (See papers with first authors of J. Chappellaz, C. Genthon, C. Lorius, J. R. Petit, and D. Raynaud.)
1986–1988	Louis A. Frank and his colleagues report images of the entire aurora oval from space using the *Dynamics Explorer 1* satellite.
1987	Hilary V. Cane, Neil R. Sheeley Jr., and Robert A. Howard show that strong interplanetary shocks are associated with fast coronal mass ejections that move at speeds greater than 500 km s^{-1}. They used data from the third *International Sun-Earth Explorer (ISEE-3)* low-frequency ($<$ 1 MHz) radio instrument and the Solwind coronagraph.
1987	Bruce T. Tsurutani and Walter D. Gonzalez show that Alfvén waves in the high-speed (coronal hole) solar wind streams can cause periodic magnetic reconnection, frequent magnetic substorms and aurora activity during the declining phase of the 11-year solar cycle, near solar minimum. They used the data gathered in 1978–1979 by the third *International Sun-Earth Explorer*, abbreviated *ISEE-3*.
1987, 1991	George C. Reid demonstrates that the globally averaged sea surface temperature over the past 130 years shows a very significant correlation with the envelope of the 11-year sunspot cycle. He explained the correlation by a variation in the Sun's total irradiance of Earth (the solar constant) over the same time interval.
1988	Donald V. Reames proposes that the abundances of solar energetic particles arriving at Earth, observed from the *International Sun-Earth Explorer* spacecraft over an 8.5-year period, imply two distinct populations of particles of separate origin. The ^{3}He and electron-rich events are attributed to impulsive solar flares; the other populations, which has lower helium and electron abundances and is responsible for most large proton events seen at Earth, are supposed to be accelerated by coronal mass ejections or interplanetary shocks.
1988–1993	Karin Labitzke and Harold Van Loon discover an association between the variability of the Sun and winds in the middle atmosphere. The winter storms in the stratosphere follow an 11-year pattern of low-pressure systems over the North Atlantic Ocean, matching the solar cycle in both period and phase.

Date	Event
1988–1992	Richard C. Willson and Hugh S. Hudson use radiometric data taken with the Active Cavity Radiometer Irradiance Monitor (ACRIM) instrument on the *Solar Maximum Mission (SMM)* to demonstrate that the total solar irradiance of Earth (the solar constant) varies in step with the 11-year cycle of solar activity. It has a total decline and rise of about 0.1%. They also used ACRIM data to show that the intense magnetic fields of sunspots produce a short-term (days) decrease of several tenths of a percent in the solar constant when the sunspots cross the visible solar disk. Douglas V. Hoyt, John R. Hickey, and their colleagues confirmed this solar cycle variation in the solar constant using data from the Earth Radiation Budget (ERB) experiment on the *Nimbus 7* satellite from 1978 to 1991.
1989	J. Randy Jokipii and Joseph Kóta argue that Alfvén waves streaming out of the Sun's polar regions may block the incoming cosmic rays.
1990	Jürg Beer and colleagues show that concentrations of the beryllium isotope ^{10}Be in polar ice can be used to study variations in solar activity, particularly the 11-year activity cycle, in time periods that pre-date historical records.
1990	Charles Lindsey and Douglas C. Braun propose that heliosiemic imaging could be used to produce seismic maps of magnetic regions on the far side of the Sun.
1990	Peter V. Foukal and Judith Lean show that the observed changes of the total solar irradiance of the Earth (the solar constant) during the previous 11-year cycle of magnetic activity can be explained by the increased emission of photosphere magnetic structures, called faculae. Their increased area and brightening near solar activity maximum exceed the irradiance reductions caused by dark sunspots whose numbers also increase in step with solar activity.
1990	Richard R. Radick, G. Wesley Lockwood, and Sallie L. Baliunas show that main-sequence stars similar to the Sun become brighter as their magnetic activity level increases. Baliunas and Robert Jastrow speculate that the long-term brightness changes of Sun-like stars indicate that the Sun could undergo substantial luminosity variations on time scales of centuries. They could exceed by a factor of four or five the 0.1% change in the Sun's total irradiance of the Earth (the solar constant) observed during the previous 11-year solar cycle.
1990	The *Ulysses* spacecraft is launched on 6 October 1990.
1990–1993	John T. Gosling and his colleagues argue that large, non-recurrent geomagnetic storms are caused by interplanetary disturbances driven by coronal mass ejections.
1991	The *Yohkoh*, meaning *sunbeam*, spacecraft is launched on 30 August 1991.
1991, 1994	Eigil Friis-Christensen and Knud Lassen find a high correlation between the variable period of the "11-year" sunspot cycle and the long-term variations of the land air temperature in the Northern Hemisphere for the past 130 years, and perhaps for the past five centuries.
1992	Uri Feldman reports that element abundances in the upper solar atmosphere are similar in nature to those in the solar wind and solar energetic particles, but different from abundances in the underlying photosphere.
1992, 1993, 1995	Steven W. Kahler and John T. Gosling reason that large, non-recurrent geomagnetic storms, interplanetary shock waves, and energetic interplanetary particle events are all mainly due to coronal mass ejections rather than solar flares. Hugh Hudson, Bernhard Haisch, and Keith T. Strong argue that both flares and coronal mass ejections result from solar eruptions that can have terrestrial consequences.

Date	Event
1994	Volker Bothmer and Rainer Schwenn provide evidence that eruptive prominences on the Sun are sources of magnetic clouds detected in the solar wind.
1994	David M. Rust notes that the helical magnetic fields in solar filaments and interplanetary magnetic clouds display a similar helicity, obeying a solar-hemisphere sign preference, and argues that they shed helical magnetic fields spawned within the Sun.
1994	David F. Webb and Russell A. Howard find that the coronal mass ejection rate tracks the sunspot number in amplitude and timing over the solar activity cycle.
1994	The *Wind* spacecraft is launched on 1 November 1994. It provides nearly continuous, direct, in-situ measurements of the solar wind, magnetic fields, and energetic particles arriving at the Earth's magnetosphere. *Wind* has also investigated the shocks generated by coronal mass ejections, examined the characteristics of magnetic clouds, and with other spacecraft measured long, steady reconnection layers in the solar wind near Earth.
1994	Qizhou Zhang and colleagues use observations of solar-type stars to demonstrate that their brightness increases with magnetic activity and determine possible brightness variations of the Sun in past centuries. This suggests that the solar brightness has increased 0.2–0.6% as the magnetic activity increased from the Maunder Minimum (1645–1715) to the 1980s.
1995	Johannes Geiss, George Gloeckler and Rudolf Von Steiger use *Ulysses* ion composition measurements to suggest that the fast solar wind originates in a region of low electron temperature and that the slow solar wind originates in a region of high electron temperature.
1995	Louis J. Lanzerotti and George M. Simnett and their co-workers discover unexpected, recurrent enhancements of electrons at high solar latitudes using *Ulysses* data.
1995	The *SOlar and Heliospheric Observatory*, abbreviated *SOHO*, is launched on 2 December 1995. The Large Angle and Spectometric Coronagraph, or LASCO for short, aboard *SOHO* carried out more than a decade of investigations of coronal mass ejections. *SOHO's* Extreme-ultraviolet Imaging Telescope, abbreviated EIT, observed solar flares and the aftermath of coronal mass ejections during the same period. The UltraViolet Coronagraph Spectrometer, or UVCS for short, aboard *SOHO* has also been used to study coronal mass ejections, as well as the solar wind.
1995–1998	Judith Lean, Jürg Beer, and Raymond Bradley use a comparison of the reconstructed solar irradiance variation and the estimated northern hemispheric surface temperature changes to show that the Sun's varying brightness correlated best with the climate in the pre-industrial period between 1610 and 1800. They also found that solar forcing could have driven about half of the 0.55 °C global warming since 1860, and that the Sun might explain about one-third of the temperature increase since 1970. Thomas J. Crowley and Kwang-Yul Kim provide a similar comparison and arrive at a similar conclusion in 1996. In 1998, Claus Fröhlich and Judith Lean presented the first reliable composite of nearly two decades of measurements of the Sun's total irradiance of Earth, or the solar constant, concluding that the Sun contributed little to global warming during the previous decade.
1997	The *Advanced Composition Explorer*, or *ACE* for short, was launched on 25 August 1997. From a vantage point just outside the Earth, *ACE* monitors the solar wind, with its charged particles and magnetic fields, and observes high-energy particles accelerated at the Sun, within the solar wind, or in the galactic regions beyond the heliosphere.

Date	Event
1997	Henrik Svensmark and Eigil Friis-Christensen show that global cloud cover during the previous solar cycle was strongly correlated with the cosmic ray flux, which is itself inversely correlated with solar activity. The observed association of global temperature variations and the solar cycle length might be explained by this solar-driven fluctuation in cloud cover.
1997	Warren B. White, Judith Lean, Daniel R. Cayan, and Michael D. Dettinger show that the global-average sea surface temperatures are warming and cooling by up to $0.1\,°C$ in time with the 11-year solar cycle for the past 50 years. These temperature changes could be related to increases and decreases in the solar irradiance of Earth.
1997, 1999	Alphonse C. Sterling and Hugh S. Hudson, and then Richard C. Canfield, Hugh S. Hudson, and David E. McKenzie, demonstrate a strong correlation between the appearance of large sigmoid (S or inverted S) shapes, in *Yohkoh* soft X-ray images, and the likelihood of a coronal mass ejection. The ejections were detected by transient X-ray activity, with arcades or cusp signatures, in the same region a few days after the twisted warning signatures.
1998	Gisela Dreschhoff and Edward J. Zeller report the discovery of unexpectedly high nitrate concentrations in Greenland ice cores, attributed to individual solar proton events injected into the winter polar stratosphere during the past 415 years, suggesting that such ice cores can be used to measure solar energetic particles arriving at the Earth in pre-historic times.
1998–2000	Michael E. Mann, Raymond E. Bradley, and Malcolm K. Hughes, and independently Thomas Crowley, examine the causes of climate change over the past 1,000 years, concluding that greenhouse gases released by human activity are responsible for global warming in the late 20th century.
1999	Michael Lockwood, Richard Stamper, and Matthew N. Wild report a doubling of the Sun's coronal magnetic field during the past 100 years.
1999–2000	Richard C. Canfield, Hugh S. Hudson, and David McKenzie find that active regions with an obvious soft X-ray sigmoid running through them are more likely to produce coronal mass ejections than active regions showing no large-scale sigmoid structures. Alphonse C. Sterling reports that a pre-eruption sigmoid pattern is present in over half the coronal mass ejections.
1999–2000	Jean Robert Petit and colleagues report the climate and atmospheric history of the past 420,000 years form the deep ice core extracted from Vostok, Antarctica. The results, also summarized by Bernhard Stauffer in 2000, indicate that the long-term increases and decreases in the air temperature are associated with the rise and fall of the atmospheric concentration of the three principal greenhouse gases – carbon dioxide, methane, and nitrous oxide.
1999–2002	Säm Krucker and colleagues, Dennis K. Haggerty and Edmond C. Roelof, and George M. Simnett and co-workers compare the solar release time of near-relativistic (40–300 keV) electrons arriving at Earth, and observed from *Wind* and *ACE*, with the onset of associated radio or soft X-ray radiation, suggesting delayed injection or propagation delays of about 10 min for some of these electrons. In 2006, Stephen Kahler reports that most near-relativistic electron events were accelerated in solar flares, but that relativistic electrons with energies above 300 keV are often accelerated by CME-driven shocks.
2000	Hilary Cane, Ian G. Richardson, and O. Christopher St. Cyr study the relationship between coronal mass ejections, abbreviated CMEs, and geomagnetic storms, concluding that only about half of frontside halo CMEs, observed from *SOHO's* LASCO, encounter Earth, that three quarters of such CMEs do not result in even moderate geomagnetic activity, and that the goeeffectiveness of a CME depends strongly on its southward magnetic field strength.

Date	Event
2000	Joseph E. Mazur and his co-workers use impulsive flare particles to demonstrate the mixing of interplanetary magnetic fields.
2000	John C. Raymond and colleagues present the first ultraviolet spectral evidence for a shock wave driven by a coronal mass ejection. These results were enhanced and extended by Angela Ciaravella and co-workers in 2005–2006.
2000	Udo Schüle and colleagues use the Solar Ultraviolet Measurements of Emitted Radiation, abbreviated SUMER, instrument aboard *SOHO* to show that quiet-Sun radiance variations occur during the 11-year solar activity cycle.
2000	Sheela Shodhan and her colleagues use the third *International Sun-Earth Explorer*, abbreviated *ISEE 3*, the eighth *Interplanetary Monitoring Platform*, abbreviated *IMP 8*, and *Wind* observations of counter-streaming electrons to infer the topology of interplanetary magnetic clouds associated with coronal mass ejections.
2000	Sami K. Solanki and co-workers estimate and model the long-term changes of the Sun's large-scale magnetic field back to the Maunder Minimum in the late 1600s, reproducing the previously reported doubling of the interplanetary magnetic field in the past 100 years.
2001	Douglas C. Braun and Charles Lindsey apply phase-sensitive helioseismic holography to *SOHO* MDI data to demonstrate seismic imaging of the far hemisphere of the Sun.
2001	Leonard F. Burlaga and co-workers identify two classes of fast solar ejecta moving past the Earth, the magnetic clouds with a flux-rope magnetic structure and complex ejecta with disordered magnetic fields. Only about one-third of the ejecta observed at Earth are magnetic clouds, and the other complex ejecta may be the result of the interaction of multiple coronal ejections moving at different speeds. Geomagnetic storms seem to be preferentially produced by magnetic clouds.
2001	Natchimuthuk Gopalswamy and co-workers detect intense radio emission following an interplanetary type II burst, at the time a fast coronal mass ejection overtakes a slow one, describing the interaction as coronal mass ejection "cannibalism" that might enhance the acceleration of solar energetic particles in the slow coronal mass ejection.
2001	Jie Zhang and co-workers use instruments aboard *SOHO* to study the temporal relationship between coronal mass ejections and solar flares, describing a three-phase speed profile for coronal mass ejections – the initiation, impulsive acceleration, and propagation phases. The initiation phase is characterized by a slow rise with a speed of about $80 \, km \, s^{-1}$ for a period of tens of minutes. The subsequent acceleration phase, which coincides with the flare impulsive rise and lasts a few to tens of minutes, ceases near the peak of the soft X-ray emission, and is followed by the propagation phase.
2002	The *Ramaty High Energy Solar Spectroscopic Imager*, abbreviated *RHESSI*, is launched on 5 February 2002, to investigate high-energy acceleration processes close to the Sun by observing flare radiation from soft X-rays to gamma rays.
2002	Thomas Woods and Gary Rottman review the solar irradiance variability and provide reference spectra from 0.1 to 200 nm for studies of planetary atmospheres.
2002	David A. Falconer, Ronald L. Moore, and Gilmer Allen Gary demonstrate that the coronal mass ejection productivity of solar active regions is correlated with their global, non-potentiality, or overall shear and twist, detected in vector magnetograms.

Date	Event
2003	George Gloeckler uses the Solar Wind Ion Composition Spectrometer, abbreviated SWICS, instruments on *Ulysses* and the *Advanced Composition Explorer*, abbreviated *ACE*, to detect ubiquitous high-speed, suprathermal ions moving from two to 50 times the solar wind speed.
2003	Angelos Vourlidas and colleagues use observations and numerical simulations of a unique coronal mass ejection seen with *SOHO's* LASCO, without the classical three-part structure, to infer direct detection of a shock driven by the ejection.
2004	Nancy U. Crooker and her colleagues use *Wind* observations to describe large-scale magnetic field inversions at solar-wind sector boundaries.
2004	Igor V. Sokolov, Ilia I. Roussev, and co-workers model solar proton acceleration by coronal-mass-ejection driven shocks, showing how protons with GeV energy can be accelerated by these shock waves within 6 solar radii.
2004	Gerard Thuillier and colleagues provide reference spectra for the solar irradiance from 0.5 to 2,400 nm.
2004	Jarmo Torsti, Esa Riihonen, and Leon Kocharov use an instrument on *SOHO* to monitor protons while the spacecraft was inside an interplanetary magnetic cloud associated with a coronal mass ejection, or CME. They conclude that the magnetic flux-rope structure of the CME provides a "highway" for the transport of solar energetic protons with a parallel mean free path of at least 10 AU.
2004–2006	Peter Foukal and co-workers conclude that solar irradiance variations are unlikely to have had a significant influence on global warming since the 17th century.
2005	Jack T. Gosling and his colleagues provide direct evidence for magnetic reconnection in the solar wind near the Earth's orbit at 1 AU.
2005	Carolus J. Schrijver and colleagues use *TRACE* and *SOHO* MDI observations to show that non-potential, coronal magnetic fields in active regions emerge from below.
2005	Allan J. Tylka and colleagues propose that the interplay between shock geometry and a compound seed population accounts for the variable composition of large, gradual solar energetic particle events.
2005	Yi-Ming Wang, Judith L. Lean, and Neil R. Sheeley Jr. model the Sun's magnetic field and irradiance since 1713. Their model suggests much smaller irradiance changes for the Maunder Minimum period in the late 1600s than previous estimates that use stellar variability results.
2005–2006	Phil Chamberlin develops the Flare Irradiance Spectral Model, abbreviated FISM.
2006	Peter Foukal, Claus Fröhlich, Hendrik Spruit, and Thomas M. L. Wigley report that variations of the Sun's total energy output measured from spacecraft since 1978 are too small to have contributed appreciably to accelerated global warming over the past 30 years, and that brightening of the Sun is unlikely to have had a significant influence on global warming since the 17th century.
2006	The Japanese *Hinode*, meaning *sunrise*, spacecraft is launched on 23 September 2006, to investigate how magnetic interactions and related processes generate the solar atmosphere and solar activity, including how magnetic energy is converted into intense ultraviolet and X-ray radiation and how magnetic interactions cause solar flares and coronal mass ejections.

Date	Event
2006	The twin spacecraft of the *Solar TErrestrial RElations Observatory*, abbreviated *STEREO A* and *B*, are launched on 25 October 2006, to obtain a three-dimensional, stereoscopic view of coronal mass ejections from their onset at the Sun to the orbit of the Earth, and to thereby investigate the origin, evolution, mechanisms, and interplanetary propagation of coronal mass ejections.
2006	David F. Webb and colleagues report the successful test flight of the Solar Mass Ejection Imager, abbreviated SMEI, aboard the Air Force's *Coriolus* spacecraft, launched on 6 January 2003. During the first 1.5 years of operation, 130 coronal mass ejections were observed traveling through the inner heliosphere, as far as the Earth's orbit and beyond, and at last 30 of these events were associated with major geomagnetic storms on Earth; most of these were observed as front-side halo events by *SOHO's* LASCO.
2006–2007	In 2006, Tai D. Phan and his colleagues report three-spacecraft observations that show magnetic reconnection can occur over extended regions in the solar wind, at least 390 times the Earth in size. In 2007, John T. Gosling and co-workers report five-spacecraft observations of oppositely directed exhaust jets from a magnetic reconnection X-line extending 4.26×10^6 km in the solar wind at 1 AU.
2007	Arik Posner demonstrates that solar energetic electrons, traveling at nearly the velocity of light, always arrive at the Earth's orbit ahead of the solar energetic protons and other ions, permitting up to 1-h forecasting of radiation hazards from the solar energetic ion events.

Appendix

I. Solar Space Missions

The internet addresses for the Solar Space Missions discussed in this book are listed alphabetically below. The internet addresses for instruments aboard these spacecraft are given in Chap. 1, and usually on the spacecraft home pages given below.

ACE:

http://www.srl.caltech.edu/ACE/

Hinode:

http://www.isas.jaxa.jp/home/solar/
http://solarb.msfc.nasa.gov/
http://solar-b.nao.ac.jp/index_e.shtml

RHESSI:

http://hesperia.gsfc.nasa.gov/hessi/
http://hessi.ssl.berkeley.edu/
http://rhessidatacenter.ssl.berkeley/
http://hesperia.gsfc.nasa.gov/rhessidatacenter/

SOHO:

http://sohowww.nascom.nasa.gov/

STEREO:

http://stereo.jhuapl.edu/
http://stereo.gsfc.nasa.gov/

Ulysses:

http://ulysses.jpl.nasa.gov/
http://helio.estec.esa.nl/ulysses/

Wind:

http://wind.nasa.gov
http://pwg.gsfc.nasa.gov/windnrt/

Yohkoh:

http://solar.physics.montana.edu/ylegacy/
http://www.lmsal.com/SXT/
http://ydac.mssl.ucl.ac.uk/ydac/

II. Helioseismology

http://soi.stanford.edu
http://sohowww.nascom.nasa.gov/
http://golfwww.medoc-ias.u-psud.fr
http://gong.nso.edu
http://solarphysics.livingreviews.org/
http://soi.stanford.edu/data/full_farside/

III. Space Weather

http://solar.sec.noaa.gov/
http://www.spaceweather.com
http://www.windows.ucar.edu/spaceweather
http://www.esa-spaceweather.net/
http://solarphysics.livingreviews.org/

IV. Solar Observatories and Groups

http://www.gong.nso.edu/
http://www.hao.ucar.edu/
http://www.nso.edu/
http://www.solarphysics.kva.se/
http://www.lmsal.com
http://hea-www.harvard.edu/SSXG/
http://solar.physics.montana.edu/ylegacy/
http://sspg1.bnsc.rl.ac.uk/Share/sol.html
http://sun.stanford.edu/

V. Virtual Observatories

http://vho.nasa.gov/
http://vspo.gsfc.nasa.gov
http://umbra.nascom.nasa.gov/vso
http://vso.nso.gov/
http://vso.stanford.edu

VI. Educational

http://ase.tufts.edu/cosmos/
http://solar-center.stanford.edu/
http://helios.gsfc.nasa.gov/
http://istp.gsfc.nasa.gov/istp/outreach/
http://istp.gsfc.nasa.gov/exhibit/
http://www.lmsal.com/YPOP/

VII. NASA

http://nasascience.nasa.gov/
http://sec.gsfc.nasa.gov
http://umbra.nascom.nasa.gov/solar_connections.html
http://solarscience.msfc.nasa.gov/

References

A

Abdurashitov, J. N., et al. (2002): Measurement of the solar neutrino capture rate by the Russian-American gallium solar neutrino experiment during one half of the 22 year cycle of solar activity. *Journal of Experimental and Theoretical Physics* **95**, 181–193.

Abramenko, V. I. (2005a): Multifractal analysis of solar magnetograms. *Solar Physics* **228**, 29–42.

Abramenko, V. I. (2005b): Relationship between magnetic power spectrum and flare productivity in solar active regions. *Astrophysical Journal* **629**, 1141–1149.

Abramenko, V. I., Longscope, D. W. (2005): Distribution of the magnetic flux in elements of the magnetic field in active regions. *The Astrophysical Journal* **619**, 1160–1166.

Abramenko, V. I., Pevtsov, A. A., Romano, P. (2006): Coronal heating and photospheric turbulence parameters: observational aspects. *The Astrophysical Journal (Letters)* **646**, L81.

Acton, L. W., et al. (1982): Chromospheric evaporation in a well-observed compact flare. *Astrophysical Journal* **263**, 409–422.

Acton, L. W., et al. (1992a): The *Yohkoh* mission for high-energy solar physics. *Science* **258**, 618–625.

Acton, L. W., et al. (1992b): The morphology of 20×10^6 K plasma in large non-impulsive solar flares. *Publications of the Astronomical Society of Japan* **44**, L71–L75.

Acuña, M. H., et al. (1995): The global geospace science program and its investigations. *Space Science Reviews* **71**, 5–21.

Acuña, M. H., et al. (2008): The STEREO/IMPACT magnetic field experiment. *Space Science Reviews* **136**, Issue 1–4, 203–226.

Adhémar, J. A. (1842): *Révolutions de la Mer*. Paris: Deluges Périodiques.

Aelig, M. R., et al. (1997): Solar wind iron charge states observed with high time resolution with *SOHO*/CELIAS/CTOF. In: *Proceedings of the fifth SOHO workshop. The corona and solar wind near minimum activity. ESA SP-101.* Noordwijk: ESA Publications Division, pp. 157–161.

Aharmin, B., et al. (2005): Electron energy spectra, fluxes, and day-night asymmetries of ^{8}B solar neutrinos from measurements with NaCl dissolved in the heavy-water detector at the Sudbury Neutrino Observatory. *Physical Review C* **72**, 055502–055549.

Ahmad, Q. R., et al. (2001): Measurement of the rate of the electron neutrino plus deuterium interactions produced by ^{8}B solar neutrinos at the Sudbury Neutrino Observatory. *Physical Review Letters* **87**, 071301–071307.

Ahmad, Q. R., et al. (2002a): Direct evidence for neutrino flavor transformation from neutral-current interactions in the Sudbury Neutrino Observatory. *Physical Review Letters* **89**, 011301–011302.

Ahmad, Q. R., et al. (2002b): Measurement of day and night neutrino energy spectra at SNO and constraints on neutrino mixing parameters. *Physical Review Letters* **89**, 011303–011307.

Ahmed, S. N., et al. (2004): Measurement of the total active ^{8}B solar neutrino flux at the Sudbury Neutrino Observatory with enhanced neutral current sensitivity. *Physical Review Letters* **92**, 181301.

Aiouaz, T., Peter, H., Lemaire, P. (2005): The correlation between coronal Doppler shifts and the supergranular network. *Astronomy and Astrophysics* **435**, 713–721.

Akasofu, S.-I. (1964): The development of the auroral substorm. *Planetary and Space Science* **12**, 273–282.

Akasofu, S.-I. (1968): *Polar and magnetospheric substorms*. Dordrecht: Reidel.

Akasofu, S.-I., Chapman, S. (1972): *Solar-terrestrial physics*. Oxford: Oxford University Press.

Akasofu, S.-I. (1977): *Physics of magnetospheric substorms*. Dordrecht: Reidel.

Akasofu, S.-I. (1981): Energy coupling between the solar wind and the magnetosphere. *Space Science Reviews* **28**, 121–190.

Akasofu, S.-I. (1989): Substorms. *EOS American Geophysical Union* **70**, 529–532.

Alazraki, G., Couterier, P. (1971): Solar wind acceleration caused by the gradient of Alfvén wave pressure. *Astronomy and Astrophysics* **13**, 380–389.

Alexander, D., Richardson, I. G., Zurbuchen, T. H. (2006): A brief history of CME science. *Space Science Reviews* **123**, 3–11.

Alexander, P. (1992): History of solar coronal expansion studies. *EOS Transactions of the American Geophysical Union* **73**(41), 433, 438.

Alfvén, H. (1942): The existence of electromagnetic-hydrodynamic waves. *Nature* **150**, 405.

Alfvén, H. (1947): Granulation, magneto-hydrodynamic waves, and the heating of the solar corona. *Monthly Notices of the Royal Astronomical Society* **107**, 211–219.

Alfvén, H. (1950): *Cosmical Electrodynamics*. Oxford: Clarendon Press.

Alfvén, H. (1975): Electric current structure of the magnetosphere. In: *Physics of the Hot Plasma in the Magnetosphere* (Eds. B. Hultqvist and L. Stenflo). New York: Plenum 1975, pp. 1–22.

Alfvén, H. (1977): Electric currents in cosmic plasmas. *Reviews of Geophysics and Space Physics* **15**, 271–284.

Alfvén, H., Carlqvist, P. (1967): Currents in the solar atmosphere and a theory of solar flares. *Solar Physics* **1**, 220–228.

Alfvén, H., Herlofson, N. (1950): Cosmic radiation and radio stars. *Physical Review* **78**, 616. Reproduced in: *A source Book in Astronomy and Astrophysics 1900–1975* (Eds. K. R. Lang and O. Gingerich). Cambridge: Harvard University Press 1977.

Aliani, P., et al. (2004): Neutrino mass parameters from KamLAND, SNO, and other solar evidence. *Physical Review* D **69**, 013005–013012.

Allen, C. W. (1947): Interpretation of electron densities from corona brightness. *Monthly Notices of the Royal Astronomical Society* **107**, 426–432.

Altschuler, M. D., Newkirk, G. (1969): Magnetic fields and the structure of the solar corona. I. Methods in calculating coronal fields. *Solar Physics* **9**, 131–149.

Amari, T., et al. (2000): A twisted flux rope model for coronal mass ejections and two ribbon flares. *Astrophysical Journal (Letters)* **529**, L49–L52.

Anderson, C. D. (1932a): Energies of cosmic-ray particles. *Physical Review* **41**, 405–421.

Anderson, C. D. (1932b): The apparent existence of easily deflectable positives. *Science* **76**, 238–239.

Anderson, C. D. (1933): The positive electron. *Physical Review* **43**, 491–494. Reproduced in: *Cosmic Rays* (Ed. A. M. Hillas). New York: Pergamon Press 1972.

Anderson, C. D., Neddermeyer, S. H. (1936): Cloud chamber observations of cosmic rays at 4300 meters elevation and near sea-level. *Physical Review* **50**, 263–271.

Anderson, K. A., Winckler, J. R. (1962): Solar flare X-ray burst on September 28, 1961. *Journal of Geophysical Research* **67**, 4103–4117.

Anderson, L. S., Athay, R. G. (1989): Chromospheric and coronal heating. *The Astrophysical Journal* **336**, 1089–1091.

Anderson, L. S., Athay, R. G. (1989): Model solar chromosphere with prescribed heating. *The Astrophysical Journal* **346**, 1010–1018.

Anderson, P. C., et al. (1998): Energetic auroral electron distributions derived from global X-ray measurements and comparison with in-situ particle measurements. *Geophysical Research Letters* **25**, 4105–4108.

Ando, H., Osaki, Y. (1975): Nonadiabatic nonradial oscillations – an application to the five-minute oscillation of the Sun. *Publications of the Astronomical Society of Japan* **27**, 581–603.

Andreev, V. E., et al. (1997): Characteristics of coronal Alfvén waves deduced from *Helios* Faraday rotation measurements. *Solar Physics* **176**, 387–402.

Antia, H. M. (1998): Estimate of solar radius from f-mode frequencies. *Astronomy and Astrophysics* **330**, 336–340.

Antia, H. M., Basu, S. (2000): Temporal variations of the rotation rate in the solar interior. *The Astrophysical Journal* **541**, 442–448.

Antia, H. M., Basu, S. (2001): Temporal variations of the solar rotation rate at high latitudes. *The Astrophysical Journal (Letters)* **559**, L67–L70.

Antia, H. M., Basu, S. (2005): The discrepancy between solar abundances and helioseismology. *The Astrophysical Journal (Letters)* **620**, L129–L132.

Antia, H. M., Chitre, S. M. (1995): Helioseismic bounds on the central temperature of the Sun. *The Astrophysical Journal* **442**, 434–445.

Antia, H. M., Chitre, S. M., Kale, D. M. (1978): Overstabilization of acoustic modes in a polytropic atmosphere. *Solar Physics* **56**, 275–292.

Antia, H. M., et al. (2001): Solar-cycle variation of the sound speed asphericity from GONG and MDI data 1995–2000. *Monthly Notices of the Royal Astronomical Society* **327**, 1029–1040.

Antiochos, S. K. (1996): A model for coronal mass ejections. *Bulletin of the American Astronomical Society* **28**, 1346.

Antiochos, S. K. (1998): The magnetic topology of solar eruptions. *The Astrophysical Journal (Letters)* **502**, L181.

Antiochos, S. K., De Vore, C.R., Klimchuk, J.A. (1999): A model for coronal mass ejections. *Astrophysical Journal* **510**, 485–493.

Antiochos, S. K., et al. (2003): Constraints on active region coronal heating. *The Astrophysical Journal* **590**, 54.

Antiochos, S. K., Noci, G. (1986): The structure of the static corona and transition region. *The Astrophysical Journal* **301**, 440–447.

Antonucci, E. (2006): Wind in the solar corona: dynamics and composition. *Space Science Reviews* **124**, 35–50.

Antonucci, E., Abbo, L., Dodero, M. A. (2005): Slow wind and magnetic topology in the solar minimum corona in 1996–1997. *Astronomy and Astrophysics* **435**, 699–711.

Antonucci, E., Abbo, L., Teloni, D. (2006): Oxygen abundance and energy deposition in the slow coronal wind. *The Astrophysical Journal* **643**, 1239–1244.

Antonucci, E., Dennis, B. R. (1983): Observations of chromospheric evaporation during the *Solar Maximum Mission*. *Solar Physics* **86**, 67–77.

Antonucci, E., Dodero, M. A., Giordano, S. (2000): Fast solar wind velocity in a polar coronal hole during solar minimum. *Solar Physics* **197**, 115.

Antonucci, E., et al. (1982): Impulsive phase of flares in soft X-ray emission. *Solar Physics* **78**, 107–123.

Antonucci, E., Gabriel, A. H., Dennis, B. R. (1984): The energetics of chromospheric evaporation in solar flares. *The Astrophysical Journal* **287**, 917–925.

Antonucci, E., Rosner, R. Tsinganos, K. (1986): On magnetic field stochasticity and nonthermal line broadening in solar flares. *The Astrophysical Journal* **301**, 975–980.

Anzer, U., Pneuman, G. W. (1982): Magnetic reconnection and coronal transients. *Solar Physics* **79**, 129–147.

Appleton, E. V. (1932): Wireless studies of the ionosphere. *Proceedings of the Institute of Electrical Engineers* **71**, 642–650.

Appleton, E. V., Barnett, M. A. F. (1925a): Local reflection of wireless waves from the upper atmosphere. *Nature* **115**, 333–334.

Appleton, E. V., Barnett, M. A. F. (1925b): On some direct evidence for downward atmospheric reflection of electric rays. *Proceedings of the Royal Society of London A* **109**, 621–641.

Appleton, E. V., Hey, J. S. (1946): Solar radio noise. *Philosophical Magazine* **37**, 73–84.

Araki, T., et al. (2005): Measurement of neutrino oscillation with KamLAND: Evidence of spectral distortion. *Physical Review Letters* **94**, 081801.

Aran, A., et al. (2007): Modeling and forecasting solar energetic particle events at Mars: the event on 6 March 1989. *Astronomy and Astrophysics* **469**, 1123–1134.

Armstrong, A. H., Harrison, F. B., Heckman, H. H., Rosen, L. (1961): Charged particles in the inner Van Allen radiation belt. *Journal of Geophysical Research* **66**, 351.

Arndt, M. B., Habbal, S. R., Karovska, M. (1994): The discrete and localized nature of the variable emission from active regions. *Solar Physics* **150**, 165–178.

Arnoldy, R. L. (1971): Signature in the interplanetary medium for substorms. *Journal of Geophysical Research* **76**, 5189–5201.

Arnoldy, R. L., Kane, S. R., Winckler, J. R. (1967): A study of energetic solar flare X-rays. *Solar Physics* **2**, 171–178.

Arnoldy, R. L., Kane, S. R., Winckler, J. R. (1968): Energetic solar flare X-rays observed by satellite and their correlation with solar radio and energetic particle emission. *Astrophysical Journal* **151**, 711–736.

Arrhenius, S. (1896): On the influence of carbonic acid in the air upon the temperature of the ground. *Philosophical Magazine and Journal of Science* **41**, 237–268.

Aschwanden, M. J. (1999): Do EUV nanoflares account for coronal heating? *Solar Physics* **190**, 233–247.

Aschwanden, M. J. (2001a): Revisiting the determination of the coronal heating function from *Yohkoh* data. *The Astrophysical Journal (Letters)* **559**, L171–L174.

Aschwanden, M. J. (2001b): An evaluation of coronal heating models for active regions based on *Yohkoh, SOHO* and *TRACE* observations. *The Astrophysical Journal* **560**, 1035–1044.

Aschwanden, M. J. (2002): The differential emission measure distribution in the multiloop corona. *The Astrophysical Journal (Letters)* **580**, L79–L83.

Aschwanden, M. J. (2004, 2006): *Physics of the Solar Corona: An Introduction.* New York: Springer-Verlag 2004, Second Edition 2006.

Aschwanden, M. J. (2006a): Particle acceleration in solar flares and escape into interplanetary space. In: *Solar Eruptions and Energetic Particles: Geophysical Monograph Series 165.* (Eds. N. Gopalswamy, R. Mewaldt and J. Torsti) Washington: American Geophysical Union, pp. 189–197.

Aschwanden, M. J. (2006b): The localization of particle acceleration sites in solar flares and CMEs. *Space Science Reviews* **124**, 361–372.

Aschwanden, M. J. (2008a): An observational test that disproves coronal nanoflare heating models. *The Astrophysical Journal (Letters)* **672**, L135–L138.

Aschwanden, M. J. (2008b): Understanding the overpressure in cooling coronal loops. *The Astrophysical Journal* – submitted.

Aschwanden, M. J., Acton, L. W. (2001): Temperature tomography of the soft x-ray corona: Measurements of electron densities, temperatures and differential emission measure distributions above the limb. *The Astrophysical Journal* **550**, 475–492.

Aschwanden, M. J., Alexander, D. (2001): Flare plasma cooling from 30 MK down to 1 MK modeled from *Yohkoh, GOES,* and *TRACE* observations of the Bastille Day event (14 July 2000). *Solar Physics* **204**, 91–120.

Aschwanden, M. J., Benz, A. O. (1997): Electron densities in solar flare loops, chromospheric evaporation upflows, and acceleration sites. *Astrophysical Journal* **480**, 825–839.

Aschwanden, M. J., Charbonneau, P. (2002): Effects of temperature bias on nanoflare statistics. *The Astrophysical Journal (Letters)* **566**, L59–L62.

Aschwanden, M. J., De Pontieu, B., Schrijver, C.J., Title, A.M. (2002): Transverse oscillations in coronal loops observed with *TRACE* – II Measurements of geometric and physical parameters. *Solar Physics* **206**, 99–132.

Aschwanden, M. J., et al. (1996a): Electron time-of-flight distances and flare loop geometries compared from *CGRO* and *Yohkoh* observations. *Astrophysical Journal* **468**, 398–417.

Aschwanden, M. J., et al. (1996b): Electron time-of-flight measurements during the Masuda flare, 1992 January 13. *Astrophysical Journal* **464**, 985–998.

Aschwanden, M. J., et al. (1998): The scaling law between electron time-of-flight distances and loop lengths in solar flares. *Astrophysical Journal* **470**, 1198–1217.

Aschwanden, M. J., et al. (1999a): Coronal loop oscillations observed with the Transition Region and Coronal Explorer. *Astrophysical Journal* **520**, 880–894.

Aschwanden, M. J., et al. (1999b): Quadrupolar magnetic reconnection in solar flares three-dimensional geometry inferred from *Yohkoh* observations. *The Astrophysical Journal* **526**, 1026–1045.

Aschwanden, M. J., et al. (1999c): Three-dimensional stereoscopic analysis of solar active region loops I. *SOHO*/EIT observations at temperatures of $(1.0–1.5) \times 10^6$ K. *The Astrophysical Journal* **515**, 842–867.

Aschwanden, M. J., et al. (1999d): Time variability of the "quiet" sun observed with *TRACE*. II. Physical parameters, temperature evolution, and energetics of euv nanoflares. *The Astrophysical Journal* **535**, 1047–1065.

Aschwanden, M. J., et al. (2000): Three-dimensional stereoscopic analysis of solar active region loops II. *SOHO*/EIT observations of temperatures of 1.5–2.5 MK. *The Astrophysical Journal* **531**, 1129–1149.

Aschwanden, M. J., et al. (2006): Theoretical modeling for the *STEREO* mission. *Space Science Reviews* **123**, 127.

Aschwanden, M. J., Lyndsay, F., Schrijver, C.J., David, A. (1999): Coronal loop oscillations observed with the *Transition Region And Coronal Explorer. The Astrophysical Journal* **520**, 880–894.

Aschwanden, M. J., Nightingale, R. W. (2005): Elementary loop structures of the solar corona analyzed from *TRACE* triple-filter images. *The Astrophysical Journal* **633**, 499–517.

Aschwanden, M. J., Nightingale, R. W., Alexander, D. (2000): Evidence for nonuniform heating of coronal loops inferred from multithread modeling of *TRACE* data. *The Astrophysical Journal* **541**, 1059–1077.

Aschwanden, M. J., Nitta, N., Wuelser, J.-P., Lement, J. (2008): First 3D reconstructions of coronal loops with the STEREO A + B spacecraft II. Electron density and temperature measurements. *The Astrophysical Journal* **680**, 1477–1495.

Aschwanden, M. J., Poland, A., Rabin, D. M. (2001): The new solar corona. *Annual Review of Astronomy and Astrophysics* **39**, 175–210.

Aschwanden, M. J., Schrijver, C. J., David, A. (2001): Modeling of coronal EUV loops observed with *TRACE* I. Hydrostatic solutions with nonuniform heating. *The Astrophysical Journal* **550**, 1036–1050.

Aschwanden, M. J., Tsiklauri, D. (2008): The non-equilibrium scaling law of cooling coronal loops. *The Astrophysical Journal* – submitted.

Aschwanden, M. J., Winebarger, A., Tsiklauri, D., Peter, H. (2007): The coronal heating paradox. *The Astrophysical Journal* **659**, 1673–1681.

Ashie, Y., et al. (2004): Evidence for an oscillatory signature in atmospheric neutrino oscillations. *Physical Review Letters* **93**, 101801.

Asplund, M., et al. (2004): Line formation in solar granulation IV. [O I], O I and OH lines and the photospheric O abundance. *Astronomy and Astrophysics* **417**, 751–768.

Aston, F. W. (1920): The mass-spectra of chemical elements. *Philosophical Magazine and Journal of Science* **39**, 611–625.

Aström, E. (1950): On waves in an ionized gas. *Arkiv Fur Fysik* **2**, 443.

Athay, R. G., Moreton, G. E. (1961): Impulsive phenomena of the solar atmosphere I. Some optical events associated with flares showing explosive phase. *The Astrophysical Journal* **133**, 935–945.

Athay, R. G., White, O. R. (1978): Chromospheric and coronal heating by sound waves. *The Astrophysical Journal* **226**, 1135–1139.

Athay, R. G., White, O. R. (1979): Chromospheric oscillations observed with *OSO 8* IV. Power and phase spectra for CIV. *The Astrophysical Journal* **229**, 1147–1162.

Atkinson, R. d'E. (1931): Atomic synthesis and stellar energy I, II. *The Astrophysical Journal* **73**, 250–295, 308–347. Reproduced in *A Source Book in Astronomy and Astrophysics 1900–1975* (Eds. K. R. Lang and O. Gingerich). Cambridge, Massachusetts: Harvard University Press 1979, 303–308.

Atkinson, R. d'E., Houtermans, F. G. (1929): Zur frage de aufbaumöglichkeit der elements in sternen. *Zeitschrift für Physik* **54**, 656.

Aulanier, G., et al. (2000): The topology and evolution of the Bastille Day flare. *Astrophysical Journal* **540**(2), 1126–1142.

Aulanier, G., et al. (2007): Slipping magnetic reconnection in coronal loops. *Science* **318**, 1588–1590.

Avignon, Y., Martres, M. J., Pick, M. (1966): Etude de la "composante lentement variable" en relation avec la structure des centres d'activité solaire associés. *Annales d'Astrophysique* **29**, 33–42.

Axford, W. I. (1960): The modulation of galactic cosmic rays in the interplanetary medium. *Planetary and Space Science* **13**, 115–130.

Axford, W. I. (1962): The interaction between the solar wind and the Earth's magnetosphere. *Journal of Geophysical Research* **67**, 3791–3796.

Axford, W. I. (1972): The interaction of the Solar wind with the interstellar medium. In: *Solar Wind: Proceedings of the Second International Conference* (Eds. C.P. Sonnett, P.j. Coleman, Jr. and J.M. Wilcox). NASA SP-308., p. 609.

Axford, W. I. (1980): Very hot plasmas in the solar system. *Highlights of Astronomy* **5**, 351–359.

Axford, W. I. (1985): The solar wind. *Solar Physics* **100**, 575–586.

Axford, W. I. (1994): The good old days. *Journal of Geophysical Research* **99**, 19199–19212.

Axford, W. I., Dessler, A. J., Gottlieb, B. (1963): Termination of solar wind and solar magnetic field. *The Astrophysical Journal* **137**, 1268–1278.

Axford, W. I., et al. (1999): Acceleration of the high speed solar wind in coronal holes. *Space Science Reviews* **97**, 25–41.

Axford, W. I., Hines, C. O. (1961): A unifying theory of high-latitude geophysical phenomena and geomagnetic storms. *Canadian Journal of Physics* **39**, 1433–1464.

Axford, W. I., Mc Kenzie, J. F. (1997): Solar wind. In: *Cosmic Winds and the Heliosphere* (Eds. J. R. Jokipii, C. P. Sonett and M. S. Giampapa). Tucson: University of Arizona Press 1997, pp. 31–66.

Axford, W. I., McKenzie, J. (1992): The origin of high speed solar wind streams. In: *Solar Wind Seven* (Eds. E. Marsch and R. Schwenn). New York: Elsevier, p. 1.

Ayres, T. R. (1997): Evolution of the solar ionizing flux. *Journal of Geophysical Research* **102**, 1641–1652.

B

Baade, W., Zwicky, F. (1934): Cosmic rays from super-novae. *Proceedings of the National Academy of Sciences (Washington)* **20**, 259.

Babcock, H. W. (1947): Zeeman effect in stellar spectra. *The Astrophysical Journal* **105**, 105–119.

Babcock, H. W. (1961): The topology of the sun's magnetic field and the 22-year cycle. *The Astrophysical Journal* **133**, 572–587.

Babcock, H. W., Babcock, H. D. (1955): The sun's magnetic field, 1952-1954. *The Astrophysical Journal* **121**, 349–366.

Bahall, J. N., Davis, R., Wolfenstein, L. (1988): Solar neutrinos: a field in transition. *Nature* **334**, 487–493.

Bahcall, J. N. (1964): Solar neutrinos I. Theoretical. *Physical Review Letters* **12**, 300–302. Reproduced in *A Source Book in Astronomy and Astrophysics, 1900–1975* (Eds. K. R. Lang, O. Gingerich). Cambridge, Massachusetts: Harvard University Press 1979, 389–395.

Bahcall, J. N. (1978): Solar neutrino experiments. *Review of Modern Physics* **50**, 881–903.

Bahcall, J. N. (1996): Solar neutrinos: where we are, where we are going. *The Astrophysical Journal* **467**, 475–484.

Bahcall, J. N., Basu, S., Kumar, P. (1997): Localized helioseismic constraints on solar structure. *The Astrophysical Journal (Letters)* **485**, L91–L94.

Bahcall, J. N., Basu, S., Pinsonneault, M., Serenelli, A. M. (2005): Helioseismological implications of recent solar abundance determinations. *The Astrophysical Journal* **618**, 1049–1056.

Bahcall, J. N., Basu, S., Serenelli, A. M. (2005): What is the neon abundance of the Sun? *The Astrophysical Journal* **631**, 1281–1285.

Bahcall, J. N., Bethe, H. A. (1990): A solution to the solar neutrino problem. *Physical Review Letters* **65**, 2233–2235.

Bahcall, J. N., et al. (1995): Progress and prospects in neutrino astrophysics. *Nature* **375**, 29–34.

Bahcall, J. N., et al. (1997): Are standard solar models reliable? *Physical Review Letters* **78**, 171–174.

Bahcall, J. N., Gonzalez-Garcia, M. C., Pena-Garay, C. (2003): Does the Sun shine by pp or CNO fusion reactions? *Physical Review Letters* **90**, 131301–131305.

Bahcall, J. N., Pinsonneault, M. H. (2004): What do we (not) know theoretically about solar neutrino fluxes? *Physical Review Letters* **92**, 121301.

Bahcall, J. N., Pinsonneault, M. H., Basu, S. (2001): Solar models: current epoch and time dependences, neutrinos and helioseismological properties. *The Astrophysical Journal* **555**, 990–1012.

Bahcall, J. N., Pinsonneault, M. H., Basu, S., Christensen-Dalsgaard, J. (1997): Are standard solar models reliable? *Physical Review Letters* **78**, 171–174.

Bahcall, J. N., Serenelli, A. M. (2005): How do uncertainties in the surface chemical composition of the Sun affect the predicted solar neutrino fluxes? *The Astrophysical Journal* **626**, 530–542.

Bahcall, J. N., Serenelli, A. M., Basu, S. (2005): New solar opacities, abundances, helioseismology, and neutrino fluxes. *The Astrophysical Journal (Letters)* **621**, L85–L88.

Bahcall, J. N., Serenelli, A. M., Basu, S. (2006): 10,000 standard solar models: a Monte Carlo simulation. *Astrophysical Journal Supplement Series* **165**, 400–431.

Bailey, D. K. (1957): Disturbances in the lower ionosphere observed at VHF following the solar flare of 23 February 1956 with particular reference to auroral zone absorption. *Journal of Geophysical Research* **62**, 431–463.

Bailey, F. (1842): Some remarks on the total eclipse of the Sun, on July 8th 1842. *Monthly Notices of the Royal Astronomical Society* **5**, 208–214.

Baker, D. N. (2000): The occurrence of operational anomalies in spacecraft and their relationship to space weather. *IEEE Transactions of Plasma Science* **28**, 2007–2016.

Baker, D. N., Carovillano, R. (1997): *IASTP* and solar-terrestrial physics. *Advances in Space Research* **20**, 531–538.

Baker, D. N., et al. (1987): Deep dielectric charging effects due to high energy electrons in Earth's outer magnetosphere. *Journal of Electrostatics* **20**, 3.

Baker, D. N., et al. (2001): The global efficiency of relativistic electron production in the Earth's magnetosphere. *Journal of Geophysical Research* **106**, 19169.

Baker, D. N., et al. (Eds., 2006): Solar dynamics and its effects on the heliosphere and earth. *Space Science Reviews* **124**, issue 1–4, 1–372. Reprinted by: Springer and the International Space Science Institute.

Balantekin, A. B., Yuksel, H. (2003): Constraints on neutrino parameters from neutral-current solar neutrino measurements. *Physical Review* D **68**, 113002–113007.

Balasubrahmanyan, V. K., Serlemitsos, A. T. (1974): Solar energetic particle event with ^{3}He/^{4}He greater than 1. *Nature* **252**, 460–462.

Baliunas, S. L. (1991): The past, present and future of solar magnetism: stellar magnetic activity. In: *The Sun in Time* (Eds. C. P. Sonett, M. S. Giampapa and M. S. Matthews). Tucson: The University of Arizona Press, pp. 809–831.

Baliunas, S. L., et al. (1995): Chromospheric variations in main-sequence stars. II. *The Astrophysical Journal* **438**, 269–287.

Baliunas, S. L., et al. (1996): A dynamo interpretation of stellar activity cycles. *The Astrophysical Journal* **460**, 848–854.

Baliunas, S. L., Jastrow, R. (1990): Evidence for long-term brightness changes of solar-type stars. *Nature* **348**, 520–523.

Baliunas, S. L., Soon, W. (1995): Are variations in the length of the activity cycle related to changes in brightness in solar-type stars? *The Astrophysical Journal* **450**, 896–901.

Baliunas, S. L., Vaughan, A. H. (1985): Stellar activity cycles. *Annual Review of Astronomy and Astrophysics* **23**, 379–412.

Balogh, A. (1998): Magnetic fields in the inner heliosphere. *Space Science Reviews* **83**, 93–104.

Balogh, A., et al. (1992): The magnetic field investigation on the *Ulysses* mission – Instrumentation and preliminary scientific results. *Astronomy and Astrophysics Supplement Series* **92**, No. 2, 221–236.

Balogh, A., et al. (1995): The heliospheric magnetic field over the south polar region of the sun. *Science* **268**, 1007–1010.

Balogh, A., et al. (1999): The solar origin of corotating interaction regions and their formation in the inner heliosphere. *Space Science Reviews* **89**, 141–178.

Balogh, A., et al. (Eds., 1999): Co-rotating interaction regions. *Space Science Reviews* **89**, 1–410. Reprinted by: Kluwer Academic Publishers and the International Space Science Institute.

Balogh, A., Marsden, R. G., Smith, E. J. (Eds., 2001): *The Heliosphere Near Solar Minimum: The Ulysses Perspective*. New York: Springer-Verlag 2001.

Bame, S. J., Asbridge, J. R., Gosling, J. T. (1977): Evidence for a structure-free state at high solar wind speeds. *Journal of Geophysical Research* **82**, 1487.

Bame, S. J., et al. (1974): The quiet corona: temperature and temperature gradient. *Solar Physics* **35**, 137–152.

Bame, S. J., et al. (1975): Solar wind heavy ion abundances. *Solar Physics* 43: 463–473.

Bame, S. J., et al. (1976): Solar cycle evolution of high-speed solar wind streams. *The Astrophysical Journal* **207**, 977–980.

Bame, S. J., et al. (1977a): A search for a general gradient in the solar wind speed at low solar latitudes. *Journal of Geophysical Research* **82**, 173–176.

Bame, S. J., et al. (1977b): Evidence for a structure-free state at high solar wind speeds. *Journal of Geophysical Research* **82**, 1487–1492.

Bame, S. J., et al. (1992): The *Ulysses* solar wind plasma experiment. *Astronomy and Astrophysics Supplement Series* **92**, No. 2, 237–265.

Bame, S. J., et al. (1993): *Ulysses* observations of a recurrent high speed solar wind stream and the heliomagnetic streamer belt. *Geophysical Research Letters* **20**(21), 2323–2326.

Bame, S. J., Hundhausen, A. J., Asbridge, J. R., Strong, I. B. (1968): Solar wind ion composition. *Physical Review Letters* **20**, 393.

Bamert, K., et al. (2004): Hydromagnetic wave excitation upstream of an interplanetary traveling shock. *The Astrophysical Journal (Letters)* **601**, L99–L102.

Bard, E., et al. (1997): Solar modulation of cosmogenic nuclide production over the last millennium: comparison between ^{14}C and ^{10}Be records. *Earth and Planetary Science Letters* **150**, 453–462.

Barnes, A. (1975): Plasma processes in the expansion of the solar wind and in the interplanetary medium. *Reviews of Geophysics and Space Physics* **13**(3), 1049–1053.

Barnes, A. (1992): Acceleration of the solar wind. *Reviews of Geophysics* **30**, 43–55.

Barnes, A., Gazis, P. R., Phillips, J. L. (1995): Constraints on solar wind acceleration mechanisms from *Ulysses* plasma observations: the first polar pass. *Geophysical Research Letters* **22**(23), 3309–3311.

Barnett, T. P. (1989): A solar-ocean relation: fact or fiction? *Geophysical Research Letters* **16**, 803–806.

Barnola, J. M., et al. (1987): Vostok ice core provides 160,000-year record of atmospheric CO_2. *Nature* **329**, 408–414.

Bartels, J. (1932): Terrestrial-magnetic activity and its relations to solar phenomena. *Terrestrial Magnetism and Atmospheric Electricity* **37**, 1–52.

Bartels, J. (1934): Twenty-seven day recurrences in terrestrial magnetic and solar activity, 1923–1933. *Terrestrial Magnetism and Atmospheric Electricity* **39**, 201–202.

Bartels, J. (1940): Solar radiation and geomagnetism. *Terrestrial Magnetism and Atmospheric Electricity* **45**, 339–343.

Bartels, J. (1963): Discussion of time-variations of geomagnetic activity indices K_P and A_P, 1932–1961. *Annales de Géophysique* **19**, 1–20.

Bastian, T. S., Benz, A. O., Gary, D. E. (1998): Radio emission from solar flares. *Annual Review of Astronomy and Astrophysics* **36**, 131–188.

Basu, S., Antia, H. M. (2003): Changes in solar dynamics from 1995 to 2002. *The Astrophysical Journal* **585**, 553–565.

Basu, S., Antia, H. M., Bogart, R. S. (2007): Structure of the near-surface layers of the Sun: Asphericity and time variation. *The Astrophysical Journal* **654**, 1146–1165.

Basu, S., et al. (1996): The Sun's hydrostatic structure from LOWL data. *The Astrophysical Journal* **460**, 1064–1070.

Battaglia, M., Grigis, P. C., Benz, O. (2006): Size dependence of solar X-ray flare properties. *Astronomy and Astrophysics* **439**, 737–747.

Baumback, M. M., Kurth, W. S., Gurnett, D. A. (1976): Direction-finding measurements of type III radio bursts out of the ecliptic plane. *Solar Physics* **48**, 361–380.

Beck, J. G., Giles, P. (2005): Helioseismic determination of the solar rotation axis. *The Astrophysical Journal (Letters)* **621**, L153–L156.

Beck, J. G., Gizon, L., Duvall, T. L. Jr. (2002): A new component of solar dynamics: North-south diverging flows migrating toward the equator with an 11 year period. *The Astrophysical Journal (Letters)* **575**, L47–L50.

Beckers, J. M. (2007): Effects of foreshortening on shallow sub-surface flows observed with local helioseismology. *Solar Physics* **240**, 3–7.

Beer, J., et al. (1988): Information on past solar activity and geomagnetism from [10]Be in the Camp Century ice core. *Nature* **331**, 675–679.

Beer, J., et al. (1990): Use of [10]Be in polar ice to trace the 11-year cycle of solar activity: information on cosmic ray history. Nature **347**, 164–166.

Belcher, J. W. (1971): Alfvénic wave pressures and the solar wind. *The Astrophysical Journal* **168**, 509–524.

Belcher, J. W., Davis, L. Jr. (1971): Large-amplitude Alfvén waves in the interplanetary medium, 2. *Journal of Geophysical Research* **76**, 3534–3563.

Belcher, J. W., Davis, L. Jr., Smith, E. J. (1969): Large-amplitude Alfvén waves in the interplanetary medium: *Mariner 5*. *Journal of Geophysical Research* **74**, 2303–2308.

Belcher, J. W., Olbert, S. (1975): Stellar winds driven by Alfvén waves. *The Astrophysical Journal* **200**, 369–382.

Bell, B., Glazer, H. (1958): Some sunspot and flare statistics. *Smithsonian Contributions to Astrophysics* **3**, 25–38.

Bemporad, A., et al. (2006): Current sheet evolution in the aftermath of a CME event. *The Astrophysical Journal* **638**, 1110–1128.

Bentley, R. D., et al. (1994): The correlation of solar flare hard X-ray bursts with Doppler blueshifted soft X-ray flare emission. *The Astrophysical Journal (Letters)* **421**, L55–L58.

Bentley, R. D., Mariska, J. T. (Eds., 1996): *Magnetic Reconnection in the Solar Atmosphere*. San Francisco: Astronomical Society of the Pacific Conference Series, p. 111.

Benz, A. O. (1986): Millisecond radio spikes. *Solar Physics* **104**, 99–110.

Benz, A. O. (2008): Flare Observations. *Living Reviews in Solar Physics* **5**, 1.

Benz, A. O., Bernold, T. E. X., Dennis, B. R. (1983): Radio blips and hard X-rays in solar flares. *Astrophysical Journal* **271**, 355–366.

Benz, A. O., Csillaghy, A., Aschwanden, M. J. (1996): Metric spikes and electron acceleration in the solar corona. *Astronomy and Astrophysics* **309**, 291–300.

Benz, A. O., et al. (1981): Solar radio blips and X-ray kernals. *Nature* **291**, 210–211.

Benz, A. O., et al. (1992): Electron beams in the low corona. *Solar Physics* **141**, 335–346.

Benz, A. O., et al. (1994): Particle acceleration in flares. *Solar Physics* **153**, 33–53.

Benz, A. O., Krucker, S. (1998a): Energy distribution of heating processes in the quiet solar corona. *The Astrophysical Journal (Letters)* **501**, L213–L216.

Benz, A. O., Krucker, S. (1998b): Heating events in the quiet solar corona. *Solar Physics* **182**, 349–363.

Benz, A. O., Krucker, S. (1999): Heating events in the quiet solar corona: Multiwavelength correlations. *Astronomy and Astrophysics* **341**, 286–295.

Berger, A. (1977): Support for the astronomical theory of climatic change. *Nature* 2 **68**, 44–45.

Berger, A. (1978a): Long-term variations of caloric insolation resulting from the Earth's orbital elements. *Quaternary Research* 9, 139–167.

Berger, A. (1978b): Long-term variations of daily insolation and Quaternary climatic changes. *Journal of Atmospheric Science* **35**(2), 2362–2367.

Berger, A. (1980): The Milankovitch astronomical theory of paleoclimates: a modern review. *Vistas in Astronomy* **24**, 103–122.

Berger, A. (1988): Milankovitch theory and climate. *Review of Geophysics* 26, 624–657.

Berger, A. (1991): Long-term history of climate ice ages and Milankovitch periodicity. In: *The Sun in Time* (Eds. C. P. Sonett, M. S. Giampapa and M. S. Mathews). Tucson: The University of Arizona Press, pp. 498–510.

Berger, T. E., et al. (2004): Solar magnetic elements at 0.1 arcsec resolution. General appearance and magnetic structure. *The Astronomy and Astrophysics* **428**, 613–628.

Berghmans, D., Clette, F. (1999): Active region EUV transient brightenings – first results by EIT of *SOHO* JOP 80. *Solar Physics* **186**, 207–229.

Bertaux, J. L., et al. (1995): SWAN: A study of Solar Wind Anisotropies on *SOHO* with Lyman alpha sky mapping. *Solar Physics* **162**, 403–439.

Bertaux, J. L., et al. (1997a): First results from SWAN Lyman-α solar wind mapper on *SOHO*. *Solar Physics* **175**, 737–770. Reprinted in: *The First Results From SOHO* (Eds. B. Fleck and Z. Svestka). Boston: Kluwer Academic Publishers, 737–770.

Bertaux, J. L., et al. (1997b): The first 1.5 year of observations from SWAN Lyman- alpha solar wind mapper on *SOHO*. In: *Proceedings of the Fifth SOHO Workshop. The Corona and Solar Wind Near Minimum Activity. ESA SP-404.* Noorwidjk: ESA Publications Division, pp. 29–36.

Bethe, H. A. (1939): Energy production in stars. *Physical Review* **55**, 434–456. Reproduced in *A Source Book in Astronomy and Astrophysics 1900–1975* (Eds. K. R. Lang and O. Gingerich). Cambridge, Massachusetts: Harvard University Press 1979, 320–338.

Bhattacharjee, A. (2004): Impulsive magnetic reconnection in the earth's magnetotail and the solar corona. *Annual Review of Astronomy and Astrophysics* **42**, 365–384.

Bieber, J. W., Rust, D. M. (1995): The escape of magnetic flux from the sun. *Astrophysical Journal* **453**, 911.

Biermann, L. F. (1948): Über die Ursache der chromosphärischen Turbulenz und des UV-Exzesses der Sonnenstrahlung. *Zeitschrift für Astrophysik* **25**, 161–177.

Biermann, L. F. (1951): Kometenschweife und solare Korpuskularstrahlung. *Zeitschrift für Astrophysik* **29**, 274–286.

Biermann, L. F. (1953): Physical processes in comet tails and their relation to solar activity. *La Physique des cometes. IAU Colloquium No. 4*, 251–262.

Biermann, L. F. (1957): Solar corpuscular radiation and the interplanetary gas. *Observatory* **77**, 109–110. Reproduced in: *A Source Book in Astronomy and Astrophysics, 1900–1975* (Eds.

K. R. Lang and O. Gingerich). Cambridge, Massachusetts: Harvard University Press 1979, pp. 147–148.

Biermann, L. F., Haxel, O., Schlüter, A. (1951): Neutrale ultrastrahlung von der sonne. *Zeitschrift für Naturforschung* **A6**, 47–48.

Bigazzi, A., Biferale, L., Gama, S. M. A., Velli, M. (2006): Small-scale anisotropy and intermittence in high- and low-latitude solar wind. *The Astrophysical Journal* **638**, 499–507.

Bigelow, F. H. (1889): *The Solar Corona Discussed by Spherical Harmonics*. Washington: The Smithsonian Institution 1889.

Bigelow, F. H. (1890a): *Further Study of the Solar Corona*, New Haven.

Bigelow, F. H. (1890b): The solar corona. *Sidereal Messenger* **9**, 93.

Bilenko, I. A. (2002): Longitudinal distribution of coronal holes during 1976–2002. *Solar Physics* **221**, 261–282.

Billings, D. E. (1959): Distribution of matter with temperature in the emission corona. *The Astrophysical Journal* **130**, 961–971.

Binns, W. R., et al. (2005): Cosmic-ray neon, wolf-rayet stars, and the superbubble origin of galactic cosmic rays. *The Astrophysical Journal* **634**, 351–364.

Birch, A. C., Kosovichev, A. G. (1998): Latitudinal variation of solar subsurface rotation inferred from *p*-mode frequency splittings measured with SOI-MDI and GONG. *The Astrophysical Journal (Letters)* **503**, L187–L190.

Bird, M. K., Edenhofer, P. (1990): Remote sensing observations of the solar corona. In: *Physics of the Inner Heliosphere I. Large-Scale Phenomena* (Eds. R. Schewenn and E. Marsch). Berlin: Springer-Verlag, pp. 13–87.

Bird, M. K., et al. (1992): The coronal-sounding experiment. *Astronomy and Astrophysics Supplement* **92**(2), 425–430.

Birkeland, K. (1896): Sur les rayons cathodiques sons l'action de forces magnetiques. *Archives des Sciences Physiques et Naturelles* **1**, 497.

Birkeland, K. (1908, 1913): *The Norwegian Aurora Polaris Expedition, 1902–1903, Vol. I., On the Cause of Magnetic Storms and the Origin of Terrestrial Magnetism* Christinania, Denmark: H. Aschehoug & Co. 1908, 1st Section; 1913, 2nd Section.

Blackett, P. M. S., Occhialini, G. P. S. (1933): Some photographs of the tracks of penetrating radiation. *Proceedings of the Royal Society of London A* **139**, 699–718.

Blackwell, D. E. (1960): The zodiacal light and its interpretation. *Endeavor* **19**, 14–19.

Blackwell, D. E., Ingham, M. F. (1961): Observations of the zodiacal light from a very high altitude station. *Monthly Notices of the Royal Astronomical Society* **122**, 129–141.

Bochsler, P., et al. (2000): Determination of the abundance of aluminum in the solar wind with *SOHO*/CELIAS/MTOF. *Journal of Geophysical Research* **105**(A6), 12659–12666.

Bochsler, P., Geiss, J., Maeder, A. (1990): The abundance of ^{3}He in the solar wind – a constraint for models of solar evolution. *Solar Physics* **128**, 203–215.

Bogdan, T. J. (2000): Sunspot oscillations: a review. *Solar Physics* **192**, 373–394.

Bogdan, T. J., et al. (2003): Waves in the magnetized solar atmosphere II. Waves from localized sources in magnetic flux concentrations. *The Astrophysical Journal* **599**, 626–660.

Bohlin, J. D., Sheeley, N. R. Jr. (1978): Extreme ultraviolet observations of coronal holes. *Solar Physics* **56**, 125–151.

Boischot, A. (1957): Caracteres d'un type d'émission hertzienne associé a certaines éruptions chromosphériques. *Comptes Rendus de l'Academie des Sciences* **244**, 1326–1329.

Boischot, A. (1958): Etude du rayonnement radioélectrique solaire sur 169 MHz a l'aide d'un grand interférometre a réseau. *Annales d'Astrophysique* **21**, 273–344.

Boischot, A., Denisse, J.-F. (1957): Les émissions de type IV et l'origine des rayons cosmiques associés aux éruptions chromosphériques. *Comptes Rendus de l'Academie des Sciences* **245**, 2194–2197.

Bond, G., et al. (1993): Correlations between climate records from north Atlantic sediments and Greenland ice. *Nature* **365**, 143–147.

Bond, G., et al. (2001): Persistent solar influence on north Atlantic climate during the holocene. *Science* **294**, 2130–2136.

Bondi, H. (1952): On spherically symmetrical accretion. *Monthly Notices of the Royal Astronomical Society* **112**, 195–204.

Bondi, H., Hoyle, F. (1944): On the mechanism of accretion by stars. *Monthly Notices of the Royal Astronomical Society* **104**, 273–282.

Bone, N. (1991): *The Aurora, Sun-Earth Interactions*. New York: Ellis Norwood.

Bonetti, A., et al. (1963): *Explorer 10* plasma measurements. *Journal of Geophysical Research* **68**, 4017–4062.

Boothroyd, A. I., Sackmann, I.-J., Fowler, W. A. (1991): Our Sun. II. Early mass loss of $0.1M$ and the case of the missing lithium. *The Astrophysical Journal* **377**, 318–329.

Borrini, G., et al. (1982): Helium abundance enhancements in the solar wind. *Journal of Geophysical Research* **87**, 7370–7378.

Boschler, P., Geiss, J. (1989): Composition of the solar wind. In: *Solar System Plasma Physics Geophysical Monograph 54* (Eds. J. H. Waite Jr., J. L. Burch and R. L. Moore). Washington, D. C.: American Geophysical Union 1989, 133–141.

Boteler, D. H., Pirjola, R. J., Nevanlinna, H. (1998): The effects of geomagnetic disturbances on electrical systems at the Earth's surface. *Advances in Space Research* **26**, 17–27.

Bothe, W., Kolhörster, W. (1929): The nature of the high-altitude radiation. *Zeitschrift für Physik* **56**, 751–777. Reproduced in English in: *Cosmic Rays* (Ed. A. M. Hillas). New York: Pergamon Press 1972.

Bothmer, V., Daglis, I. A. (2006): *Space Weather – Physics and Effects*. New York: Springer.

Bothmer, V., et al. (1996): *Ulysses* observations of open and closed magnetic field lines within a coronal mass ejection. *Astronomy and Astrophysics* **316**, 493–498.

Bothmer, V., et al. (1997): Solar energetic particle events and coronal mass ejections: new insights from *SOHO*. In: *31st ESLAB Symposium*. Noordwijk: ESA/ESTEC, pp. 207–216.

Bothmer, V., Schwenn, R. (1994): Eruptive prominences as sources of magnetic clouds in the solar wind. *Space Science Reviews* **70**, 215–220.

Bothmer, V., Schwenn, R. (1998): The structure and origin of magnetic clouds in the solar wind. *Annales Geophysicae* **16**, 1–24.

Bougeret, J. -L., et al. (2008): S/WAVES: The radio and plasma wave investigation on the STEREO mission. *Space Science Reviews* **136**, No. 1–4, 487–528.

Bracewell, R. N. (1956): Strip integration in radio astronomy. *Australian Journal of Physics* **9**, 198.

Bradley, R. S., Jones, P. D. (1993): "Little Ice Age" summer temperature variations: their nature and relevance to recent global warming trends. *Holocene* 3, 367–376.

Bradt, H. L., Peters, B. (1948): Investigation of the primary cosmic radiation with nuclear photographic emulsions. *Physical Review* **74**, 1828–1837.

Bradt, H. L., Peters, B. (1950): The heavy nuclei of the primary cosmic radiation. *Physical Review* **77**, 54–70.

Brandenburg, A. (2005): The case for a distributed solar dynamo shaped by near surface shear. *The Astrophysical Journal* **625**, 539–547.

Brandt, J. C., et al. (1969): Interplanetary gas. A calculation of angular momentum of the solar wind. *The Astrophysical Journal* **156**, 1117–1124.

Braun, D. C., Duvall, T. L. Jr., Labonte, B. J. (1987): Acoustic absorption by sunspots. *The Astrophysical Journal (Letters)* **319**, L27–L31.

Braun, D. C., Duvall, T. L. Jr., La Bonte, B. J. (1988): The absorption of high- degree p-mode oscillations in and around sunspots. *The Astrophysical Journal* **335**, 1015–1025.

Braun, D. C., Fay, Y. (1998): Helioseismic measurements of the subsurface meridional flow. *The Astrophysical Journal (Letters)* **508**, L105–L108.

Braun, D. C., Lindsey, C. (1999): Helioseismic images of an active region complex. *The Astrophysical Journal (Letters)* **513**, L79–L82.

Braun, D. C., Lindsey, C. (2000): Phase-sensitive holography of solar activity. *Solar Physics* **192**, 307–319.

Braun, D. C., Lindsey, C. (2001): Seismic imaging of the far hemisphere of the sun. *The Astrophysical Journal (Letters)* **560**, L189–L192.

Bravo, S., Stewart, G. A. (1997): Fast and slow wind from solar coronal holes. *The Astrophysical Journal* **489**, 992–999.

Bray, R. J., Loughhead, R. E., Durrant, C. J. (1967): *The Solar Granulation*. First edition London: Chapman and Hall 1967 Second edition New York: Cambridge University Press 1984.

Breen, A.R., et al. (1997) Ground and Space-based studies of solar wind acceleration. In: *The Corona and Solar Wind Near Minimum Activity. Proceedings of the Fifth SOHO Workshop. ESA SP-404*. Noordwijk, The Netherlands: ESA Publications, pp. 223–228.

Breit, G., Tuve, M. A. (1926): A test of the existence of the conducting layer. *Physical Review* **28**, 554–575.

Brekke, P., Hassler, D. M., Wilhelm, K. (1997a): Doppler shifts in the quiet-Sun transition region and corona observed with SUMER on *SOHO*. *Solar Physics* **175**, 349–374. Reprinted in: *The First Results From SOHO* (Eds. B. Fleck and Z. Svestka). Boston: Kluwer Academic Publishers, pp. 349–374.

Brekke, P., Hassler, D. M., Wilhelm, K. (1997b): Systematic redshifts in the quiet Sun transition region and corona observed with SUMER on *SOHO*. In: *The Corona and Solar Wind Near Minimum Activity. Proceedings of the Fifth SOHO Workshop. ESA SP-404*. Noordwijk: ESA Publications Division, pp. 229–234.

Bridge, H. S., et al. (1962): Direct observations of the interplanetary plasma. *Journal of the Physical Society of Japan* **17**, Supplement A-II, 553–559.

Broecker, W. S., Denton, G. H. (1990): What drives glacial cycles? *Scientific American* **262**, 49–56.

Brooks, D. H., et al. (2007): *Hinode* EUV imaging spectrometer observations of active region loop morphology: Implications for static heating models of coronal emission. *Publications of the Astronomical Society of Japan* **59**, S691–S697.

Brosius, J. W., Holman, G. D. (2007): Chromospheric evaporation in a remote solar flare-like transient observed at high time resolution with *SOHO's* CDS and *RHESSI*. *The Astrophysical Journal (Letters)* **659**, L73–L76.

Brosius, J. W., Phillips, K. J. H. (2004): Extreme-ultraviolet and X-ray spectroscopy of a solar flare loop observed at high time resolution: A case study in chromospheric evaporation. *Astrophysical Journal* **613**, 580–591.

Brosius, J. W., White, S. M. (2006): Radio measurements of the height of strong coronal magnetic fields above sunspots at the solar limb. *Astrophysical Journal (Letters)* **641**, L69–L72.

Brown, B. P., Haber, D. A., Hindman, B. W., Toomre, J. (2004): Variations of solar subsurface weather in the vicinity of active regions. In: *Helio- and Asteroseismology: Towards a Golden Future* (Ed. D. Dansey). ESA SP-559, 345.

Brown, D. S., et al. (2003): Observations of rotating sunspots from *TRACE*. *Solar Physics* **216**, 79–108.

Brown, J. C. (1971): The deduction of energy spectra of non-thermal electrons in flares from the observed dynamic spectra of hard X-ray bursts. *Solar Physics* **18**, 489–502.

Brown, J. C. (1972a): The decay characteristics of models of solar hard X-ray bursts. *Solar Physics* **25**, 158–177.

Brown, J. C. (1972b): The directivity and polarization of thick target X-ray bremsstrahlung from solar flares. *Solar Physics* **26**, 441–459.

Brown, J. C. (1973): Thick target X-ray bremsstrahlung from partially ionized targets in solar flares. *Solar Physics* **28**, 151–158.

Brown, J. C. (1975): The interpretation of spectra, polarization, and directivity of solar hard X-rays. In: *Solar gamma-, X-, and EUV radiation. Proceedings of IAU Symposium No. 68* (Ed. S. R. Kane). Boston: D. Reidel, pp. 245–282.

Brown, J. C. (1991): Energetic particles in solar flares: theory and diagnostics. *Philosophical Transactions of the Royal Society (London)* A**336**, 413–424.

Brown, J. C., Emslie, A. G. (1989): Self-similar Lagrangian hydrodynamics of beam-heated solar flare atmospheres. *The Astrophysical Journal* **339**, 1123–1131.

Brown, J. C., et al. (1990): Beam heating in solar flares: electrons or protons? *The Astrophysical Journal Supplement* **73**, 343–348.

Brown, T. M. (1985): Solar rotation as a function of depth and latitude. *Nature* **317**, 591–594.

Brown, T. M., Christensen-Dalsgaard, J. (1998): Accurate determination of the solar photospheric radius. *The Astrophysical Journal (Letters)* **500**, L195–L198.

Brown, T. M., et al. (1989): Inferring the Sun's internal angular velocity from observed p-mode frequency splittings. *The Astrophysical Journal* **343**, 526–546.

Brown, T. M., Gilliland, R. L. (1994): Astroseismology. *Annual Reviews of Astronomy and Astrophysics* **32**, 37–82.

Brown, T. M., Morrow, C. A. (1987): Depth and latitude dependence of solar rotation. *The Astrophysical Journal (Letters)* **314**, L21–L26.

Brueckner, G. E. (1974): The behavior of the outer solar corona ($3R_\odot$ to $10\,R_\odot$ during a large solar flare observed from *OSO-7* in white light. In: *Coronal Disturbances, IAU Symposium No. 57* (Ed. G. Newkirk, Jr.). Boston: Reidel, pp. 333–334.

Brueckner, G. E., Bartoe, J.-D. F. (1983): Observations of high-energy jets in the corona above the quiet sun, the heating of the corona, and the acceleration of the solar wind. *The Astrophysical Journal* **272**, 329–348.

Brueckner, G. E., et al. (1995): The Large Angle Spectroscopic Coronagraph (LASCO). *Solar Physics* **162**, 357–402.

Brun, A. S., Miesch, M. S., Toomre, J. (2004): Global-scale turbulent convection and magnetic dynamo action in the solar envelope. *The Astrophysical Journal* **614**, 1073–1098.

Bruner, E. C. Jr. (1978): Dynamics of the solar transition zone. *The Astrophysical Journal* **226**, 1140–1146.

Bruner, E. C. Jr. (1981): *OSO 8* observational limits to the acoustic coronal heating mechanism. *The Astrophysical Journal* **247**, 317–324.

Bruno, R., Carbone, V. (2005): The solar wind as a turbulence laboratory. *Living Reviews in Solar Physics* **2**–4.

Bruno, R., et al. (1986): In-situ observations of the latitudinal gradients of the solar wind parameters during 1976 and 1977. *Solar Physics* **104**, 431–445.

Bruzek, A. (1964): On the association between loop prominences and flares. *The Astrophysical Journal* **140**, 746–759.

Bryant, D. A., et al. (1962): Explorer 12 observations of solar cosmic rays and energetic storm particles after the solar flare of September 28, 1961. *Journal of Geophysical Research* **67**, 4983.

Budyko, M. I. (1969): Effect of solar radiation variations on the climate of Earth. *Tellus* **21**, 611–620.

Bumba, V. (1958): Relation between chromospheric flares and magnetic fields of sunspot groups. *Bulletin of the Crimean Astrophysical Observatory* **19**, 105–114.

Bumba, V., Howard, R. (1965): Large-scale distribution of solar magnetic fields. *The Astrophysical Journal* **141**, 1502–1512.

Bunsen, R. (1859): Letter to H. E. Roscoe in November 1859. Quoted by Roscoe in: *The Life and Experiences of Sir Henry Enfield Roscoe*, London 1906, p. 71. Reproduced by A. J. Meadows in: The origins of astrophysics, found in *The General History of Astronomy, Vol. 1. Astrophysics and Twentieth-Century Astronomy to 1950, part A* (Ed. O. Gingerich). New York: Cambridge University Press 1984, p. 5.

Burchfield, J. D. (1990): *Lord Kelvin and The Age of the Earth*. Chicago: University of Chicago Press.

Bürgi, A., Geiss, J. (1986): Helium and minor ions in the corona and solar wind: dynamics and charge states. *Solar Physics* **103**, 347–383.

Burkepile, J. T., et al. (2004): Role of projection effects on solar coronal mass ejection properties: 1. A study of CMEs associated with limb activity. *Journal of Geophysical Research* **109**, A03103.

Burkepile, J. T., St. Cyr, O. C. (1993): *A revised and expanded catalogue of mass ejections observed by the Solar Maximum Mission coronagraph.* NCAR/TN-369 + STR. Boulder, Colorado: National Center for Atmospheric Research.

Burlaga, L. F. (1971): Hydromagnetic waves and discontinuities in the solar wind. *Space Science Reviews* **12**, 600–657.

Burlaga, L. F. (1983): Understanding the heliosphere and its energetic particles. *Proceedings of the 18th International Conference of Cosmic Rays*, **12**, 21–60.

Burlaga, L. F. (1984): MHD processes in the outer heliosphere. *Space Science Reviews* **39**, 255–316.

Burlaga, L. F. (1988): Magnetic clouds and force-free fields with constant alpha. *Journal of Geophysical Research* **93**, 7217–7224.

Burlaga, L. F. (1990): Magnetic clouds. In: *Physics of the Inner Heliosphere II. Particles, Waves and Turbulence* (Eds. R. Schwenn and E. Marsch). New York: Springer-Verlag, pp. 1–22.

Burlaga, L. F. (1991): Magnetic clouds. In *Physics of the Inner Heliosphere, Vol. II: Particles, Waves and Turbulence* (Eds. R. Schwenn and E. Marsch). New York: Springer, pp. 1–22.

Burlaga, L. F. (1995): *Interplanetary Magnetohydrodynamics, Vol. 3 of International Series on Astronomy and Astrophysics*. New York, Oxford University Press.

Burlaga, L. F., et al. (1978): Sources of magnetic fields in recurrent interplanetary streams. *Journal of Geophysical Research* **83**, 4177–4185.

Burlaga, L. F., et al. (1981): Magnetic loop behind an interplanetary shock: *Voyager, Helios*, and *IMP 8* observations. *Journal of Geophysical Research* **86**, 6673–6684.

Burlaga, L. F., et al. (1982): A magnetic cloud and a coronal mass ejection. *Geophysical Research Letters* 9, 1317–1320.

Burlaga, L. F., Behannon, K. W., Klein, L. W. (1987): Compound streams, magnetic clouds and major magnetic storms. *Journal of Geophysical Research* **92**, 5725–5734.

Burlaga, L. F., et al. (1998): A magnetic cloud containing prominence material: January 1997. *Journal of Geophysical Research* **103**(A1), 277–285.

Burlaga, L. F., et al. (2001): Fast ejecta during the ascending phase of solar cycle 23: *ACE* observations, 1998–1999. *Journal of Geophysical Research* 106, A10, 20957–20977.

Burlaga, L. F., et al. (2005a): *Voyager 2* observations related to the October–November 2003 solar events. *Geophysical Research Letters* **32**, L03S05.

Burlaga, L. F., et al. (2005b): Crossing the termination shock into the heliosheath: magnetic fields. *Science* **309**, 2027–2029.

Burlaga, L. F., et al. (2008): Magnetic fields at the termination shock by *Voyager 2*. *Nature* **454**, 75–77.

Burlaga, L. F., King, J. H. (1979): Intense interplanetary magnetic fields observed by geocentric spacecraft during 1963–1975. *Journal of Geophysical Research* **84**, 6633–6640.

Burlaga, L. F., Lepping, R. P. (1977): The causes of recurrent geomagnetic storms. *Planetary and Space Science* **25**, 1151–1160.

Burlaga, L. F., Ness, N. F., Belcher, J. W., Whang, Y. C. (1996): Pickup protons and pressure-balance structures from 39 to 43 AU: *Voyager 2* observations during 1993 and 1994. *Journal of Geophysical Research* **101**(A7), 15523–15254.

Burlaga, L. F., Sittler, E., Mariani, F., Schwenn, R. (1981): Magnetic loop behind an interplanetary shock – *Voyager, Helios* and *IMP 8* observations. *Journal of Geophysical Research* **86**, 6673–6684.

Burlaga, L. F., Viñas, A.-F. (2005): Tsallis distributions of the large-scale magnetic field strength fluctuations in the solar wind from 7 to 87 AU. *Journal of Geophysical Research* **110**(A7), A07110.

Burnight, T. R. (1949): Soft X-radiation in the upper atmosphere. *Physical Review* **76**, 165.

Burroughs, W. J. (1992): *Weather Cycles: Real or Imaginary*. New York: Cambridge University Press.

Burton, M. E., et al. (1996): *Ulysses* out-of-ecliptic observations of interplanetary shocks. *Astronomy and Astrophysics* **316**, 313–322.

Burton, R. K., Mcpherron, R. L., Russell, C. T. (1975): An empirical relationship between interplanetary conditions and Dst. *Journal of Geophysical Research* **80**, 4204–4214.

Byram, E. T., Chubb, T. A., Friedman, H. (1953): The contribution of solar X- rays to E-layer ionization. *Physical Review* **92**, 1066–1067.

Byram, E. T., Chubb, T. A., Friedman, H. (1954): Solar X-ray emission. *Physical Review* **96**, 860.

Byram, E. T., Chubb, T. A., Friedman, H. (1956): The solar X-ray spectrum and the density of the upper atmosphere. *Journal of Geophysical Research* **61**, 251–263.

C

Cahill, L. J., Patel, V. L. (1967): The boundary of the geomagnetic field, August to November 1961. *Planetary and Space Science* **15**, 997–1033.

Cane, H. V. (1985a): The evolution of interplanetary shocks. *Journal of Geophysical Research* **90**, 191–197.

Cane, H. V. (1985b): The evolution of interplanetary transients, type II bursts and interplanetary shocks. *Astronomy and Astrophysics* **140**, 205–209.

Cane, H. V. (1997): The current status of our understanding of energetic particles, coronal mass ejections and flares. In: *Coronal Mass Ejections. Geophysical Monograph 99* (Eds. N. Crooker, J. A. Joselyn and J. Feynman). Washington: American Geophysical Union, pp. 205–215.

Cane, H. V., Erickson, W. C. (2005): Solar type II radio bursts and IP type II events. *The Astrophysical Journal* **623**, 1180–1194.

Cane, H. V., Lario, D. (2006): An introduction to CMEs and energetic particles. *Space Science Reviews* **123**, 45–56.

Cane, H. V., McGuire, R. E., Von Rosenvinge, T. T. (1986): Two classes of solar energetic particle events associated with impulsive and long duration soft X- ray events. *The Astrophysical Journal* **301**, 448–459.

Cane, H. V., Reames, D. V. (1988a): Soft X-ray emissions, meter-wavelength radio bursts, and particle acceleration in solar flares. *The Astrophysical Journal* **325**, 895–900.

Cane, H. V., Reames, D. V. (1988b): Some statistics of solar radio bursts of spectral types II and IV. *The Astrophysical Journal* **325**, 901–904.

Cane, H. V., Richardson, I. G., Cyr, O. C. St. (2000): Coronal mass ejections, interplanetary eject and geomagnetic storms. *Geophysical Research Letters* **27**, Issue 21, 3591–3594.

Cane, H. V., Sheeley, N. R. Jr., Howard, R. A. (1987): Energetic interplanetary shocks, radio emission, and coronal mass ejections. *Journal of Geophysical Research* **92**, 9869–9874.

Canfield, R. C., et al. (2007): *Yohkoh* SXT full-resolution observations of sigmoids: Structure, formation, and eruption. *Astrophysical Journal (Letters)* **671**, L81–L84.

Canfield, R. C., Hudson, H. S., McKenzie, D. E. (1999): Sigmoidal morphology and eruptive solar activity. *Geophysical Research Letters* **26**(6), 627–630.

Canuto, V. M., et al. (1983): The young Sun and the atmosphere and photochemistry of the early Earth. *Nature* **305**, 281–286.

Cargill, P. J. (1994): Some implications of the nanoflare concept. *The Astrophysical Journal* **422**, 381–393.

Cargill, P. J., Klimchuk, J. A. (1997): A nanoflare explanation for the heating for coronal loops observed by *Yohkoh*. *The Astrophysical Journal* **478**, 799–806.

Cargill, P. J., Klimchuk, J. A. (2004): Nanoflare heating of the corona revisited. *The Astrophysical Journal* **605**, 911–920.

Cargill, P. J., Priest, E. R. (1983): The heating of post-flare loops. *The Astrophysical Journal* **266**, 383–389.

Carloqwicz, M., Lopez, R. (2002): *Storms from the Sun: The Emerging Science of Space Weather*. Washington, D.C.: Joseph Henry Press.

Carlqvist, P. (1969): Current limitation and solar flares. *Solar Physics* 7, 377–392.

Carlsson, M., et al. (2007): Can high frequency acoustic waves heat the quiet sun chromosphere? *Publications of the Astronomical Society of Japan* **59**, S663–S668.

Carlsson, M., Judge, P. G., Wilhelm, K. (1997): SUMER observations confirm the dynamic nature of the quiet solar outer atmosphere: the internetwork chromosphere. *The Astrophysical Journal (Letters)* **486**, L63.

Carlsson, M., Stein, R. F. (1992): Non-LTE radiating acoustic shocks and Ca II K2V bright points. *The Astrophysical Journal (Letters)* **397**, L59–L62.

Carlsson, M., Stein, R. F. (1995): Does a nonmagnetic solar chromosphere exist? *The Astrophysical Journal (Letters)* **440**, L29–L32.

Carlsson, M., Stein, R. F. (1997): Formation of solar calcium H and K bright grains. *The Astrophysical Journal* **481**, 500.

Carlsson, M., Stein, R. F. (2002): Dynamic hydrogen ionization. *The Astrophysical Journal* **572**, 626–635.

Carmichael, H. (1964): A process for flares. In: *AAS-NASA Symposium on the Physics of Solar Flares NASA SP-50* (Ed. W. N. Hess). Washington: National Aeronautics and Space Administration, pp. 451–456.

Carovillano, R. L., Siscoe, G. L. (1969): Co-rotating structure in the solar wind. *Solar Physics* 8, 401–414.

Carrington, R. C. (1858): On the distribution of the solar spots in latitude since the beginning of the year 1854. *Monthly Notices of the Royal Astronomical Society* **19**, 1–3. Reproduced in: *Early Solar Physics* (Ed. A. J. Meadows). Oxford: Pergamon Press 1970, pp. 169–172.

Carrington, R. C. (1860): Description of a singular appearance seen in the Sun on September 1, 1859. *Monthly Notices of the Royal Astronomical Society* **20**, 13–15. Reproduced in: *Early Solar Physics* (Ed. A. J. Meadows). Oxford, England: Pergamon Press 1970, pp. 181–183.

Carrington, R. C. (1863): *Observations of the Spots on the Sun*. London: Williams and Norgate.

Cavendish, H. (1790): On the height of the luminous arch which was seen on Feb. 23, 1784. *Philosophical Transactions of the Royal Society (London)* **80**, 101–105.

Celsius, A. (1747): Bemerkungen über der Magnetnadel Stündliche Veränderungen in ihrer Abweichung. *Svenska Ventensk. Handl.* **8**, 296.

Cess, R. D., Ramanathan, V., Owen, T. (1980): The Martian paleoclimate and enhanced atmospheric carbon dioxide. *Icarus* **41**, 159–165.

Chae, J., et al. (1998): Chromospheric upflow events associated with transition region explosive events. *The Astrophysical Journal (Letters)* **504**, L123–L126.

Chae, J., et al. (1998): Photospheric magnetic field changes associated with transition region explosive events. *The Astrophysical Journal Letters* **497**, L109.

Chamberlain, J. W. (1960): Interplanetary gas II. Expansion of a model corona. *The Astrophysical Journal* **131**, 47–56.

Chamberlain, J. W. (1963): Planetary coronae and atmospheric evaporation. *Planetary and Space Science* **11**, 901–960.

Chamberlin, P., Woods, T. N., Eparvier, F. G. (2006): Flare Irradiance Spectral Model (FISM) use for space weather applications. *Proceedings of the ILWS Workshop* (Eds. N. Gopalswamy and A. Bhattacharyya), p. 153.

Chamberlin, T. C. (1899): An attempt to frame a working hypothesis of the cause of the glacial periods on an atmospheric basis. *The Journal of Geology* **7**, 545–584.

Chandra, S. (1991): The solar UV related changes in total ozone from a solar rotation to a solar cycle. *Geophysical Research Letters* **18**, 837–840.

Chandran, B. D. G. (2004): A review of the theory of incompressible MHD turbulence. *Astrophysics and Space Science* **292**, 17–28.

Chandran, B. D. G. (2005): Weak compressible magnetohydrodynamic turbulence in the solar corona. *Physical Review Letters* **95**, 265004–265300.

Chapman, G. A. (1984): On the energy balance of solar active regions. *Nature* **308**, 252–254.

Chapman, G. A. (1987): Variations of solar irradiance due to magnetic activity. *Annual Review of Astronomy and Astrophysics* **25**, 633–667.

Chapman, G. A., Cookson, A. M., Dobias, J. J. (1996): Variations in total solar irradiance during solar cycle 22. *Journal of Geophysical Research* **101**, 13541–13548.

Chapman, G. A., Cookson, A. M., Dobias, J. J. (1997): Solar variability and the relation of facular to sunspot areas during solar cycle 22. *The Astrophysical Journal* **482**, 541–545.

Chapman, G. A., et al. (1984): Solar luminosity fluctuations and active region photometry. *The Astrophysical Journal (Letters)* **282**, L99–L101.

Chapman, S. (1918): An outline of a theory of magnetic storms. *Proceedings of the Royal Society of London A* **95**, 61–83.

Chapman, S. (1918): The energy of magnetic storms. *Monthly Notices of the Royal Astronomical Society* **79**, 70–83.

Chapman, S. (1929): Solar streams of corpuscles – their geometry, absorption of light and penetration. *Monthly Notices of the Royal Astronomical Society* **89**, 456–470.

Chapman, S. (1950): Corpuscular influences upon the upper atmosphere. *Journal of Geophysical Research* **55**, 361–372. Reproduced in: *A Source Book in Astronomy and Astrophysics, 1900–1975* (Eds. K. R. Lang and O. Gingerich). Cambridge: Harvard University Press 1979, pp. 125–131.

Chapman, S. (1954): The viscosity and thermal conductivity of a completely ionized gas. *The Astrophysical Journal* **120**, 151–155.

Chapman, S. (1957): Notes on the solar corona and the terrestrial atmosphere. *Smithsonian Contributions to Astrophysics* **2**(1), 1–14.

Chapman, S. (1958): Thermal diffusion in ionized gases. *Proceedings of the Physical Society of London* **72**, 353–362.

Chapman, S. (1959a): Interplanetary space and the earth's outermost atmosphere. *Proceedings of the Royal Society (London)* **A253**, 462–481.

Chapman, S. (1959b): The outermost ionosphere. *Journal of Atmospheric and Terrestrial Physics* **15**, 43–47.

Chapman, S., Bartels, J. (1940): *Geomagnetism*. Oxford: Clarendon Press.

Chapman, S., Ferraro, V. C. A. (1929): The electrical state of solar streams of corpuscles. *Monthly Notices of the Royal Astronomical Society* **89**, 470–479.

Chapman, S., Ferraro, V. C. A. (1931): A new theory of magnetic storms, I, The initial phase. *Terrestrial Magnetism and Atmospheric Electricity* **36**, 77–97, 171–186; **37**, 147–156, 421–429 (1932).

Chapman, S., Ferraro, V. C. A. (1933): A new theory of magnetic storms, II, The main phase. *Terrestrial Magnetism and Atmospheric Electricity* **38**, 79–86.

Chapman, S., Ferraro, V. C. A. (1940): A theory of the first phase of geomagnetic storms. *Terrestrial Magnetism and Atmospheric Electricity* **45**, 245–268.

Chappellaz, J., et al. (1990): Ice-core record of atmospheric methane over the past 160,000 years. *Nature* **345**, 127–131.

Chappellaz, J., et al. (1993): Synchronous changes in atmospheric CH_4 and Greenland climate between 40 and 8 kyr BP. *Nature* **366**, 443–445.

Charbonneau, P. (2005): Dynamo models of the solar cycle. *Living Reviews in Solar Physics* **2**, 2.

Charlson, R. J., Wigley, T. M. L. (1994): Sulfate aerosol and climate change. *Scientific American* **270**, 48–57.

Chen, J. (1989): Effects of toroidal forces in current loops embedded in a background plasma. *The Astrophysical Journal* **338**, 453–470.

Chen, J. (1996): Theory of prominence eruption and propagation: interplanetary consequences. *Journal of Geophysical Research* **101**(A12), 27499–27519.

Chen, J. (2001): Physics of coronal mass ejections: a new paradigm of solar eruptions. *Space Science Reviews* **95**, 165–190.

Chen, J., et al. (1995): Simulation of geomagnetic storms during the passage of magnetic clouds. *Geophysical Research Letters* **22**(13), 1749–1752.

Chen, J., et al. (1997): Evidence of an erupting magnetic flux rope: LASCO coronal mass ejection of 1997 April 13. *The Astrophysical Journal (Letters)* **490**, L191–L194.

Chen, J., et al. (2000): Magnetic geometry and dynamics of the fast coronal mass ejection of 1997 September 9. *The Astrophysical Journal* **533**, 481–500.

Chen, J., et al. (2006): The flux-rope scaling of the acceleration of coronal mass ejections and eruptive prominences. *The Astrophysical Journal* **649**, 452–463.

Chen, J., Garren, D. A. (1993): Interplanetary magnetic clouds: topology and driving mechanism. *Geophysical Research Letters* **20**(21), 2319–2322.

Chen, Y., Esser, R., Strachan, L., Hu, Y. (2004): Stagnated outflow of O^5 ions in the source region of the slow solar wind at solar minimum. *The Astrophysical Journal* **602**, 415–421.

Chen, Y., Li, X. (2004): An ion-cyclotron resonance-driven three-fluid model of the slow wind near the Sun. *The Astrophysical Journal (Letters)* **609**, L41–L44.

Cheng, C.-C., Doschek, G. A., Feldman, U. (1979): The dynamical properties of the solar corona from intensities and line widths of EUV forbidden lines. *The Astrophysical Journal* **227**, 1037–1046.

Cheng, C.-C., et al. (1981): Spatial and temporal structures of impulsive bursts from solar flares observed in UV and hard X-rays. *The Astrophysical Journal (Letters)* **248**, L39–L43.

Cheng, C.-C., Pallavicini, R. (1987): Analysis of ultraviolet and X-ray observations of three homologous solar flares from SMM. *The Astrophysical Journal* **318**, 459–473.

Chiang, W. H., Foukal, P. V. (1985): The influence of faculae on sunspot heat blocking. *Solar Physics* **97**, 9–20.

Chitre, S. M., Gokhale, M. H. (1975): The five-minute oscillations in the solar atmosphere. *Solar Physics* **43**, 49–55.

Chou, D. -Y., Dai, D. -C. (2001): Solar cycle variations of subsurface meridional flows in the Sun. *The Astrophysical Journal (Letters)* **559**, L175–L178.

Chou, D.-Y., et al. (1995): Taiwan oscillation network. *Solar Physics* **160**, 237–243.

Chou, D.-Y., Ladenkov, O. (2005): Evolution of solar subsurface meridional flows in the declining phase of cycle 23. *The Astrophysical Journal* **630**, 1206–1212.

Christensen-Dalsgaard, J. (2002): Helioseismology. *Reviews of Modern Physics* **74**, 1073–1129.

Christensen-Dalsgaard, J., et al. (1985): Speed of sound in the solar interior. *Nature* **315**, 378–382.

Christensen-Dalsgaard, J., et al. (1996): The current state of solar modeling. *Science* **272**, 1286–1292.

Christensen-Dalsgaard, J., Gough, D. O., Thompson, M. J. (1991): The depth of the solar convection zone. *The Astrophysical Journal* **378**, 413–437.

Christensen-Dalsgaard, J., Gough, D. O., Toomre, J. (1985): Seismology of the Sun. *Science* **229**, 923–931.

Christiansen, W. N. (1984): The first decade of solar radio astronomy in Australia. In: *The Early Years of Radio Astronomy* (Ed. W. T. Sullivan III). New York: Cambridge University Press.

Chubb, T. A., Friedman, H., Kreplin, R. W. (1960): Measurements made of high-energy X-rays accompanying three class 2+ solar flares. *Journal of Geophysical Research* **65**, 1831–1832.

Chupp, E. L. (1971): Gamma ray and neutron emissions from the Sun. *Space Science Reviews* **12**, 486–525.

Chupp, E. L. (1976): *Gamma-ray Astronomy*. Dordrecht: D. Reidel.

Chupp, E. L. (1984): High energy neutral radiation from the Sun. *Annual Review of Astronomy and Astrophysics* **22**, 359–387.

Chupp, E. L. (1987): High-energy particle production in solar flares (SEP, gamma-ray and neutron emissions). *Physica Scripta* **T18**, 5–19.

Chupp, E. L. (1990): Emission characteristics of three intense solar flares observed in cycle 21. *The Astrophysical Journal Supplement* **73**, 213–226.

Chupp, E. L. (1990): Transient particle acceleration with solar flares. *Science* **250**, 229–236.

Chupp, E. L., Benz, A. O. (Eds., 1994): *Particle Acceleration Phenomena in Astrophysical Plasmas. Proceedings of the International Astronomical union (IAU) Colloquium 142. The Astrophysical Journal Supplement* **90**, 511–983.

Chupp, E. L., et al. (1973): Solar gamma ray lines observed during the solar activity of August 2 to August 11, 1972. *Nature* **241**, 333–334.

Chupp, E. L., et al. (1982): A direct observation of solar neutrons following the 0118 UT flare on 1980 June 21. *Astrophysical Journal (Letters)* **263**, L95–L99.

Chupp, E. L., et al. (1987): Solar neutron emissivity during the large flare on 1982 June 3. *Astrophysical Journal* **318**, 913–929.

Ciaravella, A., et al. (1997): Ultraviolet coronagraph spectrometer observation of the 1996 December 23 coronal mass ejection. *The Astrophysical Journal (Letters)* **491**, L59.

Ciaravella, A., et al. (2002): Elemental abundances and post-coronal mass ejection current sheet in a very hot active region. *Astrophysical Journal* **575**, 1116–1130.

Ciaravella, A., et al. (2005): Detection and diagnostics of a coronal shock wave driven by a partial-halo coronal mass ejection on 2000 June 28. *Astrophysical Journal* **621**, 1121–1128.

Ciaravella, A., Raymond, J. C., Kahler, S. W. (2006): Ultraviolet properties of halo coronal mass ejections: Doppler shifts, angles, shocks, and bulk morphology. *Astrophysical Journal* **652**, 774–792.

Cinicastagoli, G., et al. (1984): Solar cycles in the last centuries in ^{10}Be and O^{18} in polar ice and in thermoluminescence signals of a sea sediment. *Nuovo Cimento* **7C**, 235–244.

Cinicastagoli, G., Lal, D. (1980): Solar modulation effects in terrestrial production of carbon-14. *Radiocarbon* **22**, 133–158.

Cirtain, J. W., et al. (2007): Evidence for Alfvén waves in solar polar jets. *Science* **318**, 1580–1582.

Cirtain, J. W., et al. (2007a): Active region loops: Temperature measurements as a function of time from joint *TRACE* and *SOHO* CDS observations. *The Astrophysical Journal* **655**, 598–605.

Cirtain, J. W., et al. (2007b): Evidence for Alfvén waves in solar polar jets. *Science* **318**, 1580–1582.

Claverie, A., et al. (1979): Solar structure from global studies of the five-minute oscillation. *Nature* **282**, 591–594.

Clay, J. (1927): Penetrating radiation. *Koninklijke Nederlandse Akademie van Wetenschappen te Amsterdam, Proceedings of the Section of Sciences* **30**, 1115.

Clay, J. (1932): The Earth-magnetic effect and the corpuscular nature of (cosmic) ultra-radiation. IV. *Koninklijke Nederlandse Akademie van Wetenschappen te Amsterdam, Proceedings of the Section of Sciences* **35**, 1282–1290. Reproduced in: *Cosmic Rays* (Ed. A. M. Hillas). New York: Pergamon Press 1972.

Cleveland, B. T., et al. (1998): Measurement of the solar electron neutrino flux with the Homestake chlorine detector. *The Astrophysical Journal* **496**, 505–526.

Cline, T. L., McDonald, F. B. (1968): Relativistic electrons from solar flares. *Solar Physics* **5**, 507–530.

Cliver, E. W. (1994a): Solar activity and geomagnetic storms: the corpuscular hypothesis. *EOS* **75**, 609, 612–613.

Cliver, E. W. (1994b): Solar activity and geomagnetic storms: the first 40 years. *EOS* **75**, 569, 574–575.

Cliver, E. W. (1995a): Solar activity and geomagnetic storms: from M regions and flares to coronal holes and CMEs. *EOS* **76**, 75, 83–84.

Cliver, E. W. (1995b): Solar flare nomenclature. *Solar Physics* **157**, 285–293.

Cliver, E. W., Boriakoff, V., Feynman, J. (1998): Solar variability and climate change: geomagnetic a index and global surface temperature. *Geophysical Research Letters* **25**, 1035–1038.

Cliver, E. W., Cane, H. V. (2002): Gradual and impulsive solar energetic particle events. *EOS Transactions of the American Geophysical Union* **83**, 61–68.

Cliver, E. W., et al. (1989): Solar flare nuclear gamma-rays and interplanetary proton events. *The Astrophysical Journal* **343**, 953–970.

Cliver, E. W., Kahler, S. W., Reames, D. V. (2004): Coronal shocks and solar energetic proton events. *The Astrophysical Journal* **605**, 902–910.

Cliver, E. W., Webb, D. F., Howard, R. A. (1999): On the origin of solar metric type II bursts. *Solar Physics* **187**, 89–114.

Close, R. M., Parnell, C. E., Mackay, D. M., Priest, E. R. (2003): Statistical flux-tube properties of 3D magnetic carpet fields. *Solar Physics* **212**, 251–275.

Cocconi, G., et al. (1958): The cosmic ray flare effect. *Nuovo Cimento Supplement Series* **8**(2), 161–168.

Cohen, C. M. S., et al. (1999): New observations of heavy-ion-rich solar particle events from *ACE*. *Geophysical Research Letters* **26**(17), 2697–2700.

Colburn, D. S., Sonett, C. P. (1966): Discontinuities in the solar wind. *Space Science Reviews* **5**, 439–506.

Coleman, P. J. Jr. (1968): Turbulence, viscosity, and dissipation in the solar wind plasma. *Astrophysical Journal* **153**, 371–388.

Coleman, P. J. Jr., et al. (1966): Measurements of magnetic fields in the vicinity of the magnetosphere and in interplanetary space: preliminary results from *Mariner 4*. *Space Research* **6**, 907–928.

Coles, W. A., et al. (1980): Solar cycle changes in the polar solar wind. *Nature* **286**, 239–241.

Coles, W. A., Rickett, B. (1976): IPS observations of solar wind speed out of the ecliptic. *Journal of Geophysical Research* **77**, 4797–4799.

Compton, A. H. (1932): Variation of the cosmic rays with latitude. *Physical Review* **41**, 111–113.

Compton, A. H. (1933): A geographic study of the cosmic rays. *Physical Review* **43**, 387–403.

Contarino, L., Romano, P., Zuccarello, F. (2006): *RHESSI* and *TRACE* observations of an M 2.5 flare: A direct application of the Kopp and Pneuman model. *Astronomy and Astrophysics* **458**, 297–300.

Cook, W. R., et al. (1979): Elemental composition of solar energetic particles in 1977 and 1978. *Proceedings of the 16th International Cosmic Ray Conference* **12**, 265.

Cook, W. R., Stone, E. C., Vogt, R. (1984): Elemental composition of solar energetic particles. *Astrophysical Journal* **279**, 827–838.

Coplan, M. A., Ogilvie, K. W., Bochsler, P., Geiss, J. (1984): Interpretation of ^{3}He abundance variations in the solar wind. *Solar Physics* **93**, 415–434.

Corbard, T., Thompson, J. (2002): The subsurface radial gradient of solar angular velocity from MDI f-mode observations. *Solar Physics* **205**, 211–229.

Cormack, A. M. (1963): Representation of a function by its line integrals, with some radiological applications. *Journal of Applied Physics* **34**, 2722–2727.

Corti, G., et al. (1997): Physical parameters in plume and interplume regions from UVCS observations. In: *The Corona and Solar Wind Near Minimum Activity. Proceedings of the Fifth SOHO Workshop. ESA SP-404*. Noordwijk: ESA Publications Division, pp. 289–294.

Couvidat, S., et al (2003a): The rotation of the deep solar layers. *The Astrophysical Journal (Letters)* **597**, L77–L79.

Couvidat, S., Turck-Chièze, S., Kosovichev, A. G. (2003b): Solar seismic models and the neutrino predictions. *The Astrophysical Journal* **599**, 1434–1448.

Covington, A. E. (1951): Some characteristics of 10.7 cm solar noise. *Journal of the Royal Astronomical Society of Canada* **45**, 15–22.

Covington, A. E., Harvey, G. A. (1958): Impulsive and long-enduring sudden enhancements of solar radio emission at 10-cm wave-length. *Journal of the Royal Astronomical Society of Canada* **52**, 161–166.

Cowan, C. L. Jr., et al. (1956): Detection of the free neutrino: a confirmation. *Science* **124**, *103*.

Cox, A. (1969): Geomagnetic reversals. *Science* **163**, 237–245.

Cox, A. N., Livingston, W. C., Matthews, M. S. (Eds., 1991): *Solar Interior and Atmosphere*. Tucson: The University of Arizona Press.

Craig, I. J. D., McClymont, A. N., Underwood, J. H. (1978): The temperature and density structure of active region coronal loops. *Astronomy and Astrophysics* **70**, 1–11.

Cranmer, S. R. (2000): Ion cyclotron wave dissipation in the solar corona: the summed effect of more than 2000 ion species. *Astrophysical Journal* **532**, 1197–1208.

Cranmer, S. R. (2002): Coronal holes and the high-speed solar wind. *Space Science Reviews* **101**, 229–294.

Cranmer, S. R. (2004): New views of the solar wind with the Lambert W function. *American Journal of Physics* **72**, 1397–1403.

Cranmer, S. R., et al. (1999): An empirical model of a polar coronal hole at solar minimum. *The Astrophysical Journal* **511**, 481–501.

Cranmer, S. R., Field, G. B., Kohl, J. L. (1998): Spectroscopic constraints on models of ion-cyclotron resonance heating in the polar solar corona and fast solar wind. *EOS Transactions AGU* **79**, F722.

Cranmer, S. R., Van Ballegooijen, A. A. (2003): Alfvénic turbulence in the extended solar corona: kinetic effects and proton heating. *The Astrophysical Journal* **594**, 573–591.

Cranmer, S. R., Van Ballegooijen, A. A. (2005): On the generation, propagation and reflection of Alfvén waves from the photosphere to the distant heliosphere. *Astrophysical Journal Supplement Series* **156**, 265–293.

Cranmer, S. R., Van Ballegooijen, A. A., Edgard, R. J. (2007): Self consistent coronal heating and solar wind acceleration from anisotropic magnetohydrodynamic turbulence. *Astrophysical Journal Supplement* **171**, 520–551.

Crawford, H. J., et al. (1975): Solar flare particles: energy-dependent composition and relationship to solar composition. *The Astrophysical Journal* **195**, 213–221.

Crawford, H. J., Price, P., Sullivan, J. D. (1972): Composition and energy spectra of heavy nuclei with .5 less than E less than 40 MeV per nucleon in the 1971 January 24 and September 1 solar flares. *The Astrophysical Journal (Letters)* **175**, L149.

Croll, J. (1875): *Climate and Time in their Geological Relations.* New York: Appleton and London: David Bogue (1876).

Crommelynck, D., et al. (1995): First realization of the space absolute radiometric reference (SARR) during the *ATLAS* 2 flight period. *Advances in Space Research* **16**(8), 17–23.

Crooker, N. U. (2000): Solar and heliospheric geoeffective disturbances. *Journal of Atmospheric and Solar-Terrestrial Physics* **62**, 1071–1085.

Crooker, N. U., Cliver, E. W. (1994): Postmodern view of M-regions. *Journal of Geophysical Research* **99**, 23383–23390.

Crooker, N. U., et al. (1999): CIR morphology, turbulence, discontinuities and energetic particles. *Space Science Reviews* **89**, 179–220.

Crooker, N. U., et al. (2004): Heliospheric plasma sheets. *Journal of Geophysical Research* **109**(A3), A03107.

Crooker, N. U., Goslling, J. T., Kahler, S. W. (2002): Reducing heliospheric magnetic flux from coronal mass ejections without disconnection. *Journal of Geophysical Research (Space Physics)* **107**(A2), SSH 3–1.

Crooker, N. U., Joselyn, J. A., Feynman, J. (Eds., 1997): *Coronal Mass Ejections, Geophysical Monograph Series 99.* Washington: American Geophysical Union.

Crooker, N. U., Kahler, S. W., Larson, D. E., Lin, R. P. (2004): Large-scale magnetic field inversions at sector boundaries. *Journal of Geophysical Research* **109**(A3), A03108.

Crooker, N. U., Kahler, S. W., Larson, D. E., Lin, R. P. (2004): Large-scale magnetic field inversions at sector boundaries. *Journal of Geophysical Research* **109**, A3, A03108.

Crosby, N. B., Aschwanden, M. J., Dennis, B. R. (1993): Frequency distributions and correlations of solar X-ray flare parameters. *Solar Physics* **143**, 275–299.

Crowley, T. J. (1983): The geologic record of climatic change. *Review of Geophysics and Space Physics* **21**, 828–877.

Crowley, T. J. (2000): Causes of climate change over the past 1000 years. *Science* **289**, 270–277.

Crowley, T. J., Kim, K.-Y. (1996): Comparison of proxy records of climate change and solar forcing. *Geophysical Research Letters* **23**, 359–362.

Crowley, T. J., Kim, K.-Y. (1999): Modeling the temperature response to forced climate change over the last six centuries. *Geophysical Research Letters* **26**, 1901–1904.

Crutzen, P. J., Isaksen, I. S. A., Reid, G. C. (1975): Solar proton events: stratospheric sources of nitric oxide. *Science* **189**, 457–459.

Cubasch, U., et al. (1997): Simulation of the influence of solar radiation variations on the global climate with an ocean-atmosphere general circulation model. *Climate Dynamics* **13**, 757–767.

Culhane, J. L., et al. (1991): The Bragg Crystal Spectrometer for *SOLAR-A*. *Solar Physics* **136**, 89–104.

Culhane, J. L., et al. (2007): *Hinode* EUV study of jets in the Sun's south polar corona. *Publications of the Astronomical Society of Japan* **59**, S751–S756.

Culhane, J. L., et al. (2007): The UV imaging spectrometer for *Solar-B*. *Solar Physics* **243**, 19–61.

Culhane, J. L., Jordan, C. (Eds., 1991): *The Physics of Solar Flares.* London: The Royal Society.

Cummings, A. C., Stone, E. C., Webber, W. R. (1993): Estimate of the distance to the solar wind termination shock from gradients of anomalous cosmic ray oxygen. *Journal of Geophysical Research* **98**, 15165–15168.

Currie, R. G. (1974): Solar cycle signal in surface air temperature. *Journal of Geophysical Research* **79**, 5657–5660.

Cushman, G. W., Rense, W. A. (1976): Evidence of outward flow of plasma in a coronal hole. *The Astrophysical Journal (Letters)* **207**, L61–L62.

Czaykowska, A., De Pontieu, B., Alexander, D., Rank, G. (1999): Evidence for chromospheric evaporation in the late gradual flare phase from *SOHO*/CDS observations. *The Astrophysical Journal (Letters)* **521**, L75–L78.

D

Daglis, I. A., et al. (1999): The terrestrial ring current: origin, formation, and decay. *Review of Geophysics* **37**, 407–438.

Dahlburg, R. B., Klimchuk, J. A., Antiochos, S. K. (2005): An explanation of the "switch-on" nature of magnetic energy release and its application to coronal heating. *The Astrophysical Journal* **622**, 1191–1201.

Damon, P. E., Jirikowic, J. L. (1994): Solar forcing of global climate change. In: *The Sun as a Variable Star* (Eds. J. Pap, H. Hudson and S. Solanki). New York: Cambridge University Press, pp. 301–314.

Damon, P. E., Peristykh, A. N. (1999): Solar cycle length and twentieth century northern hemisphere warming. *Geophysical Research Letters* **26**, 2469–2472.

Damon, P. E., Sonett, C. P. (1991): Solar and terrestrial components of the atmospheric ^{14}C variation spectrum. In: *The Sun in Time* (Eds. C. P. Sonett, M. S. Giampapa and M. S. Matthews). Tucson: University of Arizona Press, pp. 360–388.

Danesy, D. (Ed., 2004): *SOHO-14/GONG 2004: Helio- and Asteroseismology: Towards a Golden Future*. ESA SP-559 2004.

Dansgaard, W., et al. (1980): Climatic record revealed by the Camp Century ice core. In: *The Late-Glacial Ages* (Ed. K. Turekian). New Haven: Yale University Press, pp. 37–46.

Dansgaard, W., et al. (1984): North Atlantic climate oscillations revealed by deep Greenland ice cores. In: *Climate Processes and Climate Sensitivity. American Geophysical Union Geophysical Monograph 29* (Eds. J. E. Hansen and T. Takahashi). Washington: American Geophysical Union, pp. 288–298.

Dansgaard, W., et al. (1989): A new Greenland deep ice core. *Science* **218**, 1273–1277.

Dasso, S., Milano, L. J., Matthaeus, W. H., Smith, C. W. (2005): Anisotropy in fast and slow wind fluctuations. *The Astrophysical Journal (Letters)* **635**, L181–L184.

David, C., et al. (1998): Measurement of the electron temperature gradient in a solar coronal hole. *Astronomy and Astrophysics* **336**, L90–L94.

David, C., Gabriel, A. H., Bely-Dubau, F. (1997): Temperature structure in coronal holes. In: *The Corona and Solar Wind Near Minimum Activity Proceedings of the Fifth SOHO Workshop. ESA SP-404*. Noordwijk: ESA Publications Division, pp. 319–322.

Davis, L. Jr. (1955): Interplanetary magnetic fields and cosmic rays. *Physical Review* **100**, 1440–1444.

Davis, L. Jr. (1972): The interplanetary magnetic field. In: *Solar Wind: The Proceedings of a Conference Sponsored by the National Aeronautics and Space Administration. NASA SP-308* (Eds. C. P. Sonett, P. J. Coleman, Jr., J. M. Wilcox) Washington: NASA, pp. 73–103.

Davis, R. Jr. (1964): Solar neutrinos II. Experimental. *Physical Review Letters* **12**, 303 305.

Davis, R. Jr., Harmer, D. S., Hoffman, K. C. (1968): Search for neutrinos from the Sun. *Physical Review Letters* **20**, 1205–1209. Reproduced in: *A Source Book in Astronomy and Astrophysics 1900–1975* (Eds. K. R. Lang and O. Gingerich). Cambridge: Harvard University Press 1979, pp. 389–395.

Davis, J. A., et al. (2008): First observation of a CME from the Sun to a distance greater than 1 AU using the Heliospheric Imager aboard STEREO A. Geophysical Research Letters – submitted.

De Forest, C. E. (2004): High-frequency waves detected in the solar atmosphere. *The Astrophysical Journal (Letters)* **617**, L89–L92.

De Forest, C. E., et al. (1997): Polar plume anatomy: results of a coordinated observation. *Solar Physics* **175**, 393–410. Reproduced in: *The First Results from SOHO* (Eds. B. Fleck and Z. Svestka). Dordrecht: Kluwer Academic Publishers, pp. 393–410.

De Forest, C. E., Gurman, J. B. (1998): Observation of quasi-periodic compressive waves in solar polar plumes. *The Astrophysical Journal (Letters)* **501**, L217–L220.

De Jager, C. (1986): Solar flares and particle acceleration. *Space Science Reviews* **44**, 43–90.

De Jager, C., Kundu, M. R. (1963): A note on bursts of radio emission and high energy ($> 20\,\text{keV}$) X-rays from solar flares. *Space Research* 3, 836–838.

De Jong, A. F. M., Mook, W. G., Becker, B. (1979): Confirmation of the Suess wiggles. *Nature* **280**, 48–49.

De Keyser, J., Roth, M., Forsyth, R., Reisenfeld, D. (2000): *Ulysses* observations of sector boundaries at aphelion. *Journal of Geophysical Research* **105**(A7), 15689–15698.

De Mairan, J. J. D. (1733): *Traité Physique et Historique de l'Aurore Boréale*. Paris: Imprimerie Royale.

De Moortel, I., Hood, A. W., Ireland, J., Walsh, R. W. (2002): Longitudinal intensity oscillations in coronal loops observed with *TRACE* II. Discussion of measured parameters. *Solar Physics* **209**, 89–108.

De Moortel, I., Ireland, J., Walsh, R. W. (2000): Observation of oscillations in coronal loops. *Astronomy and Astrophysics* **355**, L23–L26.

De Pontieu, B., Berger, T. E., Schrijver, C. J., Title, A. M. (1999): Dynamics of transition region moss at high time resolution. *Solar Physics* **190**, 419–435.

De Pontieu, B., Erdélyi, R., De Moortel, I. (2005): How to channel photospheric oscillations into the corona. *The Astrophysical Journal (Letters)* **624**, L61–L64.

De Pontieu, B., Erdélyi, R., Stewart, J. P. (2004): Solar chromospheric spicules from the leakage of photospheric oscillations and flows. *Nature* **430**, 536–539.

De Pontieu, B., et al. (2007a): A tale of two spicules: the impact of spicules on the magnetic chromosphere. *Publications of the Astronomical Society of Japan* **59**, S655–S662.

De Pontieu, B., et al. (2007b): Chromospheric Alfvenic waves strong enough to power the solar wind. *Science* **318**, 1574–1576.

De Pontieu, B., Martens, P. C. H., Hudson, H. S. (2001): Chromospheric damping of Alfvén waves. *The Astrophysical Journal* **558**, 859–871.

De Pontieu, B., Tarbell, T., Erdelyi, R. (2003): Correlations on arcsecond scales between chromospheric and transition region emission in active regions. *The Astrophysical Journal* **590**, 502–518.

De Rosa, M. L., Gilman, P. A., Toomre, J. (2002): Solar multiscale convection and rotation gradients studied in shallow spherical shells. *The Astrophysical Journal* **581**, 1356–1374.

De Rosa, M. L., Toomre, J. (2004): Evolution of solar supergranulation. *The Astrophysical Journal* **616**, 1242–1260.

De Vorkin, D. H. (1992): *Science with a Vengeance: How the Military Created the US Space Sciences after World War II*. New York: Springer-Verlag.

De Vries, H. L. (1958): Variation in concentration of radiocarbon with time and location on Earth. *Proceedings Koninlijke Nederlandse Akademie Wetenschappen* B, **61**, 94–102.

Dearborn, D. S. P., Blake, J. R. (1980a): Is the Sun constant? *The Astrophysical Journal* **237**, 616–619.

Dearborn, D. S. P., Blake, J. R. (1980b): Magnetic fields and the solar constant. *Nature* **287**, 365–366.

Dearborn, D. S. P., Blake, J. R. (1982): Surface magnetic fields and the solar luminosity. *The Astrophysical Journal* **257**, 896–900.

Debrunner, H., et al. (1983): The solar cosmic ray neutron event on June 3, 1982. *Proceedings 18th International Cosmic Ray Conference* **4**, 75–78.

Debrunner, H., Flückiger, E. O., Lockwood, J. A. (1990): Signature of the solar cosmic-ray event on 1982 June 3. *The Astrophysical Journal Supplement* **73**, 259–262.

Decker, R. B., et al. (2005): *Voyager 1* in the foreshock, termination shock, and heliosheath. *Science* **309**, 2020–2024.

Decker, R. B., et al. (2008): Mediation of the solar wind termination shock by non-thermal ions. *Nature* **454**, 67–70.

Delaboudinière, J. -P., et al. (1995): EIT: Extreme-ultraviolet Imaging Telescope for the *SOHO* mission. *Solar Physics* **162**, 291–312.

Dellinger, J. H. (1935): A new radio transmission phenomenon. *Physical Review* **48**, 705.

Dellinger, J. H. (1937): Sudden ionospheric disturbances. *Terrestrial Magnetism and Atmospheric Electricity* **42**, 49–53.

Delmas, R. J., et al. (1992): 1000 years of explosive volcanism recorded at South Pole. *Tellus* **44B**, 335–350.

Denisse, J. F. (1984): The early years of radio astronomy in France. In: *The Early Years of Radio Astronomy* (Ed. W. T. Sullivan III). New York: Cambridge University Press, pp. 303–316.

Denisse, J. F., Boischot, A., Pick, M. (1960): Propiétes deséruptions chromospheriques assoiciées a la production de rayons cosmiques par le soliel. In: *Space Research, Proceedings of the First International Space Science Symposium.* Amsterdam: North Holland Publishing Co., pp. 637–648.

Dennis, B. R. (1985): Solar hard X-ray bursts. *Solar Physics* **100**, 465–490.

Dennis, B. R. (1988): Solar flare hard X-ray observations. *Solar Physics* **118**, 49–94.

Dennis, B. R., Zarro, D. M. (1993): The Neupert effect: what can it tell us about the impulsive and gradual phases of solar flares? *Solar Physics* **146**, 177–190.

Dennison, P. A., Hewish, A. (1967): The solar wind outside the plane of the ecliptic. *Nature* **213**, 343–346.

Denskat, K. U., Neubauer, F. M. (1983): Observations of hydromagnetic turbulence in the solar wind. In: *Solar Wind Five* (Ed. M. Neugebauer). Washington, D.C.: NASA CP-2280, pp. 81–91.

Dere, K. P. (1994): Explosive events, magnetic reconnection, and coronal heating. *Advances in Space Research* **14**(4), 13–22.

Dere, K. P., et al. (1991): Explosive events and magnetic reconnection in the solar atmosphere. *Journal of Geophysical Research* **96**, 9399–9407.

Dere, K. P., et al. (1997): EIT and LASCO observations of the initiation of a coronal mass ejection. *Solar Physics* **175**, 601–612. Reproduced in: *The First Results from SOHO* (Eds. B. Fleck and Z. Svestka). Dordrecht: Kluwer Academic Publishers, pp. 601–612.

Dere, K. P., et al. (1999): LASCO and EIT observations of helical structure in coronal mass ejections. *Astrophysical Journal* **516**, 465–474.

Dere, K. P., et al. (2007): The structure and dynamics of the quiet corona from observations with the extreme ultraviolet imaging spectrometer on *Hinode*. *Publications of the Astronomical Society of Japan* **59**, S721–S726.

Dere, K. P., Mason, H. E. (1993): Nonthermal velocities in the solar transition zone observed with the high-resolution telescope and spectrograph. *Solar Physics* **144**, 217–242.

Dere, K. P., Wang, D., Howard, R. (2005): Three-dimensional structure of coronal mass ejections from LASCO polarization measurements. *The Astrophysical Journal (Letters)* **620**, L119–L122.

Desai, M. I., et al. (2003): Evidence for a suprathermal seed population of heavy ions accelerated by interplanetary shocks near 1 AU. *The Astrophysical Journal* **588**, 1149–1162.

Desai, M. I., et al. (2006): Heavy-ion elemental abundances in large solar energetic particle events and their implications for the seed population. *Astrophysical Journal* **649**, 470–489.

Desai, M. I., et al. (2006): The seed population for energetic particles accelerated by CME-driven shocks. *Space Science Reviews* **124**, 261–275.

Dessler, A. J. (1967): Solar wind and interplanetary magnetic field. *Reviews of Geophysics* **5**, 1–41.

Deubner, F. -L. (1975): Observations of low wave number nonradial eigenmodes of the Sun. *Astronomy and Astrophysics* **44**, 371–375.

Deubner, F. -L., Ulrich, R. K., Rhodes, E. J. Jr. (1979): Solar p-mode oscillations as a tracer of radial differential rotation. *Astronomy and Astrophysics* **72**, 177–185.

Deubner, F.-L., Gough, D. (1984): Helioseismology – oscillations as a diagnostic of the solar interior. *Annual Review of Astronomy and Astrophysics* **22**, 593–619.

Dicke, R. H. (1978): Is there a chronometer hidden deep in the Sun? *Nature* **276**, 676–680.

Dickinson, R. E. (1975): Solar variability and the lower atmosphere. *Bulletin of the American Meteorological Society* **56**, 1240–1248.

Dickinson, R. W., Cicerone, R. J. (1986): Future global warming from atmospheric trace gases. *Nature* **319**, 109–115.

Dietrich, W. F. (1973): The differential energy spectra of solar-flare ^{1}H, ^{3}He and ^{4}He. *The Astrophysical Journal* **180**, 955–973.

Dikpati, M. (2005): Solar magnetic fields and the dynamo theory. *Advances in Space Research* **35**(3), 322–328.

Dikpati, M., De Toma, G., Gilman, P. A. (2006): Predicting the strength of solar cycle 24 using a flux-transport dynamo-based tool. *Geophysical Research Letters* **33**, L5102.

Dikpati, M., Gilman, P. A. (2005): A shallow-water theory for the Sun's active longitudes. *The Astrophysical Journal (Letters)* **635**, L193–L196.

Dikpati, M., Gilman, P. A., Mac Gregor, K. B. (2005): Constraints on the applicability of an interface dynamo to the Sun. *The Astrophysical Journal* **631**, 647–652.

Dilke, F. W. W., Gough, D. O. (1972): The solar spoon. *Nature* **240**, 262–264, 293, 294.

Dirac, P. A. M. (1931): Quantized singularities in the electromagnetic field. *Proceedings of the Royal Society of London A* **133**, 60–72.

Dobrzycka, D., et al. (2003): Ultraviolet spectroscopy of narrow coronal mass ejections. *The Astrophysical Journal* **588**, 586–595.

Dobrzycka, D., Raymond, J. C., Cranmer, S. R. (2000): Ultraviolet spectroscopy of polar coronal jets. *The Astrophysical Journal* **538**, 922–931.

Dobson, G. M. B. (1968): Forty years' research on atmospheric ozone at Oxford: a history. *Applied Optics* **7**, 387–405.

Dodson, H. W., Hedeman, E. R. (1970): Major Hα flares in centers of activity with very small or no spots. *Solar Physics* **13**, 401–419.

Dodson, H. W., Hedeman, E. R., Owren, L. (1953): Solar flares and associated 200 Mc/sec radiation. *The Astrophysical Journal* **118**, 169–196.

Domínguez Cerdeña, I., Kneer, F., Sánchez Almeida, J. (2003): Quiet-Sun magnetic fields at high spatial resolution. *The Astrophysical Journal (Letters)* **582**, L55–L58.

Domingo, V., Fleck, B., Poland, A. I. (1995): The *SOHO* mission: An overview. *Solar Physics* **162**, 1–37.

Donea, A.-C., Lindsey, C. (2005): Seismic emission from the solar flares of 2003 October 28 and 29. *Astrophysical Journal* **630**, 1168–1183.

Dones, L., et al. (2004): Oort cloud formation and dynamics. In: *Comets II* (Eds. M. C. Festou, H. U. Keller and H. A. Weaver). Tucson: University of Arizona Press, pp. 153–174. Also in *Astronomical Society of the Pacific Conference Proceedings* **323**, 371.

Donnelly, R. F. (1967): The solar flare radiations responsible for sudden frequency deviations. *Journal of Geophysical Research* **72**, 5247–5256.

Doschek, G. A. (1983a): Solar flare X-ray spectra from the *P78–1* spacecraft. *Solar Physics* **86**, 49–58.

Doschek, G. A. (1983b): Solar instruments on the *P78–1* spacecraft. *Solar Physics* **86**, 9–17.

Doschek, G. A. (1990): Soft X-ray spectroscopy of solar flares – an overview. *Astrophysical Journal Supplement* **73**, 117–130.

Doschek, G. A. (1991): High-temperature plasma in solar flares. *Philosophical Transactions of the Royal Society (London)* **A336**, 451–460.

Doschek, G. A., et al. (1980): High-resolution X-ray spectra of solar flares III. General spectral properties of X1 –X5 type flares. *The Astrophysical Journal* **239**, 725–737.

Doschek, G. A., et al. (1993): The 1992 January 5 flare at 13.3 UT: observations from *Yohkoh*. *The Astrophysical Journal* **416**, 845–856.

Doschek, G. A., et al. (2007a): Nonthermal velocities in solar active regions observed with the extreme-ultraviolet imaging spectrometer on *Hinode*. *The Astrophysical Journal (Letters)* **667**, L109–L112.

Doschek, G. A., et al. (2007b): The temperature and density structure of an active region observed with the extreme-ultraviolet imaging spectrometer on *Hinode*. *Publications of the Astronomical Society of Japan* **59**, S707–S712.

Doschek, G. A., Feldman, U., Bohlin, J. D. (1976): Doppler wavelength shifts of transition zone lines measured in *Skylab* solar spectra. *The Astrophysical Journal (Letters)* **205**, L177–L180.

Doschek, G. A., Kreplin, R. W., Feldman, U. (1979): High-resolution solar flare X-ray spectra. *The Astrophysical Journal (Letters)* **233**, L157–L160.

Doschek, G. A., Mariska, J. T., Sakao, T. (1996): Soft X-ray flare dynamics. *The Astrophysical Journal* **459**, 823–835.

Doschek, G. A., Strong, K. T., Tsuneta, S. (1995): The bright knots at the tops of soft X-ray flare loops. Quantitative results from *Yohkoh*. *The Astrophysical Journal* **440**, 370–385.

Doschek, G. A., Warren, H. P. (2005): Chromospheric evaporation in solar flares revisited. *Astrophysical Journal* **629**, 1150–1163.

Dowdy, J. F. Jr., Rabin, D., Moore, R. L. (1986): On the magnetic structure of the quiet transition region. *Solar Physics* **105**, 35–45.

Dröge, F. (1977): Millisecond fine-structures of solar burst radiation in the range 0.2–1.4 GHz. *Astronomy and Astrophysics* **57**, 285–290.

Dröge, W., Kartavykh, Y. Y., Klecker, B., Mason, G. M. (2006): Acceleration and transport modeling of solar energetic particle charge sates for the event of 1998 September 9. *The Astrophysical Journal* **645**, 1516–1524.

Drake, J. J., Testa, P. (2005): The 'solar model problem' solved by the abundance of neon in nearby stars. *Nature* **436**, 525–528.

Dreschhoff, G., Zeller, E. J. (1998): Ultra-high nitrate in polar ice as indicator of past solar activity. *Solar Physics* **177**, 365–374.

Dryer, M. (1974): Interplanetary shock waves generated by solar flares. *Space Science Reviews* **15**, 403–468.

Dryer, M. (1982): Coronal transient phenomenon. *Space Science Reviews* **33**, 233–275.

Dryer, M. (1994): Interplanetary studies: propagation of disturbances between the Sun and the magnetosphere. *Space Science Reviews* **67**, 363–419.

Dryer, M. (1996): Comments on the origins of coronal mass ejections. *Solar Physics* **169**, 421–429.

Dryer, M. (1998): Multidimensional, magnetohydrodynamic simulation of solar generated disturbances: space weather forecasting of geomagnetic storms. *American Institute of Aeronautics and Astronautics Journal* **36**(3), 365–370.

Dryer, M., et al. (2004): Real-time shock arrival predictions during the "Halloween 2003 epoch". *Space Weather* **2**, S09001.

Dryer, M., Jones, D. L. (1968): Energy deposition in the solar wind by flare-generated shock waves. *Journal of Geophysical Research, Space Physics* **73**, 4875–4881.

Dryer, M., McIntosh, P. S. (1972): Preface to solar activity and predictions. *American Institute of Aeronautics and Astronautics Proceedings* **30**, 1–3.

Dryer, M., Wu, C.-C., Smith, Z. K. (1997): Three-dimensional MHD simulation of the April 14, 1994 interplanetary coronal mass ejection and its propagation to Earth and *Ulysses*. *Journal of Geophysical Research* **102**(A7), 14065–14074.

D'silva, S. (1998): Computing travel time in time-distance helioseismology. *The Astrophysical Journal (Letters)* **498**, L79–L82.

Duijveman, A., Hoyng, P., Machado, M. E. (1982): X-ray imaging of three flares during the impulsive phase. *Solar Physics* **81**, 137–157.

Dulk, G. A. (1985): Radio emission from the Sun and other *stars*. *Annual Review of Astronomy and Astrophysics* **23**, 169–224.

Dungey, J. W. (1953): Conditions for the occurrence of electrical discharges in astrophysical systems. *Philosophical Magazine* **44**, 725–738.

Dungey, J. W. (1958): The neutral point discharge theory of solar flares. A reply to Cowling's crit-
icism. In: *Electromagnetic Phenomena in Cosmical Physics. Proceedings of IAU Symposium
No. 6* (Ed. B. Lehnert). Cambridge: Cambridge at the University Press 1958, pp. 135–140.

Dungey, J. W. (1961): Interplanetary magnetic field and the auroral zones. *Physical Review Letters*
6, 47–48.

Dungey, J. W. (1979): First evidence and early studies of the Earth's bow shock. *Nuovo Cimento
C* **2**, 655–660.

Dungey, J. W. (1994): Memories, maxims, and motives. *Journal of Geophysical Research* **99**,
19189–19197.

Duvall, T. L. Jr. (1979): Large-scale solar velocity fields. *Solar Physics* **63**, 3–15.

Duvall, T. L. Jr. (1980): The equatorial rotation rate of the supergranulation cells. *Solar Physics*
66, 213–221.

Duvall, T. L. Jr. (1982): A dispersion law for solar oscillations. *Nature* **300**, 242–243.

Duvall, T. L. Jr., Birch, A. C., Gizon, L. (2006): Direct measurement of travel-time kernels for
helioseismology. *The Astrophysical Journal* **646**, 553–559.

Duvall, T. L. Jr., et al. (1984): Internal rotation of the Sun. *Nature* **310**, 22–25.

Duvall, T. L. Jr., et al. (1996): Downflows under sunspots detected by helioseismic tomography.
Nature **379**, 235–237.

Duvall, T. L. Jr., et al. (1997): Time-distance helioseismology with the MDI instrument. *Solar
Physics* **170**, 63–73. Reprinted in: *The First Results from SOHO* (Eds. B. Fleck and Z. Svestka).
Boston: Kluwer Academic Publishers, pp. 63–73.

Duvall, T. L. Jr., Harvey, J. W. (1983): Observations of solar oscillations of low and intermediate
degree. *Nature* **302**, 24–27.

Duvall, T. L. Jr., Harvey, J. W. (1984): Rotational frequency splitting of solar oscillations. *Nature*
310, 19–22.

Duvall, T. L. Jr., Harvey, J. W., Kosovichev, A. G., Svestka, Z. (Eds., 2000) SOHO 9: Helioseismic
Diagnostics of Solar Convection and Activity. *Solar Physics* **192**, Nos. 21–22, 1–478.

Duvall, T. L. Jr., Harvey, J. W., Pomerantz, M. A. (1986): Latitude and depth variation of solar
rotation. *Nature* **321**, 500–501.

Duvall, T. L. Jr., Jeffries, S. M., Harvey, J. W., Pomerantz, M. A. (1993): Time-distance
helioseismology. *Nature* **362**, 430–432.

Dziembowski, W. A., Goode, P. R. (2004): Helioseismic probing of solar variability: The
formation and simple assessments. *The Astrophysical Journal* **600**, 464–479.

E

Earl, J. A. (1961): Cloud-chamber observations of primary cosmic-ray electrons. *Physical Review
Letters* **6**, 125–128.

Eather, R. H. (1980): *Majestic Lights. The Aurora in Science, History, and the Arts.* Washington:
American Geophysical Union.

Eddington, A. S. (1920): The internal constitution of the stars. *Nature* **106**, 14–20.

Eddy, J. A. (1974): A nineteenth-century coronal transient. *Astronomy and Astrophysics* **34**,
235–240.

Eddy, J. A. (1976): The Maunder minimum. The reign of Louis XIV appears to have been a time
of real anomaly in the behavior of the sun. *Science* **192**, 1189–1202.

Eddy, J. A. (1977a): Climate and the changing Sun. *Climate Change* **1**, 173–190.

Eddy, J. A. (1977b): The case of the missing sunspots. *Scientific American* **236**, 80–95, May.

Eddy, J. A. (1977c): Historical evidence for the existence of the solar cycle. In: *The Solar Output
and Its Variation* (Ed. ORAN R. WHITE). Boulder: Colorado Associated University Press,
pp. 51–71.

Eddy, J. A. (1979): *A New Sun: The Solar Results From Skylab*. Washington: National Aeronautics and Space Administration SP-402 1979.

Eddy, J. A. (1983a): The Maunder minimum: a reappraisal. *Solar Physics* **89**, 195–207.

Eddy, J. A. (1983b): Keynote address: an historical review of solar variability, weather, and climate. In: *Weather and Climate Responses to Solar Variations* (Ed. B. M. McCormac). Boulder: Colorado Associated University Press, pp. 1–23.

Eddy, J. A. (1990): Some thoughts on Sun-weather relations. *Philosophical Transactions of the Royal Society of London A* **330**, 543–545.

Eddy, J. A., Gilliland, R. L., Hoyt, D. V. (1982): Changes in the solar constant and climatic effects. *Nature* **300**, 689–693.

Eddy, J. A., Stephenson, F. R., Yau, K. K. C. (1989): On pre-telescopic sunspot records. *Quarterly Journal of the Royal Astronomical Society* **30**, 60–73.

Edlén, B. (1994): An attempt to identify the emission lines in the spectrum of the solar corona. *Arkiv för Matematik, Astronomi och Fysik* **28B**, 1–4. Reproduced in: *A Source Book in Astronomy and Astrophysics, 1900-1975* (Eds. K. R. Lang and O. Gingerich). Cambridge, MA: Harvard University Press 1975, pp. 120–124.

Edlén, B. (1942): Die Deutung der Emissionslinien im Spektrum der Sonnenkorona. *Zeitschrift für Astrophysik* **22**, 30–64.

Edlén, B. (1945): The identification of the coronal lines. *Monthly Notices of the Royal Astronomical Society* **105**, 323–333.

Eguci, K., et al. (2003): First results from KamLAND: evidence for reactor anti-neutrino disappearance. *Physical Review Letters* **90**, 021802.

Einaudi, G., et al. (1999): Formation of the slow solar wind in a coronal streamer. *Journal of Geophysical Research* **104**(A1), 521–534.

Einaudi, G., et al. (2001): Plasmoid formation and acceleration in the solar streamer belt. *The Astrophysical Journal* **547**, 1167–1177.

Einstein, A. (1905): Ist die Trägheit eines Körpers von seinem Energieinhalt abhängig? (Does the inertia of a body depend upon its energy content?), *Annalen der Physik* **18**, 639–641. English translation given in *A Source Book in Astronomy and Astrophysics 1900–1975* (Eds. K. R. Lang and O. Gingerich). Cambridge, Massachusetts: Harvard University Press 1979, pp. 276–278.

Einstein, A. (1906): Das Prinzip von der Erhaltung der Schwerpunktsbewegung und die Trägheit der Energie. (The principle of the conservation of the motion of the center of gravity and the inertia of energy), *Annalen der Physik* **20**, 627–633. English translation given in *A Source Book in Astronomy and Astrophysics 1900–1975* (Eds. K. R. Lang and O. Gingerich). Cambridge, Massachusetts: Harvard University Press 1979, pp. 276–280.

Elliott, H. A., McComas, D. J., Riley, P. (2003): Latitudinal extent of large-scale structures in the solar wind. *Annales Geophysicae* **21**(6), 1331–1339.

Elliott, J. R., Kosovichev, A. G. (1998): The adiabatic exponent in the solar core. *The Astrophysical Journal (Letters)* **500**, L199–L202.

Ellison, M. A. (1942): Some studies of the motions of hydrogen flocculi by Doppler displacements of the Hα line. *Monthly Notices of the Royal Astronomical Society* **102**, 11–21.

Ellison, M. A. (1946): Visual and spectrographic observations of a great solar flare 1946 July 25. *Monthly Notices of the Royal Astronomical Society* **106**, 500–508.

Ellison, M. A. (1948): Distinction between flares and prominences. *Observatory* **68**, 69–70.

Ellison, M. A. (1949): Characteristic properties of chromospheric flares. *Monthly Notices of the Royal Astronomical Society* **109**, 1–27.

Elphinstone, R. E., Murphree, J. S., Cogger, L. L. (1996): What is a global auroral substorm? *Reviews of Geophysics* **34**, 169–232.

Elsworth, Y. P., et al. (1990): Variation of low-order acoustic solar oscillations over the solar cycle. *Nature* **345**, 322–324.

Elsworth, Y. P., et al. (1995): Slow rotation of the Sun's interior. *Nature* **376**, 669–672.

Emiliani, C. (1966): Isotopic paleotemperatures. *Science* **154**, 851–857.

Emslie, A. G., et al. (2004): Energy partition in two solar flare/CME events. *Journal of Geophysical Research* **109**(A10), A10104.

Endeve, E., et al. (2005): Release of helium from closed-field regions of the Sun. *The Astrophysical Journal* **624**, 02.

Evans, J. V. (1982): The Sun's influence on the Earth's atmosphere and interplanetary space. *Science* **216**, 467–474.

Evenson, P. Meyer, P. Pyle, K. R. (1983): Protons from the decay of solar flare neutrons. *The Astrophysical Journal* **274**, 875–882.

Evershed, J. (1948): Spectrum lines in chromospheric flares. *Observatory* **68**, 67–68.

F

Fabian, P., Pyle, J. A., Wells, R. J. (1979): The August 72 proton event and the atmospheric ozone layer. *Nature* **277**, 458–460.

Fainberg, J., Stone, R. G. (1974): Satellite observations of type III radio bursts at low frequencies. *Space Science Reviews* **16**, 145–188.

Fairfield, D. H., Cahill, L. J. Jr. (1966): Transition region magnetic field and polar magnetic disturbances. *Journal of Geophysical Research* **71**, 155–169.

Falconer, D. A., et al. (1997): Neutral-line magnetic shear and enhanced coronal heating in solar active regions. *The Astrophysical Journal* **482**, 519–534.

Falconer, D. A., Moore, R. L., Gary, G. A. (2002): Correlation of the coronal mass ejection productivity of solar active regions with measures of their global nonpotentiality from vector magnetograms: baseline results. *Astrophysical Journal* **569**, 1016–1025.

Falconer, D. A., Moore, R. L., Porter, J. G., Hathaway, D. H. (1998): Network coronal bright points: coronal heating concentrations found in the solar magnetic network. *The Astrophysical Journal* **501**, 386–396.

Fan, Y. (2004): Magnetic fields in the solar convection zone. *Living Reviews in Solar Physics* **1**, 1.

Fan, Y. (2005): Coronal mass ejections as loss of confinement of kinked magnetic flux ropes. *Astrophysical Journal* **630**, 543–551.

Fan, Y., Gibson, S. E. (2004): Numerical simulations of three-dimensional coronal magnetic fields resulting from the emergence of twisted magnetic flux tubes. *Astrophysical Journal* **609**, 1123–1133.

Fan, Y., Gibson, S. E. (2006): On the nature of the x-ray bright core in a stable filament channel. *Astrophysical Journal (Letters)* **641**, L149–L152.

Fan, Y., Gibson, S. E. (2007): Onset of coronal mass ejections due to loss of confinement of coronal flux ropes. *The Astrophysical Journal* **668**, 1232–1245.

Farrugia, C. J., Burlaga, L. F., Lepping, R. P. (1997): Magnetic clouds and the quiet storm effect at Earth. In: *Magnetic Storms: Geophysical Monograph 98* (Eds. B. T. Tsurutani, et al.). Washington: American Geophysical Union, pp. 91–106.

Farrugia, C. J., et al. (1993): The Earth's magnetosphere under continued forcing: substorm activity during the passage of an interplanetary magnetic cloud. *Journal of Geophysical Research* **98**, 7657–7671.

Feldman, U. (1992): Elemental abundances in the upper solar atmosphere. *Physica Scripta* **46**, 202–220.

Feldman, U., et al. (1980): High-resolution X-ray spectra of solar flares IV. General spectral properties of M-type flares. *The Astrophysical Journal* **241**, 1175–1185.

Feldman, U., et al. (1994): The morphology of the 10^7 plasma in solar flares I. Nonimpulsive flares. *Astrophysical Journal* **424**, 444–458.

Feldman, U., et al. (1996a): Electron temperature, emission measure, and X-ray flux in A2 to X2 X-ray class solar flares. *Astrophysical Journal* **460**, 1034–1041.

Feldman, U., Widing, K. G. (1993): Elemental abundances in the upper solar atmosphere of quiet and coronal hole regions (Te is approximately equal to 4.3×10 exp 5 K). *Astrophysical Journal* **414**, 381–388.

Feldman, W. C., et al. (1976): High-speed solar wind parameters at 1 AU. *Journal of Geophysical Research* **81**, 5054–5060.

Feldman, W. C., et al. (1977): Plasma and magnetic fields from the Sun. In: *The Solar Output and its Variations* (Ed. O. R. White). Boulder: Colorado Associated University Press, pp. 351–381.

Feldman, W. C., et al. (1981): The solar origins of solar wind interstream flows: near-equatorial coronal streamers. *Journal of Geophysical Research* **86**(A7), 5408–5416.

Feldman, W. C., et al. (1996b): Constraints on high-speed solar wind structure near its coronal base: a *Ulysses* perspective. *Astronomy and Astrophysics* **316**, 355–367.

Feldman, W. C., et al. (1997): Experimental constraints on pulsed and steady state models of the solar wind near the Sun. *Journal of Geophysical Research* **102**, 26, 905.

Feng, L. B., et al. (2007): First stereoscopic coronal loop reconstructions from STEREO SECCHI images. *The Astrophysical Journal Letters* **671**, 205.

Fermi, E. (1934): Versuch einer theorie de beta-strahlen I. (An attempt at the theory of beta-rays.), *Zeitschrift fur Physik* **88**, 161. English translation in *American Journal of Physics* **36**, 1150 (1968).

Fichtel, C. E., Guss, D. E. (1961): Heavy nuclei in solar cosmic rays. *Physical Review* **6**, 495–497.

Fireman, E. L. (1980): Solar activity during the past 10,000 years from radionuclides in lunar samples. In: *The Ancient Sun:Fossil Record in the Earth, Moon and Meteorites* (Eds. R. O. Pepin, J. A. Eddy and R. B. Merrill). New York: Pergamon Press, pp. 365–386.

Fischer, H., et al. (1999): Ice core records of atmospheric CO_2 around the last three glacial terminations. *Science* **283**, 1712–1714.

Fishman, G. J., et al. (1994): Discovery of intense gamma-ray flashes of atmospheric origin. *Science* **264**, 1313–1316.

Fisk, L. A. (1978): He-3 rich flares – a possible explanation. *The Astrophysical Journal* **224**, 1048–1055.

Fisk, L. A. (1996): Motion of the footpoints of heliospheric magnetic field lines at the Sun: Implications for recurrent energetic particle events at high heliographic latitudes. *Journal of Geophysical Research*, **101**, A7, 15547–15554.

Fisk, L. A. (2001): On the global structure of the heliospheric magnetic field. *Journal of Geophysical Research* **106**(A8), 15849–15858.

Fisk, L. A. (2003): Acceleration of the solar wind as a result of the reconnection of open magnetic flux with coronal loops. *Journal of Geophysical Research* **108**, A4, SSH 7-1, 1157.

Fisk, L. A. (2005): The open magnetic flux of the Sun 1. Transport by reconnections with coronal loops. *The Astrophysical Journal* **626**, 563–573.

Fisk, L. A., et al. (2003): Acceleration of the solar wind as a result of the reconnection of open magnetic flux with coronal loops. *American Institute of Physics Proceedings* **679**, 287–292.

Fisk, L. A., Gloeckler, G. (2006): The common spectrum for accelerated ions in the quiet-time solar wind. *The Astrophysicla Journal (Letters)* **640**, L79–L82.

Fisk, L. A., Jokipii, J. R. (1999): Mechanisms for latitudinal transport of energetic particles in the heliosphere. *Space Science Reviews* **89**, 115–124.

Fisk, L. A., Kozlovsky, B., Ramaty, R. (1974): An interpretation of the observed oxygen and nitrogen enhancements in low-energy cosmic rays. *Astrophysical Journal (Letters)* **190**, L35–L37.

Fisk, L. A., Lee, M. A. (1980): Shock acceleration of energetic particles in co-rotating interaction regions in the solar wind. *The Astrophysical Journal* **237**, 620–626.

Fisk, L. A., Schwadron, N. A. (2001a): Origin of the solar wind theory. *Space Science Reviews* **97**, 221.

Fisk, L. A., Schwadron, N. A. (2001b): The behavior of open magnetic field of the sun. *The Astrophysical Journal* **560**, 425–438.

Fisk, L. A., Schwadron, N. A., Zurbuchen, T. H. (1999): Acceleration of the fast solar wind by the emergence of new magnetic flux. *Journal of Geophysical Research* **104**, A9, 19765–19772.

Fisk, L. A., Schwadron, N. A., Zuruchen, T. H. (1998): On the slow solar wind. *Space Science Reviews* **86**, 51–60.

Fisk, L. A., Zurbuchen, T. H. (2006): Distribution and properties of open magnetic flux outside of coronal holes. *Journal of Geophysical Research* **111**, A09115.

Fisk, L. A., Zurbuchen, T. H., Schwadron, N. A. (1999): On the coronal magnetic field: consequences of large-scale motion. *The Astrophysical Journal* **521**, 868–877.

Fitzenreiter, R. J., et al. (2003): Modification of the solar wind electron velocity distribution at interplanetary shocks. *Journal of Geophysical Research* **108**(A12), SSH 1–1.

Fitzgerald, G. F. (1892): Sunspots and magnetic storms. *The Electrician* **30**, 48.

Fitzgerald, G. F. (1900): Sunspots, magnetic storms, comets' tails, atmospheric electricity and aurorae. *The Electrician* **46**, 249, 287–288.

Fleck, B. (2005): Eight years of *SOHO*: some highlights. In: *Solar Magnetic Phenomena Astrophysics and Space Science Library* **320**, (Eds. A. Hanslmeier, A. Veronig, M. Messerotti) New York: Springer, 139–166.

Fleck, B., Domingo, V., Poland, A. I. (Eds., 1995): The *SOHO* mission. *Solar Physics* **162**, Nos. 1, 2. Boston: Kluwer Academic Publishers.

Fleck, B., Gurman, J. B., Hochedez, J. -F., Robbrecht, E. (Eds., 2008): SOHO-20: transient events on the sun and in the heliosphere. *Annales Geophysicae*.

Fleck, B., Noci, G., Poletto, G. (Eds., 1994): SOHO-2: Mass supply and flows in the solar corona. *Space Science Reviews* **70**(1–2).

Fleck, B., Svestka, Z., (Eds., 1997): *The First Results from SOHO. Solar Physics* **170**(1), **175**(2). Boston: Kluwer Academic Publishers.

Fleck, B., Zurbuchen, T. H. (Eds., 2005): *Solar Wind 11/SOHO 16: Connecting Sun and Heliosphere.* ESA SP-592 2005 Noordwijk, the Netherlands.

Fleishman, G. D., Nita, G. M., Gary, D. E. (2005): Evidence for resonant transition radiation in decimetric continuum solar bursts. *Astrophysical Journal* **620**, 506–516.

Fletcher, K., Thompson, M. (Eds., 2006): *SOHO-18/GONG 2006: Beyond the Spherical Sun.* ESA SP-624 2006.

Fletcher, L. Hannah, I. G., Hudson, H. S., Metcalf, T. R. (2007): A *TRACE* white light and *RHESSI* hard X-ray study of flare energetics. *Astrophysical Journal* **656**, 1187–1196.

Fletcher, L., Huber, M. C. E. (1997): O^{5+} acceleration by turbulence in polar coronal holes. In: *The Corona and Solar Wind Near Minimum Activity. Proceedings of the Fifth SOHO Workshop. ESA SP-404*. Noordwijk: ESA Publications Division, pp. 379–384.

Fletcher, L., Hudson, H. (2001): The magnetic structure and generation of EUV flare ribbons. *Solar Physics* **204**, 69–89.

Fletcher, L., Hudson, H. (2002): Spectral and spatial variations of flare hard x-ray footpoints. *Solar Physics* **210**, 307–321.

Fletcher, L., Pollock, J. A., Potts, H. E. (2004): Tracking of *TRACE* ultraviolet flare footpoints. *Solar Physics* **222**, 279–298.

Fludra, A., et al. (1989): Turbulent and directed plasma motions in solar flares. *The Astrophysical Journal* **344**, 991–1003.

Fludra, A., et al. (1997): Active regions observed in extreme ultraviolet light by the coronal diagnostic spectrometer on *SOHO*. *Solar Physics* **175**, 487–509.

Fludra, A., Ireland, J. (2003): Inversion of the intensity-magnetic field relationship in solar active regions. *Astronomy and Astrophysics* **398**, 297–303.

Fogli, G. L., et al. (2003a): Addendum to "solar neutrino oscillation parameters after first KamLAND results. *Physical Review D* **69**, 017301.

Fogli, G. L., et al. (2003b): Solar neutrino oscillation parameters after first KamLAND results. *Physical Review D* **67**, 073002–073012.

Foley, C. R., Culhane, J. L., Acton, L. W. (1997): *Yohkoh* soft X-ray determination of plasma parameters in a polar coronal hole. *The Astrophysical Journal* **491**, 933–938.

Foley, C. R., et al. (2001): Eruption of a flux rope on the disk of the Sun: evidence for the coronal mass ejection trigger? *The Astrophysical Journal (Letters)* **560**, L91–L94.

Folland, C. K., Karl, T. R., Vinikov, K. Y. (1990): Observed climate variations and change. In: *Climate Change, the IPCC Scientific Assessment* (Eds. J. T. Houghton, G. J. Jenkins and J. J. Ephraums). Cambridge: Cambridge University Press, pp. 195–218.

Forbes, T. G. (1986): Fast-shock formation in line-tied magnetic reconnection models of solar flares. *The Astrophysical Journal* **305**, 553–563.

Forbes, T. G. (2000) A review on the genesis of coronal mass ejections. *Journal of Geophysical Research* **105**, A10, 23153–23166.

Forbes, T. G., Acton, L. W. (1996): Reconnection and field line shrinkage in solar flares. *Astrophysical Journal* **459**, 330–341.

Forbes, T. G., et al. (2006): CME theory and models. *Space Science Reviews* **123**, 251–302.

Forbes, T. G., Isenberg, P. A. (1991): A catastrophe mechanism for coronal mass ejections. *The Astrophysical Journal* **373**, 294–307.

Forbes, T. G., Malherbe, J. M., Priest, E. R. (1989): The formation of flare loops by magnetic reconnection and chromospheric ablation. *Solar Physics* **120**, 285–307.

Forbes, T. G., Priest, E. R. (1987): A comparison of analytical and numerical models for steadily driven reconnection. *Review of Geophysics* **25**, 1583–1607.

Forbush, S. E. (1937): On diurnal variation in cosmic-ray intensity. *Terrestrial Magnetism and Atmospheric Electricity* **42**, 1–16.

Forbush, S. E. (1938a): On cosmic-ray effects associated with magnetic storms. *Terrestrial Magnetism and Atmospheric Electricity* **43**, 203–218.

Forbush, S. E. (1938b): On the world-wide changes in cosmic ray intensity. *Physical Review* **54**, 975–988.

Forbush, S. E. (1946): Three unusual cosmic-ray increases possibly due to charged particles from the Sun. *Physical Review* **70**, 771–772.

Forbush, S. E. (1950): Cosmic-ray intensity variations during two solar cycles. *Journal of Geophysical Research* **63**, 651–669.

Forbush, S. E. (1954): World-wide cosmic-ray variations, 1937-1952. *Journal of Geophysical Research* **59**, 525–542.

Forbush, S. E., Stinchcomb, T. B., Schein, M. (1950): The extraordinary increase of cosmic-ray intensity on November 19, 1949. *Physical Review* **79**, 501–504.

Forrest, D. J., Chupp, E. L. (1983): Simultaneous acceleration of electrons and ions in solar flares. *Nature* **305**, 291–292.

Forrest, D. J., et al. (1985): Neutral pion production in solar flares. *Proceedings of the 19th International Cosmic Ray Conference* **4**, 146–149.

Forsyth, R. J., et al. (1996): The heliospheric magnetic field at solar minimum: *Ulysses* observations from pole to pole. *Astronomy and Astrophysics* **316**, 287–295.

Forsyth, R. J., et al. (2006): ICMEs in the inner heliosphere: origin, evolution and propagation effects. *Space Science Reviews* **123**, 383–416.

Fossat, E., Grec, G., Pomerantz, M. A. (1981): Solar pulsations observed from the geographic south pole – initial results. *Solar Physics* **74**, 59–63.

Fossum, A., Carlsson, M. (2005a): High-frequency acoustic waves are not sufficient to heat the solar chromosphere. *Nature* **435**, 919–921.

Fossum, A., Carlsson, M. (2005b): Response functions of the ultraviolet filters of *TRACE* and the detection of high-frequency acoustic waves. *The Astrophysical Journal* **625**, 556–562.

Foukal, P. (1990a): Solar luminosity variations over timescales of days to the past few solar cycles. *Philosophical Transactions of the Royal Society of London A* **330**, 591–599.

Foukal, P. (1990b): The variable Sun. *Scientific American* **262**, 34–41.

Foukal, P. (1994): Stellar luminosity variations and global warming. *Science* **264**, 238–239.

Foukal, P., Fröhlich, C., Spruit, H., Wigley, T. M. L. (2006): Variations in solar luminosity and their effect on the earth's climate. *Nature* **443**, 161–166.

Foukal, P., Lean, J. (1986): The influence of faculae on total irradiance and luminosity. *The Astrophysical Journal* **302**, 826–835.

Foukal, P., Lean, J. (1988): Magnetic modulation of solar luminosity by photospheric activity. *The Astrophysical Journal* **328**, 347–357.

Foukal, P., Lean, J. (1990): An empirical model of total solar irradiance variation between 1874 and 1988. *Science* **247**, 556–558.

Foukal, P., North, G., Wigley, T. (2004): A stellar view on solar variations and climate. *Science* **306**, 68–69.

Frank, L. A., Craven, J. D. (1988): Imaging results from *Dynamics Explorer 1. Review of Geophysics and Space Physics* **26**, 249–283.

Frank, L. A., et al. (1986): The theta aurora. *Journal of Geophysical Research* **91**, 3177–3224.

Fraunhofer, J. (1814–1815): Bestimmung des Brechungs - und Farbenzerstreutuungs – Vermögens Verschiedener Glassarten. *Denkschriften* (Munich Academy of Sciences) **5**, 195, 202. Auszug davon in Gilb. *Annalen der Physik* **56**, 264 (1817).

Frazier, E. N. (1968): A spatio-temporal analysis of velocity fields in the solar photosphere. *Zeitschrift für Astrophysik* **68**, 345–358.

Frazin, R. A., Cranmer, S. R., Kohl, J. L. (2003): Empirically determined anisotropic velocity distributions and outflows of O^{5+} ions in a coronal streamer at solar minimum. *Astrophysical Journal* **597**, 1145–1157.

Freier, P., et al. (1948a): Evidence for heavy nuclei in the primary cosmic radiation. *Physical Review* **74**, 213–217.

Freier, P., et al. (1948b): The heavy component of primary cosmic rays. *Physical Review* **74**, 1818–1827.

Freier, P., Ney, E. P., Winckler, J. R. (1959): Balloon observations of solar cosmic rays on March 26, 1958. *Journal of Geophysical Research* **64**, 685–688.

Frick, P., et al. (1997): Wavelet analysis of stellar chromospheric activity variations. *The Astrophysical Journal* **483**, 426–434.

Friedlander, M. W. (1989): *Cosmic Rays*. Cambridge, MA: Harvard University Press.

Friedman, H. (1961): X-ray and extreme ultraviolet observations of the Sun. In: *Space Research II* (Eds. H. C. Van De Hulst, C. De Jager and A. F. Moore). Amsterdam: North-Holland Pub. Co. Reproduced in: *A Source Book in Astronomy and astrophysics 1900–1975* (Eds. K. R. Lang and O. Gingerich). Cambridge: Harvard University Press 1979, pp. 56–61.

Friedman, H. (1963): Solar X-ray emission. In: *The Solar Corona. Proceedings of International Astronomical Union Symposium No. 16* (Ed. J. W. EVANS). New York: Academic Press, pp. 45–48.

Friedman, H. (1986): *Sun and Earth*. New York: Scientific American Library.

Friedman, H., Lichtman, S. W., Byram, E. T. (1951): Photon counter measurements of solar X-rays and extreme ultraviolet light. *Physical Review* **83**, 1025–1030.

Friis-Christensen, E., Lassen, K. (1991): Length of the solar cycle: an indicator of solar activity closely associated with climate. *Science* **254**, 698–700.

Friis-Christensen, E., Lassen, K. (1994): Solar activity and global temperature. In: *The Sun as a Variable Star: Solar and Stellar Luminosity Variations* (Eds. J. Pap, C. Fröhlich, H. Hudson, S. K. Solanki). New York: Cambridge University Press, pp. 339–347.

Fröhlich, C. (1987): Variability of the solar "constant" on time scales of minutes to years. *Journal of Geophysical Research* **92**, 796–800.

Fröhlich, C. (2006): Solar irradiance variability since 1978. *Space Science Reviews* **125**, 53–65.

Fröhlich, C., et al. (1995): VIRGO: Experiment for helioseismology and solar irradiance monitoring. *Solar Physics* **162**, 101–128.

Fröhlich, C., et al. (1997a): First results from VIRGO, the experiment for helioseismology and solar irradiance monitoring in SOHO. *Solar Physics* **170**, 1–25. Reprinted in: *The First Results from SOHO* (Eds. B. FLECK and Z. SVESTKA). Boston: Kluwer Academic Publishers 1997, pp. 1–25.

Fröhlich, C., et al. (1997b): In-flight performances of VIRGO solar irradiance instruments on SOHO. *Solar Physics* **175**, 267–286.

Fröhlich, C., Lean, J. (1998): The sun's total irradiance: cycles, trends and related climate change uncertainties since 1976. *Geophysical Research Letters* **25**, 4377–4380.

Fröhlich, C., Lean, J. (1999): Total solar irradiance variations: The construction of a composite and its comparison with models. In: *New Eyes to See Inside the Sun and Stars. IAU Symposium 185* (Eds. F. L. Deubner, J. Christensen-Dalsgaard and D. Kurtz). Dordrecht: Kluwer Academic Publications, pp. 89–102. *Space Science Reviews* **88**(3–4).

Fröhlich, C., Lean, J. (2002): Solar irradiance variability and climate. *Astronomische Nachrichten* **323**, 203–212.

Fröhlich, C., Lean, J. (2004): Solar radiative output and its variability: evidence and mechanisms. *Astronomy and Astrophysics Review* **12**(4), 273–320.

Fukuda, Y., et al. (1996): Solar neutrino data covering solar cycle 22, *Physical Review Letters* **77**, 1683–1686.

Fukuda, Y., et al. (1998): Evidence for oscillation of atmospheric neutrinos. *Physical Review Letters* **81**, 1562–1567.

Fukuda, Y., et al. (2001): Solar boron 8 and hep neutrino measurements from 1258 days of super-kamiokande data. *Physical Review Letters* **86**, 5651–5655.

Furth, H. P., Kileen, J., Rosenbluth, M. N. (1963): Finite-resistivity instabilities of a sheet pinch. *Physics of Fluids* **6**(4), 459–484.

Fürst, E., Benz, A. O., Hirth, W. (1982). About the relation between radio and soft X-ray emission in case of very weak solar activity. *Astrnomy and Astrophysics* **107**, 178–185.

G

Gabriel, A. H. (1971): Measurements on the Lyman alpha corona. *Solar Physics* **21**, 392–400.

Gabriel, A. H. (1976): A magnetic model of the solar transition region. *Philosophical Transactions of the Royal Society (London)* **A281**, 339–352.

Gabriel, A. H. (1977): Structure of the quiet chromosphere and corona. In: *The Energy Balance and Hydrodynamics of the Solar Chromosphere and Corona. Proceedings of IAU Colloquium No. 36* (Eds. R.-M. Bonnet and P. H. Delache). Paris: G. de Bussac Clermont-Ferrand, pp. 375–399.

Gabriel, A. H., Bely-Dubau, F., Lemaire, P. (2003): The contribution of polar plumes to the fast solar wind. *The Astrophysical Journal* **589**, 623–634.

Gabriel, A. H., et al. (1971): Rocket observations of the ultraviolet solar spectrum during the total eclipse of 1970 March 7. *The Astrophysical Journal* **169**, 595–614.

Gabriel, A. H., et al. (1995): Global Oscillations at Low Frequency from the *SOHO* mission (GOLF). *Solar Physics* **12**, 61–99.

Gabriel, A. H., et al. (1997): Performance and early results from the GOLF instrument flown in the *SOHO* mission. *Solar Physics* **175**, 207–226. Reprinted in: *The First Results from SOHO* (Eds. B. Fleck and Z. Svestka). Boston: Kluwer Academic Publishers, pp. 207–226.

Gabriel, A. H., et al. (2002): A search for solar g modes in the golf data. *Astronomy and Astrophysics* **390**, 111–113.

Gabriel, A. H., et al. (2005): Solar wind outflow in polar plumes form 1.05 to 2.4 Rsolar. *Astrophysical Journal (Letters)* **635**, L185–188.

Galileo G. (1613): *Istoria e Dimostrazioni Intorno alle Macchie Solari e Loro Accidenti*. Rome.

Gallagher, P. T., et al. (2002): *RHESSI* and *TRACE* observations of the 21 April 2002 X1.5 flare. *Solar Physics* **210**, 312–356.

Galloway, D. J., Weiss, N. O. (1981): Convection and magnetic fields in stars. *The Astrophysical Journal* **243**: 945–953.

Galvin, A. B., et al. (2008): The Plasma and Suprathermal Ion Composition (PLASTIC) investigation on the STEREO observatories. *Space Science Reviews*, **136**, No. 1–4, 437–486.

Gamow, G. (1928): Zur quanten theorie der atomzertrümmerung (On the quantum theory of the atomic nucleus). *Zeitschrift fur Physik* **52**, 510.

García, R. A., et al. (2007): Tracking solar gravity modes: The dynamics of the solar core. *Science* **316**, 1591–1597.

García, H. A. (2000): Thermal-spatial analysis of medium and large solar flares, 1976 to. 1996. *The Astrophysical Journal Supplement* **127**, 189–210.

Garrett, H. B. (1981): The charging of spacecraft surfaces. *Review of Geophysics* **19**, 577.

Gary, D. E., Hurford, G. J. (1994): Coronal temperature, density, and magnetic field maps of a solar active region using the Owens Valley Solar Array. *The Astrophysical Journal* **420**, 903–912.

Gary, D. E., Keller, C. U. (2004): S*olar and Space Weather Radiophysics – Current Status and Future Development*s. Dordrecht: Kluwer.

Gary, G. A. (2001): Plasma beta above a solar active region: Rethinking the paradigm. *Solar Physics* **203**, 71–86.

Gary, G. A., et al. (1987): Nonpotential features observed in the magnetic field of an active region. *The Astrophysical Journal* **314**, 782–794.

Gary, G. A., Moore, R. L. (2004): Eruption of a multiple-turn helical magnetic flux tube in a large flare: Evidence for external and internal reconnection that fits the breakout model of solar magnetic eruption. *Astrophysical Journal* **611**, 545–556.

Gauss, C. F. (1841): *Allgemeine Theorie des Erdmagnetismus, Resultate aus den Beobachtungen des Magnetischen Verein im Jarhre*. Translated by Mrs. Sabine, revised by Sir John Herschel in *Scientific Memoirs Selected From Transactions of Foreign Academies and Learned Societies and From Foreign Journals* **2**, 184–251.

Gavaghan, H. (1998): *Something New Under the Sun: Satellites and the Beginning of the Space Age*. New York: Springer-Verlag.

Gazis, P. R., et al. (2006): ICMEs at high latitudes and in the outer heliosphere. *Space Science Reviews* **123**, 417–451.

Geiss, J., Bochsler, P. (1991): Long time variations in solar wind properties – Possible causes versus observations". In: *The Sun in Time* (Eds. C. P. Sonett, M. S. Giampapa and M. S. Matthews). Tucson, University of Arizona Press, pp. 98–117.

Geiss, J., et al. (1970): *Apollo 11* and *12* solar wind composition experiments: fluxes of He and Ne isotopes. *Journal of Geophysical Research* **75**, 5972–5979.

Geiss, J., et al. (1995b): The southern high-speed stream – results from the SWICS instrument on *Ulysses*. *Science* **268**, 1033–1036.

Geiss, J., Gloeckler, G. (2001): Heliospheric and interstellar phenomena deduced from pickup ion observations. *Space Science Reviews* **97**, 169–181.

Geiss, J., Gloeckler, G., Fisk, L. A., von Steiger, R. (1995): C+ pickup ions in the heliosphere and their origin. *Journal of Geophysical Research* **100**(A12), 23373–23378.

Geiss, J., Gloeckler, G., Von Steiger, R. (1996a): Origin of C^+ ions in the heliosphere. *Space Science Reviews* **78**, 43–52.

Geiss, J., Gloeckler, G., Von Steiger, R. (1996b): Origin of the solar wind from composition data. *Space Science Reviews* **72**, 49–60.

Geiss, J., Reeves, H. (1972): Cosmic and solar system abundances of deuterium and helium-3. *Astronomy and Astrophysics* **18**, 126–132.

Geiss, J., Witte, M. (1996): Properties of the interstellar gas inside the heliosphere. *Space Science Reviews* **78**, 229–238.

Genthon, C., et al. (1987): Vostok ice core: climatic response to CO_2 and orbital forcing changes over the last climatic cycle. *Nature* **329**, 414–418.

Georges, T. M. (1967): Ionospheric effects of atmospheric waves. *ESSA Technical Report IER 57-ITSA 54*. Institute for telecommunication and aeronomy, Boulder, Colorado, p. 3.

Giacalone, J., Jokipii, J. R. (2004): Magnetic footpoint diffusion at the Sun and its relation to the heliospheic magnetic field. *The Astrophysical Journal* **616**, 573–577.

Gibson, S. E., et al. (2004): Observational consequences of a magnetic flux rope emerging into the corona. *The Astrophysical Journal* **617**, 600–613.

Gibson, S. E., et al. (2006): The evolving sigmoid: evidence for magnetic flux ropes in the corona before, during, and after CMEs. *Space Science Reviews* **124**, 131–144.

Gibson, S. E., Fan, Y. (2006): Coronal prominence structure and dynamics: a magnetic flux rope interpretation. *Journal of Geophysical Research* **111**, A12103.

Gibson, S. E., Low, B. C. (1998): A time-dependent three-dimensional magnetohydrodynamic model of coronal mass ejection. *Astrophysical Journal* **493**, 460.

Gilbert, W. (1600): *De Magnete, Magneticisque Corporibus, et de Magno Magnete Tellure: Physiologia Nova, Plurimis and Argumentis, and Experimentis Demonstrata*. London. English translation by P. Fleury Mottelay, *William Gilbert of Colchester...on the great magnet of the earth*. Ann Arbor 1893 and Silvanus P. Thompson, reprinted from the 1900 edition by Basic Books, New York 1958.

Gilbert, W.: *De Magnete, Magneticisque Corporibus, et de Magno Magnete Tellure: Physiologia Nova, Plurimis and Argumentis, and Experimentis Demonstrata* London 1600. English translation by P. Fleury Mottelay, *William Gilbert of Colchester... on the great magnet of the earth*. Ann Arbor 1893 and Silvanus P. Thompson, reprinted from the 1900 edition by Basic Books, New York 1958.

Giles, P. M., Duvall, T. L. Jr., Scherrer, P. H., Bogart, R. S. (1997): A subsurface flow of material from the Sun's equator to its pole. *Nature* **390**, 52–54.

Gille, J. C., Smythe, C. M., Heath, D. F. (1984): Observed ozone response to variations in solar ultraviolet radiation. *Science* **225**, 315–317.

Gilliland, R. L. (1980): Solar luminosity variations. *Nature* **286**, 838–839.

Gilliland, R. L. (1981): Solar radius variations over the past 265 years. *The Astrophysical Journal* **248**, 1144–1155.

Gilliland, R. L. (1982): Modeling solar variability. *The Astrophysical Journal* **253**, 399–405.

Gilliland, R. L. (1989): Solar evolution. *Palaeogeography, Palaeoclimatology, Palaeoecology* **75**, 35–55.

Gilman, P. A. (1974): Solar rotation. *Annual Review of Astronomy and Astrophysics* **12**, 47–70.

Gilman, P. A. (2000): Fluid dynamics and MHD of the solar convection zone and tachocline: Current understanding and unresolved problems. *Solar Physics* **192**, 27–48.

Gilman, P. A., Miesch, M. S. (2004): Limits to penetration of meridional circulation below the solar convection zone. *The Astrophysical Journal* **611**, 568–574.

Ginzburg, V. L. (1946): On solar radiation in the radio spectrum. *Comptes Rendus (Doklady) de l'Académie des Sciences de l' URSS* **52**, 487.

Ginzburg, V. L. (1956): The nature of cosmic radio emission and the origin of cosmic rays. *Nuovo Cimento Supplement* **3**, 38–48. Reproduced in: *A Source Book in Astronomy and Astrophysics 1900–1975*. Cambridge: Harvard University Press 1977, pp. 677–684.

Giordano, S., et al. (2000): Identification of he coronal sources of the fast solar wind. *The Astrophysical Journal (Letters)* **531**, L79–L82.

Giovanelli, R. G. (1939): The relations between eruptions and sunspots. *The Astrophysical Journal* **89**, 555–567.

Giovanelli, R. G. (1940): Solar eruptions. *The Astrophysical Journal* **91**, 344–349.

Giovanelli, R. G. (1946): A theory of chromospheric flares. *Nature* **158**, 81–82.

Giovanelli, R. G. (1947): Magnetic and electric phenomena in the Sun's atmosphere associated with sunspots. *Monthly Notices of the Royal Astronomical Society* **107**, 338–355.

Giovanelli, R. G. (1948): Chromospheric flares. *Monthly Notices of the Royal Astronomical Society* **108**, 163–176.

Giovanelli, R. G. (1949): A note on heat transfer in the upper chromosphere and corona. *Monthly Notices of the Royal Astronomical Society* **109**, 372.

Giovanelli, R. G., McCabe, M. K. (1958): The flare-surge event. *Australian Journal of Physics* **11**, 191–200.

Giovanni, L., Knoll, D. A. (2005): Effect of a converging flow at the streamer cusp on the genesis of the slow solar wind. *The Astrophysical Journal* **624**, 1049–1056.

Gizon, L. (2004): Helioseismology of time-varying flows through the solar cycle. *Solar Physics* **224**, 217.

Gizon, L., Birch, A. C. (2002): Time-distance helioseismology: the forward problem for random distributed sources. *The Astrophysical Journal* **571**, 966–986.

Gizon, L., Birch, A. C. (2004): Time-distance helioseismoogy: noise estimation. *The Astrophysical Journal* **614**, 472–489.

Gizon, L., Birch, A. C. (2005): Local helioseismology. *Living Reviews in Solar Physics* **2**, 6.

Gizon, L., Duvall, T. L. Jr., Schou, J. (2003): Wave-like properties of solar supergranulations. *Nature* **421**, 43–44.

Gleeson, L. J., Axford, W. I. (1968): Solar modulation of galactic cosmic rays. *The Astrophysical Journal* **154**, 1011–1019.

Gleissberg, W. (1943): Predictions for the coming sunspot-cycle. *Terrestrial Magnetism and Atmospheric Electricity* **48**, 243–244.

Gleissberg, W. (1958): The eighty-year sunspot cycle. *Journal of the British Astronomical Association* **68**, 148–152.

Gleissberg, W. (1965): The 80-year solar cycle in auroral frequency number. *Journal of the British Astronomical Association* **75**, 227–231.

Gleissberg, W. (1966): Ascent and descent in the eighty-year cycles of solar activity. *Journal of the British Astronomical Association* **76**, 265–270.

Gloeckler, G. (1999): Observation of injection and pre-acceleration processes in the slow solar wind. *Space Science Reviews* **89**, 91–104.

Gloeckler, G. (2003): Ubiquitous suprathermal tails on the soar wind and pickup ion distributions. *Astronomical Society of the Pacific Conference Proceedings* **679**, 583–588.

Gloeckler, G. et al. (1999): Unusual composition of the solar wind in the 2–3 May 1998 CME observed with SWICS on *ACE*. *Geophysical Research Letters* **26**, 157–160.

Gloeckler, G., et al. (1986): Solar wind carbon, nitrogen and oxygen abundances measured in the Earth's magnetosheath with *AMPTE/CCE*. *Geophysical Research Letters* **13**, 793–796.

Gloeckler, G., et al. (1989): Heavy ion abundances in coronal hole solar wind flows. *EOS* **70**, 424.

Gloeckler, G., et al. (1992): The solar wind ion composition spectrometer. *Astronomy and Astrophysics Supplement* **92**(2), 267–289.

Gloeckler, G., et al. (1993): Detection of interstellar pick-up hydrogen in the solar system. *Science* **261**, 70–73.

Gloeckler, G., et al. (1994): Acceleration of interstellar pickup ions in the disturbed solar wind observed on *Ulysses*. *Journal of Geophysical Research* **99**(A9), 17637–17643.

Gloeckler, G., et al. (1999): Unusual composition of the solar wind in the 2–3 May 1998 CME observed with SWICS on ACE. *Geophysical Research Letters* **26**, 157–160.

Gloeckler, G., et al. (2000): Interception of comet Hyakutake's ion tail at a distance of 500 million kilometers. *Nature* **404**, 576–578.

Gloeckler, G., et al. (2004a): Cometary ions trapped in a coronal mass ejection *The Astrophysical Journal (Letters)* **604**, L121–L124.

Gloeckler, G., et al. (2004b): Observations of the helium focusing cone with pickup ions. *Astronomy and Astrophysics* **426**, 845–854.

Gloeckler, G., Fisk, L. A., Geiss, J. (1997): Anomalously small magnetic field in the local interstellar cloud. *Nature* **386**, 374–377.

Gloeckler, G., Fisk, L. A., Zurbuchen, T. H., Schwadron, N. A. (2000): Acceleration and transport of energetic particles observed in the heliosphere. *American Institute of Physics Conference Proceedings* **528**, 221–228.

Gloeckler, G., Geiss, J. (1989): The abundances of elements and isotopes in the solar wind. *American Institute of Physics Conference Proceedings* **183**, 49–71.

Gloeckler, G., Geiss, J. (1996): Abundance of ^{3}He in the local interstellar cloud. *Nature* **381**, 210–212.

Gloeckler, G., Geiss, J. (1998): Interstellar and inner source pickup ions observed with SWICS on *Ulysses*. *Space Science Reviews* **86**, 127–159.

Gloeckler, G., Geiss, J. (1998): Measurement of the abundance of helium-3 in the Sun and in the local interstellar cloud with SWICS on *Ulysses*. *Space Science Reviews* **84**, 275–284.

Gloeckler, G., Geiss, J. (2004): Composition of the local interstellar medium as diagnosed with pickup ions. *Advances in Space Research* **34**(1), 53–60.

Gloeckler, G., Zurbuchen, T. H., Geiss, J. (2003): Implications of the observed anticorrelation between solar wind speed and coronal electron temperature. *Journal of Geophysical Research (Space Physics)* **108**(A4), SSH 8–1.

Gold, T. (1955): Discussion of shock waves and rarefied gases. In: *Gas Dynamics of Cosmic Clouds* (Eds. J. C. van de Hulst and J. M. BURGERS). New York: North-Holland 1955, p. 103.

Gold, T. (1959a): Magnetic field in the solar system. *Nuovo Cimento Supplemento* **13**, 318–323.

Gold, T. (1959b): Plasma and magnetic fields in the solar system. *Journal of Geophysical Research* **64**, 1665–1674.

Gold, T. (1960): Energetic particle fluxes in the solar system and near the Earth. *The Astrophysical Journal Supplement* **4**, 406–426.

Gold, T. (1962): Magnetic storms. *Space Science Review* **1**, 100–114.

Gold, T. (1964): Magnetic energy shedding in the solar atmosphere. In: *AAS-NASA Symposium on the Physics of Solar Flares NASA SP-50* (Ed. W. N. Hess). Washington, DC: National Aeronautics and Space Administration 1964, pp. 389–395.

Gold, T., Hoyle, F. (1960): On the origin of solar flares. *Monthly Notices of the Royal Astronomical Society* **120**, 89–105.

Goldreich, P. Murrray, N., Kumar, P. (1994): Excitation of solar p modes. *The Astrophysical Journal* **424**, 466–479.

Goldreich, P., Keeley, D. A. (1977a): Solar seismology I. The stability of the solar p-modes. *The Astrophysical Journal* **211**, 934–942.

Goldreich, P., Keeley, D. A. (1977b): Solar seismology II. The stochastic excitation of the solar p-modes by turbulent convections. *The Astrophysical Journal* **212**, 243–251.

Goldreich, P., Kumar, P. (1990): Wave generation by turbulent convection. *The Astrophysical Journal* **363**, 694–704.

Goldstein, B. E. (1993): The solar wind as we know it today. *EOS Transactions of the American Geophysical Union* **74**(20), 229.

Goldstein, B. E., et al. (1996): *Ulysses* plasma parameters: latitudinal, radial, and temporal variations. *Astronomy and Astrophysics* **316**, 296–303.

Goldstein, M. L., Roberts, D. A., Matthaeus, W. H. (1995): Magnetohydrodynamic turbulence in the solar wind. *Annual Review of Astronomy and Astrophysics* **33**, 283–326.

Goldstein, M. L., Roberts, D. A., Matthaeus, W. H. (1997): Magnetohydrodynamic turbulence in cosmic winds. In: *Cosmic Winds and the Heliosphere* (Eds. J. R. Jokipii, C. P. Sonett and M. S. Giampapa). Tuscon, University of Arizona Press, pp. 521–580.

Golub, L., et al. (1974): Solar x-ray bright points. *Astrophysical Journal (Letters)* **189**, L93–L97.

Golub, L., et al. (1980): Magnetic fields and coronal heating. *Astrophysical Journal* **238**, 343–348.

Golub, L., et al. (2007): The X-ray telescope (XRT) for the *Hinode* mission. *Solar Physics* **243**, 63–86.

Golub, L., Krieger, A. S., Harvey, J. W., Vaiana, G. S. (1977): Magnetic properties of X-ray bright points. *Solar Physics* **53**, 111–121.

Golub, L., Krieger, A. S., Vaiana, G. S. (1976a): Distribution of lifetimes for coronal soft X-ray bright points. *Solar Physics* **49**, 79–90.

Golub, L., Krieger, A. S., Vaiana, G. S. (1976b): Observation of spatial and temporal variations in X-ray bright point emergence patterns. *Solar Physics* **50**, 311–327.

Golub, L., Pasachoft J. M. (1997): The Solar Corona. New York: Cambridge University Press.

Gonzalez, W. D., et al. (1994): What is a geomagnetic storm? *Journal of Geophysical Research* **99**, 5771–5792.

Gonzalez, W. D., et al. (2004): Prediction of peak-Dst from halo CME/magnetic cloud speed observations. *Journal of Atmospheric and Solar-Terrestrial Physics* **66**, 161–165.

Gonzalez, W. D., Tsurutani, B. T. (1987): Criteria of interplanetary parameters causing intense magnetic storms (Dst < −100nT). *Planetary and Space Science* **35**, 1101–1108.

Gonzalez, W. D., Tsurutani, B. T., de Gonzalez, A. L. C. (1999): Interplanetary origin of geomagnetic storms. *Space Science Reviews* **88**, 529–582.

Goode, P. R., et al. (1991): What we know about the Sun's internal rotation from oscillations. *Astrophysical Journal* **367**, 649–657.

Gopalswamy, et al. (1998): Origin of coronal and interplanetary shocks – A new look with *Wind* spacecraft data. *Journal of Geophysical Research* **103**, 307.

Gopalswamy, N. (2006): Properties of interplanetary coronal mass ejections. *Space Science Reviews* **124**, 145–168.

Gopalswamy, N., et al. (1999): Microwave enhancement and variability in the elephant's trunk coronal hole: Comparison with *SOHO* observations. *Journal of Geophysical Research* **104**(A5), 9767–9779.

Gopalswamy, N., et al. (2001): Radio signatures of coronal mass ejection interaction: coronal mass ejection cannibalism? *Astrophysical Journal (Letters)* 548, L91–L94.

Gopalswamy, N., et al. (2002): Interacting coronal mass ejections and solar energetic particles. *Astrophysical Journal (Letters)* **572**, L103–L107.

Gopalswamy, N., et al. (2003): Coronal mass ejection activity during solar cycle 23. In: *Solar Variability as an Input to the Earth's Environment. International Solar Cycle Studies Symposium. ESA SP-535* (Ed. A. Wilson), pp. 403–414.

Gopalswamy, N., et al. (2005): Coronal mass ejections and other extreme characteristics of the 2003 October–November solar eruptions. *Journal of Geophysical Research* **110**(A9), A09S15.

Gopalswamy, N., et al. (2006): The pre-CME Sun. *Space Science Reviews* **123**, 303–339.

Gopalswamy, N., Hanaoka, Y. (1998): Coronal dimming associated with a giant prominence eruption. *Astrophysical Journal (Letters)* **498**, L179–L182.

Gopalswamy, N., Mewaldt, R., Torsti, J. (Eds., 2006): *Solar Eruptions and Energetic Particles: Geophysical Monograph Series 165*, Washington: American Geophysical Union 2006.

Gorney, D. J. (1990): Solar cycle effects on the near-earth space environment. *Reviews of Geophysics* **28**, 315–336.

Gosling, J. T. (1990): Coronal mass ejections and magnetic flux ropes in interplanetary space. In: *Physics of Magnetic Flux Ropes: Geophysical Monograph 58* (Eds. C. T. Russell et al.). Washington, DC: American Geophysical Union 1990, pp. 343–364.

Gosling, J. T. (1993): The solar flare myth. *Journal of Geophysical Research* **98**, 18937–18949.

Gosling, J. T. (1994): The solar flare myth in solar-terrestrial physics. In: *Solar System Plasmas in Space and Time. Geophysical Monograph 84* (Eds. J. L. Burch and J. H. Waite, Jr.) Washington, DC: American Geophysical Union 1994, pp. 65–69.

Gosling, J. T. (1996): Corotating and transient solar wind flows in three dimensions. *Annual Review of Astronomy and Astrophysics* **34**, 35–74.

Gosling, J. T. (1997): Coronal mass ejections – an overview. In: *Coronal Mass Ejections Geophysical Monograph 99* (Eds. N. Crooker, J. A. Joselyn and J. Feynman). Washington, DC: American Geophysical Union 1997, pp. 9–16.

Gosling, J. T., et al. (1974): Mass ejections from the Sun: A view from *Skylab. Journal of Geophysical Research* **79**, 4581–4587.

Gosling, J. T., et al. (1975): Direct observations of a flare related coronal and solar wind disturbance. *Solar Physics* **40**, 439–448.

Gosling, J. T., et al. (1976a): Solar wind speed variations 1962–1974. *Journal of Geophysical Research* **81**, 5061–5070.

Gosling, J. T., et al. (1976b): The speeds of coronal mass ejection events. *Solar Physics* **48**, 389–397.

Gosling, J. T., et al. (1980): Observations of large fluxes of He+ in the solar wind following an interplanetary shock. *Journal of Geophysical Research* **85**, 3431–3434.

Gosling, J. T., et al. (1981): Coronal streamers in the solar wind at 1 AU. *Journal of Geophysical Research* **86**, A7, 5438–5448.

Gosling, J. T., et al. (1987): Bidirectional solar wind electron heat flux events. *Journal of Geophysical Research* **92**, 8519–8535.

Gosling, J. T., et al. (1990): Coronal mass ejections and large geomagnetic storms. *Geophysical Research Letters* **17**, 901–904.

Gosling, J. T., et al. (1991): Geomagnetic activity associated with Earth passsage of interplanetary shock disturbances and coronal mass ejections. *Journal of Geophysical Research* **96**, 7831–7839.

Gosling, J. T., et al. (1994): Solar wind corotating stream interaction regions out of the ecliptic plane: *Ulysses. Space Science Reviews* **72**, 99–104.

Gosling, J. T., et al. (1994): The speeds of coronal mass ejections in the solar wind at mid heliographic latitudes: *Ulysses. Geophysical Research Letters* **21**(12), 1109–1112.

Gosling, J. T., et al. (1995): Coronal mass ejections at high heliographic latitudes: *Ulysses*. *Space Science Reviews* **72**, 133–136.

Gosling, J. T., et al. (2004): Dispersionless modulations in low-energy solar electron bursts and discontinuous changes in the solar wind electron strahl. *Journal of Geophysical Research* **10**, A05102.

Gosling, J. T., et al. (2006): Petschek-type reconnection exhausts in the solar wind well beyond 1 AU. *Ulysses*. *Astrophysical Journal* **644**, 613–621.

Gosling, J. T., et al. (2007): Direct evidence for prolonged magnetic reconnection at a continuous x-line within the heliospheric current sheet. *Geophysical Research Letters* **34**, L06102.

Gosling, J. T., et al. (2007b): Five spacecraft observations of oppositely directed exhaust jets from a magnetic reconnection x-line extending more than 4.26 million kilometers in the solar wind at 1 AU. *Geophysical Research Letters* **34**, L20108.

Gosling, J. T., Hundhausen, A. J. (1977): Waves in the solar wind. *Scientific American* **236**, 36–43 – March.

Gosling, J. T., Mc Comas, D. J., Skooug, R. M., Forsyth, R. J. (2001): Stream interaction regions at high heliographic latitudes during *Ulysses* 12/22/2004 6:25pm second polar orbit. *Space Science Reviews* **97**, 189–192.

Gosling, J. T., Pizzo, V. (1999): Formation and evolution of corotating interaction regions and their three dimensional structure. *Space Science Reviews* **89**, 21–52.

Gosling, J. T., Pizzo, V., Bame, S. J. (1973): Anomalously low proton temperatures in the solar wind following interplanetary shock waves – evidence for magnetic bottles? *Journal of Geophysical Research* **78**, 2001–2009.

Gosling, J. T., Riley, P., Mccomas, D. J., Pizzo, V. J. (1998): Overexpanding coronal mass ejections at high heliographic latitudes – observations and simulations. *Journal of Geophysical Research* **103**, 1941–1954.

Gosling, J. T., Roelof, E. C. (1974): A comment on the detection of closed magnetic structures in the solar wind. *Solar Physics* **39**, 405–408.

Gosling, J. T., Skoug, R. M. (2002): On the origin of radial magnetic fields in the heliosphere. *Journal of Geophysical Research (Space Physics)* **107**(A10 SSH), 19–1, 1327.

Gosling, J. T., Skoug, R. M., Mc Comas, D. J. (2004): Low-energy solar electron bursts and solar wind stream structure at 1 AU. *Journal of Geophysical Research* **109**, A04104.

Gosling, J. T., Skoug, R. M., Mccomas, D. J., Smith, C. W. (2005a): Direct evidence for magnetic reconnection in the solar wind near 1 AU. *Journal of Geophysical Research* **110**, A1, A01107.

Gosling, J. T., Skoug, R. M., Mccomas, D. J., Smith, C. W. (2005b): Magnetic disconnection form the Sun: observations of a reconnection exhaust in the solar wind at the heliospheric current sheet. *Geophysical Research Letters* **32**, 5. L05105.

Gough, D. O. (1976): Random remarks on solar hydrodynamics. In: *The Energy Balance and Hydrodynamics of the Solar Chromosphere and Corona*. Proceedings of the International Astronomical Union Colloquium No. 36 (Eds. R. -M. Bonnet and P. H. Delache). Paris: G. De Bussac, Clermont-Ferrand 1976, pp. 3–36.

Gough, D. O. (1981): Solar interior structure and luminosity variations. *Solar Physics* **74**, 21–34.

Gough, D. O. (1989): Deep roots of solar cycles. *Nature* **336**, 618–619.

Gough, D. O., Leibacher, J. W., Scherrer, P. H., Toomre, J. (1996): Perspectives in helioseismology. *Science* **272**, 1281–1284.

Gough, D. O., Toomre, J. (1991): Seismic observations of the solar interior. *Annual Review of Astronomy and Astrophysics* **29**, 627–684.

Graham, G. (1724): An account of observations made of the variation of the horizontal needle at London, in the latter part of the year 1722 and beginning of 1723. *Philosophical Transactions of the Royal Society (London)* **33**(383), 96–107.

Grall, R. R., et al. (1996): Rapid acceleration of the polar solar wind. *Nature* **379**, 429–432.

Gray, D. F., Livingston, W. C. (1997): Monitoring the solar temperature: Spectroscopic temperature variations of the Sun. *Astrophysical Journal* **474**, 802–809.

Greaves, W. M. H., Newton, H. W. (1929): On the recurrence of magnetic storms. *Monthly Notices of the Royal Astronomical Society* **89**, 641–646.

Grec, G., Fossat, E., Pomerantz, M. A. (1980): Solar oscillations – full disk observations from the geographic south pole. *Nature* **288**, 541–544.

Grec, G., Fossat, E., Pomerantz, M. A. (1983): Full-disk observations of solar oscillations from the geographic south pole – latest results. *Solar Physics* **82**, 55–66.

Green, C. A., Kosovichev, A. G. (2007): Magnetic effect on wavelike properties of solar supergranulation. *Astrophysical Journal (Letters), Part 2* **665**, L75–L78.

Greenland Ice-Core Project (GRIP) Members (1993): Climate instability during the last interglacial period recorded in the GRIP ice core. *Nature* **364**, 103–207.

Gribov, V. N., Pontecorvo, B. M. (1969): Neutrino astronomy and lepton charge. *Physics Letters B* **28**, 493–496.

Gringauz, K. I. (1961): Some results of experiments in interplanetary space by means of charged particle traps on Soviet space probes. *Space Research* **2**, 539–553.

Gringauz, K. I., et al. (1960): A study of the interplanetary ionized gas, high-energy electrons, and corpuscular radiation from the Sun by means of the three-electrode trap for charged particles on the second Soviet cosmic rocket. *Soviet Physics (Doklady)* **5**, 361–364.

Grip Members (1993): Climate instability during the last interglacial period recorded in the GReenland Ice-core Project (GRIP) ice core. *Nature* **364**, 203–207.

Grotrian, W. (1934): Über das Fraunhofersche Spektrum der Sonnenkorona. *Zeitschrift für Astrophysik* **8**, 124–146.

Grotrian, W. (1939): On the question of the significance of the lines in the spectrum of the solar corona. *Naturwissenschaften* **27**, 214. English translation in: *A Source Book in Astronomy and Astrophysics 1900–1975* (Eds. K. R. Lang and O. Gingerich). Cambridge, MA: Harvard University Press 1979, pp. 120–122.

Gruen, E., et al. (1992): The *Ulysses* dust experiment. *Astronomy and Astrophysics Supplement* **92**(2), 411–423.

Güdel, M. (2007): The Sun in time: Activity and environment. *Living Reviews in Solar Physics* **4**, 3.

Güdel, M., Guinan, E. F., Skinner, S. L. (1997): The X-ray Sun in time: A study of the long-term evolution of coronae of solar-type stars. *The Astrophysical Journal* **483**, 947–960.

Güdel, M., Schmitt, J. H. M. M., Benz, A. O (1994): Discovery of microwave emission from four nearby solar-type G stars. *Science* **265**, 933–935.

Gudiksen, B. V., Nordlund, A. (2002): Bulk heating and slender magnetic loops in the solar corona. *The Astrophysical Journal (Letters)* **a572**, L113–L116.

Gudiksen, B. V., Nordlund, A. (2005): An ab initio approach to the solar coronal heating problem. *The Astrophysical Journal* **618**, 1020–1030.

Guenther, D. B., Demarque, P. (1997): Seismic tests of the Sun's interior structure, composition, and age, and implications for solar neutrinos. *The Astrophysical Journal* **484**, 937–959.

Guenther, D. B., Jaffe, A., Demarque, P. (1989): The standard solar model: Composition, opacities, and seismology. *Astrophysical Journal* **345**, 1022–1033.

Guhathakurta, M., Fisher, R. (1998): Solar wind consequences of a coronal hole density profile: Spartan 201–03 coronagraph and *Ulysses* observations from 1.15 R_o to 4 AU. *Astrophysical Journal (Letters)* **499**, L215–L218.

Gurnett, D. A., Baumback, M. M., Rosenbauer, H. (1978): Stereoscopic direction finding analysis of a type III solar radio burst: Evidence for emission at $2f_p$ *Journal of Geophysical Research* **83**, 616–622.

Gurnett, D. A., et al. (1993): Radio emission from the heliopause triggered by an interplanetary shock. *Science* **262**, 199–203.

Gurnett, D. A., Kurth, W. S. (2008): Intense plasma waves observed at and near the solar wind termination shock. *Nature* **454**, 78–80.

Guzxik, J. A., Watson, I. S., Cox, A. N. (2005): Can enhanced diffusion improve helioseismic agreement for solar models with revised abundances? *Astrophysical Journal* **627**, 1049–1056.

H

Habbal, S. R., et al. (1997): Origins of the slow and the ubiquitous fast solar wind. *Astrophysical Journal (Letters)* **489**, L103–L106.

Haber, D. A., et al. (2002): Evolving submerged meridional circulation cells within the upper convection zone revealed by ring-diagram analysis. *The Astrophysical Journal* **570**, 855–864.

Haber, D. A., et al. (2004): Organized subsurface flows near active regions. *Solar Physics* **220**, 371–380.

Hagenaar, H. J., Schrijver, C. J., Title, A. M. (1997): The distribution of cell sizes of the solar chromospheric network. *Astrophysical Journal* **481**, 988–995.

Haggerty, D. K., Roelof, E. C. (2002): Impulsive near-relativistic solar electron events: delayed injection with respect to solar electromagnetic emission. *Astrophysical Journal* **579**, 841–853.

Haigh, J. D, Blackburn, M., Day, R. (2005): The response of tropospheric circulation to perturbations in lower-stratospheric temperature. *Journal of Climate* **18**, 3672–3685.

Haigh, J. D. (1994): The role of stratospheric ozone in modulating the solar radiative forcing of climate. *Nature* **370**, 544–546.

Haigh, J. D. (1996): The impact of solar variability on climate. *Science* **272**, 981–984.

Haigh, J. D. (2001): Climate variability and the influence of the Sun. *Science* **294**, 2109–2111.

Haigh, J. D. (2007): The Sun and the Earth's climate. *Living Reviews in Solar Physics* **4**, 2.

Haisch, B. M., Rodono, M. (Eds., 1989): Solar and stellar flares. *Proceedings of IAU Colloquium No. 104. Solar Physics* **121**(1, 2). Reprinted Boston: Kluwer 1989.

Haisch, B., Strong, K. T., Rodono, M. (1991): Flares on the Sun and other stars. *Annual Review of Astronomy and Astrophysics* **29**, 275–324.

Hale, G. E. (1892a): A remarkable solar disturbance. *Astronomy and Astrophysics* **11**, 611–613.

Hale, G. E. (1892b): On the condition of the Sun's surface in June and July, 1892, as compared with the record of terrestrial magnetism. *Astronomy and Astrophysics* **11**, 917–925.

Hale, G. E. (1908a): On the probable existence of a magnetic field in Sun-spots. *Astrophysical Journal* **28**, 315–343. Reproduced in: *A Source Book in Astronomy and Astrophysics 1900–1975* (Eds. K. R. Lang and O. Gingerich). Cambridge: Harvard University Press 1979, pp. 96–105.

Hale, G. E. (1908b): Solar vortices. *The Astrophysical Journal* **28**, 100–116.

Hale, G. E. (1908c): The Zeeman effect in the Sun. *Publications of the Astronomical Society of the Pacific* **20**, 287–288.

Hale, G. E. (1926): Visual observations of the solar atmosphere. *Proceedings of the National Academy of Science* **12**, 286–295.

Hale, G. E. (1929): The spectrohelioscope and its work. Part I. History, instruments, adjustments, and methods of observation. *Astrophysical Journal* **70**, 265–311.

Hale, G. E. (1931): The spectrohelioscope and its work. Part III. Solar eruptions and their apparent terrestrial effects. *Astrophysical Journal* **73**, 379–412.

Hale, G. E., et al. (1919): The magnetic polarity of sun-spots. *Astrophysical Journal* **49**, 153–178.

Halley, E. (1716): An account of the late surprising appearance of the lights seen in the air. *Philosophical Transactions of the Royal Society (London)* **29**, 406–428.

Hammer, C. U. (1977): Past volcanism revealed by Greenland ice sheet impurities. *Nature* **270**, 482–486.

Hampel, W., et al. (1999): GALLEX solar neutrino observations: Results for GALLEX IV. *Physics Letters* **44**, 127–133.

Hanaoka, Y. (1994): A flare caused by interacting coronal loops. *Astrophysical Journal (Letters)* **420**, L37–L40.

Hanaoka, Y. (1996): Flares and plasma flow caused by interacting coronal loops. *Solar Physics* **165**, 275–301.

Hanaoka, Y. (1997): Double-loop configuration of solar flares. *Solar Physics* **173**, 319–346.

Hanasoge, S. M., et al. (2006): Computational acoustics in spherical geometry: Steps toward validating helioseismology. *Astrophysical Journal* **648**, 1268–1275.

Handy, B. N., et al. (1999): The *Transition Region and Coronal Explorer. Solar Physics* **187**, 229–260.

Handy, B. N., Schrijver, C. J. (2001): On the evolution of the solar photospheric and coronal magnetic field. *Astrophysical Journal* **547**, 1100–1108.

Hansen, J. E., Lacis, A. A. (1990): Sun and dust versus greenhouse gases: An assessment of their relative roles in global climate change. *Nature* **346**, 713–719.

Hansen, J. E., Lebedeff, S. (1987): Global trends of measured surface air temperature. *Journal of Geophysical Research* **92**, 13345–13372.

Hansen, J. E., Lebedeff, S. (1988): Global surface air temperatures: Update through 1987. *Geophysical Research Letters* **15**, 323–326.

Hansen, J. F., Bellan, P. M. (2001): Experimental demonstration of how strapping fields can inhibit solar prominence eruptions. *Astrophysical Journal (Letters)* **563**, L183–L186.

Hansen, R. T., Garcia, C. J., Hansen, S. F., Yasukawa, E. (1974): Abrupt depletions of the inner corona. *Publications of the Astronomical Society of the Pacific* **86**, 500–515.

Hara, H., et al. (1994): Temperatures of coronal holes observed with *Yohkoh* SXT. *Publications of the Astronomical Society of Japan* **46**, 493–502.

Harmon, J. K., Coles, W. A. (2005): Modeling radio scattering and scintillation observations of the inner solar wind using oblique Alfvén/ion cyclotron waves. *Journal of Geophysical Research* **110**, A03101.

Harra, L. K., et al. (2007): Coronal dimming observed with *Hinode*: Outflows related to a coronal mass ejection. *Publications of the Astronomical Society of Japan* **59**, S801–S806.

Harra, L. K., Sterling, A. C. (2003): Imaging and spectroscopic investigations of a solar coronal wave: Properties of the wave front and associated erupting material. *Astrophysical Journal* **587**, 429–438.

Harrison, R. A. (1986): Solar coronal mass ejections and flares. *Astronomy and Astrophysics* **162**, 283–291.

Harrison, R. A. (1991): Coronal mass ejection. *Philosophical Transactions of the Royal Society (London)* **A336**, 401–412.

Harrison, R. A. (1994): A statistical study of the coronal mass ejection phenomenon. *Advances in Space Research* **14**(4), 23–28.

Harrison, R. A. (1997a): CME onset studies. In: *The Corona and Solar Wind Near Minimum Activity. Proceedings of the Fifth SOHO Workshop ESA SP-404.* Noordwijk: ESA Publications Division 1997, pp. 85–91.

Harrison, R. A. (1997b): EUV blinkers – the significance of variations in the extreme ultraviolet quiet Sun. *Solar Physics* **175**, 467–485. Reprinted in: *The First Results from SOHO* (Eds. B. Fleck and Z. Svestka). Boston: Kluwer Academic Publishers 1997, pp. 467–485.

Harrison, R. A. (1997c): One year of CDS: Highlights from observations using the coronal diagnostic spectrometer on *SOHO*. In: *The Corona and Solar Wind Near Minimum Activity. Proceedings of the Fifth SOHO Workshop. ESA SP-404.* Noorwijk: ESA Publications Division 1997, pp. 7–16.

Harrison, R. A., Bryans, P., Simnett, G. M., Lyons, M. (2003): Coronal dimming and the coronal mass ejection onset. *Astronomy and Astrophysics* **400**, 1071–1083.

Harrison, R. A., et al. (1990): The launch of solar coronal mass ejections: Results from the coronal mass ejection onset program. *Journal of Geophysical Research* **95**, 917–937.

Harrison, R. A., et al. (1995): The Coronal Diagnostic Spectrometer for the *Solar and Heliospheric Observatory. Solar Physics* **162**, 233–290.

Harrison, R. A., et al. (2008): First imaging of coronal mass ejections in the heliosphere viewed from outside the Sun-Earth line. *Solar Physics* **247**, 171–193.

Harrison, R. A., Lyons, M. (2000): A spectroscopic study of coronal dimming associated with a coronal mass ejection. *Astronomy and Astrophysics* **358**, 1097–1108.

Hart, A. B. (1954): Motions in the Sun at the photospheric level IV. The equatorial rotation and possible velocity fields in the photosphere. *Monthly Notices of the Royal Astronomical Society* **114**, 17–38.

Hart, A. B. (1956): Motions in the Sun at the photospheric level VI. Large scale motions in the equatorial region. *Monthly Notices of the Royal Astronomical Society* **116**, 38–55.

Hartle, R. E., Sturrock, P. A. (1968): Two-fluid model of the solar wind. *Astrophysical Journal* **151**, 1155–1170.

Hartmann, L. W., Noyes, R. W. (1987): Rotation and magnetic activity in main-sequence stars. *Annual Review of Astronomy and Astrophysics* **25**, 271–301.

Hartz, T. R. (1964): Solar noise observations from the *Alouette* satellite. *Annales d'Astrophysique* **27**, 831–836.

Hartz, T. R. (1969): Type III solar radio noise bursts at hectometer wavelengths. *Planetary and Space Science* **17**, 267–287.

Harvey, J. W. (1995): Helioseismology. *Physics Today* **48**, 32–38 – October.

Harvey, J. W., et al. (1996): The Global Oscillation Network Group (GONG) project. *Science* **272**, 1284–1286.

Harvey, J. W., Kennedy, J. R., Leibacher, J. W. (1987): GONG – to see inside our Sun. *Sky and Telescope* **74**, 470–476 – November.

Harvey, K. L., et al. (1999): Does magnetic flux submerge at flux cancellation sites? *Solar Physics* **190**, 35–44.

Harvey, K. L., Martin, S. F. (1973): Ephemeral active regions. *Solar Physics* **32**, 389–402.

Harvey, K. L., Recely, F. (2002): Polar coronal holes during cycles 22 and 23. *Solar Physics* **211**, 31–52.

Haselgrove, B., Hoyle, F. (1959): Main-sequence stars. *Monthly Notices of the Royal Astronomical Society* **119**, 112–120.

Hassler, D. M., et al. (1997): Observations of polar plumes with the SUMER instrument on *SOHO*. *Solar Physics* **175**, 375–391. Reprinted In: *The First Results from SOHO* (Eds. B. Fleck and Z. Svestka). Boston: Kluwer Academic Publishers 1997, pp. 375–391.

Hassler, D. M., et al. (1999): Solar wind outflow and the chromospheric magnetic network. *Science* **283**, 810–813.

Hastings, D. E., Garret, H. (1996): *Spacecraft–Environment Interactions. Cambridge Atmospheric and Space Science Series*. New York: Cambridge University Press 1996.

Hathaway, D. H. (1996): Doppler measurements of the Sun's meridional flow. *Astrophysical Journal* **460**, 1027–1033.

Hathaway, D. H., et al. (1996): GONG observations of solar surface flows. *Science* **272**, 1306–1309.

Hathaway, D. H., et al. (2000): The photospheric convection spectrum. *Solar Physics* **193**, 299–312.

Hathaway, D. H., et al. (2002): Radial flows in supergranules. *Solar Physics* **205**, 25–38.

Hathaway, D. H., et al. (2003): Evidence that a deep meriodional flow sets the sunspot cycle period. *Astrophysical Journal* **589**, 665–670.

Hathaway, D. H., Williams, P. E., Cuntz, M. (2006): Supergranule superrotation identified as a projection effect. *The Astrophysical Journal* **644**, 598–602.

Havnes, O. (1971): Abundances and accelerations mechanisms of cosmic rays. *Nature* **229**, 548–549.

Hays, J. D., Imbrie, J., Shackleton, N. J. (1976): Variations in the Earth's orbit: pacemaker of the ice ages. *Science* **194**, 1121–1132.

Heath, D. F., Krueger, A. J., Crutzen, J. (1977): Solar proton events: Influence on stratospheric ozone. *Science* 197: 888–889.

Heaviside, O. (1902): Telegraphy In: *Encyclopedia Britannica* **113**, 215.

Henderson-Sellers, A. (1979): Clouds and the long term stability of the earth's atmosphere and climate. *Nature* **279**, 786–788.

Hernandez, I. Gonzalez, et al. (2006): Meridional circulation variability from large aperture ring-diagram analysis of global oscillation network group and Michelson Doppler imager data. *Astrophysical Journal* **638**, 576–583.

Herschel, W. (1801): Observations tending to investigate the nature of the Sun, in order to find the causes or symptoms of its variable emission of light and heat; with remarks on the use that may possibly be drawn from solar observations. *Philosophical Transactions of the Royal Society of London* **91**, 265–318.

Hess, V. F. (1912): Concerning observations of penetrating radiation on seven free balloon flights. *Physikalishe Zeitschrift* **13**, 1084–1091. English translation in: *A Source Book in Astronomy and Astrophysics 1900–1975* (Eds. K. R. Lang and O. Gingerich). Cambridge, MA: Harvard University Press 1979, pp. 13–20.

Hewish, A. (1955): The irregular structure of the outer regions of the solar corona. *Proceedings of the Royal Society (London)* **228A**, 238–251.

Hewish, A. (1958): The scattering of radio waves in the solar corona. *Monthly Notices of the Royal Astronomical Society* **118**, 534–546.

Hey, J. S. (1946): Solar radiations in the 4–6 metre radio wavelength band. *Nature* **157**, 47–48.

Hey, J. S. (1973): *The Evolution of Radio Astronomy*. New York: Science History Publications, Neale Watson Academic Publications 1973.

Heyvaerts, J., Priest, E. R. (1983): Coronal heating by phase-mixed shear Alfvén waves. *Astronomy and Astrophysics* **117**, 220.

Heyvaerts, J., Priest, E. R. (1984): Coronal heating by reconnection in DC current systems. A theory based on Taylor's hypothesis. *Astronomy and Astrophysics* **137**, 63–78.

Heyvaerts, J., Priest, E. R., Rust, D. M. (1977): An emerging flux model for the solar flare phenomenon. *Astrophysical Journal* **216**, 123–137.

Hick, P., et al. (1995): Synoptic IPS and *Yohkoh* soft X-ray observations. *Geophysical Research Letters* **22**(5), 643–646.

Hickey, J. R., et al. (1981): Solar variability indications from *Nimbus 7* satellite data. In: *Variations of the Solar Constant* (Ed. S. SOFIA). Washington: NASA CP-2191, pp. 59–72.

Hickey, J. R., et al. (1988a): Observation of total solar irradiance variability from *Nimbus* satellites. *Advances in Space Research* **9**(7), 5–10.

Hickey, J. R., et al. (1988b): Total solar irradiance measurements by ERB/*Nimbus-7*: a review of nine years. *Space Science Reviews* **48**, 321–342.

Hiei, E., Hundhausen, A. J., Sime, D. G. (1993): Reformation of a coronal helmet streamer by magnetic reconnection after a coronal mass ejection. *Geophysical Research Letters* **20**, 2785–2788.

Higdon, J. C., Lingenfelter, R. E. (2003): The superbubble origin of ^{22}Ne in cosmic rays. *Astrophysical Journal* **590**, 822–832.

Higdon, J. C., Lingenfelter, R. E. (2006): The superbubble origin for galactic cosmic rays. *Advances in Space Research* **37**, 1913–1917.

Higdon, J. C., Lingenfelter, R. E., Ramaty, R. (1998): Cosmic-ray acceleration from supernova ejecta in superbubbles. *The Astrophysical Journal (Letters)* **509**, L33–L36.

Hilchenbach, M., et al. (1998): Detection of 55–80 keV hydrogen atoms of heliospheric origin by CELIAS/HSTOF on *SOHO*. *Astrophysical Journal* **503**, 916.

Hildner, E., et al. (1975): The sources of material comprising a mass ejection coronal transient. *Solar Physics* **45**, 363–376.

Hill, F. (1988): Rings and trumpets – three-dimensional power spectra of solar oscillations. *The Astrophysical Journal* **333**, 996–1013.

Hill, R. (1989): Solar oscillation ring diagrams and large-scale flows. *Astrophysical Journal (Letters)* **343**, L69–L71.

Hill, T. W., Dessler, A. J. (1991): Plasma motions in planetary magnetospheres. *Science* **252**, 410–415 – April.

Hillas, A. M. (1972): *Cosmic Rays*. New York: Pergamon Press 1972.

Hindman, B. W., et al. (2004): Comparison of solar subsurface flows assessed by ring and time-distance analyses. *The Astrophysical Journal* **613**, 1253–1262.

Hines, C. O. (1974): A possible mechanism for the production of sun-weather correlations. *Journal of Atmospheric Science* **31**, 589–591.

Hioter, O. P. (1747): Om Magnet-nalens Atskillige andreingar. *Kongle Swen Wetenskaps Acad. Handlgar*, 27–43.

Hirayama, T. (1974): Theoretical model of flares and prominences I. Evaporating flare model. *Solar Physics* **34**, 323–338.

Hirshberg, J., Bame, S. J., Robbins, D. E. (1972): Solar flares and solar wind helium enrichments: July 1965–July 1967. *Solar Physics* **23**, 467–486.

Hirshberg, J., Colburn, D. S. (1969): Interplanetary field and geomagnetic variations – a unified view. *Planetary and Space Science* **17**, 1183–1206.

Hodgson, R. (1860): On a curious appearance seen in the Sun. *Monthly Notices of the Royal Astronomical Society* **20**, 15–16. Reproduced in: *Early Solar Physics* (Ed. A. J. Meadows). Oxford: Pergamon Press 1970, p. 185.

Hoeksema, J. T. (1994): The large-scale structure of the heliospheric current sheet during the *Ulysses* epoch. *Space Science Reviews* **72**, 137–148.

Hoeksema, J. T., Domingo, V., Fleck, B, Battrick, B. (Eds) (1995): *SOHO-4: Helioseismology.* ESA SP-376.

Hoeksema, J. T., Scherrer, P. H. (1987): Rotation of the coronal magnetic field. *Astrophysical Journal* **318**, 428–436.

Hoeksema, J. T., Wilcox, J. M., Scherrer, P. H. (1982): Structure of the heliospheric current sheet in the early portion of sunspot cycle 21. *Journal of Geophysical Research* **87A**, 10, 331–10, 338.

Hofer, M. Y., et al. (2003): Transition to solar minimum at high solar latitudes: Energetic particles from corotating interaction regions. *Geophysical Research Letters* **30**(19), ULY 8–1.

Hollweg, J. V. (1972): Alfvénic motions in the solar atmosphere. *Astrophysical Journal* **177**, 255–259.

Hollweg, J. V. (1973): Transverse Alfvén waves in the solar wind. Wave pressure, Poynting flux, and angular momentum. *Journal of Geophysical Research* **78**, 3643–3652.

Hollweg, J. V. (1975): Waves and instabilities in the solar wind. *Reviews of Geophysics* **13**, 263–289.

Hollweg, J. V. (1978): Some physical processes in the solar wind. *Reviews of Geophysics* **16**, 689–720.

Hollweg, J. V. (1984): Resonances of coronal loops. *Astrophysical Journal* **277**, 392–403.

Hollweg, J. V. (1986): Transition region, corona and solar wind in coronal holes. *Journal of Geophysical Research* **91**, 4111.

Hollweg, J. V. (1990): MHD waves on solar magnetic flux tubes – tutorial review. In: *Physics of Magnetic Flux Ropes. Geophysical Monograph 58* (Eds. C. T. Russell, E. R. Priest and L. C. Lee). Washington, DC: American Geophysical Union 1990, pp. 23–31.

Hollweg, J. V. (2006): The solar wind: then and now. Recurrent magnetic storms: Co-rotating solar wind streams. *Geophysical Monograph Series* **167**, 19–30.

Hollweg, J. V., et al. (1982): Possible evidence for coronal Alfvén waves. *Journal of Geophysical Research* **97**, 1–8.

Hollweg, J. V., Isenberg, P. A. (2002): Generation of the fast solar wind: a review with emphasis on the resonant cyclotron interaction. *Journal of Geophysical Research (Space Physics)* **107**, A7, 1147, SSH 12–1.

Hollweg, J. V., Johnson, W. (1988): Transition regions, corona, and solar wind in coronal holes: some two-fluid models. *Journal of Geophysical Research* **87**, 1.

Holman, G. D. (2005): Energetic electrons in solar flares as viewed in X-rays. *Advances in Space Research* **35**, 1669–1674.

Holzer, T. E. (1977): Effects of rapidly diverging flow, heat addition, and momentum addition in the solar wind and stellar winds. *Journal of Geophysical Research* **82**, 23–35.

Holzer, T. E. (1989): Interaction between the solar wind and the interstellar medium *Annual Review of Astronomy and Astrophysics* **27**, 199–234.

Holzer, T. E., Axford, W. I. (1970): The theory of stellar winds and related flows. *Annual Review of Astronomy and Astrophysics* **8**, 31–60.

Holzer, T. E., Leer, E. (1997): Coronal hole structure and the high speed solar wind. In: *The Corona and Solar Wind Near Minimum Activity. Proceedings of the Fifth SOHO Workshop. ESA SP-404.* (Eds. O. Kjeldseth-Moe, A. Wilson) Noordwijk, The Netherlands: ESA Publications Division 1997, pp. 65–74.

Hood, A. W., Priest, E. R. (1979): Kink instability of solar coronal loops as the cause of solar flares. *Solar Physics* **64**, 303–321.

Hood, L. L. (1987): Solar ultraviolet radiation induced variations in the stratosphere and mesosphere. *Journal of Geophysical Research* **92**, 876–888.

Horbury, T. S., Tsurutani, B. (2001): *Ulysses* measurements of waves, turbulence and discontinuities. In: *The Heliosphere Near Solar Minimum: The Ulysses perspective* (Eds. A. Balogh, R. G. Marsden and E. J. Smith). New York: Springer, Praxis 2001, pp. 167–227.

Hosaka, J., et al. (2006): Solar neutrino measurements in super-kamiokande-1. *Physical Review D* **73**, 112001–112007.

Hovestadt, D., et al. (1978): The nuclear and ionic charge distribution particle experiments on the *ISEE-1* and *ISEE-C* spacecraft. *IEEE Transactions Geoscience Electronics* Vol. **GE–16**, 166–175.

Hovestadt, D., et al. (1981): Singly charged energetic helium emitted in solar flares. *Astrophysical Journal (Letters)* **246**, L81–L84.

Hovestadt, D., et al. (1984): Survey of He(+)/He(2+) abundance ratios in energetic particle events. *Astrophysical Journal (Letters)* **282**, L39–L42.

Hovestadt, D., et al. (1995): CELIAS – Charge, Element and Isotope Analysis System for *SOHO*. *Solar Physics* **162**, 441–481.

Hovestadt, D., Vollmer, O., Gloeckler, G, Fan, C.-Y. (1973): Measurement of elemental abundance of very low energy solar cosmic rays. *Proceedings of the 13th International Cosmic Ray Conference*, 1498–1503.

Hovestadt, D., Vollmer, O., Gloeckler, G., Fan, C.-Y. (1973): Differential energy spectra of low-energy (less than 8.5 MeV per nucleon) heavy cosmic rays during solar quiet times. *Physical Review Letters* **31**, 650–653.

Howard, R. (1974): Studies of solar magnetic fields. *Solar Physics* **38**, 283–299.

Howard, R. (1979): Evidence for large-scale velocity features on the Sun. *The Astrophysical Journal (Letters)* **228**, L45–L50.

Howard, R. (1985): Eight decades of solar research at Mount Wilson. *Solar Physics* **100**, 171–187.

Howard, R. A., et al. (1982): The observation of a coronal transient directed at Earth. *Astrophysical Journal (Letters)* **263**, L101–L104.

Howard, R. A., et al. (1985): Coronal mass ejections: 1979–1981. *Journal of Geophysical Research* **90**, 8173–8191.

Howard, R. A., et al. (1997): Observations of CMEs from *SOHO*/LASCO. In: *Coronal Mass Ejections. Geophysical Monograph 99* (Eds. N. Crooker, J. A. Joselyn and J. Feynman). Washington, DC: American Geophysical Union 1997, pp. 17–26.

Howard, R. A., et al. (2008): Sun Earth Connection Coronal and Heliospheric Investigation (SECCHI). *Space Science Reviews* **136**, Issue 1–4, 67–115.

Howard, R. A., Koomen, M. J. (1974): Observation of sectored structure in the outer solar corona: Correlation with interplanetary magnetic field. *Solar Physics* **37**, 469–475.

Howard, R. A., Sheeley, N. R. Jr., Michels, D. J., Koomen, M. J. (1985): Coronal mass ejections 1979–1981. *Journal of Geophysical Research* **90**, 8173–8191.

Howard, R., Labonte, B. J. (1980): The Sun is observed to be a torsional oscillator with a period of 11 years. *The Astrophysical Journal (Letters)* **239**, L33–L36.

Howard, R., Labonte, B. J. (1981): Surface magnetic fields during the solar activity cycle. *Solar Physics* **74**, 131–145.

Howe, R. (2003): The internal rotation of the Sun. In: *Proceedings of SOHO 12/GONG + 2002. Local and Global Helioseismology: The Present and Future. ESA SP-517* (Ed. H. Sawaya-Lacoste). Noordwijk, the Netherlands pp. 81–86.

Howe, R., et al. (2000a): Deeply penetrating banded zonal flows in the solar convection zone. *The Astrophysical Journal (Letters)* **533**, L163–L166.

Howe, R., et al. (2000b): Dynamic variations at the base of the solar convections zone. *Science* **287**, 2456–2460.

Howe, R., et al. (2004): Convection-zone dynamics from GONG and MDI, 1995–2004. In *Helio- and Asterosismology: Towards a Golden Future ESA SP-559* (Ed. D. Danesy). p. 472.

Howe, R., et al. (2006a): Large-scale zonal flows near the solar surface. *Solar Physics* **235**, 1–15.

Howe, R., et al. (2006b): Solar convection zone dynamics: How sensitive are inversion to subtle dynamo features? *The Astrophysical Journal* **649**, 1155–1168.

Hoyle, F. (1949): *Some Recent Researches in Solar Physics*. Cambridge, England: Cambridge at the University Press 1949.

Hoyle, F. (1958): Remarks on the computation of stellar evolution tracks. In: *Stellar Populations* (Ed. J. K. O'CONNELL). Vatican City: Specola Vaticana, pp. 223–226.

Hoyle, F., Bates, D. R. (1948): The production of the E-layer. *Terrestrial Magnetism and Atmospheric Electricity* **53**, 51–62.

Hoyng, P., et al. (1981): Origin and location of the hard X-ray emission in a two-ribbon flare. *Astrophysical Journal (Letters)* **246**, L155–L159.

Hoyt, D. V, Eddy, J. A., Hudson, H. S. (1983): Sunspot areas and solar irradiance variations during 1980. *Astrophysical Journal* **275**, 878–888.

Hoyt, D. V. et al. (1992): The *Nimbus 7* solar total irradiance: a new algorithm for its derivation. *Journal of Geophysical Research* **97**, 51–63.

Hoyt, D. V., Schatten, K. H. (1993): A discussion of plausible solar irradiance variations, 1700–1992. *Journal of Geophysical Research* **98**, 18895–18906.

Hoyt, D. V., Schatten, K. H. (1997): *The Role of the Sun in Climate Change*. New York: Oxford University Press 1997.

Hoyt, D. V., Schatten, K. H., Nesmes-Ribes, E. (1994): The one hundredth year of Rudolf Wolf's death: Do we have the correct reconstruction of solar activity? *Geophysical Research Letters* **21**, 2067–2070.

Hsieh, K. C., Simpson, J. A. (1970): The relative abundances and energy spectra of ^{3}He and ^{4}He from solar flares. *Astrophysical Journal (Letters)* **162**, L191–L196.

Hu, Q., et al. (2003): Double flux-rope magnetic cloud in the solar wind at 1 AU. *Geophysical Research Letters* **30**(7), 38–1.

Hu, Q., Smith, C. W., Ness, N. F., Skoug, R. M. (2004): Multiple flux rope magnetic ejecta in the solar wind. *Journal of Geophysical Research* **109**, A03102.

Hudson, H. S. (1972): Thick-target processes and white-light flares. *Solar Physics* **24**, 414–428.

Hudson, H. S. (1987): Solar flare discovery. *Solar Physics* **113**, 1–12.

Hudson, H. S. (1988): Observed variability of the solar luminosity. *Annual Review of Astronomy and Astrophysics* **26**, 473–508.

Hudson, H. S. (1991): Solar flares, microflares, nanoflares and coronal heating. *Solar Physics* **133**, 367–369.

Hudson, H. S. (1997): The solar antecedents of geomagnetic storms. In: *Magnetic Storms* (Eds. B. T. Tsurutani, W. D. Gonzales and Y. Kamide). Washington, DC: American Geophysical Union 1997, 37–44.

Hudson, H. S., Acton, L. W., Freeland, S. L. (1996): A long-duration solar flare with mass ejection and global consequences. *Astrophysical Journal* **470**, 629–635.

Hudson, H. S., Bougeret, J.-L., Burkepile, J. (2006a): Coronal mass ejections: Overview of observations. *Space Science Reviews*, **123**, 13–30.

Hudson, H. S., Cliver, E.W, (2001): Observing coronal mass ejections without coronagraphs. *Journal of Geophysical Research* **106**(A11), 251199–25214.

Hudson, H. S., Silva, S., Woodard, M. (1982): The effect of sunspots on solar irradiance. *Solar Physics* **76**, 211–219.

Hudson, H. S., et al. (1994): Impulsive behavior in solar soft X-radiation. *Astrophysical Journal (Letters)* **422**, L25–L27.

Hudson, H. S., et al. (1998): X-ray coronal changes during halo CMEs. *Geophysical Research Letters* **25**, 2481–2484.

Hudson, H. S., Haisch, B. M., Strong, K. T. (1995): Comment on 'The solar flare myth' by J. T. Gosling. *Journal of Geophysical Research* **100**, 3473–3477.

Hudson, H. S., Ryan, J. (1995): High-energy particles in solar flares. *Annual Review of Astronomy and Astrophysics* **33**, 239–282.

Hudson, H. S., Warmuth, A. (2004): Coronal loop oscillations and flare shock waves. *Astrophysical Journal (Letters)* **614**, L85–L88.

Hudson, H. S., Webb, D. F. (1997): Soft X-ray signatures of coronal ejections. In: *Coronal Mass Ejections. Geophysical Monograph 99* (Eds. N. Crooker, J. A. Joselyn and J. Feynman). Washington, DC: American Geophysical Union 1997, pp. 27–38.

Hudson, H. S., Wolfson, C. J., and Metcalf, T. R. (2006b): White-light flares: A *TRACE/RHESSI* overview. *Solar Physics* **234**, 79–93.

Hudson, M. K., et al. (2004): 3D modeling of shock-induced trapping of solar energetic particles in the Earth's magnetosphere. *Journal of Atmospheric and Terrestrial Physics* **66**, 1389–1397.

Hufbauer, K. (1991): *Exploring the Sun: Solar Science Since Galileo*. Baltimore, Maryland: Johns Hopkins University Press 1991.

Hulburt, E. O. (1938): Photoelectric ionization in the ionosphere. *Physical Review* **53**, 344–351.

Humboldt, F. W. H. A. von (1799–1804): *Voyage aux régions équinoxiales du Nouveau Continent, fait en 1799, 1800, 1801, 1802, 1803, et 1804 par Al [exandre] de Humboldt et A [imé] Bonpland*. Paris, 1805–1834.

Humboldt, F. W. H. A. Von (1845, 1847): *Kosmos*. Cotta, Stuttgart 1845, 1847.

Hundhausen, A. J. (1972a): *Coronal Expansion and Solar Wind*. New York: Springer-Verlag.

Hundhausen, A. J. (1972b): Interplanetary shock waves and the structure of solar wind disturbances. In: *Solar Wind: NASA SP-308* (Eds. C. P. Sonett, P. J. Coleman and J. M. Wilcox). Washington: NASA 1972, pp. 393–417.

Hundhausen, A. J. (1973): Nonlinear model of high-speed solar wind streams. *Journal of Geophysical Research* **78**, 1528–1542.

Hundhausen, A. J. (1977): An interplanetary view of coronal holes. In: *Coronal Holes and High Speed Wind Streams* (Ed. J. Zirker). Boulder: Colorado Associated University Press 1977, pp. 225–329.

Hundhausen, A. J. (1987): The origin and propagation of coronal mass ejections. In: *Solar Wind Six* (Eds. V. J. Pizzo, T. Holzer and D. G. Sime). Boulder, Colorado: National Center for Atmospheric Research 1987, p. 181.

Hundhausen, A. J. (1993): Sizes and locations of coronal mass ejections: *SMM* observations from 1980 and 1984–1989. *Journal of Geophysical Research* **98**, 13177–13200.

Hundhausen, A. J. (1997): An introduction. In: *Coronal Mass Ejections: Geophysical Monograph 99* (Eds. N. Crooker, J. A. Joselyn and J. Feynman). Washington: American Geophysical Union, pp. 1–7.

Hundhausen, A. J. (1997): Coronal mass ejections. In: *Cosmic Winds and the Heliosphere* (Eds. J. R. Jokipii, C. P. Sonett and M. S. Giampapa). Tucson, Arizona: University of Arizona Press 1997, pp. 259–296.

Hundhausen, A. J., Bame, S. J., Montgomery, M. D. (1970): Large-scale characteristics of flare-associated solar wind disturbances. *Journal of Geophysical Research* **75**, 4631–4642.

Hundhausen, A. J., Burkepile, J. T., St. Cyr, O. C. (1994): Speeds of coronal mass ejections: *SMM* observations from 1980 and 1984–1989. *Journal of Geophysical Research* **99**, 6543–6552.

Hundhausen, A. J., Gosling, J. T. (1976): Solar wind structure at large heliocentric distances: an interpretation of *Pioneer 10* observations. *Journal of Geophysical Research* **81**, 1436–1440.

Hundhausen, A. J., Stanger, A. L., Serbicki, S. A. (1994): Mass and energy contents of coronal mass ejections: *SMM* results from 1980 and 1984–1989. In: *Solar Dynamical Phenomena and Solar Wind Consequences. Proceedings of the Third SOHO Workshop. ESA SP-373*. Noordwijk: ESA Publications Division, pp. 409–412.

Hunt, J. J., Domingo, V. (Eds., 1994): *SOHO-3: Solar Dynamic Phenomena and Solar Wind Consequences. ESA SP-373* 1994.

Hurford, G. J., et al. (2002): The *RHESSI* imaging concept. *Solar Physics* **210**, 61–86.

Hurford, G. J., et al. (2003): First gamma-ray images of a solar flare. *Astrophysical Journal (Letters)* **595**, L77–L80.

Hurford, G. J., et al. (2006): Gamma-ray imaging of the 2003 October/November solar flares. *Astrophysical Journal (Letters)* **644**, L93–L96.

Hurford, G. J., Mewaldt, R. A., Stone, E. C., Vogt, R. E. (1975): Enrichment of heavy nuclei in He-3-rich flares. *Astrophysical Journal (Letters)* **201**, L95–L97.

Hurley, K., et al. (1992): The solar X-ray/cosmic gamma-ray burst experiment aboard *Ulysses*. *Astronomy and Astrophysics Supplement* **92**(2), 401–410.

Huttunen, E. C., et al. (2008): *STEREO* and *Wind* observations of a magnetic cloud on May 21–23, 2007. *Astrophysical Journal* – submitted.

Huttunen, K. E. J., et al. (2002): April 2000 magnetic storm: Solar wind driver and magnetospheric response. *Journal of Geophysical Research* **107**, 1440.

Huttunen, K. E. J., et al. (2005): Properties and geoeffectiveness of magnetic clouds in the rising, maximum and early declining phases of solar cycle 23. *Annales Geophysicae* **23**, 625–641.

Huttunen, K. E. J., Koskinen, H. E. J. (2004): Importance of post-shock streams and sheath region as drivers of intense magnetospheric storms and high-latitude activity. *Annales Geophysicae* **22**, 1729–1738.

I

Ichimoto, K., et al. (2007a): Twisting motions of sunspot penumbral filaments. *Science* **318**, 1597–1599.

Iichimoto, K., et al. (2007b): Fine-scale structures of the Evershed effect observed by the Solar Optical Telescope aboard *Hinode*. *Publications of the Astronomical Society of Japan* **59**, S593–S599.

Illing, R. M. E., Hundhausen, A. J. (1983): Possible observation of a disconnected magnetic structure in a coronal transient. *Journal of Geophysical Research* **99**, 10210–10214.

Imada, S., et al. (2007): Discovery of a temperature-dependent upflow in the plage region during a gradual phase of the X-class flare. *Publications of the Astronomical Society of Japan* **59**, S793–S799.

Imbrie, J. (1982): Astronomical theory of the Pleistocene ice ages. A brief historical. *Icarus* **50**, 408–432.

Imbrie, J., Imbrie, K. P. (1979): *Ice ages – Solving the Mystery*. Short Hills, New Jersey: Enslow Publishers 1979.

Imbrie, J., Imbrie, J. Z. (1980): Modeling the climatic response to orbital variations. *Science* **207**, 943–953.

Imbrie, J., Imbrie, K. P. (1986): *Ice ages – Solving the Mystery. Second Edition*. Cambridge, Massachusetts: Harvard University Press 1986.

Imbrie, J., et al. (1984): The orbital theory of Pleistocene climate: Support from a revised chronology of the marine δ^{18}O record. In: *Milankovitch and Climate, Part 1* (Eds.L. Berger et al.). Dordrecht, The Netherlands: Reidel 1984, pp. 269–305.

Imbrie, J., et al. (1992): On the structure and origin of major glaciation cycles 1. Linear responses to Milankovich forcing. *Paleoceanography* **7**, 701–738.

Innes, D. E, Inhester, B., Axford, W. I., Wilhelm, K. (1997): Bi-directional plasma jets produced by magnetic reconnection on the Sun. *Nature* **386**, 811–813.

Insley, J. E., More, V., Harrison, R. A. (1995): The differential rotation of the corona as indicated by coronal holes. *Solar Physics* **160**, 1–18.

Intergovernmental Panel On Climate Change (2001, 2007): *The Scientific Basis. Contribution of Working Group 1 to the Third and the Fourth Assessment Report of the Intergovernmental Panel on Climate Change*. New York: Cambridge University Press 2001, 2007.

Intriligator, D. S., et al. (2005): From the Sun to the outer heliosphere: Modeling and analyses of the interplanetary propagation of the October/November (Halloween) 2003 solar events. *Journal of Geophysical Research* **110**, A09S10.

Ionson, J. A. (1978): Resonant absorption of Alfvénic surface waves and the heating of solar coronal loops. *Astrophysical Journal* **226**, 650–673.

Ireland, J., Will-Davey, M., Walsh, R. W. (1999): Coronal heating events in high cadence *TRACE* data. *Solar Physics* **190**, 207–232.

Isenberg, P. A. (1983): Acceleration of heavy ions in the solar wind. In: *Solar Wind Five* (Ed. M. Neugebauer). Washington: NASA, pp. 655.

Isenberg, P. A. (1990): Investigations of a turbulent-driven solar wind model. *Journal of Geophysical Research* **95**, 6437.

Isenberg, P. A. (1991): The solar wind. *Geomagnetism* **4**, 1–85.

Isenberg, P. A. (2001): Heating of coronal holes and generation of the solar wind by ion-cyclotron resonance. *Space Science Reviews* **95**, 119.

Isenberg, P. A. (2003): The kinetic shell model of coronal heating and acceleration by ion cyclotron waves: 3. The proton halo and dispersive waves. *Journal of Geophysical Research* **109**, A03101.

Isenberg, P. A., Forbes, T. G. (1993): Catastrophic evolution of a force-free flux rope: A model for eruptive flares. *Astrophysical Journal* **417**, 368–386.

Isenberg, P. A., Forbes, T. G. (2007): A three-dimensional line-tied magnetic filed model for solar eruptions. *Astrophysics Journal* **670**, 1453–1466.

Isobe, H., et al. (2007): Flare ribbons observed with G-band and Fe I 6302 Å filters of the Solar Optical Telescope on board *Hinode*. *Publications of the Astronomical Society of Japan* **59**, S807–S813.

Izmodenov, V., Gloeckler, G., Malama, Y. (2003): When will *Voyager 1* and *2* cross the termination shock? *Geophysical Research Letters* **30**(7) 3–1.

J

Jackman, C. H., et al. (2005): Neutral atmospheric influences of the solar proton events in October–November 2003. *Journal of Geophysical Research* **110**, 9.

James, I. N., James, P. M. (1989): Ultra-low-frequency variability in a simple atmospheric circulation model. *Nature* **342**, 53–55.

Janssen, P. J. C. (1872): Observations of the solar eclipse of 12 December 1871. *Nature* 5, 249. Reproduced In: *Early Solar Physics* (Ed. A. J. Meadows). New York: Pergamon Press 1970, pp. 223–224.

Jefferies, S. M., et al. (2006): Magnetoacoustic portals and the basal heating of the solar chromosphere. *The Astrophysical Journal (Letters)* **648**, L151–L155.

Jensen, J. M., Pijpers, F. P., Thompson, M. J. (2006): Time-distance measurements of cross-correlation asymmetries around NOAA AR 10486. *Astrophysical Journal (Letters)* **648**, L75–L78.

Jing, J., et al. (2006): The statistical relationship between the photospheric magnetic parameters and the flare productivity of active regions. *The Astrophysical Journal* **38**, 259.

Jockers, K. (1970): Solar wind models based on exospheric theory. *Astronomy and Astrophysics* **6**, 215–239.

Johnson, T. H. (1938): Nature of primary cosmic radiation. *Physical Review* 54, 385–387.

Jokipii, J. R. (1966): Cosmic-ray propagation I. Charged particles in a random magnetic field. *Astrophysical Journal*, **146**, 480.

Jokipii, J. R. (1971): Propagation of cosmic rays in the solar wind. *Review of Geophysics and Space Physics* **9**, 27–87.

Jokipii, J. R., Davis, L. Jr. (1969): Long-wavelength turbulence and the heating of the solar wind. *The Astrophysical Journal* **156**, 1101–1106.

Jokipii, J. R., et al. (1995): Interpretation and consequences of large-scale magnetic variances observed at high heliographic latitude. *Geophysical Research Letters* **22**(23), 3385–3388.

Jokipii, J. R., Giacalone, J. (2004): Radial streaming anisotropies of charged particles accelerated at the solar wind termination shock. *The Astrophysical Journal* **605**, L145–L148.

Jokipii, J. R., Giacalone, J., Kóta, J. (2004): Transverse streaming anisotropies of charged particles accelerated at the solar wind termination shock. *Astrophysical Journal (Letters)* **611**, L141–L144.

Jokipii, J. R., Kóta, J. (1989): The polar heliospheric magnetic field. *Geophysical Research Letters* **16**, 1–4.

Jokipii, J. R., Levy, E. H. (1977): Effects of particle drifts on the solar modulation of galactic cosmic rays. *The Astrophysical Journal (Letters)* **213**, L85–L88.

Jokipii, J. R., Mc Donald, F. B. (1995): Quest for the limits of the heliosphere. *Scientific American* **272**, 58–63 – April.

Jokipii, J. R., Sonett, C. P., Giampapa, M. S. (Eds., 1997): *Cosmic Winds and the Heliosphere.* Tucson, Arizona: University of Arizona Press 1997.

Jones, G. H., Balogh, A. (2003): The global heliospheric magnetic field polarity distribution as seen at *Ulysses. Annales Geophysicae* **21**(6), 1377–1382.

Jones, P. D., Wigley, T. M. L., Wright, P. B. (1986): Global temperature variations between 1861 and 1984. *Nature* **322**, 430–434.

Joselyn, J. A., McIntosh, P. S. (1981): Disappearing solar filaments: a useful predictor of geomagnetic activity. *Journal of Geophysical Research* **86**, 4555–4564.

Jouzel, J., et al. (1987): Vostok ice core: A continuous isotope temperature record over the last climatic cycle (160,000 years). *Nature* **329**, 402–408.

Jouzel, J., et al. (1993): Extending the Vostok ice core record of palaeoclimate to the penultimate glacial period. *Nature* **364**, 407–412.

Joy, A. H., Humason, M. L. (1949): Observations of the faint dwarf star L726–8. *Publications of the Astronomical Society of the Pacific* **61**, 133–134.

Juckett, D. A. (2006): Long period (0.9–5.5 year) oscillations in surface spherical harmonics of sunspot longitudinal distributions. *Solar Physics* **237**, 351–364.

Judge, P. G., Tarbell, T. D., Klaus, W. (2001): A study of chromospheric oscillations using the *SOHO* and *TRACE* spacecraft. *The Astrophysical Journal* **554**, 424–444.

K

Kahler, S. W. (1977): The morphological and statistical properties of solar X-ray events with long decay times. *The Astrophysical Journal* **214**, 891–897.

Kahler, S. W. (1982): The role of the big flare syndrome in correlations of solar energetic proton fluxes and associated microwave burst parameters. *Journal of Geophysical Research* **87**, 3439–3448.

Kahler, S. W. (1987): Coronal mass ejections. *Reviews of Geophysics* **25**, 663–675.

Kahler, S. W. (1992): Solar flares and coronal mass ejections. *Annual Review of Astronomy and Astrophysics* **30**, 113–141.

Kahler, S. W. (2001): The correlation between solar energetic particle peak intensities and speeds of coronal mass ejections: Effects of ambient particle intensities and energy spectra. *Journal of Geophysical Research* **106**, 20947–20956.

Kahler, S.W. (2007): Solar sources of heliospheric energetic events: shocks or flares? *Space Science Reviews* **129**, 359–390.

Kahler, S. W., Sheeley, N. R. Jr., Liggett, M. (1989): Coronal mass ejections and associated X-ray flare durations. *Astrophysical Journal* **344**, 1026–1033.

Kahler, S. W., et al. (1984): Associations between coronal mass ejections and solar energetic proton events. *Journal of Geophysical Research* **89**, 9683–9693.

Kahler, S. W., et al. (1986): Solar filament eruptions and energetic particle events. *Astrophysical Journal* **302**, 504–510.

Kahn, F. D. (1961): Sound waves trapped in the solar atmosphere. *Astrophysical Journal* **134**, 343–346.

Kaiser, M. L. (2008): The *STEREO* mission: An introduction. *Space Science Reviews*, **136**, No. 1–4, 5–16.

Kakinuma, T. (1977): Observations of interplanetary scintillation: solar wind velocity measurements. In: *Study of Traveling Interplanetary Phenomena* (Eds. M. A. Shea, D. F. Smart and S. T. Wu). Dordrecht: D. Reidel 1977, pp. 101–118.

Kalkofen, W. (2008): Heating and dynamics of the quiet chromosphere. *Proceedings of the International Astronomical Union Symposium 247*, 93–98.

Kallenbach, R., et al. (1998): Fractionation of Si, Ne, and Mg isotopes in the solar wind as measured by *Soho*/Celias/Mtof. *Space Science Reviews* **85**, 357–370.

Kallenbach, R., Geiss, J., Gloeckler, G., Von Steiger, R. (2000): Pick-up ion measurements in the heliosphere – A review. *Astrophysics and Space Science* **274**, 97–114.

Kallenrode, M.-B. (1998): *Space Physics. An Introduction to Plasmas and Particles in the Heliosphere and Magnetospheres*. New York: Springer-Verlag 1998.

Kamio, S., et al. (2007): Velocity structure of jets in a coronal hole. *Publications of the Astronomical Society of Japan* **59**, S757–S762.

Kanbach, G., et al. (1993): Detection of a long-duration solar gamma-ray flare on June 11, 1991 with EGRET on *COMPTON-GRO*. *Astronomy and Astrophysics Supplement Series* **97**, 349–353.

Kane, R. P. (2006): The idea of space weather – a historical perspective. *Advances in Space Research* **37**, 1261–1264.

Kane, S. R. (1974): Impulsive (flash) phase of solar flares: Hard X-ray, microwave, EUV and optical :emissions. In: *Coronal Disturbances. Proceedings of IAU Symposium No. 57* (Ed. G. Newkirk, Jr.) Boston: D. Reidel 1974, pp. 105–141.

Kane, S. R. (Ed., 1975): *Solar Gamma-, X-, and EUV Radiation. Proceedings of IAU Symposium No. 68*. Boston: D. Reidel 1975.

Kane, S. R., et al. (1980): Impulsive phase of solar flares. In: *Solar Flares: A Monograph from Skylab Solar Workshop II* (Ed. P. A. Sturrock). Boulder, Colorado: Colorado Associated University Press 1980, pp. 187–229.

Kane, S. R., et al. (1986): Rapid acceleration of energetic particles in the 1982 February 8 solar flare. *Astrophysical Journal (Letters)* **300**, L95–L98.

Kane, S. R., et al. (1995): Energy release and dissipation during giant solar flares. *Astrophysical Journal (Letters)* **446**, L47–L50.

Kano, R., Tsuneta, S. (1995): Scaling law of solar coronal loops obtained with *Yohkoh*. *Astrophysical Journal* **454**, 934–944.

Kano, R., Tsuneta, S. (1996): Temperature distributions and energy scaling law of solar coronal loops obtained with *Yohkoh*. *Publications of the Astronomical Society of Japan* **48**, 535–543.

Kappenman, J. G. (1996): Geomagnetic storms and their impact on power systems. *IEEE Power Engineering Review* **16**, 5–8.

Karlén, W., Kuylenstierna, J. (1996): Evidence from the Scandinavian tree lines since the last ice age. In: *The Global Warming Debate* (Ed. J. Emsley). London: European Science and Environment Forum 1996, pp. 192–204.

Karpen, J. T., Antiochos, S. K., Klimchuk, J. A. (2006): The origin of high speed motions and threads in prominences. *Astrophysical Journal* **637**, 531–540.

Karpen, J. T., et al. (1998): Dynamic responses to magnetic reconnection in solar arcades. *Astrophysical Journal* **495**, 491.

Kasting, J. F. (1989): Long-term stability of the Earth's climate. *Palaeogeography, Palaeoclimatology, Palaeoecology* **75**, 83–95.

Kasting, J. F., Ackerman, T. P. (1986): Climatic consequences of very high carbon dioxide levels in Earth's early atmosphere. *Science* **234**, 1383–1385.

Kasting, J. F., Catling, D. (2003): Evolution of a habitable planet. *Annual Reviews of Astronomy and Astrophysics* **41**, 429–463.

Kasting, J. F., Grinspoon, D. H. (1991): The faint young sun problem. In: *The Sun in Time* (Eds. C. P. Sonett, M. S. Giampapa and M. S. Matthews). Tucson, Arizona: The University of Arizona Press 1991, pp. 447–462.

Kasting, J. F., Toon, O. B. (1989): Climate evolution on the terrestrial planets. In: *Origin and Evolution of Planetary and Satellite Atmospheres*, (Eds.S. K. Alrya, J. B. Pollack, M. S. Matthews). Tucson, University of Arizona Press, pp. 423–449.

Katsukawa, Y., et al. (2007a): Formation process of a light bridge revealed with the *Hinode* Solar Optical Telescope. *Publications of the Astronomical Society of Japan* **59**, S577–S584.

Katsukawa, Y., et al. (2007b): Small-scale jetlike features in penumbral chromospheres. *Science* **318**, 1594–1596.

Katsukawa, Y., Tsuneta, S. (2005): Magnetic properties at footpoints of hot and cool loops. *Astrophysical Journal* **621**, 498–511.

Kawabata, K. (1960): The relationship between post-burst increases of solar microwave radiation and sudden ionospheric disturbances. *Report of Ionosphere and Space Research in Japan* **14**, 405–426.

Keating, G. M., et al. (1986): Detection of stratospheric HNO_3 and NO_2 response to short-term solar ultraviolet variability. *Nature* **322**, 43–46.

Keating, G. M., et al. (1987): Response of middle atmosphere to short-term ultraviolet variations 1. Observations. *Journal of Geophysical Research* **92**, 889–902.

Keeling, C. D. (1960): The concentration and isotopic abundance of carbon dioxide in the atmosphere. *Tellus* **12**, 200–203.

Kellogg, P. J. (1962): Flow of plasma around the earth. *Journal of Geophysical Research* **67**, 3805–3811.

Kelly, P. M. (1977): Solar influence on North Atlantic mean sea level pressure. *Nature* **269**, 320–322.

Kelly, P. M., Wigley, T. M. L. (1990): The influence of solar forcing trends on global mean temperature since 1861. *Nature* **347**, 460–462.

Kelly, P. M., Wigley, T. M. L. (1992): Solar cycle length, greenhouse forcing and global climate. *Nature* **360**, 328–330.

Kelvin, Lord: see Thomson, W. (Baron Kelvin) (1892): Presidential address to the Royal Society on November 30, 1892. In: *Popular Lectures and Addresses by Sir William Thomson Baron Kelvin. Volume II. Geology and General Physics*. London: Macmillan and Company 1894, 508–529.

Kennelly, A. E. (1902): On the elevation of the electrically-conducting strata in the Earth's atmosphere. *Electrical World and Engineer* **39**, 473.

Keppler, E., et al. (1992): The *Ulysses* energetic particle composition experiment EPAC. *Astronomy and Astrophysics Supplement* **92**(2), 317–331.

Kiepenheuer, K. O. (1950): Cosmic rays and radio emission from our galaxy. *Physical Review* **79**, 738–739. Reproduced In: *A Source Book in Astronomy and Astrophysics 1900–1975* (Eds. K. R. Lang and O. Gingerich). Cambridge, Massachusetts: Harvard University Press 1977, pp. 677–679.

Kim, S., et al. (2007): Two-step reconnections in a C3.3 flare and its preflare activity observed by *Hinode* XRT. *Publications of the Astronomical Society of Japan* **59**, S831–S836.

Kim, Y.-H., et al. (2007): Small-scale X-ray/EUV jets seen in *Hinode* XRT and *TRACE*. *Publications of the Astronomical Society of Japan* **59**, S763–S769.

King, D. B., et al. (2003): Propagating EUV disturbances in the solar corona: Twowavelength observations. *Astronomy and Astrophysics* **404**, L1–L4.

Kirchhoff, G. (1861): On the chemical analysis of the solar atmosphere. *Philosophical Magazine and Journal of Science* **21**, 185–188. Reproduced in: *Early Solar Physics* (Ed. A. J. Meadows). New York: Pergamon Press 1970, pp. 103–106.

Kirchhoff, G., Bunsen, R. (1860): Chemical analysis of spectrum – observations. *Philosophical Magazine and Journal of Science* **20**, 89–109, **22**, 329–249, 498–510 (1861).

Kitai, R., et al. (2007): Umbral fine structures in sunspots observed with *Hinode* Solar Optical Telescope. *Publications of the Astronomical Society of Japan* **59**, S585–S591.

Kivelson, M. G., Russell, C. T. (Eds., 1997):*Introduction to Space Physics*. Cambridge, England: Cambridge University Press 1997.

Kjeldseth-Moe, O., Brekke, P. (1998): Time variability of active region loops observed with the coronal diagnostic spectrometer on *SOHO*. *Solar Physics* **182**, 73–95.

Kjeldseth-Moe, O., Wilson, A. (Eds., 1997): *SOHO-5: The Corona and Solar Wind Near Minimum Activity*. ESA SP-404 Noordwijk, Netherlands 1997.

Klecker, B., et al. (2006): Energetic particle observations. *Space Science Reviews* **123**, 217–250.

Klecker, B., Möbius, E., Popecki, M. A. (2007): Ionic charge states of solar energetic particles. A clue to the source. *Space Science Reviews* **130**, 273–282.

Kleim, B., Dammasch, J. E., Curdt, W., Wilhelm, K. (2002): Correlated dynamics of hot and cool plasmas in the main phase of a solar flare. *Astrophysical Journal (Letters)* **568**, L61–L65.

Kleim, B., Titov, V. S., Török, T. (2003): Formation of current sheets and sigmoidal structure by the kink instability of a magnetic loop. *Astronomy and Astrophysics* **413**, L23–L26.

Klein, J., et al. (1980): Radiocarbon concentrations in the atmosphere: 8000 year record of variations in tree rings. *Radiocarbon* **22**, 950–961.

Klein, K.-L. (2003): Introduction. In: *Energy Conversion and Particle Acceleration in the Solar Corona, Lecture Notes in Physics Vol. 612* (Ed. K.-L. Klein). New York: Springer-Verlag 2003, pp. 1–6.

Klein, L. W., Burlaga, L. F. (1982): Interplanetary magnetic clouds at 1 AU. *Journal of Geophysical Research* **87**, 613–624.

Klimchuk, J. A. (2000): Cross-sectional properties of coronal loops. *Solar Physics* **193**, 53–75.

Klimchuk, J. A. (2001): Theory of coronal mass ejections. *Space Weather, Geophysical Monograph* **125**, 143–157.

Klimchuk, J. A. (2006): On solving the coronal heating problem. *Solar Physics* **234**, 41–77.

Klimchuk, J. A., Cargill, P. J. (2001): Spectroscopic diagnostics of nanoflare- heated loops. *The Astrophysical Journal* **553**, 440–448.

Klimchuk, J. A., Gary, D. E. (1995): A comparison of active region temperatures and emission measures observed in soft X-rays and microwaves and implications for coronal heating. *The Astrophysical Journal* **448**, 925–937.

Klimchuk, J. A., Porter, L. J. (1995): Scaling of heating rates in solar coronal loops. *Nature* **377**, 131–133.

Klimchuk, J. A., Tanner, S. E. M., De Moortel, I. (2004): Coronal seismology and the propagation of acoustic waves along coronal loops. *Astrophysical Journal* **616**, 1232–1241.

Ko, Y.-K., et al. (1997): An empirical study of the electron temperature and heavy ion velocities in the south polar coronal hole. *Solar Physics* **171**, 345–361.

Ko, Y.-K., et al. (2003): Dynamical and physical properties of a post-coronal mass ejection current sheet. *Astrophysical Journal* **594**, 1068–1084.

Kocharov, L., Torsti, J. (2003): The origin of high-energy ^{3}He-rich solar particle events. *The Astrophysical Journal* **586**, 1430–1435.

Kohl, J. L., Cranmer, S. R. (Eds., 1999): SOHO-7: Coronal holes and solar wind acceleration. *Space Science Reviews* **87** (1–2), 1–368.

Kohl, J. L., et al. (1980): Measurement of coronal temperatures from 1.5 to 3 solar radii. *The Astrophysical Journal (Letters)* **241**, L117–L121.

Kohl, J. L., et al. (1995): *Spartan 201* coronal spectroscopy during the polar passes of *Ulysses*. *Space Science Reviews* **72**, 29–38.

Kohl, J. L., et al. (1995): The Ultraviolet Coronagraph Spectrometer for the *Solar and Heliospheric Observatory*. *Solar Physics* **162**, 313–356.

Kohl, J. L., et al. (1997): First results from the *SOHO* ultraviolet coronagraph spectrometer. *Solar Physics* **175**, 613–644. Reprinted in: *The First Results from SOHO* (Eds. B. Fleck and Z. Svestka). Boston: Kluwer Academic Publishers 1997, pp. 613–644.

Kohl, J. L., et al. (1998): UVCS/*SOHO* empirical determinations of anisotropic velocity distributions in the solar corona. *Astrophysical Journal (Letters)* **501**, L127–L131.

Kohl, J. L., et al. (1999): EUV spectral line profiles in polar coronal holes from 1.3 to 3.0 R solar. *The Astrophysical Journal (Letters)* **510**, L59–L62.

Kohl, J. L., Noci, G., Cranmer, R., Raymond, J. C. (2006): Ultraviolet spectroscopy of the extended solar corona. *Astronomy and Astrophysics Review* 13, 31–157.

Kohl, J. L., Withbroe, G. L. (1982): EUV spectroscopic plasma diagnostics for the solar wind acceleration region. *Astrophysical Journal* 256, 263–270.

Kojima, M., et al. (2004): Fast solar wind after the rapid acceleration. *Journal of Geophysical Research* 109(A4), A04103.

Kojima, M., Kakinuma, T. (1987): Solar cycle evolution of solar wind speed structure between 1973 and 1985 observed with the interplanetary scintillation method. *Journal of Geophysical Research* 92, 7269–7279.

Kolhörster, W. (1913): Messungen der durchdringenden Strahlung im Freiballon im größeren Höhen. *Physikalishe Zeitschrift* 14, 1153–1156.

Kolmogoroff, A. N. (1941a): Dissipation of energy in the locally isotropic turbulence. *Compt. Rend (Dokl.) Acad. Sci. (SSSR)* 32, 16.

Kolmogoroff, A. N. (1941b): The local structure of turbulence in incompressible viscous fluids for very large Reynolds numbers. *Compt. Rend (Dokl.) Acad. Sci (SSSR)* 30, 301.

Kominz, M. A., Pisias, N. G. (1979): Pleistocene climate: Deterministic or stochastic. *Science* 204, 171–173.

Komm, R., et al. (2004): Solar subsurface fluid dynamics descriptors derived from Global Oscillation Network Group and Michelson Doppler Imager data. *Astrophysical Journal* 605, 554–567.

Komm, R., et al. (2007): Divergence and vorticity of solar subsurface flows derived from ring-diagram analysis of MDI and GONG data. *The Astrophysical Journal* 667, 571–584.

Kopp, G., Lawrence, G., Rottman, G. (2005): The Total Irradiance Monitor (TIM): Science results. *Solar Physics* 230, 129–139.

Kopp, R. A., Holzer, T. E. (1976): Dynamics of coronal hole regions 1.: Steady polytropic flows with multiple critical points. *Solar Physics* 49, 43–56.

Kopp, R. A., Kuperus, M. (1968): Magnetic fields and the temperature structure of the chromosphere-corona interface. *Solar Physics* 4, 212–223.

Kopp, R. A., Pneuman, G. W. (1976): Magnetic reconnection in the corona and the loop prominence phenomenon. *Solar Physics* 50, 85–98.

Korzennik, S., Wilson, A. (Eds., 1998): *SOHO-6/GONG 98: Structure and Dynamics of the Interior of the Sun and Sun-like Stars.* ESA SP-418 1998.

Kosovichev, A. G., Duvall, T. L. Jr. (2006): Active region dynamics. *Space Science Reviews* 124, 1–12. Also in *Solar Dynamics and Its Effects on the Heliosphere and Earth, Space Science Series of ISSI Volume 22.* New York: Springer 2007, pp. 1–12.

Kosovichev, A. G., Duvall, T. L. Jr., Scherrer, P. H. (2000): Time-distance inversion methods and results – invited review. *Solar Physics* 192, 159–176.

Kosovichev, A. G., et al. (1997): Structure and rotation of the solar interior: Initial results from the MDI medium-L program. *Solar Physics* 170, 43–61. Reprinted in: *The First Results From SOHO* (Eds. B. Fleck and Z. Svestka). Boston: Kluwer Academic Publishers 1997, pp. 43–61.

Kosovichev, A. G., Schou, J. (1997): Detection of zonal shear flows beneath the Sun's surface from f-mode frequency splitting. *The Astrophysical Journal (Letters)* 482, L207–L210.

Kosovichev, A. G., Zharkova, V. V. (1998): X-ray flare quakes the Sun. *Nature* 393, 317–318.

Kosugi, T., et al. (1991): The Hard X-ray Telescope (HXT) for the *SOLAR-A* mission. *Solar Physics* 136, 17–36.

Kosugi, T., et al. (1992): The Hard X-ray Telescope (HXT) onboard Yohkoh: Its performance and some initial results. *Publications of the Astronomical Society of Japan* 44, L45–L49.

Kosugi, T., et al. (2007a): The *Hinode* (*Solar-B*) mission: An overview. *Solar Physics* 243, 3–17.

Kotoku, J., et al. (2007b): Magnetic feature and morphological study of X-ray bright points with *Hinode. Publications of the Astronomical Society of Japan* 59, S735–S743.

Koutchmy, S., et al. (2004): The August 11th, 1999 CME. *Astronomy and Astrophysics* 420, 709–718.

Koutchmy, S., Livshits, M. (1992): Coronal streamers. *Space Science Reviews* 61, 393–417.

Kozlovsky, B., Ramaty, R. (1977): Narrow lines from alpha-alpha reactions. *Astrophysical Letters* **19**, 19–24.

Krall, J., et al. (2001): Erupting solar magnetic flux ropes: Theory and observation. *Astrophysical Journal* **562**, 1045–1057.

Krall, J., et al. (2006): Flux rope model of the 2003 October 28–30 coronal mass ejection and interplanetary coronal mass ejections. *Astrophysical Journal* **642**, 541–553.

Kreplin, R. W. (1961): Solar X-rays. *Annales de Géophysique* **17**, 151–161.

Kreplin, R. W., Chubb, T. A., Friedman, H. (1962): X-ray and Lyman-alpha emission from the Sun as measured from the Nrl *Sr-1* satellite. *Journal of Geophysical Research* **67**, 2231–2253.

Krieger, A. S., et al. (1974): X-ray observations of coronal holes and their relation to high velocity solar wind streams. In: *Solar Wind Three* (Ed. C. T. Russell). Los Angeles, California: Institute of Geophysics and Planetary Physics, UCLA 1974, 132–139.

Krieger, A. S., Timothy, A. F., Roelof, E. C. (1973): A coronal hole and its identification as the source of a high velocity solar wind stream. *Solar Physics* **29**, 505–525.

Krijger, J. M., et al. (2001): Dynamics of the solar chromosphere. III. Ultraviolet brightness oscillations from *TRACE*. *Astronomy and Astrophysics* **379**, 1052–1082.

Krimigis, S. M., et al. (2003): *Voyager 1* exited the solar wind at a distance of approximately 85 AU from the Sun. *Nature* **426**, 45–48.

Krucker, S., Benz, A. O. (1998): Energy distribution of heating processes in the quiet solar corona. *Astrophysical Journal (Letters)* **501**, L213–L216.

Krucker, S., Benz, A. O., Bastian, T. S., Acton, L. W. (1997): X-ray network flares of the quiet sun. *Astrophysical Journal* **488**, 499–505.

Krucker, S., et al. (2002): Hard X-ray microflares down to 3 keV. *Solar Physics* **210**, 445–456.

Krucker, S., Hurford, G. J., Lin, R. P. (2003): Hard X-ray source motions in the 2002 July 23 gamma-ray flare. *Astrophysical Journal (Letters)* **595**, L103–L106.

Krucker, S., Larson, D. E., Lin, R. P., Thompson, B. J. (1999): On the origin of impulsive electron events observed at 1 AU. *The Astrophysical Journal* **519**, 864–875.

Krucker, S., Lin, R. P. (2000): Two classes of solar proton events derived from onset time analysis. *Astrophysical Journal (Letters)* **542**, L61–L64.

Krüger, A. (1979): *Introduction to Solar Radio Astronomy and Radio Physics*. Dordrecht D. Reidel 1979.

Kubo, M., et al. (2007a): *Hinode* observations of a vector magnetic field change associated with a flare on 2006 December 13. *Publications of the Astronomical Society of Japan* **59**, S779–S784.

Kubo, M., et al. (2007b): Formation of moving magnetic features and penumbral magnetic fields with *Hinode*/SOT. *Publications of the Astronomical Society of Japan* **59**, S607–S612.

Kucera T. A. (2006): Ultraviolet observations of prominence activation and coronal loop dynamics. *Astrophysical Journal* **645**, 1525–1536.

Kucera, T. A., Tovar, M., De Pontieu, B. (2003): Prominence motions observed at high cadences in temperatures from 10 000 to 250 000 K. *Solar Physics* **212**, 81–97.

Kucharek, H., et al. (2003): On the source and acceleration of energetic He$^+$: A long term observation with *ACE*/SEPICA. *Journal of Geophysical Research* **108**(A10), LIS 15–1.

Kuhn, J. R., Bush, R. I., Scherrer, P., Scheick, S. (1998): The sun's shape and brightness. *Nature* **392**, 155–157.

Kuhn, J. R., Kasting, J. F. (1983): The effects of increased CO_2 concentrations of surface temperature of the early Earth. *Nature* **301**, 53–55.

Kuhn, J. R., Libbrecht, K. G., Dicke, R. H. (1988): The surface temperature of the Sun and changes in the solar constant. *Science* **242**, 908–911.

Kukla, G. (1975): Missing link between Milankovitch and climate. *Nature* **253**, 600–603.

Kukla, G., et al. (1981): Orbital signature of interglacials. *Nature* **290**, 295–300.

Kumar, A., Rust, D. M. (1996): Interplanetary magnetic clouds, helicity conservation and current-core flux-ropes. *Journal of Geophysical Research* **101**(A7), 15667–15684.

Kundu, M. R. (1961): Bursts of centimeter-wave emission and the region of origin of X-rays from solar flares. *Journal of Geophysical Research* **66**, 4308–4312.

Kundu, M. R. (1965): *Solar Radio Astronomy*. New York: Wiley Interscience.

Kundu, M. R. (1982): Advances in solar radio astronomy. *Reports on Progress in Physics* **45**, 1435–1541.

Kundu, M. R., et al. (1995): Microwave and hard X-ray observations of footpoint emission from solar flares. *Astrophysical Journal* **454**, 522–530.

Kundu, M. R., Lang, K. R. (1985): The sun and nearby stars: Microwave observations at high resolution. *Science* **228**, 9–15.

Kundu, M. R., Vlahos, L. (1982): Solar microwave bursts – a review. *Space Science Reviews* **32**, 405–462.

Kundu, M. R., Woodgate, B., Schmahl, E. J. (Eds., 1989): *Energetic Phenomena on the Sun*. Boston: Kluwer Academic Publishers.

Kunow, H., et al. (1999): Co-rotating interaction regions at high latitudes. *Space Science Reviews* **89**, 221–268.

Kunow, H., et al. (Eds., 2006): Coronal mass ejections, *Space Science Reviews* **123**, 1–484. Reprinted by Springer Verlag and the International Space Science Institute.

Kuperus, M., Ionson, J. A., Spicer, D. S. (1981): On the theory of coronal heating mechanisms. *Annual Review of Astronomy and Astrophysics* **19**, 7–40.

Kyle, H. L., Hoyt, D. V., Hickey, J. R. (1994): A review of the *Nimbus 7* ERB solar data set. In: *The Sun as a Variable Star: Solar and Stellar Luminosity] Variations* (Eds. J. M. Pap, C. Fröhlich, H. S. Hudson and S. K. Solanki). New York: Cambridge University Press, pp. 9–12.

L

Labitzke, K. (1987): Sunspots, the QBO and the stratospheric temperature in the north polar region. *Geophysical Research Letters* **14**, 535–537.

Labitzke, K., van Loon, H. (1988): Associations between the 11-year solar cycle, the QBO, and the atmosphere. Part 1. The troposphere and stratosphere in the northern hemisphere winter. *Journal of Atmospheric and Terrestrial Physics* **50**, 197–206.

Labitzke, K., van Loon, H. (1990): Associations between the 11-year solar cycle, the quasi-biennial oscillation and the atmosphere: a summary of recent work. *Philosophical Transactions of the Royal Society (London)* **A330**, 557–589.

Labitzke, K., van Loon, H. (1992): On the association between the QBO and the extratropical stratosphere. *Journal of Atmosphere and Terrestrial Physics* **54**, 1453–1463.

Labitzke, K., van Loon, H. (1993): Some recent studies of probable connections between solar and atmospheric variability. *Annales Geophysicae* **11**, 1084–1094.

Labitzke, K., van Loon, H. (1995): Connection between the troposphere and stratosphere on a decadal scale. *Tellus A* **47**, 275–286.

Labonte, B. J., Howard, R. (1982): Solar rotation measurements at Mount Wilson. III – meridional flow and limbshift. *Solar Physics* **80**, 361–372.

Lacis, A. A., Carlson, B. E. (1992): Global warming: Keeping the Sun in proportion. *Nature* **360**, 297.

Lacoste, H. (Ed., 2006) *SOHO-17: 10 Years of SOHO and Beyond*. ESA SP-617 2006.

Laitinen, T. V., et al. (2005): The magnetotail reconnection region in a global MHD simulation. *Annales Geophysicae* **23**, 3753.

Laitinen, T. V., et al. (2006): On the characterization of magnetic reconnection in MHD simulations. *Annales Geophysicae* **24**, 3059–3069.

Lamb, H. H. (1965): The early Medieval warm epoch and its sequel. *Palaeogeography, Palaeoclimatology, Palaeoecology* **1**, 13–37.

Lamb, H. H. (1977): *Climate: Present, Past and Future: Climate History and the Future*. London: Methuen.

Lamb, H. H. (1982): *Climate History and the Modern World*. London: Methuen.

Landi, E., Feldman, U., Doschek, G. A. (2007): Neon and oxygen absolute abundances in the corona. *The Astrophysical Journal* **659**, 743–749.

Lane, J. H. (1870): On the theoretical temperature of the sun; under the hypothesis of a gaseous mass maintaining its volume by its internal heat, and depending on the laws of gases as known to terrestrial experiment. *American Journal of Science and Arts* (2nd series) **50**, 57–74. Reproduced in: *Early Solar Physics* (Ed. A. J. Meadows). New York: Pergamon Press 1970, pp. 257–276.

Lang, K. R. (1994): Radio evidence for nonthermal particle acceleration on stars of late spectral type. *Astrophysical Journal Supplement* **90**, 753–764.

Lang, K. R. (1996): Unsolved mysteries of the Sun – Part 1, 2. *Sky and Telescope* **92**(2), 38–42, August, **92**(3), 24–28, September.

Lang, K. R. (1997): *SOHO* reveals the secrets of the sun. *Scientific American* **276**(3), 32–47, March. Updated in: *Magnificent Cosmos*, a Scientific American Publication (1998), March. New York City, Scientific American.

Lang, K. R. (1999a): The Sun. In: *The New Solar System* (Eds. J. Kelly Beatty, C. C. Petersen, and A. Chaikin). New York: Cambridge University Press 1999, pp. 23–38.

Lang, K. R. (1999b): *Astrophysical Formulae. Vol. I. Radiation, Gas Processes and High Energy Astrophysics*. New York: Springer Verlag.

Lang, K. R. (1999c): *Astrophysical Formulae. Vol. II. Space, Time, Mass and Cosmology*. New York: Springer Verlag.

Lang, K. R. (2001): *The Cambridge Encyclopedia of the Sun*. Cambridge, England, Cambridge University Press.

Lang, K. R. (2006): *Sun, Earth and Sky*, 2nd edition. New York: Springer Verlag.

Lang, K. R., Gingerich, O. (Eds., 1979): *A Source Book in Astronomy and Astrophysics 1900–1975*. Cambridge: Harvard University Press.

Lang, K. R., et al. (1993): Magnetospheres of solar active regions inferred from spectral-polarization observations with high spatial resolution. *Astrophysical Journal* **419**, 398–417.

Langer, S. H., Petrosian, V. (1977): Impulsive solar X-ray bursts. III. Polarization, directivity, and spectrum of the reflected and total bremsstrahlung radiation from a beam of electrons directed toward the photosphere. *The Astrophysical Journal* **215**, 666–676.

Lanzerotti, L. J. (2001a): Space weather effects on communications. In: *Space Storms and Space Weather Hazards, NATO Science Series II, Vol. 38*. (Ed. I. A. Daglis). Boston: Kluwer, pp. 313–334.

Lanzerotti, L. J. (2001b): Space weather effects on technologies. In: *Space Weather* (Eds. P. Song, H. J. Singer and G. L. Siscoe). Washington, D. C: American Geophysical Union, p. 11.

Lanzerotti, L. J., et al. (1992): Heliosphere instrument for spectra, composition and anisotropy at low energies. *Astronomy and Astrophysics Supplement* **92**(2), 349–363.

Lanzerotti, L. J., et al. (1995): Over the southern solar pole: low-energy interplanetary charged particles. *Science* **268**, 1010–1013.

Lapenta, G., Knoll, D. A. (2005): Effect of a converging flow at the streamer cusp on the genesis of the slow solar wind. *The Astrophysical Journal* **624**, 1049–1056.

Lario, D., et al. (2000): Energetic proton observations at 1 and 5 AU: 2. Rising phase of the solar cycle 23. *Journal of Geophysical Research* **105**(A8), 18251–18274.

Larsen, D. E., et al. (1997): Tracing the topology of the October 18–20, 1995, magnetic cloud with 0.1 to 100 keV electrons. *Geophysical Research Letters* **24**, 1911–1914.

Lassen, K., Friis-Christensen, E. (1995): Variability of the solar cycle length during the past five centuries and the apparent association with terrestrial climate. *Journal of Atmospheric and Terrestrial Physics* **57**, 835–845.

Lassen, K., Friis-Christensen, E. (1996): A long-term comparison of sunspot cycle length and temperature change from Zurich observatory. In: *The Global Warming Debate* (Ed. J. Emsley). London: European Science and Environment Forum, pp. 224–232.

Lattes, C. M. G., et al. (1947): Processes involving charged mesons. *Nature* **159**, 694–697.

Lattes, C. M. G., Occhialini, G. P. S., Powell, C. F. (1947): Observations on the tracks of slow mesons in photographic emulsions. *Nature* **160**, 453–456, 492. Reproduced in: HILLAS (1972).

Laut, P. (2003): Solar activity and terrestrial climate: an analysis of some purported correlations. *Journal of Atmospheric and Solar-Terrestrial Physics* **65**, 801–812.

Laut, P., Gundermann, J. (2000): Solar cycle lengths and climate: a reference revisited. *Journal of Geophysical Research* **105**, 27489–27492.

Leamon, R. J., Canfield, R. C., Bleh, M. Z., Pevtsov, A. A. (2003): What is the role of the kink instability in solar coronal eruptions? *Astrophysical Journal (Letters)* **596**, L255–L258.

Leamon, R. J., Canfield, R. C., Pevtsov, A. A. (2002): Properties of magnetic clouds and geomagnetic storms associated with eruption of coronal sigmoids. *Journal of Geophysical Research (Space Physics)* **107**(A9), SSH 1–1.

Lean, J. (1987): Solar uv irradiance variation: A review. *Journal of Geophysical Research* **92**, 839–868.

Lean, J. (1989): Contribution of ultraviolet irradiance variations to changes in the Sun's total irradiance. *Science* **244**, 197–200.

Lean, J. (1991): Variations in the Sun's radiative output. *Reviews of Geophysics* **29**, 505–535.

Lean, J. (1997): The Sun's variable radiation and its relevance for Earth. *Annual Review of Astronomy and Astrophysics* **35**, 33–67.

Lean, J. (2000): Evolution of the Sun's spectral irradiance since the Maunder Minimum. *Geophysical Research Letters* **27**, 2425–2428.

Lean, J., Beer, J., Bradley, R. (1995): Reconstruction of solar irradiance since 1610: implications for climate change. *Geophysical Research Letters* **22**, 3195–3198.

Lean, J., et al. (1995): Correlated brightness variations in solar radiative output from the photosphere to the corona. *Geophysical Research Letters* **22**, 655–658.

Lean, J., et al. (1998): Magnetic sources of the solar irradiance cycle. *Astrophysical Journal* **492**, 390–401.

Lean, J., Foukal, P. (1988): A model of solar luminosity modulation by magnetic activity between 1954 and 1984. *Science* **240**, 906–908.

Lean, J., Rind, D. (1994): Solar variability: Implications for global change. *EOS* **75**(1), 1–6.

Lean, J., Skumanich, A., White, O. (1992): Estimating the Sun's radiative output during the Maunder minimum. *Geophysical Research Letters* **19**, 1591–1594.

Lean, J., Wang, Y.M., Sheeley, N. Jr. (2002): The effect of increasing solar activity on the Sun's total and open magnetic flux during multiple cycles: implications for solar forcing of climate. *Geophysical Research Letters* **29**(24), 2224.

Lee, R. B., et al. (1995): Long-term solar irradiance variability during sunspot cycle 22. *Journal of Geophysical Research* **100**, 1667–1675.

Leer, E., Holzer, T. E. (1980): Energy addition in the solar wind. *Journal of Geophysical Research* **85**, 4681–4688.

Lefebvre, S., Kosovichev, A. G. (2005): Changes in the subsurface stratification of the Sun with the 11-year activity cycle. *The Astrophysical Journal (Letters)* **633**, L149–L152.

Leibacher, J. W., Stein, R. F. (1971): A new description of the solar five-minute oscillation. *Astrophysical Letters* **7**, 191–192.

Leibacher, J. W., Van Driel-Gesztelyi, L., Gizon, L., Cally, P. (Eds., 2008): *SOHO-19/GONG 2007: Seismology of Magnetic Activity. Solar Physics.* – submitted.

Leighton, R. B. (1961): Considerations on localized velocity fields in stellar atmospheres: Prototype – The solar atmosphere. In: *Aerodynamic Phenomena in Stellar Atmospheres. Proceedings of the Fourth Symposium on Cosmical Gas Dynamics. Supplemento del Nuovo Cimento* **22**, 321–325.

Leighton, R. B. (1963): The solar granulation. *Annual Review of Astronomy and Astrophysics* **1**, 19–40.

Leighton, R. B. (1964): Transport of magnetic fields on the sun. *The Astrophysical Journal* **140**, 1547–1562.

Leighton, R. B. (1969): A magneto-kinematic model or the solar cycle. *The Astrophysical Journal* **156**, 1–26.

Leighton, R. B., Noyes, R. W., Simon, G. W. (1962): Velocity fields in the solar atmosphere I. Preliminary report. *The Astrophysical Journal* **135**, 474–499.

Leka, K. D., Canfield, R. C., Mc Clymont, A. N., Van Driel-Gesztelyi, L. (1996): Evidence for current-carrying emerging flux. *The Astrophysical Journal* **462**, 547.

Leka, K. D., Fan, Y. L., Barnes, G. (2005): On the availability of sufficient twist in solar active regions to trigger the kink instability. *The Astrophysical Journal* **626**, 1091–1095.

Lemon, R. J., Mcintosh, S. W. (2007): Empirical solar wind forecasting from the chromosphere. *Astrophysical Journal, Issue* **659**, 738–742.

Lenz, D. D., et al. (1999): Temperature and emission-measure profiles along long-lived solar coronal loops observed with the *Transition Region and Coronal Explorer. The Astrophysical Journal (Letters)* **517**, L155–L158.

Lepping, R. P., et al. (1991): The interaction of a very large interplanetary magnetic cloud with the magnetosphere and with cosmic rays. *Journal of Geophysical Research* **96**, 9425–9438.

Lepri, S. T., Zurbuchen, T. H. (2004): Iron charge state distributions as an indicator of hot ICMEs: Possible sources and temporal and spatial variations during solar maximum. *Journal of Geophysical Research* **109**(A1), A01112.

Letaw, J. R., Silberberg, R., Tsao, C. H. (1987): Radiation hazards on space missions. *Nature* **330**, 709–710.

Le Treut, H, Ghil, M. (1983): Orbital forcing, climatic interactions, and glaciation cycles. *Journal of Geophysical Research* **99**, 5167–5190.

Levine, J. S., Hays, P. B., Walker, J. C. G. (1979): The evolution and variability of atmospheric ozone over geological time. *Icarus* **39**, 295–309.

Levine, R. H. (1974): Acceleration of thermal particles in collapsing magnetic regions. *The Astrophysical Journal* **190**, 447–456.

Levine, R. H., Altschuler, M. D., Harvey, J. W. (1977): Solar sources of the interplanetary magnetic field and solar wind. *Journal of Geophysical Research* **82**, 1061–1065.

Li, B., Li, X., Hu, Y.-Q., Habbal, S. R. (2004): A two-dimensional Alfvén wave-driven solar wind model with proton temperature anisotropy. *Journal of Geophysical Research* **109**, A07103.

Li, H., et al. (2007): Response of the solar atmosphere to magnetic flux emergence from *Hinode* observations. *Publications of the Astronomical Society of Japan* **59**, S643–S648.

Li, X. (2002): Heating in coronal funnels by ion cyclotron waves. *Astrophysical Journal (Letters)* **571**, L67–L70.

Li, X. (2003): Transition region, coronal heating and the fast solar wind. *Astronomy and Astrophysics* **406**, 345–356.

Li, X. (2004): Variations of 0.7–6.0 MeV electrons at geosynchronous orbit as a function of solar wind. *Space Weather* **2**(3), S03006.

Li, X., et al. (1998): The effect of temperature anisotropy on observations of Doppler dimming and pumping in the inner corona. *The Astrophysical Journal (Letters)* **501**, L133.

Li. X., Habbal, S. R. (2005): Hybrid simulation of ion cyclotron resonance in the solar wind. Evolution of velocity distribution functions. *Journal of Geophysical Research* **110**, A10, A10109.

Li, Y., et al. (2008): The solar magnetic field and coronal dynamics of the eruption on 19 May 2007. *Astrophysical Journal Letters* **681**, L37–L39.

Libbrecht, K. G. (1989): Solar p-mode frequency splittings. *The Astrophysical Journal* **336**, 1092–1097.

Libbrecht, K. G., Woodard, M. F. (1990): Solar-cycle effects on solar oscillation frequencies. *Nature* **345**, 779–782.

Libby, W. F. (1955): *Radiocarbon Dating.* Chicago: The University of Chicago Press.

Liewer, P. C., Neugebauer, M., Zurbuchen, T. (2004): Characteristics of active-region sources of solar wind near solar maximum. *Solar Physics* **223**, 209–229.

Liewer, P., et al. (2008): Stereoscopic analysis of STEREO/EUVI observations of May 29, 2007 erupting filament. *Astrophysical Journal* – submitted.

Lighthill, M. J. (1952): On sound generated aerodynamically: I. General theory. *Proceedings of the Royal Society of London A* **211**, 564.

Lighthill, M. J. (1954): On sound generated aerodynamically: II. Turbulence as a source of sound. *Proceedings of the Royal Society of London A* **222**, 1.

Lin, J. (2007): Observational features of large-scale structures as revealed by the catastrophe model of solar eruptions. *Chinese Journal of Astronomy and Astrophysics* **7**(4), 457–476.

Lin, J., et al. (2005): Direct observations of the magnetic reconnection site of an eruption on 2003 November 18. *The Astrophysical Journal* **622**, 1251–1264.

Lin, J., et al. (2007): Features and properties of coronal mass ejection/flare current sheets. *The Astrophysical Journal (Letters)* **658**, L123–L126.

Lin, J., Forbes, T. G. (2000): Effects of reconnection on the coronal mass ejection process. *Journal of Geophysical Research* **105**(A2), 2375–2392.

Lin, J., Forbes, T. G., Isenberg, P. A., Demoulin, P. (1998): The effect of curvature on flux-rope models of coronal mass ejections. *The Astrophysical Journal* **504**, 1006.

Lin, J., Mancuso, S., Vourlidas, A. (2006): Theoretical investigation of the onsets of type II radio bursts during solar eruptions. *The Astrophysical Journal* **649**, 1110–1123.

Lin, J., Raymond, J. C., Van Ballegooijen, A. A. (2004): The role of magnetic reconnection in the observable features of solar eruptions. *The Astrophysical Journal* **602**, 422–435.

Lin, J., Soon, W. (2004): Evolution of morphological features of CMEs deduced from catastrophe model of solar eruptions. *New Astronomy* **9**, 611–628.

Lin, J., Soon, W., Baliunas, S. L. (2003): Theories of solar eruptions: A review. *New Astronomy Reviews* **47**, 53–84.

Lin, R. P. (1970): The emission and propagation of 40 keV solar electrons, I. The relationship of 40 keV electron to energetic proton and relativistic electron emission by the Sun. *Solar Physics* **12**, 266–303.

Lin, R. P. (1985): Energetic solar electrons in the interplanetary medium. *Solar Physics* **100**, 537–561.

Lin, R. P. (1987): Solar particle acceleration and propagation. *Reviews of Geophysics* **25**, 676–684.

Lin, R. P. (2005): Relationship of solar flare accelerated particles to solar energetic particles (SEPs) observed in the interplanetary medium. *Advances in Space Research* **35**, 1857–1863.

Lin, R. P. (2006): Particle acceleration by the Sun: Electrons, hard X-rays/gamma-rays. *Space Science Reviews* **124**, 233–248.

Lin, R. P., et al. (1981): A new component of hard X-rays in solar flares. *Astrophysical Journal (Letters)* **251**, L109–L114.

Lin, R. P., et al. (1984): Solar hard X-ray microflares. *The Astrophysical Journal* **283**, 421–425.

Lin, R. P., et al. (2002): The Reuven Ramaty High-Energy Solar Spectroscopic Imager (RHESSI). *Solar Physics* **210**, 3–32.

Lin, R. P., et al. (2008): The *STEREO* IMPACT Suprathermal Electron (STE) instrument *Space Science Reviews* Issue 1–4, 241–255.

Lin, R. P., Evans, L. G., Fainberg, J. (1973): Simultaneous observations of fast solar electrons and type III radio burst emission near 1 AU. *Astrophysical Letters* **14**, 191–198.

Lin, R. P., Hudson, H. S. (1976): Non-thermal processes in large solar flares. *Solar Physics* **50**, 153–178.

Lindemann, F. A. (1919): Note on the theory of magnetic storms. *Philosophical Magazine* **38**, 669–684.

Lindsey, C., Braun, D. C. (1990): Helioseismic imaging of sunspots at their antipodes. *Solar Physics* **126**, 101–115.

Lindsey, C., Braun, D. C. (1997): Helioseismic holography. *The Astrophysical Journal* **485**, 895–503.

Lindsey, C., Braun, D. C. (2000a): Basic principles of solar acoustic holography. *Solar Physics* **192**, 261–284.

Lindsey, C., Braun, D. C. (2000b): Seismic images of the far side of the Sun. *Science* **287**, 1799–1801.

Lindsey, C., Braun, D. C. (2005a): The acoustic showerglass II. Imaging active region subphoto-spheres. *The Astrophysical Journal* **620**, 1118–1131.

Lindsey, C., Braun, D. C. (2005b): The acoustic showerglass I. Seismic diagnostics of photospheric magnetic fields. *The Astrophysical Journal* **620**, 1107–1117.

Lingenfelter, R. E. (1969): Solar flare optical, neutron, and gamma-ray emission. *Solar Physics* **8**, 341–347.

Lingenfelter, R. E., et al. (1965): High-energy solar neutrons 1. Production in flares. *Journal of Geophysical Research* **70**, 4077–4086.

Lingenfelter, R. E., Ramaty, R. (1967): High energy nuclear reactions in solar flares. In: *High Energy Nuclear Reactions in Astrophysics* (Ed. B. Shen). New York: W. A. Benjamin 1967, pp. 99–158.

Linker, J. A., Mikic, Z. (1995): Disruption of a helmet streamer by photospheric shear. *Astrophysical Journal (Letters)* **38**, L45–L48.

Linsky, J. L. (1980): Stellar chromospheres. *Annual Review of Astronomy and Astrophysics* **18**, 439–488.

Lionello, R., Riley, P., Linker, J. A., Mikic, Z. (2005): The effects of differential rotation on the magnetic structure of the solar corona: Magnetohydrodynamic simulations. *Astrophysical Journal* **625**, 463–473.

Lites, B. W., Hansen, E. R. (1977): Ultraviolet brightenings in active regions as observed from *OSO-8*. *Solar Physics* **55**, 347–358.

Lites, B., et al. (2007): *Hinode* observations of horizontal quiet Sun magnetic flux and the "hidden turbulent magnetic flux". t Astrophysical Journal **634**, 651–662.*Publications of the Astronomical Society of Japan* **59**, S571–S576.

Litwin, C., Rosner, R. (1993): On the structure of solar and stellar coronae – Loops and loop heat transport. *Astrophysical Journal* **412**, 375.

Livingston, W., Wallace, L., White, O. R. (1988): Spectrum line intensity as a surrogate for solar irradiance variations. *Science* **240**, 1765–1767.

Lockwood, G. W., et al. (1984): The photometric variability of solar-type stars. IV. Detection of rotational modulation among Hyades stars. *Publications of the Astronomical Society of the Pacific* **96**, 714–722.

Lockwood, G. W., et al. (1992): Long-term solar brightness changes estimated from a survey of sun-like stars. *Nature* **360**, 653–655.

Lockwood, G. W., Skiff, B. A., Radick, R. R. (1997): The photometric variability of sun-like stars: Observations and results, 1984–1995. *The Astrophysical Journal* **485**, 789–811.

Lockwood, M., Forsyth, R. B., Balogh, A., Mc Comas, D. J. (2004): Open solar flux estimates from near-Earth measurements of the interplanetary magnetic field: Comparison of the first two perihelion passes of the *Ulysses* spacecraft. *Annales Geophysicae* **22**, 1395–1405.

Lockwood, M., Frölich, C. (2007): Recent oppositely directed trends in solar climate forcings and the global mean surface air temperature. *Proceedings of the Royal Society* **463**, 2447–2460.

Lockwood, M., Stamper, R., Wild, M. N. (1999): A doubling of the Sun's coronal magnetic field during the past 100 years. *Nature* **399**, 37–439.

Lockyer, J. N. (1869): Spectroscopic observations of the Sun. III, IV. *Proceedings of the Royal Society* **17**, 350–356, 415–418. Reproduced in: *Early Solar Physics* (Ed. A. J. Meadows). New York: Pergamon Press 1970, pp. 193–202, 233–236.

Lockyer, J. N. (1874): *Contributions to Solar Physics*. London: Macmillan

Lodders, K. (2003a): Abundances and condensation temperatures of the elements. *Meteoritics and Planetary Science* **38**, 5272.

Lodders, K. (2003b): Solar system abundances and condensation temperatures of the elements. *Astrophysical Journal* **591**, 1220–1247.

Lodge, O. (1900): Sun spots, magnetic storms, comet tails, atmospheric electricity, and aurorae. *The Electrician* **46**, 249–250, 287–288.

Lomb, N. R., Andersen, A. P. (1980): The analysis and forecasting of the Wolfsunspot numbers. *Monthly Notices of the Royal Astronomical Society* **190**, 723–732.

Longcope, D. W. (2005): Topological methods for the analysis of solar magnetic fields. *Living Reviews in Solar Physics* **2**, 7.

Longcope, D. W., Brown, D. S., Priest, E. R. (2003): On the distribution of magnetic null points above the solar photosphere. *Physics of Plasmas* **10**, 3321–3334.

Longcope, D. W., et al. (2005): Observations of separator reconnection to an emerging active region. *The Astrophysical Journal* **630**, 596–614.

Longcope, D. W., Kankelborg, C. C. (1999): Coronal heating by collision and cancellation of magnetic elements. *The Astrophysical Journal* **524**, 483–495.

Loomis, E. (1860): On the geographical distribution of auroras in the northern hemisphere. *American Journal of Science and Arts* **30**, 89.

Loomis, E. (1864): The aurora borealis, or polar light: Its phenomena and laws. *Smithsonian Institute Annual Report* 1864 Washington, D.C. US Government Printing Office 1865, 208–248.

Loomis, E. (1866–1871): Notices of auroras extracted from the meteorological journals of Reverend Ezra Stiles. *Transactions of the American Academy of Arts and Sciences* **1**, 155.

Lorius, C., et al. (1985): A 150,000-year climatic record from Antarctic ice. *Nature* **316**, 591–596.

Lorius, C., et al. (1988): Antarctic ice core: CO_2 and climatic change over the last climatic cycle. *EOS* **69**, 681, 683–684.

Lorius, C., et al. (1990): The ice-core record: climate sensitivity and future greenhouse warming. *Nature* **347**, 139–147.

Lovelock, J. E. (1979): *Gaia, a New Look at Life on Earth*. Oxford: Oxford University Press.

Lovelock, J. E. (1988): *The Ages of Gaia*. New York: Norton.

Lovelock, J. E., Margulis, L. (1973): Atmospheric homeostasis by and for the biosphere: the gaia hypothesis. *Tellus* **26**, 1–9.

Lovelock, J. E., Whitfield, M. (1982): Life span of the biosphere. *Nature* **296**, 561–563.

Low, B. C. (1996): Solar activity and the corona. *Solar Physics* **167**, 217–265.

Lu, E. T., Hamilton, R. J. (1991): Avalanches and the distribution of solar flares. *Astrophysical Journal (Letters)* **380**, L89–L92.

Lu, Q. M., Wu, C. S., Wang, S. (2006): The nearly isotropic velocity distributions of energetic electrons in the solar wind. *The Astrophysical Journal* **638**, 1169–1175.

Lucek, E. A., Balogh, A. (1998): The identification and characterization of Alfvénic fluctuations in *Ulysses* data at midlatitudes. *The Astrophysical Journal* **507**, 984–900.

Lugaz, N., Manchester, W. B. IV, Gombosi, T. I. (2005): Numerical simulation of the interaction of two coronal mass ejections from Sun to Earth. *The Astrophysical Journal* **634**, 651–662.

Luhmann, J. G., et al. (2007): *STEREO* IMPACT investigation goals, measurements, and data products overview. *Space Science Reviews*, **136**, No. 1–4, 117–184.

Lundquist, L. L., et al. (2007): Interaction between emerging flux and large-scale loop systems observed with *Hinode* XRT. Presented at the 30 May 2007 Meeting of the Solar Physics Division, American Astronomical Society, Honolulu, Hawaii.

Lüst, R., Schlüter, A. (1954): Kraftfreie magneticfelder. *Zeitschrift für Astrophysik* **34**, 263–282.

Lynch, B. J., et al. (2004): Observable properties of the breakout model for coronal mass ejections. *The Astrophysical Journal* **617**, 589–599.

Lyons, L. R. (1992): Formation of auroral arcs via magnetosphere-ionosphere coupling. *Reviews of Geophysics* **30**, 93–112.

Lyons, L. R., Williams, D. J. (1984): *Quantitative Aspects of Magnetospheric Physics*. Boston, D. Reidel.

Lyot, B. (1930): La couronne solair etudie en dehors des eclipses. *Comptes Rendus de l'Academie des Sciences Paris* **191**, 834.

M

Machado, M. E., et al. (1988): The observed characteristics of flare energy release I. Magnetic structure at the energy release site. *The Astrophysical Journal* **326**, 425–450.

Mackay, D. H., Van Ballegooijen, A. A. (2005): New results in modeling the hemispheric pattern of solar filaments. *Astrophysical Journal (Letters)* **621**, L77–L80.

Macklin, R. J. Jr., Neugebauer, M. M. (Eds., 1966): *The Solar Wind: Proceedings of a Conference held at the California Institute of Technology, Pasadena, California, April 1–4, 1964, and Sponsored by the Jet Propulsion Laboratory.* Oxford: Pergamon Press.

Maclennan, C. G., Lanzerotti, L. J., Gold, R. E. (2003): Low energy charged particles in the high latitude heliosphere: Comparing solar maximum and solar minimum. *Geophysical Research Letters* **30**(19), ULY 7–1.

Mac Low, M. -M., McCray, R. (1988): Superbubbles in disk galaxies. *The Astrophysical Journal* **324**, 776–785.

Mac Neice, et al. (2004): A numerical study of the breakout model for coronal mass ejection initiation. *The Astrophysical Journal* **614**, 1028–1041.

Mac Queen, R. M. (1980): Coronal transients: A summary. *Philosophical Transactions of the Royal Society (London)* **A297**, 605–620.

Mac Queen, R. M., et al. (1974): The outer solar corona as observed from *Skylab*: Preliminary results. *Astrophysical Journal (Letters)* **187**, L85–L88.

Mac Queen, R. M., et al. (1976): Initial results from the high altitude observatory white light coronagraph on *Skylab* – a progress report. *Philosophical Transactions of the Royal Society (London)* **A281**, 405–414.

Maher, K. A., Stevenson, D. J. (1988): Impact frustration of the origin of life. *Nature* **331**, 612–614.

Mairan, J. J.: *Traité Physique et Historique de l'Aurorae Boréale,* Paris: L'Imprimerie Royale 1733 (1st edition), 1754 (2nd revised edition).

Mall, U., Fichtner, H., Rucinski, D. (1996): Interstellar atom and pick-up ion fluxes along the *Ulysses* flight-path. *Astronomy and Astrophysics* **316**, 511–518.

Maltoni, M., et al. (2003): Status of three-neutrino oscillations after the SNO-salt data. *Physical Review* **D68**, 113010–113028.

Manchester, W. B., et al. (2004): Modeling a space weather event from the Sun to the Earth: CME generation and interplanetary propagation. *Journal of Geophysical Research* **109**, A02107.

Manchester, W. B., et al. (2005): Coronal mass ejection shock and sheath structures relevant to particle acceleration. *Astrophysical Journal* **622**, 1225–1239.

Mancuso, S., Spangler, S. R. (1999): Coronal Faraday rotation observations: measurements and limits on plasma inhomogeneities. *The Astrophysical Journal* **525**, 195–208.

Mandrini, C. H., Démoulin, P, Klimchuk, J. A. (2000): Magnetic field and plasma scaling laws: Their implications for corona heating models. *The Astrophysical Journal* **530**, 999–1015.

Mann, M. E. (2000): Climate change: lessons for a new millennium. *Science* **289**, 253–254.

Mann, M. E., Bradley, R. S., Hughes, M. K. (1998): Global-scale temperature patterns and climate forcing over the past six centuries. *Nature* **392**, 779–787.

Mann, M. E., Bradley, R. S., Hughes, M. K. (1999): Northern hemisphere temperatures during the past millennium: inferences, uncertainties, and limitations. *Geophysical Research Letters* **26**, 759–762.

Mann, M. E., Park, J., Bradley, R. S. (1995): Global interdecadal and century- scale climate oscillations during the last five centuries. *Nature* **378**, 266–270.

Manoharan, P. K., et al. (1996): Evidence for large-scale solar magnetic reconnection from radio and X-ray measurements. *Astrophysical Journal (Letters)* **468**, L73–L76.

Marconi, G. (1899): Wireless telegraphy. *Proceedings of the Institution of Electrical Engineers* **28**, 273.

Margulis, L., Lovelock, J. E. (1974): Biological modulation of the Earth's atmosphere. *Icarus* **21**, 471–489.

Mariska, J. T. (1986): The quiet solar transition region. *Annual Review of Astronomy and Astrophysics* **24**, 23–28.

Mariska, J. T. (1992): *The Solar Transition Region*. New York: Cambridge University Press.

Mariska, J. T. (1994): Flare plasma dynamics observed with the Yohkoh Bragg crystal. *The Astrophysical Journal* **434**, 756.

Mariska, J. T., Doschek, G. A., Bentley, R. D. (1993): Flare plasma dynamics observed with the Yohkoh Bragg crystal spectrometer I. Properties of the Ca XIX resonance line. *Astrophysical Journal* **419**, 418–425.

Mariska, J. T., Feldman, U., Doschek, G. A. (1978): Measurements of extreme- ultraviolet emission-line profiles near the solar limb. *Astrophysical Journal* **226**, 698–705.

Markovskii, S. A., et al. (2006): Dissipation of the perpendicular turbulent cascade in the solar wind. *Astrophysical Journal* **639**, 1177–1185.

Markovskii, S. A., Hollweg, J. V. (2004): Intermittent heating of the solar corona by heat flux-generated ion cyclotron waves. *Astrophysical Journal* **609**, 1112–1122.

Markson, R. (1978): Solar modulation of atmospheric electrification and possible implications for the Sun-weather relationship. *Nature* **273**, 103–109.

Marsch, E. (1991): MHD turbulence in the solar wind. In: *Physics of the Inner Heliospere, Vol. II* (Eds. R. Schwenn and E. Marsch). Heildelberg: Springer-Verlag, pp. 159–241.

Marsch, E. (1997): Working group 3: Coronal hole structure and high speed solar wind. In: *The Corona and Solar Wind Near Minimum Activity. Proceedings of the Fifth SOHO Workshop. ESA SP-404*. Noordwijk: ESA Publications Division, pp. 135–140.

Marsch, E. (2006): Kinetic physics of the solar corona and solar wind. *Living Reviews in Solar Physics* **3**, 1.

Marsch, E., Goertz, C. K., Richter, K. (1982): Wave heating and acceleration of solar wind ions by cyclotron resonance. *Journal of Geophysical Research* **87(A7)**, 5030–5044.

Marsch, E., Tu, C.-Y. (1990): On the radial evolution of MHD turbulence in the inner heliosphere. *Journal of Geophysical Research* **95**, 8211–8229.

Marsch, E., Tu, C. -Y. (1997): The effects of high-frequency Alfvén waves on coronal heating and solar wind acceleration. *Astronomy and Astrophysics* **319**, L17–L20.

Marsch, E., Tu, C.-Y. (1997): Solar wind and chromospheric network. *Solar Physics* **176**, 87–106.

Marsch, E., Tu, C.-Y. (2001): Evidence for pitch angle diffusion of solar wind protons in resonance with cyclotron waves. *Journal of Geophysical Research* **106**, 8357.

Marsden, R. G. (Ed., 1995): *The High Latitude Heliosphere*. New York: Kluwer Academic Publishers.

Marsden, R. G. (Ed., 2001): The 3-D heliosphere at solar maximum. *Space Science Reviews* **97**, 1–429. Reprinted by: Kluwer Academic Publishers, Dordrecht, the Netherlands.

Marsden, R. G., et al. (1987): *ISEE 3* observations of low-energy proton bidirectional events and their relation to isolated interplanetary magnetic structures. *Journal of Geophysical Research* **92**, 11009–11019.

Marsden, R. G., et al. (1996): *Ulysses* at high heliographic latitudes: An introduction. *Astronomy and Astrophysics* **316**, 279–286.

Marsden, R. G., Smith, E. J. (1996): *Ulysses*: Solar sojourner. *Sky and Telescope* **91**, 24–30, March.

Marsh, K. A. (1978): Ephemeral region flares and the diffusion of the network. *Solar Physics* **59**, 105–113.

Marsh, K. A., Hurford, G. J. (1980): VLA maps of solar bursts at 15 and 23 GHz with arcsecond resolution. *Astrophysical Journal (Letters)* **240**, L111–L114.

Marsh, M. S., Walsh, R. W., De Moortel, I., Ireland, J. (2003): Joint observations of propagating oscillations with *SOHO*/CDS and *TRACE*. *Astronomy and Astrophysics* **404**, L37–L41.

Marsh, N. D., Svensmark, H. (2000): Low cloud properties influenced by cosmic rays. *Physical Review Letters* **85**, 5004–5007.

Marsh, N. D., Svensmark, H. (2003): Solar influence on Earth's climate. *Space Science Reviews* **107**, 317–325.

Martens, P. C. H., Cauffman, D. (Eds., 2002): *Multi-wavelength Observations of Coronal Structure and Dynamics Yohkoh 10th Anniversary Meeting*. New York: Pergamon.

Martens, P. C. H., Kuin, N. P. M. (1989): A circuit model for filament eruptions and two-ribbon flares. *Solar Physics* **122**, 263–302.

Martens, P. C. H., Zwaan, C. (2001): Origin and evolution of filament-prominence systems. *Astrophysical Journal* **558**, 872–877.

Martin, S. F. (1998): Conditions for the formation and maintenance of filaments (invited review). *Solar Physics* **183**, 107–137.

Martin, S. F., Bilimoria, R., Tracadas, P. W. (1994): Magnetic field configurations basic to filament channels and filaments. In: *Solar Surface Magnetism* (Eds. R. J. Rutten and C. J. Schrijver). NATO Series C433, Kluwer Academic Publisher, Dordrecht the Netherlands, p. 303.

Martinson, D. G., et al. (1987): Age dating and the orbital theory of the ice ages: Development of a high-resolution 0 to 300,000–year chronostratigraphy. *Quaternary Research* **27**, 1–29.

Martyn, D. F. (1946): Temperature radiation from the quiet Sun in the radio spectrum. *Nature* **158**, 632–633.

Mason, D., et al. (2006): Flares, magnetic fields, and subsurface vorticity: a survey of GONG and MDI data. *Astrophysical Journal* **645**, 1543–1553.

Mason, G. M. (2001): Heliospheric lessons for galactic cosmic-ray acceleration. *Space Science Reviews* **99**, 119–133.

Mason, G. M., et al. (1986): The heavy-ion compositional signature in He-3-rich solar particle events. *Astrophysical Journal* **303**, 849–860.

Mason, G. M., et al. (1999): Origin, injection and acceleration of CIR particles: observations report of working group 6. *Space Science Reviews* **89**, 327–367.

Mason, G. M., et al. (2007): The Suprathermal Ion Telescope (SIT) for the IMPACT/SEP investigation. *Space Science Reviews* **136**, No. 1–4, 257–284.

Mason, G. M., Mazur, J. E., Dwyer, J. R. (1999): 3He enhancements in large solar energetic particle events. *Astrophysical Journal (Letters)* **525**, L133–L136.

Mason, G. M., Sanderson, T. R. (1999): CIR associated energetic particles in the inner and middle heliosphere. *Space Science Reviews* **89**, 77–90.

Masuda, S. (1994): *Hard X-ray Sources and the Primary Energy Release Site in Solar Flares.* Ph. D. Thesis. Mitaka: University of Tokyo, the Yohkoh HXT group, National Astronomical Observatory.

Masuda, S., et al. (1994): A loop-top hard X-ray source in a compact solar flare as evidence for magnetic reconnection. *Nature* **371**, 495–497.

Masuda, S., et al. (1995): Hard X-ray sources and the primary energy-release site in solar flares. *Publications of the Astronomical Society of Japan* **47**, 677–689.

Masuda, S., Kosugi, T., Tsuneta, S., Hara, H. (1996): Discovery of a loop-top hard X-ray source in impulsive solar flares. *Advances in Space Research* **17**, 4–5, 63–66.

Matsuzaki, K., et al. (2007): Hot and cool loops composing the corona of the quiet Sun. *Publications of the Astronomical Society of Japan* **59**, S683–S689.

Matthaeus, W. H., et al. (2005): Spatial correlation of solar-wind turbulence from two point measurements. *Physical Review Letters* **95**, 231101.

Matthaeus, W. H., Lamkin, S. L. (1986): Turbulent magnetic reconnection. *Physics of Fluids* **29**, 2513.

Matthes, K., et al. (2003): Improved 11-year solar signal in the Freie Universität Berlin climate middle atmosphere model (fub-cmam). *Journal of Geophysical Research* **109**, D06101.

Mattok, C. (Ed., 1992): *SOHO-1: Coronal Steamers, Coronal Loops, and Coronal and Solar Wind Compositions.* ESA SP-348 1992 Noordwijk, the Netherlands.

Maunder, E. W. (1890): Professor Spoerer's researches on sunspots. *Monthly Notices of the Royal Astronomical Society* **50**, 251–252.

Maunder, E. W. (1894): A prolonged sunspot minimum. *Knowledge* **17**(106), 173–176.

Maunder, E. W. (1905): Magnetic disturbances, 1882 to 1903, as recorded at the royal observatory, greenwich, and their association with sun-spots. *Monthly Notices of the Royal Astronomical Society* **65**, 2–34.

Maunder, E. W. (1922): The prolonged sunspot minimum, 1645–1715. *Journal of the British Astronomical Association* **32**, 140–145.

Maxwell, A., Swarup, G. (1958): A new spectral characteristic in solar radio emission. *Nature* **181**, 36–38.

Maxwell, J. C. (1860): Illustrations of the dynamical theory of gases: Part I. On the motions and collisions of perfectly elastic spheres. *Philosophical Magazine* **19**, 19.

Mazur, J. E., et al. (1992): The energy spectra of solar flare hydrogen, helium, oxygen, and iron: Evidence for stochastic acceleration. *Astrophysical Journal* **401**, 398–410.

Mazur, J. E., et al. (2000): Interplanetary magnetic field line mixing deduced from impulsive solar flare properties. *Astrophysical Journal (Letters)* **532**, L79–L82.

Mazur, J. E., et al. (2002): Charge states of energetic particles from co-rotating interaction regions as constraints on their source. *Astrophysical Journal* **566**, 555–561.

Mc Allister, A. H., Crooker, N. U. (2002): Coronal mass ejections, corotating interaction regions, and geomagnetic storms. In: *Coronal Mass Ejections. Geophysical Monograph 99* (Eds. N. Crooker, J. A. Joselyn and J. Feynman). Washington, DC: American Geophysical Union 1997, pp. 279–290.

Mc Ateer, et al. (2004): Ultraviolet oscillations in the chromosphere of the quiet Sun. *Astrophysical Journal* **602**, 436–445.

Mc Comas, D. J., et al. (1998a): *Ulysses'* rapid crossing of the polar coronal hole boundary. *Journal of Geophysical Research* **103**, 1955.

Mc Comas, D. J., et al. (1998b): *Ulysses'* return to the slow solar wind. *Geophysical Research Letters* **25**, 1–4.

Mc Comas, D. J., et al. (2000): Solar wind observations over *Ulysses'* first full polar orbit. *Journal of Geophysical Research* **105**(A5), 10419–10434.

Mc Comas, D. J., et al. (2002): *Ulysses'* second fast-latitude scan: Complexity near solar maximum and the reformation of polar coronal holes. *Geophysical Research Letters* **29**(9), 4–1.

Mccomas, D. J., et al. (2003): The three-dimensional solar wind around solar maximum. *Geophysical Research Letters* **30**(10), 1517.

McComas, D. J., et al. (2007): Understanding coronal heating and solar wind acceleration: Case for in situ near-Sun measurements. *Reviews of Geophysics* **45**, 1–26.

Mc Crea, W. H. (1929): The hydrogen chromosphere. *Monthly Notices of the Royal Astronomical Society* **89**, 483–497.

Mc Crea, W. H. (1956): Shock waves in steady radial motion under gravity. *Astrophysical Journal* **124**, 461–468.

Mc Donald, A. B., et al. (2001): First neutrino observations from the Sudbury Neutrino Observatory. *Nuclear Physics B – Proceedings Supplements* **91**, 21–28.

Mc Hargue, L. R., Damon, P. E. (1991): The global beryllium 10 cycle. *Reviews of Geophysics* **29**, 141–158.

McIntosh, P. S., Dryer, M. (Eds., 1970): *Progress in Astronautics and Aeronautics, Vol. 30.* Cambridge: MIT Press, Preface.

Mcintosh, S. W., et al. (2001): An observational manifestation of magnetoatmospheric waves in internetwork regions of the chromosphere and transition region. *Astrophysical Journal (Letters)* **548**, L237–L241.

Mcintosh, S. W., Fleck, B., Tarbell, T. D. (2004): Chromospheric oscillations in an equatorial coronal hole. *Astrophysical Journal (Letters)* **609**, L95–L98.

McIntosh, S. W., Jefferies, S. M. (2006): Observing the modification of the acoustic cutoff frequency by field inclination angle. *The Astrophysical Journal (Letters)* **647**, L77–L81.

Mcintosh, S. W., Judge, P. O. (2001): On the nature of magnetic shadows in the solar chromosphere. *Astrophysical Journal* **561**, 420–426.

Mcintosh, S. W., Leamon, R. J. (2005): Is there a chromospheric footprint of the solar wind? *Astrophysical Journal (Letters)* **624**, L117–L120.

McKenzie, D. E. (2002): Signatures of reconnection in eruptive flares. In *Multi Wavelength Observations of Coronal Structure and Dynamics* (Eds. P. C. H. Martens and D. P. Cauffman).

McKenzie, D. E., Canfield, R. C. (2008): *Hinode* XRT observations of a long-lasting coronal sigmoid. *Astronomy and Astrophysics Letters*. **481**, L65–L68.

Mc Kenzie, D. E., Hudson, H. S. (1999): X-ray observations of motions and structure above a solar flare arcade. *Astrophysical Journal (Letters)* **519**, L93–L96.

Mc Kenzie, J. F., Axford, W. I., Banaszkiewicz, M. (1997): The fast solar wind. *Geophysical Research Letters* **24**(22), 2877–2880.

Mc Kenzie, J. F., Banaszkiewicz, M., Axford, W. I. (1995): Acceleration of the high speed solar wind. *Astronomy and Astrophysics* **303**, L45–L46.

Mc Kenzie, J. F., Bornatici, M. (1974): Effect of sound waves, Alfvén waves, and heat flow on interplanetary shock waves. *Journal of Geophysical Research* **79**, 4589–4594.

Mc Kibben, R. B., et al. (1996): Observations of galactic cosmic rays and the anomalous helium during *Ulysses* passage from the south to the north solar pole. *Astronomy and Astrophysics* **316**, 547–554.

Mc Kibben, R. B., et al. (2003): *Ulysses* COSPIN observations of cosmic rays and solar energetic articles from the South Pole to the North Pole of the Sun during solar maximum. *Annales Geophysicae* **21**, 1217–1228.

Mc Kibben, R. B., Lopate, C., Zhang, M. (2001): Simultaneous observations of solar energetic particle events by *IMP 9* and the *Ulysses* COSPIN high energy telescope at high solar latitudes. *Space Science Reviews* **97**, 257–262.

Mc Lean, D. J. (2005): Metrewave solar radio bursts. In: *Solar Radiophysics* (Eds. D. J. Mc Lean and N. R. Labrum). Cambridge: Cambridge University Press 1985, pp. 37–52.

Mc Lean, D. J., Labrum, N. R. (Eds., 1985): *Solar Radiophysics*. Cambridge: Cambridge University Press.

Mc Lennan, J. C., Shrum, G. M. (1926): On the origin of the auroral green line 5577 Å and other spectra associated with the aurora borealis. *Proceedings of the Royal Society (London)* **A108**, 501.

Mc Mulllin, D. R., et al. (2004): Heliospheric conditions that affect the interstellar gas inside the heliosphere. *Astronomy and Astrophysics* **426**, 885–895.

McDonald, A. B. (2001): First results from the Sudbury Neutrino Observatory explain the missing solar neutrinos and reveal new neutrino properties. *News Release of the Sudbury Neutrino Observatory on 18 June 2001*. Sudbury Neutrino Observatory, Sudbury, Ontario, Canada.

McDonald, A. B. (2005): Sudbury Neutrino Observatory results. *Physica Scripta T* **121**, 29–32.

Meadows, A. J. (1970): *Early Solar Physics*. Oxford: Pergamon Press.

Meadows, A. J. (1975): A 100 years of controversy over sunspots and weathers. *Nature* **256**, 95–97.

Meadows, A. J., Kennedy, J. E. (1982): The origin of solar-terrestrial studies. *Vistas in Astronomy* **25**, 419–426.

Melrose, D. B. (1995): Current paths in the corona and energy release in solar flares. *Astrophysical Journal* **451**, 391–401.

Melrose, D. B. (1997): A solar flare model based on magnetic reconnection between current-carrying loops. *Astrophysical Journal* **486**, 521–533.

Mewaldt, R. A. (2006): Solar energetic particle composition, energy spectra and space weather. *Space Science Reviews* **124**, 303–316.

Mewaldt, R. A., et al. (2005): Solar-particle energy spectra during the large events of October–November 2003 and January 2005. *29th International Cosmic Ray Conference Pune*, **1**, 101–104.

Meyer, J.-P. (1981): A tentative ordering of all available solar energetic particle abundance observations. *Proceedings of the 17th International Cosmic Ray Conference* **3**, 145–152.

Meyer, J.-P. (1985a): Solar-stellar outer atmospheres and energetic particles, and galactic cosmic rays. *Astrophysical Journal Supplement Series* **57**, 173–204.

Meyer, J.-P. (1985b): The baseline composition of solar energetic particles. *Astrophysical Journal Supplement Series* **57**, 151–171.

Meyer, J.-P. (1991): Diagnostic methods for coronal abundances. *Advances in Space Research* **11**(1), 269–280.

Meyer, P., Parker, E. N., Simpson, J. A. (1956): Solar cosmic rays of February, 1956 and their propagation through interplanetary space. *Physical Review* **104**, 768–783.

Meyer, P., Simpson, J. A. (1954): Changes in amplitude of the cosmic-ray 27–day intensity variation with solar activity. *Physical Review* **96**, 1085–1088.

Meyer, P., Simpson, J. A. (1955): Changes in the low-energy particle cutoff and primary spectrum of cosmic radiation. *Physical Review* **99**, 1517–1523.

Meyer, P., Simpson, J. A. (1957): Changes in the low-energy particle cutoff and primary spectrum of cosmic rays. *Physical Review* **106**, 568–571.

Meyer, P., Vogt, R. (1961): Electrons in the primary cosmic radiation. *Physical Review Letters* **6**, 193–196.

Meyer, P., Vogt, R. (1962): High-energy electrons of solar origin. *Physical Review Letters* **8**, 387–389.

Meyer-Vernet, N. (1999): How does the solar wind blow? A simple kinetic model. *European Journal of Physics* **20**, No. 3, 167–176.

Michel, F. C., Dessler, A. J. (1965): Physical significance of inhomogeneities in polar cap absorption events. *Journal of Geophysical Research* **70**, 4305–4311.

Miesch, M. S. (2003): Numerical modeling of the solar tachocline II. Forced turbulence with imposed shear. *Astrophysical Journal* **586**, 663–684.

Miesch, M. S. (2005): Large-scale dynamics of the convection zone and tachocline. *Living Reviews in Solar Physics* **2**, 1.

Miesch, M. S., Gilman, P. A. (2004): Thin-shell magnetohydrodynamic equations for the solar tachocline. *Solar Physics* **220**, 287–305.

Mikheyev, S. P., Smirnov, A. Y. (1985): Resonance enhancement of oscillations in matter and solar neutrino spectroscopy. *Soviet Journal of Nuclear Physics* **42**, 913–917.

Mikic, Z., Lee, M. A. (2006): An introduction to theory and models of CMEs, shocks and solar energetic particles *Space Science Reviews* **123**, 57–80.

Mikic, Z., Linker, J. A. (1994): Disruption of coronal magnetic field arcades. *Astrophysical Journal* **430**, 898–912.

Milankovitch, M. M. (1920): *Théorie Mathématique des Phénomenes Thermiques Produits par la Radiation Solarie*. Académie Yugoslave des Sciences et des Arts de Zagreb. Paris: Gauthier-Villars.

Milankovitch, M. M. (1941): Kanon der Erdbestrahlung une sei Eiszeitenproblem (Canon of insolation and the ice-age problem). *Königliche Serbische Akademie, Beograd, Publication 132, Section of Mathematics and Natural Science* **33**, 1941. (English translation by the Israel Program for Scientific Translations and published for the U. S. Department of Commerce and the National Science Foundation, Jerusalem 1970).

Miller, J. A., et al. (1997): Critical issues for understanding particle acceleration in impulsive solar flares. *Journal of Geophysical Research* **102**, 14631–14659.

Millikan, R. A. (1926): High frequency rays of cosmic origin. *Proceedings of the National Academy of Sciences* **12**, 48–55.

Millikan, R. A., Cameron, G. H. (1926): High frequency rays of cosmic origin III. Measurements in snow-fed lakes at high altitudes. *Physical Review* **28**, 851–868.

Minton, D. A., Malhotra, R. (2007): Assessing the massive young Sun hypothesis to solve the warm young Earth puzzle. *Astrophysical Journal* **660**, 170–1706.

Miralles, M. P., Cranmer, S. R., Kohl, J. L. (2001): Ultraviolet coronagraph spectrometer observations of a high-latitude coronal hole with high oxygen temperatures and the next solar cycle polarity. *Astrophysical Journal (Letters)* **560**, L193–L196.

Miralles, M. P., Cranmer, S. R., Kohl, J. L. (2004): Low-latitude coronal holes during solar maximum. *Advances in Space Research* **33**, 696.

Miralles, M. P., Cranmer, S. R., Kohl, J. L. (2005): Solar cycle variations of coronal hole properties. *American Geophysical Union Spring Meeting*, Abstract SP51B-07.

Miralles, M. P., et al. (2001): Comparison of empirical models for polar and equatorial coronal holes. *Astrophysical Journal (Letters)* **549**, L257–L260.

Miranda, O. G., et al. (2004): Constraining the neutrino magnetic moment with antineutrinos from the Sun. *Physical Review Letters* **93**, 051304.

Mitalas, R., Sills, K. R. (1992): On the photon diffusion time scale for the Sun. *The Astrophysical Journal* **401**, 759.

Mitchell, J. F. B. (1989): The "greenhouse" effect and climate change. *Reviews of Geophysics* **27**, 115–139.

Mizuno, D. R., et al. (2005): Very high altitude aurora observations with the solar mass ejection imager. *Journal of Geophysical Research* **110**, A07230.

Möbius, E., et al. (1985): Direct observation of He ($+$) pick-up ions of interstellar origin in the solar wind. *Nature* **318**, 426–429.

Möbius, E., et al. (2002): Charge state of energetic (about 0.5 MeV/nucleon) ions in co-rotating interaction regions at 1 AU and implications on source populations. *Geophysical Research Letters* **29**(2), 1016.

Möbius, E., et al. (2004): Synopsis of interstellar He parameters from combined neutral gas, pickup ion and UV scattering observations and related consequences. *Astronomy and Astrophysics* **426**(3), 897–907.

Molina, M. J., Rowland, F. S. (1974): Stratospheric sink for chlorofluoromethanes. chlorine atomic-atalyzed destruction of ozone. *Nature* **249**, 810–812.

Montgomery, M. D., et al. (1974): Solar wind electron temperature depressions following some interplanetary shock waves: Evidence for magnetic merging? *Journal of Geophysical Research* **79**, 3103–3123.

Moon, Y.-J., et al. (2002): A statistical study of two classes of coronal mass ejections. *Astrophysical Journal* **581**, 694–702.

Moon, Y.-J., et al. (2007): Hinode sp vector magnetogram of AR 10930 and its cross comparison with MDI. *Publications of the Astronomical Society of Japan* **59**, S625–S630.

Moore, R. L. (1988): Evidence that magnetic energy shedding in solar filament eruptions is the drive in accompanying flares and coronal mass ejections. *Astrophysical Journal* **324**, 1132–1137.

Moore, R. L., et al. (1980): The thermal X-ray flare plasma. In: *Solar Flares. Skylab Solar Workshop II* (Ed. P. A. Sturrock). Boulder, Colorado: Colorado Associated University Press 1980, pp. 341–409.

Moore, R. L., Schmieder, B., Hathaway, D. H., Tarbell, T. D. (1997): 3–d magnetic field configuration late in a large two-ribbon flare. *Solar Physics* **176**, 153–169.

Moore, R. L., Sterling, A. C. (2006): Initiation of coronal mass ejections. In *Solar* Eruptions and Energetic Particles. Geophysical Monograph Series 165. (Ed. N. Gopalswamy, R. Mewaldt and J. Torsti). Washington: American Geophysical Union.

Moore, R. L., Sterling, A. C. (2007): The coronal-dimming footprint of a streamer puff coronal mass ejection: Confirmation of the magnetic-arch-blowout scenario. *Astrophysical Journal* **661**, 543–550.

Moore, R. L., Sterling, A. C., Hudson, H. S., Lemen, J. R. (2001): Onset of the magnetic explosion in solar flares and coronal mass ejections. *Astrophysical Journal* **552**, 833–848.

Moore, R. L., Sterling, A. C., Suess, S. T. (2007): The width of a solar coronal mass ejection and the source of the driving magnetic explosion: A test of the standard scenario for CME production. *Astrophysical Journal* **668**, 1221–1231.

Moore, R. L., Suess, S. T., Musielak, Z. E., An, C. -H. (1991): Alfvén wave trapping, network microflaring and heating in coronal holes. *The Astrophysical Journal* **378**, 347–359.

Moran, T. G., Davila, J. M. (2004): Three-dimensional polarimetric imaging of coronal mass ejections. *Science* **305**, 66–70.

Moreton, G. E. (1960): Hα observations of flare-initiated disturbances with velocities ≈ 1000 km/sec. *The Astronomical Journal* **65**, 494–495.

Moreton, G. E. (1961): Fast-moving disturbances on the Sun. *Sky and Telescope* **21**, 145–147.

Moreton, G. E. (1964): Ha shock wave and winking filaments with the flare of 20 September 1963. *The Astronomical Journal* **69**, 145.

Moreton, G. E., Severny, A. B. (1968): Magnetic fields and flares in the region cmp 20 September 1963. *Solar Physics* **3**, 282–297.

Morrison, P. (1954): Solar-connected variations of the cosmic rays. *Physical Review* **95**, 646.

Morrison, P. (1958): On gamma-ray astronomy. *Nuovo Cimento* **7**, 858–865.

Moses, D. et al. (1997): EIT observations of the extreme ultraviolet Sun. *Solar Physics* **175**, 571–599. Reprinted in: *The First Results From SOHO* (Eds. B. Fleck and Z. Svestka). Boston: Kluwer Academic Publishers 1997, pp. 571–599.

Müller-Mellin, R., et al. (1995): COSTEP – Comprehensive Suprathermal and Energetic Particle Analyser. *Solar Physics* **162**, 483–504.

Mullen, E. G., et al. (1991): A double-peaked inner radiation belt: Cause and effect as seen on CRRES. *IEEE Transactions on Nuclear Science* **38**, 1713–1717.

Munro, R. H., et al. (1979): The association of coronal mass ejection transients with other forms of solar activity. *Solar Physics* **61**, 201–215.

Munro, R. H., Jackson, B. V. (1977): Physical properties of a polar coronal hole from 2 to 5 solar radii. *The Astrophysical Journal* **213**, 874–886.

Murphy, N., Smith, E. J., Schwadron, N. A. (2002): Strongly underwound magnetic fields in co-rotating interaction regions: Observations and implications. *Geophysical Research Letters* **29**(22), 2066.

Murphy, R. J., Dermer, C. D., Ramaty, R. (1987): High-energy processes in solar flares. *Astrophysical Journal Supplement* **63**, 721–748.

Murphy, R. J., et al. (1985): Solar flare gamma-ray line spectroscopy. *Proceedings 19th International Cosmic Ray Conference (La Jolla)* **4**, 253–256.

Murphy, R. J., et al. (1991): Solar abundances from gamma-ray spectroscopy: Comparisons with energetic particle, photospheric, and coronal abundances. *The Astrophysical Journal* **371**, 793–803.

Murphy, R. J., Ramaty, R. (1984): Solar flare neutrons and gamma rays. *Advances in Space Research* **4**(7), 127–136.

N

Nagashima, K., et al. (2007): Observations of sunspot oscillations in G band and Ca II H line with Solar Optical Telescope on Hinode. *Publications of the Astronomical Society of Japan* **59**, S631–S636.

Nakamura, M., et al. (1998): Reconnection event at the dayside magnetopause on January 10, 1997. *Geophysical Research Letters* **25**, 2529–2532.

Nakariakov, V. M., et al. (1999): TRACE observations of damped coronal loop oscillations: Implications for coronal heating. *Science* **285**, 862–864.

Nakariakov, V. M., Verwichte, E. (2005): Coronal waves and oscillations. *Living Reviews in Solar Physics* **2**, 3.

Narain, U., Ulmschneider, P. (1990): Chromospheric and coronal heating mechanisms. *Space Science Reviews* **54**, 377–445.

Narain, U., Ulmschneider, P. (1996): Chromospheric and coronal heating mechanisms II. *Space Science Reviews* **75**, 453–509.

Nash, A. G., Sheeley, N. R., Wang, Y. -M. (1988): Mechanisms for the rigid rotation of coronal holes. *Solar Physics* **117**, 359–389.

National Research Council: Solar influences on global change. Washington, DC: National Academy Press 1994.

Neftel, A., et al. (1982): Ice core sample measurements give atmospheric CO_2 content during the past 40,000 years. *Nature* **295**, 220–223.

Neftel, A., et al. (1985): Evidence from polar ice cores for the increase in atmospheric CO_2 in the past two centuries. *Nature* **315**, 45–47.

Neftel, A., Oeschger, H., Suess, H. E. (1981): Secular non-random variations of cosmogenic carbon-14 in the terrestrial atmosphere. *Earth and Planetary Science Letters* **56**, 127–147.

Neidig, D. F. (1989): The importance of solar white-light flares. *Solar Physics* **121**, 261–269.

Nerney, S., Suess, S. T. (2005): Stagnation flow in thin streamer boundaries. *The Astrophysical Journal* **624**, 378–391.

Nesme-Ribes, E., Baliunas, S. L., Sokoloff, D. (1996): The stellar dynamo. *Scientific American* **275**, 46–52 – August.

Nesme-Ribes, E., et al. (1993): Solar dynamics and its impact on solar irradiance and the terrestrial climate. *Journal of Geophysical Research* **98**, 18923–18935.

Nesme-Ribes, E., Mangeney, A. (1992): On a plausible physical mechanism linking the Maunder Minimum to the Little Ice Age. *Radiocarbon* **34**(2), 263–270.

Nesme-Ribes, E., Sokoloff, D., Sadourny, R.: Solar rotation, irradiance changes and climate. In: *The Sun as a Variable Star* (Eds. J. Pap, H. Hudson and S. Solanki). New York: Cambridge University Press 1994, pp. 244–251.

Ness, N. F. (1965): The Earth's magnetic tail. *Journal of Geophysical Research* **70**, 2989–3005.

Ness, N. F. (1968): Observed properties of the interplanetary plasma. *Annual Review of Astronomy and Astrophysics* **6**, 79–114.

Ness, N. F. (1996): Pioneering the swinging 1960s into the 1970s and 1980s. *Journal of Geophysical Research* **101**(A5), 10497–10509.

Ness, N. F., Hundhausen, A. J., Bame, S. J. (1971): Observations of the interplanetary medium: Vela 3 and IMP 3, 1965–1967. *Journal of Geophysical Research* **76**, 6643–6660.

Ness, N. F., Scearce, C. S., Seek, J. B. (1964): Initial results of the *IMP 1* magnetic field experiment. *Journal of Geophysical Research* **69**, 3531–3569.

Ness, N. F., Wilcox, J. M. (1964): Solar origin of the interplanetary magnetic field. *Physical Review Letters* **13**, 461–464.

Ness, N. F., Wilcox, J. M. (1966): Extension of the photospheric magnetic field into interplanetary space. *Astrophysical Journal* **143**, 23–31.

Neugebauer, M. (1981): Observations of solar-wind helium. *Fundamentals of Cosmic Physics* **7**, 131–199.

Neugebauer, M. (1992): Knowledge of coronal heating and solar wind acceleration obtained from observations of the solar wind near 1 AU. In: *Solar Wind Seven* (Eds. E. Marsch and R. Schwenn). Oxford: Pergamon 1992, p. 69.

Neugebauer, M. (1997): Pioneers of space physics: a career in the solar wind. *Journal of Geophysical Research* **102**, A12, 26,887–26,894.

Neugebauer, M., et al. (2002): Sources of the solar wind at solar activity maximum. *Journal of Geophysical Research* **107**(A12), 1488.

Neugebauer, M., Snyder, C. W. (1962): The mission of *Mariner II* – preliminary observations. Solar plasma experiment. *Science* **138**, 1095–1096.

Neugebauer, M., Snyder, C. W. (1966): *Mariner 2* observations of the solar wind. *Journal of Geophysical Research* **71**, 4469–4484.

Neugebauer, M., Snyder, C. W. (1967): *Mariner 2* observations of the solar wind 2. Relation of plasma properties to the magnetic field. *Journal of Geophysical Research* **72**, 1823–1828.

Neupert, W. M. (1968): Comparison of solar X-ray line emission with microwave emission during flares. *The Astrophysical Journal (Letters)* **153**, L59–L64.

Neupert, W. M. (1989): Transient coronal extreme ultraviolet emission before and during the impulsive phase of a solar flare. *The Astrophysical Journal* **344**, 504–512.

Neupert, W. M., et al. (1967): Observation of the solar flare X-ray emission line spectrum of iron from 1.3 to 20 Å. *Astrophysical Journal (Letters)* **149**, L79–L83.

Neupert, W. M., et al. (1998): Obsevations of coronal structures above an active region by EIT and implications for coronal energy deposition. *Solar Physics* **183**, 305–321.

Neupert, W. M., Pizzo, V. (1974): Solar coronal holes as sources of recurrent geomagnetic disturbances. *Journal of Geophysical Research* **79**, 3701–3709.

Newell, N. E., et al. (1989): Global marine temperature variation and the solar magnetic cycle. *Geophysical Research Letters* **16**, 311–314.

Newell, P. T., Meng, C.-I., Wing, S. (1998): Relation to solar activity of intense aurorae in sunlight and darkness. *Nature* **393**, 342–345.

Newkirk, G. Jr. (Ed. 1974): *Coronal Disturbances: Proceedings of IAU Symposium No. 57.* Boston: D. Reidel 1974.

Newkirk, G. Jr., Harvey, J. (1968): Coronal polar plumes. *Solar Physics* **3**, 321–343.

Newman, M. J., Rood, R. T. (1977): Implications of solar evolution for the Earth's early atmosphere. *Science* **198**, 1035–1037.

Newton, H. W. (1930): An active region of the Sun on 1930 August 12. *Monthly Notices of the Royal Astronomical Society* **90**, 820–825.

Newton, H. W. (1932): The 27-day period in terrestrial magnetic disturbances. *Observatory* **55**, 256–261.

Newton, H. W. (1935): Note on two allied types of chromospheric eruptions. *Monthly Notices of the Royal Astronomical Society* **95**, 650–665.

Newton, H. W. (1942): Characteristic radial motions of Hα absorption markings seen with bright eruptions on the Sun's disc. *Monthly Notices of the Royal Astronomical Society* **102**, 2–10.

Newton, H. W. (1943): Solar flares and magnetic storms. *Monthly Notices of the Royal Astronomical Society* **103**, 244–257.

Newton, H. W., Nunn, M. L. (1951): The Sun's rotation derived from sunspots 1934 –1944 and additional results. *Monthly Notices of the Royal Astronomical Society* **111**, 413–421.

Ney, E. P. (1959): Cosmic radiation and the weather. *Nature* **183**, 451–452.

Ng, C. K., Reames, D. V., Tylka, A. J. (1999): Effect of proton-amplified waves on the evolution of solar energetic particle composition in gradual events. *Geophysical Research Letters* **26**(14), 2145–2148.

Nightingale, R. W., et al. (2002): Concurrent rotating sunspots, twisted coronal fans, sigmoid structures and coronal mass ejections. In: *Multi-wavelength Observations of Coronal Structure and Dynamics – Yohkoh 10th Anniversary Meeting* (Eds. P. C. H. Martens and D. Cauffman). New York: Elsevier Science 2002, p. 149.

Nisbet, E. G. (2000): The realms of Archaean life. *Nature* **405**, 625–626.

Nisbet, E. G., Sleep, N. H. (2001): The habitat and nature of early life. *Nature* **409**, 1083–1091.

Nishio, M., et al. (1997): Magnetic field configuration in impulsive solar flares inferred from coaligned microwave/X-ray images. *The Astrophysical Journal* **489**, 976.

Nitta, N. V., et al. (2006): Solar sources of impulsive solar energetic particle events and their magnetic field connection to the Earth. *The Astrophysical Journal*, **650**(1), 438–450.

Noci, G. (1973): Energy budget in coronal holes. *Solar Physics* **28**, 403–407.

Noci, G. (2002): The temperature of the solar corona. *Memoires Societa Astronomia Italiana* **74**, 704.

Noci, G., et al. (1997): First results from UVCS/SOHO. *Advances in Space Research* **20**(12), 2219.

Noci, G., et al. (1997): The quiescent corona and slow solar wind. In: *The Corona and Solar Wind near Minimum Activity Proceedings of the Fifth SOHO Workshop. ESA SP-404 (Eds. O. Kjeldseth-Moe, A. Wilson).* Noordwijk, The Netherlands: ESA Publications Division 1997, pp. 75–84.

Noci, G., Kohl, J. L., Withbroe, G. L. (1987): Solar wind diagnostics from Doppler-enhanced scattering. *The Astrophysical Journal* **315**, 706–715.

Nolte, J. T., et al. (1976): Coronal holes as sources of solar wind. *Solar Physics* **46**, 303–322.

November, L. J. (1989): The vertical component of the supergranular convection. *The Astrophysical Journal* **344**, 494–503.

November, L. J., Koutchmy, S. (1996): White-light coronal dark threads and density fine structure. *The Astrophysical Journal* **466**, 512–528.

Noyes, R. W. (1971): Ultraviolet studies of the solar atmosphere. *Annual Review of Astronomy and Astrophysics* **9**, 209–236.

Noyes, R. W., Baliunas, S. L., Guinan, E. F.: What can other stars tell us about the Sun? In: *The Solar Interior and Atmosphere* (Eds. A. N. Cox, W. C. Livingston and M. S. Matthews). Tucson, Arizona: University of Arizona Press 1991, pp. 1161–1186.

Noyes, R. W., et al. (1984): Rotation, convection, and magnetic activity in lower main sequence stars. *The Astrophysical Journal* **279**, 763–777.

Noyes, R. W., Leighton, R. B. (1963): Velocity fields in the solar atmosphere II. The oscillation field. *The Astrophysical Journal* **138**, 631–647.

Noyes, R. W., Weiss, N. O., Vaughan, A. H. (1984): The relation between stellar rotation rate and activity cycle periods. *The Astrophysical Journal* **287**, 769–773.

O

Ofman, L. (2005): MHD waves and heating in coronal holes. *Space Science Reviews* **120**, 67–94.

Ofman, L., Davila, J. M., Shimizu, T. (1996): Signatures of global mode Alfvén resonance heating in coronal loops. *The Astrophysical Journal (Letters)* **459**, L39–L42.

Ofman, L., Gary, S. P., Vinas, A. (2002): Resonant heating and acceleration of ions in coronal holes by cyclotron resonant spectra. *Journal of Geophysical Research* **107**(A12), 1461, SSH 9–1.

Ogawara, Y., et al. (1991): The *SOLAR-A* mission: An overview. *Solar Physics* **136**, 1–16.

Ogilvie, K. W., Coplan, M. A. (1995): Solar wind composition. *Reviews of Geophysics Supplement* **33**, 615–622.

Ogilvie, K. W., Desch, M. D. (1997): The Wind spacecraft and its early scientific results. In: *Results of the IASTP Program* (Ed. C. T. Russell). New York: Elsevier, p. 559.

Okamoto, T. J., et al. (2007): Coronal transverse magnetohydrodynamic waves in a solar prominence. *Science* **318**, 1577–1578.

Oort, J. H. (1950): The structure of the cloud of comets surrounding the solar system and a hypothesis concerning its origin. *Bulletin of the Astronomical Institutes of the Netherlands* **11**, 91–110. Reproduced in: *A Source Book in Astronomy and Astrophysics 1900–1975* (Eds. K. R. Lang and O. Gingerich). Cambridge: Harvard University Press 1979, pp. 132–137.

Oranje, B. J. (1983): The Ca II emission from the Sun as a star II. The plage emission profile. *Astronomy and Astrophysics* **124**, 43–49.

Orrall, F. Q., Rottman, G. J., Klimchuk, J. A. (1983): Outflow from the Sun's polar corona. *The Astrophysical Journal (Letters)* **266**, L65–L68.

Oster, L., Sofia, S., Schatten, K. (1982): Solar irradiance variations due to active regions. *The Astrophysical Journal* **256**, 768–773.

Osterbrock, D. E. (1961): The heating of the solar chromosphere, plages, and corona by magnetohydrodynamic waves. *The Astrophysical Journal* **134**, 347–388.

Otsuji, K., et al. (2007): Small-scale magnetic-flux emergence observed with Hinode Solar Optical Telescope. *Publications of the Astronomical Society of Japan* **59**, S649–S654.

Owen, T., Cess, R. D., Ramanathan, V. (1979): Enhanced CO_2 greenhouse to compensate for reduced solar luminosity on early Earth. *Nature* **277**, 640–642.

P

Paillard, D. (1998): The timing of Pleistocene glaciations from a simple multiple-state climate model. *Nature* **391**, 378–381.

Pallavicini, R., Serio, S., Vaiana, G. S. (1977): A survey of solar X-ray limb flare images: The relation between their structure in the corona and other physical parameters. *Astrophysical Journal* **216**, 108–122.

Pallé, P. L., Régulo, C., Roca-Cortés, T. (1989): Solar cycle induced variations of the low l solar acoustic spectrum. *Astronomy and Astrophysics* **224**, 253–258.

Parker, E. N. (1955): Dynamics of the interplanetary gas and magnetic fields. *The Astrophysical Journal* **125**, 668–676.

Parker, E. N. (1957a): Sweet's mechanism for merging magnetic fields in conducting fluids. *Journal of Geophysical Research* **62**, 509–520.

Parker, E. N. (1957b): Acceleration of cosmic rays in solar flares. *Physical Review* **107**, 830–836.

Parker, E. N. (1958a): Dynamical instability in an anisotropic ionized gas of low density. *Physical Review* **109**, 1874–1876.

Parker, E. N. (1958b): Dynamics of the interplanetary gas and magnetic fields. *Astrophysical Journal* **128**, 664–676.

Parker, E. N. (1958c): Interaction of the solar wind with the geomagnetic field. *The Physics of Fluids* **1**, 171–187.

Parker, E. N. (1958d): Cosmic-ray modulation by solar wind. *Physical Review* **110**, 1445–1449. Reproduced in: *Cosmic Rays* (Ed., A. M. Hillas). New York: Pergamon Press 1972.

Parker, E. N. (1959): Extension of the solar corona into interplanetary space. *Journal of Geophysical Research* **64**, 1675–1681.

Parker, E. N. (1960): The hydrodynamic theory of solar corpuscular radiation and stellar winds. *The Astrophysical Journal* **132**, 821–866.

Parker, E. N. (1961): Sudden expansion of the corona following a large solar flare and the attendant magnetic field and cosmic-ray effects. *The Astrophysical Journal* **133**, 1014–1033.

Parker, E. N. (1963): *Interplanetary Dynamical Processes*. New York: John Wiley.

Parker, E. N. (1963): The solar flare phenomenon and the theory of reconnection and annihilation of magnetic fields. *Astrophysical Journal Supplement* **8**(77), 177–211.

Parker, E. N. (1964): Dynamical properties of stellar coronas and stellar winds I. Integration of the momentum equation. *The Astrophysical Journal* **139**, 72–92.

Parker, E. N. (1965): Dynamical theory of the solar wind. *Space Science Reviews* **4**, 666–708.

Parker, E. N. (1972): Topological dissipation and the small-scale fields in turbulent gases. *Astrophysical Journal* **174**, 499–510.

Parker, E. N. (1973): Convection and magnetic fields in an atmosphere with constant temperature gradient. I. Hydrodynamic flows. *Astrophysical Journal* **186**, 643–664.

Parker, E. N. (1979): Sunspots and the physics of magnetic flux tubes. I. The general nature of the sunspot. II. Aerodynamic drag. *Astrophysical Journal* **230**, 905–913.

Parker, E. N. (1983a): Magnetic fields in the cosmos. *Scientific American* **249**, 44–65.

Parker, E. N. (1983b): Magnetic neutral sheets in evolving fields II. Formation of the solar corona. *The Astrophysical Journal* **264**, 642–647.

Parker, E. N. (1988): Nanoflares and the solar x-ray corona. *The Astrophysical Journal* **330**, 474–479.

Parker, E. N. (1991): Heating solar coronal holes. *Astrophysical Journal* **372**, 719–727.

Parker, E. N. (1996): Something stirs under the Sun. *Nature* **379**, 209–210.

Parker, E. N. (1997a): Reflections on macrophysics and the Sun. *Solar Physics* **176**, 219–247.

Parker, E. N. (1997b): Mass ejection and a brief history of the solar wind concept. In: *Cosmic Winds and the Heliopshere* (Eds. J. R. Jokipii, C. P. Sonett, and M. S. Giampapa). Tucson: University of Arizona Press 1997.

Parker, E. N. (1998): Reflection on macrophysics and the Sun (Special historical review). *Solar Physics* **176**, 2, 219–247.

Parker, E. N. (2001): A history of early work on the heliospheric magnetic field. *Journal of Geophysical Research* **106**, A8, 15797–15802.

Parker, E. N. (2002): A history of the solar wind concept. In: *The Century of Space Science I* (Eds. J.A.M. Bleeker, J. Geiss, M. Huber). Dordrecht, the Netherlands, Kluwer 2002, 225–255.

Parkinson, J. H., Morrison, L. V., Stephenson, F. V. (1980): The constancy of the solar diameter over the past 250 years. *Nature* **288**, 548–551.

Parnell, C. E., Jupp, P. E. (2000): Statistical analysis of the energy distribution of nanoflares in the quiet sun. *The Astrophysical Journal* **529**, 554–569.

Parnell, C. E., Priest, E. R., Golub, L. (1994): The three-dimensional structures of X-ray bright points. *Solar Physics* **151**, 57–74.

Paschmann, G. (1997): Observational evidence for transfer of plasma across the magnetopause. *Space Science Reviews* **80**, 217–234.

Patsourakos, S., Klimchuk, J. A., Mac Neice, P. J. (2004): The instability of steady-flow models to explain the extreme-ultarviolet coronal loops. *Astrophysical Journal* **603**, 322–329.

Pauli, W. (1930): Les theories quantities du magnetisive l'electron magnetique (The theory of magnetic quantities: The magnetic electron. *Septiene Cousieil Phys. Solvay*, Proceedings of the Sixth Solvay Conference 1930, Bruxelles, Gauthier-Vill 183–186.

Pauli, W. (1933): Remarks at the seventh solvay conference. Reproduced in the original French in: *Collected Scientific Papers of Wolfgang Pauli, Vol. 2* (Eds. R. Kronig and V. F. Weiskopf). New York: Wiley Interscience 1964.

Pawsey, J. L. (1946): Observation of million degree thermal radiation from the sun at a wave-length of 1.5 meters. *Nature* **158**, 633–634.

Payne-Scott, R., Little, A. G. (1952): The position and movement on the solar disk of sources of radiation at a frequency of 97 Mc/s. III–Outbursts. *Australian Journal of Scientific Research* **A5**, 32–46.

Payne-Scott, R., Yabsley, D. E., Bolton, J. G. (1947): Relative times of arrival of bursts of solar noise on different radio frequencies. *Nature* **160**, 256–257.

Pecker, J. C., Runcorn, S. K. (Eds., 1990): The Earth's climate and variability of the Sun over recent millennia: Geophysical, astronomical and archaeological aspects. *Philosophical Transactions of the Royal Society (London)* **A330**, 395–687. New York: Cambridge: University Press.

Peres, G., et al. (2000): The Sun as an X-Ray star. II. Using the Yohkoh/Soft X-ray Telescope-derived solar emission measure versus temperature to interpret stellar X-ray observations. *Astrophysical Journal* **528**, 537–551.

Perrin, J. B. (1920): Atomes et lumiere. *La Revue du Mois* **21**, 113–166.

Peterson, L. E. (1963): The 0.5-Mev gamma-ray and low-energy gamma-ray spectrum to 6 grams per square centimeter over Minneapolis. *Journal of Geophysical Research* **68**, 979–987.

Peterson, L. E., Winckler, J. R. (1959): Gamma ray burst from a solar flare. *Journal of Geophysical Research* **64**, 697–707.

Petit, J. R., et al. (1999): Climate and atmospheric history of the past 420,000 years from the Vostok ice core, Antarctica. *Nature* **399**, 429–436.

Petschek, H. E. (1964): Magnetic field annihilation. In: *AAS-NASA Symposium on the Physics of Solar Flares NASA SP-50* (Ed. W. N. Hess). Washington, DC: National Aeronautics and Space Administration, pp. 425–439.

Petschek, H. E., Thorne, R. M. (1967): The existence of intermediate waves in neutral sheets. *Astrophysical Journal* **147**, 1157–1163.

Pevtsov, A. A. (2000): Transequatorial loops in the solar atmosphere. *The Astrophysical Journal* **531**, 553–560.

Pevtsov, A. A., Canfield, R. C. (1999): Helicity of the photospheric magnetic field. In: *Magnetic Helicity in Space and Laboratory Plasmas. Geophysical Monograph 111*. (Eds.. M. R. Brown, R. C. Canfield, and A. A. Pevtsov). Washington: American Geophysical Union, p. 103.

Pevtsov, A. A., Canfield, R. C., Metcalf, T. R. (1995): Latitudinal variation of helicity of photospheric magnetic fields. *Astrophysical Journal (Letters)* **440**, L109–L112.

Pevtsov, A. A., Canfield, R. C., Zirin, H. (1996): Reconnection and helicity in a solar flare. *Astrophysical Journal* **473**, 533–538.

Pevtsov, A. A., et al. (1997): On the subphotospheric origin of coronal electric currents. *Astrophysical Journal* **481**, 973.

Pevtsov, A. A., et al. (2008): On the solar cycle variation of the hemispheric helicity rule. *Astrophysical Journal* **677**, 719–722.

Phan, T. D., et al. (2005): Magnetopause processes. *Space Science Reviews* **118**, 367–424.

Phan, T. D., et al. (2006): A magnetic reconnection X-line extending more than 390 Earth radii in the solar wind. *Nature* **439**, 175–178.

Phillips, J. L., et al. (1995a): *Ulysses* solar wind plasma observations at high southerly latitudes. *Science* **268**, 1030–1033.

Phillips, J. L., et al. (1995): *Ulysses* solar wind plasma observations from pole to pole. *Geophysical Research Letters* **22**(23), 3301–3304.

Phillips, J. L., et al. (1995b): Sources of shocks and compressions in the high-latitude solar wind: *Ulysses. Geophysical Research Letters* **22**(23), 3305–3308.

Phillips, K. J. H. (1991): Spectroscopy of high-temperature solar flare plasmas. *Philosophical Transactions of the Royal Society (London)* **A336**, 461–470.

Pick, M., et al. (2006): Multi-wavelength observations of CMEs and associated phenomena. *Space Science Reviews* **123**, 341–382.

Pick, M., Van den Oord, G. H. J. (1990): Observations of beam propagation. *Solar Physics* **130**, 83–99.

Piddington, J. H. (1956): Solar atmospheric heating by hydromagnetic waves. *Monthly Notices of the Royal Astronomical Society* **116**, 314.

Piddington, J. H. (1958): Interplanetary magnetic field and its control of cosmic ray variations. *Physical Review* **112**, 589–596.

Pike, C. D., Mason, H. E. (2002): EUV spectroscopic observations of spray ejecta from an X2 flare. *Solar Physics* **206**, 359–381.

Pizzo, V. J., Gosling, J. T. (1994): Three-dimensional simulation of high-latitude interaction regions: comparison with *Ulysses* results. *Geophysical Research Letters* **21**, 2063–2066.

Pneuman, G. W., Kopp, R. A. (1971): Gas-magnetic field interactions in the solar corona. *Solar Physics* **18**, 258–270.

Poincaré, H. (1896): Remarques sur une experience de M. Birkeland. *Comptes Rendus de l'Academie des Sciences* **123**, 530–533.

Poland, A. I., et al. (1981): Coronal transients near sunspot maximum. *Solar Physics* **69**, 169–175.

Poletto, G., et al. (2002): Low-latitude solar wind during the fall 1998 *SOHO-Ulysses* quadrature. *Journal of Geophysical Research (Space Physics)* **107**(A10), SSH 9–1.

Pollack, H., Huang, S., Shen, P. Y. (1998): Climate change revealed by subsurface temperatures: A global perspective. *Science* **282**, 279–281.

Pomerantz, M. A., Duggal, S. P. (1973): Record-breaking cosmic ray storm stemming from solar activity in August 1972. *Nature* **241**, 331–333.

Pontecorvo, B. M. (1967): Neutrino experiments and the problem of conservation of leptonic charge *zhurnal eksperimental'noi i teoreticheskoi fiziki* **53**, 1717–1725. *Soviet Physics JETP* **26**, 984 (1968).

Popescu, M. D., Doyle, J. G., Xia, L. D. (2004): Network boundary origins of fast solar wind seen in the low transitions regions? *Astronomy and Astrophysics* **421**, 339–348.

Porter, J. G., Dere, K. P. (1991): The magnetic network location of explosive events observed in the solar transition region. *The Astrophysical Journal* **370**, 775–778.

Porter, J. G., et al. (1987): Microflares in the solar magnetic network. *The Astrophysical Journal* **323**, 380–390.

Porter, J. G., Fontenla, J. M., Simnett, G. M. (1995): Simultaneous ultraviolet and X-ray observations of solar microflares. *Astrophysical Journal* **438**, 472–479.

Porter, J. G., Klimchuk, J. A. (1995): Soft X-ray loops and coronal heating. *The Astrophysical Journal* **454**, 499–511.

Porter, J. G., Moore, R. L. (1988): Coronal heating by microflares. In: *Solar and Stellar Coronal Structure and Dynamics* (Ed. R. C. Altrock). Sunspot: National Solar Observatory, Sacramento Peak, pp. 125–129.

Porter, J. G., Toomre, J., Gebbie, K. B. (1984): Frequent ultraviolet brightenings observed in a solar active regions with *Solar Maximum Mission. The Astrophysical Journal* **283**, 879–886.

Porter, L. J., Klimchuk, J. A. (1995): Soft X-ray loops and coronal heating. *Astrophysical Journal* **454**, 499–511.

Porter, L. J., Klimchuk, J. A., Sturrock, P. A. (1994a): The possible role of high-frequency waves in heating solar coronal loops. *The Astrophysical Journal* **435**, 502–514.

Porter, L. J., Klimchuk, J. A., Sturrock, P. A. (1994b): The possible role of MHD waves in heating the solar corona. *The Astrophysical Journal* **435**, 482–501.

Posner, A. (2007): Up to one-hour forecasting of radiation hazards from solar energetic ion events with relativistic electrons. *Space Weather* **5**, S05001–S05029.

Posner, A., et al. (2001): Nature of the boundary between open and closed magnetic field line regions at the Sun revealed by composition data and numerical models. *Journal of Geophysical Research* **106**(A8), 15869–15880.

Pres, P. (1999): The magnetic association of coronal bright points. *The Astrophysical Journal (Letters)* **510**, L73–L76.

Priem, H. N. A. (1997): CO_2 and climate: A geologist's view. *Space Science Reviews* **81**, 173–198.

Priest, E. R. (1978): The structure of coronal loops. *Solar Physics* **58**, 57–87.

Priest, E. R. (1982): *Solar Magnetohydrodynamics*. Boston: D. Reidel.

Priest, E. R. (1991): The magnetohydrodynamics of energy release in solar flares. *Philosophical Transactions of the Royal Society (London)* **A336**, 363–380.

Priest, E. R. (1996): Coronal heating by magnetic reconnection. *Astrophysics and Space Science* **237**, 49–73.

Priest, E. R. (1999): How is the solar corona heated? *Solar and Stellar Activity: Similarities and Differences. Asp Conference Series* **158**, 321–333.

Priest, E. R., Foley, C. R., Heyvaerts, J., Arber, T. D., Culhane, J. L., Acton, L. W. (1998): Nature of the heating mechanism for the diffuse solar corona. *Nature* **393**, 545–547.

Priest, E. R., Forbes, T. G. (1986): New models for fast steady-state magnetic reconnection. *Journal of Geophysical Research* **91**, 5579–5588.

Priest, E. R., Forbes, T. G. (1990): Magnetic field evolution during prominence eruptions and two-ribbon flares. *Solar Physics* **126**, 319–350.

Priest, E. R., Forbes, T. G. (2002): The magnetic nature of solar flares. *Astronomy and Astrophysics Review* **10**(4), 313–377.

Priest, E. R., Heyvaerts, J. F., Title, A. M. (2002): A flux-tube tectonics model for solar coronal heating driven by the magnetic carpet. *The Astrophysical Journal* **576**, 533–551.

Priest, E. R., Parnell, C. E., Martin, S. F. (1994): A converging flux model of an X-ray bright point and an associated canceling magnetic feature. *The Astrophysical Journal* **427**, 459–474.

Priest, E. R., Schrijver, C. J. (1999): Aspects of three-dimensional magnetic reconnection (invited review). *Solar Physics* **190**, 1–24.

Pudovkin, M. I., Veretenenko, S. V. (1996): Variations of the cosmic rays as one of the possible links between the solar activity and the lower atmosphere. *Advances in Space Research* **17**(11), 161–164.

Pulkiinen, T. I., et al. (2001): The Sun–Earth connection on time scales from years to decades and centuries. *Space Science Reviews* **95**, 625–637.

Pulkkinen, T. I. (2007): Space weather: terrestrial perspective. *Living Reviews in Solar Physics* **4**, 1.

Purcell, J. D., Tousey, R., Watanabe, K. (1949): Observations at high altitudes of extreme ultraviolet and X-rays from the Sun. *Physical Review* **76**, 165–166.

Q

Qiu, J., et al. (2007): On the magnetic flux budget in low-corona magnetic reconnection and interplanetary coronal mass ejections. *The Astrophysical Journal* **659**, 758–772.

Qiu, J., Wang, H., Cheng, C. Z., Gary, D. E. (2004): Magnetic reconnection and mass acceleration in flare-coronal mass ejections events. *Astrophysical Journal* **604**, 900–905.

Quinn, T. J., Fröhlich, C. (1999): Accurate radiometers should measure the output of the Sun. *Nature* **401**, 841.

R

Radick, R. R., Lockwood, G. W., Baliunas, S. L. (1990): Stellar activity and brightness variations: a glimpse at the Sun's history. *Science* **247**, 39–44.

Ragot, B. R. (2006a): Mean cross-field displacement of magnetic field lines: Full nonlinear calculation and comparison with generalized quasi-linear results in the solar wind. *Astrophysical Journal* **644**, 622–630.

Ragot, B. R. (2006b): Nonresonant pitch-angle scattering of low-energy electrons in the solar wind. *The Astrophysical Journal* **642**, 1163–1172.

Ragot, B. R. (2006c): Distributions of magnetic field orientations in the turbulent solar wind. *The Astrophysical Journal* **651**, 1209–1218.

Ragot, B. R. (2006d): Lengths of wandering magnetic field lines in the turbulent solar wind. *The Astrophysical Journal* **653**, 1493–1498.

Raisbeck, G. M., et al. (1981): Cosmogenic ^{10}Be concentrations in Antarctic ice during the past 30,000 years. *Nature* **292**, 825–826.

Raisbeck, G. M., et al. (1985): Evidence for an increase in cosmogenic ^{10}Be during a geomagnetic reversal. *Nature* **315**, 315–317.

Raisbeck, G. M., et al. (1987): Evidence for two intervals of enhanced ^{10}Be deposition in Antarctic ice during the last glacial period. *Nature* **326**, 273–277.

Raisbeck, G. M., et al. (1990): ^{10}Be and ^{2}H in polar ice cores as a probe of the solar variability's influence on climate. *Philosophical Transactions of the Royal Society (London)* **A300**, 463–470.

Raisbeck, G. M., Yiou, F. (1988): ^{10}Be as a proxy indicator of variations in solar activity and geomagnetic field intensity during the last 10,000 years. In: *Secular, Solar and Geomagnetic Variations in the Last 10,000 Years* (Eds. F. R. Stephenson and W. Wolfendale). Drodrecht: Kluwer, pp. 287–296.

Ramanathan, V. (1988): The greenhouse theory of climate change. *Science* **240**, 293–299.

Ramanathan, V., et al. (1989): Cloud-radiative forcing and climate: Results from the Earth Radiation Budget experiment. *Science* **243**, 57–62.

Ramaty, R. (1969): Gyrosynchrotron emission and absorption in a magnetoactive plasma. *Astrophysical Journal* **158**, 753–770.

Ramaty, R., et al. (1983): Implications of high-energy neutron observations from solar flares. *Astrophysical Journal Letters* **273**, L41–L45.

Ramaty, R., et al. (1993): Acceleration in solar flares: Interacting particles versus interplanetary particles. *Advances in Space Research* **13**(9), 275–284.

Ramaty, R., et al. (1994): Gamma-ray and millimeter-wave emissions from the 1991 June X-class flares. *The Astrophysical Journal* **436**, 941–949.

Ramaty, R., Kozlovsky, B., Lingenfelter, R. E. (1979): Nuclear gamma rays from energetic particle interactions. *Astrophysical Journal Supplement* **40**, 487–526.

Ramaty, R., Lingenfelter, R. E. (1966): Galactic cosmic-ray electrons. *Journal of Geophysical Research* **71**, 3687–3703.

Ramaty, R., Lingenfelter, R. E. (1979): γ-Ray line astronomy. *Nature* **278**, 127–132.

Ramaty, R., Lingenfelter, R. E. (1983): Gamma-ray line astronomy. *Space Science Reviews* **36**, 305–317.

Ramaty, R., Mandzhavidze, N. (1994): Theoretical models for high-energy solar flare emissions. In: *High Energy Solar Phenomena – A New Era of Spacecraft Measurements* (Eds. J. M. Ryan and W. T. Vestrand). New York: American Institute of Physics, pp. 26–44.

Ramaty, R., Murphy, R. J. (1987): Nuclear processes and accelerated particles in solar flares. *Space Science Reviews* **45**, 213–268.

Ramaty, R., Petrosian, V. (1972): Free-free absorption of gyrosynchrotron radiation in solar microwave bursts. *The Astrophysical Journal* **178**, 241–249.

Ramsay, W. (1901): The inert constituents of the atmosphere. *Nature* **65**, 161–164.

Ramsey, H. E., Smith, S. F. (1966): Flare-initiated filament oscillations. *Astronomical Journal* **71**, 197–199.

Rao, U. R. (1972): Solar modulation of galactic cosmic radiation. *Space Science Reviews* **12**, 719–809.

Raouafi, N. E., Harvey, J. W., Solanki, S. K. (2007): Properties of solar polar coronal plumes constrained by ultraviolet coronagraph spectrometer data. *The Astrophysical Journal* **658**, 643–656.

Rappazzo, A. F., Velli, M., Einaudi, G., Dahlburg, R. B. (2005): Diamagnetic and expansion effects on the observable properties of the slow solar wind in a coronal streamer. *The Astrophysical Journal* **633**, 474–488.

Raymond, J. C. (1999): Composition variations in the solar corona and solar wind. *Space Science Reviews* 87, 55–66.

Raymond, J. C. (2004): Enhanced: Imaging the Sun's eruptions in three dimensions. *Science* **305**, 49–50.

Raymond, J. C., et al. (1997): Composition of coronal streamers from the ultraviolet coronagraph spectrometer. *Solar Physics* **175**, 645–655. Reprinted in *The First Results From SOHO* (Eds. B. Fleck and Z. Svestka). Boston: Kluwer Academic Publishers, pp. 645–665.

Raymond, J. C., et al. (2000): *SOHO* and radio observations of a CME shock wave. *Geophysical Research Letters* **27**(10), 1439–1442.

Raymond, J. C., et al. (2003): Far-ultraviolet spectra of fast coronal mass ejections associated with X-class flares. *Astrophysical Journal* **597**, 1106–1117.

Raynaud, D., et al. (1993): The ice record of greenhouse gases. *Science* **259**, 926–934.

Reale, F., et al. (2007): Fine thermal structure of a coronal active region. *Science* **318**, 1582–1584.

Reale, F., Peres, G. (2000): *TRACE*-derived temperature and emission measure profiles along long-lived coronal loops: The role of filamentation. *Astrophysical Journal (Letters)* **528**, L45–L48.

Reames, D. V. (1988): Bimodal abundances in the energetic particles of solar and interplanetary origin. *Astrophysical Journal (Letters)* **330**, L71–L75.

Reames, D. V. (1990): Energetic particles from impulsive solar flares. *Astrophysical Journal Supplement Series* **73**, 235–251.

Reames, D. V. (1993): Non-thermal particles in the interplanetary medium. *Advances in Space Research* **13**, 331–339.

Reames, D. V. (1995): Solar energetic particles: A paradigm shift. *Reviews of Geophysics Supplement* **33**, 585–589.

Reames, D. V. (1997): Energetic particles and the structure of coronal mass ejections. In: *Coronal Mass Ejections. Geophysical Monograph 99* (Eds. N. Crooker, J. A. Joselyn, J. Feynman). Washington: American Geophysical Union, pp. 217–226.

Reames, D. V. (1999): Particle acceleration at the Sun and in the heliosphere. *Space Science Reviews* **90**, 413–491.

Reames, D. V. (2002): Magnetic topology of impulsive and gradual solar energetic particle events. *Astrophysical Journal (Letters)* **571**, L63–L66.

Reames, D. V., Barbier, L. M., Ng, C. K. (1996): The spatial distribution of particles accelerated by coronal mass ejection-driven shocks. *Astrophysical Journal* **466**, 473.

Reames, D. V., Kahler, S. W., Ng, C. K. (1997): Spatial and temporal invariance in the spectra of energetic particles in gradual solar events. *Astrophysical Journal* **491**, 414.

Reames, D. V., Ng, C. K. (2004): Heavy-element abundances in solar energetic particle events. *Astrophysical Journal* **610**, 510–522.

Reames, D. V., Richardson, I. G., Wenzel, K. P. (1992): Energy spectra of ions from impulsive solar flares. *Astrophysical Journal* **387**, 715–725.

Reedy, R. C., Arnold, J. R. (1972): Interaction of solar and galactic cosmic-ray particles with the Moon. *Journal of Geophysical Research* **77**, 537–555.

Reedy, R. C., Arnold, J. R., Lal, D. (1983): Cosmic-ray record in solar system matter. *Science* **219**, 127–135.

Reeves, K. K., Warren, H. P. (2002): Modeling the cooling of postflare loops. *Astrophysical Journal* **578**, 590–597.

Reid, G. C. (1976): Influence of ancient solar-proton events on the evolution of life. *Nature* **259**, 177–179.

Reid, G. C. (1987): Influence of solar variability on global sea surface temperatures. *Nature* **329**, 142–143.

Reid, G. C. (1991): Solar irradiance variations and global ocean temperature. *Journal of Geomagnetism and Geoelectricity* **43**, 795–801.

Reid, G. C. (1991): Solar total irradiance variations and the global sea surface temperature record. *Journal of Geophysical Research* **96**, 2835–2844.

Reid, G. C., Leinbach, H. (1959): Low-energy cosmic-ray events associated with solar flares. *Journal of Geophysical Research* **64**, 1801–1805.

Reinard, A. (2005): Comparison of interplanetary CME charge state composition with CME-associated flare magnitude. *Astrophysical Journal* **620**, 501–505.

Reiner, M. J. et al. (1998): On the origin of radio emissions associated with the January 6–11, 1997 CME. *Geophysical Research Letters* **25**, 2493–2496.

Reiner, M. J., Fainberg, J., Stone, R. G. (1995): Large-scale interplanetary magnetic field configuration revealed by solar radio bursts. *Science* **270**, 461–464.

Reines, F., Cowan, C. L. Jr. (1953): Detection of the free neutrino. *Physical Review* **92**, 830–831.

Reines, F., Cowan, C. L. Jr. (1956): The neutrino. *Nature* **178**, 446.

Reisenfeld, D. B., et al. (2003): Properties of high-latitude CME-driven disturbances during *Ulysses* second northern polar passage. *Geophysical Research Letters* **30**(19), ULY 5–1.

Revelle, R., Suess, H. E. (1957): Carbon dioxide exchange between atmosphere and ocean and the question of an increase in atmospheric carbon dioxide during the past decades. *Tellus* **9**, 18–27.

Rhodes, E. J. Jr., et al. (1997): Measurements of frequencies of solar oscillation for the MDI medium-*l* program. *Solar Physics* **175**, 287–310. Reprinted in: *The First Results From SOHO* (Eds. B. Fleck and Z. Svestka). Boston: Kluwer Academic Publishers, pp. 287–310.

Rhodes, E. J. Jr., Ulrich, R. K., Simon, G. W. (1977): Observations of nonradial *p*-mode oscillations on the Sun. *Astrophysical Journal* **218**, 901–919.

Ribes, E. (1990): Astronomical determinations of the solar variability. *Philosophical Transactions of the Royal Society (London)* **A330**, 487–497.

Ribes, J. C., Nesmeribes, E. (1990): The solar sunspot cycle in the Maunder minimum Ad 1645 to Ad 1715. *Astronomy and Astrophysics* **276**, 549–563.

Rice, J. B., Strassmeier, K. G. (2001): Doppler imaging of stellar surface structure.XVII. The solar-type Pleiades star HII 314 = V1038 Tauri. *Astronomy and Astrophysics* **377**, 264–272.

Richard, O., Theado, S., Vauclair, S. (2004): Updated Toulouse solar models including the diffusion-circulation coupling and the effect of μ-gradients. *Solar Physics* **220**, 234–259.

Richardson, I. G. (1997): Using energetic particles to probe the magnetic topology of ejecta. In: *Coronal Mass Ejections, Geophysical Monograph 99* (Eds. N. Crooker, J. A. Joselyn, J. Feynman). Washington: American Geophysical Union, p. 189.

Richardson, I. G. (2004): Energetic particles and co-rotating interaction regions in the solar wind. *Space Science Reviews* **111**, 3, 267–376.

Richardson, I. G., Cane, H. V. (1995): Regions of abnormally low proton temperature in the solar wind (1965–1991) and their association with ejecta. *Journal of Geophysical Research* **100**, 23397–23412.

Richardson, I. G., Cane, H. V. (2004): Identification of interplanetary coronal mass ejections at 1 AU using multiple solar wind plasma composition anomalies. *Journal of Geophysical Research* **109**, A09104.

Richardson, J. D., et al. (2008): Cool heliosheath plasma and deceleration of the upstream solar wind at the termination shock. *Nature* **454**, 63–66.

Richardson, J. D., Wang, C., Burlaga, L. F. (2003): Correlated solar wind speed density, and magnetic field changes at *Voyager 2*. *Geophysical Research Letters* **30**(23), 2007.

Richardson, R. S. (1939): Intensity changes in bright chromospheric disturbances. *Astrophysical Journal* **90**, 368–377.

Richardson, R. S. (1944): Solar flares versus bright chromospheric eruptions: A question of terminology. *Publications of the Astronomical Society of the Pacific* **56**, 156–158.

Richardson, R. S. (1951): Characteristics of solar flares. *Astrophysical Journal* **114**, 356–366.

Rickett, B. J., Coles, W. A. (1980): Solar cycle changes in the high latitude solar wind. In: *Study of the Solar Cycle From Space. Nasa Conference Publication CP-2098*. (Eds. G. Newkirk, J. B. Zirker) Washington: National Aeronautics and Space Administration 1980, pp. 233–243.

Rickett, B. J., Coles, W. A. (1982): Solar cycle evolution of the solar wind in the three dimensions. In: *Solar Wind Five. NASA Conference Publication CP-2280*. (Ed. M. Neugebauer) Washington: NASA 1983, pp. 315–321.

Rickett, B. J., Coles, W. A. (1991): Evolution of the solar wind structure over a solar cycle: Interplanetary scintillation velocity measurements compared with coronal observations. *Journal of Geophysical Research* **96**, A2, 1717–1736.

Riley, P. (2007): An alternative interpretation of the relationship between the inferred open solar flux and the interplanetary magnetic field. *Astrophysical Journal (Letters)* **667**, L97–L100.

Riley, P., Crooker, N. U. (2004): Kinematic treatment of coronal mass ejection evolution in the solar wind. *Astrophysical Journal* **600**, 1035–1042.

Riley, P., et al. (2007): "Bursty" reconnection following solar eruptions: MHD simulations and comparison with observations. *Astrophysical Journal* **655**, 591–597.

Riley, P., Gosling, J. T., Crooker, N. U. (2004): *Ulysses* observations of the magnetic connectivity between coronal mass ejections and the sun. *Astrophysical Journal* **608**, 1100–1105.

Robbins, D. E., Hundhausen, A. J., Bame, S. J. (1970): Helium in the solar wind. *Journal of Geophysical Research* **75**, 1178–1187.

Robbrecht, E., et al. (2001): Slow magnetoacoustic waves in coronal loops: EIT and TRACE. *Astronomy and Astrophysics* **370**, 591–601.

Roberts, D. A. (1989): Interplanetary observational constraints on Alfvén wave acceleration of the solar wind. *Journal of Geophysical Research* **94**, 6899–6905.

Roberts, D. A., Goldstein, M. L. (1991): Turbulence and waves in the solar wind. *Review of Geophysics Supplement* **29**, 932–943.

Roddier, F. (1975): Principe de realisation d'un hologramme acoustique de la surface du Soleil. (Procedure to form an acouostical hologram of the solar surface.) *Comptes rendus de l'Académie des Sciences Paris B* **281**, 93–95.

Roelof, E. C. (1974): Coronal structure and the solar wind. In: *Solar Wind Three* (Ed. C. T. Russell). Los Angeles: Institute of Geophysics and Planetary Physics UCLA, pp. 98–131.

Roelof, E. C., et al. (1992): Low-energy solar electrons and ions observed at *Ulysses* February–April, 1991 – The inner heliosphere as a particle reservoir. *Geophysical Research Letters* **19**(12), 1243–1246.

Rohen, G., et al. (2005): Ozone depletion during the solar proton events of October/November 2003 as seen by Sciamachy. *Journal of Geophysical Research* **110**, A09S39.

Romano, P., Contarino, L., Zuccarello, F. (2005): Observational evidence of the primary role played by photospheric motions in magnetic helicity transport before a filament eruption. *Astronomy and Astrophysics* **433**, 683–690.

Rompolt, B. (1975): Spectral features to be expected from rotational and expansional motions in fine solar structures. *Solar Physics* **41**, 329–348.

Rompolt, B. (1990): Small scale structure and dynamics of prominences. *Hvar Observatory Bulletin* **14**, 37–102.

Rosenbauer, H. R., et al. (1977): A survey of initial results of the *Helios* plasma experiment. *Journal of Geophysics* **42**, 561–580.

Rosenberg, H. (1976): Solar radio observations and interpretations. *Philosophical Transactions of the Royal Society (London)* **A281**, 461–471.

Rosenberg, R. L. (1970): Unified theory of the interplanetary magnetic field. *Solar Physics* **15**, 72–78.

Rosenberg, R. L., Coleman, P. J. Jr. (1969): Heliographic latitude dependence of the dominant polarity of the interplanetary magnetic field. *Journal of Geophysical Research* **74**, 5611–5622.

Rosenthal, C. S., et al. (2002): Waves in the magnetized solar atmosphere I. Basic processes and internetwork oscillations. *The Astrophysical Journal* **564**, 508–524.

Rosner, R., Tucker, W. H., Vaiana, G. S. (1978): Dynamics of the quiescent solar corona. *The Astrophysical Journal* **220**, 643–665.

Rossi, B. (1991): The interplanetary plasma. *Annual Review of Astronomy and Astrophysics* **29**, 1–8.

Rostoker, G., Fälthammar, C.-G. (1967): Relationship between changes in the interplanetary magnetic field and variations in the magnetic field at the Earth's surface. *Journal of Geophysical Research* **72**, 5853–5863.

Rottman, G. J. (1981): Rocket measurements of the solar spectral irradiance during solar minimum. *Journal of Geophysical Research* **86**, 6697–6705.

Rottman, G. J., Orrall, F. Q., Klimchuk, J. A. (1982): Measurements of outflow from the base of solar coronal holes. *Astrophysical Journal* **260**, 326–337.

Rouillard, A. P., et al. (2008): First imaging of co-rotating solar wind flows and their source regions using the *STEREO* spacecraft. *Geophysical Research Letters* **35**, L10110.

Roussev, I. I., et al. (2003a): A three-dimensional flux rope model for coronal mass ejections based on a loss of equilibrium. *Astrophysical Journal (Letters)* **588**, L45–L48.

Roussev, I. I., et al. (2003b): A three-dimensional model of the solar wind incorporating solar magnetogram observations. *Astrophysical Journal (Letters)* **595**, L57–L61.

Roussev, I. I., et al. (2004): A numerical model of a coronal mass ejection: shock development with implications for the acceleration of GeV protons. *Astrophysical Journal (Letters)* **605**, L73–L76.

Roussev, I. I., Sokolov, I. S. (2006): Models of solar eruptions: Recent advances from theory and simulations. In *Solar Eruptions and Energetic Particles, Geophysical Monograph Series* **165**, 89–102

Roussev, I., et al. (1999): Modeling of explosive events in the solar transition region. *Romanian Astronomical Journal* **9**, 57.

Roussev, I., et al. (2001): Modeling of explosive events in the solar transition region in a 2D environment Ii. Various MHD experiments. *Astronomy and Astrophysics* **375**, 228–242.

Roussev, I., et al. (2001): Modeling of solar explosive events in 2D environments. III. Observable consequences. *Astronomy and Astrophysics* **380**, 719–726.

Rucinski, D., Bzowski, M., Fahr, H. J. (2003): Imprints from the solar cycle on the helium atom and helium pickup ion distributions. *Annales Geophysicae* **21**(6), 1315–1330.

Russell, C. (2008): *The STEREO Mission. Space Science Reviews* **136**, Issues 1–4. New York, Springer.

Russell, C. T., (Ed., 1995): The global geospace mission. *Space Science Reviews* **71**, 1–878.

Russell, C. T., (Ed., 1997): Results of the IASTP program. *Advances in Space Research* **20**, 523–1107.

Russell, C. T., McPherron, R. L. (1973): Semiannual variation of geomagnetic activity. *Journal of Geophysical Research* **78**, 92–108.

Rust, D. M. (1976): An active role for magnetic fields in solar flares. *Solar Physics* **47**, 21–40.

Rust, D. M. (1982): Solar flares, proton showers, and the space shuttle. *Science* **216**, 939–946.

Rust, D. M. (1983): Coronal disturbances and their terrestrial effects. *Space Science Reviews* **34**, 21–36.

Rust, D. M. (1994): Spawning and shedding helical magnetic fields in the solar atmosphere. *Geophysical Research Letters* **21**(4), 241–244.

Rust, D. M. (2001): A new paradigm for solar filament eruptions. *Journal of Geophysical Research* **106**(A11), 25075–25088.

Rust, D. M., et al. (2005): Comparison of interplanetary disturbances at the *NEAR* spacecraft with coronal mass ejections at the Sun. *Astrophysical Journal* **621**, 524–536.

Rust, D. M., Hildner, E. (1976): Expansion of an X-ray coronal arch into the outer corona. *Solar Physics* **48**, 381–387.

Russell, C. (2008): The *STEREO Mission Space Science Reviews*, **138** Issues 1–4. New York Springer.

Rust, D. M., Kumar, A. (1996): Evidence for helically kinked magnetic flux ropes in solar eruptions. *Astrophysical Journal (Letters)* **464**, L199–L202.

Rust, D. M., Labonte, B. J. (2005): Observational evidence of the kink instability in solar filament eruptions and sigmoids. *Astrophysical Journal (Letters)* **622**, L69–L72.

Rust, D. M., Nakagawa, Y., Neupert, W. M. (1975): EUV emission, filament activation and magnetic fields in a slow-rise flare. *Solar Physics* **41**, 397–414.

Rust, D. M., Svestka, Z. (1979): Slowly moving disturbances in the X-ray corona. *Solar Physics* **63**, 279–295.

Rust, D. M., Webb, D. F. (1977): Soft X-ray observations of large-scale active region brightenings. *Solar Physics* **54**, 403–417.

Ruzmaikin, A. A., et al. (1996): Spectral properties of solar convection and diffusion. *Astrophysical Journal* **471**, 1022–1029.

Ryan, J., et al. (1994): Neutron and gamma-ray measurements of the solar flare of 1991 June 9. In: *High-Energy Solar Phenomena – A New Era of Spacecraft Measurements. AIP Conference Proceedings 294* (Eds. J. M. Ryan and W. T. Vestrand). New York: American Institute of Physics, p. 89.

Rye, R., Kuo, P. H., Holland, H. D. (1995): Atmospheric carbon dioxide concentrations before 2.2 billion years ago. *Nature* **378**, 603–605.

Ryutova, M., Tarbell, T. (2003): MHD shocks and the origin of the solar transition region. *Physical Review Letters* **90**, 191101.

S

Saba, J. L. R., Strong, K. T. (1991): Nonthermal broadening. *Astrophysical Journal* **375**, 789–799.

Sabine, E. (1852): Letter to John Herschel 16 March 1852. *Herschel Letters No. 15.235. (Royal Society)*. In: The origin of solar-terrestrial studies (Quoted by A. J. Meadows and J. E. Kennedy). *Vistas in Astronomy* **25**, 419–426 (1982).

Sabine, E. (1852): On periodical laws discoverable in the mean effects of the larger magnetic disturbances. *Philosophical Transactions of the Royal Society (London)* **142**, 103–124.

Sackmann, I.-J., Boothroyd, A. I. (2003): Our Sun. V. A bright young Sun consistent with helioseismology and warm temperatures on ancient Earth and Mars. *Astrophysical Journal* **583**, 1024–1039.

Sackmann, I.-J., Boothroyd, A. I., Kraemer, K. E. (1993): Our Sun III. Present and future. *Astrophysical Journal* **418**, 457–468.

Sagan, C., Chyba, C. (1997): The early faint Sun paradox: organic shielding of ultraviolet-labile greenhouse gases. *Science* **276**, 1217–1221.

Sagan, C., Mullen, G. (1972): Earth and Mars: evolution of atmospheres and surface temperatures. *Science* **177**, 52–56. .

Sagdeev, R. Z., Kennel, C. F. (1991): Collisionless shock waves. *Scientific American* **264**, 106–113, April.

Saito, T. (1975): Two-hemisphere model of the three-dimensional magnetic structure of the interplanetary space. *Science Reports of the Tohoku University*, Series 5, **26**, 37–54.

Sakai, J.-I., De Jager, C. (1996): Solar flares and collisions between current-carrying loops. *Space Science Reviews* **77**, 1–192.

Sakao, T. (1994): *Characteristics of Solar Flare Hard X-ray Sources as Revealed with the Hard X-ray Telescope aboard the Yohkoh Satellite.* Ph. D. Thesis. Mitaka: University of Tokyo, the Yohkoh HXT group, National Astronomical Observatory.

Sakao, T., et al. (2007): Continuous plasma outflows from the edge of a solar active region as a possible source of solar wind. *Science* **318**, 1585–1587.

Sakurai, T. (1976): Magnetohydrodynamic interpretation of the motion of prominences. *Publications of the Astronomical Society of Japan* **28**, 177–198.

Sakurai, T. (1991): Observations from the *Hinotori* mission. *Philosophical Transactions of the Royal Society (London)* **A336**, 339–347.

Sakurai, T., Spangler, S. R. (1994): The study of coronal plasma structures and fluctuations with Faraday rotation measurements. *Astrophysical Journal* **434**, 773–785.

Sandbaek, O., Leer, E. (1995): Coronal heating and solar wind energy balance. *Astrophysical Journal* **454**, 486–498.

Sandbaek, O., Leer, E., Hansteen, V. H. (1994): On the relation between coronal heating, flux tube divergence, and the solar wind proton flux and flow speed. *Astrophysical Journal* **436**, 390–399.

Sanderson, T. R., et al. (1983): Correlated particle and magnetic field observations of a large-scale magnetic loop structure behind an interplanetary shock. *Geophysical Research Letters* **10**, 916–919.

Sarabhai, V. (1963): Some consequences of nonuniformity of solar wind velocity. Journal of Geophysical Research **68**, 1555–1557.

Sauvaud, J.-A., et al. (2007): The IMPACT Solar Wind Electron Analyzer (SWEA). *Space Science Reviews* **136**, Issue 1–4, 227–239.

Savcheva, A., et al. (2007): A study of polar jet parameters based on *Hinode* XRT observations. *Publications of the Astronomical Society of Japan* **59**, S771–S778.

Sawaya-Lacoste, H. (Ed., 2003): *SOHO-12/GONG 2002: Local and Global Helioseismology: The Present and Future.* ESA SP-517 2003.

Scafetta, N., West, B. J. (2006): Phenomenological solar signature in 400 years of reconstructed northern hemisphere temperature record. *Geophysical Research Letters* **33**, L17718.

Schatten, K. H. (1988): A model for solar constant secular changes. *Geophysical Research Letters* **15**, 121–124.

Schatten, K. H., et al. (1985): The importance of improved facular observations in understanding solar constant variations. *Astrophysical Journal* **294**, 689–696.

Schatten, K. H., Leighton, R. B., Howard, R., Wilcox, J. M. (1972): Large scale photospheric magnetic field: The diffusion of active region fields. *Solar Physics* **26**, 283–289.

Schatten, K. H., Wilcox, J. M., Ness, N. F. (1969): A model of interplanetary and coronal magnetic fields. *Solar Physics* **6**, 442–455.

Schatzman, E. (1949): The heating of the solar corona and chromosphere. *Annales d'Astrophysique* **12**, 203–218.

Schein, M., Jesse, W. P., Wollan, E. O. (1941): The nature of the primary cosmic radiation and the origin of the mesotron. *Physical Review* **59**, 615.

Scherb, F. (1964): Velocity distributions of the interplanetary plasma detected by *Explorer 10*. *Space Research* **4**, 797–818.

Scherrer, P. H., et al. (1995): The Solar Oscillations Investigation – Michelson Doppler Imager. *Solar Physics* **162**, 129–188.

Schlesinger, M. E., Ramankutty, N. (1992): Implications for global warming of intercycle solar irradiance variations. *Nature* **360**, 330–333.

Schmahl, E., Hildner, E. (1977): Coronal mass-ejections-kinematics of the 19 December 1973 event. *Solar Physics* **55**, 473–490.

Schmelz, J. T., et al. (2005): All coronal loops are the same: Evidence to the contrary. *Astrophysical Journal (Letters)* **627**, L81–L84.

Schmidt, A. (1924): Das erdmagnetische Aussenfeld. *Zeitschrift Geophysikalische* **1**, 3–13.

Schmidt, W. K. H., et al. (1980): On temperature and speed of He^{++} and O^{6+} ions in the solar wind. *Geophysical Research Letters* **7**, 697–700.

Schöll, M., et al. (2007): Long-term reconstruction of the total solar irradiance based on neutron monitor and support data. *Advances in Space Research* **40**, 996–999.

Schou, J. (1999): Migration of zonal flows detected using Michelson Doppler Imager f mode frequency splittings. *The Astrophysical Journal (Letters)* **523**, L181–L184.

Schou, J. (2003): Wavelike properties of solar supergranulation detected in Doppler shift data. *Astrophysical Journal (Letters)* **596**, L259–L262.

Schou, J., et al. (1997): Determination of the Sun's seismic radius from the *SOHO* Michelson Doppler Imager. *Astrophysical Journal (Letters)* **489**, L197–L200.

Schou, J., et al. (1998): Helioseismic studies of differential rotation in the solar envelope by the solar oscillations investigation using the Michelson Doppler Image. *The Astrophysical Journal* **505**, 390–417.

Schove, D. J. (1955): The sunspot cycle 649 BC to 2000 AD. *Journal of Geophysical Research* **60**, 127–145.

Schrijver, C. J. (1997): Working group 6: Magnetic fields, coronal structure and phenomena. In: *The Corona and Solar Wind Near Minimum Activity. Proceedings of the Fifth SOHO Workshop. ESA SP-404*. Noordwijk: Esa Publications Division, pp. 149–153.

Schrijver, C. J. (2001): Catastrophic cooling and high-sped downflow in quiescent solar coronal loops observed with *TRACE*. *Solar Physics* **198**, 325–345.

Schrijver, C. J. (2007): Braiding-induced interchange reconnection of the magnetic field and the width of solar coronal loops. *Astrophysical Journal (Letters)* **662**, L119–L122.

Schrijver, C. J., Aschwanden, M. J., Title, A. M. (2002): Transverse oscillations in coronal loops observed with *TRACE* – I. An overview of events, movies, and a discussion of common properties and required conditions. *Solar Physics* **206**, 69–98.

Schrijver, C. J., Brown, D. S. (2000): Oscillations in the magnetic field of the solar corona in response to flares near the photosphere. *Astrophysical Journal (Letters)* **537**, L69–L72.

Schrijver, C. J., De Rosa, M. L. (2003): Photospheric and heliospheric magnetic fields. *Solar Physics* **212**, 165–200.

Schrijver, C. J., De Rosa, M. L., Title, A. M. (2002): What is missing from our understanding of long-term solar and heliospheric activity? *Astrophysical Journal* **577**, 1006–1012.

Schrijver, C. J., De Rosa, M. L., Title, A.M., Metcalf, T. R. (2005): The non-potentiality of active-region coronae and the dynamics of the photospheric magnetic field. *Astrophysical Journal* **628**, 501–513.

Schrijver, C. J., et al. (1989): Relations between the photospheric magnetic field and the emission from the outer atmospheres of cool stars. I. The solar Ca II K line core emission. *Astrophysical Journal* **337**, 964–976.

Schrijver, C. J., et al. (1996): Dynamics of the chromosphric network: Mobility, dispersal, and diffusion coefficients. *Astrophysical Journal* **468**, 921.

Schrijver, C. J., et al. (1997a): Sustaining the quiet photospheric network: The balance of flux emergence, fragmentation, merging, and cancellation. *The Astrophysical Journal* **487**, 424–436.

Schrijver, C. J., et al. (1997b): The dynamic quiet solar corona: 4 days of joint observing with MDI and EIT. In: *The Corona and Solar Wind Near Minimum Activity. Proceedings of the Fifth SOHO Workshop. ESA SP-404*. Noordwijk: ESA Publications 1997, pp. 669–674.

Schrijver, C. J., et al. (1998): Large-scale coronal heating by the small-scale magnetic field of the Sun. *Nature* **394**, 152–154.

Schrijver, C. J., et al. (1999): A new view of the solar outer atmosphere by the *Transition Region and Coronal Explorer*. *Solar Physics* **187**, 261–302.

Schrijver, C. J., Hagenaar, H. J., Title, A. M. (1997): On the patterns of the solar granulation and supergranulation. *Astrophysical Journal* **475**, 328.

Schrijver, C. J., Lemen, J. R., Mewe, R. (1989): Coronal activity in F-, G-, and K type stars. IV. Evidence for expanding loop geometries in stellar coronae. *Astrophysical Journal* **341**, 484–492.

Schrijver, C. J., Mewe, R., Walter, F. M. (1984): Coronal activity in F-, G-, and K-type stars. Ii. Coronal structure and rotation. *Astronomy and Astrophysics* **138**, 258–266.

Schrijver, C. J., Title, A. M. (2001): On the Formation of polar spots in Sun-like stars. *The Astrophysical Journal* **551**, 1099–1106.

Schrijver, C. J., Title, A. M. (2002): The topology of a mixed-polarity potential field, and inferences for the heating of the quiet solar corona. *Solar Physics* **207**, 223–240.

Schrijver, C. J., Title, A. M. (2003): The magnetic connection between the solar photosphere and the corona. *The Astrophysical Journal (Letters)* **597**, L165–L168.

Schrijver, C. J., Van Ballegooijen, A. A. (2005): Is the quiet-sun corona a quasi steady, force-free environment? *Astrophysical Journal* **630**, 552–560.

Schrijver, C. J., Zwann, C. (2000): *Solar and Stellar Magnetic Activity. Cambridge Astrophysics Series 34*, Cambridge: Cambridg University Press.

Schröder, W. (1988): Aurorae during the Maunder Minimum. *Meteorology and Atmospheric Physics* **38**, 246–251.

Schröder, W. (1994): Behavior of auroras during the Spörer minimum (1450–1550). *Annales Geophysicae* **12**, 808–809.

Schröder, W. (1997): Some aspects of the earlier history of solar-terrestrial physics. *Planetary and Space Science* **45**, 395–400.

Schröter, E. H. (1985): The solar differential rotation: Present status of observations. *Solar Physics* **100**, 141–160.

Schröter, E. H., Wöhl, H. (1975): Differential rotation, meridional and random motions of the solar Ca$^+$ network. *Solar Physics* **42**, 3–16.

Schuck, P. W. (2005): Local correlation tracking and the magnetic induction equation. *Astrophysical Journal (Letters)* **632**, L53–L56.

Schuck, P. W. (2006): Tracking magnetic footpoints with the magnetic induction equation. *Astrophysical Journal* **646**, 1358–1391.

Schüle, U., et al. (2000): Radiance variations of the quiet Sun at far-ultraviolet wavelengths. *Astronomy and Astrophysics* **354**, L71–L74.

Schultz, M. (1973): Interplanetary sector structure and the heliomagnetic equator. *Astrophysics and Space Science* **24**, 371–383.

Schumacher, J., Kliem, B. (1997): Coalescence of magnetic islands including anomalous resistivity. *Physics of Plasmas* **4**(10), 3533.

Schumacher, J., Kliem, B. (1997): Transient fast reconnection in dynamic current sheets with anomalous resistivity. *Advances in Space Research* **19**, 1797.

Schüssler, M., et al. (1996): Distribution of starspots on cool stars. I. Young and main sequence stars of $1\,M_\odot$. *Astronomy and Astrophysics* **314**, 503–512.

Schüssler, M., Solanki, S. K. (1992): Why rapid rotators have polar spots. *Astronomy and Astrophysics* **264**, L13–L16.

Schuster, A. (1911): The origin of magnetic storms. *Proceedings of the Physical Society (London)* **A85**, 61.

Schwabe, S. H. (1844): Sonnen-Beobachtungem im Jahre 1843. *Astronomische Nachrichten* **21**, No. 495, 233–236. Reprinted in: *Kosmos* (Ed. A. Von Humboldt). English translation in: Solar observations during 1843 in *Early Solar Physics* (Ed. A. J. Meadows). Oxford: Pergamon Press 1970, pp. 95–98.

Schwadron, N. A. (2002): An explanation for strongly underwound magnetic field in co-rotating rarefaction regions and its relationship to footpoint motion on the Sun. *Geophysical Research Letters* **29**(14), 1663.

Schwadron, N. A., et al. (2000): Inner source distributions; Theoretical interpretation, implications, and evidence for inner source protons. *Journal of Geophysical Research* **105**(A4), 7465–7472.

Schwadron, N. A., et al. (2005): Solar wind from the coronal hole boundaries. *Journal of Geophysical Research* **110**(A4), A04104.

Schwadron, N. A., Mc Comas, D. J. (2003): Solar wind scaling law. *Astrophysical Journal* **599**, 1395–1403.

Schwadron, N. A., Mc Comas, D. J. (2005): The sub-Parker spiral structure of the heliospheric magnetic field. *Geophysical Research Letters* **32**(3), L03112.

Schwadron, N. A., Mc Comas, D. J., De Forest, C. (2006): Relationship between solar wind and coronal heating: Scaling laws from solar X-rays. *The Astrophysical Journal* **642**, 1173–1178.

Schwarzschild, M. (1948): On noise arising from the solar granulation. *The Astrophysical Journal* **107**, 1–5.

Schwarzschild, M. (1958): *Structure and Evolution of the Stars*. Princeton: Princeton University Press, p. 207.

Schwenn, R. (1981): Solar wind and its interactions with the magnetosphere: Measured parameters. *Advances in Space Research* **1**, 3–17.

Schwenn, R. (1983): Direct correlations between coronal transients and interplanetary disturbances. *Space Science Reviews* **34**, 85–99.

Schwenn, R. (1986): Relationship of coronal transients to interplanetary shocks: 3D aspects. *Space Science Reviews* **44**(1–2), 139–186.

Schwenn, R. (1990): Large-scale structure of the interplanetary medium. In: *Physics of the Inner Heliosphere I. Large-Scale Phenomena* (Eds. R. Schwenn and E. Marsch). New York: Springer-Verlag 1990, pp. 99–181.

Schwenn, R. (2006a): Space weather: the solar perspective. *Living Reviews in Solar Physics* **3**, 2.

Schwenn, R. (2006b): Solar wind sources and their variations over the solar cycle. *Space Science Reviews* **124**, 51–76.

Schwenn, R., et al. (2005): The association of coronal mass ejections with their effects near the Earth. *Annales Geophysicae* **23**, 1033–1059.

Schwenn, R., et al. (2006): Coronal observations of CMEs. *Space Science Reviews* **123**, 127–176.

Schwenn, R., Marsch, E. (Eds., 1990): *Physics of the Inner Heliosphere 1. Large- scale Phenomena, 2 Particles, Waves and Turbulence*. New York: Springer-Verlag.

Schwenn, R., Rosenbauer, H., Muehlhaeuser, K. -H. (1980): Singly-ionized helium in the driver gas of interplanetary shock wave. *Geophysical Research Letters* **7**, 201–204.

Seehafer, N. (1990): Electric current helicity in the solar atmosphere. *Solar Physics* **125**, 219–232.

Seppälä, A., et al. (2006): Destruction of the tertiary ozone maximum during a solar proton event. *Geophysical Research Letters* **33**, L07804.

Serebryanskiy, A., Chou, D.-Y. (2005): Comparison of solar cycle variations of solar p-mode frequencies from GONG and MDI. *Astrophysical Journal* **633**(1), 1187–1190.

Severny, A. B. (1958): The appearance of flares in neutral points of the solar magnetic field and the pinch-effect. *Bulletin of the Crimean Astrophysical Observatory* **20**, 22–51.

Shackleton, N. J. (1977): The oxygen isotope stratigraphic record of the late Pleistocene. *Philosophical Transactions of the Royal Society (London)* **B280**, 169–182.

Share, G. H., et al. (2003). Directionality of flare-accelerated alpha particles at the sun. *Astrophysical Journal (Letters)* **595**, L89–L92.

Share, G. H., et al. (2004): *RHESSI* e + -e− annihilation radiation observations: Implications for conditions in the flaring solar chromosphere. *Astrophysical Journal (Letters)* **615**, L169–L172.

Share, G. H., Murphy, R. J. (1997): Intensity and directionality of flare-accelerated α-particles at the sun. *Astrophysical Journal* **485**, 409–418.

Share, G. H., Murphy, R. J., Ryan, J. (1997): Solar and stellar gamma ray observations with *COMPTON*. In: *Proceedings of the Fourth Compton Symposium* (Eds. C. D. Dermer, M. S. Strickman and J. D. Kurfess). New York: American Institute of Physics, pp. 17–36.

Shaviv, N. J. (2003): Toward a solution to the early faint Sun paradox: A lower cosmic ray flux from a stronger solar wind. *Journal of Geophysical Research* **108**, 3.

Shea, M. A., Smart, D. F. (1990): A summary of major solar proton events. *Solar Physics* **127**, 297–320.

Sheeley, N. R. Jr. (2005): Surface of the sun's magnetic field: A historical review of the flux-transport mechanism. *Living Reviews in Solar Physics* **2**, 5.

Sheeley, N. R. Jr., et al. (1975): Coronal changes associated with a disappearing filament. *Solar Physics* **45**, 377–392.

Sheeley, N. R. Jr., et al. (1977): A pictorial comparison of interplanetary magnetic field polarity, solar wind speed, and geomagnetic disturbances index during the sunspot cycle. *Solar Physics* **52**, 485–495.

Sheeley, N. R. Jr., et al. (1983): Associations between coronal mass ejections and interplanetary shocks. In *Solar Wind Five, NASA Conference Publication Vol. 2280*. Washington, DC: NASA, pp. 693–702.

Sheeley, N. R. Jr., et al. (1983): Associations between coronal mass ejections and soft X-ray events. *Astrophysical Journal* **272**, 349–354.

Sheeley, N. R. Jr., et al. (1984): Associations between coronal mass ejections and metric type II bursts. *Astrophysical Journal* **279**, 839–847.

Sheeley, N. R. Jr., et al. (1985): Coronal mass ejections and interplanetary shocks. *Journal of Geophysical Research* **90**, 163–175.

Sheeley, N. R. Jr., et al. (1997b): Measurements of flow speeds in the corona between 2 and 30 solar radii. *Astrophysical Journal* **484**, 472–478.

Sheeley et al. (1997)

Sheeley, N. R. Jr., et al. (1999): Continuous tracking of coronal outflows: Two kinds of coronal mass ejections. *Journal of Geophysical Research* **104**(A11), 24739–24768.

Sheeley, N. R. Jr., Harvey, J. W., Feldman, W. C. (1976): Coronal holes, solar wind streams, and recurrent geomagnetic disturbances: 1973–1976. *Solar Physics* **49**, 271–278.

Sheeley, N. R. Jr., Knudson, T. N., Wang, Y.-M. (2001): Coronal inflows and the Sun's nonaxisymmetric open flux. *Astrophysical Journal (Letters)* **546**, L131–L135.

Sheeley, N. R. Jr., Nash, A. G., Wang, Y.-M. (1987): The origin of rigidly rotating magnetic field patterns on the Sun. *Astrophysical Journal* **319**, 481–502.

Sheeley, N. R. Jr., Wang, Y.-M. (1991): Magnetic field configurations associated with fast solar wind. *Solar Physics* **131**, 165–186.

Sheeley, N. R. Jr., Wang, Y.-M. (2001): Coronal inflows and sector magnetism. *Astrophysical Journal (Letters)* **562**, L107–L110.

Sheeley, N. R. Jr., Wang, Y.-M. (2002): Characteristics of coronal inflows. *Astrophysical Journal* **579**, 874–887.

Sheeley, N. R. Jr., Wang, Y. -M., Phillips, J. L. (1997a): Near-Sun magnetic fields and the solar wind. In: *Cosmic Winds and the Heliosphere* (Eds. J. R. Jokipii, C. P. Sonett and M. S. Giampapa). Tucson: University of Arizona Press 1997, pp. 459–483.

Sheeley, N. R., Jr., Warren, H. P., Wang, Y.-M. (2004): The origin of postflare loops. *Astrophysical Journal* **616**, 1224–1231.

Shevgaonkar, R. K., Kundu, M. R. (1984): Three-dimensional structures of two solar active regions from Vla observations at 2, 6 and 20 centimeters wavelength. *Astrophysical Journal* **283**, 413–420.

Shevgaonkar, R. K., Kundu, M. R. (1985): Dual frequency observations of solar microwave bursts using the VLA. *Astrophysical Journal* **292**, 733–751.

Shibahashi, H. (2007): Meridional circulation and differential rotation in the solar convection zone. *Astronomische Nachrichten* **328**, 264.

Shibata, K. (1996): New observational facts about solar flares from *Yohkoh* studies – Evidence of magnetic reconnection and a unified model of flares. *Advances in Space Research* **17**(4/5), 9–18.

Shibata, K. (1997): Rapidly time variable phenomena: Jets, explosive events, and flares. In: *The Corona and Solar Wind NearMminimum Activity. Proceedings of the Fifth SOHO Workshop. ESA SP-404.* Noordwijk: Esa Publications Division, pp. 103–112.

Shibata, K., et al. (1992): Observations of X-ray jets with the *Yohkoh* soft X-ray telescope. *Publications of the Astronomical Society of Japan* **44**, L173–L179.

Shibata, K., et al. (1994): A gigantic coronal jet ejected from a compact active region in a coronal hole. *The Astrophysical Journal (Letters)* **431**, L51.

Shibata, K., et al. (1995): Hot plasma ejections associated with compact-loop solar flares. *Astrophysical Journal (Letters)* **451**, L83–L85.

Shibata, K., et al. (2007a): New solar physics with *Solar-B* mission. *Astronomical Society of the Pacific Conference Series* **369**, 1–593.

Shibata, K., et al. (2007b): Chromospheric anemone jets as evidence of ubiquitous reconnection. *Science* **318**, 1591–1593.

Shimizu, T. (1995): Energetics and occurrence rate of active-region transient brightenings and implications for the heating of the active-region corona. *Publications of the Astronomical Society of Japan* **47**, 251–263.

Shimizu, T. (Ed., 1996): *Yohkoh Views the Sun – The First Five Years.* Tokyo: The Institute of Space and Astronautical Science, National Astronomical Observatory, *Yohkoh* Group.

Shimizu, T., Tsuneta, S. (1997): Deep survey of solar nanoflares with *Yohkoh*. *Astrophysical Journal* **486**, 1045–1057.

Shimizu, T., Tsuneta, S., Acton, L. W., Lemen, J. R., Uchida, Y. (1992): Transient brightenings in active regions observed by the Soft X-ray Telescope on *Yohkoh*. *Publications of the Astronomical Society of Japan* **44**, L147–L153.

Shimojo, M., et al. (2007): Fine structures of solar X-ray jets observed with the X-ray telescope aboard *Hinode*. *Publications of the Astronomical Society of Japan* **59**, S745–S750.

Shimojo, M., Shibata, K. (1999): Occurrence rate of microflares in an X-ray bright point within an active region. *Astrophysical Journal* **516**, 934–938.

Shimojo, M., Shibata, K. (2000): Physical parameters of solar X-ray jets. *Astrophysical Journal* **542**, 1100–1108.

Shindell, D. T., et al. (2001): Solar forcing of regional climate change during the maunder minimum. *Science* **294**, 2149–2152.

Shindell, D. T., et al. (2003): Volcanic and solar forcing of climate change during the pre-industrial era. *Journal of Climate* **16**, 4094–4107.

Shine, R. A., Simon, G. W., Hulburt, N. E. (2000): Supergranule and mesogranule evolution. *Solar Physics* **193**, 313–331.

Shklovskii, I. S., Moroz, V. I., Kurt, V. G. (1960): The nature of the Earth's third radiation belt. *Soviet Astronomy AJ* **4**, 871–873.

Shodhan, S., et al. (2000): Counterstreaming electrons in magnetic clouds. *Journal of Geophysical Research* **105**(A12), 27261–27268.

Silverman, S. M. (1992): Secular variation of the aurora for the past 500 years. *Reviews of Geophysics* **30**(4), 333–351.

Simnett, G. M. (1973): Relativistic electrons in space. *Space Research* **13**, 745–762.

Simnett, G. M. (1991): Energetic particle production in flares. *Philosophical Transactions of the Royal Society (London)* **A336**, 439–450.

Simnett, G. M., Roelof, E. C., Haggerty, D. K. (2002): The acceleration and release of near-relativistic electrons by coronal mass ejections. *Astrophysical Journal* **579**, 854–862.

Simon, G. W. (1967): Observations of horizontal motions in solar granulation: Their relation to supergranulation. *Zeitschrift für Astrophysick* **65**, 345–363.

Simon, G. W., Leighton, R. B. (1964): Velocity fields in the solar atmosphere. III. Large-scale motions, the chromospheric network, and magnetic fields. *Astrophysical Journal* **140**, 1120–1147.

Simon, G. W., Title, A. M., Weiss, N. O. (1995): Kinematical models of supergranular diffusion on the Sun. *Astrophysical Journal* **442**, 886–897.

Simon, G. W., Title, A. M., Weiss, N. O. (2001): Sustaining the Sun's magnetic network with emerging bipoles. *Astrophysical Journal* **561**, 427–434.

Simon, T., Herbig, G., Boesgaard, A. M. (1985): The evolution of chromospheric activity and the spin-down of solar-type stars. *Astrophysical Journal* **293**, 551–574.

Simpson, J. A. (1954): Cosmic-radiation intensity-time variations and their origin III. The origin of 27-day variations. *Physical Review* **94**, 426–440.

Simpson, J. A. (1983): Elemental and isotopic composition of the galactic cosmic rays. *Annual Review of Nuclear Particle Science* **33**, 323–381.

Simpson, J. A. (1998): Brief history of recurrent solar modulation of the galactic cosmic rays (1937–1990). *Space Science Reviews* **83**, 169–176.

Simpson, J. A., Connell, J. J. (2001): Cosmic ray isotopic composition studies with the *Ulysses* high energy telescope: Implications for origin and distribution in the Galaxy. *Space Science Reviews* **97**, 337–341.

Simpson, J. A., et al. (1992): The *Ulysses* Cosmic Ray and Solar Particle Investigation. *Astronomy and Astrophysics Supplement Series* **92**, No. 2, 365–399.

Simpson, J. A., et al. (1995a): Cosmic ray and solar particle investigations over the south polar regions of the Sun. *Science* **268**, 1019–1023.

Simpson, J. A., et al. (1995b): The latitude gradients of galactic cosmic ray and anomalous helium fluxes measured on *Ulysses* from the Sun's south polar region to the equator. *Geophysical Research Letters* **22**(23), 3337–3340.

Singer, S. F. (1957): A new model of magnetic storms and aurorae. *EOS* **38**, 175–190.

Siscoe, G. (2000): The space-weather enterprise: Past, present, and future. *Journal of Atmospheric and Solar-Terrestrial Physics* **62**, 1223–1232.

Siscoe, G. L. (1976): Three-dimensional aspects of interplanetary shock waves. *Journal of Geophysical Research* **81**, 6235–6241.

Siscoe, G. L. (1980): Evidence in the auroral record for secular solar variations. *Review of Geophysics and Space Physics* **18**, 647–658.

Skoug, R. M., et al. (2004): Extremely high speed solar wind: 29–30 October 2003. *Journal of Geophysical Research* **109**(A9), A09102.

Skumanich, A., et al. (1984): The Sun as a star: Three-component analysis of chromospheric variability in the calcium K line. *Astrophysical Journal* **282**, 776–783.

Slottje, C. (1978): Millisecond microwave spikes in a solar flare. *Nature* **275**, 520–521.

Smith, D. M., et al. (2002): The *RHESSI* spectrometer. *Solar Physics* **210**, 33–60.

Smith, D. M., et al. (2003): High-resolution spectroscopy of gamma-ray lines from the X class solar flare of 2002 July 23. *Astrophysical Journal (Letters)* **595**, L81–L84.

Smith, D. M., et al. (2005): Terrestrial gamma-ray flashes observed up to 20 MeV. *Science* **307**, 1085–1088.

Smith, E. J. (1962): A comparison of *Explorer 6* and *Explorer 10* magnetometer data. *Journal of Geophysical Research* **67**, 2045–2049.

Smith, E. J., Balogh, A. (1995): *Ulysses* observations of the radial magnetic field. *Geophysical Research Letters* **22**(23), 3317–3320.

Smith, E. J., et al. (1995): *Ulysses* observations of Alfvén waves in the southern and northern solar hemispheres. *Geophysical Research Letters* **22**(23), 3381–3384.

Smith, E. J., et al. (2000): Recent observations of the heliospheric magnetic field at *Ulysses*: Return to low latitude. *Advances in Space Research* **26**(5), 823–832.

Smith, E. J., et al. (2003): The Sun and heliosphere at solar maximum. *Science* **302**, 1165–1169.

Smith, E. J., Marsden, R. G. (1995): *Ulysses* observations from pole-to-pole: An introduction. *Geophysical Research Letters* **22**(23), 3297–3300.

Smith, E. J., Marsden, R. G. (1998): The *Ulysses* mission. *Scientific American* **278**, 74–79, January.

Smith, E. J., Marsden, R. G., Page, D. E. (1995): *Ulysses* above the Sun's south pole – an introduction. *Science* **268**, 1005–1006.

Smith, E. J., Sonett, C. P., Dungey, J. W. (1964): Satellite observation of the geomagnetic field during magnetic storms. *Journal of Geophysical Research* **69**, 2669–2688.

Smith, E. J., Tsurutani, B. T., Rosenberg, R. L. (1978): Observations of the interplanetary sector structure up to heliographic latitudes of 16 degrees by *Pioneer 11*. *Journal of Geophysical Research* **83**, 717–724.

Smith, E. J., Wolfe, J. H. (1976): Observations of interaction regions and co-rotating shocks between one and five AU: *Pioneers 10* and *11*. *Geophysical Research Letters* **3**, 137–140.

Smith, E. J., Wolfe, J. H. (1977): *Pioneeer 10, 11* observations of evolving solar wind streams and shocks beyond 1 AU. In *Study of Traveling Interplanetary Phenomena* (Eds. M. A. Shea, D. F. Smart and S. T. Wu). Dordrecht: D. Reidel, pp. 227–257.

Smy, M. B., et al. (2004): Precise measurement of the solar neutrino day-night and seasonal variation in Super-Kamiokande-I. *Physical Review D* **69**, 011104–011109.

Sno Collaboration (1999): The Sudbury neutrino observatory projects. *Nuclear Physics B – Proceedings Supplements* **77**, 43–47.

Sno Collaboration (2000): The Sudbury Neutrino Observatory. *Nuclear Instruments and Methods in Physics Research Section A: Accelerators, Spectrometers, Detectors and Associated Equipment* **449**, 172–207.

Snodgrass, H. B. (1983): Magnetic rotation of the solar photosphere. *The Astrophysical Journal* **270**, 288–299.

Snodgrass, H. B. (1985): Solar torsional oscillations: A net pattern with wavenumber 2 as artifact. *Astrophysical Journal* **291**, 339–343.

Snodgrass, H. B., Howard, R. (1985): Torsional oscillations of the sun. *Science* **228**, 945–952.

Snyder, C. W., Neugebauer, M. (1964): Interplanetary solar-wind measurements by *Mariner II*. *Space Research* **4**, 89–113.

Snyder, C. W., Neugebauer, M., Rao, U. R. (1963): The solar wind velocity and its correlation with cosmic-ray variations and with solar and geomagnetic activity. *Journal of Geophysical Research* **68**, 6361–6370.

Socas-Navarro, H., Sánchez Almeida, J. (2002): Magnetic properties of photospheric regions with very low magnetic flux. *Astrophysical Journal* **565**, 1323–1334.

Soderblom, D. R. (1985): A survey of chromospheric emission and rotation among solar-type stars in the solar neighborhood. *Astronomical Journal* **90**, 2103–2115.

Soderblom, D. R., Baliunas, S. L. (1988): The sun among the stars: What stars indicate about solar variability. In: *Secular, Solar and Geomagnetic Variations in the Last 10,000 Uears* (Eds. F. R. Stephenson and A. W. Wolfendale). Dordrecht: Kluwer, pp. 25–48.

Sokolov, I. V., et al. (2004): A new field line advection model for solar particle acceleration. *Astrophysical Journal (Letters)* **616**, L171–L174.

Solanki, S. K. (2006): The solar magnetic field. *Reports on Progress in Physics* **69**, 563–668.

Solanki, S. K., et al. (2003): Three-dimensional magnetic field topology in a region of solar coronal heating. *Nature* **425**, 692–695.

Solanki, S. K., Fligge, M. (1998): Solar irradiance since 1874 revisted. *Geophysical Research Letters* **25**, 341–344.

Solanki, S. K., Fligge, M. (1999): A reconstruction of total solar irradiance since 1700. *Geophysical Research Letters* **26**, 2465–2468.

Solanki, S. K., Inhester, B., Schussler, M. (2006): The solar magnetic field. *Reports on Progress in Physics* **69**, (3), 563.

Solanki, S. K., Schüssler, M., Fligge. M. (2000): Evolution of the Sun's large scale magnetic field since the Maunder Minimum. *Nature* **408**, 445–447.

Somov, B. V., Kosugi, T. (1997): Collisonless reconnection and high-energy particle acceleration in solar flares. *Astrophysical Journal* **485**, 859–868.

Sonett, C. P. (1984): Very long solar periods and the radiocarbon record. *Reviews of Geophysics and Space Physics* **22**(3), 239–254.

Sonett, C. P., Colburn, D. S., Davis, L. Jr., Smith, E. J., Colman, P. J. Jr. (1964): Evidence for a collision-free magnetohydrodynamic shock in interplanetary space. *Physical Review Letters* **13**, 153–156.

Sonett, C. P., et al. (1960): Current systems in the vestigial geomagnetic field: *Explorer 6. Physical Review Letters* **4**, 161–163.

Sonett, C. P., Giampapa, M. S., Matthews, M. S. (Eds., 1991): *The Sun in Time.* Tucson: University of Arizona Press.

Sonett, C. P., Suess, H. E. (1984): Correlation of bristlecone pine ring widths with atmospheric ^{14}C variations: A climate-Sun relation. *Nature* **307**, 141–143.

Song, P., Singer, H. J., Siscoe, G. L. (Eds., 2001): *Space Weather Geophysical Monograph No. 125.* Washington: American Geophysical Union.

Soon, W. H., Posmentier, E. S., Baliunas, S. L. (1996): Inference of solar irradiance variability from terrestrial temperature changes, 1880–1993: An astrophysical application of the Sun-climate connection. *Astrophysical Journal* **472**, 891–902.

Southworth, G. C. (1945): Microwave radiation from the Sun. *Journal of the Franklin Institute* **239**, 285–297.

Spadaro, D., et al. (2003): A transient heating model for coronal structure and dynamics. *Astrophysical Journal* **582**, 486–494.

Spangler, S. R., et al. (2002): Very long baseline interferometer measurements of turbulence in the inner solar wind. *Astronomy and Astrophysics* **384**, 654–665.

Spangler, S. R., Mancuso, S. (2000): Radio astronomical constraints on coronal heating by high-frequency Alfvén waves. *Astrophysical Journal* **530**, 491–499.

Spicer, D. S., Sibeck, D., Thompson, B. J., Davila, J. M. (2006): A Kopp Pneuman-like picture of coronal mass ejections. *Astrophysical Journal* **643**, 1304–1316.

Spiegel, E. A., Weiss, N. O. (1980): Magnetic activity and variations in solar luminosity. *Nature* **287**, 616–617.

Spörer, G. F. W. (1874–1976): *Beobachtungen der Sonnenflecken zu Anclam.* Leipzig.

Spörer, G. F. W. (1887): Üeber die periodicität der Sonnenflecken seit dem Jahre 1618. *Vierteljahrsschr Astronomische Gesellschaft (Leipzig)* **22**, 323–329.

Spörer, G. F. W. (1889): Üeber die periodicität der Sonnenflecken seit dem Jahre 1618. R. Leopold-Caroline Acad. Aston. Halle **53**, 283–324.

Spörer, G. F. W. (1899): Sur les différences que présentent l'Hémisphere nord el l'Hémisphere sud du Soleil. *Bulletin Astronomique* **6**, 60.

Spreybroeck, L. P. Van, Krieger, A. S., Vaiana, G. S. (1970): X-ray photographs of the Sun on March 7, 1970. *Nature* **227**, 818–822.

Spruit, H. C. (1982): Effect of spots on a star's radius and luminosity. *Astronomy and Astrophysics* **108**, 348–355.

Spruit, H. C. (1988): Influence of magnetic activity on the solar luminosity and radius. In: *Solar Radiative Output Variations* (Ed. P. V. Foukal). Cambridge: Cambridge Research and Instrumentation, pp. 254–288.

Srivastava, N., Venkatakrishnan, P. (2002): Relationship between CME speed and geomagnetic storm intensity. *Geophysical Research Letters* **29**, 1.

St. Cyr, O. C., et al. (2000): Properties of coronal mass ejections: *SOHO* LASCO observations from January 1996 to June 1998. *Journal of Geophysical Research* **105**, 18169–18186.

Stamper, R., et al. (1999): Solar causes of the long-term increase in geomagnetic activity. *Journal of Geophysical Research* **104**(A12), 28325–28342.

Stauffer, B. (2000): Long term climate records from polar ice. *Space Science Reviews* **94**, 321–336.

Stauffer, B., et al. (1998): Atmospheric Co_2 concentration and millennial-scale climate change during the last glacial period. *Nature* **392**, 59–61.

Steig, E. L., et al. (1996): Large amplitude solar modulation cycles of ^{10}Be in Antarctica: Implications for atmospheric mixing processes and interpretation of the ice core record. *Geophysical Research Letters* **23**, 523–526.

Stein, R. F., Nordlund A. (2001): Solar oscillations and convection Ii. Excitation of radial oscillations. *Astrophysical Journal* **546**, 585–603.

Stenborg, G., Cobelli, P. J. (2003): A wavelet packets equalization technique to reveal the multiple spatial-scale nature of coronal structures. *Astronomy and Astrophysics* **398**, 1185–1193.

Stenflo, J. O. (1974): Differential rotation and sector structure of solar magnetic fields. *Solar Physics* **36**, 495–515.

Sterling, A. C. (2000): Sigmoid CME source regions at the Sun; some recent results. *Journal of Atmospheric and Solar-Terrestrial Physics* **62**, 1427–1435.

Sterling, A. C., et al. (2007): *Hinode* observations of the onset stage of a solar filament eruption. *Publications of the Astronomical Society of Japan* **59**, S823–S829.

Sterling, A. C., Hudson, H. S. (1997): *Yohkoh* SXT observations of X-ray "dimming" associated with a halo coronal mass ejection. *Astrophysical Journal (Letters)* **491**, L55–L58.

Sterling, A. C., Hudson, H. S., Thompson, B. J., Zarro, D. M. (2000):*Yohkoh* SXT and *SOHO* EIT observations of sigmoid-to-arcade evolution of structures associated with halo coronal mass ejections. *Astrophysical Journal* **532**, 628–647.

Sterling, A. C., Moore, R. L. (2004): External and internal reconnection in two filament-carrying magnetic cavity solar eruptions. *Astrophysical Journal* **613**, 1221–1232.

Sterling, A. C., Moore, R. L. (2005): Slow-rise and fast-rise phases of an erupting solar filament, and flare emission onset. *Astrophysical Journal* **630**, 1148–1159.

Stern, D. P. (1989): A brief history of magnetospheric physics before the spaceflight era. *Reviews of Geophysics* **27**, 103–114.

Stix, M. (1981): Theory of the solar cycle. *Solar Physics* **74**, 79–101.

Stone, E. C., et al. (1998): The Advanced Composition Explorer. *Space Science Reviews* **86**, 1–22, 257–632.

Stone, E. C., et al. (2005): *Voyager 1* explores the termination shock region and the heliosheath beyond. *Science* **309**, 2017–2020.

Stone, E. C., et al. (2008): An asymmetric solar wind termination shock. *Nature* **454**, 71–74.

Stone, R. G., et al. (1992): The Unified Radio and Plasma wave investigation. *Astronomy and Astrophysics Supplement Series* **92**, No. 2, 291–316.

Størmer, C. (1907): Sur les trajectories des corpuscles, electrisés dans l'espace sous l'action du magnétisme terrestre avec l'application aux aurores boréales. *Archives des sciences physiques et naturelles (Geneva)* **24**, 5, 113, 221, 317; **32**, 117–123, 190–219, 277–314, 415–436, 505–509 (1911); **33**, 51–69, 113–150 (1912).

Størmer, C. (1917): Corpuscular theory of the aurora borealis. *Journal of Geophysical Research* **22**, 23–34, 97–112.

Størmer, C. (1930): Periodische electronenbahnen im fielde lines elementarmagneton und ihre awendung auf eschenhagens elementarwellen des erdmagnetismus. *Astrophysics* **1**, 237.

Størmer, C. (1955): *The Polar Aurora.* Oxford: The Clarendon Press.

Strachan, L., et al. (2002): Empirical densities, kinetic temperatures, and outflow velocities in the equatorial streamer belt at solar minimum. *Astrophysical Journal* **571**, 1008–1014.

Strassmeier, K. G., Rice, J. B. (1998): Doppler imaging of stellar surface structure. VI. HD 129333 = EK Draconis: a stellar analog of the active young Sun. *Astronomy and Astrophysics* **330**, 685–695.

Strong, K. T. (1991): Observations from the *Solar Maximum Mission. Philosophical Transactions of the Royal Society (London).* **A336**, 327–337.

Strong, K. T., et al. (1984): A multiwavelength study of a double impulsive flare. *Solar Physics* **91**, 325–344.

Strong, K. T., et al. (1992): Observations of the variability of coronal bright points by the soft X-ray telescope on *Yohkoh. Publications of the Astronomical Society of Japan* **44**, L161–L166.

Strong, K. T., et al. (Eds., 1998): *The Many Faces of the Sun. A Summary of the Results from NASA's Solar Maximum Mission.* New York: Springer Verlag.

Strömgren, B. (1932): The opacity of stellar matter and the hydrogen content of the stars. *Zeitschrift für Astrophysik* **4**, 118–152.

Stuiver, M. (1961): Variations in atmospheric carbon-14 attributed to a variable sun. *Science* **207**, 11–19.

Stuiver, M. (1961): Variations in radiocarbon concentration and sunspot activity. *Journal of Geophysical Research* **66**, 273–276.

Stuiver, M. (1980): Solar variability and climatic change during the current millennium. *Nature* **286**, 868–871.

Stuiver, M., Braziunas, T. F. (1989): Atmospheric ^{14}C and century-scale solar oscillations. *Nature* **338**, 405–408.

Stuiver, M., Braziunas, T. F. (1993): Sun, ocean, climate and atmospheric 14 co2: an evaluation of causal and spectral relationships. *Holocene* **3**(4), 289–305.

Stuiver, M., Quay, P. D. (1980): Changes in atmospheric carbon-14 attributed to a variable Sun. *Science* **207**, 11–19.

Sturrock, P. A. (1966): Model of the high-energy phase of solar flares. *Nature* **211**, 695–697.

Sturrock, P. A. (1968): A model of solar flares. In: *Structure and Development of Solar Active Regions. International Astronomical Union Symposium No. 35* (Ed. K. O. Kiepenheuer). Dordrecht: D. Reidel Publishing Co., pp. 471–477.

Sturrock, P. A. (Ed., 1980): *Solar Flares: A Monograph From Skylab Solar Workshop II.* Boulder: Colorado Associated University Press.

Sturrock, P. A. (1989): The role of eruption in solar flares. *Solar Physics* **121**, 387–397.

Sturrock, P. A., Hartle, R. E. (1966): Two-fluid model of the solar wind. *Physical Review Letters* **16**, 628–631.

Sturrock, P. A., Uchida, Y. (1981): Coronal heating by stochastic magnetic pumping. *The Astrophysical Journal* **246**, 331–336.

Sturrock, P. A., Wheatland, M. S., Acton, L. W. (1996): *Yohkoh* soft x-ray telescope images of the diffuse solar corona. *Astrophysical Journal (Letters)* **461**, L115–L117.

Su, Y., et al. (2007): Evolution of the sheared magnetic fields of two X-class flares observed by *Hinode*/XRT. *Publications of the Astronomical Society of Japan* **59**, S785–S791.

Su, Y., Golub, L., Van Ballegooijen, A. A. (2007): A statistical study of shear motion of the footpoints in two-ribbon flares. *Astrophysical Journal* **655**, 606–614.

Sudol, J. J., Harvey, J. W. (2005): Longitudinal magnetic field changes accompanying solar flares. *Astrophysical Journal* **635**, 647–658.

Suess, H. E. (1955): Radiocarbon concentration in modern wood. *Science* **122**, 415–417.

Suess, H. E. (1965): Secular variations of the cosmic-ray produced carbon 14 in the atmosphere and their interpretations. *Journal of Geophysical Research* **70**, 5937–5952.

Suess, H. E. (1968): Climate changes, solar activity, and cosmic-ray production rate of natural radiocarbon. *Meteorology Monograph* **8**, 146–150.

Suess, H. E. (1973): Natural radiocarbon. *Endeavor* **32**, 34–38.

Suess, H. E. (1980): Radiocarbon geophysics. *Endeavor* **4**, 113–117.

Suess, H. E., Linick, T. W. (1990): The ^{14}C record in bristlecone pine wood of the past 8000 years based on the dendrochronology of the late C. W. Ferguson. *Philosophical Transactions of the Royal Society (London)* **A330**, 403–412.

Suess, S. T. (1990): The heliopause. *Reviews of Geophysics* **28**, 97–115.

Suess, S. T., et al. (1996): Latitudinal dependence of the radial IMF component – interplanetary imprint. *Astronomy and Astrophysics* **316**, 304–312.

Suess, S. T., Smith, E. J. (1996): Latitudinal dependence of the radial IMF component coronal imprint. *Geophysical Research Letters* **23**(22), 3267–3270.

Sui, L., Holman, G. D. (2003): Evidence for the formation of a large-scale current sheet in a solar flare. *Astrophysical Journal (Letters)* **596**, L251–L254.

Sui, L., Holman, G. D., Dennis, B. R. (2005): Evidence for magnetic reconnection in three homologous solar flares observed by *RHESSI*. *Astrophysical Journal, Part 1*, **612**, 546–556.

Sui, L., Holman, G. D., Dennis, B. R. (2006): Enigma of a flare involving multiple-loop interactions: Emerging, colliding loops or magnetic breakout? *Astrophysical Journal* **646**, 605–614.

Sullivan, W. T. III. (Ed., 1984): *The Early Years of Radio Astronomy*. New York: Cambridge University Press.

Svalgaard, L., Duvall, T. L., Scherrer, P. H. (1978): The strength of the Sun's polar fields. *Solar Physics* **58**, 225–239.

Svalgaard, L., et al. (1975): The Sun's sector structure. *Solar Physics* **45**, 83–91.

Svalgaard, L., Wilcox, J. M. (1975): Long-term evolution of solar sector structure. *Solar Physics* **41**, 461–475.

Svalgaard, L., Wilcox, J. M. (1976): Structure of the extended solar magnetic field and the sunspot cycle variation in cosmic ray intensity. *Nature* **262**, 766–768.

Svalgaard, L., Wilcox, J. M. (1978): A view of solar magnetic fields, the solar corona, and the solar wind in three dimensions. *Annual Review of Astronomy and Astrophysics* **16**, 429–443.

Svalgaard, L., Wilcox, J. M., Duvall, T. L. (1974): A model combining the polar and the sector structured solar magnetic fields. *Solar Physics* **37**, 157–172.

Svensmark, H. (1998): Influence of cosmic rays on Earth's climate. *Physical Review Letters* **81**, 5027–5030.

Svensmark, H., et al. (2007): Experimental evidence for the role of ions in particle nucleation under atmospheric conditions. *Proceedings of the Royal Society* **463**, 385–396.

Svensmark, H., Friis-Christensen, E. (1997): Variation of cosmic ray flux and global cloud coverage – a missing link in solar-climate relationships. *Journal of Atmospheric and Solar-Terrestrial Physics* **59**, 1225–1232.

Svensmark, H., Friss-Christensen, E. (2007): Reply to Lockwood and Frölich: The persistent role of the Sun in climate forcing. Danish National Space Center: Scientific Report 3/2007.

Svestka, Z. (1976): *Solar Flares*. Norwell: Kluwer.

Svestka, Z. (1995): On "the solar flare myth" postulated by Gosling. *Solar Physics* **160**, 153–156.

Svestka, Z., Cliver, E. W. (1992): History and basic characteristics of eruptive flares. In: *Eruptive Solar Flares. Proceedings of International Astronomical Union Colloquium No. 133* (Eds. Z. Svestka, B. V. Jackson and M. E. Machado). New York: Springer-Verlag, pp. 1–14.

Svestka, Z., et al. (1982): Observations of a post-flare radio burst in X-rays. *Solar Physics* **75**, 305–329.

Svestka, Z., et al. (1987): Multi-thermal observations of newly formed loops in a dynamic flare. *Solar Physics* **108**, 237–250.

Sweet, P. A. (1958): The neutral point theory of solar flares. In: *Electromagnetic Phenomen in Cosmical Physics. International Astronomical Union Symposium No. 6* (Ed. B. Lehnert). Cambridge: Cambridge at the University Press 1958, pp. 123–134.

Sweet, P. A. (1958): The production of high energy particles in solar flares. *Nuovo Cimento* (**10**)**8**, Suppl. 2, 188–196., Suppl. 2, 188–196.

Sweet, P. A. (1958b): The neutral point theory of solar flares. In: *Electromagnetic Phenomen in Cosmical Physics. International Astronomical Union Symposium No. 6* (Ed. B. Lehnert). Cambridge: Cambridge University Press 1958, pp. 123–134.

Sweet, P. A. (1969): Mechanisms of solar flares. *Annual Review of Astronomy and Astrophysics* **7**, 149–176.

Syrovatskii, S. I. (1981): Pinch sheets and reconnection in astrophysics. *Annual Review of Astronomy and Astrophysics* **19**, 163–229.

T

Takakura, T. (1961): Acceleration of electrons in the solar atmosphere and type IV radio outbursts. *Publications of the Astronomical Society of Japan* **13**, 166–172.

Takakura, T. (1967): Theory of solar bursts. *Solar Physics* **1**, 304–353.

Takakura, T. (1995): Imaging spectra of hard X-rays from the foot points of impulsive loop flares. *Publications of the Astronomical Society of Japan* **47**, 355–364.

Takakura, T., et al. (1993): Time variation of the hard X-ray image during the early phase of solar impulsive bursts. *Publications of the Astronomical Society of Japan* **45**, 737–753.

Takakura, T., Kai, K. (1966): Energy distribution of electrons producing microwave impulsive bursts and X-ray bursts from the Sun. *Publications of the Astronomical Society of Japan* **18**, 57–76.

Tanaka, K. (1987): Impact of X-ray observations from the *Hinotori* satellite on solar flare research. *Publications of the Astronomical Society of Japan* **39**, 1–45.

Tanaka, K. (1991): Studies on a very flare-active delta group - peculiar delta spot evolution and inferred subsurface magnetic rope structure. *Solar Physics* **136**, 133–149.

Tanaka, K., et al. (1982): High-resolution solar flare X-ray spectra obtained with rotating spectrometers on the *Hinotori* satellite. *The Astrophysical Journal (Letters)* **254**, L59–L63.

Tandberg-Hanssen, E., Emslie, A. G. (1988): *The Physics of Solar Flares*. New York: Cambridge University Press.

Tandon, J. N., Das, M. K. (1982): The effect of a magnetic field on solar luminosity. *Astrophysical Journal* **260**, 338–341.

Tang, F. (1981): Rotation rate of high-latitude sunspots. *Solar Physics* **69**, 399–404.

Telleschi, A., et al. (2005): Coronal evolution of the Sun in time: High-resolution X Ray spectroscopy of solar analogs with different ages. *Astrophysical Journal* **622**, 653–679.

Tett, S. F. B., et al. (1999): Causes of twentieth-century temperature change near the Earth's surface. *Nature* **399**, 569–572.

Thernisien, A., et al. (2008): Three-dimensional reconstruction of CMEs from SECCHI observations. *Astrophysical Journal*, Submitted.

Thomas, B. T., Smith, E. J. (1980): The Parker spiral configuration of the interplanetary magnetic field between 1 and 8.5 AU. *Journal of Geophysical Research* **85**, 6861–6867.

Thomas, B. T., Smith, E. J. (1981): The structure and dynamics of the heliospheric current sheet. *Journal of Geophysical Research* **86**, 11105–11110.

Thomas, J. H., Cram, L. E., Nye, A. H. (1982): Five-minute oscillations as a probe of sunspot structure. *Nature* **297**, 485–487.

Thompson, B. J., et al. (1998): *SOHO*/EIT observations of an earth-directed coronal mass ejection on May 12, 1997. *Geophysical Research Letters* **25**, 2465–2468.

Thompson, B. J., et al. (1999): *SOHO*/EIT observations of the 1997 April 7 coronal transient: Possible evidence of coronal Moreton waves. *Astrophysical Journal (Letters)* **517**, L151–L154.

Thompson, M. J., et al. (1996): Differential rotation and dynamics of the solar interior. *Science* **272**, 1300–1305.

Thompson, M. J., et al. (2003): The internal rotation of the sun. *Annual Review of Astronomy and Astrophysics* **41**, 599–643.

Thomsen, M. F., et al. (1998): The magnetospheric response to the CME passage of January 10–11, 1997, as seen at geosynchronous orbit. *Geophysical Research Letters* **25**, 2545–2548.

Thomson, W. (Baron Kelvin) (1892): Presidential address to the Royal Society on November 30, 1892. In: *Popular Lectures and Addresses by Sir William Thomson Baron Kelvin. Volume II. Geology and General Physics*. London: Macmillan and Company 1894, pp. 508–529.

Thuillier, G., et al. (2004) Solar irradiance reference spectra for two solar active levels. *Advances in Space Research* **34**, 256–261.

Timothy, A. F., Krieger, A. S., Vaiana, G. S. (1975): The structure and evolution of coronal holes. *Solar Physics* **42**, 135–156.

Tinsley, B. A. (1988): The solar cycle and the QBO influences on the latitude of storm tracks in the North Atlantic. *Geophysical Research Letters* **15**, 409–412.

Tinsley, B. A. (1994): Solar wind mechanism suggested for weather and climate change. *EOS Transactions of the American Geophysical Union* **75**(32), 369–376.

Titov, V., Demoulin, P. (1999): Basic topology of twisted magnetic configurations in solar flares. *Astronomy and Astrophysics* **351**, 701–720.

Tomczyk, S., et al. (2007): Alfvén waves in the solar corona. *Science* **317**, 1192–1196.

Tomczyk, S., Schou, J., Thompson, M. J. (1995): Measurement of the rotation rate in the deep solar interior. *The Astrophysical Journal (Letters)* **448**, L57–L60.

Toomre, J. (2002): Order amidst turbulence. *Science* **296**, 64–65.

Topka, K., Moore, R., Labonte, B. J., Howard, R. (1982): Evidence for a poleward meridional flow on the Sun. *Solar Physics* **79**, 231–245.

Török, T., Kliem, B. (2005): Confined and ejective eruptions of kink-unstable flux ropes. *Astrophysical Journal (Letters)* **630**, L97–L100.

Torsti, J., et al. (1995): Energetic particle experiment ERNE. *Solar Physics* **162**, 505–531.

Torsti, J., et al. (1998): Energetic ($\sim$ 1 to 50 MeV) protons associated with Earth- directed coronal mass ejections. *Geophysical Research Letters* **25**, 2525–2528.

Torsti, J., et al. (2002): Solar particle event with exceptionally high ^{3}He enhancement in the energy range up to 50 MeV nucleon. *Astrophysical Journal (Letters)* **573**, L59–L63.

Torsti, J., Laivola, J., Kocharov, L. (2003): Common overabundance of ^{3}He in high-energy solar particles. *Astronomy and Astrophysics* **408**, L1–L4.

Torsti, J., Riihonen, E., Kocharov, L. (2004): The 1998 May 2–3 magnetic cloud: an interplanetary "highway" for solar energetic particles observed with *SOHO*/ERNE. *Astrophysical Journal (Letters)* **600**, L83–L86.

Toth, G., et al. (2007): Sun to thermosphere simulation of the October 28–30, 2003 storm with the space weather modeling framework. *Space Weather* **5**, S06003.

Totsuka, Y. (1991): Recent results on solar neutrinos from Kamiokande. *Nuclear Physics B* **19**, 69–76.

Tousey, R. (1963): The extreme ultraviolet spectrum of the Sun. *Space Science Review* **2**, 3–69.

Tousey, R. (1967): Some results of twenty years of extreme ultraviolet solar research. *Astrophysical Journal* **149**, 239–252.

Tousey, R. (1973): The solar corona. *Space Research* **13**, 713–730.

Tousey, R. (1976): Eruptive prominences recorded by the X u.v. spectroheliograph on *Skylab*. *Philosophical Transactions of the Royal Society (London)* **A281**, 359–364.

Tousey, R., et al. (1946): The solar ultraviolet spectrum from a V-2 rocket. *The Astronomical Journal* **52**, 158–159.

Tousey, R., et al. (1973): A preliminary study of the extreme ultraviolet spectroheliograms from *Skylab*. *Solar Physics* **33**, 265–280.

Trattner, K. J., et al. (1996): *Ulysses* COSPIN/LET: latitudinal gradients of anomalous cosmic ray O, N and Ne. *Astronomy and Astrophysics* **316**, 519–527.

Tsuneta, S. (1995): Particle acceleration and magnetic reconnection in solar flares. *Publications of the Astronomical Society of Japan* **47**, 691–697.

Tsuneta, S. (1996): Interacting active regions in the solar corona. *Astrophysical Journal (Letters)* **456**, L63–L65.

Tsuneta, S. (1996): Structure and dynamics of magnetic reconnection in a solar flare. *Astrophysical Journal* **456**, 840–849.

Tsuneta, S. (1997): Moving plasmoid and formation of the neutral sheet in a solar flare. *Astrophysical Journal* **483**, 507.

Tsuneta, S., et al. (1983): Vertical structure of hard X-ray flare. *Solar Physics* **86**, 313–321.

Tsuneta, S., et al. (1991): The Soft X-ray Telescope for the *SOLAR-A* mission. *Solar Physics* **136**, 36–67.

Tsuneta, S., et al. (1992): Global restructuring of the coronal magnetic fields observed with the *Yohkoh* Soft X-ray Telescope. *Publications of the Astronomical Society of Japan* **44**, L211–L214.

Tsuneta, S., et al. (1992): Observation of a solar flare at the limb with the *Yohkoh* Soft X-ray Telescope. *Publications of the Astronomical Society of Japan* **44**, L63–L69.

Tsuneta, S., et al. (2008): The Solar Optical Telescope (SOT) for the *Solar-B* mission, *Solar Physics,* **249**, 167–196.

Tsurutani, B. T., et al. (1988): Origin of interplanetary southward magnetic fields responsible for major magnetic storms near solar maximum (1978–1979). *Journal of Geophysical Research* **93**, 8519–8531.

Tsurutani, B. T., et al. (1990): Interplanetary Alfvén waves and auroral (substorm) activity: *IMP 8. Journal of Geophysical Research* **95**, 2241–2252.

Tsurutani, B. T., et al. (1994): The relationship between interplanetary discontinuities and Alfvén waves: *Ulysses* observations. *Geophysical Research Letters* **21**(21), 2267–2270.

Tsurutani, B. T., et al. (1995a): Interplanetary origin of geomagnetic activity in the declining phase of the solar cycle. *Journal of Geophysical Research* **100**, 21717–21733.

Tsurutani, B. T., et al. (1995b): Large amplitude IMF fluctuations in co-rotating interaction regions: *Ulysses* at midlatitudes. *Geophysical Research Letters* **22**(23), 3397–3400.

Tsurutani, B. T., et al. (1996): Interplanetary discontinuities and Alfvén waves at high heliographic latitudes: *Ulysses. Journal of Geophysical Research* **101**, 11027–11038.

Tsurutani, B. T., Gonzalez, W. D. (1987): The cause of high-intensity, long-duration continuous AE activity (HILDCAAs): interplanetary Alfvén wave trains. *Planetary and Space Science* **35**, 405–412.

Tsurutani, B. T., Gonzalez, W. D. (1997): The interplanetary causes of magnetic storms: A review. In: *Magnetic Storms: Geophysical Monograph 98* (Eds. B. T. Tsurutani, W. D. Gonzalez, Y. Kamide and J. K. Arbailo). Washington, DC: American Geophysical Union, pp. 77–89.

Tsurutani, B. T., Gonzalez, W. D., Kamide, Y., Arballo, J. K. (Eds., 1997): *Magnetic Storms. Geophysics Monograph 98.* Washington, DC: American Geophysical Union.

Tsurutani, B. T., Lin, R. P. (1985): Acceleration of greater than 47 keV ions and greater than 2 keV electrons by interplanetary shocks at 1 AU. *Journal of Geophysical Research* **90**, 1–11.

Tu, C. -Y., et al. (2005): Solar wind origin in coronal funnels. *Science* **308**, 519–523.

Tu, C.-Y., Marsch, E. (1992): The evolution of MHD turbulence in the solar wind. In: *Solar Wind Seven,* (Eds. E. Marsch, R. Schwenn) Oxford, Pergamon, pp. 549–554.

Tu, C.-Y., Marsch, E. (1995): MHD structures, waves and turbulence in the solar wind: Observations and theories. *Space Science Reviews* **73**, 1–210.

Tu, C.-Y., Marsch, E. (2001): On cyclotron wave heating and acceleration of solar wind ions in the outer corona. *Journal of Geophysical Research* **106**(A5), 8233–8252.

Tu, C.-Y., Marsch, E., Qin, Z.-R. (2004): Dependence of the proton beam drift velocity on the proton core plasma beta in the solar wind. *Journal of Geophysical Research* **109**(A5), A05101.

Tu, C.-Y., Marsch, E., Wilhelm, K., Curdt, W. (1998): Ion temperatures in a solar polar coronal hole observed by SUMER on *SOHO*. *Astrophysical Journal* **503**, 475.

Turck-Chièze, S., et al. (2001): Solar neutrino emission deduced from a seismic model. *The Astrophysical Journal (Letters)* **555**, L69–L73.

Turck-Chieze, S, et al. (1988): Revisiting the solar model. *Astrophysical Journal* **335**, 415–424.

Turck-Chieze, S., et al. (1997): First results of the solar core from GOLF acoustic modes. *Solar Physics* **175**, 247–265. Reprinted in: *The First Results From SOHO* (Eds. B. Fleck and Z. Svestka). Boston: Kluwer Academic Publishers, pp. 247–265.

Turck-Chieze, S., et al. (2004): Looking for gravity-mode multiplets with the GOLF experiment aboard *SOHO*. *Astrophysical Journal* **604**, 455–468.

Tylka, A. J., et al. (2002): Flare- and shock-accelerated energetic particles in the solar events of 2001 April 12 and 15. *Astrophysical Journal (Letters)* **581**, L119–L123.

Tylka, A. J., et al. (2005): Shock geometry, seed populations, and the origin of variable elemental composition at high energies in large gradual solar particle events. *Astrophysical Journal* **625**, 474–495.

Tyndall, J. (1861): On the absorption and radiation of heat by gases and vapors, and on the physical connection of radiation, absorption, and conduction. *Philosophical Magazine and Journal of Science* **22A**, 276–277.

Tzedakis, P. C., et al. (1997): Comparison of terrestrial and marine records of changing climate of the last 500,000 years. *Earth and Planetary Science Letters* **150**, 171–176.

U

Uchida, Y. (1963): An effect of the magnetic field in the shock wave heating theory of the solar corona. *Publications of the Astronomical Society of Japan* **15**, 376–399.

Uchida, Y. (1968): Propagation of hydromagnetic disturbances in the solar corona and Moreton's wave phenomenon. *Solar Physics* **4**, 30.

Uchida, Y. (1974): Behavior of flare-produced coronal mhd wavefront and the occurrence of type II radio bursts. *Solar Physics* **39**, 431–449.

Uchida, Y., Altschuler, M. D., Newkirk, G. Jr. (1973): Flare-produced coronal MHD-fast-mode wavefronts and Moreton's wave phenomenon. *Solar Physics* **28**, 495–516.

Uchida, Y., Canfield, R. C., Watanabe, T., Hiei, E. (Eds., 1991): *Flare Physics in Solar Activity Maximum 22*. New York: Springer-Verlag.

Uchida, Y., et al. (1992): Continual expansion of the active-region corona observed by the *Yohkoh* Soft X-ray Telescope. *Publications of the Astronomical Society of Japan* **44**, L155–L160.

Uchida, Y., et al. (Eds., 1994): *X-ray Solar Physics From Yohkoh*. Tokyo: University Academy Press.

Uchida, Y., Kosugi, T., Hudson, H. S. (Eds., 1996): *Magnetodynamic Phenomena in the Solar Atmosphere - Prototypes of Stellar Magnetic Activity. IAU Colloquium No. 153*. Boston: Kluwer Academic Publishers.

Ugarte-Urra, I., Warren, H. P., Winebarger, A. R. (2007): The magnetic topology of coronal mass ejection sources. *Astrophysical Journal* **662**, 1293–1301.

Ulmschneider, P., Priest, E. R., Rosner, R. (Eds., 1991): *Mechanisms of Chromospheric and Coronal Heating*. New York: Springer-Verlag.

Ulrich, R. K. (1970): The five-minute oscillations on the solar surface. *Astrophysical Journal* **162**, 993–1002.

Ulrich, R. K. (1970): The five-minute oscillations on the solar surface. *The Astrophysical Journal* **162**, 993–1002.

Ulrich, R. K. (1975): Solar neutrinos and variations in the solar luminosity. *Science* **190**, 619–624.

Ulrich, R. K., Bertello, L. (1995): Solar-cycle dependence of the sun's apparent radius in the neutral iron spectral line at 525 nm. *Nature* **377**, 214–215.

Ulrich, R. K., Boyden, J. E. (2005): The solar surface toroidal magnetic field. *Astrophysical Journal (Letters)* **620**, L123–L127.

Ulrich, R. K., Rhodes, E. J. Jr. (1977): The sensitivity of nonradial *p* mode eigenfrequencies to solar envelope structure. *Astrophysical Journal* **218**, 521–529.

Underwood, J. H., et al. (1976): Preliminary results from S-056 X-ray telescope experiment aboard the *Skylab*-Apollo Telescope Mount. *Progress in Astronautics and Aeronautics* **48**, 179–195.

Unsöld, A. (1928): Über die Struktur der Fraunhoferschen Linien und die quantitative Spektral-analyse der Sonnenatmosphäre. *Zeitschrift für Physik* **46**, 765.

Ushida, Y., et al. (2003): *TRACE* observation of an arcade flare showing evidence supporting quadrupole magnetic source model for arcade flares. *Publications of the Astronomical Society of Japan* **55**(1), 305–312.

Usmanov, A. V., Dryer, M. (1995): A global 3-d simulation of interplanetary dynamics in June, 1991. *Solar Physics* **159**, 347–370.

Uzdensky, D. A. (2007): Fast collisionless reconnection condition and self- organization of solar coronal heating. *Astrophysical Journal* **671**, 2139–2153.

V

Vaiana, G. S., et al. (1968): X-ray structures of the Sun during the importance 1n flare of 8 June 1968. *Science* **161**, 564–567.

Vaiana, G. S., et al. (1973a): X-ray observations of characteristic structures and time variations from the solar corona: Preliminary results from *Skylab*. *The Astrophysical Journal (Letters)* **185**, L47–L51.

Vaiana, G. S., Krieger, A. S., Timothy, A. F. (1973b): Identification and analysis of structures in the corona from X-ray photography. *Solar Physics* **32**, 81–116.

Vaiana, G. S., Rosner, R. (1978): Recent advances in coronal physics. *Annual Review of Astronomy and Astrophysics* **16**, 393–428.

Vainio, R., Schlickeiser, R. (1999): Self-consistent Alfvén-wave transmission and test-particle acceleration at parallel shocks. *Astronomy and Astrophysics* **343**, 303–311.

Van Allen, J. A. (1975): Interplanetary particles and fields. *Scientific American* **233**, 160–162, September.

Van Allen, J. A., Fennell, J. F., Ness, N. F. (1971): Asymmetric access of energetic solar protons to the Earth's north and south polar caps. *Journal of Geophysical Research* **76**, 4262–4275.

van Allen, J. A., Krimigis, S. M. (1965): Impulsive emission of ≈ 40-keV electrons from the Sun. *Journal of Geophysical Research* **70**, 5737–5751.

van Allen, J. A., McIlwain, C. E., Ludwig, G. H. (1959): Radiation observations with satellite 1958ε. *Journal of Geophysical Research* **64**, 271–286. Reproduced in:*A Source Book in Astronomy and Astrophysics 1900–1975* (Eds. K. R. Lang and O. Gingerich). Cambridge: Harvard University Press 1979, pp. 149–151.

Van Allen, J. A., Ness, N. F. (1969): Particle shadowing by the Moon. *Journal of Geophysical Research* **74**, 91–93.

Van Ballegooijen, A. A., et al. (1998): Dynamics of magnetic flux elements in the solar photosphere. *Astrophysical Journal* **508**, 435–447.

Van Ballegooijen, A. A., Martens, P. C. H. (1989): Formation and eruption of solar prominences. *Astrophysical Journal* **343**, 971–984.

Van De Hulst, H. C. (1947): Zodiacal light in the solar corona. *Astrophysical Journal* **105**, 471–488.

Van Driel-Gesztelyi, L., et al. (1996): X-ray bright point flares due to magnetic reconnection. *Solar Physics* **163**, 145–170.

van Loon, H., Labitzke, K. (1988): Association between the 11-year solar cycle, the QBO and the atmosphere, Part II. Surface and 700 mb on the northern hemisphere in winter. *Journal of Climate* **1**, 905–920.

van Loon, H., Labitzke, K. (1990): Association between the 11-year solar cycle and the atmosphere. Part IV. The stratosphere, not grouped by the phase of the QBO. *Journal of Climate* **3**, 827–837.

Van Speybroeck, L. P., Krieger, A. S., Vaiana, G. S. (1970): X-ray photographs of the Sun on March 7, 1970. *Nature* **227**, 818–822.

Vandegriff, J., Wagstaff, K., Ho, G., Plauger, J. (2005): Forecasting space weather: predicting interplanetary shocks using neural networks. *Advances in Space Research* **36**(12), 2323–2327.

Vasyliunas, V. M. (1975): Theoretical models of magnetic field line merging. I. *Reviews of Geophysics and Space Physics* **13**, 303–336.

Vaughan, A. H., Preston, G. W. (1980): A survey of chromospheric Ca II H and K emission in field stars of the solar neighborhood. *Publications of the Astronomical Society of the Pacific* **92**, 385–391.

Veck, N. J., Parkinson, J. H. (1981): Solar abundances form X-ray flare observations. *Monthly Notices of the Royal Astronomical Society* **197**, 41–55.

Vegard, L. (1913): On spectra of the aurora borealis. *Physikalishe Zeitschrift* **14**, 677.

Veronig, A. M., et al. (2005): Physics of the Neupert effect: Estimates of the effects of source energy, mass transport, and geometry using *RHESSI* and *GOES* data. *Astrophysical Journal* **621**, 482–497.

Vestrand, W. T. (1991): High-energy flare observations from the *Solar Maximum Mission*. *Philosophical Transactions of the Royal Astronomical Society* **A336**, 349–362.

Vial, J. C., Bocchialini, K., Boumier, P. (Eds., 1999): *Space Solar Physics. Theoretical and Observational Issues in the Context of the SOHO Mission. Lecture Notes in Physics No. 507.* Heidelberg: Springer Verlag.

Vial, J. C., Kaldeich-Schürmann (Eds., 1999): *SOHO-8: Plasma Dynamics and Diagnostics in the Solar Transition Region and Corona.* ESA SP-446 1999.

Voitenko, Y., Goossens, M. (2004): Cross-field heating of coronal ions by low frequency Alfvén waves. *Astrophysical Journal (Letters)* **605**, L149–L152.

Völk, H. J. (1975): Cosmic ray propagation in interplanetary space. *Review of Geophysics and Space Physics* **13**, 547–566.

von Humboldt, F. W. H. A. (1799–1804): *Voyage aux régions équinoxiales du Nouveau Continent, fait en 1799, 1800, 1801, 1802, 1803, et 1804 par Al [exandre] de Humboldt et A [imé] Bonpland.* Paris, 1805–1834.

Von Steiger, R., et al. (1992): Variable carbon and oxygen abundances in the solar wind as observed in Earth's magnetosheath by *AMPTE*/CCE. *Astrophysical Journal* **389**, 791–799.

Von Steiger, R., et al. (2000): Composition of quasi-stationary solar wind flows from *Ulysses*/solar wind ion composition spectrometer. *Journal of Geophysical Research* **105**, A12, 27217–27238.

Von Steiger, R., Lallement, R., Lee, M. A. (Eds., 1996): The heliosphere in the local interstellar medium. *Space Science Reviews* **78**, 1–399. Reprinted by: Kluwer Academic Publishers, Dordrecht, the Netherlands and the International Space Science Institute, Bern, Switzerland.

Vorontsov, S. V., et al. (2002): Helioseismic measurement of solar torsional oscillations. *Science* **296**, 101–103.

Vourlidas, A., et al. (2003): Direct detection of a coronal mass ejection-associated shock in large angle and spectrometeric coronagraph experiment white-light images. *Astrophysical Journal* **598**, 1392–1402.

Vourlidas, A., Subramanian, P., Dere, K. P., Howard, R. A. (2000): Large angle spectrometric coronagraph measurements of the energetics of coronal mass ejections. *Astrophysical Journal* **534**, 456–467.

Vrsnak, B., Ruzjak, V., Romplt, B. (1991): Stability of prominences exposing helical-like patterns. *Solar Physics* **136**, 151–167.

W

Wagner, W. J. (1984): Coronal mass ejections. *Annual Review of Astronomy and Astrophysics* **22**, 267–289.

Wagner, W, J., Mac Queen, R. M. (1983): The excitation of type II radio bursts in the corona. *Astronomy and Astrophysics* **120**, 136–138.

Waldmeier, M. (1938): Chromosphärische eruptionen I. *Zeitschrift für Astrophysik* **16**, 276–290.

Waldmeier, M. (1940): Chromosphärische eruptionen II. *Zeitschrift für Astrophysik* **20**, 46–66.

Waldmeier, M. (1951): Spektralphotometrische klassifikation der protuberanzen. *Zeitschrift für Astrophysik* **28**, 208–218.

Waldmeier, M. (1951, 1957): *Die Sonnenkorona I, II.* Basel: Birkhäuser 1951, 1957.

Waldmeier, M. (1961): *The Sunspot Activity in the Years 1610–1960.* Zurich: Schulthess.

Waldmeier, M. (1975): The coronal hole at the 7 March 1970 eclipse. *Solar Physics* **40**, 351–358.

Waldmeier, M. (1981): Cyclic variations of the polar coronal hole. *Solar Physics* **70**, 251–258.

Wallerstein, G. (1988): Mixing in stars. *Science* **240**, 1743–1750.

Walsh, R. W., Ireland, J. (2003): The heating of the solar corona. *The Astronomy and Astrophysics Review* **12**, 1–41.

Walsh, R. W., Ireland, J., Danesy, D., Fleck, B. (Eds., 2004): *SOHO-15: Coronal Heating.* ESA SP-575 2004.

Wang, H., Chang, L., Deng, Y, Zhang, H. (2005): Reevaluation of the magnetic structure and evolution associated with the Bastille Day flare on 2000 July 12. *Astrophysical Journal* **627**, 1031–1039.

Wang, H., et al. (2002): Rapid changes of magnetic fields associated with six X-class flares. *Astrophysical Journal* **576**, 497–504.

Wang, H., Qiu, J., Jing, J., Zhang, H. (2003): Study of ribbon separation of a flare associated with a quiescent filament eruption. *Astrophysical Journal* **593**, 564–570.

Wang, T. J., et al. (2002): Doppler shift oscillations of hot solar coronal plasma seen by SUMER: A signature of loop oscillations? *Astrophysical Journal (Letters)* **574** L101–L104.

Wang, T. J., et al. (2003): Hot coronal loop oscillations observed with SUMER: Examples and statistics. *Astronomy and Astrophysics* **406**, 1105–1121.

Wang, Y.-M. (1993): Flux-tube divergence, coronal heating, and the solar wind. *Astrophysical Journal (Letters)* **410**, L123–L126.

Wang, Y.-M. (1994a): Polar plumes and the solar wind. *Astrophysical Journal (Letters)* **435**, L153–L156.

Wang, Y.-M. (1994b): Two types of slow solar wind. *Astrophysical Journal (Letters)* **437**, L67–L70.

Wang, Y.-M. (1998): Network activity and the evaporative formation of polar plumes. *Astrophysical Journal (Letters)* **501**, L145–L150.

Wang, Y.-M., et al. (1997a): Solar wind stream interactions and the wind speed- expansion factor relationship. *Astrophysical Journal (Letters)* **488**, L51–L54.

Wang, Y.-M., et al. (1997b): The green line corona and its relation to the photospheric magnetic field. *Astrophysical Journal* **485**, 419–429.

Wang, Y.-M., et al. (1998a): Coronagraph observations of inflows during high solar activity. *Geophysical Research Letters* **26**, 1203–1206.

Wang, Y.-M., et al. (1998b): Origin of streamer material in the outer corona. *Astrophysical Journal (Letters)* **498**, L165–L168.

Wang, Y. -M., et al. (1998): Observations of correlated white-light and extreme ultraviolet jets from polar coronal holes. *Astrophysical Journal* **508**, 899–907.

Wang, Y. -M., et al. (2000): The dynamical nature of coronal streamers. *Journal of Geophysical Research* **105**, A11, 25133–25142.

Wang, Y. -M., et al. (2007): The solar eclipse of 2006 and the origin of raylike features in the white-light corona. *Astrophysical Journal* **660**, 882–892.

Wang, Y. -M., Nash, A. G., Sheeley, N. R. Jr. (1989): Magnetic flux transport on the Sun. *Science* **245**, 712–718.

Wang, Y. -M., Sheeley, N. R. Jr. (1990a): Solar wind speed and coronal flux-tube expansion. *Astrophysical Journal* **355**, 726–732.

Wang, Y. -M., Sheeley, N. R. Jr. (1990b): Magnetic flux transport and the sunspot- cycle evolution of coronal holes and their wind streams. *Astrophysical Journal* **365**, 372–386.

Wang, Y. -M., Sheeley, N. R. Jr. (1992): On potential field models of the solar corona. *Solar Physics* **392**, 310–319.

Wang, Y. -M., Sheeley, N. R. Jr. (2004): Footprint switching and the evolution of coronal holes. *Astrophysical Journal* **612**, 1196–1205.

Wang, Y.-M., et al. (1999a): Coronagraph observations of inflows during high solar activity. *Geophysical Research Letters* **26**, 1203–1206.

Wang, Y.-M., et al. (1999b): Streamer disconnection events observed with the LASCO coronagraph. *Geophysical Research Letters* **26**, 1349–1352.

Wang, Y.-M., Hawley, S. H., Sheeley, N. R. Jr. (1996): The magnetic nature of coronal holes. *Science* **271**, 464–469.

Wang, Y.-M., Lean, J. L., Sheeley, N. R. Jr. (2005) Modeling the Sun's magnetic field and irradiance since 1713. *Astrophysical Journal* **625**, 522–538.

Wang, Y.-M., Pick, M., Mason, G. M. (2006): Coronal holes, jets, and the origin of ^{3}He particle events. *Astrophysical Journal* **639**, 495–509.

Wang, Y.-M., Sheeley, N. R. Jr. (1991a): Magnetic flux transport and the Sun's dipole moment: New twists to the Babcock-Leighton model. *Astrophysical Journal* **375**, 761–770.

Wang, Y.-M., Sheeley, N. R. Jr. (1991b): Why fast solar wind originates from slowly expanding coronal flux tubes. *Astrophysical Journal (Letters)* **372**, L45–L48.

Wang, Y.-M., Sheeley, N. R. Jr. (1997): The high-latitude solar wind near sunspot maximum. *Geophysical Research Letters* **24**(24), 3141–3144.

Wang, Y.-M., Sheeley, N. R. Jr. (2002a): Coronal white-light jets near sunspot maximum. *Astrophysical Journal* **575**, 542–552.

Wang, Y.-M., Sheeley, N. R. Jr. (2002b): Observations of core fallback during coronal mass ejections. *Astrophysical Journal* **567**, 1211–1224.

Wang, Y.-M., Sheeley, N. R. Jr. (2002c): Sunspot activity and the long-term variation of the Sun's open magnetic flux. *Journal of Geophysical Research* **107**(A10), 1302.

Wang, Y.-M., Sheeley, N. R. Jr. (2003): On the topological evolution of the coronal magnetic field during the solar cycle. *Astrophysical Journal* **599**, 1404–1417.

Wang, Y.-M., Sheeley, N. R. Jr. (2006): Observations of flux rope formation in the outer corona. *Astrophysical Journal* **650**, 1172–1183.

Wang, Y.-M., Sheeley, N. R. Jr., Nash, A. G. (1990): Latitudinal distribution of solar-wind speed from magnetic observations of the Sun. *Nature* **347**, 439–444.

Wang, Y.-M., Sheeley, N. R. Jr., Nash, A. G. (1991): A new solar cycle model including meridional circulation. *Astrophysical Journal* **383**, 431–442.

Warren, H. P. (2000): Fine structure in solar flares. *Astrophysical Journal (Letters)* **536**, L105–L108.

Warren, H. P., et al. (1997): Doppler shifts and nonthermal broadening in the quiet solar transition region: O VI. *Astrophysical Journal (Letters)* **484**, L91–L94.

Warren, H. P., et al. (2007): Observations of transient active region heating with *Hinode*. *Publications of the Astronomical Society of Japan* **59**, S675–S681.

Warren, H. P., Winebarger, A. R., Hamilton, P. S. (2002): Hydrodynamic modeling of active region loops. *The Astrophysical Journal (Letters)* **579**, L41–L44.

Watko, J. A., Klimchuk, J. A. (2000): Width variations along coronal loops observed by *TRACE*. *Solar Physics* **193**, 77–92.

Webb, D. F. (1992): The solar sources of coronal mass ejections. In: *Eruptive Solar Flares* (Eds. Z. Svestka, B. V. Jackson and M. E. Machado). Berlin: Springer-Verlag, pp. 234–247.

Webb, D. F. (1995): Coronal mass ejections: The key to major interplanetary and geomagnetic disturbances. *Reviews of Geophysics, Supplement* **33**, 577–583.

Webb, D. F., Burkepile, J., Forbes, T. G., Riley, P. (2003): Observational evidence of new current sheets trailing coronal mass ejections. *Journal of Geophysical Research* **108**(A12), 1440.

Webb, D. F., et al. (1998): The solar origin of the January 1997 coronal mass ejection, magnetic cloud and geomagnetic storm. *Geophysical Research Letters* **25**, 2469–2472.

Webb, D. F., et al. (2000): Relationship of halo coronal mass ejections, magnetic clouds, and magnetic storms. *Journal of Geophysical Research* **105**(A4), 7491–7508.

Webb, D. F., et al. (2005): Commission 49: Interplanetary plasma and heliosphere. *Proceedings of the International Astronomical Union* **1**, 103–120.

Webb, D. F., et al. (2006): Solar mass ejection imager (SMEI) observations of coronal mass ejections (CMEs) in the heliosphere. *Journal of Geophysical Research* **111**, A12101.

Webb, D. F., et al. (2008): Studying geoeffective icmes between the Sun and Earth: space weather implications of smei observations. Submitted to *Space Weather*.

Webb, D. F., Howard, R. A. (1994): The solar cycle variation of coronal mass ejections and the solar wind mass flux. *Journal of Geophysical Research* **99**, 4201–4220.

Webb, D. F., Hundhausen, A. J. (1987): Activity associated with the solar origin of coronal mass ejections. *Solar Physics* **108**, 383–401.

Webb, D. F., Krieger, A. S., Rust, D. M. (1976): Coronal X-ray enhancements associated with Hα filament disappearances. *Solar Physics* **48**, 159–186.

Weber, E. J., Davis, L. Jr. (1967): The angular momentum of the solar wind. *Astrophysical Journal* **148**, 217–227.

Wedemeyer-Böhm, S., Steiner, O., Bruls, J., Rammacher, W. (2007): What is heating the quiet-sun chromosphere? *Physics of the Chromospheric Plasmas ASP Converence Series* **368**, 93.

Weiss, J. E., Weiss, N. O. (1979): Andrew Marvell and the Maunder minimum. *Quarterly Journal of the Royal Astronomical Society* **20**, 115–118.

Weiss, N. O. (1990): Periodicity and aperiodicity in solar magnetic activity. *Philosophical Transactions of the Royal Society (London)* **A330**, 617–625.

Weizsäcker, C. F. Von (1938): Über Elementumwandlungen in Innern der Sterne II (Element transformation inside stars II), *Physikalische Zeitschrift* **39**, 633–646. English translation in *A Source Book in Astronomy and Astrophysics 1900–1975* (Eds. K. R. Lang and O. Gingerich). Cambridge: Harvard University Press 1979, pp. 309–319.

Whang, Y. C., Burlaga, L. F. (1994): Interaction of global merged interaction region shock with the heliopause and its relation to the 2- and 3-kHz radio emissions. *Journal of Geophysical Research* **99**(A11), 21457–21466.

Wheatland, M. S., Sturrock, P. A., Acton, L. W. (1997): Coronal heating and the vertical temperature structure of the quiet corona. *Astrophysical Journal* **482**, 510–518.

White, O. R. (Ed., 1997): *The Solar Output and Its Variation*. Boulder: Colorado Associated University Press.

White, O. R., Livingston, W. C. (1981): Solar luminosity variation III. Calcium K variation from solar minimum to maximum in cycle 21. *Astrophysical Journal* **249**, 798–816.

White, O. R., Livingston, W. C., Wallace, L. (1987): Variability of chromospheric and photospheric lines in solar cycle 21. *Journal of Geophysical Research* **92**, 823–827.

White, W. B., et al. (1997): Response of global upper ocean temperature to changing solar irradiance. *Journal of Geophysical Research* **102**, 3255–3266.

Wiegelmann, T., et al. (2005): Comparing magnetic field extrapolations with measurements of magnetic loops. *Astronomy and Astrophysics* **433**, 701–705.

Wiegelmann, T., Xia, L.-D., Marsch, E. (2005): Links between magnetic fields and plasma flows in a coronal hole. *Astronomy and Astrophysics* 432, L1–L4.

Wigley, T. M. L. (1976): Spectral analysis: Astronomical theory of climatic change. *Nature* **264**, 629–631.

Wigley, T. M. L., Kelly, P. M. (1990): Holocene climatic change, ^{14}C wiggles and variations in solar irradiance. *Philosophical Transactions of the Royal Society (London)* **A330**, 547–560.

Wigley, T. M. L., Raper, S. C. B. (1990): Climatic change due to solar irradiance changes. *Geophysical Research Letters* **17**, 2169–2172.

Wilcox, J. M. (1968): The interplanetary magnetic field, solar origin and terrestrial effects. *Space Science Reviews* **8**, 258–328.

Wilcox, J. M., Ness, N. F. (1965): Quasi-stationary co-rotating structure in the interplanetary medium. *Journal of Geophysical Research* **70**, 5793–5805.

Wild, J. P. (1950): Observations of the spectrum of high-intensity solar radiation at meter wavelengths. II – Outbursts, III – Isolated Bursts. *Australian Journal of Scientific Research* **A3**, 399–408, 541–557.

Wild, J. P. (1963): Fast phenomena in the solar corona. In: *The Solar Corona. Proceedings of IAU Symposium No. 16* (Ed. J. W. Evans). New York: Academic Press, pp. 115–127.

Wild, J. P., Mc Cready, L. L. (1950): Observations of the spectrum of high-intensity solar radiation at meter wavelengths. I – The apparatus and spectral types. *Australian Journal of Scientific Research* **A3**, 387–398.

Wild, J. P., Murray, J. D., Rowe, W. C. (1953): Evidence of harmonics in the spectrum of a solar radio outburst. *Nature* **172**, 533–534.

Wild, J. P., Roberts, J. A., Murray, J. D. (1954): Radio evidence of the ejection of very fast particles from the Sun. *Nature* **173**, 532–534.

Wild, J. P., Sheridan, K. V., Neylan, A. A. (1959): An investigation of the speed of the solar disturbances responsible for type III radio bursts. *Australian Journal of Physics* **12**, 369–398.

Wild, J. P., Smerd, S. F. (1972): Radio bursts from the solar corona. *Annual Review of Astronomy and Astrophysics* **10**, 159–196.

Wild, J. P., Smerd, S. F., Weiss, A. A. (1963): Solar bursts. *Annual Review of Astronomy and Astrophysics* **1**, 291–366.

Wilhelm, K. (2006): Solar coronal-hole plasma densities and temperatures. *Astronomy and Astrophysics* **455**, 697–708.

Wilhelm, K., et al. (1995): SUMER – Solar Ultraviolet Measurements of Emitted Radiation. *Solar Physics* **162**, 189–232.

Wilhelm, K., et al. (1998): The solar corona above polar coronal holes as seen by SUMER on *SOHO*. *Astrophysical Journal* **500**, 1023–1038.

Wilhelm, K., et al. (2000): On the source regions of the fast solar wind in polar coronal holes. *Astronomy and Astrophysics* 353, 749–756.

Wills-Davey, M. J., Thompson, D. J. (1999): Observations of a propagating disturbance in *TRACE*. *Solar Physics* **190**, 467–483.

Willson, R. C, et al. (1981): Observations of solar irradiance variability. *Science* **211**, 700–702.

Willson, R. C. (1982): Solar irradiance variations and solar activity. *Journal of Geophysical Research* **87**, 4319–4324.

Willson, R. C. (1984): Measurements of solar total irradiance and its variability. *Space Science Reviews* **38**, 203–242.

Willson, R. C. (1991): The Sun's luminosity over a complete solar cycle. *Nature* **351**, 42–44.

Willson, R. C. (1997): Total solar irradiance trend during solar cycles 21 and 22. *Science* **277**, 1963–1965.

Willson, R. C., et al. (1986): Long-term downward trend in total solar irradiance. *Science* **234**, 1114–1117.

Willson, R. C., Hudson, H. S. (1988): Solar luminosity variations in solar cycle 21. *Nature* **332**, 810–812.

Willson, R. C., Hudson, H. S. (1991): The Sun's luminosity over a complete solar cycle. *Nature* **351**, 42–44.

Willson, R. C., Mordvinov, A. V. (2003): Secular total solar irradiance trend during solar cycles 21–23. *Geophysical Research Letters* **30**, 1199.

Willson, R. F., Lang, K. R. (1984): Very large array observations of solar active regions IV. Structure and evolution of radio bursts from 20 centimeter loops. *Astrophysical Journal* **279**, 427–437.

Willson, R. F., Lang, K. R., Gary, D. E. (1993): Particle acceleration and flare triggering in large-scale magnetic loops joining widely separated active regions. *Astrophysical Journal* **418**, 490–495.

Wilson, A. (Ed., 2002): *SOHO-11: From Solar Min to Max: Half a Solar Cycle with SOHO*. ESA SP-508 2002.

Wilson, A., Pallé, P. L. (Eds., 2001): *SOHO-10/GONG 2000: Helio- and Asteroseismology at the Dawn of the Millenium*. ESA SP-464 2001.

Wilson, J. W., et al. (2004): Deep space environments for human exploration. *Advances in Space Research* 34, 1281–1287.

Wilson, O. C. (1966a): Stellar chromospheres. *Science* 151, 1487–1498.

Wilson, O. C. (1966b): Stellar convection zones, chromospheres, and rotation. *Astrophysical Journal* 144, 695–708.

Wilson, O. C. (1978): Chromospheric variations in main-sequence stars. *Astrophysical Journal* 226, 379–396.

Wilson, O. C., Wooley, R. (1970): Calcium emission intensities as indicators of stellar age. *Monthly Notices of the Royal Astronomical Society* 148, 463–475.

Wilson, P. R., McIntosh, P. S. (1991): The reversal of the solar polar magnetic fields. *Solar Physics* 136, 221–237.

Wilson, P. R., McIntosh, P. S., Snodgrass, H. B. (1990): The reversal of the solar polar magnetic fields I. The surface transport of magnetic flux. *Solar Physics* 127, 1–9.

Wilson, R. M., Hildner, E. (1984): Are interplanetary magnetic clouds manifestations of coronal transients at 1 AU? *Solar Physics* 91, 169–180.

Wimmer-Schwingruber, R. F. (2006): Coronal mass ejections. *Space Science Reviews* 123(1–3), 471–480.

Winebarger, A. R., et al. (2002): Steady flows detected in extreme-ultraviolet loops. *The Astrophysical Journal (Letters)* 567, L89–L92.

Winebarger, A. R., Warren, H. P. (2005): Cooling active region loops observed with SXT and TRACE. *Astrophysical Journal* 626, 543–550.

Winebarger, A. R., Warren, H. P., Mariska, J. T. (2003): *Transition Region and Coronal Explorer* and Soft X-ray Telescope active region loop observations: Comparisons with static solutions of the hydrodynamic equations. *The Astrophysical Journal* 587, 439–449.

Winograd, I. J., et al. (1988): A 250,000-year climatic record from great basin vein calcite: Implications for Milankovitch theory. *Science* 242, 1275–1280.

Withbroe, G. L. (1981): Activity and outer atmosphere of the sun. In: *Activity and Outer Atmospheres of the Sun and Stars, 11th Advanced Course of the Swiss National Academy of Sciences*. Sauverny: Observatoire de Geneve.

Withbroe, G. L. (1988): The temperature structure, mass, and energy flow in the corona and inner solar wind. *Astrophysical Journal* 325, 442–467.

Withbroe, G. L. (1989): The solar wind mass flux. *Astrophysical Journal (Letters)* 337, L49–L52.

Withbroe, G. L., et al. (1982): Probing the solar wind acceleration region using spectroscopic techniques. *Space Science Reviews* 33, 17–52.

Withbroe, G. L., Feldman, W. C., Ahluwalia, H. S. (1991): The solar wind and its coronal origins. In: *Solar Interior and Atmosphere* (Eds. A. N. Cox, W. C. Livingston and M. S. Matthews). Tucson: University of Arizona Press, pp. 1087–1106.

Withbroe, G. L., Noyes, R. W. (1977): Mass and energy flow in the solar chromosphere and corona. *Annual Review of Astronomy and Astrophysics* 15, 363–387.

Witte, M., et al. (1992): The interstellar neutral-gas experiment on *Ulysses*. *Astronomy and Astrophysics Supplement* 92(2), 333–348.

Woch, J., et al. (1997): SWICS/*Ulysses* observations: The three-dimensional structure of the heliosphere in the declining/minimum phase of the solar cycle. *Geophysical Research Letters* 24(22), 2885–2888.

Wolfenstein, L. (1978): Neutrino oscillations in matter. *Physical Review D* 17, 2369–2374.

Wolff, C. L. (1972a): Free oscillations of the Sun and their possible stimulation by solar flares. *The Astrophysical Journal* 176, 833–842.

Wolff, C. L. (1972b): The five-minute oscillations as nonradial pulsations of the entire Sun. *Astrophysical Journal (Letters)* 177, L87–L92.

Wolfson, R. (1983): The active solar corona. *Scientific American* 248, 104–119, February.

Wollaston, W. H. (1802): A method of examining refractive and dispersive power by prismatic reflection. *Philosophical Transactions of the Royal Society of London* **92**, 365–380.

Woo, R. (2006): Ultra-fine-scale filamentary structures in the outer corona and the solar magnetic field. *Astrophysical Journal (Letters)* **639**, L95–L98.

Woo, R., et al. (1995): Fine-scale filamentary structure in coronal streamers. *Astrophysical Journal (Letters)* **449**, L91–L94.

Woo, R., Habbal, S. R. (1997): Extension of coronal structure into interplanetary space. *Geophysical Research Letters* **24**(10), 1159–1162.

Woo, R., Habbal, S. R. (2005): Origin and acceleration of the slow solar wind. *Astrophysical Journal (Letters)* **629**, L129–L132.

Woo, R., Habbal, S. R., Feldman, U. (2004): Role of closed magnetic fields in solar wind flow. *Astrophysical Journal* **612**, 1171–1174.

Wood, B. E., et al. (1999): Comparison of two coronal mass ejections observed by EIT and LASCO with a model of an erupting magnetic flux rope. *Astrophysical Journal* **512**, 484–495.

Wood, B. E., et al. (2005): New mass-loss measurements from atmospheric Lyman alpha absorption. *Astrophysical Journal Letters* **628**, L143–L146.

Woodard, M. F. (1997): Implications of localized, acoustic absorption for heliotomographic analysis of sunspots. *The Astrophysical Journal* **485**, 890–894.

Woodard, M. F. (2002): Solar subsurface flow inferred directly from frequency wavenumber correlations in the seismic velocity field. *The Astrophysical Journal* **565**, 634–639.

Woodard, M. F., Hudson, H. (1983a): Frequencies, amplitudes and line widths of solar oscillations from total solar irradiance observations. *Nature* **305**, 589–593.

Woodard, M. F., Hudson, H. (1983b): Solar oscillations observed in the total irradiance. *Solar Physics* **82**, 67–73.

Woodard, M. F., Noyes, R. C. (1985): Change of the solar oscillation eigenfrequencies with the solar cycle. *Nature* **318**, 449–450.

Woods, T. N., et al. (2004): Solar irradiance variability during the October 2003 solar storm period. *Geophysical Research Letters* **31**, L10802.

Woods, T. N., et al. (2005): The *Solar Euv Experiment (SEE)*: mission overview and first results. *Journal of Geophysical Research* **110**, A01312.

Woods, T. N., Kopp, G., Chamberlin, P. C. (2006): Contributions of the solar ultraviolet irradiance to the total solar irradiance during large flares. *Journal of Geophysical Research* **111**, A10S14.

Woods, T. N., Rottman, G. J. (2002): Solar ultraviolet variability over time periods of aeronomic interest. In: *Comparative Aeronomy in the Solar System. Geophysical Monograph Series 130* (Eds. M. Mendillo, A. Nagy, J. Hunter Waite Jr.) Washington: American Geophysical Union, pp. 221–234.

Wu, S. T., et al. (1989): Flare energetics. In: *Energetic Phenomena on the Sun* (Eds. M. R. Kundu, B. Woodgate and E. J. Schmahl). Boston: Kluwer Academic Publishers, pp. 377–492.

X

Xia, L.-D., Marsch, E., Wilhelm, K. (2004): On the network structures in solar equatorial coronal holes – Observations of SUMER and MDI on *SOHO*. *Astronomy and Astrophysics* **424**, 1025–1037.

Xie, H., Ofman, L., Viñas, A. (2004): Multiple ions resonant heating and acceleration by Alfvén/cyclotron fluctuations in the corona and the solar wind. *Journal of Geophysical Research* **109**, A08103.

Y

Yamada, M. (1999): Review of controlled laboratory experiments on physics of magnetic reconnection. *Journal of Geophysical Research* **104**(A7), 14529–14541.

Yamamoto, T. T., et al. (2005): Magnetic helicity injection and sigmoidal coronal loops. *Astrophysical Journal* **624**, 1072–1079.

Yamauchi, Y., et al. (2004): The magnetic structure of hydrogen alpha macrospicules in solar coronal holes. *Astrophysical Journal* **605**, 511–520.

Yamauchi, Y., Suess, S. T., Steinberg, J. T., Sakurai, T. (2004): Differential velocity between solar wind protons and alpha particles in pressure balance structures. *Journal of Geophysical Research* **109**, A03104.

Yan, Y., et al. (2007): Diagnostics of radio fine structures around 3 GHz with *Hinode* data in the impulsive phase of an X3.4/4B flare event on 2006 December 13. *Publications of the Astronomical Society of Japan* **59**, S815–S821.

Yashiro, S., et al. (2004): A catalog of white light coronal mass ejections observed by the *SOHO* spacecraft. *Journal of Geophysical Research* **109**(A7), A07105.

Yiou, F., et al. (1985): ^{10}Be in ice at Vostok Antarctica during the last climatic cycle. *Nature* **316**, 616–617.

Yokoyama, T., et al. (2001): Clear evidence of reconnection inflow of a solar flare. *Astrophysical Journal (Letters)* **546**, L69–L72.

Yokoyama, T., Shibata, K. (1995): Magnetic reconnection as the origin of x-ray jets and Hα surges on the Sun. *Nature* **375**, 42–44.

Yoshida, T., Tsuneta, S. (1996): Temperature structure of solar active regions. *Astrophysical Journal* **459**, 342–346.

Yoshimori, M. (1989): Observational studies of gamma-rays and neutrons from solar flares. *Space Science Reviews* **51**, 85–115.

Yoshimori, M., et al. (1983): Gamma-ray observations from *Hinotori*. *Solar Physics* **86**, 375–382.

Yoshimori, M., et al. (1991): The Wide Band Spectrometer on the *SOLAR-A*. *Solar Physics* **136**, 69–88.

Young, C. A (1896): *The Sun*. New York: Appleton.

Young, C. A. (1869): On a new method of observing contacts at the Sun's limb, and other spectroscopic observations during the recent eclipse. *American Journal of Sciences and Arts* **48**, 370–378. Reproduced in: *Early Solar Physics* (Ed. A. J. Meadows). Oxford: Pergamon Press 1970, pp. 125–134.

Young, P. R., et al. (2007): Solar transition region features observed with *Hinode*/EIS. *Publications of the Astronomical Society of Japan* **59**, S727–S733.

Yurchyshyn, V. B., et al. (2001): Orientation of the magnetic fields in interplanetary flux ropes and solar filaments. *Astrophysical Journal* **563**, 381–388.

Yurchyshyn, V., et al. (2005): Statistical distributions of speeds of coronal mass ejections. *Astrophysical Journal* **619**, 599–603.

Yurchyshyn, V., Wang, H., Abramenko, V. (2004): Correlation between speeds of coronal mass ejections and the intensity of geomagnetic storms. *Space Weather* 2, S02001.

Z

Zaatri, A., et al. (2006): North-south asymmetry of zonal and meridional flows determined from ring diagram analysis of GONG data. *Solar Physics* **236**, 227–244.

Zank, G. P., Gaisser, T. K. (Eds., 1992): *Particle Acceleration in Cosmic Plasmas*. New York: American Institute of Physics.

Zarro, D. M., Canfield, R. C. (1989): H-alpha redshifts as a diagnostic of solar flare heating. *Astrophysical Journal (Letters)* **338**, L33–L36.

Zarro, D. M., et al. (1999): *SOHO* EIT observations of extreme-ultraviolet "dimming" associated with a halo coronal mass ejection **520**, L139–L142.

Zhang, G., Burlaga, L. F. (1988): Magnetic clouds, geomagnetic disturbances, and cosmic ray decreases. *Journal of Geophysical Research* **93**, 2511–2518.

Zhang, J., Dere, K. P. (2006): A statistical study of main and residual accelerations of coronal mass ejections. *Astrophysical Journal* **649**, 1100–1109.

Zhang, J., Dere, K., Howard, R. A. (2001): Relationship between coronal mass ejections and flares. *Eos Transactions American Geophysical Union* **82**, 47.

Zhang, J., et al. (2001): On the temporal relationship between coronal mass ejections and flares. *Astrophysical Journal* **559**, 452–462.

Zhang, J., et al. (2004): A study of the kinematic evolution of coronal mass ejections. *Astrophysical Journal* **604**, 420–432.

Zhang, K., et al. (2003): A three-dimensional spherical nonlinear interface dynamo. *Astrophysical Journal* **596**, 663–679.

Zhang, M. (2006): Helicity observations of weak and strong fields. *Astrophysical Journal (Letters)* **646**, L85–L88.

Zhang, M., Golub, L. (2003): The dynamical morphologies of flares associated with the two types of solar coronal mass ejections. *Astrophysical Journal* **595**, 1251–1258.

Zhang, M., Low, B. C. (2003): Magnetic flux emergence into the solar corona III. The role of magnetic helicity conservation. *Astrophysical Journal* **584**, 479–496.

Zhang, M., Low, B. C. (2005): The hydromagnetic nature of solar coronal mass ejections. *Annual Review of Astronomy and Astrophysics* **43**, 103–137.

Zhang, Q., et al. (1994): A method of determining possible brightness variations of the Sun in past centuries from observations of solar-type stars. *Astrophysical Journal (Letters)* **427**, L111–L114.

Zhang, T. X. (2003): Preferential heating of particles by h-cyclotron waves generated by a global magnetohydrodynamic mode in solar coronal holes. *Astrophysical Journal (Letters)* **597**, L69–L72.

Zhao, J., Kosovichev, A. G. (2003): Helioseismic observations of the structure and dynamics of a rotating sunspot beneath the solar surface. *The Astrophysical Journal* **591**, 446–453.

Zhao, J., Kosovichev, A. G. (2004): Torsional oscillation, meridional flows, and vorticity inferred in the upper convection zone of the sun by time-distance helioseismology. *The Astrophysical Journal* **603**, 776–784.

Zhao, J., Kosovichev, A. G. (2006): Surface magnetism effects in time-distance helioseismology. *Astrophysical Journal* **643**, 1317–1324.

Zhao, J., Kosovichev, A. G., Duvall, T. L. Jr. (2001): Investigation of mass flows beneath a sunspot by time-distance helioseismology. *The Astrophysical Journal* **557**, 384–388.

Zhao, X. P., Hoeksema, J. T. (1997): Is the geoeffectiveness of the 6 January 1997CME predictable from solar observations? *Geophysical Research Letters* 24, 2965–2968.

Zhou, G. P., Wang, J. X., Zhang, J. (2006): Large-scale source regions of Earth-directed coronal mass ejections. *Astronomy and Astrophysics* **445**, 1133–1141.

Zhou, X., Tsurutani, B. T. (2001): Interplanetary shock triggering of nightside geomagnetic activity: Substorms, pseudo breakups, and quiescent events. *Journal of Geophysical Research* 106, 18957–18968.

Zirin, H., Moore, R., Walters, J. (1976): Proceedings of the workshop: The solar constant and the Earth's atmosphere. *Solar Physics* **46**, 377–409.

Zirker, J. B. (1977): Coronal holes – an overview. In: *Coronal holes and High Speed Wind Streams* (Ed. J. B. Zirker). Boulder: Colorado Associated University Press, pp. 1–26.

Zirker, J. B. (1993): Coronal heating. *Solar Physics* **148**, 43–60.

Zirker, J. B. (Ed., 1977): *Coronal Holes and High Speed Wind Streams*. Boulder: Colorado Associated University Press.

Zirker, J. B., Engvold, O., Martin, S. F. (1998): Counter-streaming gas flows in solar prominences as evidence for vertical magnetic fields. *Nature* **396**, 440–441.

Zurbuchen, T. H. (2007): A new view of the coupling of the Sun and heliosphere. *Annual Review of Astronomy and Astrophysics* **45**, 297–338.

Zurbuchen, T. H., et al. (1999): The transition between fast and slow solar wind from composition data. *Space Science Reviews* **87**, 353–356.

Zurbuchen, T. H., et al. (2004): On the fast coronal mass ejections in October/November 2003: ACE-SWICS results. *Geophysical Research Letters* **31**(11), L11805.

Zurbuchen, T. H., Fisk, L. A., Gloeckler, G., Von Steiger, R. (2002): The solar wind composition throughout the solar cycle: A continuum of dynamic states. *Geophysical Research Letters* **29**(9), 66–1.

Zurbuchen, T. H., Richardson, I. G. (2006): In-situ solar wind and magnetic field signatures of interplanetary coronal mass ejections. *Space Science Reviews* **123**, 31–43.

Zurbuchen, T. H., Schwadron, N. A., Fisk, L. A. (1997): Direct observational evidence for a heliospheric magnetic field with large excursions in latitude. *Journal of Geophysical Research* **102**(A11), 24175–24182.

Author Index

Subject Index

The Stars By E. L. Schatzman and F. Praderie

Modern Astrometry 2nd Edition
By J. Kovalevsky

**The Physics and Dynamics of Planetary
Nebulae** By G. A. Gurzadyan

Galaxies and Cosmology By F. Combes,
P. Boissé, A. Mazure and A. Blanchard

Observational Astrophysics 2nd Edition
By P. Léna, F. Lebrun and F. Mignard

Physics of Planetary Rings Celestial
Mechanics of Continuous Media
By A. M. Fridman and N. N. Gorkavyi

Tools of Radio Astronomy 4th Edition,
Corr. 2nd printing
By K. Rohlfs and T. L. Wilson

Tools of Radio Astronomy Problems and
Solutions 1st Edition, Corr. 2nd printing
By T. L. Wilson and S. Hüttemeister

Astrophysical Formulae 3rd Edition
(2 volumes)
Volume I: Radiation, Gas Processes
and High Energy Astrophysics
Volume II: Space, Time, Matter
and Cosmology
By K. R. Lang

Galaxy Formation 2nd Edition
By M. S. Longair

Astrophysical Concepts 4th Edition
By M. Harwit

Astrometry of Fundamental Catalogues
The Evolution from Optical to Radio
Reference Frames
By H. G. Walter and O. J. Sovers

Compact Stars. Nuclear Physics, Particle
Physics and General Relativity 2nd Edition
By N. K. Glendenning

The Sun from Space By K. R. Lang

Stellar Physics (2 volumes)
Volume 1: Fundamental Concepts
and Stellar Equilibrium
By G. S. Bisnovatyi-Kogan

Stellar Physics (2 volumes)
Volume 2: Stellar Evolution and Stability
By G. S. Bisnovatyi-Kogan

Theory of Orbits (2 volumes)
Volume 1: Integrable Systems
and Non-perturbative Methods
Volume 2: Perturbative
and Geometrical Methods
By D. Boccaletti and G. Pucacco

Black Hole Gravitohydromagnetics
By B. Punsly

Stellar Structure and Evolution
By R. Kippenhahn and A. Weigert

Gravitational Lenses By P. Schneider,
J. Ehlers and E. E. Falco

Reflecting Telescope Optics (2 volumes)
Volume I: Basic Design Theory and its
Historical Development. 2nd Edition
Volume II: Manufacture, Testing, Alignment,
Modern Techniques
By R. N. Wilson

Interplanetary Dust
By E. Grün, B. Å. S. Gustafson, S. Dermott
and H. Fechtig (Eds.)

The Universe in Gamma Rays
By V. Schönfelder

Astrophysics. A New Approach 2nd Edition
By W. Kundt

Cosmic Ray Astrophysics
By R. Schlickeiser

Astrophysics of the Diffuse Universe
By M. A. Dopita and R. S. Sutherland

The Sun An Introduction. 2nd Edition
By M. Stix

Order and Chaos in Dynamical Astronomy
By G. J. Contopoulos

Astronomical Image and Data Analysis
2nd Edition By J.-L. Starck and F. Murtagh

The Early Universe Facts and Fiction
4th Edition By G. Börner

Printed in the United States
By Bookmasters